D1632855

b
ɟ6

Live Cell

A   ᴼᴿᴬᵀᴼᴿʸ ᴹᴾᶜᴬᴹᴾᵁˢ

Imaging

Live Cell

A LABORATORY MANUAL

Imaging

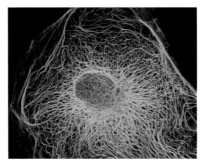

EDITED BY

ROBERT D. GOLDMAN
Northwestern University Medical School

DAVID L. SPECTOR
Cold Spring Harbor Laboratory

COLD SPRING HARBOR LABORATORY PRESS
Cold Spring Harbor, New York

Live Cell Imaging
A LABORATORY MANUAL

Publisher	John Inglis
Acquisition Editors	John Inglis and Kaaren Janssen
Project Manager and Developmental Editor	Kaaren Janssen
Developmental Editors	Tracy Kuhlman, Siân Curtis, and Michael Zierler
Project Coordinator	Mary Cozza
Permissions Coordinator	Maria Falasca
Production Manager	Denise Weiss
Production Editor	Rena Steuer
DVD Production Editor	Mala Mazzullo
Compositors	Compset, Inc. and Stephanie Collura
Cover Designer	Michael Albano

Front cover artwork: A living PtK2 rat kangaroo epithelial cell transfected with EGFP-Keratin 18 and stained with DAPI to highlight the DNA (*blue*) within the nucleus. Note the extensive array of keratin-containing tonofibrils in the cytoplasm. This confocal image was prepared from a Z-stack of 40 sections (0.2 μm focal steps) using a Zeiss LSM 510 microscope equipped with both an Ar laser and a UV laser. Photo provided by Satya Khuon and Robert D. Goldman, Northwestern University Medical School.

Back cover artwork: Human U2OS 2-6-3 cell (deconvolved projection using a DeltaVision system) showing an actively transcribing genetic locus (*bright green region in nucleus*), the mRNPs derived from the locus (*green particles in nucleoplasm*), as well as the protein product that is directed to peroxisomes (*blue cytoplasmic structures*). Photo provided by Susan M. Janicki and David L. Spector, Cold Spring Harbor Laboratory; see Chapter 26.

Library of Congress Cataloging-in-Publication Data

Live cell imaging : a laboratory manual / edited by Robert D. Goldman, David Spector.
    p. cm.
 Includes bibliographical references (p.).
 ISBN 0-87969-683-4 (pbk. : alk. paper) -- ISBN 0-87969-682-6 (hardcover : alk. paper)
 1. Cells--Laboratory manuals. 2. Microscopy--Laboratory manuals. I. Goldman, Robert D.,
1939- II. Spector, David L.

  QH583.2.L485 2004
  571.6'078--dc22

                        2004012095

10 9 8 7 6 5 4 3

*We wish to dedicate this book to our wives
Anne Goldman and Mona Spector
for their constant support, devotion, and understanding
throughout the preparation of this book*

Contents

DVD Contents

Chapter 5

Chapter 9

Chapter 13

Preface

M ANY OF THE REMARKABLE IMAGES AND ANALYSES of the dynamic properties of molecules in living cells described in this book stem from technical advances of the early 1970s. This was the period when commercially available epifluorescence microscopes equipped with efficient dichroic filter systems were introduced. These instruments permitted cell biologists to take full advantage of indirect immunofluorescence techniques and, for the first time, to photograph cells with 15–60-second exposure times, albeit on 35 mm film. Epifluorescence also made it possible to use the highest numerical aperture oil immersion objectives as both condenser and objective lenses. Accompanying these developments in microscopy, advances in the production and use of antibodies permitted the widespread use of double or even triple label indirect immunofluorescence, thereby providing investigators with the opportunity to carefully examine the associations between different components within the same subcellular compartment. Subsequently, developments in fluorescence in situ hybridization (FISH) permitted the localization of individual genes on chromosomes and specific mRNAs. The more recent introduction of confocal, deconvolution, and multiphoton microscopes, along with continuous advances in image-capturing devices, have made it possible to improve resolution, map in significant detail the three-dimensional organization of structures within cells, and capture images with unprecedented speeds, frequently in the range of milliseconds.

The images produced using high-resolution immunofluorescence techniques have revealed the remarkable complexity of cellular architecture and provided important insights into major cellular structures, including the different cytoskeletal systems, the nucleus and its various compartments, and membranous organelles. However, images of fixed and stained cells are static, providing only a brief snapshot of the organization and properties of these cellular structures. This major drawback stimulated cell biologists and microscopists to develop methods for studying specific types of molecules or multicomponent complexes in living cells, using the sensitivity of fluorescence-based imaging. At first, this involved the tracking of the incorporation of proteins (e.g., in the case of the cytoskeletal systems, tubulin, vimentin, and actin) directly conjugated with a fluorochrome such as X-rhodamine. Such derivatized proteins are usually microinjected into cells, and over time, they are incorporated into the structures containing their endogenous counterparts. This permits the investigator to carry out time-lapse observations and undertake fluorescence recovery after photobleaching (FRAP) experiments to monitor the dynamic properties of specific cellular proteins in live cells. However, the fluorescence signal is usually weak under these conditions, and the overall photobleaching of specimens in the microscope field undermines the number of time-resolved images that can be captured.

The amazing discovery in the early 1990s that genetically encoded fluorescent proteins could be expressed in cells and organisms through the use of green fluorescent protein (GFP) represented a revolution in the field of cell biology. The use of GFP-tagged proteins has many advantages as described throughout this book, but the greatest of these is the ability to capture more time-resolved images, providing researchers with the opportunities to observe the dynamic properties of tagged molecules for extended periods of time and extract quantitative information from living cells.

The specific methods used for live cell imaging described in the 35 chapters in this book have been written by world-renowned authorities. Each chapter is accompanied by remarkable images and, in many cases, movies on the accompanying DVD that clearly demonstrate the enormous benefits derived from the coordinated use of the technical advances in microscopic imaging. These advances have made it possible for researchers to delve into the mysteries of subcellular architecture and function in unprecedented ways. As a result, cell biologists are now undertaking experiments and analyses of living cells that, until a few years ago, they never dreamed would be possible.

The book is divided into two major sections. In the first, "Detection and Approaches to Live Cell Imaging," the technical aspects of live cell imaging are presented. Here, the reader learns about GFP variants, how best to express a GFP fusion protein in cells, and the different imaging modalities that can be used in conjunction with GFP fusion proteins. In the second section of the book, "Imaging of Live Cells and Organisms," specific examples are presented. These examples utilize GFP and various live cell imaging approaches to study the dynamics of subcellular components or processes from single molecules to single cells in culture, tissues, and animals. Each chapter is written in a fashion that is extremely useful for both those who have never used live cell imaging in their research programs, as well as for the more expert microscopist. In addition, this book provides an excellent reference for students interested in learning about the way in which live cell imaging can be used to extract important functional information from cells or organisms that cannot be obtained by other more traditional biochemical, molecular, or genetic methodologies.

We wish to express our thanks and gratitude to all of the authors and their research groups throughout the world who made this volume a reality. We are grateful to the fabulous staff at Cold Spring Harbor Laboratory Press who converted our dream into an actuality. In particular, we thank our Publisher, John Inglis, for agreeing to take on the project and our Project Manager, Kaaren Janssen, for coordinating the efforts and keeping this book in the forefront of our minds. We thank Developmental Editors, Tracy Kuhlman, Siân Curtis, and Michael Zierler; Project Coordinator, Mary Cozza; Permissions Coordinator, Maria Falasca; Production Manager, Denise Weiss; Production Editor, Rena Steuer; DVD Production Editor, Mala Mazzullo; and Desktop Editor, Stephanie Collura.

ROBERT D. GOLDMAN
DAVID L. SPECTOR

SECTION 1

Detection and Approaches to Live Cell Imaging

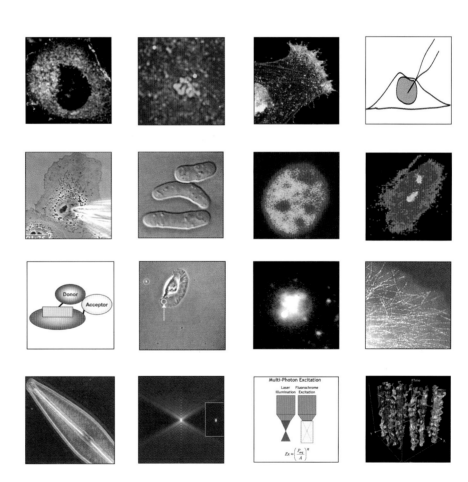

CHAPTER ONE

Fluorescent Protein Tracking and Detection

Mark A. Rizzo and David W. Piston

Department of Molecular Physiology and Biophysics, Vanderbilt University, Nashville, Tennessee 37232

INTRODUCTION

THE 1990S BROUGHT AN ABUNDANCE OF TECHNICAL ADVANCES that enabled both the widespread use of molecular cloning techniques and the measurement of cellular processes in living systems using optical microscopy. The synthesis of these two technologies has been a watershed for biologists, who can now easily and effectively film their favorite proteins in the cellular environment. The biggest star of these movies has been green fluorescent protein (GFP).

The discovery of GFP can be traced back to the early 1960s, when researchers studying the bioluminescent properties of the *Aequorea victoria* jellyfish isolated a blue-emitting, Ca^{++}-dependent bioluminescent protein that they named aequorin (Shimomura et al. 1962). Alongside this purification, another protein was found that was not luminescent, but which fluoresced green under UV light; this was eventually named "green fluorescent protein." Over the next two decades, it was determined that aequorin and GFP work together in the light organs of *A. victoria* to convert Ca^{++}-induced luminescent signals to the green fluorescence that was characteristic of the species (Shimomura 1998). Although GFP was eventually cloned in 1992 (Prasher et al. 1992), its potential as a molecular probe was not appreciated until it was used for tracking gene expression (Chalfie et al. 1994). GFP has since been engineered to produce a vast number of differently colored mutants, fusion proteins, and biosensors that are broadly referred to as fluorescent proteins (FPs). More recently, FPs from other species have been identified, resulting in further expansion of the color palette. With the rapid progression of FP technology, the usefulness of this genetically encoded fluorophore for a wide variety of applications beyond the simple tracking of FP-tagged molecules in living cells is now becoming fully appreciated.

The purposes of this chapter are to provide a description of the types of FPs that are currently available and tips on how they can be used successfully, to discuss some of the caveats and potential hazards associated with using FPs, and to identify areas of current (and sometimes much needed) development.

Features of *Aequorea victoria* GFP

One of the most important things to appreciate about GFP is that the entire 27-kD structure is essential to the development and maintenance of its fluorescence. Remarkably, the principle fluorophore is derived from just three amino acids: Ser65, Tyr66, and Gly67 (Fig. 1). Although this simple amino acid motif is commonly found throughout nature, it does not generally result in fluorescence. Unique to GFP is the location of this motif in the center of a remarkably stable barrel structure consisting of 11 β strands (Ormö et al. 1996). In the context of this special environment, a reaction occurs between the carboxyl carbon of Ser65 and the amino nitrogen of Gly67 that results in formation of an imidazolin-5-one ring. Further oxidation results in conjugation of the imidazolin ring with Tyr66 and maturation of a fluorescent species. Importantly, the native GFP fluorophore exists in two states: a predominant, protonated form with an excitation maximum at 395 nm and a less prevalent, unprotonated form with an excitation peak at ~475 nm. Regardless of the excitation wavelength, fluorescence emission is maximal at 507 nm, although the peak is not well defined.

There are two features of the fluorophore that have important implications for its use as a probe. First, the photophysics of GFP as a fluorophore are quite complex and thus, the molecule can accommodate quite a bit of modification. Much has been done to fine-tune the fluorescence of native GFP to provide a broad range of molecular probes, but the vast potential of GFP as a starting material for constructing molecular probes cannot be understated.

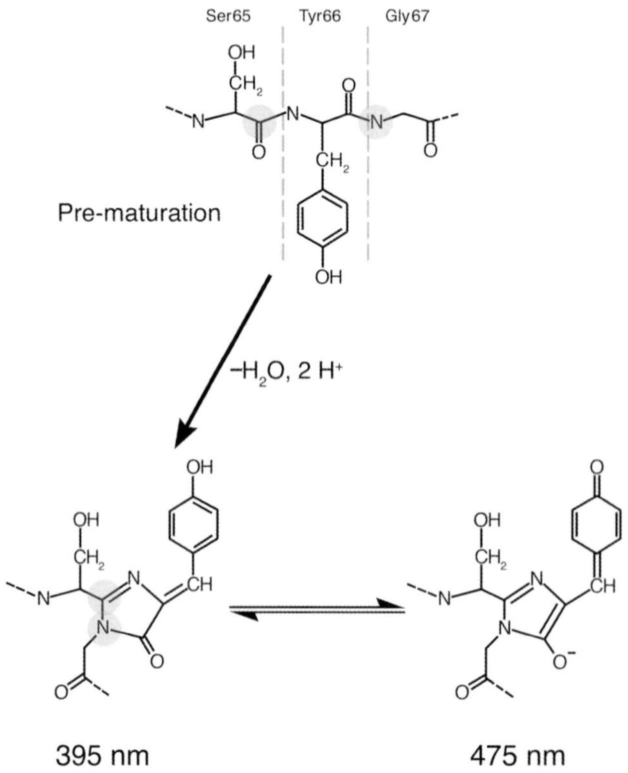

FIGURE 1. Maturation of the GFP fluorophore. Reaction of the carboxyl carbon of Ser65 and the amino nitrogen of Gly67 (highlighted) leads to formation of fluorophore. Two absorptive states are known to exist: a predominant, protonated form that absorbs at 395 nm and a less prevalent, unprotonated form that absorbs at approximately 475 nm.

Second, GFP fluorescence is very dependent on the structure surrounding the fluorophore. Denaturation of GFP destroys its fluorescence, and mutations to residues surrounding the fluorophore can greatly change the fluorescent properties of GFP. The packing of residues inside the β barrel is extremely stable, which results in a very high fluorescence quantum efficiency (up to 80%; Patterson et al. 1997). This tight protein structure also confers resistance to changes in pH, temperature, and denaturants such as urea (Ward 1998). Mutations in GFP that affect its fluorescence generally have negative effects on this stability, resulting in a reduction of quantum efficiency and greater environmental sensitivity. Although some of these defects can be overcome by additional mutations, derivative FPs are generally more sensitive to the environment than native GFP. These limitations should be considered when designing experiments.

THE FP COLOR PALETTE

A broad range of FP variants has been produced with colors that span the visible spectrum (Table 1). Adaptation of *A. victoria* GFP has resulted in FPs that range in color from blue to yellow. Red FPs from new species have been discovered and adapted for use in mammalian systems. In addition, the brightness and stability of FPs have been enhanced to improve their overall usefulness. The following sections provide an overview of the many different variants of FPs and highlight some of the more important practical considerations for their use.

Mutations that Improve Use for Mammalian Systems

To adapt GFPs for use in mammalian systems, several modifications of GFP were made and are found in all commonly used variants. First, the maturation of the fluorescence was optimized for 37°C. Maturation of the wild-type GFP fluorophore is quite efficient at 28°C, but increasing the temperature to 37°C substantially reduces overall maturation and results in decreased fluorescence (Patterson et al. 1997). Mutation of Phe64 to Leu (Cormack et al. 1996) results in improved maturation of fluorescence at 37°C that is at least equivalent to maturation at 28°C. This mutation is present in the most popular varieties of *A. victoria*-derived FPs but is not the only mutation that improves folding at 37°C. For example, α-GFP (F99S/M153T/V163A) does not contain the F64L mutation but folds slightly better at 37°C than at 28°C (Patterson et al. 1997).

In addition to improving maturation at 37°C, optimization of codon usage for mammalian expression has also improved overall brightness of GFP expressed in mammalian cells (Yang et al. 1996). Over 190 silent mutations were introduced into the coding sequence to optimize expression in human tissues. A Kozak translation initiation site was also introduced by insertion of a Val as the second amino acid. These improvements have resulted in a very useful probe for live cell imaging of mammalian cells and are common to all of the currently used FPs derived from *A. victoria* GFP.

Green FPs

Although native GFP is nicely fluorescent and wonderfully stable, the excitation maximum is close to the ultraviolet range. Because ultraviolet light requires special optical considerations, it is not well suited for most microscopy systems. In addition, ultraviolet light can damage cells and is not the best choice for live cell imaging. Fortunately, the excitation maximum of GFP can be shifted to a more amenable 488 nm by a single point mutation (S65T; Heim et al. 1995). This mutation is featured in the most popular variant of GFP, the

TABLE 1. Commonly used FPs

Species	FP	Excitation (nm)	Emission (nm)	Mutations	Reference
Aequorea victoria					
Green	wtGFP	395	475		
	EGFP	489	508	F64L, S65T, H231L	Patterson et al. 2001
	GFP²	399	511	F64L	Zimmermann et al. 2002
Blue	EBFP	383	445	F64L, S65T, Y66H, T145F	Patterson et al. 2001
Cyan	ECFP	434 (453)	477 (501)	F64L, S65T, Y66W, N146I, M153T, V163A, H231L	Patterson et al. 2001
	CGFP	458	504	ECFP mutations, T203Y	Sawano and Miyawaki 2000
Yellow	EYFP	514	527	S65G, V68L, S72A, T203Y, H231L	Patterson et al. 2001
	Citrine	516	529	EYFP + Q69M	Griesbeck et al. 2001
	Venus	515	528	EYFP + F46L, F64L, M153T, V163A, S175G	Nagai et al. 2002
Monomeric				A206K, L221K, F223R	Zacharias et al. 2002
Discosoma striata					
Red	DsRed	558	583		Patterson et al. 2001
	DsRed2	558	583	R2A, K5E, K9T, A105V, I161T, S197A	Bevis and Glick 2002
	DsRed-Express	554	586	R2A, K5E, N6D, T21S, H41T, N42Q, V44A, C117S, T217A	Bevis and Glick 2002
	RedStar	558	583	2S (insertion), R17K, V96I, F124L, M182K, P186Q, T202I	Knop et al. 2002
Monomeric	mRed1	584	607	>30 amino acid substitutions	Campbell et al. 2002

enhanced GFP (EGFP) that is sold commercially in easy-to-use vectors by BD Biosciences Clontech (see Chapter 2). Furthermore, EGFP can be imaged using commonly available fluorescein filter sets and is among the brightest of the currently available FPs. These features have made EGFP both the most popular FP and the best choice for most single-label FP experiments. The only drawbacks to the use of EGFP are slight pH sensitivity and a weak tendency to dimerize. These caveats are discussed in greater detail later in this chapter.

Other than EGFP, several other green FPs are currently in use. The best of these in terms of photostability and brightness may be the Emerald variety (S65T/S72A/N149K/M153T/I167T). Emerald is no longer commercially available, so relative to EGFP, it is seldom used. Other companies sell humanized variants that offer distinct advantages for fluorescence resonance energy transfer (FRET) experiments. PerkinElmer offers GFP² (F64L), which retains the 400-nm excitation peak and can be used as an effective partner for enhanced yellow fluorescent protein (EYFP) (Zimmermann et al. 2002). A variant of the S65C mutation (Heim et al. 1995) that has a peak excitation of 474 nm is now being sold by QbioGene and may be a more suitable FRET partner for enhanced blue fluorescent protein (EBFP) than the redshifted EGFP.

Yellow FPs

Yellow FPs were created after the crystal structure of GFP revealed that Thr203 was near the chromaphore (Ormö et al. 1996). Mutation of this residue to Tyr was introduced to stabilize the excited state dipole moment of the chromaphore and resulted in an ~20-nm redshift of both the excitation and emission spectra (Ormö et al. 1996). Further development led to EYFP, which is one of the brightest and most popular FPs. Its brightness and fluorescence spectrum make EYFP well suited for fluorescence microscopy and multicolor FP experiments. It is also useful for energy transfer experiments when paired with enhanced cyan fluorescent protein (ECFP) or GFP[2]. However, EYFP is far from perfect. It is very sensitive to acidic pH and loses 50% of its fluorescence at pH 6.5 (Patterson et al. 2001). In addition, it is also sensitive to Cl⁻ and photobleaches much more quickly than the green FPs. Recent development has solved some of these problems. The Citrine variant of yellow FP (Griesbeck et al. 2001) has been shown to be much more resistant to photobleaching, acidic pH, and other environmental effects, and Venus (Nagai et al. 2002) is the fastest maturing and brightest variant available to date. Use of these newer, more robust variants has already been shown to be particularly important when using quantitative techniques such as FRET (Takemoto et al. 2003). Although they are not yet commercially available, they are highly recommended.

Blue and Cyan FPs

The blue and cyan variants of GFP resulted from direct modification of Tyr66 in the fluorophore (Fig. 1). Conversion of this residue to His results in blue emission (450 nm) whereas conversion to Trp results in a major fluorescence peak around 480 nm along with a shoulder that peaks around 500 nm (Heim et al. 1994). Both of these varieties are weakly fluorescent and require secondary mutations to increase folding efficiency and overall brightness (Heim and Tsien 1996). Still, the so-called "enhanced" variants of this class of FP (EBFP and ECFP) are only about 25% as bright as EGFP (Patterson et al. 2001). Additionally, excitation of these FPs is best in spectral regions that are not commonly used, so special filters and/or laser sources are required. Despite these drawbacks, the widespread interest in multicolor labeling and FRET has popularized their use. This is especially true for ECFP, which can be excited off peak by an argon laser and is much more resistant to photobleaching than EBFP. In contrast to other FPs, there has not been much interest in developing better FPs in this area of the spectrum, and it seems unlikely that improved cyan and blue FPs will be developed anytime soon.

Red FPs

A major goal of fluorescent protein development has become the construction of a red FP that is just as effective as EGFP. The advantages of a good red FP range from the potential compatibility with existing confocal microscopes and filter sets to an increased ability to image whole animals, which are more transparent to red light. Because redshifting the fluorescent properties of *A. victoria* FPs beyond yellow has been unsuccessful, several groups have begun searching for a red FP among the tropical corals. The first of these coral-derived red FPs to be routinely used was derived from *Discosoma striata* and is commonly referred to as DsRed, or DrFP583 (Matz et al. 1999). Once fully matured, the emission peak of DsRed is 583 nm whereas the excitation spectrum has a major peak at 558 nm and a minor peak around 500 nm. However, maturation of DsRed fluorescence occurs slowly and proceeds through a time when it fluoresces green (Baird et al. 2000). This so-called green state has proven problematic for multiple labeling experiments with other green FPs because of the spectral overlap. Furthermore, DsRed is an obligate tetramer and can form large protein aggregates inside cells. Although these features are inconsequential for the use of DsRed as a reporter of gene expression, the usefulness of DsRed as an epitope tag is severely limited. In contrast to the *A. victoria* FPs, which have been successfully used to tag hundreds of proteins, the DsRed conjugates have proven much less successful and are often toxic.

Some of these problems have been overcome through mutagenesis. The second-generation DsRed, DsRed2, contains several mutations at the amino terminus that prevent formation of protein aggregates and reduce toxicity. In addition, the fluorophore maturation time is reduced (Bevis and Glick 2002). DsRed2 is still a tetramer, but it is more compatible with green FPs in multiple labeling experiments due to its quicker maturation. More recently, third-generation mutants have been developed that further reduce maturation time and increase brightness in terms of peak cellular fluorescence. Red fluorescence from DsRed-Express (BD Biosciences Clontech) can be observed within an hour after expression, as compared to ~11 hours for DsRed and ~6 hours for DsRed2 (Bevis and Glick 2002). A yeast-optimized variant, RedStar, has been developed that also has an improved maturation time and increased brightness (Knop et al. 2002). The presence of a green state in DsRed-Express and RedStar is not apparent, making these varieties the best choice for multiple labeling experiments. These variants are still obligate tetramers, so they are not the best choice for labeling proteins.

Several other red FPs have been isolated from reef coral. One of the first to be adapted for mammalian use was HcRed1, which was isolated from *Heteractis crispa* and is now commercially available from BD Biosciences Clontech. HcRed1 was originally derived from a nonfluorescent chromoprotein that absorbs red light (Gurskaya et al. 2001). Mutagenesis of HcRed1 resulted in a weakly fluorescent obligate dimer (588-nm excitation; 618-nm emission). Although the fluorescence spectrum of this protein is adequate for separation from DsRed, it tends to coaggregate with DsRed and is far less bright. Recently, an HcRed construct containing two molecules in tandem was made commercially available by Evrogen. In principle, dimerization between HcRed molecules preferentially occurs within the tandem pairing and produces a monomeric tag, but because the overall brightness of the probe has not yet been improved, it is not a good choice for routine application in live cell microscopy.

Monomeric Variants

In their natural states, most FPs are oligomeric. Likewise, *A. victoria* GFP is thought to participate in a tetrameric complex with aequorin (Ward 1998), but this complex has only been observed at very high protein concentrations and the tendency of *A. victoria* FPs to dimerize is generally very weak (K_d >100 μM; Zacharias et al. 2002). Dimerization of FPs has thus not generally been observed when expressed in mammalian systems, but when FPs are targeted to specific cellular compartments such as the plasma membrane, the local concentration of FPs can theoretically become high enough to permit dimerization. This is particularly a concern when performing FRET experiments in which even the slightest tendency of FP dimerization could cloud data interpretation. Therefore, a series of mutations has been found that can further reduce the tendency of *A. victoria* FPs to dimerize (Zacharias et al. 2002): A206K, L221K, and F223R. The K_d of these mutants was found to be highest for A206K, although both L221K and F223R pushed the K_d into the low-mM range.

Construction of a monomeric DsRed has been much more difficult. Greater than 30 amino acid changes to its structure were required for creation of the first-generation monomeric DsRed (Campbell et al. 2002). This initial variant is much dimmer than the native protein and photobleaches very quickly, so it is not a very robust choice compared to monomeric green and yellow FPs. Interestingly, both the excitation and emission spectra are slightly redshifted. Although the fluorescent properties of monomeric Red make it challenging to use, the progress so far is a strong indication that further development will produce more useful and important varieties of red FPs.

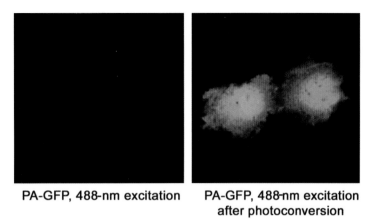

PA-GFP, 488-nm excitation

PA-GFP, 488-nm excitation
after photoconversion

FIGURE 2. Photoactivation of PA-GFP. COS 7 cells expressing PA-GFP were imaged using 488-nm excitation light before (*left*) and after (*right*) photoconversion by ~60 seconds of irradiation with a 100-W Hg^{2+} lamp through a D405/40x excitation filter and 440DCLP dichroic mirror (Chroma Technology). Fluorescence was collected using a Zeiss LSM410 confocal microscope with a 63x plan-apochromat 1.4NA objective lens. Alternatively, photoconversion can be achieved by irradiation with a laser source, such as a 413-nm krypton ion laser. (Images courtesy of George Patterson and Jennifer Lippincott-Schwartz, NIH.)

Molecular Highlighters

One of the most interesting developments in FP research has been the use of FPs as molecular highlighters (Table 2). For example, the photoactivatable GFP (PA-GFP) contains a single point mutation to the native GFP that allows photoconversion of the excitation peak from ultraviolet to blue by illumination with an ~400-nm light (Patterson and Lippincott-Schwartz 2002). Unconverted PA-GFP has an excitation peak similar to the wild-type GFP (~400 nm). After photoconversion, the excitation at 488 nm increases ~100-fold (see Fig. 2). This allows very high contrast between the unconverted and converted pools of PA-GFP and is useful for tracking the dynamics of subpopulations of molecules within a cell.

Other FPs can also be used as molecular highlighters. Three-photon excitation (<760 nm) of DsRed can convert the DsRed fluorescence from red to green (Marchant et al. 2001). This is likely due to selective photobleaching of the red chromophores in DsRed, resulting in observable fluorescence from the green state. The "Timer" variant of DsRed gradually turns from bright green (500-nm emission) to bright red (580-nm emission) over the course of several hours (Terskikh et al. 2000). The relative ratio of green to red can then be used to gather temporal data for gene expression studies. An example of how Timer can be used is shown in Figure 3. In this experiment, Timer was fused to a membrane-bound protein that is processed in the Golgi. Newly synthesized protein (green) is found only in the Golgi, whereas the plasma membrane contains protein labeled only with mature (red) Timer.

Most recently, molecular highlighters have been developed from FPs cloned from additional species. Kaede, a FP isolated from stony coral, photoconverts from green to red in the presence of ultraviolet light (Ando et al. 2002). Unlike PA-GFP, the conversion of fluorescence occurs by absorption of light that is spectrally distinct from its illumination. However, this protein is an obligate tetramer, making it less suitable for use as an epitope tag than PA-GFP. Kindling FP (KFP1) has been developed from a nonfluorescent chromoprotein from *Anemonia sulcata* (Chudakov et al. 2003) and is now commercially available from Evrogen. KFP1 is also nonfluorescent until illuminated with green light. Low-intensity light results in

TABLE 2. FP-based biosensors

Category	Variant	Comments and references
FRET-based sensors		
Ca^{++}	cameleon	K_d1 = 100 nM, K_d2 = 4.3 µM, 1.6-fold dynamic (Miyawaki et al. 1997)
	cameleon (YC6.1)	K_d = 110 nM, 2-fold dynamic range (Truong et al. 2001)
Phosphorylation	protein kinase A	Nagai et al. 2000; Zhang et al. 2001
	Src	Ting et al. 2001
	Abl	Ting et al. 2001
	EGF receptor	Ting et al. 2001
	insulin receptors	Sato et al. 2002
Miscellaneous	nitric oxide	Pearce et al. 2000
	voltage (VSFP)	Sakai et al. 2001
	cAMP	Zaccolo et al. 2000
	cGMP	Honda et al. 2001
Sensitized FPs		
Ca^{++}	camgaroo-2	Brightens with Ca^{++}, K_d =5 µM (Griesbeck et al. 2001)
	G-CaMP	Brightens with Ca^{++}, K_d = 235 nM, folds at 28°C (Nakai et al. 2001)
	pericams:	Folds at 37°C, K_d=1.2 µM (Nagai et al. 2001)
	flash	Brightens with Ca^{++}
	inverse	Dims upon binding Ca^{++}
	ratiometric	Changes excitation from 420 nm to 490 nm with Ca^{++}, 10-fold dynamic range
pH	tandem EGFP:ECFP:EYFP	Uses pH dependency of FPs to detect pH 5–8 (Llopis et al. 1998)
	pHluorins:	pKa = 7.1 (Miesenbock et al. 1998; Sankaranarayanan et al. 2000)
	ratiometric	Changes excitation from 470 nm to 410 nm in acid, 2.5-fold dynamic range
	ecliptic	Fluorescence quenches in acid
	deGFP4	pKa = 7.4, changes emission from 518 nm to 461 nm (400-nm excitation) in acid, 5-fold dynamic range (Hanson et al. 2002)
Cl$^-$	EYFP	Jayaraman et al. 2000
	tandem ECFP:EYFP	Kuner and Augustine 2000
Voltage	FlaSH	Siegel and Isacoff 1997
Molecular highlighters		
	PA-GFP	100-fold increase in fluorescence after 413-nm illumination (Patterson and Lippincott-Schwartz 2002)
	DsRed	Converts to green emission by three-photon excitation (<760 nm) (Marchant et al. 2001)
	pTimer	Converts from green (483-nm excitation; 500-nm emission) to red (558-nm excitation; 583-nm emission) over time (Terskikh et al. 2000)
	Kaede	Emission shifts from 518 nm to 582 nm (480 excitation) when exposed to ultraviolet light (Ando et al. 2002)
	KFP1	Reversible or irreversible fluorescence (580-nm excitation; 600-nm emission) using green light
		Blue light quenches reversible fluorescence (Chudakov et al. 2003)

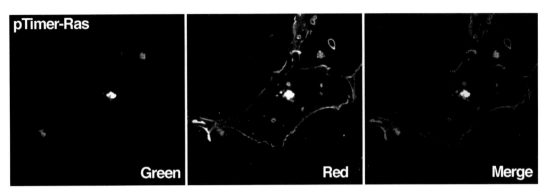

FIGURE 3. Use of Timer to track protein trafficking. Timer protein was fused to a membrane-associated protein processed in the Golgi. Newly produced protein is localized in the Golgi (green), whereas only older protein (red) is found on the plasma membrane.

a transient red fluorescence that decays over a few minutes. Illumination with blue light quenches the kindled fluorescence immediately, allowing tight control over fluorescent labeling. In contrast, high-intensity illumination results in irreversible kindling and allows for stable highlighting similar to PA-GFP. The ability to precisely control fluorescence is particularly useful when tracking particle movement in a crowded environment. For example, this approach has been successfully used to track the fate of neural plate cells in developing *Xenopus* embryos and the movement of individual mitochondria in PC12 cells (Chudakov et al. 2003).

APPLICATIONS

Common Uses of FPs in Living Cells

FPs are quite versatile and have been successfully employed in almost every biological discipline from microbiology to systems physiology. They have been extremely useful as reporters for gene expression studies in both cultured cells and whole animals. In live cells, FPs are most commonly used to track the localization and dynamics of proteins, organelles, and other cellular compartments (see Chapters 19–35). The construction and expression of GFP fusion proteins are explored in Chapter 2. In addition, methods for quantitative examination of dynamic protein localization are examined in Chapter 7. FPs can also be used to assess protein–protein interactions in living cells through the use of FRET (see Chapter 8).

The following discussion provides general tips for imaging FPs, as well as quantitative imaging of FPs and imaging several FPs at the same time. Also provided is an overview of the different types of biosensors that have been derived from FPs. Because this manual focuses primarily on live cell microscopy techniques, the discussion of FP uses is restricted to this context.

General Tips for Digital Imaging of FPs in Living Cells

The most commonly used FP, EGFP, is quite bright and thus provides a large signal-to-noise ratio. This allows flexible imaging and a high probability of success, even when the optical conditions are less than ideal. However, when the experiment demands use of one of the dimmer mutants, such as blue or cyan FPs, the signal-to-noise ratio can be considerably less and optimization of fluorescence detection is required.

Selecting an Objective Lens

Although the selection of the objective lens is often given the least amount of thought, it may be the most important consideration when preparing for an imaging experiment. The objective lens not only determines the spatial resolution, but also the amount of light collected. In addition, it may provide optical corrections that are essential (or detrimental) to the success of the experiment.

Tips for Choosing the Proper Lens for FP Imaging

1. Choose a lens with a high numerical aperture (NA). The general principle regarding any imaging collection is simple: The higher the signal-to-noise ratio, the better the image. This is particularly crucial to image collection on a laser-scanning confocal system, where the number of photons collected is often quite small. Hence, selection of the proper objective lens is particularly useful when optimizing photon collection, because the brightness of an image is proportional to the $(NA)^4/(magnification)^2$ of the lens. This principle is illustrated in Figure 4A in which a cell expressing EGFP targeted to the mitochondria is imaged over the same field of view using three different objectives: a 40x 1.3NA lens, a 63x 1.4NA lens, and a 100x 1.3NA lens. Although the number of pixels and detector conditions are identical, the cell is brightest using the 40x lens. In contrast, the cell is barely visible using the 100x lens. The 100x lens can still be used, but the detector gain must be increased, resulting in increased noise and a poorer image.

2. Avoid excessive magnification. Resolution is determined by the NA of the lens, *not* the magnification. Simply increasing the digital enlargement during image collection on a

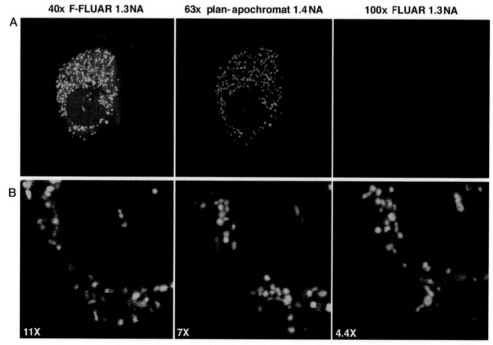

FIGURE 4. Comparison of three oil immersion objectives. (*A*) A cell expressing mitochondria-targeted EGFP is imaged with three different objective lenses and identical detection conditions on a Zeiss LSM510 confocal microscope. (*B*) The conditions in *A* are used with digital enlargement to image the same region of the cell in *A*. Note that the resolving power of the objectives is comparable, due to the similar NA of the objectives.

confocal microscope using a 40x objective results in an image equivalent to the 100x lens (Fig. 4B). The resolution of both lenses is the same because they have the same NA. This is not to suggest that increased magnification is not beneficial; in fact, choosing a high magnification objective lens can be very helpful when imaging very small objects, such as secretory granules, using a wide-field microscopy system. Because image size relative to the detector size has an important role in determining spatial sampling frequency, the optimal magnification is determined by the imaging camera (see Piston 1998 for a more in-depth discussion). Thus, the best choice of objective lens often depends on the optical requirements of the instrumentation in addition to the specific requirements of a particular experiment.

3. Avoid unnecessary correction. The third consideration concerns the optical corrections included in objectives for chromatic aberration (apochromat) and flatness of field (plan). These corrections are obtained by including more lenses inside an objective. With each additional lens, the total transmission of light decreases, which decreases the FP signal that is detected. Although these lenses are required for some specialized applications such as multicolor imaging, a good signal-to-noise ratio (and thus a good image) will be more difficult to obtain. This is especially problematic when imaging some of the dimmer fluorescent proteins, such as ECFP or HcRed. Therefore, a 40x F-FLUAR 1.3NA objective lens is preferable for imaging a single FP color. For multicolor experiments, switch to a plan-apochromat 1.4NA objective lens for the necessary color correction.

Selecting a Filter Set

When selecting a filter set to image FPs, it is helpful to keep in mind the way in which the filters will be used. For example, when using filters to distinguish fluorescence by sight, as is the case for a fluorescence dissecting microscope, the emphasis is on collecting as much light as possible from the sample. This is important because (1) the eye is relatively insensitive to incoming light, (2) it is often very helpful to be able to find the cells that express the *least* amount of a FP as possible, and (3) the eye can distinguish different colors quite accurately. Thus, the filter sets that are the most helpful for visualizing FPs by sight are very broad-passage filters that will provide abundant excitation light and fluorescence emission. For example, the Endow GFP filter set from Chroma (HQ470/40 exciter; Q495LP dichroic; HQ525/50 emitter) or a set with a long-pass emission filter is very good for the visualization of EGFP fluorescence in many microscopes.

With digital imaging, the priorities are very different. Typical wide-field and confocal microscopes have very sensitive detectors that easily detect a FP. However, these detectors do not distinguish the source of the emitted photons and detect autofluorescence in addition to FPs. Therefore, the most desirable filters for detection by charge coupled device chips and photomultipier tubes are narrow-passage filters that are specific to the emission of the FP being examined. With EGFP, for example, a narrow band-pass emission filter (HQ515/30) is very useful for filtering out contaminating autofluorescence in both wide-field and confocal experiments (Piston et al. 1999).

Quantitative Analysis of Protein Expression using FPs

Because the ratio of FP to labeled protein is 1:1, the concentration of molecules in the cell can be calculated from the total cellular fluorescence using known concentrations of recombinant FP as a reference standard. EGFP is typically used for this purpose because of its photostability and brightness. For calculation of cellular EGFP concentrations, known concentrations of EGFP can be embedded in polyacrylamide slabs and imaged alongside cells expressing EGFP-tagged constructs (Patterson et al. 1998; Piston et al. 1999). Reference standards have also been developed for calculating the density of membrane-bound proteins by binding precise quantities of His-tagged EGFP to coated beads (Chiu et al. 2001). Unfortunately, such calibration standards are not yet commercially available. As FPs are increasingly used to label molecules in whole organisms using knock-in technology, quantitative analysis of protein expression will become an increasingly useful technique.

Multicolor Imaging

The primary purpose for developing a full palette of useful FPs is to track several cellular events simultaneously. Given the number of available FPs, the best pairings may not be obvious. Recommendations here are restricted to confocal imaging (see Chapter 14), because this offers the necessary spatial precision essential for most colocalization experiments.

Imaging Multiple Colors Without Spectral Detection

Imaging multiple colors of fluorophores using filter sets requires careful consideration for filter cross talk, i.e., the spillover fluorescence of one dye into another channel. Cross talk can occur during both excitation and emission of different FPs. Cross talk between fluorophore excitations occurs toward the blue end of the spectrum, whereas cross talk between emissions occurs toward the red. For example, a green dye can often be detected through red emission filters, but a red dye cannot be seen through a green filter. For this reason, multicolor imaging proceeds with the reddest dye imaged first, using excitation wavelengths that do not cross over to the bluer dye. For example, to image EYFP and EGFP in the same cell using the Zeiss LSM510 confocal microscope, EYFP would be collected first using 514-nm excitation, which is beyond the absorption of EGFP, and then with a 530–550-nm band-pass filter. The emission filter is not very critical here, since only EYFP is excited. A second scanning should then be used to image EGFP using the 477-nm argon laser line together with a very tight 490–500-nm emission band-pass filter. This filter should be correctly positioned to exclude EYFP fluorescence and allow specific capture of EGFP. Although the peak of the 488-nm argon laser is closer to the excitation maximum of EGFP, it is too close to the emission filter band pass and reflected laser light would interfere with image collection.

   This example illustrates two points. The first is that multicolor imaging with filters requires careful consideration for cross talk at the expense of the signal-to-noise ratio (because the band-pass width of the collections is severely restricted) and temporal resolution (because two separate collection strategies are required). The second point is less obvious. For quantitative multicolor imaging, special optical designs particular to the experiment are required. Given the expense of the instrumentation and the prevalence of core facilities, the instrumentation available may not fit the needs of the experiment. Although many of the LSM510 systems are equipped with the first set of filters for collecting EYFP, the special filters for collection of EGFP in the presence of EYFP are not offered in standard systems. Therefore, the most important point when designing multicolor imaging experiments is to *ensure that the necessary equipment is available* before setting up the experiment. Although this may seem obvious, checking equipment configurations can save much time and effort, particularly when making the investment in FP experiments. This information becomes increasingly important given the rising number of available FPs and the possible number of ways of combining FPs for a given experiment. While most imaging systems are designed to be flexible, all imaging solutions cannot be accommodated on a single system.

Although the brightest FPs are green and yellow, their spectral peaks are separated by only 25 nm. Multicolor experiments using this pair are possible, but cross talk between filter sets presents a problem. As a result, green and yellow FPs are not often used together. By far, the most widely used combination is a yellow FP paired with ECFP. Although ECFP is not much brighter than EBFP, its use has two advantages: (1) excitation of ECFP with an argon laser (458 nm) is easily adapted to current confocal microscope systems and (2) ECFP is much more resistant to photobleaching than blue FPs (Patterson et al. 2001). Combining DsRed with green or yellow FPs is also a good pairing that can be easily separated, but some important considerations must be kept in mind. DsRed and DsRed-Express are recommended over first-generation DsRed because they mature significantly faster. In addition, because DsRed is an obligate tetramer, it may cause unintended effects on the biology of the protein of interest. Unfortunately, no effective rule of thumb is available to predict whether DsRed will work as a tag. Some proteins simply tolerate the DsRed tag, whereas others do not. For triple labeling using only FPs, the combination of cyan, yellow, and DsRed offers a good solution.

There are some trade-offs to using several FPs in the context of live cell imaging. For example, the rate of data acquisition slows with the addition of each color, because each color requires different collection conditions. Addressing fluorescence cross talk becomes more complicated, especially because the emission spectra of FPs tend to be quite broad. Finally, because the different FPs vary in relative brightness, each color may require different signal integration times. Cyan and blue FPs are quite dim and may thus require longer collection times that saturate the brighter green and yellow FPs. Therefore, it is important to balance the needs of the experiment (i.e., fast collection versus best separation) with the choice of FP. For example, using EYFP together with DsRed2 is a good choice for fast collection, because these two FPs can share a single excitation setting, whereas using ECFP with EYFP can provide better spectral separation.

Imaging Multiple Colors with Spectral Detection

The disadvantage of using filter sets to detect fluorescence emission is that all the photons collected by the filter are treated the same way, regardless of the source. Recently, instrumentation was developed to distinguish the source of the fluorescence based on emission spectra of the sample. Confocal imaging systems from Zeiss, Leica, and Bio-Rad are now capable of spectral imaging and separating the fluorescence emission of dyes that have overlapping spectra (see Chapter 15).

How many different colors can be used simultaneously? In theory, the number of FPs that one can spectrally separate at one time is unlimited. Figure 5 contains an example of successful separation of EBFP, ECFP, EGFP, EYFP, DsRed, and HcRed in a single cell. The successful application of linear unmixing to spectral scanning techniques changes the way in which FPs are selected for multicolor experiments. Instead of choosing combinations that enable the best separations (such as ECFP paired with EYFP or EGFP paired with dsRED), the best FPs can now be chosen. Generally, FPs should be used in the following order when designing a multicolor experiment:

1. Green (Emerald or EGFP).

2. Yellow (Venus, Citrine, or EYFP).

3. Other FPs.

As described above, the current generation of the other FPs has significant limitations. Note that in Figure 5 the images acquired for EBFP and ECFP are of a far lower quality than the EGFP and EYFP images. This is due to a low signal-to-noise ratio upon data collection

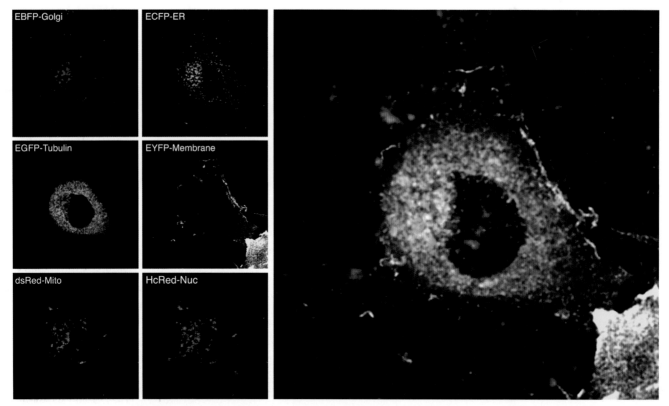

FIGURE 5. Imaging six FPs in a single cell. Six FPs that were targeted to different cellular compartments (*left*) are resolved in a single cell by using spectral detection and linear unmixing on an LSM510 META confocal microscope (Zeiss). A merged image generated from the six separated images is shown on the *right*.

and is a recurring difficulty using the existing variants of the blue and cyan FPs. The problem of hetero-oligomerization between red FPs is also demonstrated in Figure 5. Although DsRed (targeted to the mitochondria) and HcRed (targeted to the nucleus) are directed to different cellular locations, they colocalize in both the mitochondria and the nucleus. Fortunately, many of these drawbacks will be solved by development of the current set of FPs and discovery of new FPs. The number of FPs that can be used for live cell imaging will thus continue to increase.

Biosensor FPs

Basic Strategies for Development of Biosensors

A number of biosensor FPs have been developed for reporting on a wide variety of intracellular processes (see Table 2). Two strategies have been generally employed in the construction of these sensors (Fig. 6). The first strategy takes advantage of the sensitivity of FRET to changes in molecular distances and orientation (see Chapter 8). Typically, two FPs (usually ECFP and EYFP) are inserted at opposing ends of a sensor protein. Changes in the conformation of the sensor protein cause changes in the FRET between the FP pairing. A ratiometric change in the fluorescence output of the two FPs using a single excitation parameter is typically observed. The second strategy uses the structure of the GFP molecule itself. As discussed above, the fluorescent properties of a FP are greatly dependent on the interaction

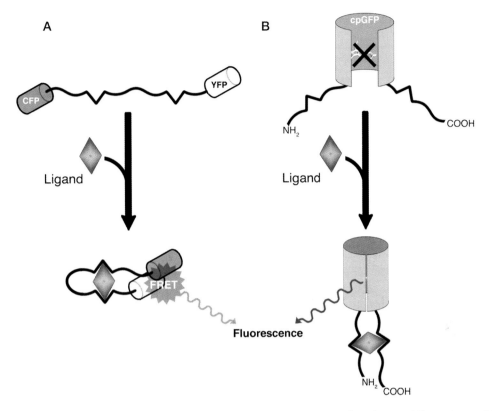

FIGURE 6. Construction of biosensors using FPs. Two basic strategies have been used to generate biosensors using FPs. In the FRET-based approach *(A)*, matched FPs (for example, CFP and YFP) are placed on the ends of a biosensor region. When the biosensor region associates with a ligand, the conformation of the probe is altered and brings the two FPs closer together in space. This association results in increased FRET. Alternatively, a sensitized FP can be constructed using a variety of approaches that tie the fluorescence of a single FP to its environment. Commonly, a biosensor region is incorporated into the structure of an FP *(B)*. Association of the biosensor region with ligand results in a change in the conformation of the FP and a change in the properties of its fluorescence.

between the fluorophore and the enveloping amino acids. Changes in the packing of the fluorophore, either physically or by mutation, can have dramatic effects on its fluorescent properties. As a result, a wide variety of sensors have been developed that manipulate the conformation of a single FP to sensitize its fluorescence to biological events. Typically, sensitized FPs are constructed either through strategic incorporation of a conformationally sensitive protein domain into the FP or through mutagenesis of the FP itself.

Ca++ Sensors

Ca++ indicators were among the first FP-derived biosensors. The first attempt, named cameleon, used a FRET-based approach (Miyawaki et al. 1997). Initially, a single *Xenopus* calmodulin and the Ca++ calmodulin-binding domain of myosin light chain kinase (M13 domain) were flanked by either EBFP:EGFP or ECFP:EYFP pairs. The M13 domain only binds the calmodulin in the presence of Ca++, resulting in increased FRET. Because of the two Ca++ binding sites on calmodulin, these cameleon probes have a biphasic response to changes in intracellular Ca++ over a large concentration range of 10 nM–100 μM. However, the maximum change in the FRET ratio is only 1.6-fold, making these probes useless for most cell biological applications. More recently, cameleon was redesigned using a structure-based approach (Truong et al. 2001). The Ca^{++} sensing domain of the original construct was replaced by a new sensing domain comprised of the calmodulin-binding domain of calmodulin-dependent protein kinase kinase inserted in the middle of calmodulin. This improves the dynamic range of the probe and allows sensing of Ca^{++} concentrations between 50 nM and 1 μM. Despite these improvements, the dynamic range is limited to a twofold maximum, which is insufficient for most experiments.

An alternative approach has yielded better results. Several sensitized FPs were generated by inserting Ca^{++}-sensitive domains at the beginning of the seventh β strand of the β-can structure (Baird et al. 1999; Greisbeck et al. 2001; Nagai et al. 2001; Nakgai et al. 2001). This particular site is quite tolerant of insertions that do not destroy maturation of fluorescence (Baird et al. 1999). The latest generation of the "camgaroos," which contains a calmodulin insertion, has a K_d for Ca^{++} ~5 μM and an approximately sixfold increase in fluorescence intensity upon binding Ca^{++} (Griesbeck et al. 2001). Another strategy uses the unique barrel structure of GFP to rearrange the ends of the FP by linking the natural termini and creating a new start codon in the middle of the FP sequence (Baird et al. 1999). The new termini of this circularly permuted FP can then be linked to biosensor elements such as calmodulin and the M13 domain (Nagai et al. 2001; Nakai et al. 2001). The first of these probes is based on the EGFP structure (G-CaMP) and requires incubation at 28°C for maturation of fluorescence (Nakai et al. 2001). Probes based on EYFP have produced better results and are more amenable for work with mammalian cells because fluorescence maturation can occur at 37°C (Nagai et al. 2001). One of these sensors, ratiometric pericam, changes its excitation peak from 420 nm to 490 nm upon binding Ca^{++} (Nagai et al. 2001). Overall, this probe seems to possess the best combination of affinity (K_d = 1.2 μM) and dynamic range (a maximum of a tenfold increase in ratios). However, the sensitivity of these probes pales in comparison to the commonly used synthetic probes such as fluo-3 (K_d ~400 nM; >40-fold increase in brightness) (Minta et al. 1989), and they are not as useful outside of the special advantages inherent to genetically encoded indicators.

pH Sensors

There are two approaches for using FPs to sense changes in pH. The first strategy uses the pH sensitivity of the second-generation FPs (Patterson et al. 2001). Because the fluorescence of EGFP (fluorescence pKa = 5.9) and EYFP (pKa = 6.5) becomes quenched in acidic pH, fusion of these FPs with a second, less pH-sensitive FP such as ECFP (pKa = 4.7) or DsRed results in the creation of a ratiometric probe that can be used to measure the pH of cellular compartments (Llopis et al. 1998). In a similar application, yellow FPs were targeted to secretory granules to effectively mark the point of vesicle fusion with the membrane. Because the lumen of secretory granules tends to be very acidic, yellow fluorescence is suppressed until a fusion pore opens. On contact with the neutral extracellular solution, the brightness increases, resulting in a burst of fluorescence (Han et al. 2002). Although the use of yellow FPs as pH sensors is useful and relatively straightforward, fluorescence quenching is not specific to pH. For example, halide ions such as Cl$^-$ also quench yellow FP fluorescence (Jayaraman et al. 2000). In fact, identical strategies have been employed to sense Cl$^-$ transients (Kuner and Augustine 2000). Therefore, this approach is better suited for detecting relative changes in pH than for quantification of pH.

The second approach is derived from the effect of chromophore protonation on the native GFP chromophore (Fig. 1). Deprotonation of the chromophore results in a shift of the excitation peak from 395 nm to 475 nm. As a result of this demonstrated effect of proton sensitivity, attempts were made to create ratiometric sensors through mutagenesis. The first successful varieties resulted in the creation of two probes with pKa of ~7.1; these are named pHluorins (Miesenbock et al. 1998; Sankaranarayanan et al. 2000). The excitation peak of ratiometric pHluorin shifts from 470 nm to 410 nm as the pH acidifies, whereas ecliptic pHluorin becomes nonfluorescent in acidic pH. More recently, a dual-emission pH sensor (deGFP4) was developed with pKa of 7.4 (Hanson et al. 2002). As the pH acidifies, the ratio of 518-nm emission to 461-nm emission decreases. The dynamic range of this probe seems to be the largest of the currently described probes and will probably be the most useful.

Phosphorylation

Unlike the biosensors described above, no currently available synthetic probe can directly report kinase activity in real time. But, the FRET-based detectors of phosphorylation are one of the more important classes of FP-based biosensors developed to date (Nagai et al. 2000; Ting et al. 2001; Zhang et al. 2001; Sato et al. 2002). These probes generally consist of a single construct containing a phosphorylation motif specific to a class of kinases and a phosphopeptide binding domain flanked by cyan and yellow FPs on the ends. When the construct is phosphorylated by a kinase, the phosphopeptide binding domain binds the phosphorylated sequence. This results in bringing the two terminal FPs very close to one another in space, allowing energy transfer. Remarkably, this simple strategy has proven to be specific, robust, and easily adaptable to a growing variety of both Ser/Thr (Nagai et al. 2000; Zhang et al. 2001) and Tyr kinases (Ting et al. 2001; Sato et al. 2002). The drawback of these sensors, however, is their small dynamic range. Similar to the other intramolecular FRET-based sensors such as cameleon (see above), the conformational change of the probes typically results in less than a 50% change in the FRET ratio. While still resolvable, the small contrast of these probes can make their use rather challenging.

Protease Cleavage Sensors

One of the best-described applications of FP-based sensors is the FRET-based protease cleavage assay. In this assay, FRET pairs are conjugated by a linker that incorporates a consensus protease cleavage site. These constructs typically display very strong energy transfer that disappears entirely upon cleavage of the linker sequence. This assay features high contrast and is generally very robust. It has been used successfully to measure the activity of a large number of proteases. In particular, a number of constructs have been generated that measures the activation of the caspase family of proteases during apoptosis using both EBFP:EGFP and ECFP:EYFP FRET pairs (Xu et al. 1998; Mahajan et al. 1999). The ECFP:EYFP pairing has been particularly popular and is now commercially available from BD Biosciences Clontech. However, a note of caution is in order when using the ECFP:EYFP pairing for making measurements in harsh conditions such as those that occur with apoptosis. A recent comparison between caspase-3 probes based on ECFP:EYFP and ECFP:Venus suggested strong disadvantages to using the ECFP:EYFP pairing (Takemoto et al. 2003). Although the construction of the linker sequences was identical, the disappearance of sensitized emission from the ECFP:EYFP pairing occurred much earlier during apoptosis than with the ECFP:Venus pairing. Although the relevance of these findings remains to be determined, the importance of considering the environmental sensitivity of FPs when designing experiments cannot be stressed enough.

Aside from the occasional caveat, the FP-based protease assay offers tremendous advantages over assays based on traditional fluorescent dyes. One of the best examples of these advantages used a protease sensor for calpain targeted to the cell membrane by fusion with a subcellular localization sequence (Vanderklish et al. 2000). This simple experiment permitted examination of enzymatic activity in distinct cellular compartments with a spatial specificity that is not possible with conventional fluorescent probes.

More Sensors

The number of other sensors using both sensitized FPs and the FRET-based approach is increasing. Despite limitations in the dynamic range of FRET-based biosensors, this strategy seems to be the most widely used so far, probably because of its inherent ratiometric design and the ease of probe construction. This approach has already been successfully used to detect a wide array of "agonists," including nitric oxide (Pearce et al. 2000), cyclic nucleotides (Zaccolo et al. 2000; Honda et al. 2001), and voltage (Sakai et al. 2001). This list will certainly grow as this proven strategy is applied to other interesting biological questions.

New strategies for developing biosensors are also being developed. For example, a novel protein–protein interaction assay has been developed by splitting EYFP into two parts (Hu et al. 2002). This so-called bimolecular fluorescence complementation assay involves attaching complementary portions of EYFP to each of two interacting proteins. Stable interaction of the target proteins results in formation of the EYFP structure and maturation of fluorescence. Although maturation occurs with a half-life on an hourly timescale, this assay has been successfully used to examine the subcellular localization of interacting transcription factor complexes in living cells. This powerful technique is yet another example of the flexibility of FPs in building molecular probes. Additional strategies for employing FPs in genetically encoded biosensors will certainly be developed.

FUTURE DIRECTIONS

Currently, the focus of FP development is centered on two basic goals. The first is to perfect the current palette of *A. victoria*-derived FPs. The second aim is to obtain a useful red FP. Progress toward these goals has been quite impressive. The latest generation of *A. victoria* variants has solved most of the deficiencies of the first generation of FPs, particularly for the yellow and green FPs. The search for a monomeric, bright, and fast-maturing red FP has introduced several new and interesting classes of FP, particularly from species of coral. Development of existing FPs, together with new technologies such as insertion of unnatural amino acids (Wang et al. 2003), will further expand the color palette. As spectral separation techniques become better developed and more commonly employed, these newer varieties will supplement the existing palette, especially in the yellow and red regions of the spectrum.

The current trend of fluorescent probes is to expand the role of dyes that fluoresce into the far red. In mammalian cells, both autofluorescence and absorption of light by tissues are greatly reduced at the red end of the spectrum. Thus, the development of far-red fluorescent probes would be extremely useful for the examination of thick specimens and whole animals. Given the success of FPs as reporters in transgenic systems, the development of far-red FPs for use in whole organisms will become increasingly important in the coming years.

In addition, the large potential for FP use in the engineering of biosensors is just now being realized. The number of biosensor constructs is rapidly growing. By using structural information, development of these probes has led to improved sensitivity and will continue to do so. The success of these endeavors certainly suggests that almost any biological parameter will be measurable using the appropriate FP-based biosensor.

REFERENCES

Ando R., Hama H., Yamamoto-Hino M., Mizuno H., and Miyawaki A. 2002. An optical marker based on the UV-induced green-to-red photoconversion of a fluorescent protein. *Proc. Natl. Acad. Sci.* **99:** 12651–12656.

Baird G.S., Zacharias D.A., and Tsien R.Y. 1999. Circular permutations and receptor insertion within green fluorescent proteins. *Proc. Natl. Acad. Sci.* **96:** 11241–11246.

———. 2000. Biochemistry, mutagenesis, and oligomerization of DsRed, a red fluorescent protein from coral. *Proc. Natl. Acad. Sci.* **97:** 11984–11989.

Bevis B.J. and Glick B.S. 2002. Rapidly maturing variants of the Discosoma red fluorescent protein (DsRed). *Nat. Biotechnol.* **20:** 83–87.

Campbell R.E., Tour O., Palmer A.E., Steinbach P.A., Baird G.S., Zacharias D.A., and Tsien R.Y. 2002. A monomeric red fluorescent protein. *Proc. Natl Acad. Sci.* **99:** 7877–7882.

Chalfie M., Tu Y., Euskirchen G., Ward W.W., and Prasher D.C. 1994. Green fluorescent protein as a marker for gene expression. *Science* **263:** 802–805.

Chiu C.S., Kartalov E., Unger M., Quake S., and Lester H.A. 2001. Single-molecule measurements calibrate green fluorescent protein surface densities on transparent beads for use with "knock-in" animals and other expression systems. *J. Neurosci. Methods* **105:** 55–63.

Chudakov D.M., Belousov V.V., Zaraisky A.G., Novoselov V.V., Staroverov D.B., Zorov D.B., Lukyanov S., and Lukyanov K.A. 2003. Kindling fluorescent proteins for precise in vivo photolabeling. *Nat. Biotechnol.* **21:** 191–194.

Cormack B., Valdivia R.H., and Falkow S. 1996. FACS-optimized mutants of the green fluorescent protein (GFP). *Gene* **173:** 33–38.

Griesbeck O., Baird G.S., Campbell R.E., Zacharias D.A., and Tsien R.Y. 2001. Reducing the environmental sensitivity of yellow fluorescent protein. *J. Biol. Chem.* **276:** 29188–29194.

Gurskaya N.G., Fradkov A.F., Terskikh A., Matz M.V., Labas Y.A., Martynov V.I., Yanushevich Y.G., Lukyanov K.A., and Lukyanov SA. 2001. GFP-like chromoproteins as a source of far-red fluorescent proteins. *FEBS Lett.* **507:** 16–20.

Han W., Li D., and Levitan E.S. 2002. A new green fluorescent protein construct for localizing and quantifying peptide release. *Ann. N.Y. Acad. Sci.* **971:** 627–633.

Hanson G.T., McAnaney T.B., Park E.S., Rendell M.E.P., Yarbrough D.K., Chu S., Xi L., Boxer S.G., Montrose M.H., and Remington S.J. 2002. Green fluorescent protein variants as ratiometric dual emission pH sensors. 1. Structural characterization and preliminary application. *Biochemistry* **41:** 15477–15488.

Heim R. and Tsien R.Y. 1996. Engineering green fluorescent protein for improved brightness, longer wavelengths and fluorescence resonance energy transfer. *Curr. Biol.* **6:** 178–182.

Heim R., Cubitt A.B., and Tsien R.Y. 1995. Improved green fluorescence. *Nature* **373:** 663–664.

Heim R., Prasher D.C., and Tsien R.Y. 1994. Wavelength mutations and posttranslational autoxidation of green fluorescent protein. *Proc. Natl. Acad. Sci.* **91:** 12501–12504.

Honda A., Adams S.R., Sawyer C.L., Lev-Ram V., Tsien R.Y., and Dostmann W.R. 2001. Spatiotemporal dynamics of guanosine 3′,5′-cyclic monophosphate revealed by a genetically encoded, fluorescent indicator. *Proc. Natl. Acad. Sci.* **98:** 2437–2442.

Hu C.D., Chinenov Y., and Kerppola T.K. 2002. Visualization of interactions among bZIP and Rel family proteins in living cells using bimolecular fluorescence complementation. *Mol. Cell* **9:** 789–798.

Jayaraman S., Haggie P., Wachter R.M., Remington S.J., and Verkman A.S. 2000. Mechanism and cellular applications of a green fluorescent protein-based halide sensor. *J. Biol. Chem.* **275:** 6047–6050.

Knop M., Barr F., Riedel C.G., Heckel T., and Reichel C. 2002. Improved version of the red fluorescent protein (drFP583/DsRed/RFP). *BioTechniques* **33:** 592–602.

Kuner T. and Augustine G.J. 2000. A genetically encoded ratiometric indicator for chloride: Capturing chloride transients in cultured hippocampal neurons. *Neuron* **27:** 447–459.

Llopis J., McCaffery J.M., Miyawaki A., Farquhar M.G., and Tsien R.Y. 1998. Measurement of cytosolic, mitochondrial, and Golgi pH in single living cells with green fluorescent proteins. *Proc. Natl. Acad. Sci.* **95:** 6803–6808.

Mahajan N.P., Harrison-Shostak D.C., Michaux J., and Herman B. 1999. Novel mutant green fluorescent protein protease substrates reveal the activation of specific caspases during apoptosis. *Chem. Biol.* **6:** 401–409.

Marchant J.S., Stutzmann G.E., Leissring M.A., LaFerla F.M., and Parker I. 2001. Multiphoton-evoked color change of DsRed as an optical highlighter for cellular and subcellular labeling. *Nat. Biotechnol.* **19:** 645–649.

Matz M.V., Fradkov A.F., Labas Y.A., Savitsky A.P., Zaraisky A.G., Markelov M.L., and Lukyanov S.A. 1999. Fluorescent proteins from nonbioluminescent Anthozoa species. *Nat. Biotechnol.* **17:** 969–973.

Miesenbock G., De Angelis D.A., and Rothman J.E. 1998. Visualizing secretion and synaptic transmission with pH-sensitive green fluorescent proteins. *Nature* **394:** 192–195.

Minta A., Kao J.P., and Tsien R.Y. 1989. Fluorescent indicators for cytosolic calcium based on rhodamine and fluorescein chromophores. *J. Biol. Chem.* **264:** 8171–8178.

Miyawaki A., Llopis J., Heim R., McCaffery J.M., Adams J.A., Ikura M., and Tsien R.Y. 1997. Fluorescent indicators for Ca2+ based on green fluorescent proteins and calmodulin. *Nature* **388:** 882–887.

Nagai T., Sawano A., Park E.S., and Miyawaki A. 2001. Circularly permuted green fluorescent proteins engineered to sense Ca2+. *Proc. Natl. Acad. Sci.* **98:** 3197–3202.

Nagai T., Ibata K., Park E.S., Kubota M., Mikoshiba K., and Miyawaki A. 2002. A variant of yellow fluorescent protein with fast and efficient maturation for cell-biological applications. *Nat. Biotechnol.* **20:** 87–90.

Nagai Y., Miyazaki M., Aoki R., Zama T., Inouye S., Hirose K., Iino M., and Hagiwara M. 2000. A fluorescent indicator for visualizing cAMP-induced phosphorylation in vivo. *Nat. Biotechnol.* **18:** 313–316.

Nakai J., Ohkura M., and Imoto K. 2001. A high signal-to-noise Ca^{2+} probe composed of a single green fluorescent protein. *Nat. Biotechnol.* **19:** 137–141.

Ormö M., Cubitt A.B., Kallio K., Gross L.A., Tsien R.Y., and Remington S.J. 1996. Crystal structure of the *Aequorea victoria* green fluorescent protein. *Science* **273:** 1392–1395.

Patterson G.H. and Lippincott-Schwartz J. 2002. A photoactivatable GFP for selective photolabeling of proteins and cells. *Science* **297:** 1873–1877.

Patterson G., Day R.N., and Piston D. 2001. Fluorescent protein spectra. *J. Cell Sci.* **114:** 837–838.

Patterson G.H., Knobel S.M., Sharif W.D., Kain S.R., and Piston D.W. 1997. Use of the green fluorescent protein and its mutants in quantitative fluorescence microscopy. *Biophys. J.* **73:** 2782–2790.

Patterson G.H., Schroeder S.C., Bai Y., Weil A., and Piston D.W. 1998. Quantitative imaging of TATA-binding protein in living yeast cells. *Yeast* **14:** 13–25.

Pearce L.L., Gandley R.E., Han W., Wasserloos K., Stitt M., Kanai A.J., McLaughlin M.K., Pitt B.R, and Levitan E.S. 2000. Role of metallothionein in nitric oxide signaling as revealed by green fluorescent fusion protein. *Proc. Natl. Acad. Sci.* **97:** 477–482.

Piston D.W. 1998. Choosing objective lenses: The importance of numerical aperture and magnification in digital optical microscopy. *Biol. Bull.* **195:** 1–4.

Piston D.W., Patterson G.H., and Knobel S.M. 1999. Quantitative imaging of the green fluorescent protein (GFP). *Methods Cell Biol.* **58:** 31–48.

Prasher D.C., Eckenrode V.K., Ward W.W., Prendergast F.G., and Cormier M.J. 1992. Primary structure of the *Aequorea victoria* green fluorescent protein. *Gene* **111:** 229–233.

Sakai R., Repunte-Canonigo V., Raj C.D., and Knopfel T. 2001. Design and characterization of a DNA-encoded, voltage-sensitive fluorescent protein. *Eur. J. Neurosci.* **13:** 2314–2318.

Sankaranarayanan S., De Angelis D., Rothman J.E., and Ryan T.A. 2000. The use of pHluorins for optical measurements of presynaptic activity. *Biophys. J.* **79:** 2199–2208.

Sato M., Ozawa T., Inukai K., Asano T., and Umezawa Y. 2002. Fluorescent indicators for imaging protein phosphorylation in single living cells. *Nat. Biotechnol.* **20:** 287–294.

Sawano A. and Miyawaki A. 2000. Directed evolution of green fluorescent protein by a new versatile PCR strategy for site-directed and semi-random mutagenesis. *Nucleic Acids Res.* **28:** E78.

Shimomura O. 1998. The discovery of green fluorescent protein. In *Green fluorescent protein: Properties, applications, and protocols* (ed. M. Chalfie and S. Kain), Chap. 1. Wiley-Liss, New York.

Shimomura O., Johnson F.H., and Saiga Y. 1962. Extraction, purification, and properties of aequorin, a bioluminescent protein from the luminous hydromedusan *Aequorea*. *J. Cell. Comp. Physiol.* **59:** 223–239.

Siegel M.S. and Isacoff E.F. 1997. A genetically encoded optical probe of membrane voltage. *Neuron* **19:** 735–741.

Takemoto K., Nagai T., Miyawaki A., and Miura M. 2003. Spatio-temporal activation of caspase revealed by indicator that is insensitive to environmental effects. *J. Cell. Biol.* **160:** 235–243.

Terskikh A., Fradkov A., Ermakova G., Zaraisky A., Tan P., Kajava A.V., Zhao X., Lukyanov S., Matz M., Kim

S., Weissman I., and Siebert P. 2000. "Fluorescent Timer": Protein that changes color with time. *Science* **290:** 1585–1588.

Ting A.Y., Kain K.H., Klemke R.L., and Tsien R.Y. 2001. Genetically encoded fluorescent reporters of protein tyrosine kinase activities in living cells. *Proc. Natl. Acad. Sci.* **98:** 15003–15008.

Truong K., Sawano A., Mizuno H., Hama H., Tong K.I., Mal T.K., Miyawaki A., and Ikura M. 2001. FRET-based in vivo Ca2+ imaging by a new calmodulin-GFP fusion molecule. *Nat. Struct. Biol.* **8:** 1069–1073.

Vanderklish P.W., Krushel L.A., Holst B.H., Gally J.A., Crossin K.L., and Edelman G.M. 2000. Marking synaptic activity in dendritic spines with a calpain substrate exhibiting fluorescence resonance energy transfer. *Proc. Natl. Acad. Sci.* **97:** 2253–2258.

Wang L., Xie J., Deniz A.A., and Schultz P.G. 2003. Unnatural amino acid mutagenesis of green fluorescent protein. *J. Org. Chem.* **68:** 174–176.

Ward W.W. 1998. Biochemical and physical properties of green fluorescent. In *Green fluorescent protein: Properties, applications, and protocols* (ed. M. Chalfie and S. Kain), Chap. 3. Wiley-Liss, New York.

Xu X., Gerard A.L., Huang B.C., Anderson D.C., Payan D.G., and Luo Y. 1998. Detection of programmed cell death using fluorescence energy transfer. *Nucleic Acids Res.* **26:** 2034–2035.

Yang T.T., Cheng L., and Kain S.R. 1996. Optimized codon usage and chromophore mutations provide enhanced sensitivity with the green fluorescent protein. *Nucleic Acids Res.* **24:** 4592–4593.

Zaccolo M., De Giorgi F., Cho C.Y., Feng L., Knapp T., Negulescu P.A., Taylor S.S., Tsien R.Y., and Pozzan T. 2000. A genetically encoded, fluorescent indicator for cyclic AMP in living cells. *Nat. Cell Biol.* **2:** 25–29.

Zacharias D.A., Violin J.D., Newton A.C., and Tsien R.Y. 2002. Partitioning of lipid-modified monomeric GFPs into membrane microdomains of live cells. *Science* **296:** 913–916.

Zhang J., Ma Y., Taylor S.S., and Tsien R.Y. 2001. Genetically encoded reporters of protein kinase A activity reveal impact of substrate tethering. *Proc. Natl. Acad. Sci.* **98:** 14997–15002.

Zimmermann T., Rietdorf J., Girod A., Georget V., and Pepperkok R. 2002. Spectral imaging and linear unmixing enables improved FRET efficiency with a novel GFP2-YFP FRET pair. *FEBS Lett.* **531:** 245–249.

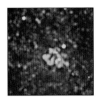

Constructing and Expressing GFP Fusion Proteins

David L. Spector[a] and Robert D. Goldman[b]

[a]Cold Spring Harbor Laboratory, One Bungtown Road, Cold Spring Harbor, New York 11724
and [b]Northwestern University Medical School, Department of Cell and
Molecular Biology, Chicago, Illinois 60611

INTRODUCTION

GFP (GREEN FLUORESCENT PROTEIN) FUSION PROTEINS have been used to address a wide range of questions in individual cells, as well as in tissues of a particular organism (see Chapters 19–35). However, investigators must take extreme care in using GFP and its derivatives—cyan fluorescent protein (CFP), yellow fluorescent protein (YFP), etc.; see Chapter 1—to ensure that the resultant fusion protein is expressed at or close to the endogenous level of the parent protein, and that it is full length, localizes correctly, and behaves normally once incorporated in the cell. As the GFP protein itself is 27 kD, one must consider the potential effects of placing such a large tag in association with a protein under investigation. This chapter discusses how these goals can be achieved and provides examples to assist the investigator in designing and implementing experiments using GFP fusion proteins.

CONSTRUCTING FUSION PROTEINS

Vectors are commercially available that contain multiple cloning sites which allow a cDNA or genomic sequence of interest to be placed in-frame at the carboxyl or amino terminus of the GFP-coding region (e.g., BD Biosciences Clontech; Qbiogene; Stratagene). The choice of which vector to use will depend on the location of critical regions of interaction or folding in the protein under study. For example, if a protein of interest has a critical interaction domain at its carboxyl terminus, a logical decision would be to place GFP at the amino terminus of the coding sequence. However, from practical experience, the logical choice is not always the correct choice and so it may be beneficial to prepare amino- and carboxy-terminal fusions simultaneously and then to assess both fusions for correct localization and associations as described below. In addition to amino- and carboxy-terminal fusions, GFP has been placed internally within a protein's coding sequence, and it has also been used in a bimolecular fluorescence complementation (BiFC) assay for determination of the locations of protein interactions in living cells (Hu et al. 2002). In the case of internal placement, the mouse major histocompatibility complex (MHC) class I allele H2L^d was labeled with GFP placed between the α3 and transmembrane domains of the native protein (Marguet et al. 1999; Rocheleau et al. 2003). GFP was flanked by amino- and carboxy-terminal linkers, each containing six amino acid residues (PGSIAT and LGMDEL, respectively); for further discussion of linkers, see below. In the BiFC approach, nonfluorescent fragments of YFP or other fluorescent proteins were identified that could reconstitute the fluorophore only when brought together by interactions between proteins covalently linked to each fragment (Hu et al. 2002; Hu and Kerppola 2003).

Commercially available vectors are usually supplied with cytomegatovirus (CMV) or simian virus-40 (SV40) promoters, both of which are strong and drive high-level expression. In particular cases when low levels of expression are required, it may be necessary to replace such promoters with weaker promoters such as SV2 (Tsukamoto et al. 2000; Janicki et al. 2004) or endogenous promoter sequences. In addition, commercial vectors are available that use the murine RNA polymerase II promoter or the murine phosphoglycerate kinase promoter for expression in ES cells (Qbiogene, Inc.). In the case of proteins whose constitutive expression may not be tolerated in cells, vectors are available that allow for inducible expression of fusion proteins (Rubinchik et al. 2000) (Qbiogene, Inc.). Inducible expression may also be useful in order to study spatial and temporal changes in localization at various time points post induction (Ahmad and Henikoff 2002).

An additional consideration in the development of GFP fusion proteins is the correct folding of both the protein of interest and GFP. Many investigators have utilized a linker sequence between GFP and the multicloning site to allow for folding of GFP and the protein of interest without steric hindrance. This approach can also be useful in cases where GFP may sterically interfere with interactions between the protein of interest and its functional partners. A variety of linker sequences have been utilized. Expression of actin GFP in *Saccharomyces cerevisiae* was accomplished by using the normal actin promoter, followed by the entire actin-coding sequence, a segment in-frame encoding ten alanine residues, followed by the GFP sequence (Doyle and Botstein 1996). Other workers have utilized ten glycine-alanine repeats fused to the ATG codon of the GFP sequence (Grava et al. 2000) or a flexible hydrophilic linker (GEGQGQGQGPGRGYAYRS) between the gene of interest and GFP (Leonhardt et al. 2000). In addition, vectors are commercially available that contain a six-amino-acid (GGGGGA) flexible linker (Qbiogene, Inc.).

EXPRESSING THE FUSION PROTEIN

GFP fusion proteins can be transiently or stably expressed. Transient expression will result in high levels of expression from plasmids that are not integrated into the cellular genome, whereas stable expression, in which genes are integrated into chromosomal DNA and maintained with the use of selectable markers (e.g., G418), can be selected to result in more native levels of expression. Introduction of plasmid sequences into cells can be accomplished by electroporation, chemical treatment (CaPO$_4$), various lipofection reagents purchased as kits (e.g., FuGENE 6, Roche Applied Science; Lipofectamine 2000, Invitrogen; DOTAP, Biontex Inc.), retroviral transformation (see Chapter 3), or microinjection (see Chapter 4). Irrespective of the approach taken, optimal conditions must be determined for individual cell types. We routinely use electroporation and have worked out conditions such that GFP fusion proteins can be detected in human U2OS cells as soon as 2 hours postelectroporation using a Gene Pulser II (Bio-Rad) operating at 170 V, 950 µF (Janicki et al. 2004), thereby allowing for low levels of protein expression. The level of expression of a particular fusion protein can vary depending on the amount of plasmid DNA transfected, the promoter strength, and the time of expression. These conditions should be optimized to obtain a low level of expression that correctly mimics the localization of the native protein and provides enough fluorescence signal to perform time-lapse studies.

Although transient expression is quick and can provide informative results, in many cases, it is beneficial and/or essential to develop stable cell lines expressing the fusion protein of interest. In addition to providing more native levels of expression, individual clones can be generated from single cells, the integration site of the plasmid mapped, and copy number determined; and, because every cell in the population is expressing the fusion protein, cell cycle analyses and biochemical fractionation are significantly easier to accomplish. The following protocol, developed by Margaret Wheelock (University of Nebraska) and modified by Paula Bubulya in the Spector laboratory, can be used for the development of stable cell lines expressing GFP fusion proteins.

DEVELOPMENT OF STABLE MAMMALIAN CELL LINES EXPRESSING GFP FUSION PROTEINS

Before beginning this protocol, the following issues must be considered and the reagents prepared and organized.

- **DNA for transfection:** Prepare DNA by the $CsCl_2$ method or use commercially available kits (i.e., Qiagen). The amount of DNA used in transfection varies depending on the transfection method used and should be optimized by transient transfection to obtain appropriate levels of protein expression prior to attempting stable transfection.

- **Antibiotic for selection of stable clones:** The appropriate concentration of antibiotic used for selection purposes will vary depending on the antibiotic and the cell type being investigated. Before beginning the stable transfection, determine the amount of antibiotic that will kill most of the cells in a dish within 2–4 days. Some antibiotics act quickly (puromycin), whereas others take several days (Geneticin/G418).

- **Screening protocol: Plan ahead.** Stable clones will need to be screened and frozen stocks will have to be prepared for approximately 4–8 weeks after transfection. For GFP fusion proteins, cells will be scored for expression by fluorescence microscopy. As it is common to obtain clones that express a truncated construct, positive clones should be further screened by immunoblotting to verify the correct size of the fusion construct.

- **Cells for transfection:** Split cells 2 days prior to the day of transfection to stimulate cell growth. Plate cells for transfection on the day before the transfection. For electroporation, aim for an 80% confluent dish of cells to use the following day. For transfection protocols that require attached cells, such as calcium phosphate or lipid-mediated transfection, aim for a 10–20% or 20–40%, respectively, confluent dish of cells to use the following day. When transfecting cells that grow very rapidly, such as baby hamster kidney (BHK) cells, use fewer cells per dish.

Transfection Procedure

Although optimal transfection procedures (e.g., calcium phosphate, electroporation, or FuGENE 6, Roche Applied Science) vary depending on cell type, this general transfection procedure has been successful for stable transfection of HeLa, A-431, U2OS, BHK, and HT1080 cells.

Day 1

1. Prepare the DNA according to manufacturers instructions for the transfection procedure.

   For electroporation of HeLa cells, this protocol has been successful with a mixture of 2 μg of target DNA and 20–40 μg of sheared salmon sperm DNA (Amresco) added directly to the cuvette.

2. Transfect the cells.

   For transfections that require attached cells: Prepare DNA according to the specific transfection protocol and add it directly to the cells. Incubate the cells for 24 hours.

   For electroporation: Trypsinize the cells for the minimum time necessary for removal from culture dishes; suspend and gently centrifuge the cells in a 15-ml conical tube to obtain a cell pellet. Each plate of 80% confluent cells can be used for four separate transfections. For example, resuspend the pellet in 800 μl of culture medium and add 200 μl of cell suspension to each cuvette containing DNA. Mix the cells and DNA. Adjust settings on the eletroporator (240 V; 950 μF for HeLa cells) and electroporate.

3. Allow the cells to sit for 2 minutes at room temperature following electroporation, and then add 1 ml of culture medium to the cuvette and mix well to break apart clusters of dead cells. Plate cells directly into a 100-mm dish with 10 ml of culture medium and incubate the cells for 24 hours at 37°C.

Day 2

4. Remove the culture medium from transfected cells. Wash the cells three times with sterile phosphate-buffered saline (PBS) and add fresh medium. Incubate the cells for 24 hours at 37°C.

   For very fast-growing cells such as BHK cells, each step of this procedure can be performed ahead of schedule to prevent the cells from becoming too confluent.

Day 3

5. Trypsinize the cells long enough to obtain a single-cell suspension. Prepare a total of ten 100-mm dishes each containing 8 ml of medium. It is a good idea to reuse the original dish, as some transfected cells will be left behind. Resuspend the pellet of transfected cells in 10 ml of medium and plate 1 ml per dish. Incubate the cells for 24 hours at 37°C.

Day 4

6. Prepare 10 ml of a 10X stock (in medium) of the appropriate antibiotic used for selection, and then add 1 ml of 10X antibiotic near the edge of each dish, swirling to mix well. Monitor the progress of the transfection every few days by observing the amount of cell death and the color of the culture medium; yellow medium will be due to cell overgrowth. If the medium becomes very dense with cell debris, change to fresh antibiotic medium after 4–7 days.

Days 14–21 (approximately)

7. Colonies of stably transfected cells should have appeared during this time. View the colonies by *carefully* holding the plate up to ceiling lights.

8. Allow the colonies to reach a diameter of ~4 mm, and then transfer them to 24-well plates. Mark the position of each colony with permanent marker on the bottom of the dish. Choose about ten colonies per plate for screening, trying to choose only colonies

that are not in contact with other colonies. Of course, the more clones screened, the better the chance of obtaining a positive clone or several clones with different levels of fusion protein expression.

> The colonies can be transferred by scraping with the tip of a pipette (using either a Gilson P-1000 or P-200 pipetman) while simultaneously releasing the plunger to draw up the cells with medium. An alternative is to use cloning rings to trypsinize each colony, but this technique is much more difficult to use. These 24-well plates are the stock plates and should always be handled with great care and kept sterile.

Days 21–28 (approximately)

9. Screen individual wells in the 24-well plates as the cells become confluent. In the case of scraped colonies, confluence would be defined as multiple large clumps of cells (4–6 mm in diameter) in several places in the well, rather than a monolayer covering the entire bottom of the well.

   a. Prepare a second set of 24-well plates to use for screening. *If screening by fluorescence microscopy, place a small round coverslip in each well.*

   b. Mark the top of each well with its corresponding clone number.

   c. Rinse confluent cells with PBS, and add 3–4 drops of trypsin (place in the incubator to loosen large clumps). Add 1 ml culture medium with serum when the cells are released and pipette gently to resuspend cells.

   d. Transfer ~25–50% of the cells to the screening wells.

   > About 25% of the remaining stock cells can be discarded if desired to avoid overgrowing the stock plates. The recommended goal is to screen the clones transferred to the screening plates before the wells in the stock plates reach confluence again.

Stable Clones

10. After stable clones are identified, transfer them to 6-well dishes and then to 100-mm dishes. From this point, the clones can be expanded to make frozen stocks.

    > If a clone contains a mixture of positive and negative cells, simple serial dilution procedures or fluorescence-activated cell sorting (FACS) can be used to isolate the positive cells.

IS THE FUSION PROTEIN LOCALIZED CORRECTLY?

Irrespective of whether the fusion protein is expressed transiently or stably, it must be determined whether it is localized correctly. If familiar with the localization of the protein under study, examine its localization using a fluorescence microscope equipped with the correct excitation/emission filter cube for the respective fluorescent protein (see Chapter 1). If the protein under study has not previously been localized, perform a side-by-side localization with an antibody that specifically recognizes the protein under study. This is easily accomplished by fixing the cells and processing for single indirect immunofluorescence. For example, in the case of cells transfected with enhanced GFP-tagged constructs, the secondary antibody could be tagged with Texas Red. Using high-resolution confocal microscopy, complete colocalization of the two fluorophores should be achieved. If an antibody is not available, double-label localization should be performed with an antibody to a partner protein that is expected to be colocalized with the protein of interest. Evaluate any suspicious or novel localization very carefully as the GFP tag can, in certain cases, interfere with the correct localization of the protein under study. In addition, the tag can alter protein folding and/or result in aggregation, most notably in the cytoplasm. Such localization, for example, with an amino-terminal fusion protein may not be observed with a carboxy-terminal fusion protein, or visa versa. Suspicious localization of the fusion protein should be taken very seriously and further studies should be placed on hold until the localization is understood. Of course, for a novel protein of unknown function, it may not be easy to distinguish between correct and aberrant localization of the fusion protein. In these cases, every attempt should be made to prepare an antibody to determine the normal localization of the protein in nontransfected cells.

IS THE FUSION PROTEIN FUNCTIONAL?

As is the case with any fusion protein, and in particular those containing the 27-kD GFP protein, one must determine if the fusion protein is functional. However, in many cases, the term "functional" is rather loosely defined. Upon development of a GFP fusion protein, the first parameters that should be investigated are (1) the protein's localization as discussed above, (2) whether the fusion protein is full length, and (3) its expression level as compared to the endogenous protein. The last two parameters can be determined simultaneously by immunoblot analysis using an antibody to the endogenous protein that will recognize both the endogenous protein and the GFP fusion protein. In this case, there will be a retardation of mobility in SDS-PAGE preparations due to the increased molecular weight of the fusion protein. If such an antibody is not available, antibodies to GFP (i.e., a monoclonal antibody available from Roche) can be used to determine the size of the expressed fusion protein. This characterization is extremely important as expression of a truncated protein can result in incorrect interpretation and if followed up can result in a significant loss of time and resources. Immunoprecipitation followed by immunoblotting can be used to determine if the expressed fusion protein has assembled into a complex with its expected binding partners as determined from studies of the native cellular protein. Such analysis can be performed either with an antibody to the endogenous protein or with anti-GFP antibodies (i.e., a monoclonal antibody available from Roche) that function in immunoprecipitation assays. If a cell line is available or can be developed in which the protein of interest is deficient, rescue by the GFP fusion protein can be examined. This approach was used in Chinese hamster ovary cells deficient in the ERCC1 DNA repair protein (Houtsmuller et al. 1999). Finally, if the specific function of the protein of interest is known, the ability of the fusion protein to replace the endogenous protein can be examined in an immunodepleted extract used for a functional assay, i.e., in vitro transcription or in vitro pre-mRNA splicing reaction. In addition, in vivo functional assays can be performed to determine if the fusion protein is recruited to a site where a particular function is inducible; ie., transcription/pre-mRNA splicing (Misteli et al. 1997; Janicki et al. 2004) or DNA repair (Essers et al. 2002; Lisby et al. 2003) or if it is recruited to functional sites at a particular time in the cell cycle, i.e., DNA replication (Leonhardt et al. 2000).

Although GFP has been extremely useful, a major drawback is its large size, 27 kD, as a protein tag. Recently, a method has been developed that permits the fluorescent labeling of transient or stably expressed proteins containing a small tetracysteine sequence CCXXCC (where X is any amino acid) (Griffin et al. 1998; Gaietta et al. 2002). Proteins carrying this motif can be labeled in vivo with the nonfluorescent, membrane-permeant biarsenical derivative of fluorescein, FlAsH-EDT2 (4′,5′-bis[1,3,2-dithioarsolan-2-yl]fluorescein-[1,2-

ethanedithiol]2). FlAsH-EDT2 binds with high affinity (K_d 10 PM) to the tetracysteine motif, resulting in strong green fluorescence, whereas ReAsH-EDT2 (a biarsenical derivative of the fluorophore resorufin), when bound to the tetracysteine tag, emits a red fluorescence (Gaietta et al. 2002). These two reagents have been used to label tetracysteine motif-tagged intracellular proteins in living cells and in combination to perform pulse-chase studies (Gaietta et al. 2002). The method is especially useful because not only is the ReAsH-EDT2 reagent fluorescent, but it can also be used, by photoconverting diaminobenzidine (DAB) to a highly insoluble reaction product, to visualize tagged proteins at high resolution by electron microscopy (Gaietta et al. 2002).

REFERENCES

Ahmad K. and Henikoff S. 2002. The histone variant H3.3 marks active chromatin by replication-independent nucleosome assembly. *Mol. Cell* **9**: 1191–1200.

Doyle T. and Botstein D. 1996. Movement of yeast cortical actin cytoskeleton visualized in vivo. *Proc. Natl. Acad. Sci.* **93**: 3886–3891.

Essers J., Houtsmuller A.B., van Veelen L., Paulusma C., Nigg A.L., Pastink A., Vermeulen W., Hoeijmakers J.H., and Kanaar R. 2002. Nuclear dynamics of RAD52 group homologous recombination proteins in response to DNA damage. *EMBO J.* **21**: 2030–2037.

Gaietta G., Deerinck T.J., Adams S.R., Bouwer J., Tour O., Laird D.W., Sosinsky G.E., Tsien R.Y., and Ellisman M.H. 2002. Multicolor and electron microscopic imaging of connexin trafficking. *Science* **296**: 503–507.

Grava S., Dumoulin P., Madania A., Tarassov I., and Winsor B. 2000. Functional analysis of six genes from chromosomes XIV and XV of *Saccharomyces cerevisiae* reveals *YOR145c* as an essential gene and *YNL059c/ARP5* as a strain-dependent essential gene encoding nuclear proteins. *Yeast* **16**: 1025–1033.

Griffin B.A., Adams S.R., and Tsien R.Y. 1998. Specific covalent labeling of recombinant protein molecules inside live cells. *Science* **281**: 269–272.

Houtsmuller A.B., Rademakers S., Nigg A.L., Hoogstraten D., Hoeijmakers J.H., and Vermeulen W. 1999. Action of DNA repair endonuclease ERCC1/XPF in living cells. *Science* **284**: 958–961.

Hu C.D. and Kerppola T.K. 2003. Simultaneous visualization of multiple protein interactions in living cells using multicolor fluorescence complementation analysis. *Nat. Biotechnol.* **21**: 539–545.

Hu C.D., Chinenov Y., and Kerppola T.K. 2002. Visualization of interactions among bZIP and Rel family proteins in living cells using bimolecular fluorescence complementation. *Mol. Cell* **9**: 789–798.

Janicki S.M., Tsukamoto T., Salghetti S.E., Tansey W.P., Sachidanandam R., Prasanth K.V., Ried T., Shav-Tal Y., Bertrand E., Singer R.H., and Spector D.L. 2004. From silencing to gene expression: Real-time analysis in single cells. *Cell* **116**: 683–698.

Leonhardt H., Rahn H.P., Weinzierl P., Sporbert A., Cremer T., Zink D., and Cardoso M.C. 2000. Dynamics of DNA replication factories in living cells. *J. Cell Biol.* **149**: 271–280.

Lisby M., Mortensen U.H., and Rothstein R. 2003. Colocalization of multiple DNA double-strand breaks at a single Rad52 repair centre. *Nat. Cell Biol.* **5**: 572–577.

Marguet D., Spiliotis E.T., Pentcheva T., Lebowitz M., Schneck J., and Edidin M. 1999. Lateral diffusion of GFP-tagged H2Ld molecules and of GFP-TAP1 reports on the assembly and retention of these molecules in the endoplasmic reticulum. *Immunity* **11**: 231–240.

Misteli T., Cáceres J.F., and Spector D.L. 1997. The dynamics of a pre-mRNA splicing factor in living cells. *Nature* **387**: 523–527.

Rocheleau J.V., Edidin M., and Piston D.W. 2003. Intrasequence GFP in class I MHC molecules, a rigid probe for fluorescence anisotropy measurements of the membrane environment. *Biophys. J.* **84**: 4078–4086.

Rubinchik S., Ding R., Qiu A.J., Zhang F., and Dong J. 2000. Adenoviral vector which delivers FasL-GFP fusion protein regulated by the tet-inducible expression system. *Gene Ther.* **7**: 875–885.

Tsukamoto T., Hashiguchi N., Janicki S.M., Tumbar T., Belmont A.S., and Spector D.L. 2000. Visualization of gene activity in living cells. *Nat. Cell Biol.* **2**: 871–878.

CHAPTER THREE

Viral Vectors for Introduction of GFP

Rusty Lansford

California Institute of Technology, Biological Imaging Center, Pasadena, California 91125

INTRODUCTION

LIVING CELLS ARE CONTINUOUSLY INTERACTING with and responding to their surrounding environments. As never before, investigators may choose to study the everchanging events of living cells using static or dynamic collection modalities. Dynamic imaging techniques offer dramatic insights into the spatial and temporal controls of biological processes. We face similar choices when we choose to follow our favorite sporting event by watching TV or by reading the newspaper. By watching the game through the dynamic modality of video, we are able to better appreciate the ebb and flow of the game, the interactions required between players to score a goal. Static snapshots of the same game viewed in a newspaper or sports journal would likely be able to show the players scoring the goal, but the reader would have no sense of the interactions between players that went into setting up that goal. The need for dynamic viewing is as required in science as it is in sport in order to begin to appreciate the true context of the event. To appreciate the intricacies and ceaseless interactions that cells constantly undergo, we should utilize equipment and reagents to watch such events using dynamic modalities. The union of the green fluorescent protein (GFP) and viral vectors affords investigators this opportunity. GFP permits the contents of cells to be illuminated and followed in a dynamic manner, whereas viral vectors permit GFP to be introduced into cells in a very specific and efficient manner.

Viruses are ubiquitous, submicroscopic agents that are able to infect throughout the three major domains of living organisms: archaea, bacteria, and eukarya. Viruses have been found globewide in all the nooks and crannies that contain living organisms. Even though viruses are notorious for inducing numerous devastating diseases in humans, animals, and plants, only a small percentage are actually pathogenic. Viruses are obligate intracellular parasites whose genetic material, either DNA or RNA, is ensconced in a protective coat and transmitted from one host cell to another. Viruses are therefore considered to be the most efficient vehicles for transferring genetic material into host cells, which has made them prime candidates for gene-transfer vectors. Likewise, viruses that encode GFP are able to introduce the GFP gene efficiently into host cells which makes them essential research tools.

Most virus particles range between 30 and 100 nm in diameter. Viral genomes are extremely diverse in that they can consist of nucleic acid molecules that are monomer or multimer, linear or circular, single-stranded or double-stranded, DNA or RNA. The size of viral DNA genomes can range from 3 to 375 kb, whereas the range of viral RNA genomes can range from 3 to 30 kb. The life cycle of a typical virus can be summarized as follows: (1) binding to and entry of the virions into the host cell, (2) uncoating of the nucleic acids, (3) replication of the nucleic acids and generation of viral RNAs, (4) expression of viral proteins, and (5) assembly of nascent virions and their release from the host cells.

Viruses adeptly replicate themselves by introducing their genomes into cells and by commandeering the cells biosynthetic machinery in order to express their proteins, produce progeny viruses, and egress from the host cell. Virus replication and gene expression can be so efficient that they interfere with the host cell's own housekeeping abilities and lead to disease. The ruthless efficacy by which viruses parasitize numerous cell types is in part what makes them such attractive gene vectors. Exogenous genes such as GFP can be incorporated into viral particles or into viral genomes in order to introduce the GFP proteins or GFP genes into cells; virions used for this purpose are called viral vectors. Viral vectors are very diverse much like the virions that they are derived from. Individual virions generally contain either an RNA or a DNA genome that is ensconced by coat proteins and a lipid bilayer. The nucleic acid genome may be single- or double-stranded. Some viral vectors integrate their genomes into host-cell chromosomes, whereas others maintain episomal stability. The transcriptional promoters and enhancers that drive and maintain RNA expression vary greatly among viruses in the different cell types in which they are active or suppressed.

Viral vectors are typically generated from viruses by replacing viral genes that are needed for the replication phase of their life cycle with foreign transgenes such as GFP. To produce viable recombinant viral vectors, the deleted viral genes can be either integrated into the genome of a packaging cell line or expressed from a plasmid or virus so that the viral gene products are provided in *trans*. The recombinant viral vectors ideally maintain their ability to infect and transduce the same cell types that they would normally have infected. Viral vectors that express GFP have been shown to be extremely useful for discerning intracellular protein trafficking and organelle movement, for understanding how viruses interact with their cellular hosts, and for experiments in neurobiology, developmental biology, and plant biology, to give only a few examples.

The most important criteria for investigators is to make sure to properly match the strengths of the chosen viral vector with their experimental requirements. It is also vital to pay close attention to the color variant of GFP (or combinations thereof) to use along with the best protein tag to direct the GFP for appropriate subcellular localization. This chapter summarizes the basic aspects of viral structure and function and describes how a few selected viruses are being used to introduce GFP into experimental cells and embryos. So many distinct GFP-expressing viral vectors have been developed during the past decade that even listing them here is well beyond the possible scope of this chapter.

CHOOSING THE RIGHT VIRAL VECTOR FOR EXPERIMENT

A myriad of choices must be considered when deciding which GFP-expressing viral vector to use for which experiment. The investigator must choose the viral system that best suits the experiment. This involves determining what is required in terms of tissue tropism, gene expression patterns and levels, genome versus episomal integration, and time delay for GFP expression. Other considerations are immunologic effects, ease and expense of virus generation and purification, ease of recombinant DNA techniques, and stability of viral genomes. Some of these issues are considered below.

Genome versus Episome

Integration of the viral vector genome into the host genome, preferably in a site-specific location, ensures that the transgene is carried with the cell in a Mendelian manner and is therefore not lost during the lifetime of the cell. However, genomic integration that is beneficial for some experiments may be deleterious for others, thus the trade-offs must be evaluated. A serious disadvantage of genome integration is that the viral DNA may integrate near a cellular proto-oncogene or some other important gene, thereby disrupting the normal expression pattern of that gene, which might induce transformation of the infected cells into cancerous cells (Coffin et al. 1997). On the other hand, episomal maintenance does not guarantee that the viral genome is maintained in the infected cell. It may be lost from the infected cell, especially from proliferating cells. Here again, the pros and cons of each viral vector must be considered in the context of the particular experimental setting.

Promoter Control

Cell- and tissue-specific gene expression is an extremely important research topic with profound implications for both basic and applied research. The most frequently used viral vector promoters are viral in origin, often derived from a different virus than the vector backbone to minimize recombination events (cytomegalovirus, simian virus-40, and Rous sarcoma virus transcription units are experimental favorites). Viral promoters have the advantages of being smaller, stronger, and better understood than most other cellular promoter sequences. This is also the case for viral enhancers with the added advantage that these elements are often located close to the promoter they govern within their viral origin. This proximity likely means that their interactions with one another are well-known. Vertebrate cell genomes are enormous compared with viral genomes, and in most cases, promoters and their enhancers (or silencers) are poorly studied, even if they have been identified and shown to be related to one another. Cellular promoter behavior is difficult to predict a priori in the context of a viral vector (Harris and Lemoine 1996). In the cell, transcriptional enhancers and silencers typically are found at sites distant from their promoter sequences and, when included within a smaller distance, may fail to operate. Insulator elements included within viral vectors have worked with mixed results to date. However, one of the drawbacks of the use of viral promoters is their tendency to be silenced, especially in transgenic animals. Viral promoters can be silenced due to positional effects related to their chromatin integration site, by the transcription complexes of genes that the virus may have integrated into or adjacent to, or by DNA methylation of their transcription elements by the host cell. Gene expression can also be down-regulated at the level of mRNA stability (Qin and Gunning 1997).

Tissue Tropism

The host range or tropism that a virus can infect is generally restricted. The surface of viruses includes many copies of one type of protein that bind to or absorb to receptor molecules on a host cell. The cellular receptors that viruses bind are very diverse. Most cells express only a specific subset of the known cellular receptors. The limited cellular host-cell range results from the fact that viruses recognize only specific receptors. Once bound to a receptor, viruses can either fuse with the cell membrane or enter via the endocytic pathway.

Considerable research has recently gone into modifying the tissue tropism of viral vectors. The vesicular stomatitis virus (VSV)-G-pseudotyped retroviruses efficiently (>99%) infect cells throughout embryonic development. The infected cells appear to follow their standard migratory routes and undergo normal differentiation. Embryonic cells that have been infected with retroviruses expressing GFP have been successfully recorded in vivo using time-lapse imaging in combination with laser-scanning confocal and 2-photon microscopy (see Chapters 14 and 15).

CHOOSING THE BEST FLUORESCENT PROTEIN

Cellular fate-mapping and migration studies are carried out with fluorescent dyes or retroviruses expressing marker genes such as lacZ (Price et al. 1987). The problem with these approaches is that they do not permit long-term dynamic observation of the developing embryo. Dyes are diluted out of dividing cells and the retrovirus-expressed marker genes are not intrinsically fluorescent, which prevents dynamic imaging. GFP has instead become an essential and favored tool for the rapid detection of gene transfer, cell tracking, and cell mechanism studies. What makes GFP such an effective experimental tool is that it can be visualized without interruption or termination of an experiment, as is required with the detection of other commonly used markers such as alkaline phosphatase and β-galactosidase.

GFP is a 238-amino-acid polypeptide isolated from the jellyfish *Aequorea victoria,* which acts as an energy-transfer acceptor deriving excitation energy from emission of blue light via aequorin. GFP emits green fluorescence when excited by blue light, which can be observed using standard light or laser microscope technology. In addition, fluorescent proteins are favored over vital fluorescent dyes because they can be readily targeted to subcellular compartments, they can be introduced into a broad range of tissues and organisms in a very specific manner, and they can be designed to respond to a great variety of biological events and signals. Furthermore, their use seldom triggers photodynamic toxicity and they are relatively photostable (Tsien 1998). However, even the task of choosing the ideal fluorescent protein can be rather complicated.

There are now numerous color variants of GFP to choose from including BlueFP, CyanFP, Cyan-GreenFP, GreenFP, YellowFP, or RedFP (Zhang et al. 2002). In addition, it is important to select a GFP encoded by a sequence whose codon usage is suited to the organisms under study (Crameri et al. 1996; Yang et al. 1996; Zolotukhin et al. 1996). The ability to localize different GFP variants to specific cellular organelles allows these structures to be specifically imaged, cell divisions to be followed, and morphological changes to be dynamically observed. For example, H2B-GFP (histone 2B-GFP) is localized to the cell nucleus and binds to chromosomal DNA. This permits the various stages of mitosis and thus cell divisions to be followed. The H2B-GFP also greatly enhances cell tracking via time-lapse videomicroscopy when compared to conventional vital dyes and untargeted GFPs since a dark GFP-free cytoplasm improves the ability to distinguish neighboring cells from one another. A few additional examples include the use of actin-GFP or actin-binding protein GFPs to follow actin-related events (such as heart looping or cell migration) and plasma-membrane-localized GFPs to follow cell migrations and changes in cell morphology. In addition, mitochondrial GFPs are used to follow cell movements since the labeled mitochondria are bright and punctate. It is also possible to view dynamically how cells and tissues intermingle and interact with one another by labeling each cell or tissue with a distinct GFP color variant and then recording the cells using multispectral laser-scanning microscope imaging techniques.

RETROVIRAL VECTORS

Retroviral vectors are most frequently derived from the Moloney murine leukemia virus (Mo-MLV). Mo-MLV vectors are only able to infect dividing cells because the preintegration nucleoprotein complex of the virus cannot traverse an intact nuclear membrane (Miller et al. 1990; Roe et al. 1993). However, lentiviral vectors, like those derived from human immuno-deficiency virus type 1 (HIV-1) are able to infect nondividing cells because they can actively transport their proviral genome across the intact nuclear membrane (Weinberg et al. 1991; Lewis et al. 1992). Retroviral vectors have the advantage that they stably integrate into the genome of the cell they infect.

Virion Properties

Retroviruses are a class of enveloped viruses containing a single-stranded RNA molecule as the genome (Coffin et al. 1997). The basic retroviral genome is approximately 10 kb and consists of the 5′ and 3′ long terminal repeats (LTRs) and the *gag*, *pol*, and *env* genes. The LTRs are located at each end of the genome and contain promoter/enhancer regions and sequences involved in integration into the host-cell genome. Following viral uptake by the host cell, the viral genome is reverse-transcribed into double-stranded DNA, which then integrates into the host genome and is expressed as proteins. In addition, there are sequences required for packaging the viral DNA (psi) and for RNA splicing. *gag* encodes three proteins that form the shell of the virion; *pol* encodes reverse transcriptase, integrase, and RNase H which are necessary for viral integration into host chromosomal DNA. The *env* gene encodes the envelope glycoprotein that extends from the lipid membrane of the virion and functions as a ligand for the cellular viral receptor.

Vector Properties

Viruses deleted for one or more of their structural genes (e.g., *gag*, *pol*, or *env*) are infectious when propagated in a viral helper cell that expresses the missing gene or genes. Retroviruses are typically limited to introducing approximately 7.5 kb of genetic information due to packaging constraints of the RNA into viral capsid core (Coffin et al. 1997). The proteins encoded by the *gag*, *pol*, and *env* genes are expressed from transfected plasmids in packaging or helper cell lines, thereby providing "in *trans*" these elements required for viral assembly. The retroviral vector is rendered replication-defective as it no longer carries the deleted structural genes required for replication, nor does its target cell. Thus, the deleted viral vector introduces and expresses the transgene, the transgene is stably maintained within the target cell genome, and the replication-defective retrovirus is incapable of establishing an active infection in the host. Usually all viral regions homologous to the vector backbone are deleted, and the structural genes are expressed by at least two transcription units to prevent viral recombination and reconstitution. Even so, replication-competent retroviruses do occur at a low frequency.

Numerous investigators have generated replication-defective retroviral vectors that express the GFP marker. Pseudotyping alters the host range of a virus by exchanging the surface antigens among both DNA and RNA viruses (Zavada 1982). The resulting VSV-G-pseudotyped retroviruses possess a broad host range and can be concentrated 1000-fold with minimal loss of biological activity (Burns et al. 1993). Retroviral vectors have the advantage that they stably integrate into the genome of the cell they infect and are replicated and segregated to daughter cells along with the rest of the chromosome into which

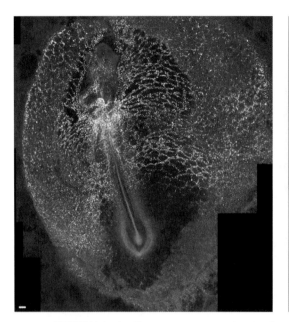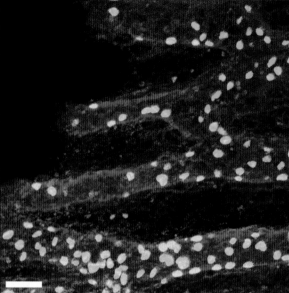

FIGURE 1. (*Left*) H2B-YFP expression in developing blood vessels 15 hours after injection of the rH2B-YFP-expressing retroviral vector into quail blood island primordia. Laser-scanning confocal microscopy reveals YFP labeling of blood vessels within the 12-somite quail embryo. (*Right*) Two populations of cells arise from blood island injections. Intraembryonic blood vessels of a 12-somite quail embryo ~15 hours after blood island primordia injection shows numerous YFP+ (*green*) cells within the QH1+ (*red*) vasculature. Bar, 25 μm.

they integrated (Fig. 1). This trait is exploited in cell-tracking and cell-fate experiments since the GFP-expressing cells can be assumed to descend from a single infected cell. The available carrying capacity for retroviral vectors is approximately 7.5 kb (Verma and Somia 1997), which is too small for some genes even if the cDNA is used. Retroviral constructs are small enough to be manipulated easily by standard cloning techniques. Numerous cloning cassettes have been generated so that it is trivial to shuttle in new gene components. A serious disadvantage of retroviruses is that they can transform the infected cells into cancerous cells by integrating near a cellular proto-oncogene thereby driving inappropriate expression from the LTR, or disrupting a tumor suppressor gene (Coffin et al. 1997).

LENTIVIRUS VECTORS

HIV-1 is the etiologic agent of acquired immunodeficiency syndrome or AIDS. The cell surface receptor CD4 found on T helper cells is the major receptor for HIV (Maddon et al. 1988). Lentiviruses are unique retroviruses in that they are capable of infecting both proliferating and nonproliferating cells (Naldini et al. 1996). HIV-based vectors have also been used to efficiently generate GFP-expressing transgenic animals (Lois et al. 2002).

Virion Properties

The approximately 9-kb HIV-1 RNA genome contains nine open reading frames (ORFs) which encode 15 proteins (Coffin et al. 1997). Three of the ORFs encode the Gag, Pol, and Env polyproteins that are common to all retroviruses. HIV-1 also encodes six additional accessory proteins—Vif, Vpr, Nef, Tat, Rev, and Vpu (Frankel and Young 1998). Vif, Vpr, and Nef are found in the viral particle, Tat and Rev provide essential gene regulatory functions, and Vpu assists in virion assembly. HIV-1 is capable of infecting nondividing cells because the Vpr proteins actively transport the proviral complex through the intact nuclear membrane (Naldini et al. 1996). Current viral helper or packaging cell lines utilize separate plasmids for a pseudotyped *env* gene and for supplying the viral structural and regulatory genes in *trans* (Zufferey et al. 1997; Dull et al. 1998).

For years, numerous investigators have attempted to generate transgenic animals using retroviruses because of their high rate of infectivity and because they are able to stably integrate into the genome of cells. However, the generation of transgenic animals with retroviruses such as the Mo-MLV proved to be impractical due to silencing of the provirus during development, resulting in low levels of transgene expression (Jaenisch et al. 1981; Jahner et al. 1982). HIV-based vectors have been used to generate transgenic mice and rats very efficiently, with the vast majority of the animals expressing high levels of GFP (Lois et al. 2002). Even more impressive, mice generated using the lentiviral vectors with muscle-specific, T-lymphocyte-specific, or ubiquitous transcription elements express high levels of GFP only in the appropriate cell types.

ADENOVIRUS VECTORS

Adenovirus infections of humans are common and cause cold-like symptoms. Adenovirus has a simple construction that comprises a capsid, fibers, a core, and associated protein(s). Neutralizing sera have been used to identify 49 different serotypes of adenovirus that are divided into six groups (A to F) on the basis of genome size, composition, homology, and organization. Most of these cause benign respiratory tract infections in humans. Subgroup-C serotypes 2 or 5 are primarily used as viral vectors. The life cycle does not usually involve integration into the host genome, rather they replicate as episomal elements in the nucleus of the host cell, and thus there is minimal risk of insertional mutagenesis. Adenovirus C can infect rodent, chicken, and primate cells (Flint et al. 2000; Fields et al. 2001).

Virion Properties

The adenovirus is a DNA virus containing a linear double-stranded DNA genome of approximately 36 kb (van Regenmortel et al. 2000; Fields et al. 2001). The adenovirus particle is nonenveloped with a diameter of 80–110 nm. The regular icosahedron capsid consists of 252 capsomers that are each 8–10 nm in diameter (12 are pentons and 240 are hexons). One or two filaments protrude from each of the 12 vertices. Each of the 12 penton bases is tightly associated with one or two fiber coat proteins that have a shaft of 9–77.5 nm length with a distal knob. Cellular uptake of adenovirus involves contact of the adenovirus fiber coat protein with a cell surface receptor, including the major histocompatibility complex (MHC) class I molecule (Hong et al. 1997), the Coxsackie B viruses and adenovirus receptor (Bergelson et al. 1997), and the integrin family of cell surface heterodimers (Wickham et al. 1993).

The double-stranded DNA genome is not segmented and encodes structural and nonstructural proteins (van Regenmortel et al. 2000). Virions consist of ten structural proteins located in the capsid, fibers, and core. Each end of the viral genome is flanked by 100 –150 bp of repeated DNA sequence (inverted terminal repeat). Viral gene expression occurs in a sequential cascade. Adenoviral genes are grouped as early (E) genes whose expression precedes viral DNA replication and the transcription of late (L) genes at 6–8 hours postinfection. The E genes encode regulatory proteins for viral replication, and the L genes encode structural proteins vital for virion assembly. The E1A and E1B products of the E1 gene are required for viral replication. The E3 gene products thwart the host immune response to adenovirus by preventing T-cell-mediated cytolysis.

Vector Properties

The advantages of using adenovirus vectors include their broad host range, their ability to infect and deliver transgenes to both dividing and nondividing cells, and their ability to replicate episomally within the host nucleus. Furthermore, the viral stock can be concentrated to high titers (10^{11} to 10^{12} pfu/ml) and produced on a large scale (Flint et al. 2000; Fields et al. 2001). Because adenoviral DNA is episomal, insertional mutagenesis is much less of a concern than it is for retroviral vectors. However, this also means that adenovirus expression is relatively short-lived (5–10 days), as the proportion of transduced cells decreases with each cell division or possibly due to immune intervention (Verma and Somia 1997). Adenovirus vectors can also induce immune responses that can cause numerous problems for clinicians and experimentalists alike.

Adenovirus vectors are designed to be replication-defective for safety purposes, in order to insert more foreign DNA into their genomes and to avoid recognition by their host's immune systems. The adenoviral genome is approximately 36 kb of which up to 30 kb can be replaced with foreign DNA (Smith et al. 1995). Numerous variant adenovirus vectors have been generated that contain deletions of the E genes within the viral genome. Deletion of the E1A and part of the E1B gene results in replication-deficient viruses that must be propagated in helper cell lines that provide the deleted E1 gene products in *trans*. The amount of DNA that can be inserted into adenovirus virions cannot exceed approximately 105% of the wild-type genome, which only permits the insertion of about 2 kb of foreign DNA into full-length adenovirus vectors, 5–6 kb of foreign DNA into E1A-deleted vectors, or 7–8 kb of foreign DNA into E1- and E3-deleted vectors. Additional vectors have been generated that contain an E2A temperature-sensitive mutant (Engelhardt et al. 1994) or an E4 deletion (Armentano et al. 1997). Vectors developed recently may only contain the inverted terminal repeat (ITR) region of the viral genome—containing the origin of replication—and a packaging sequence around the transgene with all the necessary viral genes being provided in *trans* by a helper virus (Chen et al. 1997) to minimize recognition of the adenovirus vector by the immune system. Two disadvantages of existing adenovirus vectors is the induction of T-cell-mediated response against adenoviral proteins that cause a local inflammatory reaction resulting in lysis of the transduced cells which shortens the duration of transgene expression. The transient nature of gene expression and the induction of host immune responses seriously jeopardize the efficacy of adenovirus gene delivery and will require considerable new research to bolster user confidence.

GFP-expressing Adenovirus Vectors

The previous paragraph briefly mentions some of the serious drawbacks to using adenovirus vectors. GFP-expressing adenovirus could be used to better understand intracellular and extracellular events of the adenovirus life cycle. For instance, GFP-impregnated adenovirus could be used to infect cells of interest. Intracellular events such as endosome release could be dynamically followed using low-light microscopy techniques, thereby permitting insight into subcellular organelles and structures with which adenovirus interacts to traffic through the cell. This information could facilitate the design of more efficient expression vectors.

Additionally, GFP-impregnated and -expressing adenovirus vectors could be used to determine how cells of the immune system recognize adenovirus virions. These same adenovirus vectors will continue to be useful for investigators analyzing the gene expression stability in target cells, along with determining the tissue tropism of these vectors.

ALPHAVIRUS VECTORS

Alphaviruses are neurotropic and are able to cross the blood-brain barrier (Lundstrom 1999; Dubensky et al. 2000). Alphaviruses are potentially valuable vectors because their plus-strand RNA genomes self-replicate within the host-cell cytoplasm and they can express very high levels of protein. Alphavirus expression vectors have been generated from Semliki Forest virus (SFV) (Liljestrom and Garoff 1991), and Sindbis virus (SIN) (Xiong et al. 1989).

Virion Properties

The alphavirus is an enveloped RNA virus containing a positive linear single-stranded RNA genome that is 11–12 kb in length (van Regenmortel et al. 2000; Fields et al. 2001). The RNA genome consists of two regions, one containing the replicase genes (*nsp1-4*) and the other containing the structural genes (*C, E1, E2*). Genome replication occurs entirely within the cytoplasm of the infected cell. Cloning is relatively simple in these genomes and infectious RNA can be transcribed in vivo or in vitro. In-vitro-transcribed mRNA that is transfected into host cells gives rise to viable virus. The Sindbis virus genome is 11,703 nucleotides long and has been fully sequenced (Accession number J02363). The approximately 70-nm icosahedral (T = 4) virions are enveloped by a lipid membrane. The lipids are derived from host-cell membranes. Viral membranes include phosphatidyl ethanolamine, phosphatidyl choline, phosphatidyl serine, cholesterol, and sphingomyelin (van Regenmortel et al. 2000). The virus infects arthropod and vertebrate hosts.

Vector Properties

The advantages of alphavirus systems include (1) a broad host tropism for mammalian and insect cells, (2) the release of positive-strand RNA directly into the host-cell cytoplasm, permitting quick gene expression, (3) the extremely high levels of transgene expression, and (4) the ability to infect nondividing cells (Wahlfors et al. 2000). Alphavirus replicons can express as much as 25% of the protein of a cell over a period of 72 hours. A potential disadvantage of these vectors is that they can induce cell death via apoptosis and inhibit macromolecular synthesis. However, suicide vectors may be desirable in certain experimental and clinical scenarios.

Alphavirus expression vectors are usually made as propagation-incompetent RNA replicons (Wahlfors et al. 2000). Replicons contain both the *cis* and *trans* alphavirus genetic elements required for RNA replication, as well as heterologous gene expression via the native subgenomic promoter. After in vitro transcription, replicon RNA is often transfected into cells. The replicon RNA is then translated into the four nonstructural proteins that comprise the alphaviral replicase, but not the structural proteins required for packaging into particles. This keeps expression limited to only the cells that have been transfected with the replicon. Viral replication progresses through a minus-strand RNA intermediate and subsequently generates two distinct positive-strand RNA species, corresponding to a genomic-length vector RNA and an abundant subgenomic RNA encoding the heterologous gene (Polo et al. 1999). Alternatively, the replicon RNA can be introduced directly into cells as plasmid DNA (Dubensky et al. 1996).

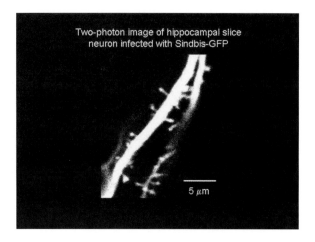

FIGURE 2. High levels of GFP expression from alphavirus vectors. Sindbis vectors that express high levels of GFP have been used to detect subtle changes in neuronal morphology in response to physiological stimuli (Maletic-Savatic et al. 1999; Chen et al. 2000). (Image kindly provided by R. Malinow.)

The structural proteins are encoded from subgenomic mRNA, which makes subcloning in foreign genes under the control of the subgenomic promoter straightforward (Lundstrom et al. 2003). Vectors can be generated by inserting transgenes downstream from the structural genes or between the structural genes and the nonstructural genes (Dubensky et al. 2000). Using this approach, the subgenomic mRNAs are transcribed for the structural proteins and for the transgene. A major drawback of this approach is that only 2000 nucleotides or less may be inserted into the genome without deleteriously affecting virus packaging. Another approach is to delete the structural genes and replace them with the desired transgene, thereby permitting a 5-kb transgene to be inserted. This approach requires the host cells to be coinfected with a helper virus that supplies the structural proteins in *trans*.

GFP-expressing Alphavirus

GFP-expressing alphaviruses are especially desirable for use in neuronal cells both in vitro and in vivo (Lundstrom 1999). The high levels of GFP protein produced from alphavirus vectors permits dynamic imaging of delicate dendrites (Chen et al. 2000). GFP expressing alphaviruses are also ideal for neuronal tracing studies, because, if the correct protein tag is placed on GFP, the copious amounts of expressed GFP could pass into connected cells. Sindbis vectors that express GFP have been used to detect subtle changes in neuronal morphology in response to physiological stimuli (see Fig. 2) (Maletic-Savatic et al. 1999; Chen et al. 2000).

Many transfection schemes used for the nervous system result in significant glial cell labeling, with minimal neuronal cell labeling. If the goal is to infect more neuronal cells than other cells, then alphaviruses seem to be ideal delivery system (Ehrengruber et al. 1999, 2001; Lundstrom 1999). For example, cortical and hippocampal primary neurons in culture are efficiently infected, resulting in 75–95% GFP-positive cells. Following the injection of SFV vectors into neuronal tissue slice cultures more than 90% of the infected cells are neurons.

POXVIRUS VECTORS

Many potentially useful viruses fall under the heading of poxvirus; the best known of these is the vaccinia virus (VV). VV was used to vaccinate hundreds of millions of humans in the successful eradication of smallpox (Fenner et al. 1989).

Virion Properties

VV contains one molecule of linear double-stranded DNA with 10-kb hairpin loops at each end (Moss 1996). The VV genome is approximately 192 kb long and has been fully sequenced (Accession number M35027, NC001559) (Goebel et al. 1990). VVs are enveloped and brick-shaped; they are about 250 nm in diameter, 250–300 nm long, and 200 nm high. The complex genome contains nearly 200 genes that encode proteins for RNA and DNA synthesis, virus assembly, and immune system suppression. A defining characteristic of VV is its ability to carry out both DNA replication and transcription within the host-cell cytoplasm. VV can do this because its genome encodes all of the necessary enzymes for replication and transcription, thereby not needing the host cell's machinery located within the nucleus. Early viral gene transcripts can be detected within 15 minutes after entry into the cell (Baldick and Moss 1993). Intermediate and late gene transcription occurs during and after viral DNA replication.

Vector Properties

Advantages of the VV system include its broad host range, quick and efficient cytoplasmic expression of transgenes, authentic protein processing, and ability to accommodate more than 25 kb of transgenic DNA. There is an extensive literature of using VV as expression vectors for various foreign genes. Disadvantages of the VV system include difficulties associated with cloning transgenes into the 200-kb genome and host-cell death. The 200-kb genome is too large to manipulate using conventional cloning methods. Thus, homologous recombination is the most widely used method of inserting foreign genes such as GFP into VV.

VV is infectious and can only be used in Biosafety containment facilities. Two highly attenuated VV vectors, NYVAC and Ankara, are safer than their wild-type counterparts; they can be grown to high titer levels, and are approved for use outside Biosafety containment facilities (Moss 1996; Pacietti 1996).

Viral promoters are required for transgene expression because VV replicates in the cytoplasm with their own transcription system. The promoters are only about 30 bp long. Vaccinia vectors are often designed to express transgenes from the early promoters that are expressed soon after infection and prior to VV cytotoxic effects; however, the early promoters tend to transcribe at relatively low levels. To obtain both early transcription and high levels, it is possible to use certain exogenous promoters, including the bacteriophage T7 and SP6 promoters.

GFP-expressing Vaccinia Virus

GFP expressed from VV will be useful for analyzing protein processing and trafficking, antigen presentation, viral transport, and protein structure and function relationships (Carroll and Moss 1997; Moss and Ward 2001; Rietdorf et al. 2001). Figure 3 shows HeLa cells that were infected with a recombinant VV that expresses GFP attached to the carboxyl terminus of the viral B5R membrane protein (Image kindly provided by Brian Ward and Bernard Moss). The infected cells were fixed, permeabilized, and stained with rhodamine-conjugated phalloidin (red) to visualize F-actin. Expression of B5R-GFP leads to the fluorescent labeling of enveloped virions (green), which collect in the periphery of the cell and can be seen on the tips of thick actin tails (arrows).

PLANT VIRUS VECTORS

Plant viruses such as tobacco mosaic virus (TMV) and cowpea chlorotic mottle virus (CCMV), can constitute as much as 10% of the dry weight of a plant. Large amounts of

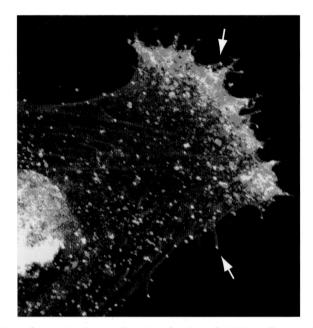

FIGURE 3. Visualization of vaccinia virus at the tips of actin tails. HeLa cells were infected with a recombinant vaccinia virus that expresses GFP attached to the carboxyl terminus of the viral B5R membrane protein (B5R-GFP). Infected cells were fixed, permeabilized, and stained with rhodamine-conjugated phalloidin (*red*) to visualize F-actin. Expression of B5R-GFP leads to the fluorescent labeling of enveloped virions (*green*), which collect in the periphery of the cell and can be seen on the tips of thick actin tails (*arrows*). (Image kindly provided by Brian M. Ward and Bernard Moss.)

pure, properly folded, biologically active proteins can be obtained from transgenic plants. Plant-virus-based vectors provide one of the most economical sources of biomass production and therefore are attractive candidates for the large-scale production of commercial products (Pogue et al. 2002). However, such production would require widespread inoculation and dissemination of many acres of plants with viral vectors. The potential spread of the engineered virus to susceptible plants and the accumulation of recombinant viruses in the environment fuels concerns about the use of plant-virus-based vectors (Pogue et al. 2002). Thus, GFP-expressing plant viruses are now being used not only to understand basic plant biology and pathology, but are to examine the efficacy and safety of plant-based biotechnology by determining the spread of plant viruses between plants and potentially between plants and animals. In addition, these viruses can be studied with respect to cell-to-cell movement, systemic infection, and the induction of disease.

TOBACCO MOSAIC VIRUS

TMV has contributed greatly to molecular biology research for more than a century (Creager et al. 1999). In 1892, TMV was shown to be the agent of tobacco mosaic disease as it could pass through a filter capable of retaining bacteria (Ivanowski 1892). TMV RNA was used in the first decisive experiments demonstrating that nucleic acids carry hereditary information and that nucleic acid alone can be sufficient for viral infectivity (Fraenkel-Conrat 1956; Gierer and Schramm 1956).

Virus Properties

The TMV genome usually consists of a monomer, or multiple molecules of plus-sense RNA that is encapsulated by identical viral coat proteins that stack in a helix around the RNA molecule (van Regenmortel et al. 2000; Fields et al. 2001). Virions usually contain one molecule of linear single-stranded RNA 6395 nucleotides (J02415; X68110) long. The genome is capped at the 5′ end and has a tRNA-like structure at the 3′ end. The genome encodes two structural virion proteins, a 26.5-kD protein of unknown function and a 17.5-kD coat protein. There are also four nonstructural proteins that encode a replicase protein, a helicase protein, and two other unknown proteins. The TMV genome likely replicates on membranous structures in the cytoplasm. The virions are nonenveloped rod-shaped helices ranging from 70 to 300 nm in length with an 18-nm-diameter and a 2-nm-diameter axial canal. The capsid is made up of 2130 identical capsomers that form in a highly ordered, right-handed helical manner around the RNA genome. TMV is found in all parts of the host plant, and it infects cells via direct contact with wounded areas on plant surfaces.

Vector Properties

The advantage of TMV-based viral vectors is that they can constitute as much as 10% of the dry weight of a plant, which potentially permits copious amounts foreign gene products to be harvested from infected plants within days of inoculation. The 6395-nucleotide TMV genome is relatively small and can easily be manipulated. Leader sequences that target a transgene to specific locations in the cell, such as the apoplast, cytosol, or endomembrane (Yusibov et al. 1999), may be used to study plant biology and virus transmission and stability or to purify desired protein products. A TMV-based transient gene expression vector, 30B, has been constructed that expresses GFP (30B-GFP) (Shivprasad et al. 1999). The 30B expression vector contains heterologous sequences from tobacco mild green mosaic virus U5, including the coat protein subgenomic mRNA promoter, the coat protein ORF, and the 3′ nontranslated region.

One of the main disadvantages of TMV-based vectors is that hybrid viruses can develop within plants by deleting the foreign gene and repeated sequences, retaining only those sequences required for optimal replication and movement (Dawson et al. 1989). The 30B-GFP vector has been used in challenge experiments that assessed the competitiveness of the recombinants with the parent virus (Rabindran and Dawson 2001). It was found that the recombinants were outcompeted by and less pathogenic than TMV, data that not only were insightful for studying viral recombination, but also addressed the concern of the persistence of recombinant viruses in the environment compared with their wild-type counterparts. It has been noted that the stability of the transgene can be increased if duplicated sequences are not included in the vector.

TURNIP CRINKLE VIRUS

The turnip crinkle virus (TCV) is a small plant virus with a broad host range that, although not yet well studied, shows promise in its usefulness as a plant vector.

Viral Properties

The TCV genome is composed of two linear, positive-sense single-stranded RNA molecules that are 4.2 kb long (Brunt et al. 1996; Qu and Morris 1999). TCV is a single-stranded RNA virus of the *Carmovirus* group (Carrington et al. 1987) (Accession code M22445). The viral RNA genome encodes two subgenomic RNAs and contains five ORFs: p28 and p88, which specify an RNA-dependent RNA polymerase; p38, which specifies the viral coat protein; and p8 and p9, which are essential for cell-to-cell movement (Qu and Morris 1999). The virions are isometric and nonenveloped with a diameter of 28 nm. Virions can be found in all parts of the host plant. TCV is thought to be transmitted by insect vectors (*Phyllotreta* and *Psylloides)* and by contact between plants (van Regenmortel et al. 2000; Fields et al. 2001).

Vector Properties

There are several advantages to developing TCV plant vectors. TCV has a small genome that can be easily manipulated using conventional cloning techniques, and it has a broad host range, including the ability to infect commonly studied plants such as tobacco and *Arabidopsis* (Qu and Morris 1999). Finally, there are several TCV-resistant lines or mutants that affect TCV replication in *Arabidopsis*. One minor drawback with TCV is that limited research has been done with it at this point.

GFP-expressing TCV

GFP-expressing TCV has been used to study viral movement and symptomatology (Cohen et al. 2000). The constructs may be used in combination with different mutants and inoculation strategies to study mechanisms underlying cell-to-cell movement, systemic infection, and disease induction. GFP-expressing TCV may provide a valuable tool to screen *Arabidopsis* for plasmodesmata mutations. Plant cells are separated by cell walls; thus, viruses have evolved to use plant cell intercellular connections termed plasmodesmata for cell-to-cell transport (McLean et al. 1997). GFP-expressing TCV could be used to visualize the mechanisms by which viral movement proteins facilitate transport of viral nucleic acids through plant plasmodesmata and into neighboring cells (Lazarowitz and Beachy 1999).

CONCLUSIONS

Viruses are provocative by nature—they are the causative agents of many annoying and other deadly diseases, yet their potential impact as indispensable tools in cell biology research and gene therapy are just beginning to be realized. Viruses are ubiquitous throughout nature and have evolved to infect a myriad of host cells using numerous infection routes in order to express their own genes and propagate themselves. More than a century after TMV was determined to be the causative agent of plant disease, the mechanisms of viral replication and infectivity are only now beginning to be understood. Virologists will continue to adapt viruses to numerous experimental and clinical objectives by better understanding how the nearly infinite number of viruses continue to infect all known life forms.

By combining high-resolution confocal or 2-photon microscopy (see Chapters 14 and 15) with GFP-expressing viruses, investigators will be able to gain dynamic insight into subcellular events that regulate cellular biology. Just imagine being able to label and track thousands of cells simultaneously to understand how cells and tissues move in tightly choreographed routines in order to form an embryo or a plant. Imagine being able to label cells with a fluorescent tag using multicolor GFP-expressing viruses, and then following these cell and tissue movements using multispectral, time-lapse fluorescence microscopy in three dimensions. Imagine analyzing the recorded data using automated cell-tracking and color discrimination software capable of distinguishing the subtle movements of thousands of individual cells and identifying a handful of genes expressed in these cells. Gene expression patterns and cell migration data collected using confocal and 2-photon laser-scanning microscopy could be integrated with MRI-collected data sets in order to compare subcellular resolution data in the anatomical context of the whole embryo in three dimensions. By combining the unique ability of viruses to efficiently introduce GFP into diverse cells with that of new imaging and image analysis platforms, our imaginations and our time will once again be our limiting factors.

REFERENCES

Armentano D., Zabner J., Sacks C., Sookdeo C.C., Smith M.P., St George J.A., Wadsworth S.C., Smith A.E., and Gregory R.J. 1997. Effect of the E4 region on the persistence of transgene expression from adenovirus vectors. *J. Virol.* **71:** 2408–2416.

Baldick C.J., Jr. and Moss B. 1993. Characterization and temporal regulation of mRNAs encoded by vaccinia virus intermediate-stage genes. *J. Virol.* **67:** 3515–3527.

Bergelson J.M., Cunningham J.A., Droguett G., Kurt-Jones E.A., Krithivas A., Hong J.S., Horwitz M.S., Crowell R.L., and Finberg R.W. 1997. Isolation of a common receptor for Coxsackie B viruses and adenoviruses 2 and 5. *Science* **275:** 1320–1323.

Brunt A., Crabtree K., Dallwitz M., Gibbs A., Watson L., and Zurcher E. 1996. Plant viruses online: Descriptions and lists from the VIDE database. At http://biology.anu.edu.au; http://image.fs.uidaho.edu/vide.

Burns J.C., Friedmann T., Driever W., Burrascano M., and Yee J.K. 1993. Vesicular stomatitis virus G glycoprotein pseudotyped retroviral vectors: Concentration to very high titer and efficient gene transfer into mammalian and nonmammalian cells. *Proc. Natl. Acad. Sci.* **90:** 8033–8037.

Carrington J.C., Morris T.J., Stockley P.G., and Harrison S.C. 1987. Structure and assembly of turnip crinkle virus. IV. Analysis of the coat protein gene and implications of the subunit primary structure. *J. Mol. Biol.* **194:** 265–276.

Carroll M.W. and Moss B. 1997. Poxvirus as expression vectors. *Curr. Opin. Biotechnol.* **8:** 573–577.

Chen B.E., Lendvai B., Nimchinsky E.A., Burbach B., Fox K., and Svoboda K. 2000. Imaging high-resolution structure of GFP-expressing neurons in neocortex in vivo. *Learn. Mem.* **7:** 433–441.

Chen H.H., Mack L.M., Kelly R., Ontell M., Kochanek S., and Clemens P.R. 1997. Persistence in muscle of an adenoviral vector that lacks all viral genes. *Proc. Natl. Acad. Sci.* **94:** 1645–1650.

Coffin J.M., Hughes S.H., and Varmus H.E. 1997. *Retroviruses.* Cold Spring Harbor Laboratory Press, Cold Spring Harbor, New York.

Cohen Y., Gisel A., and Zambryski P.C. 2000. Cell-to-cell and systemic movement of recombinant green fluorescent protein-tagged turnip crinkle viruses. *Virology* **273:** 258–266.

Crameri A., Whitehorn E.A., Tate E., and Stemmer W.P. 1996. Improved green fluorescent protein by molecular evolution using DNA shuffling. *Nat. Biotechnol.* **14:** 315–319.

Creager A.N.H., Scholthof K.G., Citovsky V. and Scholthof H.B. 1999. Tobacco mosaic virus: Pioneering Research for a Century. *Plant Cell* **11:** 301–308.

Dawson W.O., Lewandowski D.J., Hilf M.E., Bubrick P., Raffo A.J., Shaw J.J., Grantham G.L., and Desjardins P.R. 1989. A tobacco mosaic virus-hybrid expresses and loses an added gene. *Virology* **173:** 285–292.

Dubensky T.W., Polo J.M., and Jolly D.J. 2000. Alphavirus-based vectors for vaccine and gene therapy applications. In *Gene therapy: Therapeutic mechanisms and strategies* (ed. N.S. Templeton and D.D. Lasic), pp. 109–130. Marcel Dekker, New York.

Dubensky T.W., Jr., Driver D.A., Polo J.M., Belli B.A., Latham E.M., Ibanez C.E., Chada S., Brumm D., Banks T.A., Mento S.J., Jolly D.J., and Chang S.M. 1996. Sindbis virus DNA-based expression vectors: Utility for in vitro and in vivo gene transfer. *J. Virol.* **70:** 508–519.

Dull T., Zufferey R., Kelly M., Mandel R.J., Nguyen M., Trono D., and Naldini L. 1998. A third-generation lentivirus vector with a conditional packaging system. *J. Virol.* **72:** 8463–8471.

Ehrengruber M.U., Hennou S., Bueler H., Naim H.Y., Deglon N., and Lundstrom K. 2001. Gene transfer into neurons from hippocampal slices: Comparison of recombinant Semliki Forest virus, adenovirus, adeno-associated virus, lentivirus, and measles virus. *Mol. Cell. Neurosci.* **17:** 855–871.

Ehrengruber M.U., Lundstrom K., Schweitzer C., Heuss C., Schlesinger S., and Gahwiler B.H. 1999. Recombinant Semliki Forest virus and Sindbis virus efficiently infect neurons in hippocampal slice cultures. *Proc. Natl. Acad. Sci.* **96:** 7041–7046.

Engelhardt J.F., Ye X., Doranz B., and Wilson J.M. 1994. Ablation of E2A in recombinant adenoviruses improves transgene persistence and decreases inflammatory response in mouse liver. *Proc. Natl. Acad. Sci.* **91:** 6196–6200.

Fenner F., Wittek R., and Dumbell K.R. 1989. *The orthopoxvirus.* Academic Press, San Diego.

Fields B.N., Howley P.M., Griffin D.E., Lamb R.A., Martin M.A., Roizman B., Straus S., and Knipe D. 2001. *Field's virology.* Lippincott Williams & Wilkins, Philadelphia.

Flint S.J., Enquist L.W., Krug R.M., Racaniello V.R., and Skalka A.M. 2000. *Principles of virology.* ASM Press, Washington, D.C.

Fraenkel-Conrat H. 1956. The role of the nucleic acid in the reconstitution of active tobacco mosaic virus. *J. Am. Chem. Soc.* **78:** 882–883.

Frankel A.D. and Young J. 1998. HIV-1: Fifteen proteins and an RNA. *Annu. Rev. Biochem.* **67:** 1–25.

Gierer A. and Schramm G. 1956. Infectivity of ribonucleic acid from tobacco mosaic virus. *Nature* **177:** 702–703.

Goebel S.J., Johnson G.P., Perkus M.E., Davis S.W., Winslow J.P., and Paoletti E. 1990. The complete DNA sequence of vaccinia virus. *Virology* **179:** 247–266.

Harris J.D. and Lemoine N.R. 1996. Strategies for targeted gene therapy. *Trends Genet.* **12:** 400–405.

Hong S.S., Karayan L., Tournier J., Curiel D.T., and Boulanger P.A. 1997. Adenovirus type 5 fiber knob binds to MHC class I alpha2 domain at the surface of human epithelial and B lymphoblastoid cells. *EMBO J.* **16:** 2294–2306.

Ivanowski D. 1892. Über die Mosaikkrankheit der Tabakspflanze. *Bull. Acad. Imp. Sci. St. Petersb., Nouv. Sér* **3:** 67–70.

Jaenisch R., Jahner D., Nobis P., Simon I., Lohler J., Harbers K., and Grotkopp D. 1981. Chromosomal position and activation of retroviral genomes inserted into the germ line of mice. *Cell* **24:** 519–529.

Jahner D., Stuhlmann H., Stewart C.L., Harbers K., Lohler J., Simon I., and Jaenisch R. 1982. De novo methylation and expression of retroviral genomes during mouse embryogenesis. *Nature* **298:** 623–628.

Lazarowitz S.G. and Beachy R.N. 1999. Viral movement proteins as probes for intracellular and intercellular trafficking in plants. *Plant Cell* **11:** 535–548.

Lewis P.F., Hensel M., and Ernerrnan M. 1992. Human immunodeficiency virus infection of cell arrested in the cell cycle. *EMBO J.* **11:** 3053–3058.

Liljestrom P. and Garoff H. 1991. A new generation of animal cell expression vectors based on the Semliki Forest virus replicon. *BioTechnology* **9:** 1356–1361.

Lois C., Hong E.J., Pease S., Brown E.J., and Baltimore D. 2002. Germline transmission and tissue-specific expression of transgenes delivered by lentiviral vectors. *Science* **295:** 868–872.

Lundstrom K. 1999. Alphaviruses as tools in neurobiology and gene therapy. *J. Recept. Signal Transduct. Res.* **19:** 673–686.

Lundstrom K., Abenavoli A., Malgaroli A., and Ehrengruber M.U. 2003. Novel Semliki Forest virus vectors with reduced cytotoxicity and temperature sensitivity for long-term enhancement of transgene expression. *Mol. Ther.* **7:** 202–209.

Maddon P.J., McDougal J.S., Clapham P.R., Dalgleish A.G., Jamal S., Weiss R.A., and Axel R. 1988. HIV infection does not require endocytosis of its receptor, CD4. *Cell* **54:** 865–874.

Maletic-Savatic M., Malinow R., and Svoboda K. 1999. Rapid dendritic morphogenesis in CA1 hippocampal dendrites induced by synaptic activity. *Science* **283:** 1923–1937.

McLean B.G., Hempel F.D., and Zambryski P.C. 1997. Plant intercellular communication via plasmodesmata. *Plant Cell* **9:** 1043–1054.

Miller D.G., Adam M.A., and Miller A.D. 1990. Gene transfer by retrovirus occurs only in cells that are actively replicating at the time of infection. *Mol. Cell. Biol.* **10:** 4239–4242.

Moss B. 1996. Genetically engineered poxviruses for recombinant gene expression, vaccination, and safety. *Proc. Natl. Acad. Sci.* **93:** 11341–11348.

Moss B. and Ward B.M. 2001. High-speed mass transit for poxviruses on microtubules. *Nat. Cell Biol.* **3:** E245–246.

Naldini L., Blomer U., Gage F.H., Trono D., and Verma I.M. 1996. Efficient transfer, integration, and sustained long-term expression of the transgene in adult rat brains injected with a lentiviral vector. *Proc. Natl. Acad. Sci.* **93:** 11382–11388.

Pacietti E. 1996. Applications of pox virus vectors to vaccination: An update. *Proc. Natl. Acad. Sci.* **93:** 11349–11353.

Pogue G.P., Lindbo J.A., Garger S.J., and Fitzmaurice W.P. 2002. Making an ally from an enemy: Plant virology and the new agriculture. *Annu. Rev. Phytopathol.* **40:** 45–74.

Polo J.M., Belli B.A., Driver D.A., Frolov I., Sherrill S., Hariharan M.J., Townsend K., Perri S., Mento S.J., Jolly D.J., Chang S.M., Schlesinger S., and Dubensky T.W., Jr. 1999. Stable alphavirus packaging cell lines for Sindbis virus and Semliki Forest virus-derived vectors. *Proc. Natl. Acad. Sci.* **96:** 4598–4603.

Price J., Turner D., and Cepko C. 1987. Lineage analysis in the vertebrate nervous system by retrovirus-mediated gene transfer. *Proc. Natl. Acad. Sci.* **84:** 156–160.

Qin H. and Gunning P. 1997. The 3′-end of the human beta-actin gene enhances activity of the beta-actin expression vector system: Construction of improved vectors. *J. Biochem. Biophys. Methods* **36:** 63–72.

Qu F. and Morris J.T. 1999. Carmoviruses (Tombusviridae). In *Encyclopedia of virology* (ed. A. Granoff and R.G. Webster), pp. 243–247. Academic Press, San Diego.

Rabindran S. and Dawson W.O. 2001. Assessment of recombinants that arise from the use of a TMV-based transient expression vector. *Virology* **284:** 182–189.

Rietdorf J., Ploubidou A., Reckmann I., Holmstrom A., Frischknecht F., Zettl M., Zimmermann T., and Way M. 2001. Kinesin-dependent movement on microtubules precedes actin-based motility of vaccinia virus. *Nat. Cell Biol.* **3:** 992–1000.

Roe T., Reynolds T.C., Yu G., and Brown P.O. 1993. Integration of murine leukemia virus DNA depends on mitosis. *EMBO J.* **12:** 2099–2108.

Shivprasad S., Pogue G.P., Lewandowski D.J., Hidalgo J., Donson J., Grill L.K., and Dawson W.O. 1999. Heterologous sequences greatly affect foreign gene expression in tobacco mosaic virus-based vectors. *Virology* **255:** 312–323.

Smith L.C., Eisensmith R.C., and Woo S.L. 1995. Gene therapy in heart disease. *Adv. Exp. Med. Biol.* **369:** 79–88.

Tsien R.Y. 1998. The green fluorescent protein. *Annu. Rev. Biochem.* **67:** 509–544.

van Regenmortel M.H.V., Fauquet C.M., Bishop D.H.L., Carstens E.B., Estes M.K., Lemon S.M., Maniloff J., Mayo M.A., McGeoch D.J., Pringle C.R., and Wickner R.B. 2000. *Virus taxonomy: The classification and nomenclature of viruses. The Seventh Report of the International Committee on Taxonomy of Viruses.* Academic Press, San Diego.

Verma I.M. and Somia N. 1997. Gene therapy—Promises, problems and prospects. *Nature* **389:** 239–242.

Wahlfors J.J., Zullo S.A., Loimas S., Nelson D.M., and Morgan R.A. 2000. Evaluation of recombinant alphaviruses as vectors in gene therapy. *Gene Ther.* **7:** 472–480.

Weinberg J.B., Matthews T.J., Cullen B.R., and Malim M.H. 1991. Productive human immunodeficiency virus type 1 (HIV-1) infection of nonproliferating human monocytes. *J. Exp. Med.* **174:** 1477–1482.

Wickham T.J., Mathias P., Cheresh D.A., and Nemerow G.R. 1993. Integrins alpha v beta 3 and alpha v beta 5 promote adenovirus internalization but not virus attachment. *Cell* **73:** 309–319.

Xiong C., Levis R., Shen P., Schlesinger S., Rice C.M., and Huang H.V. 1989. Sindbis virus: An efficient, broad host range vector for gene expression in animal cells. *Science* **243:** 1188–1191.

Yang T.T., Cheng L., and Kain S.R. 1996. Optimized codon usage and chromophore mutations provide enhanced sensitivity with the green fluorescent protein. *Nucleic Acids Res.* **24:** 4592–4593.

Yusibov V., Shivprasad S., Turpen T.H., Dawson W., and Koprowski H. 1999. Plant viral vectors based on tobamoviruses. *Curr. Top. Microbiol. Immunol.* **240:** 81–94.

Zavada J. 1982. The pseudotypic paradox. *J. Gen. Virol.* **63:** 15–24.

Zhang J., Campbell R.E., Ting A.Y., and Tsien R.Y. 2002. Creating new fluorescent probes for cell biology. *Nat. Rev. Cell Biol.* **3:** 906–918.

Zolotukhin S., Potter M., Hauswirth W.W., Guy J., and Muzyczka N. 1996. A "humanized" green fluorescent protein cDNA adapted for high-level expression in mammalian cells. *J. Virol.* **70:** 4646–4654.

Zufferey R., Nagy D., Mandel R.J., Naldini L., and Trono D. 1997. Multiply attenuated lentiviral vector achieves efficient gene delivery in vivo. *Nat. Biotechnol.* **15:** 871–875.

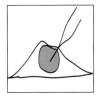

CHAPTER FOUR

Gene Delivery by Direct Injection and Facilitation of Expression by Mechanical Stretch

David A. Dean

Division of Pulmonary and Critical Care Medicine and Department of Microbiology-Immunology, Feinberg School of Medicine, Northwestern University, Chicago, Illinois 60611

INTRODUCTION

MODERN CELL BIOLOGY RELIES EVER MORE HEAVILY ON MOLECULAR TOOLS to facilitate the study of cellular processes. With the advent of recombinant DNA technology, reverse transcriptase–polymerase chain reaction (RT-PCR) to amplify almost any gene, and a number of fluorescent proteins that can be fused to any desired target protein, functional studies of the roles of proteins within cells have exploded. After creating the appropriate fusion protein plasmid (or any plasmid for that matter), the DNA must be delivered to the cell for expression and function studies. Several common approaches to DNA delivery include liposome-mediated transfection, electroporation, and direct DNA delivery by microinjection. This chapter focuses primarily on methods to directly inject expression plasmids into cells and the use of mechanical stretch to increase gene delivery and expression.

CONSIDERATIONS FOR DNA MICROINJECTION

Nuclear versus Cytoplasmic Microinjection

Direct microinjection of genetic material into cells began in the 1970s with the demonstration that mRNA isolated from one cell type could be translated in other cells after it was microinjected into the recipients (Graessmann and Graessmann 1971; Gurdon et al. 1971; Lane et al. 1971). In 1980, it was shown that when a plasmid expressing thymidine kinase was microinjected into the nuclei of thymidine-kinase-deficient mouse fibroblasts, between 50% and 100% of cells showed enzyme activity at 24 hours postinjection (Capecchi 1980). In contrast, gene expression was not detected in more than 1000 cytoplasmically injected cells during the same time frame. Similar results have been obtained in numerous systems (Graessman et al. 1989; Mirzayans et al. 1992; Thornburn and Alberts 1993; Zabner et al. 1995; Dean et al. 1999). These results demonstrate two factors:

51

First, expression of foreign genes can be obtained easily in microinjected cells. If the frequency of gene expression following nuclear injection is equal to or greater than 50%, microinjection of less than 100 cells can result in many expressing cells to study. Second, the site of injection can be critically important. Implicit in these experiments is the fact that the microinjected cells did not divide during the course of the experiments. During mitosis, the nuclear envelope breaks down, eliminating a major barrier to gene transfer. If plasmids are present in the cytoplasm, they have full access to the nuclear compartment during this stage of the cell cycle. However, in nondividing cells, the nuclear envelope provides a substantial barrier to the DNA, as is seen in the above experiments. These experiments would suggest that plasmids are incapable of entering the nuclei of nondividing cells, but this is not the case. Data from a number of labs suggest that certain DNA sequences can promote nuclear entry and gene expression even in nondividing cells in a sequence-specific manner (Graessman et al. 1989; Dean 1997; Dean et al. 1999; Vacik et al. 1999; Mesika et al. 2001). However, to obtain maximal expression of desired transgenes in cells that will not undergo mitosis during the course of the experiment, microinjection of the DNA into the nucleus is strongly recommended.

DNA Concentration

Gene expression is dependent on the copy number of the gene. Thus, in microinjected cells, expression is dependent on the number of plasmids delivered to the cell. It has been shown that as few as one to three CMV (cytomegalovirus) promoter-driven, green fluorescent protein (GFP)-expressing plasmids injected into the nucleus of a cell can result in detectable, albeit low, GFP expression in a reasonable percentage of injected cells (Dean et al. 1999). Increasing the number of plasmids delivered to the cell increases both the percentage of cells expressing the gene product and the amount of gene product (Fig. 1) (Graessman et al. 1989; Dean et al. 1999; Ludtke et al. 2002). In typical experiments, DNA should be injected at a starting concentration of about 300 ng/μl, but it can be used at concentrations between 50 ng/μl and 500 ng/μl if sufficient expression is not obtained at first. A typical microinjection system is designed to deliver less than 10% of the total cell volume, which corresponds to an injection volume of approximately 10^{-14} to 10^{-12} liters (0.01–1 pl). At a concentration of 300 ng/μl, about 5000 copies of a 6-kb plasmid would be delivered to the cell, assuming that 0.1 pl is delivered. Greater concentrations of DNA can be delivered to the cell, but two problems arise. First, gene expression saturates above

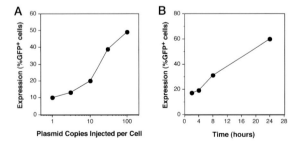

FIGURE 1. Dose and time dependency of transgene expression in microinjected cells. (*A*) Human umbilical vein endothelial cells (HUVECs) were grown on etched coverslips and microinjected with various copy numbers of pEGFP-N1 (Clontech) expressing enhanced GFP from the CMV immediate-early promoter. Eight hours later, enhanced GFP-expressing cells were counted and expressed as a percentage of cells injected. (*B*) HUVECs grown on etched coverslips were microinjected with ten copies of pEGFP-N1, and GFP expression was assessed at the indicated times following injection.

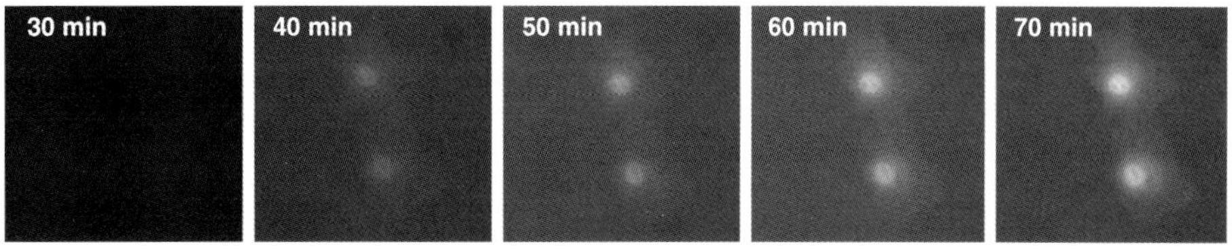

FIGURE 2. Early time course of gene expression in microinjected cells. TC7 cells (African green monkey kidney epithelial cells) were microinjected with pEGFP-N1 at 300 ng/μl and assessed for enhanced GFP expression at the indicated times. All photographs were taken with the same exposure time. As can be seen, enhanced GFP is first detected in these cells at 40 minutes postinjection, and the expression increases with time.

several hundred to a thousand copies of DNA per cell, so the benefit of delivering more plasmids will be lost. Second, at concentrations above 1 mg/ml, DNA becomes technically difficult to inject due to viscosity and aggregation. Furthermore, additional problems of injecting too much DNA also include potential toxicity of the gene product and other problems associated with overexpression. Thus, lower concentrations are a better starting point.

Timing of Gene Expression

Most studies assess gene expression at 4–24 hours postnuclear injection. Unless very low copy numbers of plasmids are injected into the cell, significant expression can be detected by the earlier time point. One parameter that can affect the timing of gene expression is promoter strength: Weak promoters usually take longer to produce sufficient protein to visualize. However, when using strong promoters, expression at early times after injection is readily detectable. Indeed, using a DNA concentration of 300 ng/μl, GFP expression from the CMV promoter (e.g., pEGFP-N1 from Clontech) can be detected in cells easily within 40 minutes of microinjection (Figs. 1 and 2). Similar results have been obtained using plasmids expressing products from the SV40 early promoter and the CMV immediate-early promoter/enhancer.

Microinjection Needles

Microinjection needles can either be pulled from glass capillaries on a pipette puller in the laboratory or be purchased premade and sterile from a number of companies. The advantage of pulling needles in the laboratory is that a variety of different needle types can be pulled, depending on the samples and cells being injected. However, once the settings are established for pulling the desired needle type, the parameters are seldom changed. The protocols for using two pipette pullers are described below, one using a high-end Flaming/Brown pipette puller from Sutter Instruments, and one using a less expensive alternative that produces fine needles, but with a little more user input. An added advantage of pulling needles in the lab is cost; once a pipette puller has been purchased, boxes of glass capillaries are inexpensive (a box of 500 is usually less than $50). The advantages to buying preformed and sterilized needles include increased uniformity of needles from one to another, ease of use (open the packet, fill the needle, and inject), high quality, and not having to invest in a pipette puller. However, premade microinjection needles can be expensive (up to $5 each at the time of publication).

Several variables must be addressed for pulling needles, including filament design, heat, pull strength (tension), and delay time between heating and pulling. There are several types of filaments that will heat the capillaries, including box (surrounds the capillary in a box) and trough (a "U" shape) filaments. On the less expensive pullers, the trough is sufficient for most needle types; however, for the high-end instruments, refer to the manufacturer for recommendations according to the desired pipettes. The heat setting will affect the length and tip size of the needle; high heat will typically produce longer needles and finer tips. The pull strength will also affect length and tip size, with greater pull strength producing longer tapered needles with finer tips. Finally, shorter delay times between heating and pulling can result in longer tapers and finer needles, but if the time is too short, the glass will form fibers resembling glass wool and no needle will be formed.

Micromanipulation and Microinjection Systems

There are two common types of microinjection systems, one of which uses a constant flow of sample and the other a pulsed flow. The former is very simple and can be accomplished on a relatively low budget. In this method, a constant flow of sample is delivered from the tip of the pipette and the amount of sample injected into the cell is determined by how long the pipette remains in the cell (Graessman and Graessman 1986). Although this means that each cell will receive a slightly different amount of sample, with practice, microinjections can become highly reproducible. A typical system is composed of a pressure regulator that can be adjusted for two pressures, back pressure and injection pressure (e.g., World Precision Instruments Pneumatic PicoPump PV830), a capillary holder, and a coarse and fine micromanipulator (e.g., Narishige, World Precision Instruments, Stoelting, etc). In this case, using a manual micromanipulator, the needle is positioned above the cell to be injected and lowered into the cell (Fig. 3). As the needle is lowered, the cell is slightly deformed because the tip is entering at an angle. Because sample is constantly flowing out of the needle, this may not be suitable for precious samples, although experience shows that even 5 μl of sample is more than enough to inject 1000 cells using this method (D.A. Dean, unpubl.). The second type of microinjection system uses a pulsed flow. The most common system of this type currently being used is the Eppendorf Femtojet injector coupled with the Eppendorf InjectMan. The advantage to these systems is that much more control is available over the injection parameters and hence, variability in injections is reduced. Another nice feature of this system is the dynamics of the injection itself. The needle is positioned over the site to be injected and when a button is pressed to inject, the needle is pulled back in the xy direction to allow a diagonal insertion of the needle into the cell, causing a direct piercing of the cell (Fig. 3). This is fast and may do less damage to the cell.

PLATING CELLS FOR MICROINJECTION

Coverslips for microinjection must be marked so that microinjected cells can be identified at time points after injection. Coverslips can be etched by the user or pre-etched coverslips can be purchased (e.g., Eppendorf CELLocate coverslips) (Fig. 4). User-etched coverslips are slightly more economical, but both types of coverslips work equally well.

Coverslip Etching

1. Place 25 × 25-mm number 1 coverslips on a solid, clean surface (benchtop or in a tissue culture).

**Traditional Micromanipulator/
Microinjection System**   **Eppendorf InjectMan System**

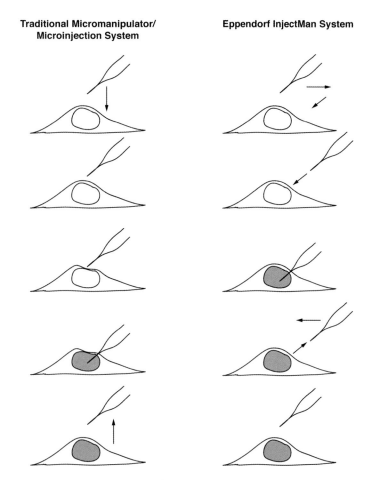

FIGURE 3. Microinjection Systems. A traditional constant pressure microinjection system using a manual micromanipulator is shown at the left. With this system, the needle is positioned over the site of injection (the nucleus), and the needle is lowered directly down into the cell. When the needle touches the cell at an angle, the membranes are distorted slightly until the needle enters the cell to deliver its contents (volume delivered depends on time inside the cell). The needle is then lifted out of the cell to complete the process. On the right is an illustration of the Eppendorf InjectMan system using a motorized micromanipulator. After the needle has been positioned, the controller pulls the needle back in the *xy* direction and then lowers the needle on a diagonal so that the tip directly pierces the cell. After the contents have been delivered (based on the time set on the Femtojet injector), the needle exits the cell on the same diagonal and returns to its original position.

2. Use a diamond pen (diamond scriber VWR 52865-005) to lightly etch an asymmetric figure onto the coverslip (Fig. 4).

   If too much pressure is applied while etching, the coverslip will snap in half or will break later in the experiment (typically after much time and energy has been spent completing all microinjections). However, if too little pressure is applied, the etched figure will be extremely difficult to detect under the microscope, both during microinjection and later when cells are being visualized.

3. Layer etched coverslips onto aluminum wrap that is cut to fit into a 100-mm glass Petri dish. About ten square coverslips can be placed onto one layer (do not overlap, or they will stick together, rendering them useless). Layer aluminum foil and coverslips to fill the dish and cover with the glass top (~15 full layers in a dish).

4. Autoclave the coverslips on dry cycle (20 minutes sterilize, 30 minutes dry).

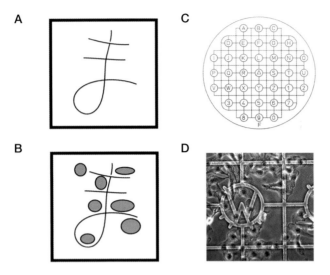

FIGURE 4. Etched coverslips for microinjection. (*A*) Sample of an asymmetric figure etched onto a coverslip; (*B*) areas for microinjection. Because of the asymmetric nature of the etching, different plasmids can be injected at these sites and identified later by their position relative to the figure. (*C*) A CELLocate coverslip; (*D*) cells growing in one area of the CELLocate coverslip.

Plating Cells

1. Place sterile, etched coverslips (or sterile CELLocate coverslips) into the wells of a 6-well plate (1 coverslip per well) and plate cells into the wells as would normally be done.

   Although most cells will adhere to the coverslips, some cells will not. Two approaches can be taken to promote cell adherence. First, the coverslips may need to be cleaned. Coverslips and glass slides may have a thin film of grease on them that can limit cell adherence. To clean etched coverslips, incubate them in 2 N NaOH<!> for 2 hours, rinse them extensively with dH$_2$O, and layer them into the Petri dish prior to autoclaving in Step 3 above. A second way to get cells to attach is to coat the coverslips with extracellular matrix proteins, such as MatriGel or rat tail collagen, after autoclaving.

2. When the cells are at the desired confluency, remove the coverslip from the 6-well dish and transfer them to a 60-mm tissue culture dish containing 5 ml of the appropriate medium. Transfer to microinjection-outfitted microscope.

   If injections and time out of the incubator are limited to less than 20–30 minutes, cells can be maintained without damage in standard medium containing or lacking serum (e.g., DMEM + 10% fetal bovine serum). However, because the lack of CO$_2$ will raise the pH of the medium, many investigators perform microinjections in a buffered medium. This is especially important if longer injection times are needed or if the cells being used are especially sensitive to pH. In this case, the medium can be buffered by the addition of 20 mM HEPES (pH 7.4).

MICROINJECTION NEEDLES

The preparation (pulling) of microinjection needles is described for two models, the Sutter Flaming/Brown Pipette Puller Model P-97 and the PUL-1 Micropipette Puller from World Precision Instruments.

Sutter Flaming/Brown Pipette Puller Model P-97

1. Push the pipette holders together into locked position.

2. Place borosilicate glass capillaries into capillary holders, in locked position.

3. Select program to be run. *For Femtotip-like microinjection pipettes*: Use 10-cm long (1.0-mm [outer diameter, OD], 0.78-mm [inner diameter, ID]) thin-wall borosilicate glass capillaries with filament (Sutter instruments BF100-78-10). The filament on the P-97 Pipette Puller should be a 2.5 × 4.5-mm Box filament (Sutter, FB245B).

Step	Heat	Pull	Velocity	Time	Pressure
1	Ramp	100	10	250	500

This program will loop three times to produce a needle with a 0.5-µm ID tip that is ~100–150 µm long.

For common, long-taper microinjection pipettes: Use 10-cm long (1.0-mm OD, 0.75-mm ID) thin-wall borosilicate glass capillaries with filament (World Precision Instruments, TW100F-4). A FB255B Box filament (2.5 × 2.5 mm) should be used to create these needles.

Step	Heat	Pull	Velocity	Time	Pressure
1	580	none	50	145	500
2	570	115	30	145	500

This program will produce a tip with a 0.5-µm ID that is ~150–200 µm long.

These parameters should be used as starting values and will most likely need to be adjusted to get the desired tips.

4. Carefully remove the needles (two will be produced) and place them in a pipette storage container (Electrode Storage Receptacle, World Precision Instruments, E210 with a 1.0-mm needle insert) with tips down.

   For best results, pull needles within several hours of use. In climates with high humidity, they should be used within an hour. Furthermore, they should be allowed to sit for at least 5 minutes before filling with DNA or protein solutions.

PUL-1 Micropipette Puller (World Precision Instruments)

1. Push the pipette holders together into locked position.

2. Place borosilicate glass capillaries into capillary holders, in locked position.

   It is very important to make sure that the capillaries are held tightly in the holder. If there is any play, the needles can slip a little and result in poor, or no, needles.

3. Set heat to setting 6 and delay to setting 3 for the initial needles using a 1.5-mm U-shaped filament. Move the tension knob on the back to the middle of the range.

4. Pull needles by pressing the Auto button.

5. Remove the needles and observe them by phase-contrast microscopy.

6. Adjust the settings to obtain a needle that has a uniform taper.

   Multiple needles will likely need to be pulled when the instrument is first set up to define the appropriate pulling parameters. However, once the settings have been established, this machine will produce highly reproducible needles for months at a time.

7. Carefully remove the needles (two will be produced) and place them in a pipette storage container (Electrode Storage Receptacle, World Precision Instruments, E210 with a 1.0-mm needle insert) with tips down.

DNA SAMPLE PREPARATION AND LOADING

Plasmid DNA is purified using standard procedures. The resulting preparation can then be delivered into the microinjection needle by either a back-filling or a forward-filling approach.

DNA Sample Preparation

1. Purify the plasmids using commercially available resin-based plasmid purification kits and resuspend DNA in 10 mM Tris-Cl (pH 8.0) and 1 mM EDTA, as described by the manufacturer (e.g., Qiagen, Promega, Invitrogen, Bio-Rad).

    Alternatively, high-quality plasmids can be purified by alkaline lysis and cesium chloride gradient centrifugation, although with the advent of inexpensive and rapid commercially available resin-based kits, the more traditional methods are not as frequently used.

2. Dilute the plasmid to a final concentration between 50 and 500 ng/μl in 0.5× phosphate-buffered saline (PBS).

    Although 5 μl of plasmid solution is sufficient to fill more than ten needles and inject several thousand cells, it is best to make 50–100 μl of DNA solution to facilitate subsequent steps. Other investigators inject DNA suspended in sterile ddH$_2$O, 1× PBS, or other buffered solutions with similar results.

3. Place the DNA solution into the cup of a 0.22-μm cellulose acetate Spin-X centrifuge tube filter (Corning 8160) and centrifuge at 16,000g in a microcentrifuge for 1 minute.

    Alternatively, if enough DNA solution has been made, any 0.22-μm syringe filter will also work. For precious samples, the low retention volume of the Spin-X tubes will prevent solution loss. For volumes less than 50 μl, prewet the Spin-X filter with 0.5× PBS to reduce sample loss.

4. Remove the filter from the Spin-X tube and centrifuge the filtrate at 16,000g for 15 minutes at 4°C to remove any particulate or aggregated matter.

5. Transfer the DNA to a clean tube.

    If extra DNA solution is stored for use at later times (i.e., days), it should always be re-centrifuged as in Step 4 prior to use.

Back-filling Microinjection Needles

1. Withdraw 5 μl of particulate-free DNA in 0.5× PBS into either a 10-μl Hamilton syringe outfitted with a 1.5–2-inch-long needle (Hamilton Microliter number 901), or an Eppendorf Microloader pipette tip (Eppendorf 5242-956.003) on a 10-μl Eppendorf pipette.

2. Carefully insert the syringe (or tip) into the back end of the microinjection needle and dispense ~0.2–0.5 μl of solution while slowly rotating or twisting the microinjection needle between index finger and thumb.

    The twisting of the needle aids in the delivery of the DNA solution and helps pull it down into the tip.

3. Remove the syringe (or tip) from the microinjection needle and place the needle into a second pipette storage container filled with 1 cm of H$_2$O.

    The humidified chamber will help to prevent evaporation of the small volumes of sample in the needles.

4. Use filled needles for microinjection within 1 hour if possible.

Front-filling Microinjection Needles

1. Withdraw 5 µl of particulate-free DNA in 0.5× PBS using a pipettor and place onto a piece of Parafilm.

2. Gently touch the tip of the microinjection needle to the surface of the drop of fluid and allow the fluid to be wicked into the needle by capillary action.

    Many investigators suggest that this approach will also aid in drawing any particulates or dust in the needle up away from the tip and to the top surface of the sample, away from the opening. Also, fewer air bubbles are usually present by front-filling. Thus, needle clogging can be reduced, but if samples are adequately filtered and centrifuged, this is not an issue.

3. Place the filled needle into a second pipette storage container filled with 1 cm of H_2O.

4. Use filled needles for microinjection within 1 hour if possible.

MICROINJECTIONS

The microinjection process is described here for each of two systems, pulsed flow or constant flow (for further details of each system, see the discussion on Micromanipulation and Microinjection Systems above and Fig. 3).

Eppendorf InjectMan and Femtojet System

1. Set starting parameters on Femtojet System:

Injection Pressure (P_i):	145 hPa
Compensation (holding) Pressure (P_c):	35 hPa
Time (t_i):	0.3 sec

    These values will change with the solutions being injected (more viscous solutions need a higher injection pressure) and the individual needles. Although each needle pulled with the same program should have the same internal diameter at the tip, even slight differences can necessitate greatly different injection pressures. The pressures and times indicated are good starting points for the two types of needles pulled above (or purchased Femtotips).

2. Press the Menu button on the Femtojet to allow a new pipette to be loaded onto the device.

3. Place the needle into the pipette holder on the micromanipulator by inserting the large end into the grip head so that the butt of the needle just goes past the compression O-ring in the grip head. Screw the grip head–needle assembly onto the Universal Capillary Holder. For Femtotips, screw the femtotip into the Femtotip adapter on the Capillary Holder.

4. Rotate the Capillary holder back to a position parallel with the micromanipulator.

5. Press the Menu button again to pressurize the system for microinjection.

    CAUTION: When the system is pressurized, it can act like a rocket launcher; if the needle is not fully secured into the grip head, it can shoot off of the capillary holder when the pressure is reapplied. Although small, the needle can cause severe damage to eyes and other tissues, and if biohazardous solutions or samples are in the needle, contamination is a serious concern.

6. Move the capillary forward until the needle tip appears above the objective, and then lower the needle into the medium. Use Fast speed for these motions, until the needle touches the medium.

7. Focus the microscope on the cell monolayer. Start with a 10X objective for microinjections. If there is not enough resolution to distinguish nuclei from cytoplasm, change to a $20\times$ for the injections themselves.

8. Switch speed to slow and continue to lower the needle until a blurry image appears above the cells. If necessary, move the needle in the *x-y* plane to position the tip slightly above the area of cells to be injected. The needle should be ~100 μm above the cells.

9. Press the Clean button to ensure that the needle is not clogged and that the sample will flow easily. Differences in the refractive index of the injected solution and the medium will allow a fan-shaped plume of solution to be seen exiting the tip of the needle.

   If the solution exiting the needle flows in an irregular pattern, the needle is more likely bad and may cause cellular damage.

10. Set the injection plane by slowly lowering the needle until it touches and then depresses and pierces a test cell. When the needle is inside the cell, press the Z-Limit button on the micromanipulator controller, and then raise the needle to just above the cells (again, about 100 μm).

11. Position the tip of the needle directly above the nucleus to be injected and press the Inject button on the top of the micromanipulator joystick to initiate the microinjection (Fig. 3).

12. Continue with injections until completed or the needle clogs (see discussion on Needle Blockage in the Troubleshooting section below).

Pneumatic PicoPump (Constant Pressure) and Manual Micromanipulator

1. Attach a tank of nitrogen gas to the pressure inlet on the PicoPump (WPI, PV820 or PV830).

2. Make sure that the injection pressure is off before attaching needle.

   Attach a foot switch (World Precision Instruments, 3260) to the PicoPump via the Remote plug to activate injection pressure. While the switch is depressed, the injection pressure will flow.

   CAUTION: When the system is pressurized, the needle can become a hazardous projectile. Make sure that the needle is secured into the holder and the injection pressure is not activated before attaching.

3. Set the holding pressure to 4 psi and the injection pressure to 40 psi as a starting point.

4. Insert the needle in the 1.0-mm micropipette holder (World Precision Instruments, MPH6S) so that the end of the needle just goes through the gasket; screw the head to the base, locking the needle into the holder.

5. Attach the needle-micropipette holder to a PicoNozzle (World Precision Instruments, 5430-10) and place the PicoNozzle onto the micromanipulator.

6. Advance and lower the needle so that it is just above the medium and over the objective.

7. Lower the needle into the medium and make sure that the hold pressure is high enough to prevent the medium from entering the needle, but low enough so that the sample is not being wasted.

8. Focus the microscope on the cell monolayer. Start with a $10\times$ objective for microinjections. If there is not enough resolution to distinguish nuclei from cytoplasm, change to a $20\times$ for the injections themselves.

9. Lower the needle to just above the cells to be injected and depress the foot switch to ensure that the needle is not clogged and that the sample will flow easily. The injection pressure may need to be increased or decreased, depending on the sample flow seen in the medium.

10. When injections are to begin, press the foot switch to start the sample flowing and then lower the needle rapidly into the cell to be injected and raise it as soon as fluid has been delivered.

> Because the time of injection is not set or uniform, leave the needle in the cell until a very slight change in refractive index of the cell or fluid delivery can be detected. It is important not to leave the needle in the cell too long or the cell will die (ideally, it should be in the cell for less than 0.5 seconds; with practice, this is easy to master). The easiest way is to turn the micromanipulator joystick one-quarter turn so that the needle goes down into the cell and almost immediately turn it one-quarter turn in the opposite direction to raise the needle. All micromanipulators can be adjusted to vary the distance traveled by one turn.

11. Continue with injections until completed or the needle clogs (see below).

TROUBLESHOOTING

Needle Blockage

The two most common problems encountered with microinjection are the ability to create "good" needles and needle clogging. Both are the same. Although user manipulation of the pipette puller parameters is needed to obtain needles with the appropriate taper and tip, with practice and limited initial effort, the parameters can be easily defined. However, once injections are begun, needles will clog and cease to inject. Even with painstaking attention to filtering samples and centrifuging them to remove particulates, clogging will occur. Two ways to overcome this are to increase the injection pressure to get the needle flowing again (i.e., blow out the blockage) and to tap the tip to try to dislodge any particulates (this can also include cell debris or extracellular matrix that becomes attached to needles during the injection process). For constant flow systems, try increasing the injection pressure while gently tapping the needle against a cell-free area on the coverslip. If this does not work, try dragging the needle across one of the etches. The drawback to this is that the needle can also break (but a broken needle and a clogged one are both equally useless). For the pulsed Eppendorf system, press the Clean button on the Femtojet to flush out the needle. If this does not work, use the Clean function on the micromanipulator to rapidly pull the needle out of the medium and return it to its original position. This approach can use the surface tension of the medium to remove attached debris. If none of these methods work, simply replace the blocked needle with a new one.

Cellular Damage

Another common problem that can arise after DNA microinjection is cell damage or death. Even though great care is taken during the microinjection process to treat the cells gently, a fraction of the injected cells will die or not express gene product. Although cells may slightly change their morphology following injection, many of these will survive and express the injected transgene. However, some cells will be injured. Three parameters can be adjusted to try to decrease this damage. First, if a microinjection needle is left in a cell too long, too much sample volume will be delivered and may result in damage. Second, if the injection pressure is too high, again, too much fluid will be injected into the cell. Another problem with too high an injection pressure is that the cell can be "blown up" or damaged by force; fluid exiting a 0.5-μm diameter is expelled at a high pressure and this can damage cellular architecture and function. Thus, by reducing the injection time and the injection pressure, damage may be averted. Another way to decrease cell injury is to adjust the depth of needle penetration in injected cells. Coverslips are uneven, and consequently, cells grow at slightly different heights across the coverslip. Thus, although the injection depth may be fine for one group of cells, it is possible that the needle could push too far into other cells, causing damage. To circumvent this, close attention should be paid to the injection Z-limit.

METHODS TO INCREASE GENE TRANSFER AND EXPRESSION

Apart from increasing the copy number of plasmids injected into cells or increasing the incubation time, there are relatively few methods to increase gene expression. However, recent experiments have shown that mechanical strain can increase gene transfer and expression (Taylor et al. 2003). A number of tissues, including the heart, lung, and vasculature, are constantly exposed to the mechanical strain that can be simulated in cell culture by application of cyclic stretch. Mechanical stretch induces numerous biological responses in cells, including alterations in the cytoskeleton, activation of cell-signaling pathways, and upregulation of transcription factors (Ingber 1997; Li and Xu 2000). Intriguingly, these responses are directly related to the process of gene delivery and expression. Exogenous DNA, either viral or nonviral, must cross the plasma membrane into the cell, travel through the cytoplasm and the cytoskeletal networks, enter the nucleus, and be transcribed in order for a transfection to be successful. By investigating the effects of cyclic stretch on transfection efficiency and subsequent gene expression, it was found that 30 minutes of cyclic stretch applied immediately following DNA delivery to cells can increase gene expression by up to 20-fold in a variety of cell types, including epithelial cells (A549 cells, murine lung epithelial cells, CHO cells, HepG2 cells), smooth muscle cells, skeletal muscle cells, and fibroblasts, among others (Taylor et al. 2003; D.A. Dean, unpubl.). Both the number of cells expressing product and the relative amount of gene expression in a given cell increase with the application of cyclic stretch (Fig. 5). Similar increases in gene expression have been detected in cells transfected either by liposomes (e.g., Lipofectin and Lipofectamine) or by electroporation (a Petripulser electrode is used to electroporate adherent cells, BTX, PP35-2P). Results suggest that cyclic stretch exerts its effects by increasing the intracellular trafficking of the DNA (e.g., increased movement from the cytoplasm to and into the nucleus).

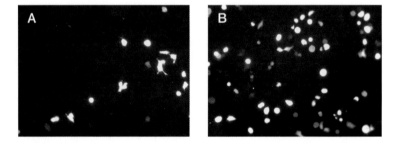

FIGURE 5. Cyclic stretch enhances transfection efficiency. A549 cells were grown under static conditions in Laminin-coated BioFlex plates and transfected with pEGFP-N1 using Lipofectin, according to the manufacturer. Immediately following lipoplex addition, the cells were grown statically for 24 hours (*A*) or stretched (10% change in basement membrane area at 1 Hz) for 30 minutes, followed by 23 hours of growth under static conditions (*B*). Twenty-four hours following DNA addition, GFP expression was observed.

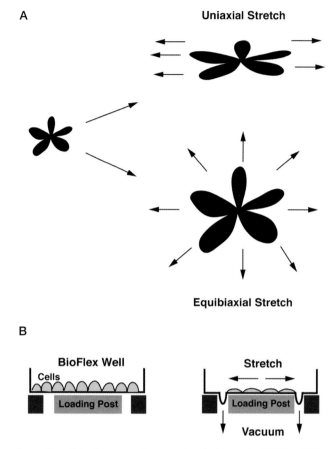

FIGURE 6. Theory and practice of device to produce mechanical stretch. (*A*) The differences between uniaxial (one direction) and equibiaxial (all directions) stretch are illustrated. (*B*) The design of a BioFlex plate is shown at rest and with an applied vacuum. As can be seen, the cells are tall at rest, but are stretched flat with vacuum.

Equibiaxial Stretching Apparatus

Two common types of stretches can be applied to cells: uniaxial (one direction) and equibiaxial (equal in all directions) (Fig. 6). The Flexercell 3000FX (Flexcell International, McKeesport, PA) device allows for both types of strains, depending on the culture plates and posts used in the machine. All of the experiments described here use equibiaxial stretch to increase gene transfer and expression. The machine consists of a base plate that holds four six-well BioFlex plates hooked to a vacuum through a controller unit that can regulate the frequency and amount of vacuum applied to the plates. Bioflex plates are the size of a standard 6-well tissue culture plate, but instead of having a polystyrene bottom, each well has a silastic membrane on which the cells are grown. Plates can be purchased with or without several different extracellular matrix proteins coating the membranes, including laminin, collagen, elastin, and fibronectin. Plates are placed on the base plate (with loading posts inserted), which is housed within a cell culture incubator, and stretch is applied according to the program being run. As vacuum is applied, the membrane pulls across the loading post and deforms at the edge to stretch the membrane and attached cells in all directions (Fig. 6). When the vacuum is released, the membrane returns to its relaxed state. Frequencies and degrees of stretch can be altered, but maximal effects on gene transfer and expression have been seen using 60 cycles/min (1 Hz) and a 10% change in surface area (10% area strain) (Taylor et al. 2003).

Enhancement of Gene Transfer and Expression by Stretch

1. Set up the Flexercell 3000FX Baseplate inside a cell culture incubator. Run vacuum lines through the access port in the incubator wall to the Flexercell 3000FX controller.

2. Plate the cells on BioFlex plates (Flexercell, BF3001U) at a desired density in the appropriate medium.

3. Grow cells for at least 24 hours prior to transfection to ensure proper attachment to the membranes.

4. Transfect cells by method of choice (liposomes, polyethyleneimine [PEI], DEAE-dextran, electroporation) according to published or manufacturer's directions.

   Alternatively, cells can be microinjected at defined areas in the center of the well as described in the preceding protocols in this chapter. To mark injection areas, use a marker to draw an asymmetric figure on the underside of the well.

5. Place a small amount of grease onto the tops of each loading post in the Flexercell Baseplate and on the gaskets.

   The grease will allow the plates to seal against the unit for a vacuum to be applied and will allow membranes to slide without resistance while they are being stretched and distorted.

6. Within 10 minutes of either electroporation or DNA complex addition, place the plates onto the base plate of the Flexercell 3000FX and make sure a seal is made.

7. Place a Plexiglas board onto the plates and place weights (extruded lead brick, Research Products International 2831, or four 1-liter bottles filled with H_2O) on top of the board to keep the plates attached to the base plate during the stretching regimen.

8. Apply cyclic stretch at 10% area stretch and 60 cycles per minute.

9. Apply stretch for between 30 minutes and 24 hours.

   If liposome-mediate transfections are performed in the absence of serum, begin stretching the cells immediately following DNA-liposome complex addition and add serum back to the cells at 2–4 hours following complex addition, according to the transfection reagent manufacturer. To do this, briefly shut off the stretch device, remove the plates to the tissue culture hood, add serum or fresh medium, and return plates to the Baseplate.

10. Remove plates at any time and observe gene expression.

Electroporation of Adherent Cells on Bioflex Plates

1. Plate the cells on Bioflex plates and grow them to the desired density in appropriate medium.

   As for all other transfection methods, subconfluent cells are much easier to transfect; once the cells become quiescent, transfection efficiency decreases dramatically.

2. Rinse the cells with PBS and add 1 ml of plasmid (10 μg) suspended in serum-free and antibiotic-free medium to each well of cells.

3. Place the PetriPulser (BTX, PP35-2P) into the dish so that the electrodes are in the DNA solution, but rest ~1–2 mm above the cell monolayer.

4. Electroporate the cells using one square wave pulse of 10-msec duration at 100 to 160 V, depending on the cell types being electroporated.

5. Remove the electrode, and add 4 ml of serum- and antibiotic-containing medium at 10 minutes postelectroporation.

REFERENCES

Capecchi M.R. 1980. High efficiency transformation by direct microinjection of DNA into cultured mammalian cells. *Cell* **22:** 479–488.

Dean D.A. 1997. Import of plasmid DNA into the nucleus is sequence specific. *Exp. Cell Res.* **230:** 293–302.

Dean D.A., Dean B.S., Muller S., and Smith L.C. 1999. Sequence requirements for plasmid nuclear entry. *Exp. Cell Res.* **253:** 713–722.

Graessmann A. and Graessmann M. 1971. The formation of melanin in muscle cells after the direct transfer of RNA from Harding-Passey melanoma cells. *Hoppe Seylers Z. Physiol. Chem.* **352:** 527–532.

Graessman M. and Graessman A. 1986. Microinjection of tissue culture cells using glass microcapillaries: Methods. In *Microinjection and organelle transplantation techniques: Methods and applications* (ed. J. E. Celis et al.), pp. 3–37. Academic Press, London.

Graessman M., Menne J., Liebler M., Graeber I., and Graessman A. 1989. Helper activity for gene expression, a novel function of the SV40 enhancer. *Nucleic Acids Res.* **17:** 6603–6612.

Gurdon J.B., Lane C.D., Woodland H.R., and Marbaix G. 1971. Use of frog eggs and oocytes for the study of messenger RNA and its translation in living cells. *Nature* **233:** 177–182.

Ingber D.E. 1997. Tensegrity: The architectural basis of cellular mechanotransduction. *Annu. Rev. Physiol.* **59:** 575–599.

Lane C.D., Marbaix G., and Gurdon J.B. 1971. Rabbit haemoglobin synthesis in frog cells: The translation of reticulocyte 9 s RNA in frog oocytes. *J. Mol. Biol.* **61:** 73–91.

Li C. and Xu Q. 2000. Mechanical stress-initiated signal transductions in vascular smooth muscle cells. *Cell. Signal.* **12:** 435–445.

Ludtke J.J., Sebestyen M.G., and Wolff J.A. 2002. The effect of cell division on the cellular dynamics of microinjected DNA and dextran. *Mol. Ther.* **5:** 579–588.

Mesika A., Grigoreva I., Zohar M., and Reich Z. 2001. A regulated, NFkappaB-assisted import of plasmid DNA into mammalian cell nuclei. *Mol. Ther.* **3:** 653–657.

Mirzayans R., Remy A.A., and Malcom P.C. 1992. Differential expression and stability of foreign genes introduced into human fibroblasts by nuclear versus cytoplasmic microinjection. *Mutat. Res.* **281:** 115–122.

Taylor W., Gokay K.E., Capaccio C., Davis E., Glucksberg M.R., and Dean D.A. 2003. Effects of cyclic stretch on gene transfer in alveolar epithelial cells. *Mol. Ther.* **7:** 542–549.

Thornburn A.M. and Alberts A.S. 1993. Efficient expression of miniprep plasmid DNA after needle microinjection into somatic cells. *BioTechniques* **14:** 356–358.

Vacik J., Dean B.S., Zimmer W.E., and Dean D.A. 1999. Cell-specific nuclear import of plasmid DNA. *Gene Ther.* **6:** 1006–1014.

Zabner J., Fasbender A.J., Moninger T., Poellinger K.A., and Welsh M.J. 1995. Cellular and molecular barriers to gene transfer by a cationic lipid. *J. Biol. Chem.* **270:** 18997–19007.

COMPANIES AND RESOURCES

Bio-Rad
1000 Alfred Nobel Drive
Hercules, CA 94547
(800) 424-6723
http://www.biorad.com

Corning
45 Nagog Park
Acton, MA 01720
(800) 492-1110
http://www.corning.com/lifesciences/

BTX Instrument Division
Harvard Apparatus, Inc.
84 October Hill Road
Holliston, MA 01746-1388
(800) 597-0580
http://www.btxonline.com/

Brinkmann-Eppendorf
One Cantiague Road
P.O. Box 1019
Westbury, NY 11590-0207
(800) 645-3050
http://www.eppendorf.com/en/

Flexcell International
500 Fifth Avenue, Suite 501
McKeesport, PA 15132
(800) 728-3714
http://www.flexcellint.com

Hamilton
P.O. Box 10030
Reno, NV 89520-0012
(800) 648-5950
http://www.hamiltoncomp.com

Invitrogen Corporation
1600 Faraday Avenue
P.O. Box 6482
Carlsbad, California 92008
(800) 955-6288
http://www.invitrogen.com

Narishige International USA, Inc.
1710 Hempstead Turnpike
East Meadow, NY 11554
(800) 445-7914
http://www.narishige.co.jp/niusa

Promega Corporation
2800 Woods Hollow Road
Madison, WI 53711
(800) 356-9526
http://www.promega.com

Qiagen
28159 Avenue Stanford
Valencia, CA 91355
(800) 426-8157
http://www1.qiagen.com/

Research Products International (RPI)
10 N. Business Center Drive
Mount Prospect, IL 60056-2190
(800) 323-9814
http://www.rpicorp.com

Stoelting
Physiology Division
620 Wheat Lane
Wood Dale, IL 60191
(630) 860-9700
http://www.stoeltingco.com/physio/

Sutter Instruments
51 Digital Drive
Novato, CA 94949
(415) 883-0128
http://www.sutter.com/

VWR
1310 Goshen Parkway
West Chester, PA 19380
(800) 932-5000
http://www.vwr.com

World Precision Instruments (WPI)
175 Sarasota Center Boulevard
Sarasota, Florida 34240
(941) 371-1003
http://www.wpiinc.com

Microinjection of Fluorophore-labeled Proteins

Yulia Komarova, John Peloquin, and Gary Borisy

Department of Cell and Molecular Biology, Northwestern University Medical School, Chicago, Illinois 60611

INTRODUCTION

VISUALIZATION OF THE DYNAMICS OF PROTEIN ASSEMBLIES in living cells is important to connect in vitro biochemical kinetics and in vivo intracellular processes. Direct-pressure microinjection with a micropipette is an essential tool for introducing a variety of impermeant substances into the cytoplasm or nucleus of plant and animal cells. Since 1951, when capillary electrodes were used for the first time to introduce small molecules into living cells, the microinjection technique has been developed and adopted for numerous studies. Cellular organelles, DNA and RNA, enzymes, structural proteins, metabolites, ions, and antibodies are just some of the molecular and cellular elements that have been delivered into living cells by needle microinjection, thereby replacing the test tube with the cell (Meredith et al. 2000; Dammermann and Merdes 2002; Becker et al. 2003; Shan et al. 2003; Wakayama et al. 2003).

Microinjection remains the most direct method to gain insight into the function and dynamics of intracellular components, to produce transgenic animals, or to overcome male infertility. Although the development of novel methods for introduction of genes into living cells by transfection using $CaPO_4$ coprecipitation, DEAE-dextran, lipofection, or electroporation and by infection with retroviral, adenoviral, or lentiviral vectors provides an alternative strategy, microinjection still permits the widest range of applications.

The advantages of microinjection are (1) efficiency: close to 100% requiring no selection process; (2) quantitativeness: the amount of substance delivered can be accurately controlled; (3) targeting: delivery can be to the cytoplasm or nucleus; (4) temporal control: injection can be precisely timed to study short-term cellular events; (5) minimal perturbation: generally no effect on short-term function (allowing immediate data collection beginning within seconds following the introduction of molecules into cells) and long-term viability; (6) versatility: no cell-type or injectant restriction; (7) thrift: only a small amount of sample is required; and (8) ability to introduce simultaneously several probes at fixed ratios. Disadvantages of microinjection include (1) small sample size: hundreds of cells (but not thousands) can be microinjected, limiting its applicability for some biochemical studies; (2) considerable expertise is required; and (3) specialized apparatus is necessary.

This chapter discusses how to set up a microinjection system with regard to the instrumentation and materials required and the commercial suppliers from which they may be obtained. It provides step-by-step descriptions of how to microinject adherent mammalian cells and helpful troubleshooting guides and considers special requirements for imaging of injected fluorophore-labeled proteins, Cy3/5-labeled tubulin, and rhodamine-labeled actin.

COMPONENTS OF A MICROINJECTION SYSTEM

A basic microinjection system requires an inverted light microscope (Fig. 1, equipment 1), micromanipulator (Fig. 1, equipment 2), micropipette holder (Fig. 2, equipment 3), gas pressure regulator (Fig. 1, equipment 3 and 4), micropipette puller, glass capillary tubing, micrometer syringe, and vibration isolation table (Fig. 1, equipment 9). We use the Narishige

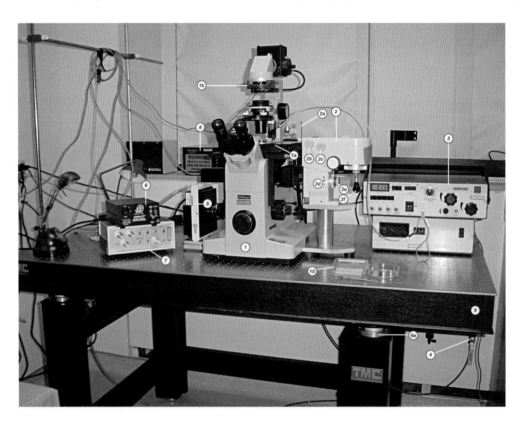

FIGURE 1. Microinjection system.
1. Nicon Diaphot inverted microscope. *1a*, Microscope stage; *1b*, shutter for transmission light.
2. Leitz micromanipulator. *2a*, Headstage; *2b*, coarse adjustment of the transverse movement; *2c*, coarse adjustment of the sagittal movement; *2d*, coarse and (*2e*) fine vertical adjustment; *2f*, joystick operator.
3. Narishige air pressure regulator, model IM-200.
4. Connections to compressed air from house supply.
5. Photometrics SenSys CCD camera.
6. UniBlitz shatter driver 1 model VMM-DI (for epifluorescence).
7. UniBlitz shatter driver 2 model SD-10 (for transmission light).
8. Power supply for mercury lamp.
9. Vibration tabletop on (*9a*) supports (TMC, Peabody, Massachusetts).
10. Custom-made container for micropipettes.

IM-200 air regulator (Narishige USA, Inc., Greenvale, New York) and the Leitz micromanipulator (Leitz, Inc., Rockleigh, New Jersey).

More elaborate systems can be assembled according to the experimental needs of the investigator. The system can be supplemented with a specimen incubator, CCD (charge-coupled device) camera (Fig. 1, equipment 5), shutter controller (Fig. 1, equipment 6 and 7), epifluorescence (Fig. 1, equipment 8), digital image processing software, and computer.

Microscope

Any sort of inverted microscope (Fig. 1, equipment 1) equipped with phase-contrast, Hoffman phase-contrast or Nomarski optics (differential interference contrast [DIC]) can be used for microinjection. A 25× or 40× dry objective with high numerical aperture (NA) is typically used for microinjection of adherent cell cultures. The lower-magnification objective (25×) is used to find and guide the micropipette tip to the surface of the cell. However, this objective does not give sufficient magnification for microinjection where a 40x objective is used. A novice learning the technique can switch from one objective to another during capillary replacement and microinjection.

Depending on size and/or geometry of the specimen, different optics may be preferred. If the specimen is flat (well spread cells in culture) phase contrast works better than DIC. When cells are rounded (early stages of cell spreading, neuronal primary cell culture) and nuclear injection is required, DIC is preferable.

Some experiments require selecting cells for microinjection that are already expressing fluorescent fusion proteins. To visualize the fluorescent signal, the microscope should be equipped with a mercury lamp, shutter controller, neutral density filter set, and a single dye filter cube. A combination of fluorescence observation and microinjection requires plan achromatic objectives (Plan-Fluor or Plan-Fluotar).

If the investigator wants to microinject and then immediately observe cells with high resolution, it is possible to inject using the 60× or 100× oil-immersion objective; however, with these objectives, it is more difficult to find the needle due to the narrower plane of focus.

Glass Capillary Tubing

Glass capillary tubing is used for micropipette fabrication. Glass tubing stock is available in different barrel and cross-section configurations and also with a wide range of diameters and wall thicknesses. Capillaries are offered in two configurations—with or without an inner filament. Capillaries with inner filaments can be easily back-loaded (see below) as the solution is drawn to the tip by capillarity. A thin-wall circular one-barrel glass capillary with an outer diameter of 1–1.5 mm is preferred for microinjection of mammalian cells. The outer diameter of the capillary and the thickness of its wall are two of the most important characteristics of glass tubing. For microinjection of small or sensitive cells, a narrower outer diameter gives a finer tip, which enables delivery of a smaller volume of injectant and helps minimize cell injury. The thickness of the tubing wall is also important because a pulled micropipette has the same ratio of outer/inner diameters at the tip as the original capillary tubing. Having a larger internal diameter for the pulled capillary improves flow characteristics and prevents frequent clogging. However, tips that have a large outer diameter will produce a large wound at the cell surface. A micropipette pulled from thin-wall tubing optimizes both the fineness of the tip and the size of its opening.

The most popular material for micropipette fabrication is borosilicate glass that provides excellent strength. For some special applications, soda glass or alumino-silicate glass tubing may be preferred. Quartz tubing is also available. However, due to its high melting point, quartz cannot be pulled using conventional pullers. The Sutter laser-based P-2000 puller is ideal for this purpose (Sutter Instrument Company, Novato, California).

> **Optional.** There is no special requirement for washing or treatment of glass capillary tubing prior to micropipette fabrication. However, if desired, micropipettes can be **sterilized** or silanized inside a multipurpose pipette storage container (World Precision Instruments, Inc). Pipettes are silanized by adding silane to the reagent well in the center of container. A container with pipettes in situ can be dry-heat-sterilized at 170°C. Of course, sterilization of micropipettes occurs when they are pulled (see below).

Pipette Puller

Micropipettes are fabricated with a micropipette puller. The main principle of pulling is application of tension from the ends of the glass tubing while the middle of tubing is heated and melted by a heating filament. The heating filament has at least one loop (circular or squared) where the capillary tubing is centered. When the heating device is turned on, a current runs through the filament causing it to heat up, thereby melting the middle of the capillary tube. The ends of the capillary tube are clamped into two arms of the machine which will electronically or gravitationally pull apart, producing two micropipettes.

A pulled micropipette consists of four parts: the tip, the shank, the shoulder, and the shaft. The shape (length and angle of the shank) and diameter of the tip opening are dependent on the current, tension of pulling, or length of gravity fall (in the case of a vertical puller). By adjusting values for the various parameters in a pulling sequence, a wide variety of pipette shapes can be achieved.

A factor that can adversely affect pipette shape is residual heat in the heating element. A platinum alloy filament helps to eliminate this problem as it has an extremely low thermal mass and cools very rapidly. The tip of the pipette pulled below the heater element is not affected by residual heat.

Numerous types of micropipette pullers are available—from very simple mechanical to sophisticated models fashioned with electronic programming. Either horizontal or vertical pullers are also available. David Kopf Instruments (Tujunga, California), is at top of the list of companies producing pipette pullers. Other brands include MicroData Instrument Inc. (Plainfield, New Jersey); Trans-Tek, Inc. (Ellington, Connecticut); Energy Beam Sciences, Inc. (Agawam, Massachusetts); HEKA Instruments Inc. (Southboro, Massachusetts); World Precision Instruments (Sarasota, Florida); Bio-Logic-Science Instruments SA (CLAIX, France); Stoelting Co. (Wood Dale, Illinois); GENEQ Inc., (Montreal, Quebec, Canada); Macropipette Making (Sydney, Austria).

Micromanipulator

A micromanipulator is an essential part of a microinjection system (Fig. 1, device 2). It is used to hold and manipulate a micropipette during microinjection. The micromanipulator consists of two main parts: headstage (2a) and remote controller (2b–f). The headstage holds and moves the micropipette, and the controller regulates pipette movement. The remote controller usually includes both coarse (2b–d) and fine (2e) control adjustments and a dynamic joystick operator (2f). The position of the micropipette tip can be moved in four ways: horizontal movement (sagittal [X] and transverse [Y]), vertical movement (Z) and tilt

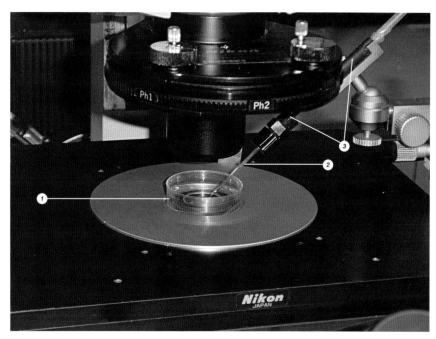

FIGURE 2. Orientation of the micropipette to the microscope stage. A dish containing cells (*1*) is placed into the special adaptor on the microscope stage. The micropipette (*2*) is inserted into the instrument collar (*3*) attached to the headstage of the micromanipulator. The micropipette forms an angle of about 30–40° to the microscope stage. The micropipette tip is lowered into cell culture medium directly above the center of lens.

angle (T). When adherent cells are microinjected, the micropipette is held at an angle of 30–40° to the microscope stage (Fig. 2). During microinjection, the tip of micropipette penetrates the cell by vertical (*z* axis) movement.

Micromanipulators can consist of two separate units (Narishige USA, Inc., Greenvale, New York; Newport Corporation, Fountain Valley, California) or be designed as one piece of equipment (Leitz, Inc., Rockleigh, New Jersey; World Precision Instruments, Inc. Sarasota, Florida). To ensure stable drift-free micromanipulation, Leitz and WPI bolt the headstage to a solid massive base with strong magnetic feet. The base also serves to elevate the manipulator to the level of the microscope stage. The joystick arm and control knobs are located at the base. When a micromanipulator includes two separate units (headstage and controller), the headstage is attached directly to the microscope stage (special adapter is required) or maintained on the magnetic stand. The remote controller should be placed on a vibration isolation table at a comfortable distance for hand manipulation.

Macromanipulators have a variety of driver mechanisms: mechanical (Leitz, Leica), oil, or water hydraulic (Narishige; Newport Corporation), and stepper motor (Zeiss, Eppendorf, WPI). More recent models, for example, Eppendorf InjectMan NI2 (Eppendorf) and motorized micromanipulator SM325 (WPI) have programmable joystick units. Each of these micromanipulators has attributes that are designed to ensure smooth and drift-free movements of the micropipette. A shorter step of the movement provides finer micromanipulation.

A micropipette holder (instrument collar; Fig. 2, device 3) is required for attaching the micropipette to the headstage. Micropipette holders are sold separately from the micromanipulator and are available in many different sizes. The size of the holder should match the outer diameter of capillary tubing to hold the micropipette tightly.

Air Pressure Regulator

An air pressure regulator is used to adjust the pressure of compressed gas that forces the outflow of a small volume of liquid from the micropipette tip. We use the Narishige IM-200 air regulator (Narishige USA, Inc.) (Fig. 1, equipment 3). It controls both the pressure and the duration of the injection. Depending on the pressure applied, a volume ranging from microliters to femtoliters can be delivered (McCaman et al. 1977; Palmer et al. 1980). Practically, it is possible to microinject a constant volume within a 50% difference among cells. For a method of measuring the injection volume based on the fluorescence intensity, see Wang (1982).

Pressure is applied pneumatically from either a tank of nitrogen or with compressed air from the house supply (Fig. 1, equipment 4). Compressed air can be used only for injection of oxygen-insensitive materials, whereas nitrogen has general application. The regulator gives a digital readout of the applied pressure in psi/kPascal. The duration of injection is adjusted using an internal clock with a digital switch.

The Narishige regulator has three auxiliary features: fill/hold, balance, and clear. The fill/hold feature is used to generate suction to fill a micropipette from the tip (see below) or to hold a cell with a second pipette. A low balance pressure is applied to generate slow flow from the tip between injections, preventing pipette clogging and inflow of the medium into the pipette. The clear feature generates a momentary application of high pressure to clear a clogged tip.

Vibration-free Environment

A vibration-free environment is required for successful microinjection. To prevent uncontrolled mechanical displacements of a micropipette tip and injected specimen, the system should be mounted on a vibration isolation table (Fig. 1, equipment 9, device 9a). The table consists of a heavy tabletop (equipment 9) on supports (device 9a) that can be inflated by compressed air. When the supports are inflated, the table rides on a cushion of air that damps floor vibrations. Newport Corporation (Fountain Valley, California) and Technical Manufacturing Company (Peabody, Massachusetts) are two brands that provide vibration isolation tables.

DIRECT PRESSURE MICROINJECTION

Preparation of Cell Culture for Microinjection

Following microinjection, live imaging or observation of injected cells after fixation and processing for immunofluorescence would be impossible if the particular cell(s) that were injected cannot be found. Photoetched grided coverslips (Bellco Glass, Vineland, New Jersey) greatly facilitate localization of injected cells. Rectangular (18 and 23 mm) or circular (25 mm in diameter) coverslips are available; each coverslip has an alphanumeric coded grid.

Coverslips can be attached to a 35-mm dish with a 20-mm diameter hole in the bottom (McKenna and Wang 1989). Either vacuum grease or hypotoxic silicone glue (e. g., Sylgard 184, Dow-Corning Corporation, Midland, Michigan) are used to seal a coverslip inside (or outside) the dish. If an experiment requires correlative microscopy and microinjected cells must be fixed and immunostained after live observation, vacuum grease is preferable because it allows easy removal of the coverslip from the dish.

Dishes with attached coverslips can be sterilized by UV illumination for 30–45 minutes prior to cell plating. There is no special sterile requirement for the room where microinjection is performed. Generally, the process of injection does not lead to cell contamination, even if cells are microinjected for an hour in a nonsterile room.

TABLE 1. Some mammalian cells used for microinjection

Name	Cell type	Origin	ATTC number
CHO-K1	epithelial	ovary; *Cricetulus griseus* (Chinese hamster)	CCL-61
CV-1	fibroblastic	kidney; normal; *Cercopithecus aethiops* (African green monkey)	CCL-70
COS-7	fibroblastic	SV40 transformed derivative of CV-1	CRL-1651
3T3-Swiss	fibroblastic	embryo; *Mus musculus* (mouse)	CCL-92
NRK	epithelial	kidney; normal; *Rattus norvegicus* (rat)	CRL-6509
PtK1	epithelial	kidney; normal; *Potorous tridactylis* (potoroo)	CRL-6493
LLCPK	epithelial	kidney; normal; porcine	

Many mammalian cells and some types of primary fibroblasts are good candidates for microinjection (see Table 1). It is important to plate cells at least 12 hours prior to microinjection (24 hours is optimal). Most cell types are very sensitive to plasma membrane damage during the early hours after replating.

Some cells, for example, LLCPK and primary cultured neuronal cells, are very sensitive to pH changes occurring while a dish is outside the CO_2 incubator. They cannot be microinjected for longer than 15 minutes unless special bicarbonate-free media is used. For microinjection of pH-sensitive cells, we recommend using Leibovitz's L-15 medium (GIBCO-Invitrogen Corporation). For microinjection of primary neuronal cells, we use L-15 medium enriched with Glucose (0.5%) and L-glutamine (0.9%). In addition, medium without bicarbonate and phenol red, but containing 20–25 mM HEPES, can be used for cells that cannot tolerate L-15 medium.

Troubleshooting Guide for Cell Culture

Problem: Cells were grown to a semiconfluent monolayer, but the grids and letters cannot be seen very well on the Bellco coverslip.

Solution: When the coverslip was attached, it was placed with the grid facing up. Cells mask the grid pattern if both are in the same focal plane. The grid can be seen only in areas that are free from cells. To relocate the cells in a semiconfluent monolayer, mount the coverslip such that the grid and cells are located on opposite sides of coverslip.

Problem: The grid can be seen using the dry objective, but not when using the oil objective.

Solution: The grid is facing down into the oil that obscures the grid pattern. Remove the oil from the coverslip with absolute ethanol and find the cells using the dry objective. Do not move the dish, then change objective.

Pulling Micropipettes

To pull micropipettes, it is best to follow the procedure described in the manual accompanying the puller. Because pullers differ, it is not possible to provide a generic step-by-step protocol. We do provide some general tips and a troubleshooting guide for micropipette fabrication. Be extremely careful when removing a pulled micropipette from the puller. Do not use the tip of a pipette if it is accidentally touched. Pulled micropipettes can be stored in a commercially available container or simply use a flat dish with lid and attach micropipettes to the bottom with a loop of double-stick tape, to keep the tips elevated. The micropipettes are usable for a couple of days, but no longer than 1 week.

Two main parameters of a micropipette tip affect the rate of the flow: the internal diameter of the tip and the angle of the shoulder (taper) to the tip. The larger the tip opening, the more material is delivered for the same applied pressure and time. A tip diameter of ~0.3 μm is optimal for the microinjection of mammalian cells in culture (e.g., CHO, PtK1, and COS-7). A 10% increase in diameter increases the delivery rate by more than 30% and can cause cell damage. A smaller tip diameter will result in frequent clogging from protein aggregates. The smaller the angle of the shoulder to the shank, the longer the shank. A 10% decrease in angle will decrease the delivery rate by about 10%. For nuclear injection, a smaller angle is needed to limit delivery of the probe into cytoplasm lying above the nucleus. By varying parameters of the puller, different shapes of micropipette pulled from the same type of tubing can be achieved.

Troubleshooting Guide for Pulling micropipettes

Problem: The micropipette has a very short shank.

Solution: Too low a current was used for micropipette fabrication. Increase the current.

Problem: No flow from the micropipette even when very high pressure is used.

Solution:

- *The tip of the micropipette is clogged by aggregates in the sample.* Raise the needle and press briefly the "clear" pushbutton on the air controller. This will apply high pressure, to clear the clogged pipette. If this procedure fails, try fabricating tips with larger openings (decrease filament current) or centrifuge the sample at high speed.

- *The tip was sealed when the micropipette was pulled.* Using too high a current will seal the tips; therefore, decrease the current. Also, we recommend selecting micropipettes with a good tip opening size using the protocol for pipette calibration (see below). It is possible to break a sealed or clogged tip by placing the "sealed" tip above a field that is free of cells. Lower the tip very slowly until a small piece of the tip is broken off. Then try to inject the cells again.

Problem: Cells explode during microinjection.

Solution: The flow from the tip is too high.

- Too high a pressure was used for microinjection; decrease the pressure.

- The internal diameter of the tip is too large; increase the current when pulling the micropipette.

Problem: The tubing cannot be broken into two micropipettes. They remain as one very long stretched piece of glass.

Solution: Exceedingly high current and solenoid power were used; decrease both parameters.

Calibration of Micropipette Tips

It is essential to estimate the internal diameter of the pulled micropipette tip when adjusting parameters for a new puller or new type of glass tubing. We provide here a modification of a protocol for use of a Leitz micromanipulator as originally described by Hagag (Hagag et al. 1990).

1. Fill a 35-mm culture dish with methanol.

2. Place the holder into the Leitz micromanipulator.

3. Attach the measuring micropipette tightly to a holder that is connected to air pressure regulator.

4. Place the culture dish on the microscope stage.

5. Turn on the Narishige air pressure regulator.

6. Set up the microinjection time for 1 minute.

7. Turn on the compressed air or nitrogen gas.

8. Position the tip of micropipette over a dish filled with methanol.

9. Lower the tip of the micropipette into the methanol. An objective need not be used to observe bubbles coming out from a tip. However, observation with a low-magnification objective may be useful.

10. Push the "injection" button.

11. Watch for bubbles. If bubbles did not appear, increase the injection pressure until bubbles begin escaping from the tip of the micropipette.

12. Record the minimal pressure at which bubbles appear. The pressure is shown on the display in kiloPascal (kPa).

13. Substitute the recorded pressure into the equation below to determine the internal micropipette tip diameter.

$$\text{diameter } (\mu m) = 70.4/[\text{pressure (kPa)}]^{1.01}$$

Preparation of a Sample

Materials for microinjection (proteins, DNA, RNA, buffer, ions, and antibodies) should be stored in small aliquots of 5–10 μl in liquid nitrogen or at −80°C. Proteins are stored in stock solutions at a concentration of 10 μg/μl. The stock can be diluted with a physiological buffer (see Step 4) if microinjection of a small amount of the fluorophore-labeled protein is desired (e.g., for speckle microscopy; see Chapter 12). DNA can simply be diluted with water.

Protein solutions should be centrifuged prior to microinjection to remove protein aggregates that could produce frequent clogging of the micropipette tip during microinjection. The sample should be centrifuged either in a microfuge at maximum speed or in an ultracentrifuge. We do not recommend centrifugation of proteins for longer than 5–10 minutes in a microfuge—to avoid warming of the sample. If an ultracentrifuge is available, for example, TL-100 ultracentrifuge (Beckman Coulter, Inc.), we suggest centrifugation at 50,000 rpm in a TLA-100 rotor (200,000g) for 20 minutes at 4°C.

Proteins tagged with HA (hemagglutinin epitope) or GST (glutathione-S-transferase) tend to aggregate and clog the micropipette tip. To avoid the frequent clogging, therefore, the tag should be cleaved prior microinjection.

The protein sample should always be kept on the ice during microinjection. Most proteins (e.g., tubulin and actin) are stable for several hours at 4°C, but denature rapidly at room temperature. We do not recommend refreezing a sample.

1. Cool the ultracentrifuge to 4°C.

> **MICROINJECTION BUFFERS**
>
> | | |
> |---|---|
> | 50 mM K-Glutamate | 100 mM PIPES |
> | 0.5 mM KCI (pH 7.0) | 1 mM $MgCl_2$ |
> | | 1 mM EGTA (pH 6.8–7.0) |
> | | |
> | (Waterman-Storer and Salmon 1997) | (Schaefer et al. 2002) |

2. Remove an aliquot of the protein from the freezer and place it on ice.

3. Allow the sample to melt.

4. If dilution is needed, dilute the sample with microinjection buffer.

5. Transfer the sample to a centrifuge tube (Beckman) for TLA-100 rotor.

6. Insert the tube into a precooled rotor (the rotor is usually kept at 4°C).

7. Balance the sample with an additional tube containing an equal volume of liquid.

8. Insert the rotor into the centrifuge.

9. Set parameters for centrifugation: speed of centrifugation 50,000 rpm; time 20 minutes; temperature +4°C.

10. After centrifugation, remove the sample tube from the rotor and place it on ice.

Loading Micropipettes with Sample

There are two ways to load micropipettes: from the tip, also known as "front-loading" or from the unpulled end or "backfilling." For backfilling, use a 10 μl Hamilton microsyringe (Fisher, VWR) or special loading plastic tips (Eppendorf).

Micropipette Loading from the Tip

1. Attach the micropipette to the holder/instrument collar that is connected to the gas pressure controller by rubber tubing.

2. Remove the tube containing the sample from the ice bucket.

3. Place the tip of the micropipette into the top of the solution (sample is drawn into the micropipette by capillary action).

4. Press the "fill" pushbutton on the gas pressure controller to apply suction of about 82 kPa. The suction is applied as long as the button is left in.

5. When a sufficient amount of solution is drawn into the tip of micropipette, release the "fill" pushbutton.

6. Return the sample to ice.

7. Place the micropipette over the culture dish.

8. Apply low balance pressure and quickly lower the micropipette into the culture media.

9. Start the microinjection (see below).

Loading of the Micropipette from the Unpulled End: Backfilling

1. Precool the microsyringe by placing it on the ice.

2. Load the microsyringe with 2–3 µl of sample (take care not to disturb the pellet) and place it back on ice.

3. Use forceps to remove a micropipette from the storage container.

4. With the other hand, insert the needle of the syringe into the barrel of the micropipette.

5. Inject a small amount of the sample (0.1 µl) into micropipette.

> The tip of a micropipette fabricated from tubing with inner filaments will fill quickly. Discard any micropipette that is not filled to the very tip.

6. Return the syringe to the ice.

7. Attach the micropipette to the holder.

8. Place the micropipette over the culture dish.

9. Apply low balance pressure, and quickly lower the micropipette into media.

10. Start the microinjection (see below).

Troubleshooting Guide for Loading Micropipettes

Problem: The micropipette is blown out of the holder when pressure is increased.

Solution: The micropipette was not attached tightly to the holder. The size of the holder does not match the size of the outer diameter of tubing used for micropipette fabrication.

Problem: The loaded micropipette becomes empty when increasing the pressure.

Solution: The micropipette tip has too large an opening. As soon as the pressure is increased, the solution is blown out.

- The micropipette tip was broken.
- The micropipette has too large internal diameter.

Problem: The micropipette is always getting clogged even before the cell is microinjected. Particles coming out from the tip cannot be seen under microscope.

Solution:

- Increase the pressure and see if it helps.
- The protein has a tendency to aggregate: Use a lower concentration of the protein for microinjection, dilute stock with microinjection buffer, or increase time for ultracentrifugation.
- The micropipette is sealed. Decrease the current for pipette fabrication.
- A bubble at the tip. Try to load sample from the tip.
- The filled tip was left too long in the air before dipping it into media. The sample dried at the tip and clogged the micropipette

Microinjection

For practicing the microinjection technique, we highly recommend the use of fluorescent-labeled dextran of high molecular weight at a concentration of 5 mg/ml (Sigma Chemical Co.). First, using a labeled substance helps with monitoring and learning how to control delivery of the material into the cell, because the fluorescent signal can be observed after microinjection, thus providing immediate feedback. Second, by looking in

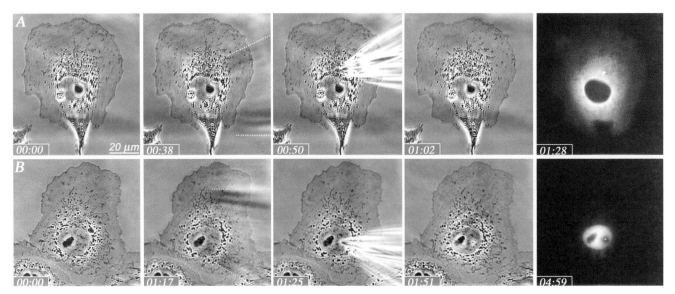

FIGURE 3. Microinjection of COS-7 cells into cytoplasm (*A*) or nucleus (*B*) with FITC-labeled dextran. The tip of the micropipette generates two shadows (each marked by a dotted line) when it is higher than the cell focal plane (in the second images of *A* and *B*). The moments of cell microinjection are shown in the third images of the panels. Delivery of dextran into the cytoplasm or nucleus is demonstrated by the fluorescent images (*right panels*). When dextran is microinjected into cytoplasm (or into nucleus) it shows only cytoplasmic (or nuclear) distribution. Times are shown in the lower left corners in minutes and seconds. Bar, 20 μm.

the appropriate fluorescent channel, it can be determined if the material is flowing properly out of the pipette tip. Third, because fluorescent-labeled dextran is an inert polymer of high molecular weight, it remains in the cell compartment into which it is injected; i.e., when injected into the cytoplasm, it remains only in the cytoplasm (Fig. 3A) or, when injected into the nucleus, only in the nucleus (Fig. 3B). Perform all steps from Step 10 rapidly to avoid drying of the sample at the tip and micropipette clogging.

1. At least 12 hours prior microinjection, plate cells into a dish with a photoetched coverslip in the bottom.

2. Pull micropipettes and store them in a container.

3. *Optional*: Calibrate the diameter of tips as described above.

4. Centrifuge a sample and keep it on ice in an ice bucket covered with a lid.

5. Set up the dish with cells on the microscope stage (Fig. 2) and find a suitable cell for injection. Always choose healthy, good looking cells.

6. Use the printed grid pattern to record the location of cells to be microinjected.

7. Turn on the Narishige air pressure controller.

8. Turn on compressed air or nitrogen gas.

9. Attach the holder to the Leitz micromanipulator.

10. Load the micropipette with protein to be microinjected.

11. Attach the micropipette tightly to the holder connected to air pressure controller.

12. Use the "balance pressure" knob to apply a small amount of pressure to the micropipette before lowering it into the dish, to prevent inflow of culture medium into the micropipette caused by capillary action.

13. Position the tip of the micropipette at a 30–40° angle to the microscope stage.

14. Lower the tip into the medium.

15. Locate the needle just above the center of the objective lens.

16. Find the micropipette tip under the microscope.

    a. First find the shadow from the tip. The shadow will not be seen without performing the lateral movements of the micropipette.

       Optional: Shift to the low-power objective while guiding the micropipette. The wider the field of view, the easier to find the tip.

    b. Use the course focus knob to slowly lower the micropipette tip and move the tip transversely (along *y* axis) with a small amplitude. The tip gives two shadows when it is above the focal plane (cells should be in focus), one from each side of the tip wall (Fig. 3, the shadows are shown in the second image on each panel, the outer side of the shadow is marked by a dotted line in the figure). The higher the tip, the larger the distance between the two shadows.

    c. Lower the micropipette with the fine adjustment until the tip is clearly seen just above the cells. From time to time, move the micropipette sagittaly (along the *x* axis). This is needed to maintain the tip in the center of field.

17. *Optional*: Check the flow before microinjection. Focus on the tip of the micropipette. Check if small particulates are being blown from the tip into the dish. If a fluorescent-labeled protein is microinjected, the material flowing out of the pipette can be seen by looking in the appropriate fluorescent channel.

18. Move the needle tip over the top of the cell that will be microinjected.

19. Increase the pressure. Typically, we apply pressure of 5–12 kPa to generate a continuous flow from the micropipette tip. When the tip touches the cell and penetrates the cortex, the solution continues to enter the cell until the micropipette is pulled out.

20. Prepare to microinject the sample.

    For microinjection into the cytoplasm (Fig. 3A; Movies 5.1 and 5.1A on the accompanying DVD):

    a. Position the needle tip above the thick part of cytoplasm (usually thicker near the nucleus).

    b. Lower the tip by fine control and penetrate the cell by gently touching the cell cortex.

       If injection occurs, a "wave" will be seen spreading from the point of injection. If the cell does not react, raise the needle and increase the pressure. Try microinjection of another cell. Do not repeatedly penetrate the same cell.

    For nuclear microinjection (Fig. 3B; Movies 5.2 and 5.2A on the accompanying DVD):

    a. Position the needle tip above the nucleus.

    b. Lower the tip by fine control and penetrate the cell (more deeply than for cytoplasmic injection). The tip has to enter the cell nucleus.

       If injection occurs, a bright zone will appear around the tip.

21. Microinject the selected cells. Keep the microinjection system to less than 30 minutes.

22. Return the dish to the incubator.

23. Leave cells in the incubator at least 30 minutes before continuing microinjection.

Troubleshooting Guide for Microinjection

Problem: The micropipette tip cannot be found.
Solution: The micropipette is not in the field of observation. Try to place the micropipette directly above the center of the objective lens. Use the low power objective to guide the micropipette. Make sagittal and transverse displacements of the micropipette with greater amplitude. Try to view the shadow. Lower the micropipette slightly and repeat the procedure.

Problem: The micropipette tip already broken when finally found.

Solution:

- Be more careful when handling the micropipette. Do not touch anything with the tip.

- The pipette tip breaks when trying to find it under the microscope. The micropipette was not moved transversely to place the end of the shadow into the center of the field. When the needle is lowered, the tip is not directly in the center. The shadow cannot be seen clearly when the needle is lowered, causing the tip to touch the coverslip away from the field of observation.

Problem: The wave can be seen when microinjecting the material into the cell, but the bright fluorescence cannot be seen inside the cell.

Solution: The low balance pressure was not applied before dipping the tip into the dish. Medium was drawn into the tip by capillary action and thus mostly medium was microinjected.

Problem: Micropipette clogs after microinjecting only a few cells.

Solution: Increase the pressure from time to time during microinjection. Use the "clear" feature of the gas regulator to unclog the tip. Centrifuge the sample longer.

Problem: Microinjection of neuronal cells was tried and all cells died.

Solution: Begin practicing the microinjection technique with CHO or COS-7 cells. They are more tolerant to microinjection. Practice for a week or two before attempting microinjection of more sensitive cells such as neurons. Do not use a freshly plated culture of neuronal cells for microinjection. Instead, primary neurons, cultured for 1 day, are recommended to start with. About 50% survival is a good result for neuronal cells.

Problem: Very high percentage of cell death occurs after microinjection. The cells permanently change morphology and die even though too much sample was not microinjected.

Solution:

- The cells that are microinjected are very sensitive. Change tubing for micropipette fabrication. Use tubing with a smaller outer diameter. Try to fabricate very fine tips. Use low pressure for microinjection.

- Check whether the probe is toxic for the cells. Check if the original protein stock (if commercial) contains any toxic component (glycerol, azide, etc.) for protein stabilization. Try to microinject buffer and see if the cells show the same percentage of survival. If the buffer contains the toxic component, the material for microinjection should be dialyzed into the appropriate carrier solution (buffer for microinjection).

Problem: Cells microinjected with a probe change morphology. Is this a specific reaction to a probe and not a result of the microinjection itself?

Solution: Check if cells change morphology after microinjection of buffer alone.

Problem: How does one know if the cell is dying?

Solution: Check if the cells change morphology gradually. Usually, good microinjection does not cause permanent changes in morphology and cells recover within 5–10 minutes. The first sign that cells are dying is that mitochondria become condensed and exhibit high contrast (with phase-contrast optics).

LIVE IMAGING OF CYTOSKELETAL PROTEINS

Considerable effort has been made to visualize dynamics of cytoskeletal components in living cells. One of the earliest attempts to study microtubule dynamics in vivo employed video-enhanced differential interference contrast microscopy and digital image processing (Cassimeris et al. 1988). However, this method did not become commonly used because the observations were strictly limited. Thus, microtubule assembly and disassembly could be recorded only in ~10–20% of the cells and only within the thin part of lamellae where individual plus ends of microtubules (MTs) were clearly seen due to the low density of cytoplasm and the absence of many intracellular components. Purification of cytoskeletal proteins, tubulin, actin, and myosin (Spudich and Watt 1971; Borisy et al. 1975; Ikebe and Hartshorne 1985) and their further cytochemical modification (Wang et al. 1982; Sammak and Borisy 1988; Wang 1989; Hyman et al. 1991; Isambert et al. 1995; Verkhovsky et al. 1995)—conjugation with fluorescent dyes—allowed an extension of in vivo kinetics studies. Although the great majority of fluorescent protein analogs are now fluorescent fusion proteins, fluorophore-labeled proteins remain the most direct and versatile method for studying cytoskeletal function.

Microinjection of fluorescent analogs of cytoskeletal proteins permitted direct observation of the dynamics and organization of the cytoskeleton in living cells. Fluorophore-tagged proteins diffuse rapidly within the cytoplasm of microinjected cells, become incorporated with endogenous proteins into polymer structures, and label them along the lattice (Saxton et al. 1984; see Movie 5.3). The fraction of probe incorporated (about 5–10% of the total cellular protein pool) is sufficient to label polymer structures uniformly (Waterman-Storer and Salmon 1998). Photoactivation of fluorescently tagged proteins leads to an emission of photons that may be detected by a cooled charge-coupled device (CCD) camera and transferred to a high-resolution digital image. Time-lapse imaging of the fluorescent signal, i.e., collection of successive frames over time, permits the analysis of the dynamic properties of fluorophore-labeled structures in living cells (Movies 5.4–5.9).

Cytoplasmic microinjection of fluorophore-tagged proteins has become a powerful approach to study the distribution and kinetics of cytoskeletal systems in living cells (Verkhovsky et al. 1995; Rodionov and Borisy 1997; Waterman-Storer 1998; Komarova et al. 2002b; Wittmann et al. 2003). Using the microinjection technique, numerous assays can be carried out including laser photobleaching (Gupton et al. 2002; Komorova et al. 2002b; see also Chapter 7), photoactivation (Yvon et al. 2001, 2002; see also Chapter 10), fluorescent speckle microscopy (FSM) (Waterman-Storer and Salmon 1998; Chang et al. 1999; Kapoor and Mitchison 2001; Salmon et al. 2002; Waterman-Storer and Danuser 2002; see also Chapter 12), and total internal reflection fluorescent microscopy (TIRFM) (Lanni et al. 1985; Sund and Axelrod 2000; see also Chapter 34).

The general principles of time-lapse imaging of directly conjugated proteins are similar to those applied for record GFP-fusion proteins (Chapter 1). Here we discuss only the specific requirements for the imaging of fluorophore-labeled proteins.

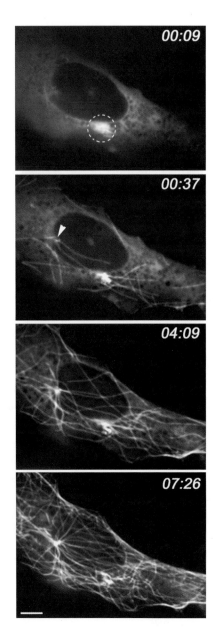

FIGURE 4. CHO cell microinjected with Cy3-labeled tubulin. Observation of incorporation of fluorophore-labeled tubulin subunit into microtubules was started 9 seconds after microinjection. Non-steady-state condition. Bright spot (circled) indicates site of microinjection. Incorporation of Cy3-labeled tubulin subunits takes place both at the plus ends of microtubules and at the centrosome (*arrowhead*). Visualization of microtubule radial array occurs while novel microtubules assemble. Times are shown in the upper right corners in minutes and seconds. Bar, 5 μm.

Steady-state Condition

Microinjection of labeled proteins such as g-actin and tubulin causes an increase in the concentration of monomer (dimer) molecules within a cell. It results in a shifting of the monomer (dimer)/polymer ratio and in an alteration of protein kinetics. Following microinjection, the cell must achieve a steady-state condition—the establishment of an equilibrium between assembly and disassembly of micromolecules. In addition, fluorescent-labeled proteins incorporate only into novel polymerized structures (Fig. 4; Movie 5.3) suggesting that complete incorporation of labeled molecules requires at least one cycle of protein turnover. Tubulin or actin turnover might vary from one cell type to another or according to different physiological conditions within the same cell type. Nevertheless, 1–1.5 hours is sufficient to achieve complete incorporation of labeled proteins and a steady-state condition. It is advisable to place the dish containing microinjected cells back into the incubator for at least 1 hour to achieve complete recovery and incorporation of labeled molecules.

Reduction of Photodamage and Photobleaching

The main principle of live imaging is similar to that of fixed immunofluorescence samples. However, specific technical problems such as photobleaching and photodamage are inherent problems related to the imaging of fluorophore-labeled molecules in living cells. For example, an excessive illumination of dye during time-lapse acquisition leads to photobleaching effects and loss of fluorescence. In addition, the emission of fluorescent light is accompanied by heat and free oxygen radical release. These cause serious problems for observation of fluorophores such as Cy3, Cy3.5, rhodamine, Texas Red, Alexa 568, etc. The number of collected images and duration of exposure are strictly limited due to photobleaching and photodamage. To eliminate these problems and improve the temporal resolution and imaging capability during live observation, cells injected with fluorophore-labeled proteins should be treated with the biocatalytic oxygen-depleting agent, Oxyrase (Oxyrase, Inc., Mansfield, Ohio) during the period of observation (Mikhailov and Gundersen 1995).

Oxyrase acts to remove dissolved oxygen from the medium. The enzyme requires a hydrogen donor (we use DL-lactic acid as a substrate) to show oxygen-depleting activity. The product of this reaction is pyruvic acid, which is natural and hypotoxic. Oxyrase is commercially available as a mixture with or without substrate. The final concentration of oxyrase should be about 0.3–0.6 units/ml, sufficient to significantly reduce photodamage and photobleaching of a sample (Waterman-Storer and Salmon 1997). We prefer to use oxyrase without substrate and add substrate directly into medium before we start live observation of the cells microinjected with fluorophore-labeled tubulin or actin. We add 9 μl of 98% DL-lactic acid and 60 μl of oxyrase (1.8 units) into a dish containing 3 ml of medium. About 3 ml of mineral oil should be layered above the medium immediately after oxyrase preparation. Mineral oil effectively "seals off" the medium from exposure to air by reducing oxygen exchange, which would saturate all the oxyrase in the medium. Mineral oil also protects the medium from evaporation. The cells should be treated with the oxyrase preparation 15–20 minutes before observation to achieve equilibrium between enzyme and substrate.

MOVIE LEGENDS

MOVIE 5.1. Microinjection of COS-7 cells into cytoplasm. Images were recorded every 2 seconds. Bar, 20 μm.

MOVIE 5.1A. Microinjection of COS-7 cells into cytoplasm. Images were recorded every 1 second. Bar, 20 μm.

MOVIE 5.2. Microinjection of COS-7 cells into nucleus. Images were recorded every 2 seconds. Bar, 20 μm.

MOVIE 5.2A. Microinjection of COS-7 cells into nucleus. Images were recorded every 1 second. Bar, 20 μm.

MOVIE 5.3. Incorporation of fluorophore-labeled tubulin into microtubules. Non-steady-state condition. Observation was started 12 seconds after microinjection of Cy-3-labeled tubulin into CHO-K1 cell. Fluorophore-labeled tubulin subunits incorporate at the plus ends of microtubules and at the centrosome (*arrowhead*). Time is shown in the upper right corner in minutes and seconds. Bar, 5 mm. Images were observed on a Nikon Diaphot 300 inverted microscope equipped with a Plan100× 1.25 NA objective using a Cy3 filter set. Images of 16-bit depth were collected with CH 350 slow scan, cooled CCD camera (Photometrics, Tucson, AZ) driven by Metamorph imaging software (Universal Imaging, Westchester, PA).

MOVIE 5.4. MT dynamics in CHO-K1 cells. In the steady state, MTs grow persistently from the centrosome toward the cell margin (MTs growing from the centrosome are in *red*). The growth is compensated by shortening from both plus (MTs shortened from the periphery back to the centrosome are in *blue*) and minus (MTs released from the centrosome and shortened from the minus ends are in *green*) ends. Time is shown in the upper right corner in minutes and seconds. Bar, 5 μm. Images were acquired on a spinning disk confocal microscope system. The system was equipped with a Nikon Eclipse TE 200 inverted microscope (Plan Fluor 100× 1.3 NA objective); spinning disk confocal scane-head (Yokogawa Electronics Corp., Tokyo, Japan); 3W argon laser and with 12-bit digital cooled CCD Cool Snap HQ CCD camera (Photometrics, Tucson, AZ).

MOVIE 5.5. MT dynamics in CHO-K1 cytoplasts with and without the centrosome. In the cytoplasts with the centrosome, minus ends of MTs (indicated with "–") are anchored at the centrosome and are stable. Plus ends of MTs (indicated with "+") are facing toward the cell cortex and are dynamic. MTs grow persistently in internal cytoplasm and show dynamic instability at the cell periphery. In a centrosome-free cytoplast, MTs spontaneously appear in the cytoplasm, then translocate by means of treadmilling. Images were collected at a 6-second intervals. Bar, 5 μm. Images were acquired using the same system as that for Movie 5.3.

MOVIE 5.6. "Speckle" analysis of treadmilled MTs. The "speckles" (indicated with *arrowheads*) remain stationary, indicating that MT translocation is driven by treadmilling, incorporation of molecules at the advancing (polymerizing) end, and disassociation from the opposite (depolymerizing) end without any net displacement of the molecules within an MT lattice. Plus and minus ends of MTs are indicated by "+" and "–," respectively. Images were collected at 3-second intervals. Bar, 5 μm. Images were acquired using the same system as that for Movie 5.3.

MOVIE 5.7. Dynamics of noncentrosomal MTs in PtK-1 cell. In PtK-1 cells, minus ends of noncentrosomal MTs (colorized in *blue* and *red*) are stable and display no dynamics. Plus ends undergo alternating phases of growth and shortening. Plus and minus ends of MTs are indicated with "+" and "–," respectively. Images were collected at 5-second intervals. Bar, 5 μm. Images were acquired using the same system as that for Movie 5.3.

MOVIE 5.8. Formation of noncentrosomal MTs in PtK-1 cell. In PtK-1 cells, noncentrosomal MTs are formed by a "de novo" mechanism (*blue*) and via the "breakage" of preexisting MT (*red*). Plus and minus ends of MTs are indicated with "+" and "–." Images were collected at 5-second intervals. Bar, 5 μm. Images were acquired using the same system as that for Movie 5.3.

MOVIE 5.9. Dynamics of actin in a locomoting fish keratocyte. In locomoting fish keratocytes, actin "speckles" (*circles*) remain stationary with respect to the substratum while the leading edge is advancing. Images were collected at 2-second intervals. Bar, 5 μm. Images were acquired using the same system as that for Movie 5.3.

REFERENCES

Becker B.E., Romney S.J., and Gard D.L. 2003. XMAP215, XKCM1, NuMA, and cytoplasmic dynein are required for the assembly and organization of the transient microtubule array during the maturation of *Xenopus* oocytes. *Dev. Biol.* **261:** 488–505.

Borisy G.G., Marcum J.M., Olmsted J.B., Murphy D.B., and Johnson K.A. 1975. Purification of tubulin and associated high molecular weight proteins from porcine brain and characterization of microtubule assembly in vitro. *Ann. N.Y. Acad. Sci.* **253:** 107–132.

Cassimeris L., Pryer N.K., and Salmon E.D. 1988. Real-time observations of microtubule dynamic instability in living cells. *J. Cell Biol.* **107:** 2223–2231.

Chang S., Svitkina T.M., Borisy G.G., and Popov S.V. 1999. Speckle microscopic evaluation of microtubule transport in growing nerve processes. *Nat. Cell. Biol.* **1:** 399–403.

Dammermann A. and Merdes A. 2002. Assembly of centrosomal proteins and microtubule organization depends on PCM-1. *J. Cell Biol.* **159:** 255–266.

Gupton S.L., Salmon W.C., and Waterman-Storer C.M. 2002. Converging populations of f-actin promote breakage of associated microtubules to spatially regulate microtubule turnover in migrating cells. *Curr. Biol.* **12:** 1891–1899.

Hagag N., Viola M., Lane B., and Randolph J.K. 1990. Precise, easy measurement of glass pipet tips for microinjection or electrophysiology. *BioTechniques* **9:** 401–406.

Hyman A., Drechsel D., Kellogg D., Salser S., Sawin K., Steffen P., Wordeman L., and Mitchison T. 1991. Preparation of modified tubulins. *Methods Enzymol.* **196:** 478–485.

Ikebe M. and Hartshorne D.J. 1985. Effects of Ca^{2+} on the conformation and enzymatic activity of smooth muscle myosin. *J. Biol. Chem.* **260:** 13146–13153.

Isambert H., Venier P., Maggs A.C., Fattoum A., Kassab R., Pantaloni D., and Carlier M.F. 1995. Flexibility of actin filaments derived from thermal fluctuations. Effect of bound nucleotide, phalloidin, and muscle regulatory proteins. *J. Biol. Chem.* **270:** 11437–11444.

Kapoor T.M. and Mitchison T.J. 2001. Eg5 is static in bipolar spindles relative to tubulin: Evidence for a static spindle matrix. *J. Cell Biol.* **154:** 1125–1133.

Komarova Y.A., Vorobjev I.A., and Borisy G.G. 2002a. Life cycle of MTs: Persistent growth in the cell interior, asymmetric transition frequencies and effects of the cell boundary. *J. Cell Sci.* **115:** 3527–3539.

Komarova Y.A., Akhmanova A.S., Kojima S., Galjart N., and Borisy G.G. 2002b. Cytoplasmic linker proteins promote microtubule rescue in vivo. *J. Cell Biol.* **159:** 589–599.

Lanni F., Waggoner A.S., and Taylor D.L. 1985. Structural organization of interphase 3T3 fibroblasts studied by total internal reflection fluorescence microscopy. *J. Cell Biol.* **100:** 1091–1102.

McCaman R.E., McKenna D.G., and Ono J.K. 1977. A pressure system for intracellular and extracellular ejections of picoliter volumes. *Brain Res.* **136:** 141–147.

McKenna N.M. and Wang Y.L. 1989. Culturing cells on the microscope stage. *Methods Cell Biol.* **29:** 195–205.

Meredith G.D., Sims C.E., Soughayer J.S., and Allbritton N.L. 2000. Measurement of kinase activation in single mammalian cells. *Nat. Biotechnol.* **18:** 309–312.

Mikhailov A.V. and Gundersen G.G. 1995. Centripetal transport of microtubules in motile cells. *Cell Motil. Cytoskelet.* **32:** 173–186.

Palmer M.R., Wuerthele S.M., and Hoffer B.J. 1980. Physical and physiological characteristics of micropressure ejection of drugs from multibarreled pipettes. *Neuropharmacology* **19:** 931–938.

Rodionov V.I. and Borisy G.G. 1997. Microtubule treadmilling in vivo. *Science* **275:** 215–218.

Salmon W.C., Adams M.C., and Waterman-Storer C.M. 2002. Dual-wavelength fluorescent speckle microscopy reveals coupling of microtubule and actin movements in migrating cells. *J. Cell Biol.* **158:** 31–37.

Sammak P.J. and Borisy G.G. 1988. Detection of single fluorescent microtubules and methods for determining their dynamics in living cells. *Cell Motil. Cytoskelet.* **10:** 237–245.

Saxton W.M., Stemple D.L., Leslie R.J., Salmon E.D., Zavortink M., and McIntosh J.R. 1984. Tubulin dynamics in cultured mammalian cells. *J. Cell Biol.* **99:** 2175–2186.

Schaefer A.W., Kabir N., and Forscher P. 2002. Filopodia and actin arcs guide the assembly and transport of two populations of microtubules with unique dynamic parameters in neuronal growth cones. *J. Cell Biol.* **158:** 139–152.

Shan J., Munro T.P., Barbarese E., Carson J.H., and Smith R. 2003. A molecular mechanism for mRNA trafficking in neuronal dendrites. *J. Neurosci.* **23:** 8859–8866.

Spudich J.A. and Watt S. 1971. The regulation of rabbit skeletal muscle contraction. I. Biochemical studies of the interaction of the tropomyosin-troponin complex with actin and the proteolytic fragments of myosin. *J. Biol. Chem.* **246:** 4866–4871.

Sund S.E. and Alexrod D. 2000. Actin dynamics at the living cell submembrane imaged by total internal reflection fluorescence photobleaching. *Biophys. J.* **79:** 1655–1669.

Tirnauer J.S., Grego S., Salmon E.D., and Mitchison T.J. 2002. EB1-microtubule interactions in *Xenopus* egg extracts: Role of EB1 in microtubule stabilization and mechanisms of targeting to microtubules. *Mol. Biol. Cell.* **13:** 3614–3626.

Verkhovsky A.B., Svitkina T.M., and Borisy G.G. 1995. Myosin II filament assemblies in the active lamella of fibroblasts: Their morphogenesis and role in the formation of actin filament bundles. *J. Cell Biol.* **131:** 989–1002.

Wakayama S., Cibelli J.B., and Wakayama T. 2003. Effect of timing of the removal of oocyte chromosomes before or after injection of somatic nucleus on development of NT embryos. *Cloning Stem Cells* **5:** 181–189.

Wang Y.L. 1989. Fluorescent analog cytochemistry: Tracing functional protein components in living cells. *Methods Cell Biol.* **29:** 1–12.

Wang Y.L., Heiple J.M., and Taylor D.L. 1982. Fluorescent analog cytochemistry of contractile proteins. *Methods Cell Biol.* **25 Pt B:** 1–11.

Waterman-Storer C.M. 1998. Microtubules and microscopes: How the development of light microscopic imaging technologies has contributed to discoveries about microtubule dynamics in living *cells. Mol. Biol. Cell* **9:** 3263–3271.

Waterman-Storer C.M. and Danuser G. 2002. New directions for fluorescent speckle microscopy. *Curr. Biol.* **12:** R633–R640.

Waterman-Storer C.M. and Salmon E.D. 1997. Actomyosin-based retrograde flow of microtubules in the lamella of migrating epithelial cells influences microtubule dynamic instability and turnover and is associated with microtubule breakage and treadmilling. *J. Cell Biol.* **139:** 417–434.

Waterman-Storer C.M. and Salmon E.D. 1998. How microtubules get fluorescent speckles. *Biophys. J.* **75:** 2059–2069.

Wittmann T., Bokoch G.M., and Waterman-Storer C.M. 2003. Regulation of leading edge microtubule and actin dynamics downstream of Rac1. *J. Cell Biol.* **161:** 845–851.

Yvon A.M., Gross D.J., and Wadsworth P. 2001. Antagonistic forces generated by myosin II and cytoplasmic dynein regulate microtubule turnover, movement, and organization in interphase cells. *Proc. Natl. Acad. Sci.* **98:** 8656–8661.

Yvon A.M., Walker J.W., Danowski B., Fagerstrom C., Khodjakov A., and Wadsworth P. 2002. Centrosome reorientation in wound-edge cells is cell type specific. *Mol. Biol. Cell* **13:** 1871–1880.

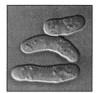

CCD Cameras for Fluorescence Imaging of Living Cells

Phong Tran

Department of Cell and Developmental Biology, University of Pennsylvania, Philadelphia, Pennsylvania 19104

INTRODUCTION

IN RECENT YEARS, BIOLOGISTS HAVE GAINED RENEWED INTEREST in applying fluorescence microscopy to monitor directly in living cells the dynamics of cell biological functions such as protein expression, protein dynamics, and protein-protein interactions. This renaissance is due to a synergistic combination of the improvements in optics and microscope performance, the improvements in camera performance, and the increased use of green fluorescent protein (GFP) and its genetically encoded variants as noninvasive fluorescent biosensors.

This chapter focuses on the application of the currently most widely utilized digital camera for fluorescence imaging of living cells: the charge-coupled device (CCD) camera. The first section provides explanations of the typical specifications of a research-grade CCD camera. The second section presents some sample calculations to illustrate the integration and optimization of the CCD camera into the imaging system. Throughout this chapter, for clarity, are images and measurements taken with the Photometric CoolSNAP$_{HQ}$ Monochrome CCD camera that point out the features of a typical research-grade CCD camera.

UNDERSTANDING THE CCD CAMERA: WHAT DOES IT ALL MEAN?

This section defines terms commonly used by manufacturers to describe the specifications of a typical research-grade CCD camera.

The CCD Sensor

A CCD chip consists of a rectangular array of square pixels. Each pixel is an individual detector of silicon photodiode coupled to an electron storage well. Incident photons hitting the photodiodes will be converted into electrons which are then stored in the potential storage wells for subsequent rapid transfer or "readout" to the amplifier. During readout, electrons in an entire row of pixels are first transferred in the vertical direction to the horizontal register. Next, the electrons in this register are transferred horizontally to the amplifier. The amplifier reads and clocks the accumulated electrons in each pixel and, through an analog-to-digital converter, converts the amount of electrons into equivalent digital signals, a process called digitization.

Two designs are commonly used to achieve rapid transfer of electrons: the frame-transfer design and the interline-transfer design. In the frame-transfer design, the rectangular array of pixels is divided into two equal halves—one half is the imaging array and the second half becomes the storage array. The imaging array receives incident photons, and the resulting accumulated electrons are then quickly shifted to the storage array after exposure or "integration" time. The storage array is masked by a metal shield and therefore cannot receive incident photons. Instead, it serves as a temporary storage area for the electrons from the imaging array. So while the electrons are being read out from the storage array to the amplifier, the imaging array can already start to collect incident photons. Drawbacks of the frame-transfer design include light smearing due to the relatively long shifting time coupled to the inefficiency in shifting the electrons in the many pixels from the imaging array to the storage array.

The interline-transfer design overcomes the drawbacks of the frame-transfer design. In the interline-transfer design, the odd columns of pixels serve as the imaging arrays, and the even columns of pixels serve as the storage arrays. The sequential organization of imaging and storage arrays gives an interline appearance, hence the name. This design allows for fast and efficient shifting of accumulated electrons from the imaging pixels to the immediately adjacent storage pixels. One drawback of this design is that the actual available surface areas of the imaging arrays are small due to the intricacy of the metal shields of individual columns of storage pixels. To increase the amount of photons reaching the available surfaces of the imaging pixels, complementary arrays of microlenses can be positioned before the imaging pixels. These microlenses serve to focus the incident photons directly into the small imaging surfaces, instead of letting them stray onto nonimaging surfaces such as the metal shields.

Both the frame-transfer and interline-transfer designs are relatively fast and efficient, with the imaging arrays and storage arrays working simultaneously to update the image in a process called "progressive scanning."

Quantum Efficiency

Ideally, for 100 photons entering the imaging pixel, 100 electrons would be generated, and 100 units of signal would be detected. This ideal situation illustrates 100% quantum efficiency (QE). The QE is also different for different wavelengths in the light spectrum. The CCD does not have perfect QE, instead, the QE ranges from 0% to 80% depending on the wavelength of the photons entering the pixels. The spectral sensitivity curve of a CCD shows the relationship between QE and wavelengths. In general, due to the coating used during construction, the CCD does not have very high QE in the blue wavelengths region (350–450 nm). However, The CCD has very high QE for photons with wavelengths in the green and red regions (450–650 nm).

The CCD Format

For each CCD array, the number of total pixels and dimensions of each pixel determine the spatial resolution of the CCD relative to the magnified projected image. A point source object can be resolved from a neighboring point source object on the xy plane if they are separated from each other by a minimum distance d_{xy}:

$$d_{xy} = \frac{0.61 \times \lambda}{NA} \tag{1}$$

where λ is the wavelength of the light being imaged and NA is the numerical aperture of the objective lens. Furthermore, to adequately view this minimum resolvable distance between two objects, a minimum sampling frequency of 2× is required. This means that for 2× sampling, 2 pixel units are required to adequately cover the magnified projected d_{xy} distance. The required size of a pixel, P_{xy}, for adequate or optimum sampling of the distance d_{xy} can be calculated as

$$P_{xy} \leq \frac{d_{xy} \times M}{F} \tag{2}$$

where d_{xy} is the minimum resolvable distance between two objects, M is the magnification power of the objective lens, and F is the desired sampling frequency.

Full-well Capacity

The maximum electron storage capacity of each pixel is called the "full-well capacity," and it is proportional to the surface area of the pixel. Beyond this full-well capacity is the saturation limit, where further accumulated electrons cannot be contained in one pixel and therefore would overflow into adjacent pixels, causing a phenomenon called "blooming." Often, antiblooming gates can be built into the CCD, but this would decrease the surface area of the pixel. The full-well capacity value is used to determine the signal-to-noise ratio and the dynamic range of the CCD.

Signal-to-Noise Ratio

The signal of the image, which originates from the fluorescent object, goes through the objective lens, reaches the CCD, and is converted into an intensity signal is expressed as

$$S = I \times QE \times \tau \tag{3}$$

where S is the signal (electrons), I is the input intensity (photons/sec) reaching the CCD, QE is the quantum efficiency of the CCD, and τ is the integration time. The maximum signal S that can be generated by the CCD, of course, is the full-well capacity of the pixel.

All signals are degraded by noise. There are three types of noise contributing to the final quality of the image signal: shot noise, read noise, and dark noise. Shot noise comes from the inherent uncertainty in counting the incoming photons, which follows Poisson statistic, and is therefore calculated as

$$N_{shot} = \sqrt{S} \tag{4}$$

where N_{shot} is the shot noise (electrons) and S is the incoming signal (electrons). In general, a higher input signal would necessarily yield a higher shot noise. This can be advantageous, as a very high shot noise would dominate the noise contribution from read and dark noises, and therefore simplify calculations of the signal-to-noise ratio and the dynamic range.

Read noise is associated with the electronics of the CCD and occurs during the readout of electrons. Read noise gives a measure of the efficiency of charge transfer. In general, a slow readout rate generates low read noise, and a fast readout rate generates high read noise. Read noise is designated N_{read} and has units of electrons. A CCD can have multiple readout rates.

Dark noise is caused by spontaneous creation and accumulation of electrons in the storage wells by thermal energy, or heat, surrounding the CCD. The rate of spontaneous electron creation is temperature-dependent. The total amount of accumulated electrons is dependent on the integration time. In general, as the CCD is cooled down to a very low temperature compared to the ambient room temperature, less dark noise is generated by the CCD. The dark noise can be calculated as

$$N_{dark} = \sqrt{C \times \tau} \tag{5}$$

where N_{dark} is the dark noise (electrons), C is the rate of spontaneous electron creation (electrons/sec), and τ is the integration time (sec). The total noise of the CCD is then is calculated as

$$N_{total} = \sqrt{N_{shot}^2 + N_{read}^2 + N_{dark}^2} \qquad (6)$$

The signal-to-noise ratio gives a measure of the sensitivity of the camera. The signal-to-noise ratio is calculated as

$$\frac{S}{N_{total}} \qquad (7)$$

where S is the incoming signal, and N_{total} is the total noise of the CCD. In general, the higher the S/N ratio the more sensitive the CCD.

Dynamic range

The dynamic range of a CCD camera represents its ability to detect fine differences in the intensity levels of the signal. The dynamic range is defined as

$$D_R = \frac{S}{\sqrt{N_{read}^2 + N_{dark}^2}} \qquad (8)$$

where D_R is the dynamics range, S is the full-well capacity of the pixel, N_{read} is the read noise, and N_{dark} is the dark noise.

The dynamic range is useful for determining the digitization requirements for a given signal. During the analog-to-digital conversion of the electrons by the CCD, the converted signal is digitized by a digitizer into *gray levels*. The digitizer can have different *bit* outputs. The total number of different gray levels generated by a digitizer is equal to $2^{(bit)}$. Therefore, an 8-bit, 10-bit, 12-bit, and 16-bit digitizer would yield gray levels of 2^8, 2^{10}, 2^{12}, and 2^{16} or 256; 1,024; 4,096; and 65,536 different gray levels, respectively. In general, the dynamic range should not be higher than total gray levels for optimum representation of the different intensity levels. Furthermore, whereas the computer can discriminate among gray levels up to 16 bits or higher, the human eyes can only detect less than 8 bits or 256 gray levels. Therefore, the monitor, through various compression algorithms used by the software, will display the higher-bit images as an 8-bit image.

Linearity

The linearity represents the ability of the pixels to respond to incident photons over a range of integration times. A perfectly linear response means that at a constant light intensity, if 1 second of exposure produces 1000 accumulated electrons, then 2 seconds of exposure would produce 2000 electrons, 3 seconds of exposure would produce 3000 electrons, and so on. Deviation from this linear response gives a measure of the nonlinearity of the CCD. In general, CCD cameras have excellent linearity, making them very useful for quantitative measurements of the signals.

Readout Speed

The readout speed refers to how fast the accumulated electrons in each pixel can be transferred and digitized into a signal. The readout speed affects the readout noise. In general, the faster the readout speed, the more read noise is generated. Readout speed has units of MHz.

USING THE CCD CAMERA: PRACTICAL SAMPLE CALCULATIONS

In this section, we point out the different parameters of the Photometric CoolSNAP$_{HQ}$ Monochromatic CCD camera (Photometrics, Roper Scientific, Inc.) and give some useful sample calculations to illustrate and clarify the different terms defined in the previous section.

What Is the CCD of the CoolSNAP$_{HQ}$?

The CoolSNAP$_{HQ}$ uses a high-quality CCD chip manufactured by Sony called ICX285. This chip has an interline-transfer design, utilizes arrays of microlenses, and is monochromatic; therefore, images are acquired in black and white or gray values.

What Is the Quantum Efficiency of the CoolSNAP$_{HQ}$?

Figure 1 shows the spectral sensitivity curve of the CoolSNAP$_{HQ}$. This particular CCD shows a respectable approximately 60% QE between 450 and 650 nm. This means that for 100 photons entering the imaging pixel, approximately 60 electrons are generated, and approximately 60 units of signal are detected.

What Is the CCD Format of the CoolSNAP$_{HQ}$?

Table 1 lists some common objective lenses and the required pixel size for optimum imaging. As an example, we calculate the d_{xy} and P_{xy} of the 100× 1.4 NA objective lens for imaging GFP fluorescence (510 nm is the fluorescent emission of GFP). From Equation (1), the minimum resolvable distance

$$d_{xy} = \frac{0.61 \times 510 \text{ nm}}{1.4} \approx 222 \text{ nm}$$

From Equation (2), at 2× sampling frequency, the required pixel size

$$P_{xy} = \frac{222 \text{ nm} \times 100}{2} \approx 11,100 \text{ nm, or } 11.1 \text{ μm} \times 11.1 \text{ μm}$$

The CoolSNAP$_{HQ}$ has an array of 1392 × 1040 pixels, with each pixel having surface dimensions of 6.45 μm × 6.45 μm. This dimension exceeds the 2× sampling frequency required for optimal imaging of diffraction-limited objects using a 100× objective lens.

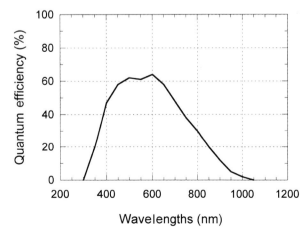

FIGURE 1. Spectral sensitivity curve of a CCD. Curve is taken from the manufacturer's specification for the CoolSNAP$_{HQ}$.

TABLE 1. Objective lens magnification, numerical aperture, pixel size, and signal brightness

Magnification	NA[a]	d_{xy} (μm)[b]	P_{xy} (μm)[c]	Brightness[d]
10	0.3	1.037	5.2	0.11
20	0.5	0.622	6.2	0.20
40	0.7	0.444	8.9	0.20
63	1.32	0.236	7.4	1.00
100	1.4	0.222	11.1	0.50

[a] The objective lenses are from Leica.
[b] d_{xy} is calculated from Equation (1).
[c] P_{xy} is calculated from Equation (2), using a sampling frequency of 2.
[d] Brightness is first calculated from Equation (10), then normalized.

What Is the Full-well Capacity of the CoolSNAP$_{HQ}$?

In general, the full-well capacity can be estimated by multiplying the surface dimensions of the pixel (in micrometers) by 10^3. For example, the CoolSNAP$_{HQ}$ is estimated to have a full-well capacity of $6.45 \times 6.45 \times 1000 = 41,000$ electrons/pixel. Due to the fact that not all of the surface area of each pixel is available to receive incident photons, the actual full-well capacity of a pixel is often much less than estimated. The CoolSNAP$_{HQ}$ has an *actual* full-well capacity of 16,000 electrons/pixel.

What Is the Signal-to-Noise Ratio of the CoolSNAP$_{HQ}$?

We can calculate the maximum signal-to-noise ratio of the CoolSNAP$_{HQ}$. The manufacturer listed the full-well capacity or maximum signal $S = 16,000$ electrons; the read noise $N_{read} = 6$ electrons at 10 MHz readout rate; and the dark noise $N_{dark} = 0.05$ electrons/pixel/sec at $-30°C$. The shot noise N_{shot} can be calculated from Equation (4) as $N_{shot} = \sqrt{16,000} = 126$ electrons. Therefore, from Equation (6), $N_{total} = \sqrt{126^2 + 6^2 + 0.05^2} \approx 126$ electrons (at 1-second integration time). The maximum signal-to-noise ratio achievable by the CoolSNAP$_{HQ}$ then becomes, from Equation (7),

$$\frac{S}{N_{total}} = \frac{16,000}{126} \approx 127$$

Often, the signal-to-noise ratio can be found expressed in decibel units, with the following relationship:

$$\left(\frac{S}{N_{total}}\right)_{db} = 20 \times \log\left(\frac{S}{N_{total}}\right) = 20 \times \log(127) \approx 42 \text{ db}$$

The CoolSNAP$_{HQ}$ also has a second readout speed at 20 MHz. This readout speed gives a slightly higher read noise $N_{read} =$ electrons at 20 MHz. For very high signals, the readout speed does not significantly affect the image quality.

In practice, very low signals would give very low signal-to-noise ratios. The signal-to-noise ratio of a very dim object can be estimated by dividing the average intensity of the object in a region of interest (ROI) over the standard deviation of the intensities in this region.

What Is the Dynamic Range of the CoolSNAP$_{HQ}$?

For the CoolSNAP$_{HQ}$, the read noise $N_{read} = 6$ electrons at 10 MHz readout rate, and the dark noise $N_{dark} = 0.05$ electrons/pixel/sec at $-30°C$. Therefore, the maximum dynamic range can be calculated from Equation (8) as

$$D_R = \frac{16,000}{\sqrt{6^2 + 0.05^2}} \approx 2667$$

With $D_R \approx 2{,}667$ gray levels, the CoolSNAP$_{HQ}$ uses a 12-bit digitizer (4096 gray levels) to efficiently sample the complete dynamic range.

What Is the Linearity of the CoolSNAP$_{HQ}$?

The CoolSNAP$_{HQ}$ has an excellent nonlinearity rating of <1%. This means that the input signal is linearly dependent on the integration time.

What Is the Readout Speed of the CoolSNAP$_{HQ}$?

The CoolSNAP$_{HQ}$ has two readout speeds: 10 MHz and 20 MHz. This means that at 10 MHz, 10,000,000 pixels/sec can be read out, and at 20 MHz, 20,000,000 pixels/sec can be read out. The readout speed is used to determine the frame rate of the CCD. One full frame of the CoolSNAP$_{HQ}$ has $1392 \times 1040 = 1{,}447{,}680$ pixels. Therefore, at 10 MHz read out, a maximum full-frame rate of

$$\frac{10{,}000{,}000 \; pixels/\sec}{1{,}447{,}680 \; pixels/full\,frame} \approx 7 \; \text{full frames/sec}$$

can be achieved, and at 20 MHz, a maximum full-frame rate of

$$\frac{20{,}000{,}000 \; pixels/\sec}{1{,}447{,}680 \; pixels/full\,frame} \approx 14 \; \text{full frames/sec}$$

can be achieved. This is the maximum full-frame rate achievable by the CoolSNAP$_{HQ}$, without considering the integration time.

The specifications of the CCD are parameters measured by the manufacturers under optimal conditions. Parameters such as quantum efficiency, noise, readout rates, and digitization bits are useful first approximations for judging the quality of the CCD camera, and thus the subsequent quality of the images to be acquired. In practice, the actual performance of a CCD is affected by the choice of software/computer and the microscope being used. A feeling for the performance of the CCD camera can be quickly gained by measuring some simple parameters. Below, we describe some measurements taken with the Photometric CoolSNAP$_{HQ}$ Monochromatic CCD camera, coupled to the Leica DMRA2 microscope with the HCX Plan Apo 100× 1.4–0.7 NA oil CS objective lens, and controlled by the MetaMorph v5.6 software.

Temperature and Noise

The CoolSNAP$_{HQ}$ operates at $-30°C$ to ensure low dark noise N_{dark}. The noise contribution can be measured at different temperatures by acquiring images at different temperatures. Figure 2 shows images acquired at 1-second integration, with the microscope light shutter

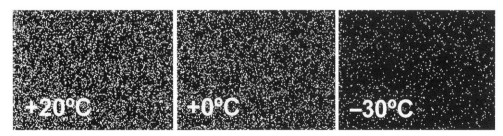

FIGURE 2. Noise decreases with decreasing operating temperature. As the operating temperature of the camera is decreased, the image noise also decreases. Images were exposed for 1 second (with the microscope shutter closed) at different temperatures.

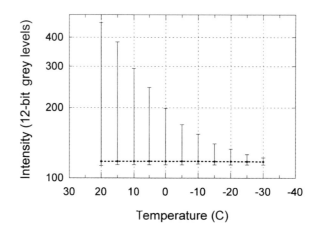

FIGURE 3. Noise decreases with decreasing operating temperature. As the operating temperature of the camera is decreased, the image noise also decreases. The average signal intensities remained unchanged over a range of temperatures. However, the noise, expressed as the range of minimum and maximum gray values, decreases significantly as the operating temperature is decreased.

closed, and at different CCD operating temperature. At +20°C, the image appears very noisy, with a lot of white speckles. At −30°C, the same image appears much less noisy, with fewer white speckles. Figure 3 shows quantitatively the noise contribution due to the temperature of the CCD. Throughout the range of temperature sampled, between −30°C and +20°C, the average intensity or gray levels among the images remained constant at 117. However, the range of gray levels is very large at high temperatures and decreases as the temperature decreases. At +20°C, the average gray level is 117.73 ± 1.07 with a minimum value of 113 and a maximum value of 462. At −30°C, the average gray level is 117.41 ± 0.74 with a minimum value of 114 and a maximum value of 122. Clearly, there is less variation in the gray levels among the pixels as the temperature of the CCD is decreased. The average gray level of about 117 also represents the *noise floor* of the CCD. The noise floor is the minimum gray level that can be detected by the CCD; below this level, the fluorescent object cannot be detected.

The noise floor changes with integration time. Figure 4 shows qualitatively the accumulation of dark noise over time. At 0.01-second exposure, the dark noise is low, with the image appearing black throughout. In contrast, at 10-second exposure, the dark noise is very evident, with the image appearing white. Figure 5 shows quantitatively the accumulation of dark noise over time. In general, the dark noise of the CCD is very low. Integration time between 0.01 and 10 seconds only increased the noise floor two gray levels, from 114 to 116. This means that for a very low-fluorescent object, the exposure time can be very long to get a brighter signal without significantly increasing the noise floor.

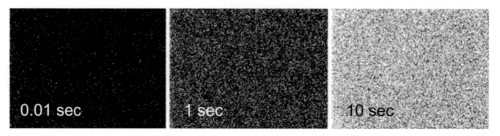

FIGURE 4. Dark noise accumulates over long integration time. As the exposure time is increased, the noise level also increases. Images were exposed with the shutter closed.

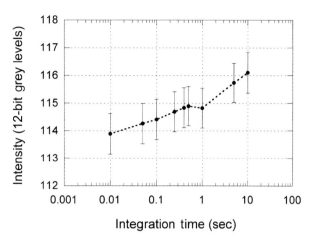

FIGURE 5. Dark noise accumulates over long integration time. As the exposure time is increased, the noise level also increases. Images were exposed with the shutter closed. Note that with a range of exposure from 10 msec to 10 sec, the average intensity due to dark noise only increases 2 gray values.

Readout Speed and Noise

The CoolSNAP$_{HQ}$ has two readout speeds, 10 MHz and 20 MHz. In general, with a relatively high signal, the readout speed does not significantly affect the image quality. However, when the signal is low, or the integration time very short, then the readout speed can affect the image quality. Figure 6 shows qualitative differences between the two readout speeds. At the same 10-msec integration time, the 10-MHz image has an average intensity of 178.80 ± 8.93 gray levels compared to the slightly less bright and noisier 20-MHz image at 175.66 ± 9.36 gray levels.

Image Time

The readout speed defines how fast the electrons are read out by the CCD, and therefore the frame rate of the CCD. In actual experiments, the "image time" may be a better representation of the speed of image acquisition. Image time represents the total time required for an actual image to be captured and displayed on the monitor. The image time takes into account integration time, charge transfer and digitization time, signal to computer time, and image display time. The integration or exposure time varies with the fluorescent signals of the

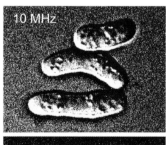

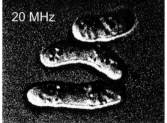

FIGURE 6. Read noise increases with higher readout rates. The images were exposed for 10 msec, and readout at 10 MHz and 20 MHz, respectively. Note that at the slower readout of 10 MHz, the image appears brighter and has a higher average intensity level of 178.80 ± 8.93 and a higher signal-to-noise ratio of 20.03. At the higher readout of 20 MHz, the same image appears darker and has a lower average intensity level of 175.66 ± 9.36 and a lower signal-to-noise ratio of 18.77.

object, ranging from several milliseconds to tens of seconds. The charge transfer and digitization time are specified for each CCD by the manufacturer. The signal to computer and the image display time vary with the imaging capture board and the speed of the computer being used. Currently, the most common mode of signal transfer is via the PCI board, with the Firewire board becoming more popular.

The image time can be increased by optimizing any of the parameters mentioned above. It can also be increased by choosing a smaller region of interest (ROI), which would yield fewer pixels to be read out, and therefore increase the frame rate. Another way to increase the image time is to do pixel "binning."

Binning

The readout method employed in the CCD allows for pixel binning. Binning combines the accumulated electrons in adjacent horizontal and vertical pixels, effectively giving a larger "pixel." This means that the full-well capacity of the new combined pixel would be larger than a single pixel, allowing for more electron accumulation and increasing the signal. However, the trade-off is that binning decreases the spatial resolution of the CCD. Binning also results in smaller image sizes. Binning of 2×2, 3×3, and 4×4 would decrease the size of the image by 4-, 9-, and 16-fold, respectively. Therefore, binning would decrease the overall image time. Binning helps the user to achieve optimization between the captured image xy spatial resolution and image signal intensity. Binning of 1×1 is the default setting and means that the individual pixel is used as is (6.45×6.45 μm), giving full xy spatial resolution. Binning of 2×2 means that the square area of 4 adjacent pixels have been combined into one larger pixel (effectively 12.90×12.90 μm). Binning of 4×4 means that the square area of 16 adjacent pixels have been combined into one larger pixel (effectively 25.80×25.80 μm), etc. Binning of 2×2 decreases the spatial resolution by one-half, but also increases the signal 4-fold. Binning of 4×4 decreases the spatial resolution by one-fourth, but also increases the signal 16-fold, etc.

Figure 7 shows qualitatively the effects of binning of the intensity of the image and the spatial resolution of the image. At the same 20-msec integration time, as the binning is increased, the image intensity appears brighter and the image size appears smaller. For

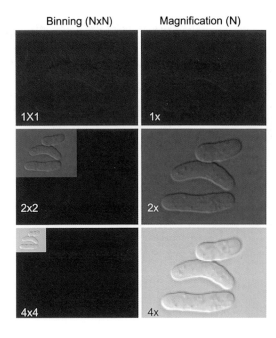

FIGURE 7. Binning increases the signal intensity and decreases the spatial resolution. All images were exposed at 20 msec. Note that the image size becomes decreasingly smaller but brighter at increasingly higher binning. When the smaller images are magnified, pixilation can be seen due to loss of spatial resolution at higher binning.

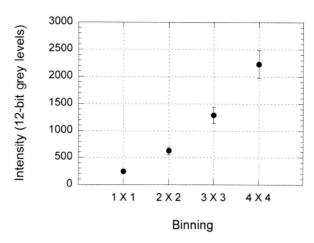

FIGURE 8. Binning increases the signal intensity. Note that binning increases the signal intensity, but not in a linear manner.

example, the original size of the 1 × 1 binning image is 420 × 318 pixels. This size decreases to 210 × 159 pixels and 140 × 106 pixels at 2 × 2 and 4 × 4 binning, respectively. The smaller image has a "pixilated" appearance when magnified to the same size as the original image due to the decrease in spatial resolution. Figure 8 shows quantitatively the effects of binning on the image intensity. For example, the average intensity of the image taken at the same 20-msec integration time increases from 243.25 ± 16.82 to 624.11 ± 66.66 to 2229.45 ± 255.7 with binning of 1 × 1 to 2 × 2 to 4 × 4, respectively.

Gain

The gain function of the CCD camera allows for electronic amplification of the *input* signal, i.e., the accumulated electrons in each pixels. This is useful because a very low signal could be amplified significantly higher than the noise floor, allowing detection. The CoolSNAP$_{HQ}$ allows for two gain settings of 1× and 2×.

Figure 9 shows qualitatively the effect of increasing integration time. As the integration time is increased, more photons will be collected by the CCD, thereby increasing the input signal and

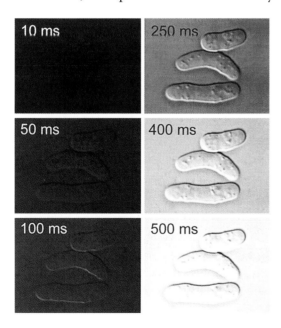

FIGURE 9. Longer integration time increases the signal intensity. As one exposes for longer time, the image appears brighter.

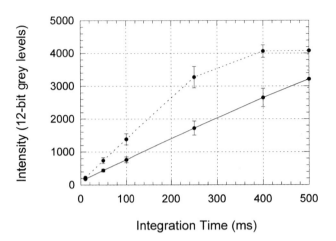

FIGURE 10. Gain function increases signal intensity of the input signal. Note that the average intensity signals are measured directly from the images of cells shown in Figure 9. Solid line is 1× gain and dotted line is 2× gain. Note the excellent linearity of the signal as the integration time is increased in the 1× gain. For 2× gain, the linearity remains excellent, except at higher integration time when the signal begins to saturate.

overall intensity of the image. Figure 10 shows qualitatively the effect of gain on the signal. At the same integration time, the 2× gain setting increases the overall image intensity by twofold.

Offset or Gamma

The offset or "gamma" is a useful feature for further amplification of the *output* signal, i.e., the gray values shown after digitization. Offset or gamma values range from 0.1 to 5. As the values go from lower to higher, the overall signal intensity increases. Changes in intensity are not linear, however. The default gamma or offset value is 1, which is assigned to the output signal. As the gamma or offset is decreased to <1, the overall image intensity would also decrease. However, more "weight" is given to the high gray levels (the bright pixels), and they decrease at a faster rate than the low gray levels (the dark pixels). As the gamma or offset is increased to >1, the overall image intensity would also increase. However, this time, more "weight" is given to the low gray levels and they increase at a faster rate than the high gray levels. Figure 11 shows the qualitative changes in an image that has been modified by the gamma or offset function.

Objective Lens Performance

In general, for fluorescence imaging, and especially low-light imaging, objective lenses with the highest possible NA are required to capture the most amount of available light. The NA is calculated as

$$NA = n \times \sin \theta \tag{9}$$

where NA is the numerical aperture of the objective lens, n is the refractive index of the imaging medium, and θ is the half angle of the cone of light reaching the objective lens. For an objective lens of $NA = 1.4$, $n = 1.515$ for the immersion oil being used, and the calculated $\theta = 67.5°$. This means the full cone of light collected by lenses of 1.4 NA is 135°. The fluorescence sample fluoresces as a three-dimensional sphere of light or 360°. This means that only 135°/360° = 37.5% of available light is being collected from the sample. When factoring in the inefficiency of various optical elements such as filters and lenses to transmit light, the actual signal reaching the CCD is relatively low. The amount of light captured by the

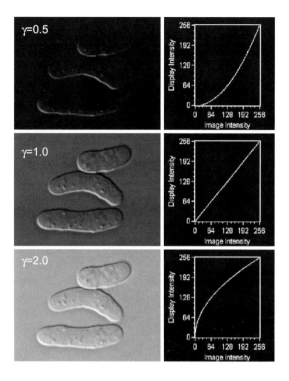

FIGURE 11. Gamma or offset modifies signal intensity of the output signal.

objective is also affected by the magnification of the objective lens. The brightness is calculated as

$$B = \frac{NA^4}{M^2} \tag{10}$$

where B is the brightness, NA is the numerical aperture of the objective lens, and M is the magnification of the objective lens. Assuming an equivalent 1.4 NA, a given signal would appear to be threefold brighter when imaged with a 60× objective lens compared to the 100× objective lens. Table 1 lists the relative brightness of a signal when captured with different objective lenses.

CONCLUSION

We have given a basic description of the specifications and features of a research-grade CCD camera. In general, in choosing a CCD, ask some simple questions such as: How bright is the sample? What temporal resolution is required? What spatial resolution is needed? The answers to these questions can help in the choice of CCD, and the simple tests as outlined above can help optimize the performance of the CCD once it is chosen.

We also recommend reading Inoué and Spring (1997), Herman and Tanke (1998), and Murphy (2001). These "classic" books promote a thorough understanding of optical microscopy, which would further aid in integrating the CCD with the microscope.

ACKNOWLEDGMENTS

I thank Drs. Anne Paoletti and Michel Bornens of the Institut Curie (Paris) for kindly providing the microscope and camera setup from which the images were taken. I also thank Drs. Ted

Salmon (University of North Carolina, Chapel Hill), Rudolf Oldenbourg, and Shinya Inoué (Marine Biological Laboratories, Woods Hole, MA) for continued guidance and support.

REFERENCES

Herman B. and Tanke H.J. 1998. *Fluorescence microscopy,* 2nd edition. Bios Scientific Publishers, Taylor and Francis Group. Abingdon, Oxfordshire, England.
Inoué S. and Spring K. 1997. *Video microscopy: The fundamentals,* 2nd edition. Plenum Press, New York.
Murphy D.B. 2001. *Fundamental of light microscopy and digital imaging.* Wiley-Liss, New York.

Photobleaching Techniques to Study Mobility and Molecular Dynamics of Proteins in Live Cells: FRAP, iFRAP, and FLIP

Gwénaël Rabut and Jan Ellenberg

Gene Expression and Cell Biology Programmes, European Molecular Biology Laboratory, D-69117 Heidelberg, Germany

INTRODUCTION

THE TECHNIQUE OF FLUORESCENCE RECOVERY after photobleaching (FRAP, also called fluorescence photobleach recovery) was introduced in the mid 1970s to study the diffusion of biomolecules in living cells (Edidin et al. 1976). For several years, it was used mainly by a small number of biophysicists who had developed their own photobleaching systems. Since the mid 1990s, FRAP has gained increasing popularity, due to the conjunction of two factors. First, photobleaching techniques are easily implemented on confocal laser-scanning microscopes (CLSMs) (McNally and Smith 2001). FRAP has therefore become available to anyone who has access to such a microscope. Second, the advent of the green fluorescent protein (GFP) has allowed easy fluorescent tagging of proteins and observation in living cells (see Chapters 1–2).

Together with FRAP's renewal and thanks to the versatility of modern CLSMs, which allow control of laser intensity at any point of the image, other photobleaching techniques have been developed. In essence, they all aim at perturbing the steady-state fluorescence distribution in a specimen by bleaching fluorescence in selected regions. After the perturbation, users can observe and then analyze how the fluorescence distribution relaxes toward the steady state. Because the photochemical bleaching of suitable fluorophores is essentially irreversible, changes of fluorescence intensity in the bleached and unbleached regions are due to the exchange of bleached and fluorescent molecules between those regions. The various photobleaching techniques differ by the size of the bleached region, the number of bleach events, and how the fluorescence relaxation is analyzed. The most commonly used photobleaching techniques are called FRAP, FLIP, and iFRAP (Fig. 1), but photobleaching is extremely flexible, and other experimental designs such as spatial analysis can be implemented to best study the process of interest.

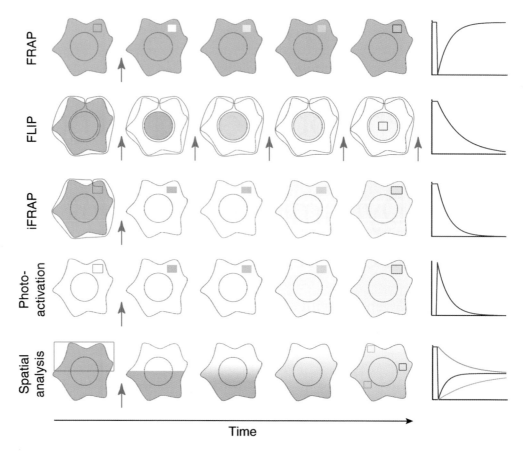

FIGURE 1. Common photobleaching/photoactivation techniques. Schematic depicting a typical cell with nucleus and cytoplasm. Bleach regions are marked by red boundaries. Arrows indicate bleaching events. Regions of interest for quantitation are denoted in the last time point in a color corresponding to the plot.

In a typical FRAP experiment (Fig. 1), a small region of the fluorescent specimen is bleached once. The images are analyzed to display the fluorescence recovery in the bleached region. FRAP experiments are very easy to set up and are thus very popular, providing qualitative information about the behavior of molecules in the bleached region ("mobile" or "immobile"). In addition, a comparison of the recovery times of different molecules or of identical molecules observed under different experimental conditions provides useful information about the system studied. For quantitative analysis of molecular dynamics, FRAP is best suited for the measurement of the apparent diffusion coefficient of molecules (see Table 1), although other parameters can also be extracted.

In an iFRAP (inverse FRAP) experiment (Fig. 1), all of the fluorescence of the specimen is bleached except for a small region of interest. The postbleach images are analyzed to display the loss of fluorescence from the unbleached region. iFRAP experiments are therefore very similar to experiments where nonfluorescent molecules are locally photoactivated or uncaged (Fig. 1) (see Chapters 10–11). Such experiments are now possible with GFP fusion proteins, thanks to the recent development of a photoactivatable GFP (Patterson and Lippincott-Schwartz 2002). Both iFRAP and photoactivation experiments provide similar qualitative information about mobility and equilibration times as do FRAP experiments. For quantitative analysis, iFRAP is limited by the time needed for bleaching large areas (which can take several seconds). It is thus mostly useful for analyzing the dissociation kinetics of molecules bound to an immobile structure for several seconds. Photoactivation, which is

TABLE 1. Applications of photobleaching methods

Application	Method[a]	Examples[b]
Descriptive[c]		
Detection of mobile and immobile fractions	FRAP, iFRAP, FLIP, photoactivation or spatial analysis	Essers et al. (2002); Houtsmuller et al. (1999); Nehls et al. (2000)
Comparison of kinetics	FRAP, iFRAP, FLIP or photoactivation	Boisvert et al. (2001); Calapez et al. (2002); Chen and Huang (2001); Cheutin et al. (2003)
Compartment connectivity	FLIP	Cole et al. (1996); Ellenberg et al. (1997); Kruhlak et al. (2000); Zaal et al. (1999)
	spatial analysis	Kohler et al. (1997)
	photoactivation	Patterson and Lippincott-Schwartz (2002)
Molecular dynamics analysis		
Diffusion coefficient	FRAP	Axelrod et al. (1976); Brown et al. (1999); Ellenberg et al. (1997); Handwerger et al. (2003); Seksek et al. (1997); Wedekind et al. (1996)
		For a review, Klonis et al. (2002)
	photoactivation	Politz (1999)
	spatial analysis	Nikonov et al. (2002); Sciaky et al. (1997); Siggia et al. (2000)
Association and dissociation kinetics	FRAP	Carrero et al. (2003)
	iFRAP	Dundr et al. (2002a,b)
	FLIP	Belgareh et al. (2001); Kruse et al. (2002); Phair and Misteli (2000)
	photoactivation	

[a]The methods listed here are described in Figure 1.
[b]This list is not meant to be exhaustive.
[c]Because this information is not linked to a model, and therefore to biophysical or biochemical properties, it is referred to as "descriptive," in contrast to molecular dynamics analysis.

much faster, can be used to analyze diffusion processes as well as rapid (and slow) dissociation kinetics.

FLIP (fluorescence loss in photobleaching) experiments differ from FRAPs and iFRAPs by the repetitive bleaching of the same region of the specimen (Fig. 1). The repetition of the bleach event constantly depletes the cell of fluorescent molecules that can move into the bleached region, thereby causing a fluorescence loss from unbleached regions. FLIP experiments are especially useful in demonstrating connectivity between different regions and compartments of a cell and for studying fluxes between them. Quantitative analyses of molecular dynamics are also possible, especially if the bleach frequency is constant and high enough not to be limiting for depletion kinetics (see the section on Bleaching).

Every photobleaching experiment is unique. The way it is performed depends on the biological molecule to be studied, the fluorophore used to tag it, the geometry of the cell, as well as the microscope setup available and how the images will be analyzed. It is therefore not possible to give a step-by-step protocol for performing a photobleaching experiment. However, there are useful guidelines and controls that should be considered for most photobleaching applications. This chapter first presents the material required for performing photobleaching experiments on a CLSM and their analysis. It then describes general considerations on how to perform photobleaching experiments as well as the necessary controls. Finally, different possible ways to analyze the data are presented. For more details, see recent reviews on photobleaching techniques (Houtsmuller and Vermeulen 2001; Lippincott-Schwartz et al. 2001; McNally and Smith 2001;

Klonis et al. 2002). For those interested in performing FRAP experiments to measure diffusion coefficients, method papers and more specialized reviews are also available (Axelrod et al. 1976; Politz 1999; Siggia et al. 2000; Klonis et al. 2002; Verkman 2002).

MATERIALS FOR PHOTOBLEACHING EXPERIMENTS AND THEIR ANALYSIS

Fluorescent Probes

An ideal fluorescent probe for photobleaching experiments should be bright and stable under low-intensity illumination and undergo a rapid irreversible photobleaching, without photo-chemical reactions, at high-intensity illumination. Depending on the biological system, the choice is between chemical fluorophores and fluorescent proteins. Fluorescent proteins can be used like large epitope tags to label almost any protein of interest fluorescently and have the great advantage that cells directly express these fusion proteins from introduced plasmid DNA or mRNA (see Chapter 2). Smaller chemical fluorophores are advantageous when fluorescent proteins appear to sterically impair the function of the protein of interest, or when studying nonproteic molecules (e.g., lipids or nucleic acids). However, using biomolecules labeled with chemical fluorophores poses several disadvantages. Before use, chemically labeled proteins must be purified, labeled with the selected fluorophore, and introduced into the cell by microinjection (see Chapter 5) or another technique, such as bead loading (McNeil and Warder 1987) or protein transduction (Morris et al. 2001). In addition to the labor-intensive nature of the labeling and the invasiveness of the delivery methods, a major disadvantage of many chemical fluorophores is that their bleaching produces free radicals, which can be toxic for the cell (see the section below on Bleaching-induced Cellular Damage).

Fluorescent Proteins

Among all the fluorescent protein variants that have been described (see Tsien 1998; Zhang et al. 2002), the enhanced GFP variant (EGFP, F64L/S65T) appears to be most suitable for studies in mammalian cultured cells at 37°C (Patterson et al. 1997). Bleaching of EGFP and similar variants is mostly irreversible, although some spontaneous reversibility is observed (for more details, see the section below on Reversibility of Bleaching: Dark States of Fluorescent Proteins). At high local concentrations, which are easily reached for overexpressed membrane proteins, EGFP and its variants have a tendency to multimerize and can distort the dynamics of the tagged protein. This property can be abolished by point mutations in the dimerization interface (Zacharias et al. 2002).

The EYFP (enhanced yellow fluorescent protein) variant (although less photostable than EGFP) can be used as an alternative for very abundant proteins that are difficult to bleach when tagged with EGFP. However, EYFP displays a higher reversibility of bleaching. The ECFP (enhanced cyan fluorescent protein) variant is less suitable for photobleaching experiments because it is excited (and bleached) at shorter, more toxic wavelengths. Although several forms of red fluorescent proteins (RFPs) have also been described (DsRed and variants, HcRed and variants; see Zhang et al. 2002), a robust monomeric RFP is still lacking (see Chapter 1); in general, RFPs have not been explored systematically for photobleaching applications.

For photoactivation, which can have significant advantages over FRAP, especially for experiments on short time scales, the photoactivatable form of GFP (PA-GFP) (Patterson and Lippincott-Schwartz 2002) is so far the best-characterized protein. Other photoactivatable mutants have been described (Ando et al. 2002; Chudakov et al. 2003), but they still need to be evaluated carefully for in vivo applications.

Chemical Fluorophores for Protein Labeling

A large variety of chemical fluorophores (including fluorescein and its derivatives, rhodamine, Texas red, and carbocyanine) have been used for photobleaching experiments. No systematic study comparing their photobleaching characteristics has been published, which makes it difficult to choose the best fluorophore. Red dyes, such as Cy3 and TRITC, are often difficult to bleach with the common HeNe lasers on CLSMs and should therefore be avoided unless strong green lasers are available. The popular Alexa dyes (e.g., Alexa488 and Alexa543) are very stable and difficult to bleach, and in our experience, they are to some degree photoactivated under certain imaging conditions. Fluorescein and its derivatives are probably the most used chemical fluorophore for photobleaching experiments. However, they tend to bleach too easily, even at low illumination intensity used for image acquisition, and their photobleaching is partly reversible (Stout and Axelrod 1995; Periasamy et al. 1996). Certain chemical fluorophores can also be used for photoactivation experiments (see Chapters 10 and 11).

Regardless of whether fluorescent proteins or chemical fluorophores are used, it is very important to test photobleaching reversibility (see the section below on Reversibility of Bleaching: Dark States of Fluorescent Proteins) and photoactivation of the chosen fluorophore under the imaging conditions to be used.

Microscope Setup

Many different microscope setups have been used for photobleaching experiments (e.g., see Verkman 2003). Today, most biologists use commercial CLSMs. Photobleaching techniques are easily implemented on these microscopes (McNally and Smith 2001), which permit a lot of flexibility in the experimental design. Their major drawbacks are limited time resolution and sensitivity, which can be problematic for studies of rapid processes and low-light fluorescence typical of living cells. If images must be acquired faster than ~10 frames per second, specialized systems are necessary, which typically combine a laser for spot bleaching in the cell with widefield image acquisition by a fast CCD (charge-coupled device) camera (Verkman 2003).

This section highlights features of CLSMs that are of special importance when performing photobleaching experiments.

Lasers for Imaging and Bleaching Fluorophores

Appropriate lasers are obviously absolutely critical for photobleaching experiments. The bleaching of a fluorophore is achieved by illuminating it with the same wavelength as used for its excitation, but with a much higher intensity. For EGFP bleaching, 488-nm Ar-ion lasers rating ~25 mW are sufficient for many applications. Stronger lasers (rated up to ~500 mW) are needed for applications requiring rapid bleaching times, or when using more photostable fluorophores (e.g., Alexa 488). We anticipate that RFPs will become increasingly popular in the future as they can be imaged with standard rhodamine or Texas red optics. However, most CLSMs use relatively weak HeNe lasers for green wavelengths, so it may be desirable to configure a system with stronger green lasers such as Nd:YAG 532 nm or Ar-Kr 546 nm if photobleaching is a major application. For PA-GFP photoactivation, a 25-mW 405-nm blue diode and an 80-mW 413-nm Kr-ion laser have been used successfully (Patterson and Lippincott-Schwartz 2002).

AOTFs for Switching between Imaging and Bleaching Modes

Photobleaching experiments require rapid switching between a low-intensity illumination mode for image acquisition and a high-intensity illumination mode for bleaching. Ideally, the imaging mode uses 1000 times less energy than the bleaching mode. Currently, the best way to achieve this switch on CLSMs—which use the same laser for both modes of illumination—is to control the laser intensity by acousto-optical tunable filters (AOTFs). These devices allow laser transmission into the specimen to be attenuated over a wide range (100% to 0.1%) within microseconds. They therefore provide flexibility (the user can freely choose the laser transmission in both imaging and bleaching mode) and allow the user to switch between imaging and bleaching mode very rapidly (critical for observing fast processes). In addition, AOTFs can be used to bleach regions of arbitrary shapes within cells by switching between imaging and bleaching intensities on a pixel-by-pixel basis during the linear scanning of the beam (Wedekind et al. 1994). AOTFs are used in the current generation of Bio-Rad, Leica, and Zeiss CLSMs.

Controlling Laser Intensity Fluctuations

CLSMs are equipped with photomultipliers to detect the emitted fluorescence as well as the transmitted light of a sample. They can, in addition, have a special detector to directly measure the laser intensity. This detector, sometimes referred to as the monitor diode, can be important when performing photobleaching experiments, because it allows the stability of the laser illumination to be independently controlled (see the sections on Laser Stability and Detector Linearity, and Practical Considerations for Image Analysis). The current generation of Zeiss CLSMs is equipped with a monitor diode.

Microscope Objectives

The objectives used for photobleaching experiments have essentially the same characteristics as the objectives used for other fluorescence applications. Most importantly, they should have high transmission and low autofluorescence and be corrected for spherical (and possibly chromatic) aberrations (e.g., Fluar or Planapochromat objectives).

The numerical aperture (NA) of the objective is the property that most influences photobleaching experiments. The NA is a measure of the opening angle of an objective. High-NA objectives focus the laser beam best along the optical axis, but do not bleach out-of-focus regions of the specimen efficiently (Fig. 2). Conversely, low-NA objectives

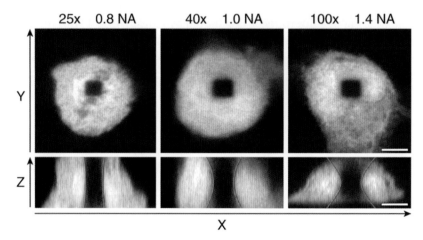

FIGURE 2. Effect of objective numerical aperture on the bleaching pattern. Fixed mitotic cells expressing EGFP were bleached with oil immersion objectives of various numerical apertures (NA). Bar, 5 μm.

bleach the sample more uniformly along the optical axis (Fig. 2), but do so less efficiently because the beam is not as well-focused. In addition, low-NA objectives collect less light and have lower resolution than high-NA objectives. When choosing an objective for photobleaching experiments, a compromise must be made between detection sensitivity, precision, efficiency, and homogeneity of bleaching along the optical axis. This compromise will depend on the thickness of the sample to be photobleached and the biological process being analyzed. For instance, a 40 × 1.0 NA oil immersion objective is a good compromise for most purposes when working with adherent mammalian fibroblasts, as it ensures homogeneous illumination over the entire height of the cell (~6 μm) while preserving relatively high bleaching efficiency and collection of fluorescent signal. To uniformly bleach mitotic fibroblasts (that can round up to ~25 μm) or thicker polarized cells (like MDCK cells), a 25 × 0.8 NA immersion objective is preferable (see Fig. 2). Conversely, to most efficiently bleach and image molecules in thin cellular extensions, use a 63 × 1.4 NA oil immersion objective.

Bleach Control Software

All of the functions of current CLSMs are computer-controlled. The microscope operating software must contain a bleaching function that allows the user to flexibly design photobleaching experiments. This function should allow users to precisely define the area to be bleached (one or multiple regions of any shape, especially useful for iFRAP and FLIP); to choose the laser transmission and scan speed, and zoom in the bleaching mode; and to choose how many times the bleached region is scanned during bleaching ("bleach iterations"). Finally, such a bleaching function needs to be incorporated in a time-lapse function, enabling the user to define when to switch between bleaching and imaging modes and how frequently the bleach should be repeated.

Currently, the microscope software allowing the best flexibility for bleaching is the LSM software operating Zeiss CLSMs. Importantly, to give users more flexibility in experimental design, some microscope operating software (Zeiss LSM, Leica LCS, Bio-Rad LaserSharp2000) offers a macro interface that allows programming of new photobleaching protocols.

Software for Data Analysis

Once a photobleaching experiment is performed, analysis of the results may require the use of additional software packages. A selection of useful programs is shown in Table 2. For more details, see the section below on Image Analysis and Kinetic Modeling.

PERFORMING A PHOTOBLEACHING EXPERIMENT

The optimal conditions for performing a photobleaching experiment on a given biological sample must be determined empirically. Once this is done, image acquisition and bleaching become routine processes that can be automated in an application protocol. Sample preparation for photobleaching experiments is similar to that for live cell imaging and will not be discussed here (see Chapter 17). When working with sensitive samples or especially toxic conditions (short wavelengths, high-intensity illuminations), the addition of antioxidants to the imaging medium can be useful to maintain cell viability (see the section below on Bleaching-induced cellular damage).

TABLE 2. Software for photobleaching experiments analysis

	URL	License	Platform	Features
Image quantitation[a]				
Image J	http://rsb.info.nih.gov/ij	free	Mac, PC	User-dedicated functions can be implemented in Java.
Metamorph	http://www.universal-imaging.com/	license	PC	
Image alignment				
TurboReg (Image J plugin)	http://bigwww.epfl.ch/thevenaz/turboreg	free	Mac, PC	Time series images are aligned to a single reference image. Nonlinear alignments are possible.
Autoaligner	http://www.bitplane.com	license	PC	Time series images are aligned to the previous image in the image stack, which does not always work efficiently. Very good manual alignment interface.
Kinetic modeling				
Virtual Cell	http://www.nrcam.uchc.edu	free for academy	Mac, PC	Easy way to write both compartmental and spatial models taking cellular geometry into account.
SAAM II	http://www.saam.com	license	PC	Easy compartmental model setup.
Berkeley Madonna	http://www.berkeleymadonna.com	license	Mac, PC	Powerful differential equation solver.
Matlab	http://www.mathworks.com	license	Mac, PC	Very good graphic output possibilities.
Gepasi	http://www.gepasi.org	free	PC	
WinSAAM	http://www.winsaam.com	free	PC	
Systems Biology Workbench	http://www.sbw-sbml.org/	free	PC	

[a] In addition, many software packages originally developed to control microscopes now include powerful image analysis features, e.g., Zeiss LSM and Axiovision, Leica LCS. This list is not meant to be exhaustive.

Parameters for Image Acquisition

Open or Closed Pinhole?

When performing photobleaching experiments with a CLSM, the user should first decide whether to use an open or closed pinhole. Closing the pinhole reduces out-of-focus blur and may appear obvious (otherwise, why use a CLSM?). However, imaging thin optical sections has drawbacks for photobleaching experiments. First, closing the pinhole typically reduces the signal, making quantitation of fluorescence intensities less accurate. This can be resolved by increasing the laser transmission in the imaging mode, reducing the scan speed, or increasing scan averaging, but all of these may cause undesired bleaching during image acquisition. Second, if using thick samples and/or high-NA objectives, out-of-focus regions are not bleached efficiently (see Fig. 2). In this case, a thin optical section gives the impression that the bleach is complete, although this is not true over the entire depth of the specimen. The result is that exchange of bleached molecules in the focal plane with unbleached out-of-focus molecules will distort the fluorescence equilibration kinetics. It is therefore frequently preferable to use a completely open pinhole as well as low-NA objectives to ensure homogeneous illumination in *z* (see the section below on Microscope Setup) when performing photobleaching experiments with CLSMs. Nevertheless, confocality can be useful for some photobleaching experiments, for example, when a large diffusely distributed pool of fluorescent molecules masks the localized pool of interest. An example is given in the section on Constructing a Compartment Model, where the protein of interest localizes to kinetochores and the cytoplasm, and only a thin optical section can reveal the kinetochore bound pool (within that section, see Fig. 6). In addition, very thick samples (e.g., whole embryos or large oocytes) cannot be bleached homogenously in *z*, even with low-NA objectives. Here, using a closed pinhole will reveal the bleached area clearly, but it may be necessary to acquire images of fluorescence recovery in three dimensions (3D) over time to estimate the contribution of out-of-focus molecules to the equilibration process.

Laser Power and Transmission

Illumination intensity (in the imaging and bleaching modes) can be controlled by (1) adjusting the strength of the laser beam by changing the output of the laser power supply and (2) modulating the laser transmission into the specimen using acousto-optical tunable filters (AOTFs) (see section on Microscope Setup). For photobleaching experiments, it is best (when possible) to use the full transmission range of AOTFs to maximize the difference between bleaching and imaging mode and thus achieve the highest bleaching efficiency and minimize undesired photobleaching during image acquisition (referred to as "acquisition photobleaching" from now on, see also the section on Controlling Acquisition Photobleaching).

Adjust the laser power in the imaging mode to the lowest value that will result in an acceptable image while using the lowest possible transmission of the AOTF. Then test in the bleaching mode the minimal laser transmission required for instantaneous and complete bleach (see the section on Bleaching). For good photobleaching experiments, the laser intensity for bleaching should be at least 100-fold higher than for imaging.

Scan Field and Zoom for Image Acquisition

CLSMs allow the user to acquire images in only a fraction of the field accessible to the scanning laser beam. It is therefore possible to zoom in on the sample by acquiring the same number of pixels from a smaller area, in addition to the optical magnification set by the objective. When imaging cultured cells, users can typically use high-magnification zooms to image only a subcellular compartment (e.g., the nucleus) or lower zooms to image multiple cells.

For photobleaching experiments, zooms that completely include the cell to be photobleached in the scan field should whenever possible be used to evaluate the contribution of all cellular pools of molecules to the equilibration after the bleach. Consider also using lower zooms and positioning the scan field to monitor neighboring control cells in the same field as the bleached cell (see the section on Controlling Acquisition Photobleaching). However, this need for relatively low zooms should be balanced with the necessity for good spatial and temporal resolution of dynamic processes, which are typically achieved at high zooms. In addition, the photobleaching efficiency (but also the undesired bleaching during imaging!) increases with the square of the zoom factor (Centonze and Pawley 1995), making high zooms more favorable for rapid bleaching.

Scan Speed, Average, and Image Acquisition Frequency

Since photobleaching is typically used to monitor dynamic processes, the scan speed should be set as fast as possible to yield an acceptable image. Frame or line averaging can be used to reduce image noise, but should be avoided to reduce undesired acquisition photobleaching and allow high time-resolution. For some experiments, for example, when studying diffusion, the time resolution might actually be much more critical than the spatial resolution. In this case, the images obtained should allow one to measure fluorescence intensities, although they do not have to meet the standards required for high-resolution three-dimensional

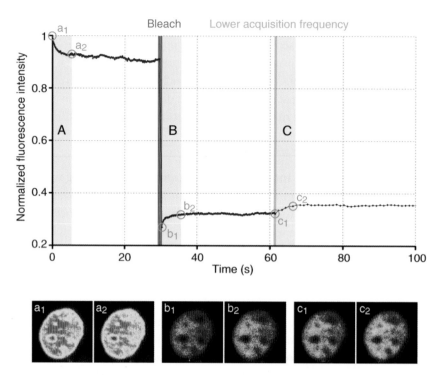

FIGURE 3. Dark state of EGFP. Total fluorescence of a living cell expressing histone H2B-EGFP. The cell was initially imaged at 7 frames per second and then bleached, and finally, the image acquisition frequency was lowered to 1 frame per second. Images of selected time points (a_1, a_2, b_1, b_2, c_1, c_2) are pseudo-colored for easier comparison. (*A*) Initially, due to high acquisition frequency, a fraction of the EGFPs are driven into a nonfluorescent dark state, resulting in a fluorescence intensity decrease from a_1 to a_2. (*B*) Photobleaching step drives more EGFPs into the dark state, but they recover fluorescence in a few seconds (b_1 to b_2). (*C*) At lower acquisition frequency, a smaller fraction of EGFPs are driven into the dark state, accounting for fluorescence increase upon lowering acquisition frequency (c_1 to c_2).

reconstruction of a fixed specimen. In extremely fast cases, reducing the number of pixels in the image to, for example, 128 × 30 instead of a standard value such as 512 × 512, will allow a substantial increase of the acquisition frequency.

The image acquisition frequency should be adjusted to resolve the dynamic phase of the recovery with good temporal resolution. A rule of thumb is to acquire at least 20 data points during the time required for half of the equilibration to occur. When using fluorescent proteins, *do not* change the acquisition frequency during the experiment, because at high-frequency illumination, a fraction of the fluorescent proteins is driven into a reversible dark state (see the section on Reversibility of Bleaching: Dark States of Fluorescent Proteins). Changing the acquisition frequency alters the fraction of nonfluorescent fluorescent proteins, leading to fluorescence variations due to photophysical changes, not molecular movements (see Fig. 3).

Detector Gain and Offset

Failure to properly adjust the gain and offset of the detector will render image quantitation impossible. The offset must be set low enough so that background pixels have a measured intensity slightly greater than zero (otherwise, some signal is lost). Similarly, the detector gain must be set so that only very few or no pixels are saturated. When working with low noise, acquiring 12-bit images (4096 gray values), rather than 8-bit images (256 gray values), will increase the effective dynamic range of the measurements.

Acquisition of Prebleach Images

Some images should be acquired prior to the photobleach. These images are used as a reference for the steady-state distribution of the fluorescent molecules. Typically, 3–10 prebleach images (depending on the noise) are sufficient to allow averaging. However, when imaging fluorescent proteins at high frequency (>1 image/second), 50–100 prebleach images must be acquired to ensure that the fraction of fluorescent proteins driven into a dark state has reached a steady state (see the section on Reversibility of Bleaching: Dark States of Fluorescent Proteins and Fig. 3). This is especially important when prebleach images are used to estimate the bleaching rate during imaging (see the section on Controlling Acquisition Photobleaching).

Bleaching

Ideally, each bleaching event should be instantaneous (no fluorescent or bleached molecules should be able to leave or enter the bleached region during the bleaching time) and complete (all fluorescent molecules in the bleached region should be bleached). This is rarely possible, and with the limited laser power available on most CLSMs (see the section on Microscope Setup), instantaneous and complete bleaching are often incompatible goals (the more complete the bleach, the longer it takes . . .). Complete bleaching yields better contrast between bleached and unbleached regions, and thus allows more precise measurement of the exchange between bleached and unbleached molecules. The bleaching time should not exceed one tenth of the half-time of the equilibration, otherwise a significant fraction of bleached molecules will have exchanged out of the bleached region.

In practice, bleaching parameters must be calibrated for each experiment in pilot photobleaching runs. One initially bleaches with full transmission of the laser. Thus, only the zoom, the scan speed, and the number of bleach iterations will affect bleaching efficiency and time. The first experiment should be carried out with a single bleach iteration, at the same scan speed and zoom used for imaging (if zoom and scan speed are changed, there will be a delay for switching between bleaching and imaging mode on most CLSMs). If a single bleach pass is sufficient to bleach the region completely, the laser transmission can be lowered for the bleach until bleaching starts to be incomplete. On the other hand, if a single iteration does not bleach the region completely, the number of bleach iterations should be increased. However, for very dynamic processes, if the laser power is insufficient, exchange processes during the bleach might limit the bleach depth that can be achieved locally. If a delay between bleaching and imaging mode is not critical for the experiment, the bleach can also use a higher zoom than the imaging mode to achieve better bleaching efficiency.

When performing repetitive bleaches in FLIP experiments, the bleaching frequency should be set just fast enough not to limit fluorescence depletion from unbleached regions, but also slow enough to allow significant exchange of fluorophores between bleached and unbleached regions during the time delay between bleaches. Typically, one increases the bleach frequency until the rate of depletion of fluorescence from an unbleached region does not increase any further.

How to Decide When to Stop the Experiment

Optimally, the total data collection after the initial bleach (or photoactivation) should be 10–50 times longer than the half-time of the observed process (Axelrod et al. 1976; Gordon et al. 1995). For initial photobleaching experiments, (1) acquire data until no noticeable fluorescence change is observed in the sample, and (2) analyze the experiments and estimate the half-time of the observed process (see the section on Parameters That Describe Fluorescence Equilibration Kinetics). Perform additional experiments for a time period that is more than ten times this estimated half-time. Careful analysis of the experiments can reveal subpopulations of molecules with different equilibration kinetics. In such cases, it may be necessary to carry out even longer experiments to resolve the kinetics of the fraction of molecules with the "slowest" exchange rate.

NECESSARY CONTROLS AND POSSIBLE ARTIFACTS

Calibrating the Bleached Region in *x-y* and *z* Dimensions

For each set of parameters affecting the bleach (i.e., objective, laser power and transmission, scan zoom and speed, pixel number, and number of bleach iterations), it is important to calibrate the part of the specimen that is affected by the bleach in all three spatial dimensions. This is best done in fixed specimens (because molecular movements have ceased) by performing a bleach and then acquiring a high-resolution *z*-stack of the bleached specimen to measure the dimensions of the bleached regions. In the *x-y* plane, the correspondence of the actual bleach boundary to the region selected before the bleaching is typically very good. This correspondence fails when very high zooms are used, because the pixel dimensions can be smaller than the diameter of the focused laser beam (see Fig. 2). Such lateral straying is problematic in FLIP experiments in which a cellular compartment (e.g., the cytoplasm) should be bleached selectively without affecting an adjacent one (e.g., the nucleus) (see Fig. 1). Along the optical axis, the bleach region will only correspond to the selected region if an objective of sufficiently low NA is used to illuminate the entire thickness of the specimen homogeneously (see the section on Microscope Setup and Fig. 2). If the NA is too high, an image of the double conical shape of the beam profile will be bleached through the depth of the fixed sample (Fig. 2), and a lower-NA objective should be used. Objectives with both very high magnification and NA (e.g., 100× NA 1.4 or higher) can be used to bleach volumes limited in *z* to a few microns, allowing rough three-dimensional bleaching patterns to be obtained (Beaudouin et al. 2002). More precise definition of the bleached volume along the optical axis can only be achieved by multiphoton excitation (Brown et al. 1999; Coscoy et al. 2002).

If the bleach region appears to scatter more in the live sample than in the fixed calibration specimen, the bleached molecules have exchanged out of the bleached region during the bleach. If enough laser power is available, decrease the bleaching time until bleached regions in live and fixed cells are similar.

Controlling Acquisition Photobleaching

Despite using low laser transmission for the imaging mode, some bleaching will occur with each image acquired to monitor fluorescence equilibration. It is essential to design photobleaching experiments in such a way that this can be corrected (see the section on Practical Considerations for Image Analysis). Possible controls include the following:

1. *Total cell intensity.* When the fluorescence from the entire cell is acquired (in 3D, by using either low-NA objectives or open pinhole or *z*-stacks), the total cell fluorescence can be used for data correction (see the section on Practical Considerations for Image Analysis). This is the method of choice, because it also corrects for laser-intensity fluctuations (see below) and for fluorescence loss due to photobleaching (for FRAPs). However, total cell fluorescence cannot be used for FLIPs (because fluorescence is regularly depleted) and may not be practical for iFRAPs (because little fluorescence is left after the bleach, resulting in a poor signal-to-noise ratio).

2. *Neighboring cell.* Alternatively, the images can contain at least one unbleached control cell in the field in addition to the bleached cell. Fluorescence of the control cell can then serve for correction of acquisition photobleaching and laser-intensity fluctuations. This is especially useful for FLIPs (and possibly iFRAPs). However, it is not always feasible: Imaging large fields lowers the time resolution; it can also be difficult to find two fluorescent cells next to each other.

3. *Prebleach series.* Finally, a large number of prebleach images can be used to estimate the rate of acquisition photobleaching, which is then used to correct the whole experiment. This is convenient for rapid imaging, when experiments are not dramatically lengthened by a prebleach series. However, under such conditions, a fraction of the fluorescent proteins might be driven into a dark state and the first data points may need to be excluded to estimate the rate (see below and Fig. 3).

If none of these is practical, separate control experiments can be performed that use settings identical to those for the photobleaching experiments (objective, zoom, pixel number, laser power, laser transmission, scan speed, averaging) but do not actually photobleach. Acquisition photobleaching can then be inferred from the control cells.

Reversibility of Bleaching: Dark States of Fluorescent Proteins

Analyses of photobleaching experiments rely on the irreversibility of the bleaching because only then does fluorescence equilibration between bleached and unbleached regions reflect exclusively molecular movements between those regions. However, all fluorophores, including FITC (Stout and Axelrod 1995; Periasamy et al. 1996) and GFP (Swaminathan et al. 1997; Levin et al. 2001), show some degree of reversible photobleaching, which must be identified to avoid misinterpretations. Photobleaching reversibility can occur by several mechanisms. Triplet-state sequestration (see Periasamy et al. 1996; Song et al. 1996; Swaminathan et al. 1997) occurs on a time scale of a few milliseconds and is therefore typically not an issue for photobleaching experiments with CLSMs. Critical are the slower mechanisms of reversibility that occur on the timescale of seconds to minutes. Fluorescent proteins (FPs) exhibit fluctuations between fluorescent and dark states (Dickson et al. 1997; Pierce et al. 1997; Haupts et al. 1998) that can be light-dependent (Garcia-Parajo et al. 2000; Schwille et al. 2000). During illumination (due to image acquisition or bleaching), a pool of the FPs is driven into a dark state that appears bleached but will recover fluorescence after a few seconds. When the illumination frequency is higher than the recovery time, a fraction of FPs is maintained in the dark state. This fraction remains in steady state, unless the illumination frequency or intensity changes (e.g., during the bleach) (see Fig. 3). Therefore, during a few seconds after bleaching, FP fluorescence variations are partly due to bleaching reversibility. With the most common FP variant, EGFP, this typically affects less than 10% of the molecules, although this reversible fraction is higher in other variants such as YFP.

Bleaching reversibility can be detected simply by measuring total cell fluorescence (again in 3D, see above) after bleaching. If fluorescence increases within seconds after the bleach, this is due to photobleaching reversibility. Photobleaching reversibility of FPs should always be measured in intact cells, as the photophysical properties of FPs can change dramatically after fixation. There is no good way to easily correct for photobleaching reversibility. If it is problematic, such as in FRAP experiments with subsecond time resolution, the use of the photoactivatable GFP will eliminate this problem. PA-GFP displays some degree of photoactivation reversibility (J. Beaudouin and J. Ellenberg, unpubl.), but this can simply be corrected as if it was bleaching due to image acquisition (see above).

Laser Stability and Detector Linearity

Laser intensities fluctuate over time, especially for gas lasers at low tube currents or when the laser is near the end of its lifetime. These fluctuations result in artifactual fluorescence variations. To correct for them, measure the excitation fluctuation using strategies 1 or 2 described in the section on Controlling Acquisition Photobleaching, because these strategies also correct for laser fluctuations. Alternatively, directly measure the laser intensity during the experiment using a monitor diode (see the section on Microscope Setup). The monitor diode readings can then be used directly to correct the fluorescence intensities (see the section on Practical Considerations for Image Analysis).

CLSMs are equipped with photomultipliers to detect the fluorescence emitted by the sample. To quantitate photobleaching experiments, the detectors need to respond linearly to the amount of fluorescence emitted by the sample. This can be checked by imaging solutions of different fluorophore concentrations under identical microscope settings (Fink et al. 1998; Piston et al. 1999).

Bleaching-induced Cellular Damage

Photobleaching experiments have often been criticized for the potential damage to a cell. The validity of photobleaching experiments has been demonstrated in various cases (e.g., see Jacobson et al. 1978; Wolf et al. 1980). Local heating created during the bleaching is limited (less than 0.5°C under typical conditions; Axelrod 1977) and is insufficient to significantly modify diffusion or chemical reactions. Nevertheless, cases of damage due to bleaching have been reported (Lepock et al. 1978; Vigers et al. 1988), and some cellular processes (such as mitosis) are extremely sensitive to radiation. For example, strong bleaching of chromosomes during prophase can delay entry into prophase or arrest cells in G_2 (J. Beaudouin and J. Ellenberg, unpubl.).

Essentially two sources of cytotoxicity are due to high-intensity illumination. First, when fluorophores are bleached, their oxidation generates radicals that can produce reactive oxygen species (ROS) if oxygen has access to the fluorophore. ROS will react quickly with neighboring molecules and can locally damage the cell. This effect is especially strong for some chemical dyes, but much less severe for FPs because their cyclic tripeptide fluorophore is shielded inside the protein barrel (Yang et al. 1996; Wachter et al. 1998; Gaietta et al. 2002; Tour et al. 2003) and radicals are more likely to react within the FP itself. FP bleaching is not influenced by removal of oxygen from the medium (Pierce et al. 1997). Second, cellular molecules (e.g., proteins, DNA, and pigments) can directly absorb the light used to image and bleach fluorophores. In general, toxicity increases dramatically at shorter wavelengths (below ~450 nm) where proteins and DNA start to absorb. However, cells containing pigments such as hematocytes (see Bloom and Webb 1984) or plant cells are also sensitive to longer wavelengths.

It is essential to control for cytotoxicity when performing photobleaching experiments, especially when working with short wavelengths and/or light-sensitive cellular processes. It is useful to monitor transmission images of the observed sample for dramatic morphological changes due to toxicity during the experiment. Whenever possible, the cellular processes investigated by photobleaching should be characterized by independent means to detect photobleaching artifacts. For example, when working with mitotic cells, the normal timing of the cell cycle and the mitotic index should be calibrated in control cells to verify that photobleaching does not inhibit cell cycle progression. If cytotoxic effects are observed, fluorophores that can be bleached at longer wavelengths should be used. Also adding antioxidants such as vitamin C (1 mg/ml, freshly dissolved in imaging medium) (Gerlich et al. 2001) or Trolox (Tsien 1998), a vitamin E analog, can be helpful.

Cell Movements and Focus Drifts

During bleaching experiments on time scales longer than a few seconds, cellular and subcellular movements lead to movements of the bleached and unbleached regions to be analyzed. In addition, for experiments lasting hours, cells can migrate out of the imaged field, and dividing cells tend to round up and leave the focus plane. In this case, automated tracking of cells or cellular structures and repositioning of the microscope stage/focus can be very useful and can be implemented on many microscopes in user-defined macros (Rabut and Ellenberg unpubl.). Alternatively, users can acquire larger fields to accommodate lateral cell movement and z-stacks to accommodate movement along the optical axis and then select the in-focus frames for analysis. In addition, over a long period of time, microscope stages tend to drift and the focus must be corrected either manually or by using an autofocusing system. Although autofocusing protocols are not yet available on CLSM operating software, it can be programmed using user-defined macros.

Protein Synthesis and Degradation

Synthesis or degradation of FP-tagged proteins can occur during long-term experiments distorting fluorescence quantitations. Proteolysis typically affects the cell homogeneously at a constant rate. Therefore, it can be corrected in the same manner as acquisition photobleaching (see the section on Controlling Acquisition Photobleaching). The rate of synthesis can be determined by completely bleaching the whole cell and observing the rate of global fluorescence increase over long times. In transient transfections, this rate can be extremely variable from cell to cell and can change over time, making straightforward corrections impossible. If new synthesis is a problem, photoactivatable GFP should be used, as newly synthesized PA-GFP is nonfluorescent and will not interfere with the measurements (Patterson and Lippincott-Schwartz 2002). Alternatively, protein synthesis inhibitors such as cycloheximide can be used, but their toxicity can create other artifacts.

IMAGE ANALYSIS AND KINETIC MODELING

Practical Considerations for Image Analysis

To evaluate and compare the results of photobleaching experiments, the observed fluorescence equilibration must be quantitated, background subtracted, corrected, and finally normalized. These four steps constitute the basic image analysis procedure. The fluorescence intensity measurement is performed by measuring the total, or more often the average, pixel values in regions of interest (ROIs) of the cell, using an image analysis software that can handle time series data. Most CLSM operating software packages (such as Zeiss LSM or Leica LCS) have such measurement tools. If not, Image J is a very powerful and constantly expanding image-processing software in the public domain that will meet the needs of most users. In addition, other commercial programs are available (see examples in Table 2). The raw intensity measurements are then imported into a spreadsheet program for further data processing (background subtraction, correction, and normalization) and graphic representation.

Image Alignment

Before performing any image analysis with constant measurement regions on an entire time series, it is essential to check that the cellular regions to be analyzed are not moving over time. If movements have occurred during the time series, first align the images either automatically or manually, or quantitate one image at a time by manually repositioning the measurement ROI. The latter method can be very time-consuming and introduces measurement errors. Alignment can be done efficiently using the Image J plugin TurboReg (Thevenaz et al. 1998) or the commercial tool Autoaligner (Bitplane AG) (see Table 2).

Background Subtraction

The average pixel value of an ROI always contains background fluorescence (from the medium, glass coverslip, objectives, and other sources). Therefore, any intensity measurement must first be background-subtracted. Background is measured in regions of the image outside the cell. The average pixel value in such an ROI is then subtracted from the average pixel values of the cellular ROIs for all time points (Fig. 4).

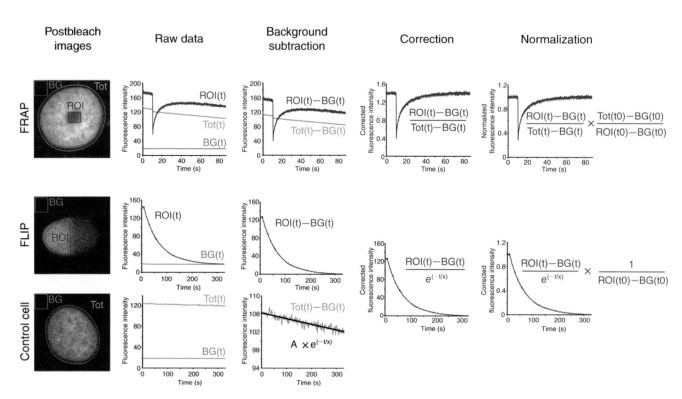

FIGURE 4. Steps of image analysis. The four steps of image analysis (raw data measurement, background subtraction, data correction and normalization) are illustrated for a FRAP and a FLIP experiment. The regions used for fluorescence intensity measurement are shown in the postbleach images. The equations for each step of image analysis are displayed next to the corresponding graph. For the FRAP experiment, the total cell intensity is used to correct for the photobleached fraction, as well as acquisition photobleaching. For the FLIP experiment, a control cell is used to correct for acquisition photobleaching. The background-subtracted fluorescence intensity in the control cell is fitted to the equation $A \times e^{(-t/x)}$, where A is the initial fluorescence value. $e^{(-t/x)}$ is then used for correcting fluorescence intensity in the bleached cell.

Corrections for Laser Fluctuations and Acquisition Photobleaching

Once background-subtracted, the fluorescence measurements should be a linear function of the fluorophore concentration in the region of interest, and intensity changes should reflect molecular movements between bleached and unbleached regions. However, acquisition photobleaching, laser-intensity fluctuations, and photobleaching reversibility will also lead to intensity changes, and thus must be corrected. Acquisition photobleaching and laser-intensity fluctuations are directly reflected in the total cell intensity. Therefore, the data can be corrected by dividing the background-subtracted fluorescence measurement by the total cell intensity at each time point. This is straightforward and also corrects for the bleached fraction in FRAP experiments (Fig. 4).

Total cell correction is not always possible, for instance, when using scan fields that do not encompass the entire cell or when performing FLIP experiments. In such cases, the data can be corrected by dividing the fluorescence measurement at each time point by a function representing the artifactual signal variations. In the case of acquisition photobleaching, this function is of the type $e^{-t/x}$. Values for x can be easily determined by fitting the total fluorescence intensity of an unbleached neighboring cell (see Fig. 4) or the gradual fluorescence decrease in the prebleach or postbleach images. In the case of laser-intensity fluctuation, the function is of the type $Diode(t) - x0$, where $Diode(t)$ is the monitor diode reading at the time t and $x0$ is a factor correcting for the nonzero offset of the diode. Values for $x0$ can be determined by measuring laser intensity as a function of the monitor diode value by varying the AOTF transmission while using the same settings as in the actual experiment. As already mentioned, there is no straightforward way to correct for photobleaching reversibility.

Normalization to Compare Different Experiments

To be able to compare different experiments, the corrected fluorescence measurements must be normalized, either to arbitrary fluorescence units or to numbers of fluorescent molecules. Obviously, normalizing to an arbitrary fluorescence unit is much easier, and generally, the first or prebleach time point is normalized to one (Fig. 4). However, using fluorophore calibration standards, it is also possible to convert fluorescence measurements into actual numbers of molecules that generated the signal (Fink et al. 1998; Hirschberg et al. 1998; Chiu et al. 2001; Dundr et al. 2002a). Knowing the numbers of molecules can be very beneficial for analyzing molecular dynamics from photobleaching experiments by kinetic modeling.

Parameters That Describe Fluorescence Equilibration Kinetics

The result of the complete image-processing routine is a kinetic plot displaying fluorescence changes in the bleached and/or unbleached regions of the cell over time. From this plot, a number of parameters can be extracted directly that describe the kinetics of fluorescence equilibration. Because these parameters are not linked to a model of molecular dynamics and do not directly relate to biophysical or biochemical properties, we refer to them as "descriptive" parameters, in contrast to "molecular dynamics," which can be extracted by kinetic modeling (see below).

- *(Im)mobile fraction.* Kinetic plots of photobleaching experiments immediately reveal whether the studied molecule has mobile and immobile fractions of fluorescent molecules within a region of the cell (Fig. 5). "Immobile" fraction is a relative definition during the time of observation, because no biological molecules are irreversibly bound to their substrates in living cells for infinitely long times. The mobile and immobile fractions are

FRAP experiment

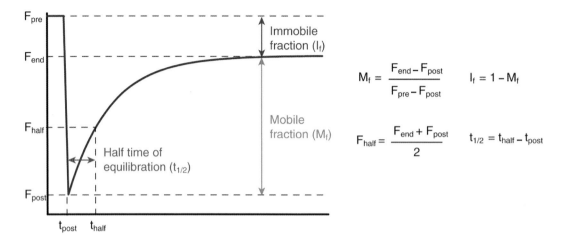

FLIP or iFRAP experiment

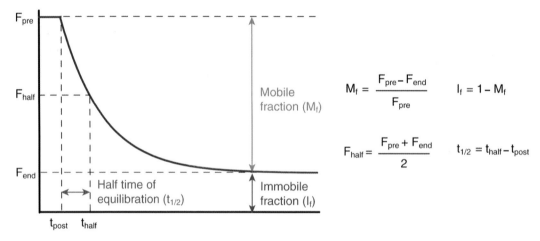

FIGURE 5. Parameters to describe photobleaching experiment kinetics. Idealized plots and equations showing how to calculate the mobile (M_f) and immobile (I_f) fractions as well as the half-time of equilibration ($t_{1/2}$) from FRAP, iFRAP, and FLIP experiments. Very similar analysis can be done with photoactivation experiments.

determined by calculating the ratios of the final to the initial fluorescence intensity (see equations in Fig. 5). For FRAPs, the immobile fraction can only be calculated if the fluorescence measurements are corrected for loss of signal due to the photobleaching step. (The uncorrected fluorescence intensity in the bleached region will never reach its initial intensity because of the loss of fraction of the total fluorescence in the photobleach.) Comparison of the mobile fraction of the same molecule under different experimental conditions can provide insight about the nature of those fractions.

- $t_{1/2}$. The time for the exchange of half of the mobile fraction between bleached and unbleached regions can be directly read from the kinetic plot (see Fig. 5) and is referred to as the halftime ($t_{1/2}$) of equilibration (or recovery for FRAPs). Comparing $t_{1/2}$ for the same molecules and process under different experimental conditions can be used to determine the steps limiting the exchange rate.

Basics of Kinetic Modeling, Simulation, and Optimization

Photobleaching experiments can also be exploited to quantitatively measure biophysical and biochemical parameters of the investigated molecular dynamics. FRAP had initially been developed to measure diffusion coefficients and flow velocity (Axelrod et al. 1976). In living cells, most proteins have binding partners; they form protein complexes and/or associate with relatively immobile cellular structures such as the cytoskeleton, chromosomes, or membranes. Therefore, photobleaching experiments contain information not just about the diffusional properties of the studied molecule, but also about its binding characteristics. Kinetic modeling and computer simulation allow this information to be extracted (Phair and Misteli 2001).

A kinetic model describes the different molecular species in a system in terms of their rates of formation and decay. Essentially, two kinds of kinetic models have been used by cell biologists: compartmental and spatial. Compartmental models (Jacquez 1996) describe a biological process using a finite number of compartments. Each compartment contains a molecule (or complex of molecules) at a cellular location, (e.g., a transcription factor, in the cytoplasm, the nucleoplasm, or bound to DNA). Within each compartment, molecular dynamics are not limited by diffusion and they are thus defined as "well mixed." Transfer of molecules between compartments represents an exchange of molecules between different locations (e.g., between the cytoplasm and the nucleus), chemical states (e.g., phosphorylation), or chemical interactions (e.g., binding to DNA). In a compartmental model, the only variable is time. Rates of transfer between compartments are either fixed or fitted to best match the experimental data. However, some biological processes are limited by diffusion even on the scale of light microscopy. For such processes, well-mixed compartments cannot be defined and space variables must be taken into account by using spatial models (see below).

Constructing a Compartment Model

To make a compartmental model, the user must first draw a list of all the states of the molecules in the system of interest from a priori knowledge or a hypothesis about the biological process under study (Fig. 6). A state represents a molecule at a given place (in Fig. 6, it is kinetochore or cytoplasm) in the cell and undergoing specific chemical interactions (in Fig. 6, the kinetochore pool has mobile and immobile components, reflecting different chemical interactions). Each molecular state is represented in the model by a separate compartment. All transfers between the compartments within the model are then defined (i.e., movements, chemical modifications, and chemical interactions). The transfers represent the inputs and outputs for each state of the model. Finally, bleaching events are represented by efflux from the bleached compartments (photoactivation would be represented by an influx).

The next step consists in defining differential equations for each state. A differential equation expresses the instantaneous change in the number of molecules for a given state. This change is equal to the number of molecules formed minus the number of molecules removed from this state at any time. The number of molecules added to a state at any time is the sum of transfers (or influxes) into the compartment. Conversely, the number of molecules removed from a certain state is given by the sum of transfers (or effluxes) removing molecules from the compartment. First, the rate laws for all individual transfers must be defined and are then combined in differential equations for each molecular state of the model. Typical examples for rate laws are (1) mass action (chemical reactions such as modifications [e.g., phosphorylation] or interactions [e.g., binding to DNA]) or (2) Fick's law (e.g., nucleocytoplasmic traffic fluxes). Since photobleaching (or photoactivation) occurs only at discrete times, time conditions need to be added to the bleaching rate constants such that the bleaching rate is zero at times when no photobleach occurred in the experiment. Finally, the

Experiment:

GFP-Nup133 localizes to the cytoplasm and the kinetochores of mitotic cells. The cytoplasm is bleached repeatedly (FLIP experiment) to measure the dissociation rate of GFP-Nup133 bound to the kinetochores

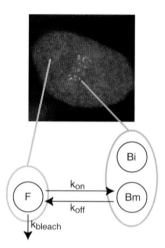

Compartment Model:

States:
GFP-Nup133 free in the cytoplasm (F)
GFP-Nup133 dynamically bound to the kinetochores (Bm)
GFP-Nup133 stably bound to the kinetochores (Bi)

Transfers:
Photobleaching of the cytoplasmic fluorescence (k_{bleach})
Dissociation of GFP-Nup133 from the kinetochores (k_{off})
Association of GFP-Nup133 to the kinetochores (k_{on})

Translation in differential equations:

Rate laws for the transfers:

Bleaching rate:	$k_{bleach} \times F$
Dissociation rate:	$k_{off} \times Bm$
Association rate:	$k_{on} \times R \times F = k_{on}^* \times F$

(R represents the free receptors of Nup133 on the kinetochores. Their concentration is unknown but is constant in steady state. Therefore $k_{on} \times R$ is constant and is represented by k_{on}^*.)

Rate laws are combined to write the differential equations of each state
$$dF/dt = k_{off} \times Bm - k_{on}^* \times F - k_{bleach} \times F$$
$$dBm/dt = k_{on}^* \times F - k_{off} \times Bm$$
$$dBi/dt = 0$$

Initial conditions:

Steady state implies that Association rate = Dissociation rate:
$$k_{on}^* \times F = k_{off} \times Bm$$

Parameter optimization and simulation:

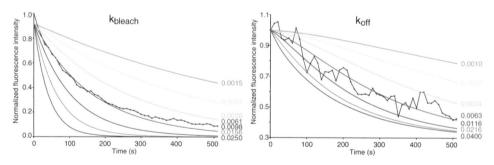

FIGURE 6. Illustration of compartmental modeling. Compartmental model of GFP-Nup133 localization in mitotic cells. See figure for details. (Reprinted, with permission, from Belgareh et al. 2001; © The Rockefeller University Press.)

model assumes the system is in steady state before the first fluorescence perturbation by writing the equilibrium equations between all transfers and determining the steady-state rate constants.

Constructing a Spatial Model

Compartmental models can be written when molecular dynamics are independent of the spatial position within the compartment. Such models have been used very successfully to model molecular fluxes (Hirschberg et al. 1998; Smith et al. 2002) and association and/or

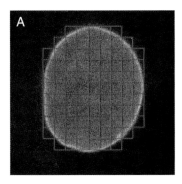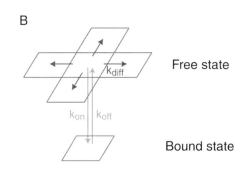

FIGURE 7. Spatial modeling. For spatial modeling, the cell is segmented into discrete subcellular domains as shown in *A*. In each domain, fluorescent molecules are found in free and bound states (*B*). Bound molecules can dissociate from their binding site (k_{off}), whereas free molecules can associate with their binding site (k_{on}) or diffuse to the neighboring spatial domains (k_{diff}).

dissociation constants from localized binding sites (Belgareh et al. 2001; Dundr et al. 2002b; Carrero et al. 2003). However, this assumption can only be made if diffusion is fast compared to the studied processes and does not limit it. If the molecular dynamics occur on time scales similar to that of diffusion, a spatial model must be constructed. A straightforward way to write a spatial model is to subdivide the cell into a finite number of spatial elements (Fig. 7A) (Schaff et al. 1997; Sciaky et al. 1997; Siggia et al. 2000). Within each element, a molecule can be present in different molecular states and transfer between those states (Fig. 7B). In addition, the different molecular states in a spatial element can exchange with neighboring elements by diffusion (Fig. 7B). Spatial models are especially relevant for studying processes such as intracellular signaling where cellular geometry has a role in controlling their spatial and temporal characteristics (Fink et al. 1999, 2000).

Computer Simulation and Optimization

Once a kinetic model (spatial or compartmental) for the studied processes has been formulated, it can be used to simulate experiments. The simulation numerically solves the system of coupled differential equations that defines the model. In the case of photobleaching experiments, the model will predict the response of the system to the fluorescence perturbation. The prediction depends on the values of the parameters of the model, i.e., the rates of transfer between molecular states. Their values are typically unknown for the first simulation and must be determined by optimization, in which the simulation tries to fit the prediction as closely as possible to the experimental data.

If the optimization process identifies one or several sets of parameter values that can properly fit the experimental data, the hypothesis about the biological mechanism of the studied process that formed the basis of the model offers a possible solution. It does not mean that the hypothesis is correct, because there are often several parameter sets or completely different models that can account for the experimental data. Typically, the simplest model that can account for the experimental data is chosen as a working model. Assuming that this model is correct, the parameters of the model will represent parameters of the studied molecular dynamics. Then the model can be used to calculate the biophysical and biochemical parameters such as residency time, steady-state fluxes, molecular abundances, and diffusion constants, which reveal more information about the system (see Hirschberg et al. 1998; Fink et al. 2000; Dundr et al. 2002b; Smith et al. 2002). Importantly, the working model can serve to make predictions about the biological system that can subsequently be tested by new experiments (see examples in Fink et al.

2000). Often, the first model fails to properly fit all the experimental data with any set of parameters. This means that the kinetic model is incorrect and often that the biological hypothesis that supports the model is wrong. In such a case, kinetic modeling can be used to invalidate theories (see examples in Hirschberg et al. 1998), and the model must be redefined to try to explain the experimental data.

Software for Kinetic Modeling

To perform kinetic modeling, a software package is needed that can solve differential equation and optimize the model parameters. SAAM II (SAAM Institute, Inc.) and Berkeley Madonna (see Table 2) allow easy drawing of compartmental models via a graphic interface. Differential equations for the fluxes between the compartments are derived automatically, and parameters of the model can be fitted to imported experimental data. Virtual Cell (Schaff et al. 1997) is the only software ready to use for developing spatial models (for a more thorough description of Virtual Cell, see Slepchenko et al. 2002), but other software such as Matlab (The MathWorks, Inc.) or Berkeley Madonna also have a flexible programming interface to construct kinetic models (compartmental or spatial), specify experimental protocols, solve systems of differential equations, and optimize model parameters.

CONCLUSION

Advances in microscopy techniques and the development of fluorescent probes now permit the development of techniques to probe molecular processes at work in living cells. The photobleaching and photoactivation techniques described in this chapter are becoming very popular because they are easy to perform and give immediate qualitative results. They are used to analyze the kinetics of biological processes in living cells. Often, these processes can be described by a kinetic model, which is useful to test theories and determine biochemical and biophysical parameters. One of the major challenges for the future will be to standardize the analysis procedures for different photobleaching applications, permitting a move from descriptive experiments to experiments that derive molecular dynamics parameters. As such parameters can often be determined by other methods, either in vivo (e.g., fluorescence correlation spectroscopy [FCS] for diffusion and interactions [Medina and Schwille 2002]) or in vitro (e.g., surface plasmon resonance for chemical interactions; McDonnell 2001), this would make it possible to directly compare results from photobleaching experiments with results from completely different experimental systems. Finally, photobleaching has applications other than kinetic analyses. Selective photobleaching can be used to enhance the contrast of dim structures in regions next to other bright objects (Presley et al. 1997). In addition, patterns can be bleached on smooth stable surfaces to analyze their deformations during various cellular processes (Daigle et al. 2001; Beaudouin et al. 2002; Gerlich et al. 2003).

REFERENCES

Ando R., Hama H., Yamamoto-Hino M., Mizuno H., and Miyawaki A. 2002. An optical marker based on the UV-induced green-to-red photoconversion of a fluorescent protein. *Proc. Natl. Acad. Sci.* **99:** 12651–12656.

Axelrod D. 1977. Cell surface heating during fluorescence photobleaching recovery experiments. *Biophys. J.* **18:** 129–131.

Axelrod D., Koppel D.E., Schlessinger J., Elson E., and Webb W.W. 1976. Mobility measurement by analysis of fluorescence photobleaching recovery kinetics. *Biophys. J.* **16:** 1055–1069.

Beaudouin J., Gerlich D., Daigle N., Eils R., and Ellenberg J. 2002. Nuclear envelope breakdown proceeds by microtubule-induced tearing of the lamina. *Cell* **108:** 83–96.

Belgareh N., Rabut G., Bai S.W., van Overbeek M., Beaudouin J., Daigle N., Zatsepina O.V., Pasteau F., Labas V., Fromont-Racine M., Ellenberg J., and Doye V. 2001. An evolutionarily conserved NPC subcomplex, which redistributes in part to kinetochores in mammalian cells. *J. Cell Biol.* **154:** 1147–1160.

Bloom J.A. and Webb W.W. 1984. Photodamage to intact erythrocyte membranes at high laser intensities: Methods of assay and suppression. *J. Histochem. Cytochem.* **32:** 608–616.

Boisvert F.M., Kruhlak M.J., Box A.K., Hendzel M.J., and Bazett-Jones D.P. 2001. The transcription coactivator CBP is a dynamic component of the promyelocytic leukemia nuclear body. *J. Cell Biol.* **152:** 1099–1106.

Brown E.B., Shear J.B., Adams S.R., Tsien R.Y., and Webb W.W. 1999. Photolysis of caged calcium in femtoliter volumes using two-photon excitation. *Biophys. J.* **76:** 489–499.

Calapez A., Pereira H.M., Calado A., Braga J., Rino J., Carvalho C., Tavanez J.P., Wahle E., Rosa A.C., and Carmo-Fonseca M. 2002. The intranuclear mobility of messenger RNA binding proteins is ATP dependent and temperature sensitive. *J. Cell Biol.* **159:** 795–805.

Carrero G., McDonald D., Crawford E., de Vries G., and Hendzel M.J. 2003. Using FRAP and mathematical modeling to determine the in vivo kinetics of nuclear proteins. *Methods* **29:** 14–28.

Centonze V. and Pawley J. 1995. Tutorial on practical confocal microscopy and use of the confocal test specimen. In *Handbook of biological confocal microscopy* (ed. J. Pawley), pp. 549–570. Plenum, New York.

Chen D. and Huang S. 2001. Nucleolar components involved in ribosome biogenesis cycle between the nucleolus and nucleoplasm in interphase cells. *J. Cell Biol.* **153:** 169–176.

Cheutin T., McNairn A.J., Jenuwein T., Gilbert D.M., Singh P.B., and Misteli T. 2003. Maintenance of stable heterochromatin domains by dynamic HP1 binding. *Science* **299:** 721–725.

Chiu C.S., Kartalov E., Unger M., Quake S., and Lester H.A. 2001. Single-molecule measurements calibrate green fluorescent protein surface densities on transparent beads for use with "knock-in" animals and other expression systems. *J. Neurosci. Methods* **105:** 55–63.

Chudakov D.M., Belousov V.V., Zaraisky A.G., Novoselov V.V., Staroverov D.B., Zorov D.B., Lukyanov S. and Lukyanov K.A. 2003. Kindling fluorescent proteins for precise in vivo photolabeling. *Nat. Biotechnol.* **21:** 191–194.

Cole N.B., Smith C.L., Sciaky N., Terasaki M., Edidin M., and Lippincott-Schwartz J. 1996. Diffusional mobility of Golgi proteins in membranes of living cells. *Science* **273:** 797–801.

Coscoy S., Waharte F., Gautreau A., Martin M., Louvard D., Mangeat P., Arpin M., and Amblard F. 2002. Molecular analysis of microscopic ezrin dynamics by two-photon FRAP. *Proc. Natl. Acad. Sci.* **99:** 12813–12818.

Daigle N., Beaudouin J., Hartnell L., Imreh G., Hallberg E., Lippincott-Schwartz J., and Ellenberg J. 2001. Nuclear pore complexes form immobile networks and have a very low turnover in live mammalian cells. *J. Cell Biol.* **154:** 71–84.

Dickson R.M., Cubitt A.B., Tsien R.Y., and Moerner W.E. 1997. On/off blinking and switching behaviour of single molecules of green fluorescent protein. *Nature* **388:** 355–358.

Dundr M., McNally J.G., Cohen J., and Misteli T. 2002a. Quantitation of GFP-fusion proteins in single living cells. *J. Struct. Biol.* **140:** 92–99.

Dundr M., Hoffmann-Rohrer U., Hu Q., Grummt I., Rothblum L.I., Phair R.D., and Misteli T. 2002b. A kinetic framework for a mammalian RNA polymerase in vivo. *Science* **298:** 1623–1626.

Edidin M., Zagyansky Y., and Lardner T.J. 1976. Measurement of membrane protein lateral diffusion in single cells. *Science* **191:** 466–468.

Ellenberg J., Siggia E.D., Moreira J.E., Smith C.L., Presley J.F., Worman H.J., and Lippincott-Schwartz J. 1997. Nuclear membrane dynamics and reassembly in living cells: Targeting of an inner nuclear membrane protein in interphase and mitosis. *J. Cell Biol.* **138:** 1193–1206.

Essers J., Houtsmuller A.B., van Veelen L., Paulusma C., Nigg A.L., Pastink A., Vermeulen W., Hoeijmakers J.H., and Kanaar R. 2002. Nuclear dynamics of RAD52 group homologous recombination proteins in response to DNA damage. *EMBO J.* **21:** 2030–2037.

Fink C., Morgan F., and Loew L.M. 1998. Intracellular fluorescent probe concentrations by confocal microscopy. *Biophys. J.* **75:** 1648–1658.

Fink C.C., Slepchenko B., Moraru I.I., Schaff J., Watras J., and Loew L.M. 1999. Morphological control of inositol-1,4,5-trisphosphate-dependent signals. *J. Cell Biol.* **147:** 929–936.

Fink C.C., Slepchenko B., Moraru I.I., Watras J., Schaff J.C., and Loew L.M. 2000. An image-based model of calcium waves in differentiated neuroblastoma cells. *Biophys. J.* **79:** 163–183.

Gaietta G., Deerinck T.J., Adams S.R., Bouwer J., Tour O., Laird D.W., Sosinsky G.E., Tsien R.Y., and Ellisman M.H. 2002. Multicolor and electron microscopic imaging of connexin trafficking. *Science* **296:** 503–507.

Garcia-Parajo M.F., Segers-Nolten G.M., Veerman J.A., Greve J., and van Hulst N.F. 2000. Real-time light-driven dynamics of the fluorescence emission in single green fluorescent protein molecules. *Proc. Natl. Acad. Sci.* **97:** 7237–7242.

Gerlich D., Beaudouin J., Gebhard M., Ellenberg J., and Eils R. 2001. Four-dimensional imaging and quantitative reconstruction to analyse complex spatiotemporal processes in live cells. *Nat. Cell Biol.* **3:** 852–855.

Gerlich D., Beaudouin J., Kalbfuss B., Daigle N., Eils R., and Ellenberg J. 2003. Global chromosome positions are transmitted through mitosis in mammalian cells. *Cell* **112:** 751–764.

Gordon G.W., Chazotte B., Wang X.F., and Herman B. 1995. Analysis of simulated and experimental fluorescence recovery after photobleaching. Data for two diffusing components. *Biophys. J.* **68:** 766–778.

Handwerger K.E., Murphy C., and Gall J.G. 2003. Steady-state dynamics of Cajal body components in the *Xenopus* germinal vesicle. *J. Cell Biol.* **160:** 495–504.

Haupts U., Maiti S., Schwille P., and Webb W.W. 1998. Dynamics of fluorescence fluctuations in green fluorescent protein observed by fluorescence correlation spectroscopy. *Proc. Natl. Acad. Sci.* **95:** 13573–13578.

Hirschberg K., Miller C.M., Ellenberg J., Presley J.F., Siggia E.D., Phair R.D., and Lippincott-Schwartz J. 1998. Kinetic analysis of secretory protein traffic and characterization of golgi to plasma membrane transport intermediates in living cells. *J. Cell Biol.* **143:** 1485–1503.

Houtsmuller A.B. and Vermeulen W. 2001. Macromolecular dynamics in living cell nuclei revealed by fluorescence redistribution after photobleaching. *Histochem. Cell Biol.* **115:** 13–21.

Houtsmuller A.B., Rademakers S., Nigg A.L., Hoogstraten D., Hoeijmakers J.H., and Vermeulen W. 1999. Action of DNA repair endonuclease ERCC1/XPF in living cells. *Science* **284:** 958–961.

Jacobson K., Hou Y., and Wojcieszyn J. 1978. Evidence for lack of damage during photobleaching measurements of the lateral mobility of cell surface components. *Exp. Cell Res.* **116:** 179–189.

Jacquez J.A. 1996. *Compartmental analysis in biology and medicine.* BioMedware, Ann Arbor.

Klonis N., Rug M., Harper I., Wickham M., Cowman A., and Tilley L. 2002. Fluorescence photobleaching analysis for the study of cellular dynamics. *Eur. Biophys. J.* **31:** 36–51.

Kohler R.H., Cao J., Zipfel W.R., Webb W.W., and Hanson M.R. 1997. Exchange of protein molecules through connections between higher plant plastids. *Science* **276:** 2039–2042.

Kruhlak M.J., Lever M.A., Fischle W., Verdin E., Bazett-Jones D.P., and Hendzel M. J. 2000. Reduced mobility of the alternate splicing factor (ASF) through the nucleoplasm and steady state speckle compartments. *J. Cell Biol.* **150:** 41–51.

Kruse C., Jaedicke A., Beaudouin J., Bohl F., Ferring D., Guttler T., Ellenberg J., and Jansen R.P. 2002. Ribonucleoprotein-dependent localization of the yeast class V myosin Myo4p. *J. Cell Biol.* **159:** 971–982.

Lepock J.R., Thompson J.E., and Kruuv J. 1978. Photoinduced crosslinking of membrane proteins by fluorescein isothiocyanate. *Biochem. Biophys. Res. Commun.* **85:** 344–350.

Levin M.H., Haggie P.M., Vetrivel L., and Verkman A.S. 2001. Diffusion in the endoplasmic reticulum of an aquaporin-2 mutant causing human nephrogenic diabetes insipidus. *J. Biol. Chem.* **276:** 21331–21336.

Lippincott-Schwartz J., Snapp E., and Kenworthy A. 2001. Studying protein dynamics in living cells. *Nat. Rev. Mol. Cell. Biol.* **2:** 444–456.

McNally J.G. and Smith C.L. 2001. Photobleaching by confocal microscopy. In *Confocal and two-photon microscopy* (ed. A. Diaspro), pp. 525–538. Wiley-Liss, New York.

McDonnell J.M. 2001. Surface plasmon resonance: Towards an understanding of the mechanisms of biological molecular recognition. *Curr. Opin. Chem. Biol.* **5:** 572–577.

McNeil P.L. and Warder E. 1987. Glass beads load macromolecules into living cells. *J. Cell. Sci.* **88(Pt 5):** 669–678.

Medina M.A. and Schwille P. 2002. Fluorescence correlation spectroscopy for the detection and study of single molecules in biology. *BioEssays* **24:** 758–764.

Morris M.C., Depollier J., Mery J., Heitz F., and Divita G. 2001. A peptide carrier for the delivery of biologically active proteins into mammalian cells. *Nat. Biotechnol.* **19:** 1173–1176.

Nehls S., Snapp E.L., Cole N.B., Zaal K.J., Kenworthy A.K., Roberts T.H., Ellenberg J., Presley J.F., Siggia E., and Lippincott-Schwartz J. 2000. Dynamics and retention of misfolded proteins in native ER membranes. *Nat. Cell Biol.* **2:** 288–295.

Nikonov A.V., Snapp E., Lippincott-Schwartz J., and Kreibich G. 2002. Active translocon complexes labeled with GFP-Dad1 diffuse slowly as large polysome arrays in the endoplasmic reticulum. *J. Cell Biol.* **158:** 497–506.

Patterson G.H. and Lippincott-Schwartz J. 2002. A photoactivatable GFP for selective photolabeling of proteins and cells. *Science* **297:** 1873–1877.

Patterson G.H., Knobel S.M., Sharif W.D., Kain S.R., and Piston D.W. 1997. Use of the green fluorescent protein and its mutants in quantitative fluorescence microscopy. *Biophys. J.* **73:** 2782–2790.

Periasamy N., Bicknese S., and Verkman A.S. 1996. Reversible photobleaching of fluorescein conjugates in air-saturated viscous solutions: Singlet and triplet state quenching by tryptophan. *Photochem. Photobiol.* **63:** 265–271.

Phair R.D. and Misteli T. 2000. High mobility of proteins in the mammalian cell nucleus. *Nature* **404:** 604–609.

———. 2001. Kinetic modelling approaches to in vivo imaging. *Nat. Rev. Mol. Cell. Biol.* **2:** 898–907.

Pierce D.W., Hom-Booher N., and Vale R.D. 1997. Imaging individual green fluorescent proteins. *Nature* **388:** 338.

Piston D.W., Patterson G.H., and Knobel S.M. 1999. Quantitative imaging of the green fluorescent protein (GFP). *Methods Cell Biol.* **58:** 31–48.

Politz J.C. 1999. Use of caged fluorochromes to track macromolecular movement in living cells. *Trends Cell Biol.* **9:** 284–287.

Presley J.F., Cole N.B., Schroer T.A., Hirschberg K., Zaal K.J., and Lippincott-Schwartz J. 1997. ER-to-Golgi transport visualized in living cells. *Nature* **389:** 81–85.

Schaff J., Fink C.C., Slepchenko B., Carson J.H., and Loew L.M. 1997. A general computational framework for modeling cellular structure and function. *Biophys. J.* **73:** 1135–1146.

Schwille P., Kummer S., Heikal A.A., Moerner W.E., and Webb W.W. 2000. Fluorescence correlation spectroscopy reveals fast optical excitation-driven intramolecular dynamics of yellow fluorescent proteins. *Proc. Natl. Acad. Sci.* **97:** 151–156.

Sciaky N., Presley J., Smith C., Zaal K.J., Cole N., Moreira J.E., Terasaki M., Siggia E., and Lippincott-Schwartz J. 1997. Golgi tubule traffic and the effects of brefeldin A visualized in living cells. *J. Cell Biol.* **139:** 1137–1155.

Seksek O., Biwersi J., and Verkman A.S. 1997. Translational diffusion of macromolecule-sized solutes in cytoplasm and nucleus. *J. Cell Biol.* **138:** 131–142.

Siggia E.D., Lippincott-Schwartz J., and Bekiranov S. 2000. Diffusion in inhomogeneous media: Theory and simulations applied to whole cell photobleach recovery. *Biophys. J.* **79:** 1761–1770.

Slepchenko B.M., Schaff J.C., Carson J.H., and Loew L.M. 2002. Computational cell biology: Spatiotemporal simulation of cellular events. *Annu. Rev. Biophys. Biomol. Struct.* **31:** 423–441.

Smith A.E., Slepchenko B.M., Schaff J.C., Loew L.M., and Macara I.G. 2002. Systems analysis of Ran transport. *Science* **295:** 488–491.

Song L., Varma C.A., Verhoeven J.W., and Tanke H.J. 1996. Influence of the triplet excited state on the photobleaching kinetics of fluorescein in microscopy. *Biophys. J.* **70:** 2959–2968.

Stout A.L. and Axelrod D. 1995. Spontaneous recovery of fluorescence by photobleached surface-adsorbed proteins. *Photochem. Photobiol.* **62:** 239–244.

Swaminathan R., Hoang C.P., and Verkman A.S. 1997. Photobleaching recovery and anisotropy decay of green fluorescent protein GFP-S65T in solution and cells: Cytoplasmic viscosity probed by green fluorescent protein translational and rotational diffusion. *Biophys. J.* **72:** 1900–1907.

Thevenaz P., Ruttimann U.E., and Unser M. 1998. A pyramid approach to subpixel registration based on intensity. *IEEE Trans. Image Process.* **7:** 27–41.

Tour O., Meijer R.M., Zacharias D.A., Adams S.R., and Tsien R.Y. 2003. Genetically targeted chromophore-assisted light activation. *Nat. Biotechnol.* **21:** 1505–1508.

Tsien R.Y. 1998. The green fluorescent protein. *Annu. Rev. Biochem.* **67:** 509–544.

Verkman A.S. 2002. Solute and macromolecule diffusion in cellular aqueous compartments. *Trends Biochem. Sci.* **27:** 27–33.

———. 2003. Diffusion in cells measured by fluorescence recovery after photobleaching. *Methods Enzymol.* **360:** 635–648.

Vigers G.P., Coue M., and McIntosh J.R. 1988. Fluorescent microtubules break up under illumination. *J. Cell Biol.* **107:** 1011–1024.

Wachter R.M., Elsliger M.A., Kallio K., Hanson G.T., and Remington S.J. 1998. Structural basis of spectral shifts in the yellow-emission variants of green fluorescent protein. *Structure* **6:** 1267–1277.

Wedekind P., Kubitscheck U., and Peters R. 1994. Scanning microphotolysis: A new photobleaching technique based on fast intensity modulation of a scanned laser beam and confocal imaging. *J. Microsc.* **176(Pt 1):** 23–33.

Wedekind P., Kubitscheck U., Heinrich O., and Peters R. 1996. Line-scanning microphotolysis for diffraction-limited measurements of lateral diffusion. *Biophys. J.* **71:** 1621–1632.

Wolf D.E., Edidin M., and Dragsten P.R. 1980. Effect of bleaching light on measurements of lateral diffusion in cell membranes by the fluorescence photobleaching recovery method. *Proc. Natl. Acad. Sci.* **77:** 2043–2045.

Yang F., Moss L.G., and Phillips G.N., Jr. 1996. The molecular structure of green fluorescent protein. *Nat. Biotechnol.* **14:** 1246–1251.

Zaal K.J., Smith C.L., Polishchuk R.S., Altan N., Cole N.B., Ellenberg J., Hirschberg K., Presley J.F., Roberts T.H., Siggia E., Phair R.D., and Lippincott-Schwartz J. 1999. Golgi membranes are absorbed into and reemerge from the ER during mitosis. *Cell* **99:** 589–601.

Zacharias D.A., Violin J.D., Newton A.C., and Tsien R.Y. 2002. Partitioning of lipid-modified monomeric GFPs into membrane microdomains of live cells. *Science* **296:** 913–916.

Zhang J., Campbell R.E., Ting A.Y., and Tsien R.Y. 2002. Creating new fluorescent probes for cell biology. *Nat. Rev. Mol. Cell. Biol.* **3:** 906–918.

FRET and Fluorescence Lifetime Imaging Microscopy

Marc Tramier, Daniele Sanvitto, Valentina Emiliani,
Christiane Durieux, and Maite Coppey-Moisan

*Institut Jacques Monod, UMR 7592 CNRS, University Paris VI and Paris VII,
Paris, France*

INTRODUCTION

RECENT STUDIES HAVE REVEALED THE EXISTENCE of more than 10,000 protein-protein interactions and several hundred multiprotein complexes in the proteome of *Saccharomyces cerevisiae* (Gavin et al. 2002; Ho et al. 2002). However, these results must be interpreted with caution. Many other interactions have likely not been detected, and among the numerous interactions already detected, some appear to have no physiological significance. Because intricate networks of interactions among proteins, RNA, and signaling molecules regulate biological processes, the determination of subcellular localization of these complexes, as well as the spatiotemporal regulation of their association and dissociation, is of prime importance for understanding their biological relevance. Fluorescence resonance energy transfer (FRET) measurements provide a useful tool to detect interactions between tagged molecules. Moreover, in the last decades, the development of high-spatial-resolution fluorescence-based microscopy coupled to advances in fluorescent protein tagging (see Chapter 1) has opened new ways to investigate subcellular processes in single living cells. More recently, fluorescence lifetime imaging microscopy (FLIM) has started to mature, and for many applications, it represents a useful alternative to fluorescence-intensity-based microscopy, especially for FRET measurements. Indeed, the determination and quantification of FRET are difficult tasks to carry out under the microscope in living cells, owing to the fluorophore concentration and light path dependence of fluorescence-intensity-based measurements, parameters that are often unknown in living cells. Moreover, processes such as photoconversion (Creemers et al. 1999; Malvezzi-Campeggi et al. 2001), photoinduction of a "dark state" (Garcia-Parajo et al. 2000), photobleaching (Patterson et al. 1997), or emissions from other fluorescent species (i.e., green-emitting immature state of DsRed), can occur, introducing pitfalls in FRET determination. This chapter presents an overview of a method for FRET imaging in living cells, which is based on time- and space-correlated single-photon counting (TSCSPC) detection. This technique is minimally invasive compared to other techniques and

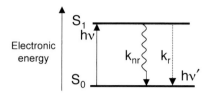

FIGURE 1. Perrin-Jablonski diagram.

provides direct quantitative measurements of intrinsic FRET efficiency and relative fraction of the bound complex.

FLUORESCENCE DECAY AND LIFETIMES

Fluorescence is a radiative process (rate constant k_r), which takes place when molecules excited by light absorption revert to their original state by light emission. Fluorescence is a very brief and transient phenomenon (typically occurring in the picosecond to nanosecond range), in which the molecules in the excited state S_1 release energy as they return to their unexcited state S_0 (Fig. 1). The kinetics of the fluorescence decay (rate constant $k = k_{nr} + k_r$) depends on the relative proportion of the various pathways for returning to ground state. Radiative (k_r) and nonradiative (k_{nr}) values thus contain information about the environment of the fluorophore in the excited state. This information is lost when measuring the steady-state fluorescence intensity, which is represented by the area under the decay curve as shown in the top part of Figure 2. For example, a similar steady-state fluorescence intensity (different curves with the same area in Fig. 2, bottom) can correspond to two different situations: high fluorophore concentration associated with a fast fluorescence decay (curve 1) or low fluorophore concentration associated with a long decay (curve 2). In principle, steady-state fluorescence intensity acquired by using a CCD (charge-coupled device) camera, as visualized in Figure 3, is insufficient to provide a direct comparison of the fluorophore concentration in different cells cultured in the same Petri dish. In the example displayed in Figure 3, which shows green fluorescent protein (GFP) tagged thymidine kinase expressed in living cells following transient transfection, the slopes of the fluorescence

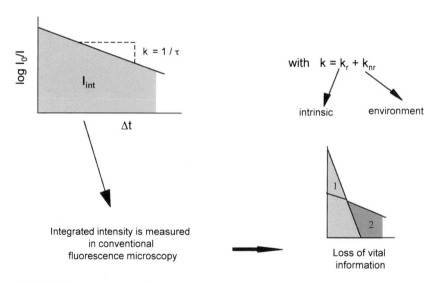

FIGURE 2. Comparison between fluorescence decay and steady-state intensity.

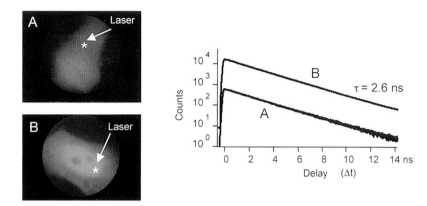

FIGURE 3. GFP fluorescence decays coming from two subcellular regions with different steady-state intensities.

decays of GFP expressed in the two cells are parallel. This means that the environment of the fluorophore is the same in the two situations, and thus the difference in the steady-state fluorescence intensity arises from a different level of protein expression. The fluorescence lifetime (τ is the inverse of the rate constant k of the various decay processes) is a measure of all deactivation pathways through which the excited fluorophore can return to the ground state and represents a sensitive probe for a variety of interesting events, such as FRET, which may occur at the subcellular level.

PHOTOPHYSICAL PRINCIPLE OF FRET AND ITS MEASUREMENT

FRET is a nonradiative process that occurs in the excited state of fluorophores when the energy is transferred from a donor fluorophore to an acceptor chromophore. The rate constant for energy transfer (k_T) depends on the distance between the two chromophores, the extent of the overlap between the donor emission and acceptor excitation spectra, the quantum yield of the donor, the fluorescence lifetime of the donor, and the relative orientation of the donor and acceptor (Fig. 4) (Förster 1948).

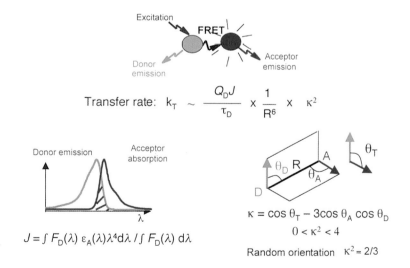

FIGURE 4. FRET process. (k_T) Transfer rate; (Q_D) quantum yield of donor; (τ_D) fluorescence lifetime of donor; (J) spectral overlap of donor emission and acceptor absorption; (R) distance between donor and acceptor; (κ^2) orientational factor.

FRET Efficiency and R_0 Determination

In practice, it is the FRET efficiency that is determined. The FRET efficiency is the ratio of the deactivation of the donor by energy transfer over all of the deactivation pathways: $E = k_T/(k_r + k_{nr} + k_T)$. If R_0 is the distance between the donor and the acceptor at which 50% of FRET efficiency takes place, then $E = 1/[1 + (R/R_0)^6]$.

R_0 can be calculated from the spectroscopic and mutual dipole orientational parameters of the donor and acceptor. Owing to the very short working range (i.e., 2–8 nm) of FRET, specific interactions between two proteins can be determined only if the fluorescent tags are close enough to each other, which is in practice at a distance close to R_0. An absence of FRET does not mean that the tagged macromolecules are not interacting. This relatively short working distance of FRET is a limitation for studying multiprotein complexes or interactions between very large proteins.

Orientation Factors

FRET is a physical process involving intermolecular long-range dipole-dipole coupling. In addition to the distance between the donor and acceptor chromophores, the other essential geometrical requirements for effective transfer are the orientations of the transition dipole of donor and acceptor, which either must be oriented favorably to each other or must have a certain degree of rotational freedom. To satisfy this latter condition for GFP-tagged proteins, because the GFP chromophore is tightly immobilized within the protein pocket, the entire GFP must be able to undergo some degree of rotational freedom relative to the tagged protein. This flexible link between the protein tag and the protein of interest can be provided by inserting additional amino acids between the two polypeptide chains during cloning. In this case, the orientation factor, κ^2, is taken as 2/3, which corresponds to a random orientation of the transition dipole of donor and acceptor. Otherwise, $0 < \kappa^2 < 4$, depending on the orientations of the tagged chromophores, which are typically unknown.

Measurement of FRET by Donor Fluorescence Lifetime

Several experimental methods for measuring FRET in living cells by fluorescence imaging microscopy are discussed below.

- *Steady-state fluorescence intensity.* This method is very popular, because it requires only conventional digital fluorescence microscopy. Indeed, it relies on the measurement of the increase of steady-state fluorescence intensity of the acceptor and the decrease of steady-state fluorescence intensity of the donor, which occur if FRET takes place when exciting the donor. Time-lapse studies of the variations of these fluorescence signals allow several biochemical processes to be followed in living cells, especially in the case of intramolecular FRET (due to the stable stoichiometry of the donor and the acceptor) (Miyawaki et al. 1997). However, intermolecular FRET must be analyzed with great care as the stoichiometry of donor and acceptor cannot be controlled and can generate false FRET signals. Moreover, if the characteristic FRET efficiency for a particular molecular interaction is not known, steady-state fluorescence intensity-based methods cannot provide quantitative measurements of FRET even if normalization procedures are carried out (Hoppe et al. 2002).

- *Photobleaching methods.* Apparent FRET efficiency can be determined by either acceptor-photobleaching FRET (Wouters et al. 1998) or donor-photobleaching FRET (Bastiaens and

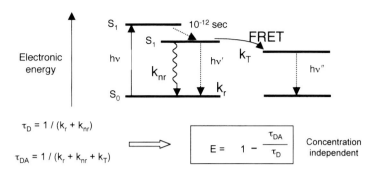

FIGURE 5. Perrin-Jablonski diagram of FRET process and characterization of FRET efficiency through the fluorescence lifetime of the donor.

Jovin 1996). The first method consists of measuring the steady-state fluorescence intensity before and after photobleaching the acceptor; the latter uses the difference between the rate of steady-state photobleaching of the donor in the presence of the acceptor and the rate of steady-state photobleaching of the donor in the absence of the acceptor.

- *Fluorescence lifetime of the donor.* This method provides a direct measurement of the characteristic FRET efficiency for the particular molecular interaction. FRET takes place in the excited state of the donor through a nonradiative process. The energy transfer rate is thus added to the total rate constant of deactivation of the excited donor. As a consequence, the fluorescence lifetime of the donor decreases relative to the one of the donor measured in the absence of FRET (Fig. 5). FRET efficiency is thus determined according to the following equation:

$$E = 1 - \tau_{DA}/\tau_D$$

τ_D and τ_{DA} are the fluorescence lifetimes of the donor in the absence and presence of the acceptor, respectively. These parameters are independent of the concentration of the donor and the acceptor. Through FLIM, it is possible to measure the fluorophore fluorescence lifetime for every pixel of a two-dimensional image.

Time-resolved images can be obtained through a direct (time domain) or indirect (frequency domain) method. In frequency domain, the phase shift and the demodulation of the fluorescence relative to the modulated excitation are measured. In the case of different lifetimes, as occurs when only a fraction of the donor is in interaction with the acceptor, a mean lifetime is obtained for each pixel. In fact, to resolve two different decay times, multifrequency experiments must be used. Nevertheless, a global analysis of the data on the totality of the pixels is sufficient in most cases to extract an image of the ratio of the bound versus unbound donor, assuming a constant FRET efficiency over the different pixels (Verveer et al. 2000). In time domain, the time-resolved decay of fluorescence after a pulsed excitation is acquired for every pixel. The fluorescence lifetimes and the proportion of fluorescence species corresponding to different lifetimes are determined from the best fit of the decays using mono-, bi-, or triexponential models. Other models can, however, be taken into account such as those based on lifetime distributions.

PRINCIPLE OF TIME-CORRELATED SINGLE PHOTON COUNTING

For the time-correlated single-photon counting (TCSPC) method (O'Connor and Phillips 1984), the time delay of a single emitted photon after a pulsed excitation is measured

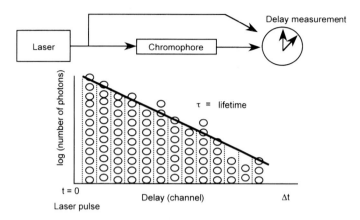

FIGURE 6. Principle of TCSPC.

sequentially and sampled into different acquisition channels (Fig. 6). The decay obtained corresponds to the return to the ground state of the fluorophore population excited by an infinitely short pulse of light. TCSPC is quite sensitive, as it requires only a very low excitation level (~1000 times lower than other approaches), which makes it well suited for studies of living cells.

In the case of two emitted photons, only the first one will be processed, which will induce artificially faster fluorescence kinetics (shorter lifetime). To measure only one photon for one excitation pulse, the count rate of the measured photons must be lower than 1% of the excitation pulse frequency. In practice, this requirement is the limiting parameter for the acquisition rate.

Excitation Source

A short pulsed excitation source is required for the TCSPC method with a pulse width that is negligible compared to the fluorescence time decay, which is usually in the nanosecond time scale. In the authors' experiments, a picosecond titanium:sapphire laser was used, with an 80-MHz repetition rate and picosecond pulses of infrared light tunable between 700 and 1000 nm. After pulse picking and frequency doubling, 350–500-nm ps pulses with a repetition rate of 4 MHz (one pulse each 250 ns) were obtained.

The TCSPC method requires that the time between pulses be long enough to allow the excited fluorophores to return to the ground state before the next pulse arrives. A pulse picker was used because an 80-MHz repetition rate corresponds to 12.5 ns between pulses, and thus for a fluorophore with a lifetime of nanoseconds, the excitation pulse is delivered when a part of the molecules are still excited.

Detectors

For the TCSPC method, photomultiplier tubes with a time resolution of tens of picoseconds are used. The authors combine the time correlation with the spatial correlation of the single photon counted. This is achieved by using a quadrant-anode detector (QA, Europhoton GmbH, Germany) (Fig. 7) (Kemnitz et al. 1997). Fluorescence decays are acquired by counting and sampling single emitted photons according to both the time delay between their arrival and the laser pulse and their *xy* coordinates. In the QA detector, the incident photon hits a photocathode, and the resulting electron is amplified by two microchannel plates that produce a cone-shaped cloud of electrons. The electrons are collected by four independent

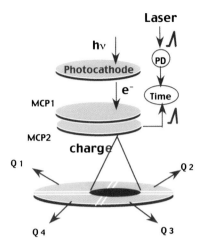

Charge ratio = xy–space

FIGURE 7. Quadrant-anode TSCSPC detector. See text for details.

quadrant-anodes and the charge ratio is used to calculate the *xy* position of the incident photon. Simultaneously, for time correlation, a time-to-amplitude converter operates between the signal coming from the second microchannel plate and the signal from a fast photodiode triggered by the pulsed excitation source.

As an alternative, it is possible to couple TCSPC detection with confocal microscopy, allowing one to use space correlation determined from scanning. This is achieved with a computer card (e.g., see Becker & Hickel, GmbH, Germany) that integrates delay time measurements (between emitted photon and excitation pulse) and scan position. However, in contrast to the TCSPC-QA detection, for which the count rate is adjusted on the whole image, the count rate of the confocal scanning-based method must be adjusted at the position of the maximum signal. Consequently, a pre-scan of the sample is required, and more importantly, the whole image is acquired at a lower average count rate. Therefore, to obtain the same image quality as in wide field conditions, a longer acquisition time is needed.

PICOSECOND FLUORESCENCE LIFETIME MICROSCOPY

Picosecond FLIM is a method based on TSCSPC that captures the fluorescence decay simultaneously in 256 × 256 pixels of a two-dimensional spatial image and in 4096 time channels.

Instrumentation Setup

A typical system for picosecond FLIM is presented in Figure 8, and includes the following.

- A mode-locked titanium sapphire laser (e.g., Millennia 5W/Tsunami 3960-M3BB-UPG kit, Spectra-Physics). This laser delivers picosecond pulses and can be tuned between 760 and 980 nm.

- A frequency doubler and pulse picker, which provide an excitation range between 380 and 490 nm, and a repetition rate of 4 MHz, respectively (e.g., Spectra-physics 3980-35).

- Optics for coupling the laser to the microscope, including a beam expander and a galvanometer.

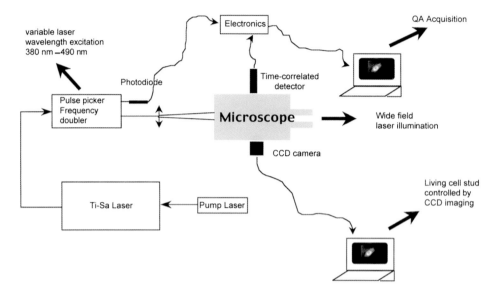

FIGURE 8. Picosecond FLIM setup.

For wide-field illumination, the width of the laser beam is expanded to fit approximately the height of the mercury lamp beam size. In this configuration, the same lenses (which are internal to the microscope in the Leica DMIRBE) required to illuminate the whole field of view can be kept for both types of excitation.

By using a beam expander and microscope optics for wide field (brightfield illumination), the resulting illuminated field contains interference fringes coming from the coherence of the laser light passing through the optical elements. To avoid these interference fringes, we add a galvanometer system just before the beam expander that averages the position of the fringes (Fig. 9).

• An inverted epifluorescence microscope (e.g., Leica DMIRBE) for high-resolution subcellular picosecond FRET imaging with a 100× objective (numerical aperture [NA] = 1.3). Supplementary magnification lenses are placed on the fluorescence exit port, 1.6× (optovar) and 1.9× (single-lens reflex [SLR]). The housing of the mercury lamp is kept for conventional illumination, and a moving mirror allows changes between laser and lamp excitation.

• A TSCSPC (QA) detector using a photocathode cooled to –20 °C by flowing nitrogen gas produced by heating liquid nitrogen. Under this condition, the background of the QA detector is 100 cps with a count rate up to 50 kHz.

• A photodiode placed on the laser beam of the residual nondoubled infrared light after the frequency doubler. The photodiode is connected to the electronic devices of the QA detector and is used to determine the reference time for measuring the time delay of the emitted photon.

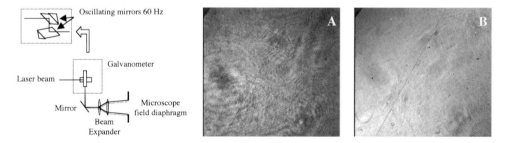

FIGURE 9. Wide-field laser illumination. The coherence of laser light gives rise to interference fringes (*A*) that are averaged by using a galvanometer system (*B*).

General Methodology

Instrument Response Function

The instrument response function (IRF) is the temporal response of the entire system and corresponds to the experimental measurement of the excitation pulse. IRF measurement is carried out by placing a reflective surface (such as a mirror) on the microscope plate and by choosing a beamsplitter as dichroic mirror. The IRF is recorded before each experiment for two reasons: (1) as a control of the setup and (2) to be used in the analysis of the experimental decays (see section on Analysis of the Decay).

The temporal response of the system corresponds more or less to the temporal response of the time-correlated photomultiplier tube (PMT), i.e., full width at half-maximum of ~100 ps in the case of a QA detector. This response is characteristic of the fluctuation in the duration of the electron avalanche in the microchannel plates. The laser pulse of ~1 ps corresponds to a subresolution time.

Care must be taken before sending the reflected laser beam onto the detector to avoid photodamage of the detector. To protect the detector, add neutral density filters up to optical density (OD) = 4.

Sample Preparation

Live adherent cells cultured on glass coverslips enables high NA objectives to be used with an inverted microscope. Even stable cell lines expressing low levels of GFP fusion proteins can be used for picosecond FLIM. The coverslips are mounted in a special holder (PeCon GmbH and LaCon, Germany) allowing the addition of medium and placed on the stage of the inverted microscope. A temperature-controlled chamber is mounted on the stage with CO_2 circulation for long-term observations of mammalian cells. A culture medium without phenol red and flavin (e.g., Dulbecco's modified Eagle's medium [DMEM] from Invitrogen) is used when acquiring the fluorescence decays (see Chapter 9).

Guidelines for Acquisition of Fluorescence Decays for FRET Measurements

1. Choose an appropriate band-pass emission filter that selects for the donor fluorescence and rejects the acceptor fluorescence.

2. Accumulate enough single-photon events to be able to analyze the decays further. The required accumulation time depends on the fluorescence intensity of the sample, the photosensitivity of the fluorescent probes, and the complexity of the decays (mono- or biexponential, the values of lifetimes, and the proportion of lifetime components). Neutral density filters usually must be added to obtain the required photon count rate (~50 kHz).

3. Acquire fluorescence photons from cells expressing donor and acceptor.

4. Acquire fluorescence photons from cells expressing donor only.

5. Acquire fluorescence photons from untransfected cells and/or from the culture medium to quantify the background contribution.

System Calibration

To calibrate the QA detector, a temporal resolution of less than 100 ps was determined by measuring the fluorescence lifetime of erythrosine in water (90 ± 4 ps). The response of the

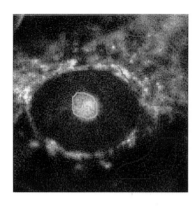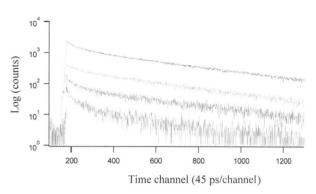

FIGURE 10. Extraction of fluorescence decays from picosecond FLIM data. Picosecond FLIM data of ethidium fluorescence in a living cell from which the fluorescence decay of the entire field (black curve) or of a different region of interest is extracted. The green, red, yellow, and blue curves correspond to the decays from cytoplasmic, nuclear, nucleolar and extracellular areas, respectively.

system to a 100-nm-diameter fluorescent bead (the diameter is less than the optical resolution) was recorded. This should yield a spatial resolution of 500 nm, which corresponds to the limitation of the numerization of the image (pixel size of ~250 nm) when using 300× magnification (objective 100×, optovar 1.6× and SLR system 1.9×).

Computer Acquisition Process

The acquisition software stores the space coordinate (x,y) and the time correlation (Δt) sequentially for each single-photon counted. This information is used to reconstruct the fluorescence image (by using the space coordinate), the fluorescence decay of the whole field (by using the time coordinate), or the fluorescence decay of a region of interest (by using both) (Fig. 10). In picosecond FLIM data, a fluorescence decay is stored at every (x,y) pixel.

ANALYSIS OF THE DECAY

Analysis of FRET data is carried out in three steps: (1) graphic comparison of the decays between donor alone and donor in the presence of the acceptor; (2) fit of the decays to determine and quantify fluorescence lifetimes; and (3) construction of lifetime images from a relevant physical model.

Visualization and Comparison of Decays

Fluorescence decays from different regions of interest are normalized and then displayed. Igor Pro software (Wavemetrics) for this step works well in our hands, but several other software packages, such as Origin, are very useful for comparison by eye. Because the fluorescence decay is independent of the fluorophore concentration, it is possible to compare the different decays coming from different cells, samples, and experiments.

Determination of Lifetimes

Fluorescence data are analyzed to gather information on the different discrete lifetimes that can be detected in the decay. To obtain lifetimes from fluorescence decays, experimental measurements are fitted by the convolution product of a multi-exponential theoretical model with the IRF (see the section above on Instrument Response Function): $i(t) = \text{IRF}(t) * \Sigma a_i\, e^{-t/\tau_i}$, where a_i corresponds to the relative contribution of the ith fluorescent species, each characterized by its fluorescence lifetime τ_i and $*$ corresponds to the convolution product. Data are analyzed by a Marquardt nonlinear least-squares algorithm using, for example, the Globals software package developed at the Laboratory for Fluorescence Dynamics at the University of Illinois at Urbana-Champaign.

Two modeling strategies are used sequentially: First, a phenomenological approach provides a set of parameters for each individual decay, parameters which are used to establish a physical modeling of FRET process.

Phenomenological Approach

Decays are fitted independently by discrete models (with discrete lifetimes) to recover both the appropriate model and the parameters (a_i, τ_i).

1. Starting with a single lifetime model, note the mean least-squares parameter χ^2.

2. Add a second lifetime in the model and compare the two χ^2 (the smaller χ^2 is the one that gives the better fit).

3. Again add another lifetime. Continue this addition process until χ^2 becomes minimal or until an abnormal solution appears, such as quasi-identical lifetimes, negative parameters, or aberrant parameters (very long or very short lifetimes). Be careful, as sometimes the fit is dependent on the initial chosen parameters .

Physical Modeling

The goal is to find a physical model that is consistent with the results of several independent fits. For example, the fluorescence decay of the donor alone is monoexponential and becomes biexponential in the presence of the acceptor; the short one is the lifetime of the donor involved in FRET and the long one is the lifetime of the donor alone. a_i, the relative contribution of the two lifetimes, corresponds to the proportion of the fluorophores engaged or not in the FRET process.

Imaging the Lifetimes and Components

At this point, data are analyzed using the physical model discussed above. Either pixel by pixel or by binning more pixels, fluorescence decays are fitted and the results are mapped in an image. Based on the chosen model, lifetimes are either fixed or left free during the fitting procedure.

As an example, Figure 11 represents the fluorescence lifetime map of ethidium in a living cell. The time-intergrated fluorescent image is analyzed (in a binned 50 × 50-pixel matrix) using a monoexponential model.

In Figure 12, the results of localization of thymidine kinase–GFP and actin-YFP (yellow fluorescent protein) in live cells are shown. A 100 × 100-pixel fitting matrix, extracted from the images of Figure 12, was analyzed using a biexponential model. The values of lifetimes of $\tau_1 = 2.40 \pm 0.04$ ns and $\tau_2 = 2.84 \pm 0.02$ ns were determined, and the relative concentrations of thymidine kinase–GFP and actin-YFP were each mapped in a two-dimensional image.

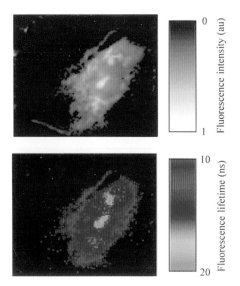

FIGURE 11. Fluorescence lifetime map of ethidium in a living cell.

EXAMPLES OF PICOSECOND FRET IMAGING MICROSCOPY

Imaging the Formation of NFE2 using GFP and DsRed

Due to the monoexponential decay of GFP, the GFP-DsRed pair is very well suited for quantitative FRET determination in living cells. The R_0 value for this donor-acceptor pair is 47 Å. The formation of the transcription factor NFE2, composed of two different subunits, one (p45) fused to GFP and the other (MafG) to DsRed, can be spatially resolved within the nucleus of a living cell by picosecond FLIM. Fluorescence decays are acquired, visualized, and analyzed as follows.

Decay Visualization and Comparison

Figure 13 shows that the fluorescence decay of GFP fused to p45 is globally faster when the other subunit (MafG) fused to the acceptor DsRed is present in the nucleus. This is expected if FRET occurs between GFP and DsRed.

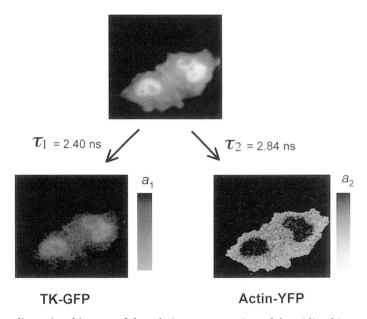

FIGURE 12. Two-dimensional images of the relative concentration of thymidine kinase–GFP and actin-YFP extracted by fitting a 100 × 100-pixel FLIM matrix with a biexponential model.

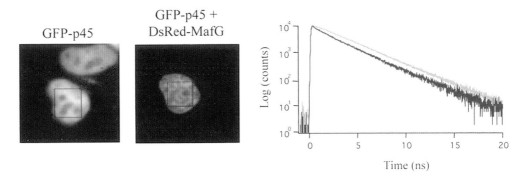

FIGURE 13. Decay comparison of GFP-p45 alone and coexpressed with DsRed-MafG. (Green curve) GFP-p45; (Red curve) GFP-p45 + DsRed-MafG.

Determination of Lifetimes

The best fit of the green curve (cell expressing only GFP-p45) is a monoexponential decay with a lifetime of 2.68 ± 0.15 ns (S.D., $n = 4$). Conversely, the best fit of the red curve (cell expressing both GFP-p45 and DsRed-MafG) is a biexponential decay with lifetime $\tau_1 = 2.82 \pm 0.17$ ns and $\tau_2 = 0.90 \pm 0.17$ ns (S.D., $n = 7$) (Tramier et al. 2002). Figure 14 shows typical results of fits with mono- and biexponential models. The value of 2.8 ns corresponds to a typical fluorescence lifetime expected for the GFP chromophore, which is not significantly different from the value found

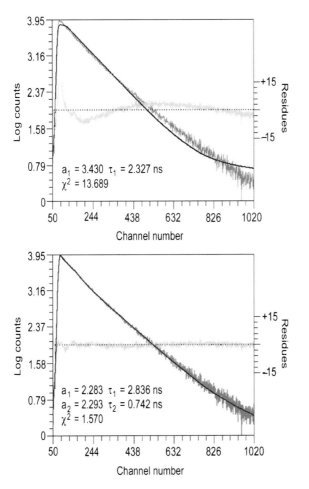

FIGURE 14. Characteristic fitting results of GFP-p45 coexpressed with DsRed-MafG using mono- (*top*) and biexponential (*bottom*) models.

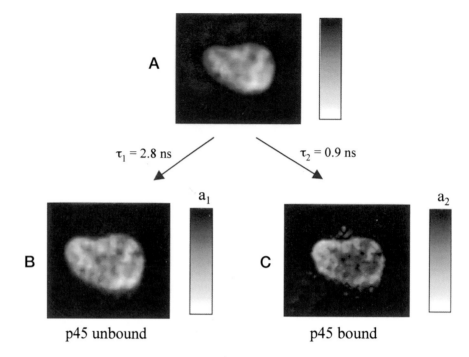

$\tau_1 = 2.8$ ns $\tau_2 = 0.9$ ns

a₁   a₂

p45 unbound   p45 bound

FIGURE 15. Two-dimensional images of the relative concentration of unbound and bound GFP-p45 with DsRed-MafG extracted by fitting a 50 × 50-pixel FLIM matrix with a biexponential model using fixed lifetimes ($\tau_1 = 2.8$ ns; $\tau_2 = 0.9$ ns). Fluorescence intensity (*A*) and preexponential factors a_1 and a_2 (*B* and *C*) are normalized between 0 (*black*) and 1 (*white*).

in monotransfected cells. The value of 0.9 ns can be interpreted as the lifetime of the GFP chromophores involved in FRET with DsRed chromophores. However, this attribution must be cross-checked with control experiments on DsRed (see the section below on Control Experiments).

Imaging the Relative Contributions of Bound and Unbound Subunits

Figure 15 presents the result of the analysis of the fluorescence decay image of a cell cotransfected with GFP-p45 and DsRed-MafG (Fig. 15A). Using a biexponential model, the data were fitted with fixed lifetimes, $\tau_1 = 2.8$ ns and $\tau_2 = 0.9$ ns, corresponding to the GFP species fused to p45 in the unbound and bound state, respectively. Figure 15B,C shows the spatial distribution (50 × 50 pixels picosecond FLIM data) of the preexponential factors for each compound. In other words, Figure 15C represents the spatially resolved localization of the heterodimeric transcription factor, which appears similar to the chromatin pattern. In contrast, the spatially resolved localization of GFP-p45, which is not in close interaction with MafG, shows a more diffuse pattern (Fig. 15B). These results suggest that the MafG subunit, via its interaction with p45, recruits p45 to the nuclear DNA.

Imaging with a CFP-YFP Pair

The aggregation and the existence of green species of DsRed can be a problem when using GFP-DsRed pairs in FRET studies. Another widely used donor-acceptor pair in cellular biology studies is cyan fluorescent protein (CFP)-YFP. However, owing to the biexponential behavior of the donor (CFP), the quantification of the FRET efficiency and the fraction of bound donor are not possible in practice (Tramier et al. 2002). Nevertheless, the existence of FRET can be revealed from the decrease of the mean fluorescence lifetime of CFP chromophores.

CONTROL EXPERIMENTS

Although the occurrence of a fluorescence lifetime shorter than that of the donor typically can be the signature of a FRET process, nevertheless it can also be due to other sources and give rise to misinterpretation of the results. Two of the sources that can yield this misleading occurrence are cellular autofluorescence and the green-emitting species of DsRed.

Certain components of cells can autofluoresce, especially lipofusion pigments, which are often present in cytoplasmic vesicles in established cell lines. This autofluorescence can occur with a short lifetime that is observed in addition to the standard GFP decay time.

The steady-state intensity of the green species of DsRed (green-emitting species of DsRed) cannot be distinguished from the green fluorescence of GFP, whether involved or not in a FRET process. However, the fluorescence decay of the green-emitting species of DsRed is characterized by two fluorescence lifetimes, $\tau_{g1} = 0.51 \pm 0.08$ ns and $\tau_{g2} = 0.034 + 0.051$ in living cells (Tramier et al. 2002), and by three lifetimes, $\tau_{g1} = 0.47$ ns, $\tau_{g2} = 0.12$ ns, and $\tau_{g3} = 0.026$ ns in purified DsRed (Cotlet et al. 2001). τ_{g1} is associated with the immature green species of DsRed, whereas the two others are associated with FRET between green- and red-emitting species of DsRed within the tetramer. Another minor species with a $\tau_r = 3.46$ ns can be observed with the green detection filter, corresponding to the blue side emission spectra of the mature red-emitting chromophore.

To distinguish between true FRET emissions and artifacts, control experiments must be carried out on mock-transfected cells and on DsRed-transfected cells to determine the contribution of the autofluorescence and the value of the lifetimes of the green-emitting species of the DsRed, respectively. The picosecond resolution of the fluorescence kinetics allows one to discriminate between close or similar lifetime values. This property of picosecond FRET imaging microscopy is very important and permits one to ascribe precisely the fluorescence lifetime of FRET between the GFP-tagged donor and the DsRed-tagged acceptor, and thus to determine the interaction between the two proteins of interest.

FUTURE DEVELOPMENTS

2-Photon Excitation

The laser used for fluorescence decay measurements can be adapted to obtain 2-photon excitation, thus providing a confocal spatial resolution. Titanium:sapphire lasers exist in two versions, picosecond and femtosecond modes. A femtosecond laser can be used to provide 2-photon excitation coupled to the TCSPC detection method. Real-time 2-photon excitation can be obtained through multifoci sytems (multilenses mounted on a rotating wheel) (Bewersdorf et al. 1998) or by using the TriMScope (LaVision Biotech GmbH, Germany). These instruments can be conveniently coupled to spatially resolved detectors such as TSC-SPC-QA or gated-CCD.

New Probe Development

Thus far, there are no perfect donor-acceptor pairs to generate FRET within living cells. With the CFP-YFP pair, the donor CFP presents two lifetime components in its decay, impeding in practice a quantitative analysis of FRET. On the other hand, owing to its monoexponential fluorescence decay, GFP (in the majority of cases) is a very good "donor." However, the only fluorescent protein acceptor that has an absorption spectra overlapping the emission spectra of GFP is DsRed. Unfortunately, the existence of an immature green species of DsRed can overlap the fluorescence of GFP-tagged proteins. Novel "red-fluorescent" proteins and maybe more interesting "green-absorbing" proteins have recently been developed from *Anemona sulcata* (Chudakov et al. 2003). If the chromophore of these nonfluorescent proteins does not reveal green fluorescence at all, this new protein would be a better acceptor for the GFP donor in the analysis of protein-protein interactions in living cells.

In addition, several organic molecules are good acceptors, such as rhodamine, Cy3, and Alexa 568. These fluorophores, however, cannot be used for endogeneous labeling of proteins expressed in living cells.

REFERENCES

Bastiaens P.I. and Jovin T.M. 1996. Microspectroscopic imaging tracks the intracellular processing of a signal transduction protein: Fluorescent-labeled protein kinase C beta I. *Proc. Natl. Acad. Sci.* **93:** 8407–8412.

Bewersdorf J., Pick R., and Hell S.W. 1998. Multifocal multiphoton microscopy. *Opt. Lett.* **23:** 655–657.

Chudakov D.M., Belousov V.V., Zaraisky A.G., Novoselov V.V., Staroverov D.B., Zorov D.B., Lukyanov S., and Lukyanov K.A. 2003. Kindling fluorescent proteins for precise in vivo photolabeling. *Nat. Biotechnol.* **21:** 191–194.

Cotlet M., Hofkens J., Habuchi S., Dirix G., van Guyse M., Michiels J., Vanderleyden J., and de Schryver F.C. 2001. Identification of different emitting species in the red fluorescent protein DsRed by means of ensemble and single-molecule spectroscopy. *Proc. Natl. Acad. Sci.* **98:** 14398–14403.

Creemers T.M.H., Lock A.J., Subramaniam V., Jovin T.M., and Völker S. 1999. Three photoconvertible forms of green fluorescent protein identified by spectral hole-burning. *Nat. Struct. Biol.* **6:** 557–560.

Förster T. 1948. Zwischenmolekulare energiewanderung und fluoreszenz. *Ann. Phys.* **2:** 55–75.

Garcia-Parajo M.F., Segers-Nolten G.M.J., Veerman J.A., Greve J., and van Hulst N.F. 2000. Real-time light-driven dynamics of the fluorescence emission in single green fluorescent protein molecules. *Proc. Natl. Acad. Sci.* **97:** 7237–7242.

Gavin A.C., Bösche M., Krause R., Grandi P., Marzioch M., Bauer A., Schultz J., Rick J.M., Michon A.M., Cruciat C.M., Remor M., Hofert C., Schelder M., Brajenovic M., Ruffner H., Merino A., Klein K., Hudak M., Dickson D., Rudi T., Gnau V., Bauch A., Bastuck S., Huhse B., Leutwein C., Heurtier M.A., Copley R.R., Edelmann A., Querfurth E., Rybin V., Drewes G., Raida M., Bouwmeester T., Bork P., Seraphin B., Kuster B., Neubauer G., and Superti-Furga G. 2002. Functional organization of the yeast proteome by systematic analysis of protein complexes. *Nature* **415:** 141–147.

Ho Y., Gruhler A., Heilbut A., Bader G.D., Moore L., Adams S.L., Millar A., Taylor P., Bennett K., Boutilier K., Yang L., Wolting C., Donaldson I., Schandorff S., Shewnarane J., Vo M., Taggart J., Goudreault M., Muskat B., Alfarano C., Dewar D., Lin Z., Michalickova K., Willems A.R., Sassi H., Nielsen P.A., Rasmussen K.J., Andersen J.R., Johansen L.E., Hansen L.H., Jespersen H., Podtelejnikov A., Nielsen E., Crawford J., Poulsen V., Sorensen B.D., Matthiesen J., Hendrickson R.C., Gleeson F., Pawson T., Moran M.F., Durocher D., Mann M., Hogue C.W., Figeys D., and Tyers M. 2002. Systematic identification of protein complexes in *Saccharomyces cerevisiae* by mass spectrometry. *Nature* **415:** 180–183.

Hoppe A., Christensen K., and Swanson J.A. 2002. Fluorescence resonance energy transfer-based stoichiometry in living cells. *Biophys. J.* **83:** 3652–3664.

Kemnitz K., Pfeifer L., Paul R., and Coppey-Moisan M. 1997. Novel detectors for fluorescence lifetime imaging on the picosecond time scale. *J. Fluoresc.* **7:** 93–98.

Malvezzi-Campeggi F., Jahnz M., Heinze K.G., Dittrich P., and Schwille P. 2001. Light-induced flickering of DsRed provides evidence for distinct and interconvertible fluorescent states. *Biophys. J.* **81:** 1776–1785.

Miyawaki A., Llopis J., Heim R., McCaffery J.M., Adams J.A., Ikura M., and Tsien R.Y. 1997. Fluorescent indicators for Ca^{2+} based on green fluorescent proteins and calmodulin. *Nature* **388:** 882–887.

O'Connor D.V. and Phillips D. 1984. *Time-correlated single photon counting*. Academic Press, New York.

Patterson G.H., Knobel S.M., Sharif W.D., Kain S.R., and Piston D.W. 1997. Use of the green fluorescent protein and its mutants in quantitative fluorescence microscopy. *Biophys. J.* **73:** 2782–2790.

Tramier M., Gautier I., Piolot T., Ravalet S., Kemnitz K., Coppey J., Durieux C., Mignotte V., and Coppey-Moisan M. 2002. Picosecond-hetero-FRET microscopy to probe protein-protein interactions in live cells. *Biophys. J.* **83:** 3570–3577.

Verveer P.J., Squire A., and Bastiaens P.I. 2000. Global analysis of fluorescence lifetime imaging microscopy data. *Biophys. J.* **78:** 2127–2137.

Wouters F.S., Bastiaens P.I., Wirtz K.W., and Jovin T.M. 1998. FRET microscopy demonstrates molecular association of non-specific lipid transfer protein (nsL-TP) with fatty acid oxidation enzymes in peroxisomes. *EMBO J.* **17:** 7179–7189.

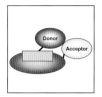

Monitoring Protein Dynamics Using FRET-based Biosensors

Teng-Leong Chew

Cell Imaging Facility and Department of Cell and Molecular Biology,
Northwestern University Feinberg School of Medicine;

Rex L. Chisholm

Department of Cell and Molecular Biology, and Center for Genetic Medicine,
University Feinberg School of Medicine,
Chicago, Illinois 60611

INTRODUCTION

THE USE OF GREEN FLUORESCENT PROTEIN (GFP) and its various derivatives permits direct observation of the dynamic properties of specific proteins in live cells by simultaneously providing maximal spatial and temporal resolution. However, the use of these fluorescent probes on their own does not provide important biochemical information regarding, for example, protein-protein interactions, enzymatic activity, and conformational changes in proteins. One of the ultimate challenges in microscopy is to devise biosensors that simultaneously combine the spatiotemporal resolution of fluorescent probes with the functional information afforded by biochemical analyses. The use of fluorescence resonance energy transfer (FRET) has produced significant progress toward this goal. FRET is a process that shifts energy from an electronically excited molecule (the donor fluorophore) to a neighboring molecule (the acceptor or quencher), returning the donor molecule to its ground state without fluorescence emission. The following are two simple criteria for choosing a fluorophore pair for FRET.

- The donor emission spectrum must overlap significantly with the excitation spectrum of the acceptor.

- The excitation light for the donor must not directly excite the acceptor.

Many chemical and protein-based fluorophore pairs fit these requirements. The most common fluorescent protein pair used in live-cell imaging is the combination of cyan fluorescent protein (CFP) as the donor and yellow fluorescent protein (YFP) as the acceptor. Table 1 lists some of the commonly used donor/acceptor pairs as well as their maximal excitation and emission wavelengths.

TABLE 1. Example of FRET fluorophore pairs

Donor	Excitation$_{Donor}$	Emission$_{Donor}$	Acceptor	Excitation$_{Acceptor}$	Emission$_{Acceptor}$
CFP	440 nm	480 nm	YFP	520 nm	535 nm
BFP	365 nm	460 nm	GFP	488 nm	535 nm
CFP	440 nm	480 nm	dsRed1	560 nm	610 nm
FITC	488 nm	535 nm	Cy3	525 nm	595 nm
Cy3	525 nm	595 nm	Cy5	633 nm	695 nm
GFP	488 nm	535 nm	Rhodamine	543 nm	595 nm

As dictated by the Förster equation (Förster 1965), FRET efficiency depends on the inverse sixth power of the distance, R, separating the donor and acceptor fluorophores: $E = R_o^6/(R_o^6 + R^6)$. R_o is the Förster distance, or the critical transfer distance. It is dependent on the FRET pair used, and can be calculated from the spectral properties of the donor and the acceptor. FRET is unique with respect to the experimentally accessible distance range (1–10 nm), which represents the typical distances involved in molecular interactions or intramolecular domains. This narrow working range makes FRET an extremely powerful tool to detect not only molecular interactions, but also large conformational changes within single molecules; an attractive property utilized in many biosensors.

TECHNICAL ADVANCES

The powerful combination of maximizing spatiotemporal resolution coupled with the biochemical information provided by FRET has resulted in the development of numerous innovative biosensors. Each of these biosensors utilizes one of two detection mechanisms to display the biological event of interest.

- The gain of FRET as the two fluorophores are brought closer to one another.
- The loss of FRET as the distance between the two fluorophores increases.

Listed in Figure 1 are some examples of FRET-based optical biosensors.

TECHNICAL CONSIDERATIONS FOR FRET

The design of a FRET-based biosensor does not necessarily guarantee successful and accurate experimental results. The most critical aspect of FRET image analysis lies in the investigator's ability to sieve through potential artifacts to obtain the precise FRET signals. Proper image analysis is discussed in detail in this section and should be given serious consideration during experimental design.

The shift of emission peaks from the donor to the acceptor as FRET increases makes it extremely difficult to obtain the absolute value of FRET in vivo. Therefore, FRET data are usually displayed as ratio images. Ratio images are constructed with a numerator image and a denominator image, wherein the ratio of the intensity value of every pixel from the two images will be obtained. Utilizing the dual emission property of most FRET biosensors, the numerator and denominator images are usually taken using a single excitation wavelength, that of the donor. The ratio of the acceptor fluorescence intensity to that of the donor is then obtained. In the case of CFP-YFP pair, the ratio would be (F_{YFP}, with CFP excitation)/(F_{CFP}, with CFP excitation).

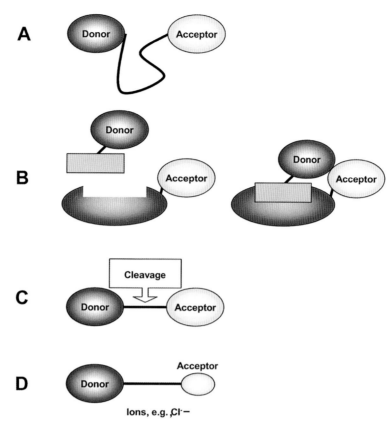

FIGURE 1. Structural design and experimental approaches of various FRET-based optical biosensors. (*A*) One of the most common, yet powerful, architectural designs of FRET sensors relies on the intramolecular conformational change in the linkers bridging the two fluorophores. The linker can consist of any effector-binding domain that will undergo significant conformational alteration upon binding to its cognate effector. This structural change in turn directly affects the efficiency of FRET (Miyawaki et al. 1997; Persechini et al. 1997; Suzuki et al. 1998; Nagai et al. 2000; Sato et al. 2000; Mochizuki et al., 2001; Chew et al. 2002). (*B*) The interaction of two proteins can be dynamically detected by tagging the two binding partners with FRET fluorophore pairs. The interaction of the two proteins will bring the fluorophores within the Förster distance, thus facilitating FRET (Adams et al. 1991; Day 1998; Llopis et al. 2000; Verveer et al. 2000; Day et al. 2001; Janetopoulos et al. 2001; Majoul et al. 2001). (*C*) In vivo protein or polynucleotide processing can be dynamically detected by flanking the cleavage site with a FRET fluorophore pair. Molecular cleavage will be translated into loss of FRET (Xu et al. 1998; Mahajan et al. 1999; Tyas et al. 2000; Vanderklish et al. 2000; Luo et al. 2001; Kohl et al. 2002). (*D*) Some fluorophores are sensitive to certain ions while their FRET partners are not. For example, YFP, but not CFP, is chloride-sensitive, and the emission intensity will decrease correspondingly in the presence of chloride, as depicted by the smaller oval. The ratio of FRET-dependent emission will change in proportion to the concentration of Cl^-, turning a CFP-YFP fusion protein into a sensitive Cl^- sensor (Kuner and Augustine 2000).

Loss of Information on Protein Concentration

The acceptor-to-donor intensity ratio thus displays the relative level of energy transfer at any given pixel. However, as shown by a representative spectral shift during FRET changes, the donor intensity will decrease with the increase in the acceptor intensity, and vice versa (Fig. 2). The simultaneous changes in intensities of both emission peaks thus preclude the possibility of relying on fluorescent intensity to determine the relative concentration of the biosensor in vivo.

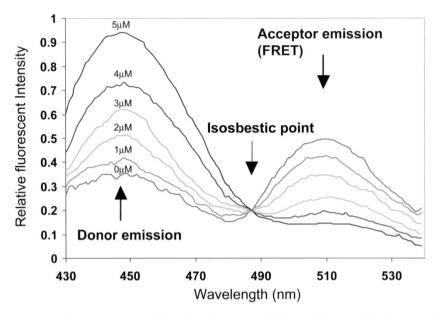

FIGURE 2. A typical FRET spectrum showing the reciprocal change between the donor and acceptor emission peaks. In this case, the capability of the FRET biosensor to detect Ca^{2+}/calmodulin concentration is determined. The purified FRET biosensor in buffer containing 1 μM calmodulin was exposed to increasing concentrations of free Ca^{2+}. Emission scans (using 380-nm excitation) were obtained at increasing $CaCl_2$ concentrations as shown above each corresponding spectrum. When the acceptor emission at 509 nm reaches its maximum (maximal FRET), the emission of the donor drops to the lowest level at 448 nm. The intensity of a FRET biosensor is therefore not a reliable gauge for protein concentration. This underscores the need to use the intensity-modulated display coupled with a ratio algorithm (described in the text) to extract the information on protein concentration from a FRET ratio image.

To include the relative protein concentration into the FRET ratio equation, a mathematical manipulation called intensity modulated display (IMD) can be used in conjunction with a different method of obtaining ratio images (Tsien and Harootunian 1990).

In the case of intramolecular FRET, such as most Ca^{2+} or GTPase biosensors, the two fluorophores are linked in a single polypeptide, and therefore will always be in equal molar ratio. The direct excitation of the donor will result in fluorescence emission intensity directly correlated with the amount of protein. Thus, a different kind of ratio image can be constructed as follows (Chew et al. 2002): (F_{YFP}, with CFP excitation)/(F_{YFP}, with YFP excitation).

The numerator image is the direct result of FRET, and the denominator is identical to acquiring an image of YFP alone. In this ratio, the intensity of the denominator will correlate directly with the amount of biosensor available at a given pixel. The higher the concentration of biosensor, the higher the value of the denominator. Using intensity modulated display, the investigator can display the ratio (which in the case of most Ca^{2+} or GTPase biosensors, this will be the biological activities of the tagged enzyme) according to the denominator intensity. This method of calculating the ratio should be avoided unless a constant stoichiometry of the donor-acceptor pair can be ensured (such as when the two fluorophores are physically linked). The ratio is usually displayed by color, and the intensity of each color denotes the concentration of the protein, as shown in the ratio bar in Figure 3.

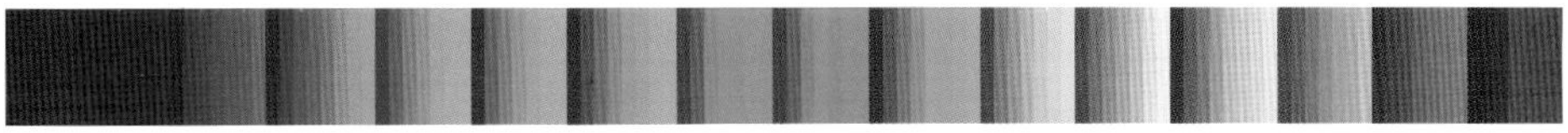

low ratio high ratio

FIGURE 3. The ratio bar is generated using the intensity modulated display in the Metamorph software. Ratio images are constructed by displaying the ratio in 16 different color hues with the lowest ratio displayed at the violet end of the ratio spectrum. Using the intensity of the denominator image (which serves as the internal control irrespective of FRET), the ratio colors were displayed in 16 different intensity levels by the intensity modulated display mode (Tsien and Harootunian 1990). Therefore, depending on the enrichment and the relative activity of the biosensor, the color denotes the relative biological activity being assayed, whereas the intensity of that ratio color represents the amount of the biosensor being localized by the cell.

Background and Autofluorescence Subtraction

Almost all fluorescence live cell imaging experiments will encounter autofluorescence problems. Autofluorescence can be due to cellular proteins or components of the cell culture medium, such as phenol red or a high concentration of riboflavin. Certain media produce much higher background fluorescence than others, regardless of whether it is free of phenol red. One common way to deal with autofluorescence is to perform background subtraction. An average value of background intensity (a region where there is no cell) can be obtained and subtracted from the acquired image. However, in FRET experiments, this practice poses the following problem: The severity of autofluorescence can vary between different channels during a FRET experiment. Investigators who rely on ratio-imaging frequently find that in the process of eliminating background fluorescence, different values must be subtracted from the numerator versus the denominator, thus inadvertently changing the actual ratio.

The first step in reducing background fluorescence is therefore to choose the best imaging medium. Many commercially available perfusion chambers for live cell imaging do not permit gaseous exchange, thereby precluding the use of CO_2 buffering. It is therefore critical to choose a medium designed for a CO_2-free environment, such as one containing HEPES buffer or Leibovitz-15 (L-15) medium. Some cell lines may not be able to tolerate a sudden change of culture condition. Under those circumstances, the mixing of regular culture medium with L-15 medium will greatly reduce the level of autofluorescence.

Recent technological advances in emission fingerprinting provide, for the first time, a way to easily overcome autofluorescence problems. The Zeiss META module is capable of recognizing the spectral properties of autofluorescence and removing it from the image without affecting the actual intensity of the fluorophores.

Motion Artifacts

Images of most multicolored live cell experiments are taken sequentially. In the case of very rapid biological processes, fluorophores may move to a different pixel over short time intervals, resulting in an artifact that causes the donor and acceptor fluorophore to misalign in the resultant ratio images. In practice, this artifact will be interpreted as a loss of FRET signal. To avoid this type of artifact, it is important to correlate live cell FRET imaging data with images of fixed cells. The FRET biosensor should show the same spectral pattern in both cases. It is also important to program the image acquisition software to alternate the sequence of capturing the numerator and denominator images. A true FRET signal should not display any alternating signals.

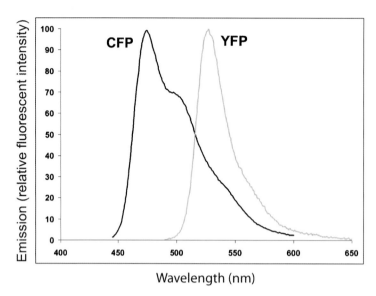

FIGURE 4. Emission spectra for CFP and YFP, a commonly used FRET fluorophore pair. It is important to note that the emission spectrum of the donor fluorophore in many instances overlaps significantly with that of the acceptor. This emission cross talk will produce artificial FRET using the FRET filter set and must be mathematically corrected (see text for details).

Emission Cross Talk

Although CFP and YFP represent excellent fluorophore pairs for energy transfer experiments, the long trailing tail of the CFP emission spectrum overlaps that of the YFP spectrum to a significant extent (Fig. 4), potentially producing "emission cross talk." This is a problem that cannot be corrected by filter design due to the extent of overlapping.

The degree of CFP "bleed-through" can be established by acquiring a CFP-expressing cell using the YFP detection system. The easiest way to accurately separate the overlapping CFP and YFP spectra would be to perform spectral unmixing based on the emission fingerprints of these two fluorophores. This can be achieved with a device such as the Zeiss META. Emission cross talk can also be determined using conventional epifluorescence microscopes and a simple algorithm developed by Youvan et al. (1997). To perform the correction, the extent of cross talk must first be determined either with cells singly expressing each of the FRET fluorophores or with fluorescent beads with the same fluorescent properties. Overall, the Youvan method can be employed using the following notation X_u^y, where X can be D, A, or F, indicating a monochrome image acquired through the Donor, Acceptor, or FRET channels. The subscript u can be substituted by d or a to represent a pixel from a donor or acceptor fluorophore. The substantial overlap of the donor and the acceptor spectra will ensure that cross-talk fluorescence will be seen in unprocessed channels. The superscript y can be replaced by b to indicate that the monochrome image has been background-subtracted. For example, F_d^b represents a background-subtracted image of the donor fluorophore taken using the FRET channel. Using this notation, the following two ratios can be established:

$F_d^b / D_d^b =$ ratio of FRET channel intensity to Donor channel intensity for a "pure donor" pixel after background subtraction on each monochrome image.

$F_a^b / A_a^b =$ ratio of FRET channel intensity to Acceptor channel intensity for a "pure acceptor" pixel after background subtraction on each monochrome image.

These ratios are essentially the fractional "bleed" of donor and acceptor, respectively, into the FRET channel. They are then used to correct the FRET channel pixel intensities. The pixel in the corrected FRET image (F^c) is then given by

$$F^c = F^b - (F_d^b / D_d^b) \times D^b - (F_a^b / A_a^b) \times A^b$$

Fluorescence cross-talk analysis can also be performed using a method devised by Gordon et al. (1998). Using the same notation as the Youvan method, the Gordon FRET correction can be expressed as

$$F^c = [F_d - (D_d \times R_D) - (A_a \times R_E)] / A_a$$

where R_D is the ratio of the detection efficiencies of the donor emission intensities through the acceptor and donor filter sets. It can be obtained from a sample that has only donor molecules by exciting at a single wavelength and dividing the total intensities detected through the acceptor and donor channels. R_E is the ratio of the extinction coefficients of the acceptor when excited at the acceptor and donor wavelengths. It can be determined by exciting a sample with only the acceptor molecules at the donor and acceptor wavelengths and dividing the total intensities detected through the acceptor channel. These two ratios are constant factors and are assumed to be spatially invariant in the image. Some commercially available software packages, such as the Zeiss FRET Analyzer Package, now include these correction methods, thus allowing users to easily correct for the resultant FRET ratio images.

FRET cross talk can also be empirically eliminated by adjusting the illumination intensity until cross talk is absent, as determined by spectral analysis. However, depending on relative protein concentrations indicated by the two fluorophores, and other intracellular conditions, this empirical method may not always be useful.

Acceptor Photobleaching

The hallmark of a true FRET spectrum requires energy transfer from the donor to the acceptor. This phenomenon dictates whether FRET occurs; the donor emission peak will drop, concomitant with an increase in the acceptor emission. These spectra will all intersect at one point referred to as the isosbestic point. To test whether the biosensor actually undergoes true FRET in vivo, a technique called acceptor photobleaching can be employed. Photobleaching will cause the acceptor to lose its capacity to absorb the energy from the donor, causing the donor to surge to the maximum as if there is no FRET. This will confirm that the emission detected by the FRET channel comes from true FRET, and not due to channel cross talk or cross-excitation of the acceptor by the donor excitation light. In fact, an apparent energy transfer efficiency $E_D(i)$ can be determined in each position i of an image as follows: $E_D(i) = 1 - F^D(i)/F_{pb}^D(i)$, where $F^D(i)$ and $F_{pb}^D(i)$ are the donor emission images before and after photobleaching. This correction is best performed in fixed biospecimens, because photobleaching requires prolonged illumination during which translocation of the donor-tagged molecules can occur, generating a motion artifact.

Differential Bleaching Rates

Different fluorophores can have different photobleaching rates, which may introduce errors into FRET calculations. Thus, rates of photobleaching of the donor and the acceptor must be established before performing acceptor photobleaching. In general though, the difference in photobleaching rates may be small enough to be omitted, especially when CFP (cyan variant of GFP) instead of BFP (blue variant of GFP) is used as the donor.

Pixel Saturation and Ratio Image Construction

Saturated pixels will greatly alter an image and therefore must be avoided. One of the easiest ways to confirm that there is a saturated pixel problem is to point the cursor to the brightest spot on the acquired image. Most image analysis software will show the corresponding pixel value as the cursor moves around the image. Alternatively, draw a line across the brightest spot, and perform a line scan. Many imaging software packages are now equipped with different modes for displaying saturated pixels. To avoid pixel saturation, reduce the exposure time, fluorescent intensity, or detector gain (for both channels), until neither the numerator nor the denominator shows any saturated pixels before the data are used for the construction of ratio images.

Molecular Quenching and Overexpression of Biosensor

Most biosensors are designed to detect the dynamic interaction of the sensor and the signaling molecule. It is therefore likely that overexpression of the biosensor may cause a rapid depletion of the signaling molecules in vivo, causing potential biological artifacts. Care must be taken to ensure that the overexpression level of the biosensor does not interfere with biological functions. It may be necessary to add a comparable amount of the signaling molecules to the experimental system to offset the sequestration of endogenous pools. For example, Persechini et al. (1997) microinjected equal molar ratios of calmodulin with the Ca^{2+}/calmodulin fluorescence indicator protein to obtain an accurate in vivo measurement of Ca^{2+}/calmodulin.

EXAMPLE OF FRET BIOSENSOR

Myosin light-chain kinase (MLCK) is a Ca^{2+}/calmodulin-dependent kinase that phosphorylates the 20-kD regulatory light chain (RLC) of myosin II. To separate the MLCK-mediated signal from the many pathways that converge on RLC, a FRET-based MLCK biosensor was created (Chew et al. 2002) based on the fluorescent indicator protein (FIP) generated by Persechini et al. (1997). In the biosensor, the MLCK molecule is linked at the carboxy-terminal end to the FIP, which consists of a donor connected to an acceptor via the Ca^{2+}/calmodulin-binding domain derived from MLCK, as shown in Figure 5. In the absence of Ca^{2+}/calmodulin, the linker adopts a coiled conformation, bringing the two fluorophores to each other to facilitate FRET. When the biosensor colocalizes with the increased intracellular pool of Ca^{2+}/calmodulin, where MLCK will be activated, Ca^{2+}/calmodulin will bind to its cognate regulatory domain in MLCK, and to the linker region. The binding of Ca^{2+}/calmodulin to the linker will alter the conformation to an open position, thus lengthening the distance between the fluorophores and disrupting FRET. MLCK activity is inversely proportional to the degree of FRET. MLCK is most active when the FRET is minimal. The biosensor therefore permits the simultaneous localization of MLCK, as well as its relative activity state in vivo.

Using the IMD image construction technique discussed above in Loss of Information on Protein Concentration, ratio images can be generated to report three critical pieces of information regarding the dynamic properties of MLCK in live cells: localization, relative activity, and local concentration of the kinase. Ratio images are constructed based on the following ratio: ($F_{acceptor}$, with donor excitation)/($F_{acceptor}$, with acceptor excitation).

The result is displayed in 16 different color hues, with the lowest ratio displayed at the blue end. Using the fluorescence intensity of the denominator (which serves as the internal

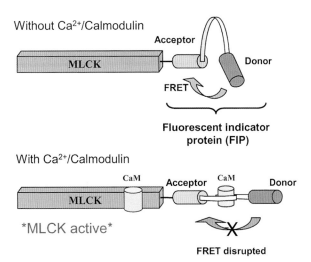

FIGURE 5. Schematic diagram of MLCK-FIP. The fluorescent indicator protein (FIP) consists of a donor fluorophore (BFP or CFP) linked to an acceptor (GFP or YFP, respectively) via the Ca²⁺/calmodulin-binding domain derived from MLCK (Chew et al. 2002). In the absence of Ca²⁺/calmodulin, the coiled Ca²⁺/calmodulin-binding domain allows FRET between the fluorophores (Persechini et al. 1997). When Ca²⁺/calmodulin (*green cylinder*) is present, it binds and activates MLCK and also binds between the fluorophores, thus disrupting FRET.

control irrespective of FRET), the ratio colors are then displayed as 16 intensity levels. Since the pixel intensity of the denominator image directly correlates with protein concentration, the IMD display reflects the relative local concentration of MLCK in vivo. Besides the obvious pattern of localization, which is apparent from a two-dimensional image, every pixel in this ratio image carries two additional types of information, as exemplified in Figure 6.

The combined information derived from the ratio images also requires a better method for displaying the data. For example, it is possible to display the ratio in 16 different colors with peak heights representing the pixel intensity of the denominator image. This type of display acts to highlight subtle molecular changes previously overlooked using a two-dimensional display. A good example is shown in Figure 7, where the transient enrichment of MLCK along actively contracting stress fibers is brought to the observer's attention by the increased peak height.

CONCLUSION

This chapter provides only a glimpse of an extremely useful tool to study molecular dynamics in live cells. The powerful capability of FRET to report protein interaction and conformational changes has led to numerous new insights in various molecular mechanisms. Although the number of examples using FRET in this manner is too numerous for a complete listing, some of the interesting findings generated by innovative designs of FRET-based sensors include the interaction of various activators and their downstream effectors (calmodulin and Ca²⁺ [Miyawaki et al. 1997; Persechini et al. 1997]); MLCK and Ca²⁺/calmodulin [Chew et al. 2002]; and GTP-bound Rac/Cdc42 with p21-activated kinase [Gardiner et al. 2002; Kraynov et al. 2000]), the activation of secondary messengers or signaling pathways (the activation of Ras [Mochizuki et al. 2001], caspases [Mahajan et al. 1999; Luo et al. 2001], and ErbB1 receptor [Verveer et al. 2000]), and intramolecular conformational changes (the lever arm of myosin [Suzuki et al. 1998]).

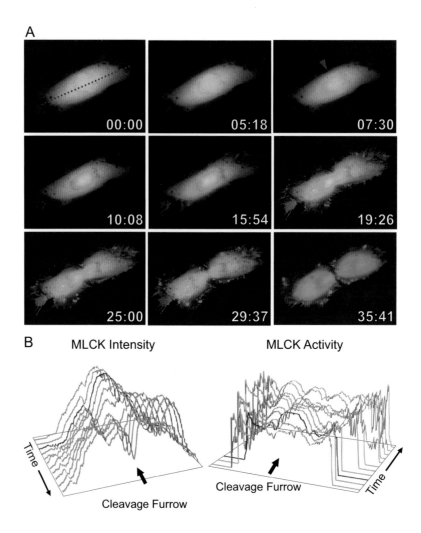

FIGURE 6. In a two-dimensional image, not only is the localization pattern apparent, but every pixel value in the ratio images constructed using IMD carries two sets of information: the biological activity reported by the change in FRET and the relative protein concentration. A line scan (*red dotted line*) performed across a cell undergoing cytokinesis allows the relative MLCK activity to be separated from the relative local concentration of the kinase. These two sets of information are then plotted against time in the lower panel *B*. Note the differing direction of the time axis in each graph. MLCK localization to the cleavage furrow decreases as cytokinesis continues, whereas MLCK activity increases. Red data lines on the graphs indicate the onset of cleavage furrow contraction. These images have not been volume-corrected. Changes in MLCK intensity may reflect changes in thickness of the cell and/or MLCK concentration. See Movie 9.1 on the accompanying DVD.

It should be emphasized that although these methods afford exciting new opportunities to delve into subcellular structure/function relationships at high resolution in live cells, there are also numerous potential biological artifacts, mathematical problems, and spectral challenges that are often overlooked during the initial phase of designing the biosensor. Some of the issues of concern are loss of information on protein concentration, autofluorescence problems, motion artifacts, and emission cross talk. This chapter summarizes the most common technical problems encountered with FRET-based biosensors and various approaches to resolving these problems.

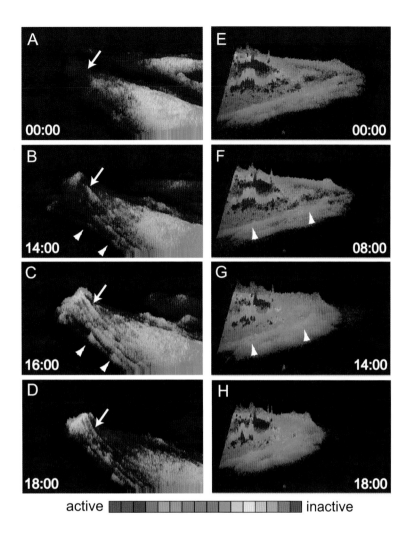

active [ratio bar] inactive

FIGURE 7. An alternative way to display IMD ratio images is to split the intensity information from the ratio color, which in most common cases denotes the biological activity. This series of ratio images shows the MLCK profile during active cellular contraction in PTK-2 cells. Live cells imaging was performed using a Bioptechs FCS2 perfusion chamber (Bioptechs, Inc., Butler, Pennsylvania) system maintained at 37°C, with a chamber thickness of 0.75 mm. To reduce the background fluorescence and to image the cells in the absence of CO_2, the cell medium was mixed with equal volume of Leibovitz-15 medium without phenol red, supplemented with 10% fetal bovine serum. Fresh medium was perfused at 0.1 ml/min into the chamber during imaging. Cell images were taken using MetaMorph software. The images were taken with a 200–300-ms exposure with an ~50-ms interval between the numerator and denominator images. Ratio images were constructed as described in this chapter, and the resulting IMD images displayed with the following modification: The color hues represent relative MLCK activity according to the ratio bar below, whereas the local concentration of MLCK is depicted by the peak height. A local enrichment of active MLCK will be shown as high peaks with blue/green color. Panels A–D and E–H show the MLCK profiles in two different contracting PTK-2 cells analyzed using this method. Arrows indicate regions where diffuse MLCK is recruited to stress fiber, and arrowheads show sites of in situ activation. In panels A–D, for example, there is a surge of MLCK activity at the left side of the cell, corresponding to the regional cellular contraction. By separating the intensity information from the ratio color, the data highlight the enrichment of MLCK protein at the region of active contraction concomitant with the increase in activity. Note the transient and highly localized recruitment of active MLCK to the region of contraction and along stress fibers. Panels E–H provide another example of transient recruitment and localized activation of MLCK during activation contraction. See Movies 9.2 and 9.3 on the accompanying DVD.

MOVIE LEGENDS

MOVIE 9.1. Distribution and relative kinase activity of myosin light chain kinase in cell undergoing cytokinesis, as monitored by FRET biosensor.

MOVIE 9.2. Recruitment and in situ activation of myosin light chain kinase along stress fiber during active concentration (Example 1).

MOVIE 9.3. Recruitment and in situ activation of myosin light chain kinase along stress fiber during active concentration (Example 2).

REFERENCES

Adams S.R., Harootunian A.T., Buechler Y.J., Taylor S.S., and Tsien R.Y. 1991. Fluorescence ratio imaging of cyclic AMP in single cells. *Nature* **349:** 694–697.

Chew T.L., Wolf W.A., Gallagher P.J., Matsumura F., and Chisholm R.L. 2002. A fluorescent resonant energy transfer-based biosensor reveals transient and regional myosin light chain kinase activation in lamella and cleavage furrows. *J. Cell Biol.* **156:** 543–553.

Day R.N. 1998. Visualization of Pit-1 transcription factor interactions in the living cell nucleus by fluorescence resonance energy transfer microscopy. *Mol. Endocrinol.* **12:** 1410–1419.

Day R.N., Periasamy A., and Schaufele F. 2001. Fluorescence resonance energy transfer microscopy of localized protein interactions in the living cell nucleus. *Methods* **25:** 4–18.

Förster T. 1965. Delocalized excitation and excitation transfer. *Modern Quant. Chem.* **3:** 93–137.

Gardiner E.M., Pestonjamasp K.N., Bohl B.P., Chamberlain C., Hahn K.M., and Bokoch G.M. 2002. Spatial and temporal analysis of Rac activation during live neutrophil chemotaxis. *Curr. Biol.* **12:** 2029–2034.

Gordon G.W., Berry G., Liang X.H., Levine B., and Herman B. 1998. Quantitative fluorescence resonance energy transfer measurements using fluorescence microscopy. *Biophys. J.* **74:** 2702–2713.

Janetopoulos C., Jin T., and Devreotes P. 2001. Receptor-mediated activation of heterotrimeric G-proteins in living cells. *Science* **291:** 2408–2411.

Kohl T., Heinze K.G., Kuhlemann R., Koltermann A., and Schwille P. 2002. A protease assay for two-photon crosscorrelation and FRET analysis based solely on fluorescent proteins. *Proc. Natl. Acad. Sci.* **99:** 12161–12166.

Kraynov V.S., Chamberlain C., Bokoch G.M., Schwartz M.A., Slabaugh S., and Hahn K.M. 2000. Localized Rac activation dynamics visualized in living cells. *Science* **290:** 333–337.

Kuner T. and Augustine G.J. 2000. A genetically encoded ratiometric indicator for chloride: Capturing chloride transients in cultured hippocampal neurons. *Neuron* **27:** 447–459.

Llopis J., Westin S., Ricote M., Wang Z., Cho C.Y., Kurokawa R., Mullen T.M., Rose D.W., Rosenfeld M.G., Tsien R.Y., Glass C.K., and Wang J. 2000. Ligand-dependent interactions of coactivators steroid receptor coactivator-1 and peroxisome proliferator-activated receptor binding protein with nuclear hormone receptors can be imaged in live cells and are required for transcription. *Proc. Natl. Acad. Sci.* **97:** 4363–4368.

Luo K.Q., Yu V.C., Pu Y., and Chang D.C. 2001. Application of the fluorescence resonance energy transfer method for studying the dynamics of caspase-3 activation during UV-induced apoptosis in living HeLa cells. *Biochem. Biophys. Res. Commun.* **283:** 1054–1060.

Mahajan N.P., Harrison-Shostak D.C., Michaux J., and Herman B. 1999. Novel mutant green fluorescent protein protease substrates reveal the activation of specific caspases during apoptosis. *Chem. Biol.* **6:** 401–409.

Majoul I., Straub M., Hell S.W., Duden R., and Soling H.D. 2001. KDEL-cargo regulates interactions between proteins involved in COPI vesicle traffic: Measurements in living cells using FRET. *Dev. Cell* **1:** 139–153.

Miyawaki A., Llopis J., Heim R., McCaffery J.M., Adams J.A., Ikura M., and Tsien R.Y. 1997. Fluorescent indicators for Ca^{2+} based on green fluorescent proteins and calmodulin. *Nature* **388:** 882–887.

Mochizuki N., Yamashita S., Kurokawa K., Ohba Y., Nagai T., Miyawaki A., and Matsuda M. 2001. Spatiotemporal images of growth-factor-induced activation of Ras and Rap1. *Nature* **411:** 1065–1068.

Nagai Y., Miyazaki M., Aoki R., Zama T., Inouye S., Hirose K., Iino M., and Hagiwara M. 2000. A fluorescent indicator for visualizing cAMP-induced phosphorylation in vivo. *Nat. Biotechnol.* **18:** 313–316.

Persechini A., Lynch J.A., and Romoser V.A. 1997. Novel fluorescent indicator proteins for monitoring free intracellular Ca^{2+}. *Cell Calcium* **22:** 209–216.

Sato M., Hida N., Ozawa T., and Umezawa Y. 2000. Fluorescent indicators for cyclic GMP based on cyclic GMP-dependent protein kinase Ialpha and green fluorescent proteins. *Anal. Chem.* **72:** 5918–5924.

Suzuki Y., Yasunaga T., Ohkura R., Wakabayashi T., and Sutoh K. 1998. Swing of the lever arm of a myosin motor at the isomerization and phosphate-release steps. *Nature* **396:** 380–383.

Tsien R.Y. and Harootunian A.T. 1990. Practical design criteria for a dynamic ratio imaging system. *Cell Calcium* **11:** 93–109.

Tyas L., Brophy V.A., Pope A., Rivett A.J., and Tavare J.M. 2000. Rapid caspase-3 activation during apoptosis revealed using fluorescence-resonance energy transfer. *EMBO Rep.* **1:** 266–270.

Vanderklish P.W., Krushel L.A., Holst B.H., Gally J.A., Crossin K.L., and Edelman G.M. 2000. Marking synaptic activity in dendritic spines with a calpain substrate exhibiting fluorescence resonance energy transfer. *Proc. Natl. Acad. Sci.* **97:** 2253–2258.

Verveer P.J., Wouters F.S., Reynolds A.R., and Bastiaens P.I. 2000. Quantitative imaging of lateral ErbB1 receptor signal propagation in the plasma membrane. *Science* **290:** 1567–1570.

Xu X., Gerard A.L., Huang B.C., Anderson D.C., Payan D.G., and Luo Y. 1998. Detection of programmed cell death using fluorescence energy transfer. *Nucleic. Acids Res.* **26:** 2034–2035.

Youvan D.C., Silva C.M., Bylina E.J., Coleman W.J., Dilworth M.R., and Yang M.M. 1997. Calibration of fluorescence resonance energy transfer in microscopy using genetically engineered GFP derivatives on nickel chelating beads. *Bio/Technology* **3:** 1–18.

Application of Light-directed Activation of Caged Biomolecules and CALI to Problems in Cell Motility

David Humphrey,[a,b] Zenon Rajfur,[b] Barbara Imperiali,[c] Gerald Marriott,[d] Partha Roy,[a] and Ken Jacobson[a,b]

[a] Department of Cell and Developmental Biology, University of North Carolina, Chapel Hill, North Carolina 27599;
[b] Lineberger Comprehensive Cancer Center, University of North Carolina, Chapel Hill, North Carolina 27599;
[c] Department of Chemistry, Massachusetts Institute of Technology, Cambridge 02139;
[d] Department of Physiology, University of Wisconsin at Madison, Madison, Wisconsin 53706

PHOTOACTIVATION OF CAGED COMPOUNDS

ACTIVATION AND INACTIVATION OF PROTEINS using light-based techniques, such as photoactivation of caged peptides or proteins and chromophore-assisted laser inactivation (CALI), offer insights into cellular dynamics not achievable using genetic means. The ability to selectively alter the activity of a specific protein at a defined time and location inside a cell allows the correlation of changes in protein activity and cellular behavior. One illustrative example, discussed below, is an experiment where photoactivation (by the release of a protective caging group) of thymosin β4, an actin monomer sequestering protein, in migrating fish keratocytes results in cell turning (Roy et al. 2001). The following pages contain a brief discussion of the background and evolution of these techniques, along with several biological examples, and an outline of the general methodologies.

A caged compound, peptide, or protein is prepared by covalently linking it to a photolabile, protecting group via a limited number of critical functional groups in the biomolecules (Marriott 1994). In most cases, the caging reagent is based on the 2-nitrobenzyl group (Marriott and Walker 1999). Under carefully defined conditions, a brief irradiation with UV light (320–400 nm), can lead to an intramolecular photoisomerization reaction that induces the cleavage of the chemical bond between the cage group and the unmodified, biologically active species, releasing a 2-nitrosobenzophenone photoproduct. It should be noted that under certain conditions (i.e., intense radiation or absence of scavengers), undesirable alternate reaction pathways can occur (McCray and Trentham 1989). The first report of a caged,

TABLE 1. Some commercially available caged small molecules and caging reagents

Nucleotides	Commercial source
ATP (NPE-caged, DMNPE-caged)	Molecular Probes
ADP (NPE-caged)	Molecular Probes
cAMP (DMNB-caged)	Molecular Probes/Calbiochem
cGMP	Calbiochem
GTP	Calbiochem
GDP	Dojindo
Neurotransmitters and second messengers	
γ-Aminobutyric acid ([CNB-caged] GABA)	Molecular Probes
Nitric oxide	Molecular Probes
Aspartic acid	Calbiochem/Molecular Probes
Glutamic acid	Calbiochem/Molecular Probes
Inositol 1,4,5-trisphosphate	Calbiochem/Molecular Probes
Calcium regulators	
Ca^{2+} (NP-EGTA)	Molecular Probes
Ca^{2+} (DM-Nitrophen)	Calbiochem
Fluorescent dyes	
Fluorescein	Molecular Probes
Fluorescein dextran	Molecular Probes
Caging reagents	
1-(bromomethyl)-2-nitro-4,5-dimethoxybenzene	Aldrich
2-nitroveratryloxychlorocarbamate (NVOC-Cl)	Fluka

biologically relevant molecule described the synthesis and use of caged ATP (Kaplan et al. 1978). Since that time, a number of compounds, including nucleotides, neurotransmitters, fluorescent dyes, metal ions, peptides, and small proteins, have been synthesized in a caged form (Shigeri et al. 2001) and used as probes of cellular dynamics (Adams and Tsien 1993; Marriott and Walker 1999). A number of caged small molecules are commercially available (for examples, see Table 1). This chapter focuses on the synthesis and implementation of caged peptides and proteins in the study of complex biological systems.

Design of Caged Peptides and Proteins

When choosing a particular molecule for photoactivation studies, it is necessary to have some structural knowledge of the molecule in order to design an appropriately caged species that will retain its biological inactivity until uncaging is effected. The following points should be considered when designing a caged biomolecule:

- The caged reagent should react with the protein of interest in aqueous solution using physiologically relevant conditions (e.g., neutral-slightly basic pH and small amounts of organic solvent).

- The bond formed between the caged group and the protein should be stable under physiological conditions. Linking the caged group to the biomolecule through a thioether, ether, carbonate, or carbamate bond is preferable to linkage via an ester, which can be cleaved inside the cell.

- The caged group should exhibit a high molar absorptivity (>1000 M^{-1} cm^{-1}) in order to increase the probability of excitation in the lamellipodium, which is usually <1 μm thick.

- The photoisomerization reaction leading to bond cleavage should proceed with a reasonable quantum yield (>0.1) and exhibit an action spectrum in the near-UV wavelength region (340–400 nm) to avoid interference with other biomolecules (McCray and Trentham 1989).

- The photoproducts of the photoactivation reaction should not react with other functional groups in the protein. In the case of the 2-nitrosobenzophenone photoproduct, a thiol-based scavenger such as dithiothreital (DTT) must be added in vitro. Scavenging of the photoproduct may be achieved in vivo by glutathione.

Given these requirements, substituted 2-nitrobenzyl-caged groups have important advantages when uncaging protein activity. First, the higher extinction of 2-nitro-4,5-dimethoxybenzene group versus unsubstituted 2-nitrobenzyl groups (5000 M^{-1} cm^{-1} vs. 500 M^{-1} cm^{-1}) increases the chance of excitation of the caged protein within the cell, which is usually less than 10 μm thick. Second, dimethoxy-substituted 2-nitrobenzyl-caged groups have a red-shifted action spectrum compared to unsubstituted 2-nitrobenzyl groups. The broad near-UV action spectrum of the 2-nitro-4,5-dimethoxybenzene group is centered at 350 nm and extends to 400 nm; this wavelength region is useful since most endogenous chromophores in a cell have excitation maxima below 350 nm. Two minor disadvantages of the 2-nitro-4,5-dimethoxybenzene-caged group are the slow progression of the photoisomerization reaction compared to certain unsubstituted 2-nitrobenzyl groups (milliseconds vs. microseconds; Walker et al. 1998) and the quantum yield is somewhat lower. However, the rate of the photoisomerization reaction is not usually an important issue for cell-based uncaging of protein activity, and the benefits of reduced photodamage using caged groups with red-shifted action spectra outweigh the problem of a slightly reduced photocleavage efficiency.

Examples of Caged Peptides/Proteins

Critical lysine residues on G-actin (monomeric form), which are essential for polymerization (including Lys-61), have been caged (Marriott 1994). Irradiation with UV light releases the cage, enabling actin polymerization to occur. Additionally, if the target protein has multiple domains, it is possible to specifically inactivate one of the domains with a caging group, without diminishing the activity or function of the others. Heavy meromyosin (HMM), a subfragment of skeletal muscle myosin II, which mechanically pushes actin filaments in a process that consumes ATP, was caged at Cys-707, which is localized in the hinge region of the protein (Marriott and Heidecker 1996). The presence of the cage inhibited actin filament motion by HMM, as observed by an in vitro motility assay, but not the ATPase activity. Thus, the caging resulted in the uncoupling of the two activities until photoactivation. Caging of individual protein domains is useful for dissecting structure-function relationships within, as well as between, proteins.

Caged molecules could prove to be extremely valuable in elucidating signal transduction pathways that regulate cell migration and other processes. Traditional approaches have relied on chemical inhibition or molecular biology techniques, involving generation of dominant active and negative mutants. However, many proteins have no known chemical inhibitors and even those that do often lack specificity. Biological approaches may lack the necessary control of activation in the desired time regime. A common theme in signal transduction pathways is the phosphorylation of serine, threonine, and tyrosine residues by kinases, which serve as molecular switches to regulate protein activity. The synthesis of caged phosphopeptides has been reported recently (Pan and Bayley 1997; Rothman et al. 2002); in the future, it may also be possible to produce caged phosphoamino acids within whole proteins (Pan and Bayley 1997). Photoreleased phosphopeptides from synthetic caged phosphopeptides can bind to downstream effectors and prevent protein activation by competing with downstream partners, thereby inhibiting signal transduction. Thus, photoreleased phosphopeptides may serve a role similar to that of dominant-negative mutants, with the added benefits of more precise temporal, and possibly spatial, control.

Synthesis of Caged Peptides/Proteins

For the study of actin dynamics and signal transduction pathways, caged peptides can be prepared by chemical modification of reactive amino acid residues within key recognition sequences (see Fig. 1). The reaction conditions differ based on the nature of the side chain

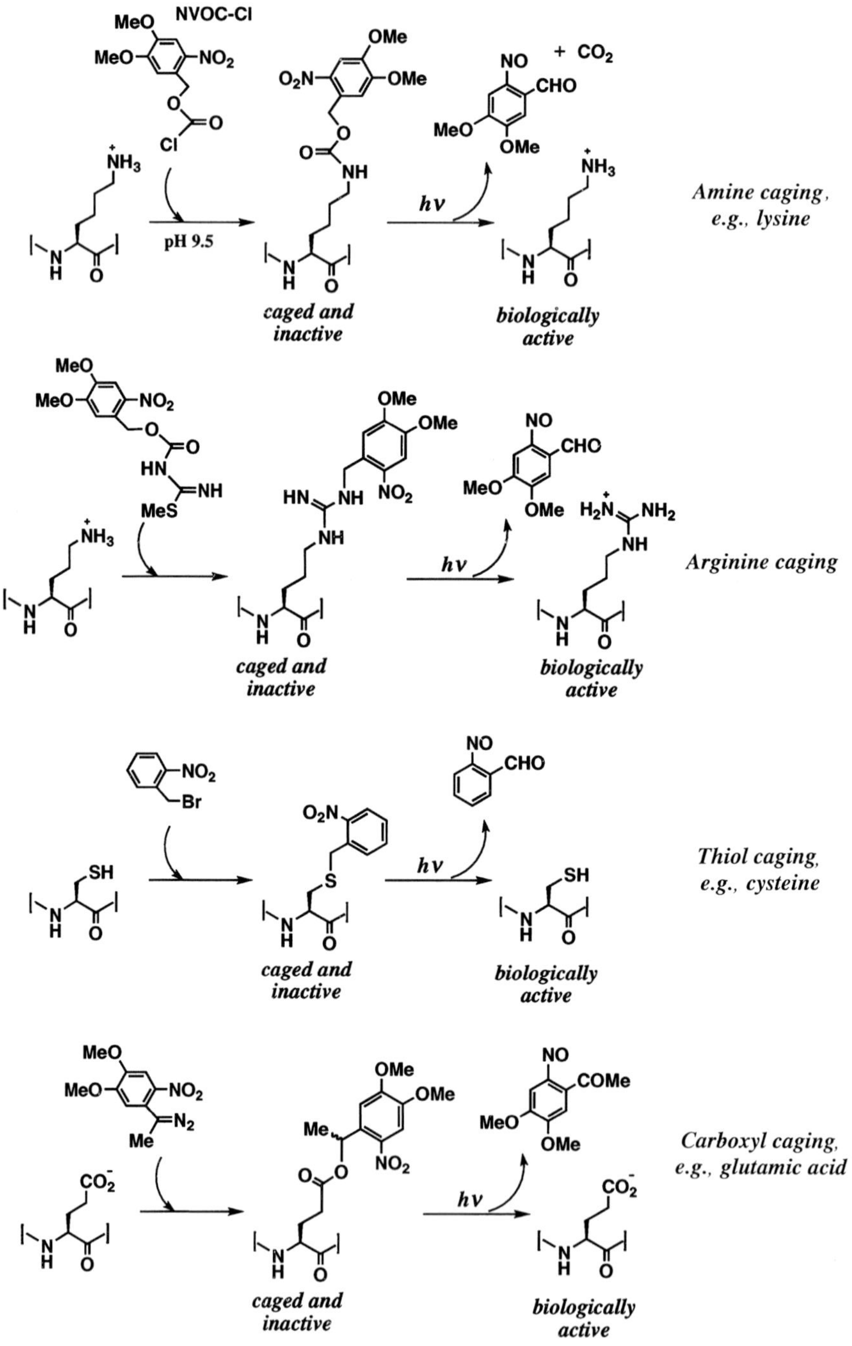

FIGURE 1. Caging and uncaging of amino acid side-chain functionality. Common caging reactions for amine, guanidine, thiol, and carboxyl side-chain moieties are given. In the case of arginine, the synthetic scheme necessitates starting with the alternate amino, ornithine; uncaging reveals the native amino acid. The designation *hν* indicates that irradiation with light (300–400 nm) will break the bond between the amino acid residue and the caging group.

to be modified and whether it is an amine, a thiol, or a carboxylic acid. For example, the modification of the lysine on G-actin is carried out at basic pH using nitroveratryloxychloroformate (NVOC-Cl). The urethane bond that is formed in this transformation can be removed by UV irradiation (320–420 nm), thereby regenerating the intact molecule (Marriott 1994). Caged arginine can be prepared by treating the corresponding ornithine-containing peptides with a guanidylating agent that includes the caging group (Wood et al. 1998). The thiol side chain of cysteine can be modified using 2-nitrobenzyl bromide (Pan and Bayley 1997), and carboxyl groups can be modified with 1-(4,5-dimethoxy-2-nitrophenyl)diazoethane. This compound can be generated in situ using a kit from Molecular Probes that includes the hydrazone precursor to the esterification agent. Bayley and coworkers have also reported a chemoenzymatic synthesis of caged phosphorothiopeptides, involving a chemical modification step (Zou et al. 2001). The advantage of using chemical modification methods to cage peptides is that the chemistry is direct and relatively easy. However, because most proteins contain multiple reactive groups, it can be difficult to control the position and number of the modified sites.

A second general method for generating caged compounds is the nonsense codon suppression technique. Originally introduced by the Schultz (Noren et al. 1989) and Chamberlain (Bain et al. 1989) groups in 1989, this method involves incorporation of a nonsense mutation into the sequence encoding the amino acid of interest. The caged protein is then synthesized using an in vitro translation system. This nonsense codon suppression technique allows photolabile amino acids to be incorporated at a desired site in any protein of interest. However, it should be noted that the method is technically challenging and may be beyond the chemical capabilities of most biological laboratories. An additional problem is that because expression must be carried out in a cell-free translation system, relatively small quantities of protein are produced.

A third technique for preparing caged peptides and small proteins is solid-phase peptide synthesis (SPPS). This method of generating polypeptides can be used to incorporate amino acids that have already been caged into polypeptides as they are synthesized on the SPPS resin (Tatsu et al. 1996, 1999; Merrifield 1997; Walker et al. 1998; Marriott and Walker 1999). SPPS consists of three separate processes: chain assembly on a resin, simultaneous or sequential cleavage and deprotection of the resin-bound, fully protected peptide, and purification and characterization of the final peptide. Caged phosphopeptides can be synthesized by SPPS using an interassembly approach (see Fig. 2) (Rothman et al. 2002).

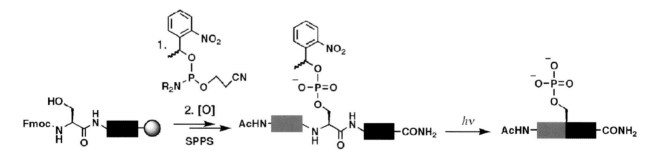

FIGURE 2. Interassembly approach for the synthesis of caged phosphopeptides. In the interassembly approach, peptides are synthesized on the solid-phase resin up to the phosphorylation site. Then, serine, threonine, or tyrosine is incorporated without side-chain protection. The caged phosphate group is introduced via phosphitylation, which introduces the trivalent phosphorous species. Peptide synthesis is then completed using standard solid-phase peptide synthesis protocols. The designation *hv* indicates that irradiation with light (300-400 nm) will break the bond between the amino acid residue and the caging group.

Characterization

Once the caged peptide has been synthesized and purified using reversed phase–high-performance liquid chromatography (RP-HPLC), it is important to fully characterize the efficiency of uncaging and the difference in protein activity before and after photoactivation and to demonstrate that no spontaneous cleavage of the bond between the cage and the protein occurs. Furthermore, it should be demonstrated that neither the caged species nor the released photoproduct participates in reactions with other proteins inside the cell.

The quantum yield for uncaging defines the efficiency of uncaging, relative to known caged molecules. To determine the quantum yield for uncaging, HPLC analysis is used to assess the photolysis of the caged compound versus that of a well-characterized molecule such as caged P_i under the same reaction conditions (Kaplan et al. 1978; Ellis-Davies and Kaplan 1994). The HPLC signals for the samples are normalized using a photochemically inert internal standard. Typical quantum yields for uncaging of o-nitrobenzyl species are in the 10–65% range (Rothman et al. 2002).

Whether uncaging can produce a local effect (on the micrometer scale) depends primarily on the local concentration of binding targets and, secondarily, on the diffusion coefficient of the uncaged reagent. Thus, even if a small, uncaged peptide diffuses rapidly, a local effect will be elicited, provided there is a sufficient density of targets in the irradiated zone to which the peptide can bind with high affinity.

Examples Utilizing Caged Compounds in Cells

There are now several specific examples of experiments in which photoactivation of caged compounds was used to gain insights into cell migration. These examples demonstrate the potential and power of photoactivation. In the first example, the uncaged peptides serve as specific inhibitors of protein function. This allows investigators to design peptides that specifically inhibit a particular region of the target protein. As a result of uncaging, the target's potential binding partners are competitively inhibited or enzymatic activity is disrupted. This makes it possible to study structure-function relationships of proteins directly and to observe their effects in cells.

Walker et al. (1998) constructed two different peptide sequences that, upon uncaging, serve as probes by specifically blocking calmodulin and myosin light-chain kinase (MLCK), respectively. The calmodulin inhibitory peptide (20 amino acids long) contained a caged tyrosine at position 5, whereas the MLCK inhibitory peptide (13 amino acids long) incorporated a caging group on Tyr-9 (both positions given relative to the amino terminus). The caged peptides were introduced into migrating eosinophil cells by microinjection. Before photolysis of the caged compounds, the eosinophils showed normal migration, demonstrating that microinjection does not alter cell behavior and that the cage was not spontaneously cleaved in the intracellular environment. Release of the cage moiety prompted a rapid (between 10 and 60 seconds) inhibition of lamellipod extension, granule flow, and net translation of cells on the substrate. In this case, global, but not local, uncaging was effective, prompting the authors to suggest that locally uncaged peptides rapidly diffuse throughout the cell, equilibrating to a concentration beneath the threshold required for activity.

Since MLCK phosphorylates only conventional myosin (II), myosin II must be involved in migratory functions. Antibodies against MLCK (Wilson et al. 1991) or myosin II (Honer et al. 1988) also disrupt vertebrate cell migration, although not as quickly as observed using photoactivated peptides.

A second example makes use of a caged version of the actin monomer-binding protein, thymosin β4, to study its role in migrating fish keratocytes (Roy et al. 2001). The caging group employed was NVOC-Cl (see Fig. 1, top) at a final labeling ratio from 2 to 4 caged (NVOC) lysine residues per thymosin β4. Lys-18 and Lys-19 residues are known to be critical for the electrostatic contacts required for proper actin binding (Vancompernolle et al. 1991, 1992; Van Troys et al. 1996). Therefore, in theory, caging either one or both of these lysine residues should prevent actin binding prior to photoactivation. In vitro, caged thymosin β4 did not alter the kinetics of actin polymerization in actin polymerization assays until the caged protein was photoactivated. After release, thymosin β4 bound actin monomers and prevented them from assembling into actin filaments. Unlike the previous example, where the uncaged peptides inhibited protein function, in this case, the protein was activated by the release of the caged moiety. The advantage of these experiments over traditional protein overexpression studies lies in the fact that the photoactivation approach permits the local and rapid alteration of intracellular protein concentration.

To study its effect in cells, caged thymosin β4 was introduced into migrating keratocytes, which have a fan-like appearance, by bead loading (see below). When the caged compound was photoactivated in the wing of the cell, this region became a pivot point, about which the cell turned (Fig. 3). Note that, in this case, the photoactivation effect was local. This appears to be due to the local availability of an actin monomer, to which the uncaged molecule rapidly binds, thereby depressing actin polymerization at a local level (Roy et al. 2001).

Condeelis and co-workers have succeeded in caging the F-actin severing factor, cofilin. The third residue from the carboxyl terminus of cofilin was mutated from serine to cysteine,

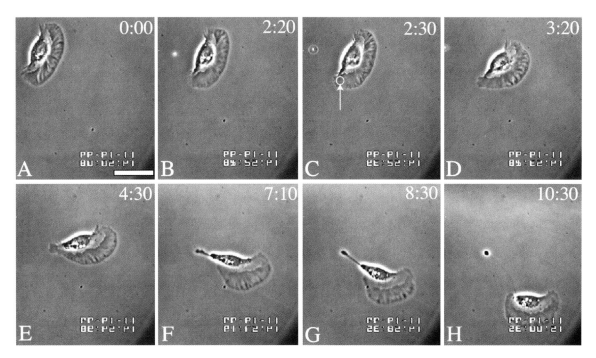

FIGURE 3. Locomoting keratocyte before and after thymosin β4 photoactivation. Time-lapse video microscopy of a locomoting keratocyte after photorelease of thymosin β4 at the wing (marked by a circle) showing dramatic turning of the cell about the site of photoactivation. *B* and *C* are the frames acquired immediately before and after photoactivation, respectively. About 8 minutes after photoactivation of thymosin β4, normal locomotion resumed (*H*). Bar, 15 μm. The experimental conditions were as follows: 3-μm beam diameter, 100X microscope objective, 10-μW laser power at the specimen plane, and 100-ms laser irradiation. (Reprinted, with permission, from Roy et al. 2001 © Rockefeller University Press.)

which could then be caged by chemical modification with α-bromo-(2-nitrophenyl) acetic acid (Ghosh et al. 2002). The caged species electrostatically resembles the inactive phosphoprotein. Upon uncaging, the cysteine-containing peptide became constitutively active because LIM-kinase could not phosphorylate the cysteine sulfhydryl. When cofilin S3 (caged)C was globally photoactivated in epidermal growth factor (EGF)-sensitive mammary carcinoma cells, the cell increased its speed of migration. Local photoactivation produced a protrusion, presumably because the newly freed barbed ends enable a burst of actin polymerization (Ghosh et al. 2004).

Using Caged Peptides/Proteins

Introduction of Peptide/Protein into Cells

In the previous sections, the design, synthesis, and characterization of caged compounds were discussed. An important aspect of these studies is the introduction of the caged peptide into cells. Three strategies can be used: cell-permeant peptide vectors, bead loading, and microinjection.

Cell-permeant peptides. Several peptide sequences can be conjugated to target peptides and proteins to facilitate their transport across the plasma membrane of intact cells. These cell-permeant peptides were originally derived from diverse protein sources, such as the DNA-binding domain of the *Drosophila* transcription factor antennepedia, the Tat transcription-activating factor of HIV-1(human immunodeficiency virus type 1), and Kaposi's sarcoma fibroblast growth factor (K-FGF) (Lindgren et al. 2000). Examples of some of the sequences that have been successfully implemented are shown in Table 2.

Bead loading. The basic idea behind the bead-loading procedure is that cells are bathed in a medium containing the peptide/protein of interest. Glass beads are then sprinkled onto the cells. The beads create temporary holes in the cell membranes, allowing some of the peptide molecules to enter cells at random (McNeil and Warder 1987). Due to the low efficiency of this technique, a large amount of protein or peptide is needed to label a relatively small number of cells. If the peptides or proteins are not fluorescent, it is useful to co-load the cells with an inert fluorescent marker, such as rhodamine dextran, to indicate successful loading. A representative protocol using fish keratocytes is given. These cells do not need to be kept at 37°C. If other cell types are used, they should be allowed to recover at 37°C and cell culture medium should be removed by washing with phosphate-buffered saline (PBS) prior to experimentation.

1. Allow fish keratocytes to migrate from 2–3 fish scales by incubating the scales on a glass coverslip or on a glass-bottomed Petri dish (MatTek Corp.) in 50 μl of protein solution (containing 4 parts caged protein solution [10–40 mg/ml] and one part indicator solution [e.g., 1 mg/ml rhodamine dextran]).

2. Sprinkle acid-washed glass beads, with a diameter of 450–600 μm (Sigma Aldrich) directly into the cells (just enough to cover the surface).

TABLE 2. Peptides that facilitate protein transport across the plasma membrane

Peptide	Sequence	Reference
Antennapedia (43–58)	RQIKIWFQNRRMKWKK	Prochiantz (1996)
TaT (48–60)	GRKKRRQRRRPPQ	Schwarze and Dowdy (2000)
K-FGF	AAVALLPAVLLALLAP	Lindgren et al. (2000)

3. Very carefully, remove the beads immediately by washing the cells twice with PBS. (PBS: 0.9 mM $CaCl_2$, 2.7 mM KCl, 1.5 mM KH_2PO_4, 0.5 mM $MgCl_2 \cdot 6H_2O$, 0.14 M NaCl, 8 mM $Na_2HPO_4 - 7H_2O$.)

> A challenging aspect of the bead-loading technique is to wash away the beads without disrupting the cells. One effective method is to carefully transfer the coverslip bearing the cells to another culture dish, thereby leaving most of the beads behind, and then performing the PBS washes.

4. After the washes, add fresh media to the cells and allow them to recover for 30 minutes before observing them under the microscope.

Microinjection protocol. The following protocol uses a semiautomated microinjection system (Eppendorf FemtoJet Microinjector along with the Eppendorf InjectMan) for the introduction of caged compounds into cells. Unlike the bead-loading approach, this technique allows the investigator to choose which cells are targeted.

1. Mix 2 μl of caged peptide (1 mM) with 0.2 μl of rhodamine dextran (2 mg/ml, adjustable according to the user's microscopy and imaging needs).

2. Pellet any debris that might clog the injection needle by centrifuging the sample in a microcentrifuge for 2–5 minutes at 10,000g.

3. Load the microinjection needle with 2 μl of the protein/rhodamine dextran mix. This volume of protein solution is enough to microinject several hundred cells.

4. Place the dish of cells to be microinjected on the microscope stand.

5. Using a 40× long-working-distance objective for visualization, gently lower the tip of the needle to 5 μm above the surface of the cells. At this point, the cells will be in focus and the needle will appear slightly out of focus.

6. Lower the capillary until the tip of the needle touches the cell as evidenced by slight cellular deformation. Set the *z* limit on the InjectMan to the desired value (here, ~100 fl).

7. Raise the needle several microns above the cell and position the tip over the region to be microinjected (nucleus or cytoplasm).

> If injecting into the cytoplasm, it is advisable to inject close to the nucleus since spreading cells flatten near their periphery, which could result in broken needles or inadequate injection.

8. Microinject the cell by pressing the "Inject" key or the button on top of the joystick.

9. If possible, return the cells to the 37°C incubator and allow them to recover for 30 minutes before proceeding to the experiments (for more detailed information see Chapter 5).

Comparison of UV Light Sources for Photoactivation

To remove the caged group from its associated compound, the sample is irradiated with UV light. This photolyzes the bond between the caging group and the compound, liberating biologically active molecules. UV light can be provided by the 365-nm line of a mercury arc lamp or a UV laser, coupled to the microscope. Both of these light sources enable photochemical uncaging inside single cells when the irradiating light is passed through pinholes of different sizes imaged onto the specimen (see below). However, a focused laser is capable of delivering a much larger light flux to the specimen resulting in shorter uncaging times (Ishihara et al. 1997).

Irradiation time required for uncaging will vary depending on the cell type and the efficiency of uncaging. For the thymosin β4 experiments, an He-Cd laser (354 nm) (or an

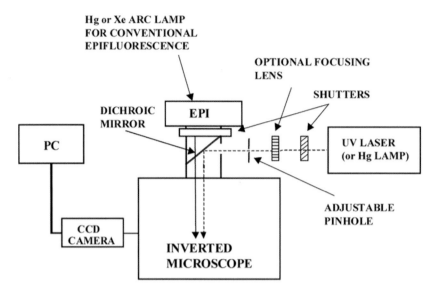

FIGURE 4. Schematic diagram of a typical wide-field microscope photoactivation setup.

argon ion multiline UV laser) was focused onto a 3-μm area on the cell surface using a 100× objective. The estimated power of the laser beam, after passing through the objective, was 10 μW, as measured using a power meter. The illumination time was 100 ms. In the experiments carried out by Walker et al. (1998), the extent of uncaging was estimated to be 45–55% by comparison with a caged fluorescein standard. The use of a caged fluorescence control, such as fluorescein, is useful not only to quantify the degree of photoactivation, but also as a control to ensure that successful uncaging is occurring with the illumination system.

Microscope and Camera Requirements

These experiments require the use of an inverted microscope that allows simultaneous epifluorescence observation and irradiation with UV light (laser or arc lamp) for photoactivation (see Fig. 4). The UV light is directed through an electromechanical shutter (Vincent Associates) to control exposure time precisely. In the case of the laser, light is weakly focused on a pinhole that is placed on a microscope image plane. The light then passes through the pinhole, which controls the width of the beam that will be brought to focus on the specimen by the microscope objective. The size of the pinhole can be varied to alter the area illuminated on the specimen (Ishihara et al. 1997). A dichroic mirror (reflection <400 nm; Chroma Technology) placed just behind the rear illumination entrance port of the microscope is used to reflect UV light into the microscope while simultaneously allowing the transmission of epifluorescence excitation light.

For epifluorescence illumination, light is passed through an electromechanical shutter and an excitation filter. In the microscope, both the UV photoactivation beam and the epifluorescence excitation light are reflected to the objective and the specimen by a dichroic mirror, which reflects all light below 500 nm. Emission is passed through an emission barrier filter before being imaged by a charged coupled device (CCD) camera. This arrangement of the filter and dichroic mirror allows the investigator to monitor the specimen and introduce the uncaging beam simultaneously. It should be noted that this type of experiment could be readily conducted with a UV laser-scanning confocal microscope that has the software to scan the beam in a small region of interest.

Environmental Chamber

To observe the effects of uncaging on cell behavior, it is necessary to keep the cells on the microscope stage for extended periods of time. During the experiments, cells must be kept under physiological conditions (37°C, 5% CO_2) using an environmental chamber (Harvard Apparatus). The chamber, which is temperature- and CO_2-controlled, is mounted on the microscope stage.

Image Analysis

Images are acquired with a cooled CCD camera (e.g., Hamamatsu model C4880) driven by image acquisition software (e.g., Metamorph, IPLab, ImagePro etc.). Changes in cell morphology or cell motility can occur over extended periods of time so the image acquisition software must allow images to be accquired at regular intervals (time-lapse acquisition module). This is a standard feature included in most advanced image acquisition software. The acquired images are then processed (e.g., increase contrast, align time lapse images) and analyzed using the image analysis software.

Design of Control Experiments

Since photoactivation involves brief irradiation of the cell with UV light, it is important to demonstrate clearly that the cell has not been damaged by the process. To do this, cells without caged peptide are irradiated for the same length of time and monitored for any differences in their behavior or morphology after UV exposure. It is also important to eliminate the possibility that the product of photoactivation, the leaving group, might be toxic to the cell. This can be done in the case of peptides, for example, by introducing an inert caged peptide into the cell. Upon photocleavage, the caging reagent and the inert peptide are released. The cell can then be tested for differences in cell morphology or behavior after uncaging, as compared to a cell that was photoactivated after receiving the uncaged inert peptide. It is also important to demonstrate that the caged biomolecule is not active either in vitro or in vivo. This is especially important in the case of caged actin (Marriott 1994).

CHROMOPHORE-ASSISTED LASER INACTIVATION

Chromophore-assisted laser inactivation (CALI) is a technique that permits targeting and specific inactivation of a protein within a mixture of other proteins either in vitro (Jay 1988; Linden et al. 1992; Liao et al. 1994) or in vivo (Chang et al. 1995; Sydor et al. 1996; Wang et al. 1996; Buchstaller and Jay 2000; Diefenbach et al. 2002; Rajfur et al. 2002). When the dye is a fluorophore and nonlaser light sources are employed for fluorophore-assisted light inactivation, the technique is known as FALI (Beck et al. 2002; Marek and Davis 2002). This technique has the potential to eliminate the activity of a protein in a localized region of the cell at a particular time. It is this precise spatiotemporal inactivation of the target protein that provides valuable information that is not attainable by other methods.

Developed by Daniel Jay and co-workers in 1988 (Jay 1988), CALI has been used to inactivate enzymes, cytoskeletal proteins, membrane proteins, transcription factors, and proteins involved in signal transduction pathways. In brief, the protein to be inactivated is either targeted by an antibody bearing a chromophore, expressed as an enhanced green fluorescent protein (EGFP) fusion protein, or expressed as a protein that will bind FlAsH (see below). To achieve inactivation, a laser (or nonlaser) source produces light at the appropriate

TABLE 3. Efficacy of CALI/FALI reagents

Reagent	Type	λ_{abs}	Effectiveness[a]
Malachite green	CALI	620 nm	++
Fluorescein	FALI	488 nm	++++
FlAsH	FALI	488 nm	++++
EGFP	FALI	488 nm	++

[a]Adapted from Surrey et al. (1998). In the case of FlAsH (Marek and Davis 2002), inactivation efficacy is assumed to be the same as fluorescein.

wavelength, which is absorbed by the chromophore. When the dye is irradiated, photo-chemical reactions produce short-lived free radicals. These highly reactive species lead to local chemical modifications that have a substantial probability of inactivating one or more functions of the protein of interest (Liao et al. 1994). If the light is focused, using a micro-scope objective, to irradiate a particular region of a cell or sample, the technique is called micro-CALI (Diamond et al. 1993).

Selection of the Protein for Inactivation

Two important considerations when designing conventional CALI experiments are the antibody and the chromophore. The antibody must be raised against an epitope of the target protein that (1) does not inhibit function upon binding and (2) is specific to the protein of interest. A potential limitation with larger proteins is that the bound anti-body could be too far away from the functional domains of the protein for inactivation—the free radicals generated upon irradiation might be unable to reach the domain of interest.

The chromophores/fluorophores that have been used for CALI/FALI to date are listed in Table 3. These molecules vary in their ability to convert absorbed energy into free rad-icals, which ultimately inactivate the target protein. In the original experiments, malachite green was the fluorophore chosen for antibody conjugation (Jay 1988). Results showed that the CALI effect was diminished in the presence of specific free radical quenchers (in the case of malachite green, hydroxyl radical quenchers). By varying the concentration of hydroxyl quenchers and correlating their effects on CALI experiments, it was calculated that the hydroxyl radicals were short–lived (10^{-12} seconds) with a half-maximal damage radius of 15 Å.

The effectiveness of malachite green, fluorescein, and GFP as fluorophores for CALI was tested directly using an in vitro β-galactosidase assay (Surrey et al. 1998). Fluorescein had a 50% higher inactivation efficiency for this enzyme than malachite green using the same anti-bodies. More recent studies show that excitation of fluorescein generates singlet oxygen with a half-maximal radius of damage of 40 Å (Beck et al. 2002). Inactivation of GFP constructs by CALI required about the same power as malachite green (Surrey et al. 1998).

Employing GFP fusion proteins has several inherent advantages over using fluorescein- and malachite-green-conjugated antibodies for CALI/FALI. The biggest advantage is that antibody is not required for GFP. In the case of cytoplasmic or nuclear targets, antibody must be introduced by bead loading, trituration loading, or microinjection. GFP constructs are relatively easy to make, are widely used, and can be introduced into cells by transfection, eliminating the need for microinjection. The main disadvantage is that the constructs may need to be overexpressed to some degree to obtain an observable effect of inactivation on cell behavior or morphology.

Examples of CALI/FALI in Cells

Genetic deletion strategies reveal that loss of a particular protein often results in cellular deficiencies. As a result of these studies, investigators know that the protein of interest is important for the trait of interest, motility, for example, but are left to determine the precise mechanistic role of the protein. CALI has proven to be a valuable technique for dissecting integrated cellular structures and processes in motile phenomena.

Due to the distinct possibility of functional redundancy and lack of isoform-specific inhibitors, mechanistic studies of the role of myosin family proteins in cells have been difficult. Jay et al. (Chang et al. 1995; Sydor et al. 1996; Wang et al. 1996; Diefenbach et al. 2002) tested the roles of various myosins, including myosin V, myosin Ic, and myosin II in neuronal growth cone motility. Using CALI to inactivate myosin II in the entire chick dorsal root ganglion, a 25% reduction in neurite outgrowth was observed. When the laser was focused to a spot covering the entire growth cone (15 μm in diameter), micro-CALI of myosin II resulted in lamellipodial retraction with no effect on filipodial dynamics. In addition, micro-CALI of myosin Ic markedly reduced retrograde F-actin flow, whereas micro-CALI of myosin II had no effect on this flow.

CALI permits selective disruption of specific proteins within a complex. Focal adhesion complexes anchor cells by providing a linkage between the extracellular matrix proteins and the actin cytoskeleton within cells. In experiments carried out by Rajfur et al. (2001), EGFP–α-actinin was expressed in Swiss-3T3 fibroblasts. When peripheral focal adhesions containing EGFP-α-actinin were irradiated with laser light, stress fibers became detached (Fig. 5). The retraction of stress fibers due to EGFP-CALI strongly suggests a functional link between α-actinin and actin-containing stress fibers at focal adhesions. In contrast, CALI of EGFP-FAK (focal adhesion kinase) in focal adhesions had no effect, although the kinase activity of FAK was diminished by CALI in vitro. This indicated that the CALI-induced stress fiber detachment was specific for EGFP α-actinin.

Studies by Surrey et al. (1998) showed the effectiveness of using fluorescein as a dye for inactivation of β-galactosidase by CALI (see Table 3). On this basis, CALI/FALI of fluorescein compared to malachite green should require a less intense source of light to achieve inactivation. In fact, using only diffuse light from a slide projector, Beck et al. (2002) tested the ability of FALI to inactivate the β1 subunit of integrin receptors of invasive cancer cells.

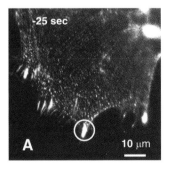 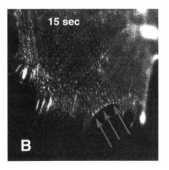 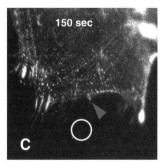

FIGURE 5. CALI of EGFP–α-actinin at focal adhesion sites. Stress fibers of Swiss-3T3 cells expressing EGFP–α-actinin detach from irradiated focal adhesion (circled in A). These stress fibers are denoted by EGFP–α-actinin densities (small bright spots on the cell image; some are indicated by arrows, (B). Irradiation took place at time 0 (between A and B). Retraction of peripheral stress fibers can be seen in C by comparing circle (site of original stress fiber connection to focal adhesion) and arrowhead. The experimental conditions: 500-mW laser power (40 mW at the specimen plane) at 488 nm for 100 ms, 2.2-μm laser beam diameter after focusing through a 100× microscope objective (Reprinted, with permission, from Rajfur et al. 2002).

FALI was performed on HT1080 cells (human fibrosarcoma cells) labeled with a fluorescein-conjugated β1 integrin antibody, migrating across a filter coated with matrigel. Invasiveness of these cells was decreased by 45% using FALI.

FlAsH (4′5′-bis(1,3,2-dithioarsolan-2-yl)fluorescein), a fluorescein derivative (Griffin et al. 1998; Gaietta et al. 2002), has recently been shown to be an effective fluorophore for FALI (Marek and Davis 2002). When placed in the cell medium, FlAsH-EDT$_2$ is nonfluorescent. However, this membrane-permeable compound becomes fluorescent inside the cell upon binding to a tetracysteine motif. Recombinant proteins to be targeted are epitope-tagged with the tetracysteine motif, so that they bind FlAsH after it has been administered extracellularly. As a result, proteins are labeled inside cells by an extracellular administration of FlAsH-EDT$_2$. The advantages of this technique are that it is noninvasive, eliminates the use of antibody constructs, and allows precise localization of the flurophore, since the tetracysteine motif is encoded within the gene for the protein of interest. Since inactivation of target protein is directly related to its distance from the fluorophore, the absence of an antibody may bring the fluorophore closer to the targeted protein, thus enhancing the likelihood of inactivation. Marek and Davis (2002) used the FlAsH-FALI strategy to analyze the role of synaptotagmin I at *Drosophila* neuromuscular junctions. Synaptotagmin had been previously shown to have a role in neurotransmitter release by analysis of gene knockouts in *Drosophila* and *Caenorhabditis elegans* (Littleton et al. 1993; Nonet et al. 1993). Electrophysiology, combined with FlAsH-FALI, of synaptotagmin demonstrated that a decrease in the release of calcium-dependent neurotransmitter at the neuromuscular junction was the result of a defect in the post-docking step of vesicle fusion.

CALI/FALI EXPERIMENTAL PROTOCOLS

In Vitro Assays for CALI/FALI

The main objective of CALI/FALI is specific inactivation of a target protein within a group of proteins. After generating the chromophore-antibody complex (or GFP construct), but before introducing it into cells, it may be beneficial to characterize the inactivation using purified proteins in vitro. First, a measure of the protein's intrinsic activity must be obtained. Next, protein activity must be assayed when it is linked to the CALI/FALI reagent. (If an antibody conjugated with malachite green or fluorescein is used, the binding of the antibody alone should not appreciably alter the protein's activity.) Finally, the decrease in activity after CALI must be measured. Although in vitro conditions are different from the cellular environment, in vitro assays may provide useful guidelines for obtaining maximal inactivation inside cells such as irradiation power and time (i.e., energy dose). To eliminate irradiation alone as the source of inactivation, samples containing antibody, but no chromophore, should also be subjected to the same laser irradiation. As a final control, irradiation of an irrelevant protein labeled with the same chromophore/fluorophore should not alter the activity of the target protein.

Introduction of Antibody/Fluorophore into Cell

CALI procedure requires the presence of the inactivating chromophore molecules in the cell. Different methods can be used to introduce the chromophore-conjugated antibody molecules into the cell e.g., bead loading, microinjection, electroporation, transfection mediated by chemical agents, and expression of protein-fluorophore chimera constructs in the case of EGFP and FlAsH. Bead loading and microinjection protocols were described previously (see section above,

Using Caged Peptides/Proteins). Electroporation utilizes an electrical pulse to create transient holes in the cell membrane, which allow the molecules of interest to pass into the cell. The method is efficient (in some cases ~80% of cells can be loaded), but some cell types are susceptible to damage caused by the electric field.

A more convenient way to introduce a chromophore into the cell (in this case, EGFP or FlaSH) is to express it inside the cell using traditional transfection techniques, mediated by chemical agents. Unlike bead loading or microinjection, this technique is noninvasive and may be less traumatic to the cells. The loading efficiency (in this case, judged from the protein expression level) can be adjusted by altering the transfection conditions.

Irradiation (Light Sources, Power)

Laser light of varying wavelengths has been used in most cases of CALI/FALI, but filtered sources, including a mercury arc lamp or even a slide projector lamp, have also been employed (Beck et al. 2002; Marek and Davis 2002). The most important criterion in choosing the light source is matching the wavelengths emitted by the source to the absorption spectrum of the CALI/FALI chromophore. To ensure optimal inactivation, when CALI/FALI requires the excitation of as many chromophores as possible in a subcellular area (less than 10 μm in diameter), it is necessary to use a focused laser, which delivers a high light energy density.

Microscope Requirements

Most CALI/FALI procedures are performed at the single cell or tissue level, and thus require the use of a microscope. An inverted microscope is convenient when CALI/FALI procedure requires the use of laser light. It is fairly easy to introduce the laser beam through an additional input/output optical port, guide it through the conventional epifluorescence path, and focus the beam on the observed sample (Fig. 6). Additional equipment, including a set of

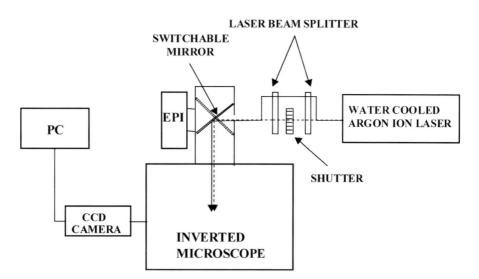

FIGURE 6. Schematic diagram of CALI/FALI experimental setup for EGFP/fluorescein. The setup is similar to that used in Figure 4, except that an argon ion laser operating at 488 nm is used. A laser beam-splitter divides the initial laser beam into two beams. One is used as a probing (or visualizing) beam and carries only 0.01% of the initial power (*solid line*). The second is the irradiation beam (*dashed line*). A fast electromechanical shutter controls the duration of irradiation. These two beams are then combined in one optical path for CALI irradiation so that the probe beam and CALI beam strike exactly the same spot on the specimen (for details, see Koppel 1979).

fast shutters and guiding mirrors, is also required. Specific descriptions of experimental setups can be found in the literature (Wang et al. 1996; Buchstaller and Jay 2000; Rajfur et al. 2002). Similarly, when the CALI/FALI procedure is performed on live cells, an environmental chamber with controlled temperature and CO_2 level is necessary to keep cells viable (see section on Environment Chamber).

Controls

Since high-intensity light (laser or lamp) may cause nonspecific (nonchromophore-mediated) effects, proper illumination controls are essential. First, nonspecific effects of CALI/FALI irradiation should be checked on cells that are not loaded with chromophore to ensure that light alone is not changing cell behavior. It is equally important to check whether cells loaded with an irradiated CALI/FALI chromophore—**not** bound to the protein of interest (the chromophore could be loaded as a dextran conjugate or expressed in a different, inconsequential protein)—do not show changes in cell behavior. The illumination controls must be performed in addition to the protein activity assays with/without antibody (see section on In Vitro Assays for CALI/FALI). Only after showing that the controls do not exhibit any changes is it safe to assume that changes, observed after CALI/FALI treatment, are caused by specific inactivation of the protein of interest in live cells.

ACKNOWLEDGMENTS

This work was supported by a National Institutes of Health Cell Migration Consortium grant (IK54 GM-64346) and a Lineberger Comprehensive Cancer Center fellowship.

REFERENCES

Adams S.R. and Tsien R.Y. 1993. Controlling cell chemistry with caged compounds. *Annu. Rev. Physiol.* **55:** 755–784.

Bain J.D., Glabe C.G., Dix T.A., and Chamberlain A.R. 1989. Biosynthetic site-specific incorporation of a non-natural amino acid into a polypeptide. *J. Am. Chem. Soc.* **111:** 8013–8014.

Beck S., Sakurai T., Eustace B.K., Beste G., Schier R., Rudert F., and Jay D.G. 2002. Fluorophore-assisted light inactivation: A high-throughput tool for direct target validation of proteins. *Proteomics* **2:** 247–255.

Buchstaller A. and Jay D.G. 2000. Micro-scale chromophore-assisted laser inactivation of nerve growth cone proteins. *Microsc. Res. Tech.* **48:** 97–106.

Chang H.Y., Takei K., Sydor A.M., Born T., Rusnak F., and Jay D.G. 1995. Asymmetric retraction of growth cone filopodia following focal inactivation of calcineurin. *Nature* **376:** 686–690.

Diamond P., Mallavarapu A., Schnipper J., Booth J., Park L., O'Connor T.P., and Jay D. G. 1993. Fasciclin I and II have distinct roles in the development of grasshopper pioneer neurons. *Neuron* **11:** 409–421.

Diefenbach T.J., Latham V.M., Yimlamai D., Liu C.A., Herman I.M., and Jay D.G. 2002. Myosin 1c and myosin IIB serve opposing roles in lamellipodial dynamics of the neuronal growth cone. *J. Cell Biol.* **158:** 1207–1217.

Ellis-Davies G.C. and Kaplan J.H. 1994. Nitrophenyl-EGTA, a photolabile chelator that selectively binds Ca^{2+} with high affinity and releases it rapidly upon photolysis. *Proc. Natl. Acad. Sci.* **91:** 187–191.

Gaietta G., Deerinck T.J., Adams S.R., Bouwer J., Tour O., Laird D.W., Sosinsky G.E., Tsien R.Y., and Ellisman M.H. 2002. Multicolor and electron microscopic imaging of connexin trafficking. *Science* **296:** 503–507.

Ghosh M., Ichetovkin I., Song X., Condeelis J.S., and Lawrence D.S. 2002. A new strategy for caging proteins regulated by kinases. *J. Am. Chem. Soc.* **124:** 2440–2441.

Ghosh M., Song X., Mouneimne G., Sidani M., Lawrence D.S., and Condeelis J.S. 2004. Cofilin promotes actin polymerization and defines the direction of cell motility. *Science* **304:** 743–746.

Griffin B.A., Adams S.R., and Tsien R.Y. 1998. Specific covalent labeling of recombinant protein molecules inside live cells. *Science* **281**: 269–272.

Honer B., Citi S., Kendrick-Jones J., and Jockusch B.M. 1988. Modulation of cellular morphology and loco-motory activity by antibodies against myosin. *J. Cell Biol.* **107**: 2181–2189.

Ishihara A., Gee K., Schwartz S., Jacobson K., and Lee J. 1997. Photoactivation of caged compounds in single living cells: An application to the study of cell locomotion. *BioTechniques* **23**: 268–274.

Jay D.G. 1988. Selective destruction of protein function by chromophore-assisted laser inactivation. *Proc. Natl. Acad. Sci.* **85**: 5454–5458.

Kaplan J.H., Forbush B., 3rd, and Hoffman J.F. 1978. Rapid photolytic release of adenosine 5′-triphosphate from a protected analogue: Utilization by the Na:K pump of human red blood cell ghosts. *Biochemistry.* **17**: 1929–1935.

Koppel D.E. 1979. Fluorescence redistribution after photobleaching. A new multipoint analysis of membrane translational dynamics. *Biophys. J.* **28**: 281–291.

Liao J.C., Roider J., and Jay D.G. 1994. Chromophore-assisted laser inactivation of proteins is mediated by the photogeneration of free radicals. *Proc. Natl. Acad. Sci.* **91**: 2659–2663.

Linden K.G., Liao J.C., and Jay D.G. 1992. Spatial specificity of chromophore assisted laser inactivation of protein function. *Biophys. J.* **61**: 956–962.

Lindgren M., Hallbrink M., Prochiantz A., and Langel U. 2000. Cell-penetrating peptides. *Trends Pharmacol. Sci.* **21**: 99–103.

Littleton J.T., Stern M., Schulze K., Perin M., and Bellen H.J. 1993. Mutational analysis of *Drosophila* synaptotagmin demonstrates its essential role in Ca(2+)-activated neurotransmitter release [comment]. *Cell* **74**: 1125–1134.

Marek K.W. and Davis G.W. 2002. Transgenically encoded protein photoinactivation (FlAsH-FALI): Acute inactivation of synaptotagmin I. *Neuron* **36**: 805–813.

Marriott G. 1994. Caged protein conjugates and light-directed generation of protein activity: Preparation, photoactivation, and spectroscopic characterization of caged G-actin conjugates. *Biochemistry* **33**: 9092–9097.

Marriott G. and Heidecker M. 1996. Light-directed generation of the actin-activated ATPase activity of caged heavy meromyosin. *Biochemistry* **35**: 3170–3174.

Marriott G. and Walker J.W. 1999. Caged peptides and proteins: New probes to study polypeptide function in complex biological systems. *Trends Plant Sci.* **4**: 330–334.

McCray J.A. and Trentham D.R. 1989. Properties and uses of photoreactive caged compounds. *Ann. Rev. Biophys. Biophys. Chem.* **18**: 239–270.

McNeil P.L. and Warder E. 1987. Glass beads load macromolecules into living cells. *J. Cell Sci.* **88**: 669–678.

Mendel D., Ellman J.A., Chang Z., Veenstra D.L., Kollman P.A., and Schultz P.G. 1992. Probing protein stability with unnatural amino acids. *Science* **256**: 1798–1802.

Merrifield B. 1997. Concept and early development of solid-phase peptide synthesis. *Methods Enzymol.* **289**: 3–13.

Nonet M.L., Grundahl K., Meyer B.J., and Rand J.B. 1993. Synaptic function is impaired but not eliminated in *C. elegans* mutants lacking synaptotagmin. *Cell* **73**: 1291–1305.

Noren C.J., Anthony-Cahill S.J., Griffith M.C., and Schultz P.G. 1989. A general method for site-specific incorporation of unnatural amino acids into proteins. *Science* **244**: 182–188.

Pan P. and Bayley H. 1997. Caged cysteine and thiophosphoryl peptides. *FEBS Lett.* **405**: 81–85.

Prochiantz A. 1996. Getting hydrophilic compounds into cells: Lessons from homeopeptides. *Curr. Opin. Neurobiol.* **6**: 629–634.

Rajfur Z., Roy P., Otey C., Romer L., and Jacobson K. 2002. Dissecting the link between stress fibres and focal adhesions by CALI with EGFP fusion proteins. *Nat. Cell Biol.* **4**: 286–293.

Rothman D.M., Vazquez M.E., Vogel E.M., and Imperiali B. 2002. General method for the synthesis of caged phosphopeptides: Tools for the exploration of signal transduction pathways. *Org. Lett.* **4**: 2865–2868.

Roy P., Rajfur Z., Jones D., Marriott G., Loew L., and Jacobson K. 2001. Local photorelease of caged thymosin beta4 in locomoting keratocytes causes cell turning. *J. Cell Biol.* **153**: 1035–1048.

Schwarze S.R. and Dowdy S.F. 2000. In vivo protein transduction: Intracellular delivery of biologically active proteins, compounds and DNA. *Trends Pharmacol. Sci.* **21**: 45–48.

Shigeri Y., Tatsu Y., and Yumoto N. 2001. Synthesis and application of caged peptides and proteins. *Pharmacol. Therapeut.* **91**: 85–92.

Surrey T., Elowitz M.B., Wolf P.E., Yang F., Nedelec F., Shokat K., and Leibler S. 1998. Chromophore-assisted light inactivation and self-organization of microtubules and motors. *Proc. Natl. Acad. Sci.* **95:** 4293–4298.

Sydor A.M., Su A.L., Wang F.S., Xu A., and Jay D.G. 1996. Talin and vinculin play distinct roles in filopodial motility in the neuronal growth cone. *J. Cell Biol.* **134:** 1197–1207.

Tatsu Y., Shigeri Y., Ishida A., Kameshita I., Fujisawa H., and Yumoto N. 1999. Synthesis of caged peptides using caged lysine: Application to the synthesis of caged AIP, a highly specific inhibitor of calmodulin-dependent protein kinase II. *Bioorgan. Med. Chem. Lett.* **9:** 1093–1096.

Tatsu Y., Shigeri Y., Sogabe S., Yumoto N., and Yoshikawa S. 1996. Solid-phase synthesis of caged peptides using tyrosine modified with a photocleavable protecting group: Application to the synthesis of caged neuropeptide Y. *Biochem. Biophys. Res. Commun.* **227:** 688–693.

Vancompernolle K., Vandekerckhove J., Bubb M.R., and Korn E.D. 1991. The interfaces of actin and Acanthamoeba actobindin. Identification of a new actin-binding motif. *J. Biol. Chem.* **266:** 15427–15431.

Vancompernolle K., Goethals M., Huet C., Louvard D., and Vandekerckhove J. 1992. G- to F-actin modulation by a single amino acid substitution in the actin binding site of actobindin and thymosin beta 4. *EMBO J.* **11:** 4739–4746.

Van Troys M., Dewitte D., Goethals M., Carlier M.-F., Vandekerckhove J., and Ampe C. 1996. The actin binding site of thymosin beta 4 mapped by mutational analysis. *EMBO J.* **15:** 201–210.

Walker J.W., Gilbert S.H., Drummond R.M., Yamada M., Sreekumar R., Carraway R.E., Ikebe M., and Fay F.S. 1998. Signaling pathways underlying eosinophil cell motility revealed by using caged peptides. *Proc. Natl. Acad. Sci.* **95:** 1568–1573.

Wang F.S., Wolenski J.S., Cheney R.E., Mooseker M.S., and Jay D.G. 1996. Function of myosin-V in filopodial extension of neuronal growth cones. *Science* **273:** 660–663.

Wilson A.K., Gorgas G., Claypool W.D., and de Lanerolle P. 1991. An increase or a decrease in myosin II phosphorylation inhibits macrophage motility. *J. Cell Biol.* **114:** 277–283.

Wood J.S., Koszelak M., Liu J., and Lawrence D.S. 1998. A caged protein kinase inhibitor. *J. Am. Chem. Soc.* **120:** 7145–7146.

Zou K., Miller W.T., Givens R.S., and Bayley H. 2001. Caged thiophosphotyrosine peptides. *Angew. Chem. Int. Ed. Engl.* **40:** 3049–3051.

Photoactivation-based Labeling and In Vivo Tracking of RNA Molecules in the Nucleus

Joan C. Ritland Politz,[a] Richard A. Tuft,[b] and Thoru Pederson[a]

[a]Program in Cell Dynamics and Department of Biochemistry and Molecular Pharmacology;
[b]Department of Physiology, University of Massachusetts Medical School,
Worcester, Massachusetts 01605

INTRODUCTION

THIS CHAPTER DESCRIBES A RECENTLY DEVELOPED METHOD for observing and measuring the movement of RNA molecules in the nucleus of living mammalian cells. RNA traffic in the cell has been studied primarily to identify the cytoplasmic sites to which certain mRNAs are delivered and the mechanisms that are involved (Richter 2001; Farina and Singer 2002; Kloc et al. 2002). But before they enter the cytoplasmic highways, RNAs are born in the crowded nucleus and must move to the nuclear pores for export (Politz and Pederson 2000). Several methods have been developed for studying the intranuclear transport of RNA in living cells (for review, see Pederson 2001), one of which (Politz 1999; Politz et al. 1999) is described in detail here. This method is presented comprehensively, including technical details, so that investigators will be able to apply it to their own particular systems.

HYBRIDIZATION OF DNA OLIGONUCLEOTIDES TO RNA IN VIVO

Given the temperature and ionic strength within mammalian cells, the well-developed principles of nucleic acid intermolecular association give little reason to doubt that short (30–43-nucleotide) oligodeoxynucleotides would hybridize to a complementary RNA target if a sufficient concentration of each were present and the particular RNA region being targeted was sufficiently unstructured and unmasked by bound proteins. Indeed, DNA oligonucleotides as short as 13 nucleotides were reported to bind complementary RNA sequences in Rous sarcoma virus (RSV)-infected chick embryo fibroblasts over 25 years ago (Zamecnik and Stephenson 1978), and much work reporting the "antisense effect," usually characterized as an inhibition of protein synthesis, was reported in ensuing years. Direct evidence that antisense DNA oligonucleotides hybridize with complementary RNA sequences in living cells was provided by the demonstration that oligo(dT)$_{43}$, taken up by cultured mammalian cells, becomes bound to sites (i.e., poly[A]) at which it can prime cDNA synthesis by reverse transcriptase

A

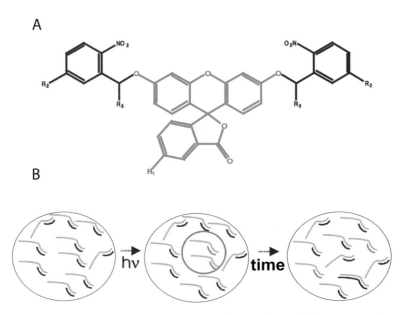

B

FIGURE 1. (*A*) Structure of caged fluorescein. R1, R2, and R3 indicate different groups in various caged fluoresceins that have been synthesized (Mitchison et al. 1998). (*B*) Cartoon showing uncaging strategy. The black lines represent caged (nonfluorescent) oligonucleotides hybridized to RNA molecules (*gray*) inside the nucleus. The blue circle represents an uncaging site in which the green lines represent uncaged oligonucleotides hybridized to RNA. These molecules move out from the uncaging site over time (*right side of diagram*). (Modified, with permission, from Politz 1999.)

(Politz et al. 1995). Although it remains difficult to predict effective hybridization targets in a given RNA based solely on computational or even empirically determined secondary structures (e.g., see Stull et al. 1992; Matthews et al. 1999), with careful application and characterization, this technique has proved to be very useful in RNA tracking studies in live cells.

THE NEXT STEP: A CAGED FLUOROPHORE

In 1988, Krafft et al. described the synthesis of caged fluorescein compounds in which the fluorescein ring system is locked into a nonfluorescent tautomer by virtue of attached *o*-nitrobenzene groups (Fig. 1A). The ether bonds that link these *o*-nitrobenzene groups to the fluorescein ring are photolabile. Irradiation with 360-nm light cleaves the ether bonds, and the fluorescein then rapidly adopts its fluorescent (xanthen-3-one) form. Mitchison (1989) described the synthesis and first use of a caged fluorescein in a biological application, a study of microtubule dynamics. The method described here uses caged fluorescein-labeled DNA oligonucleotides as hybridization tags to track the movements of RNAs in cells. The labeled oligonucleotides are introduced into living mammalian cells, where they demonstrably hybridize to complementary RNA. After site-specific photoactivation at desired sites within the cell, the RNA movements away from those sites are followed (Fig. 1B) and digitally recorded using a rapid acquisition microscopy system developed at our institution.

CELLS

The experiments upon which this method is based were carried out primarily using the rat L6 myoblast cell line growing in Dulbecco's modified Eagle's medium (DMEM) containing

10% fetal bovine serum, but HeLa cells were also used in some work. Cells were plated onto 25-mm round coverslips, placed into small, 30-mm Petri dishes, and grown overnight to about 60% confluency. The cells are then transfected with oligodeoxynucleotides, as described below.

OLIGODEOXYNUCLEOTIDE PROBES

Oligodeoxynucleotides can be obtained from a commercial vendor (i.e., Integrated DNA Technologies) with amino-modified thymidines placed at predetermined positions in the sequence. In this work, the thymidines used were modified at carbon 5 with a hexylamine group (amino-modifier C6 dT; Integrated DNA Technology or Glen Research Corp.). The length of the linker arm seems to be important for stability of the final labeled oligonucleotide/RNA hybrid in the cell; this may be because the hexylamine lies in the RNA/DNA groove and physically interferes with RNase H binding (Ueno et al. 1997). The choice of the target RNA sequence and oligonucleotide design are important determinants of experimental success and should be given due attention. Target sequences most likely to be effective can be judged by several criteria, including the availability of an RNA site to the solvent environment (e.g., chemical mapping studies may have been carried out with the RNA of interest), the presence of protein-binding sites, and the RNA secondary structure (i.e., the ribonucleoprotein structure may be available, RNA footprinting studies may have been done, or the crystal structure of the RNA may have been solved). Two other key criteria in oligonucleotide design are the G+C content of the sequence (regions that are <50% G+C should be chosen), and its uniqueness (the chosen oligonucleotide cannot be one that has the potential to cross-hybridize with other cellular RNAs). Finally, the amino-modified thymidines should be positioned approximately every ten nucleotides, partly to avoid intermolecular self-quenching of the fluorescein, but also, insofar as possible, to ensure that they are complementary to adenosines in the RNA target. In the case of an RNA target sequence that does not contain properly positioned adenosines, the modified thymidines may be placed at noncomplementary sites, but mismatches of more than two bases in the exact antisense sequence will decrease hybridization efficiency to an unacceptable level.

COUPLING OF OLIGONUCLEOTIDES WITH CAGED FLUORESCEIN

NOTE: <!> indicates hazardous materials. See Cautions Appendix.

Caged fluorescein succidimidyl esters show different rates of hydrolysis and aggregation in aqueous solution, depending on the chemistry of both the amino-reactive moieties and the caging groups. The compounds should therefore be stored desiccated, either as a solid or in anhydrous dimethylsulfoxide (DMSO), at −80°C, and suspended in aqueous solution just before the coupling reaction. The caged compounds should also be protected from light. The authors have exclusively used succinimidyl-ester-modified caged fluorochromes, which react well with amino-modified thymidines, in particular, CMNB2-AF-*N*-hydroxysuccinimide (Mitchison et al. 1998; OANB2AF in Mitchison et al. 1994). This particular reaction proceeds best in basic solution: 0.2 M sodium bicarbonate buffer (pH 8.9).

1. Dissolve a total of 100 nmoles of modified amino groups (e.g., 25 nmoles of oligonucleotide if there are four amino-modified thymidines per oligonucleotide) in sterile, deionized H_2O (~130–140 µl).

2. Dissolve 1 mg of the caged fluorescein<!> in anhydrous DMSO<!> (usually 10–20 µl) and mix this with the amino group solution (prepared in Step 1) and 40 µl of 1.0 M sodium bicarbonate buffer (pH 9.0), to give a total volume of 200 µl.

> The water solubility of caged fluoresceins varies (see Mitchison et al. 1994), so if the compound precipitates, add more DMSO to the reaction mixture. However, the volume of DMSO used should not exceed one third of the total volume (also see www.molecularprobes.com and Politz and Singer 1999). Allow the reaction to proceed for 12–24 hours at room temperature in the dark.

3. Prepare a Sephadex G-50 column in a 25-ml pipette and buffer it with 10 mM triethylammonium bicarbonate (pH 8.5). Load the reaction mixture onto the column.

4. Collect 1-ml fractions in microcentrifuge tubes and then transfer 10–20-µl aliquots from each fraction to fresh microcentrifuge tubes.

5. Determine which of the fractions contain labeled oligonucleotide by uncaging and exciting the fluorescein: Stand the aliquots in an open-bottomed tube rack and place the rack on a UV light box<!> (e.g., FOTO/UV 300 Ultraviolet Transilluminator, Fotodyne, 312-nm light). Alternatively, use a hand-held long-wavelength UV lamp<!> (e.g., Model UVGL-58 [UVP], long wavelength = 366 nm). The presence of the oligonucleotide is indicated by green fluoresence.

> Operationally, both UV sources have adequate uncaging and fluorescein excitation light for the in vitro assay of uncaging. Using this method, uncaging (to give green fluorescent color) through the UV semitranslucent polypropylene tube walls can take several seconds, but it is a convenient way to determine which fractions contain the labeled oligonucleotide.

6. Lyophilize the oligonucleotide-containing fractions and resuspend the labeled product in deionized H$_2$O (usually 100 µl).

7. Check the purity of the oligonucleotide by subjecting 2–3 µg to electrophoresis on a 10–12% (w/v) denaturing polyacrylamide<!> gel. Label the oligonucleotide of interest with standard fluorescein (as described here) and run it on the same gel as a marker.

> The standard oligonucleotide will fluoresce immediately when the gel is placed on the UV box described in Step 4. The caged oligonucleotide should run as a single band (assuming a single oligonucleotide was labeled) that is initially nonfluorescent and then becomes fluorescent over time (seconds). If two bands (or more) are present with different uncaging characteristics, it is likely that the sample is contaminated with caged fluorescein aggregates of large size. Bands that contain DNA can be identified on these gels by placing the gel on a fluor-impregnated TLC plate and visualizing the dark (absorbing) DNA bands using a short-wavelength UV hand-held lamp (e.g., Model UVGL-58 (UVP), short wavelength = 254 nm).

8. Remove contaminants by precipitating the oligonucleotide in 3 volumes of cold acetone<!> (rather than ethanol) and 0.1 volume 3 M ammonium acetate. If this is not effective, purify the oligonucleotide by preparative gel electrophoresis.

> Aggregates must be removed or they will cause spurious results in live cell experiments.

9. Uncage an aliquot of the purified oligonucleotide by transferring it to a microfuge tube and lying it directly on the UV box for 30 minutes.

> This treatment provides close to complete uncaging.

10. Determine the concentration of the oligonucleotide by measuring absorbance at 260 nm. This estimate will be high if the uncaging is incomplete. Determine the number of fluoresceins per molecule by reading the absorbance at 488 nm and then calculating the molar ratio of fluorescein to oligonucleotide (the extinction coefficient for fluorescein is 68,000 cm^{-1} M^{-1}).

11. If necessary, determine the rate and degree of uncaging over time using a fluorimeter.

INTRODUCING CAGED-FLUORESCEIN OLIGONUCLEOTIDES INTO CELLS

In most of the our studies using this method, oligonucleotides are complexed with a cationic lipid to facilitate their entry into cells. Note, however, that detectable levels will enter unaided if the oligonucleotide is added to the medium alone at an appropriate concentration (1 μM or more) (Politz et al. 1995; Politz and Singer 1999). The choice of cationic lipid is important, and the ratio of lipid to oligonucleotide must be optimized for each new oligonucleotide mix, following the manufacturer's instructions for the particular cationic lipid in use. Currently, we use Lipofectamine 2000 (GIBCO-BRL), but Tfx-50 (Promega) and Pfx-6 (discontinued when Invitrogen merged with GIBCO-BRL) have also been used with excellent success. Oligofectamine (GIBCO-BRL) gave low uptake of oligonucleotides in our initial trials.

The following conditions provide optimal levels of oligo(dT) hybridization to poly(A) RNA in the cell, as evaluated using in situ reverse transcription (Politz et al. 1995; Politz and Singer 1999) and fluorescence correlation spectroscopy (Politz et al. 1998).

1. Complex 4.5 μg of oligo(dT) or oligo(dA) 43 mer (Politz et al. 1999) or 9.75 μg of 33-mer anti-28S rRNA oligonucleotides (for sequence, see Politz et al. 2003) with 6 μl of Lipofectamine 2000 according to manufacturer instructions.

2. Add this complex to the cells (growing on coverslips in 30-mm Petri dishes) in a final volume of 1.5 ml of Opti-MEM (Optimized modified Eagle's medium, GIBCO-BRL).

3. Incubate the cells with the oligonucleotide for 2 hours and then in fresh medium (without oligonucleotide) for a further 30 minutes to 1 hour.

   This treatment will facilitate efflux of internalized oligonucleotide that has not become bound within the cell.

PHOTOACTIVATION AND TRACKING OF RNAs

There are many ways to effectively image live cells on the microscope stage (see Chapter 17). We routinely use the following technique.

1. Mount cells growing on coverslips in a chamber suitable for use on a microscope stage.

   We mount coverslips between two metal rings sealed with a gasket to form a chamber above the coverslip.

2. Heat the chamber (metal rings) with circulating water to maintain the cells at 37°C.

3. Use Leibovitz's L15 medium (GIBCO) containing 10% serum (no phenol red) or DMEM buffered with HEPES (no phenol red) at this stage of the experiments.

4. Mount the chamber on the stage of an inverted epifluorescence microscope (Fig. 2) (Rizzuto et al. 1998) that incorporates a high-speed focus drive, wide-field laser illumination<!>, and a small-format (128 × 128 pixels) frame-transfer CCD camera (MIT Lincoln Laboratory).

   All elements of the photolysis/imaging protocols (laser shutters, camera control, piezoelectric focus drive) are under the control of custom software running under Windows on a Pentium PC. The camera is thermoelectrically cooled, has ~70% quantum efficiency in the visible range, and has a readout noise of 6.5 electrons rms (root mean square).

5. Identify the nucleus to be uncaged in bright field (or use a phase-contrast lens and condenser) and move the region to be uncaged to the center of the field using standard stage controls.

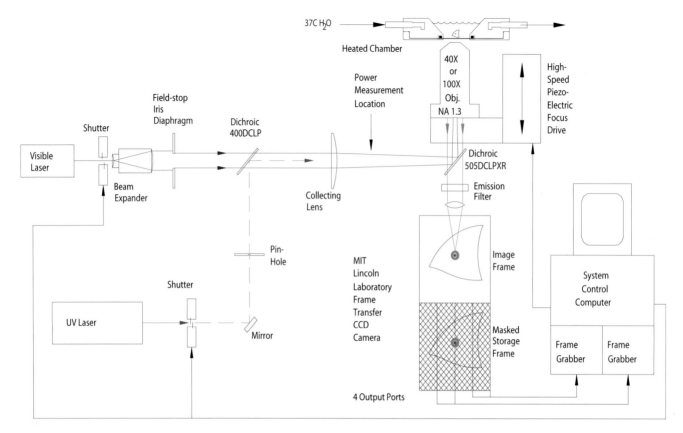

FIGURE 2. Diagram of imaging system.

If diffusion times are to be measured, it is helpful to choose a site near the center of the nucleus so that the distance the signal travels before encountering (and being slowed by) the nuclear membrane is maximal.

6. Uncage the signal.

The caged fluorescein is uncaged by the 351- and 364-nm lines of an argon laser (Coherent). The unexpanded UV beam, shuttered by a second LS3, is incident on a pinhole (Edmund Scientific, 35-μm diameter for 40× objective, 100-μm diameter for 100× objective) and combined with a 488-nm beam (see below and Fig. 2) by a 400DCLP dichroic mirror (Chroma, Inc.). The pinhole is mounted in a 3-axis positioner that allows it to be focused on the cell and centered in the field of the camera. To ensure consistency in our uncaging power, we measure (1815-C power meter, 818-UV detector; Newport Corp.) the UV power in the beam at a position between the collecting lens and the epifluorescence dichroic (see Fig. 2) several times during the course of an experiment. Uncaging is routinely carried out for 65 ms at a measured beam power of 90–120 μW (~1.6 kW/cm^2 in the focused spot on the sample).

7. Collect and integrate fluorescence images of the resulting uncaged signal over time on the image frame of the CCD using exposure times ranging from 2 to 60 ms (operator controllable).

Either 60 two-dimensional images (usually 10-ms exposure time with 500 ms between exposures) or 5 three-dimensional image stacks of 21 planes each (0.25-μm sections, 3–5 ms exposure time) are captured at 500-ms intervals after uncaging. The camera used to optimize this procedure was thermoelectrically cooled, had ~70% quantum efficiency in the visible range, and had a readout noise of 6.5 electrons rms. Uncaged fluorescein is excited by the 488-nm line of an argon-krypton laser (Coherent Inc.) that passes through a fast laser shutter (LS3, Vincent Assoc.), a beam expander, and a field-stop iris diaphragm (see Fig. 2). The beam then passes through a

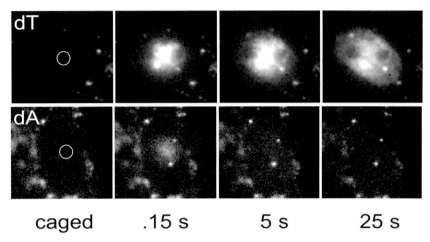

caged   .15 s   5 s   25 s

FIGURE 3. Movement of poly(A) RNA away from nuclear uncaging site. (*Top*) Distribution of uncaged oligo(dT) signal (bound to poly[A] RNA) in the L6 myoblast nucleus over time. (*Bottom*) Distribution of uncaged oligo(dA) in the L6 myoblast nucleus over time. Circles represent approximate uncaging site.

fused silica collecting lens, is reflected by a 505 DCLPXR dichroic mirror (Chroma Inc.), and is concentrated on the cell to be imaged by either a 40× NA 1.3 fluor objective or 100× NA 1.3 fluor objective (Nikon). The optical configuration images the field stop onto the cell, so that the area illuminated can be limited to one slightly larger than a single cell. After the exposure is complete, the image is transferred to the CCD storage frame in 50 μs and read out from four CCD output ports through four 2.5 Mpixel/second 12-bit analog to digital converters, and then to two frame-grabber boards, comprising a total of 8 Mbytes of dual ported memory (Bitflow, Inc.). The readout process is complete in 1.8 ms.

MOVEMENT OF POLY(A) RNA

An example of the movement of poly(A) RNA labeled with uncaged oligo(dT) is shown in Figure 3 (top). The signal moves away from the site in all directions (except into the nucleolus) with an apparent diffusion coefficient of 0.6 μm squared per second (Politz et al. 1999). Figure 3 (bottom) shows the much more rapid dispersion of the unhybridized control, oligo(dA). It is important to use a control oligonucleotide in each experiment so that the movement of a free oligonucleotide is characterized under identical conditions. Changes in cell morphology, especially retraction of lamellipodia, are an indication that the cells are being affected by the light used in imaging, and a new coverslip should be mounted. It is usually possible to image cells on a coverslip for approximately 2 hours without obvious detriment to cell health (i.e., when cells are returned to standard tissue culture conditions, they divide and grow normally for at least 2–3 days).

Acquisition of three-dimensional image stacks at sequential time points after uncaging, followed by constrained iterative deconvolution (Carrington et al. 1995; see also Chapter 15), allows higher resolution of the RNA distribution pattern. If a complementary technique is used to visualize chromosomes, detailed mapping of the signal with respect to the chromatin territories can be carried out (see Chapter 25). Hoescht 33342 (Molecular Probes), which is taken up by live cells and binds double-stranded DNA, can be added to the medium during the efflux period to a final concentration of 2.5 μM (see Chapter 28) and excited using the 360-nm line of the argon laser (Politz et al. 1999). Alternatively, stable cell lines containing green fluorescent protein (GFP)-labeled histone H2B may be used if a differently colored uncaged dye is developed (see below).

THE CURRENT AVAILABILITY OF CAGED FLUOROPHORES

The development of the method described here was enabled by a generous gift from Tim Mitchison (then at the University of California, San Francisco, now at Harvard Medical School) of a soluble caged fluorescein compound (OANBF in Mitchison et al. 1994; CMNB in Mitchison et al. 1998). The authors are currently exploring the further synthesis of this compound. A similar (but not identical) caged fluorescein is available from Molecular Probes (C-20050), but unfortunately, it has not so far proved ideal for current applications, due to aggregation problems that cause unacceptably low yields. The field of cell biology in general, and the study of intracellular nucleic acid dynamics in particular, would greatly benefit from the development of caged fluorophores that are easy to synthesize, possess high water solubility, low rates of spontaneous uncaging, rapid and high efficiencies of uncaging, and high fluorescence intensities (for a more detailed list of desirable characteristics, see Mitchison et al. 1998).

CONCLUSION: A MOVING STORY

In addition to the procedure described in this chapter, there are other methods to observe the movement of RNA in living cells. One very useful technique involves the microinjection of fluorescently labeled RNAs into cells (Jacobson and Pederson 1997; Pederson 2001). Another innovative method to study RNA dynamics is described in Chapter 35 in this volume. The use of independent methods, and the cross-confirmatory results obtained, will be important in this early stage of our understanding of RNA traffic in the cell.

ACKNOWLEDGMENTS

The authors are deeply indebted to Timothy Mitchison for his generous gift of caged fluorescein and thoughtful discussions. Without his help, this line of research would have been very difficult to pursue. The collaborative role of Robert Singer during the early phases of this method's development is warmly acknowledged, as is the key support of our late colleague Frederic Fay at the authors' institution. This work was supported by National Institutes of Health grants GM-21595 and GM-60551; early stages of this work were supported by NIH NRSA postdoctoral fellowship AR-08361 to J.C.R.P.

REFERENCES

Carrington W.A., Lynch R.M., Moore E.D., Isenberg G., Fogarty K.E., and Fay F.S. 1995. Superresolution three-dimensional images of fluorescence in cells with minimal light exposure. *Science* **268**: 1483–1487.

Farina K. and Singer R.H. 2002. The nuclear connection in RNA transport and localization. *Trends Cell Biol.* **12**: 466–472.

Jacobson M.R. and Pederson T. 1997. RNA traffic and localization reported by fluorescent molecular cytochemistry in living cells. In *Analysis of mRNA formation and function* (ed. J.D. Richter), pp. 341–359. Academic Press, New York.

Kloc M., Zearfoss N.R., and Etkin L.D. 2002. Mechanisms of subcellular mRNA localization. *Cell* **108**: 533–544.

Krafft G.A., Sutton W.R., and Cummings R.T. 1988. Photoactivable fluorophores. 3. Synthesis and photoactivation of fluorogenic difunctionalized fluoresceins. *J. Am. Chem. Soc.* **110**: 301–303.

Matthews D.H., Burkard M.E., Freier S.M., Wyatt J.R., and Turner D.H. 1999. Predicting oligonucleotide affinity to nucleic acid targets. *RNA* **4**: 1458–1469.

Mitchison T.J. 1989. Polewards microtubule flux in the mitotic spindle: Evidence from photoactivation of fluorescence. *J. Cell Biol.* **109:** 637–652.

Mitchison T.J., Sawin K.E., and Theriot J.A. 1994. Caged fluorescent probes for monitoring cytoskeleton dynamics. In *Cell biology: A laboratory handbook* (ed. J.E. Celis), vol. 3, pp. 65–74. Academic Press, New York.

Mitchison T.J., Sawin K.E., Theriot J.A., Gee K., and Mallavarapu A. 1998. Caged fluorescent probes. *Methods Enzymol.* **291:** 63–78.

Pederson T. 2001. Fluorescent RNA cytochemistry: Tracking gene transcripts in living cells. *Nucleic Acids Res.* **29:** 1013–1016.

Politz J.C. 1999. Use of caged fluorochromes to track macromolecular movement in living cells. *Trends Cell Biol.* **9:** 284–287.

Politz J.C. and Pederson T. 2000. Movement of mRNA from transcription site to nuclear pores. *J. Struct. Biol.* **129:** 252–257.

Politz J.C. and Singer R.H. 1999. In situ reverse transcription for the detection of hybridization between oligonucleotides and their intracellular targets. *Methods* **18:** 281–285.

Politz J.C., Taneja K.L., and Singer R.H. 1995. Characterization of hybridization between synthetic oligodeoxynucleotides and RNA in living cells. *Nucleic Acids Res.* **23:** 4946–4953.

Politz J.C., Tuft R.A., and Pederson T. 2003. Diffusion-based transport of nascent ribosomes in the nucleus. *Mol. Biol. Cell* **14:** 4805–4812.

Politz J.C., Browne E.S., Wolf D.E., and Pederson T. 1998. Intranuclear diffusion and hybridization state of oligonucleotides measured by fluorescence correlation spectroscopy. *Proc. Natl. Acad. Sci.* **95:** 6043–6048.

Politz J.C., Tuft R.A., Pederson T., and Singer R.H. 1999. Movement of nuclear poly(A) RNA throughout the interchromatin space in living cells. *Curr. Biol.* **9:** 285–291.

Richter D. 2001. *Cell polarity and subcellular RNA localization.* Springer Verlag, Heidelberg.

Rizzuto R., Carrington W., and Tuft R.A. 1998. Digital imaging microscopy of living cells. *Trends Cell Biol.* **8:** 288–292.

Stull R.A., Taylor L.A., and Szoka F.C., Jr. 1992. Predicting antisense oligonucleotide inhibitory efficacy: A computational approach using histograms and thermodynamic indices. *Nucleic Acids Res.* **20:** 3501–3508.

Ueno Y., Kumagai I., Haginoya N., and Matsuda A. 1997. Effects of 5-(*N*-aminohexyl)carbamoyl-2′deoxyuridine on endonuclease stability and the ability of oligodeoxynucleotide to activate RNase H. *Nucleic Acids Res.* **25:** 3777–3782.

Zamecnik P.C. and Stephenson M.L. 1978. Inhibition of Rous sarcoma virus replication and transformation by a specific oligodeoxynucleotide. *Proc. Natl. Acad. Sci.* **75:** 280–284.

CHAPTER TWELVE

Fluorescent Speckle Microscopy of Cytoskeletal Dynamics in Living Cells

Torsten Wittmann, Ryan Littlefield, and Clare M. Waterman-Storer

Department of Cell Biology, The Scripps Research Institute, La Jolla, California 92037

INTRODUCTION

FLUORESCENT SPECKLE MICROSCOPY (FSM) is a recently developed method for analyzing the dynamics of macromolecular structures in vivo and in vitro (Waterman-Storer et al. 1998). FSM is a spin-off of the classic method of fluorescent analog cytochemistry. In this method, a fluorescently labeled protein is microinjected, or expressed, in living cells and incorporated into macromolecular structures, rendering them fluorescent for dynamic analysis, by video or time-lapse fluorescence microscopy (Wang 1989; Matz et al. 2002). This classic approach is limited in its ability to report protein dynamics because of high background fluorescence, caused by unincorporated and out-of-focus incorporated fluorescent subunits, and because of the difficulty in detecting movement or turnover of subunits, caused by the uniform labeling of fluorescent structures in microscope images. These problems have been partially alleviated by the use of more technically challenging methods, such as laser photobleaching and photoactivation of fluorescence, to mark structures in limited cell areas, such that movement and subunit turnover in the marked region can be measured at steady state (Wang 1985; Wolf 1989; Theriot and Mitchison 1991; Mitchison et al. 1998; Yvon and Wadsworth 2000).

FSM utilizes a very low amount of fluorescent protein in the cell and provides information similar to that obtained using photomarking techniques. However, it delivers simultaneous kinetic information from large areas of the cell. It is capable of detecting non-steady-state molecular dynamics and measuring variation in molecular dynamics at very high spatial resolution. FSM also significantly reduces out-of-focus fluorescence, improving the visibility of fluorescently labeled structures and their dynamics in thick specimens. And because only low levels of fluorescent protein are needed in cells for FSM to work, the problems associated with protein overexpression no longer arise (Waterman-Storer and Salmon 1999). Finally, FSM can be used with various modes of fluorescence microscopy, as demonstrated recently with spinning-disk confocal FSM and total internal reflection FSM (Waterman-Storer and Danuser 2002; Adams et al. 2003). This chapter briefly reviews the theory of FSM image formation and then details design considerations for setting up an FSM imaging system, live cell specimen preparation, and FSM image acquisition. FSM image processing and recent advances in quantitative analysis are also discussed.

BASICS OF FSM IMAGE FORMATION

FSM was first discovered accidentally when it was noticed in high-magnification, high-resolution images of tissue cells injected with X-rhodamine tubulin that some fluorescent microtubules exhibited variations in fluorescence intensity along their lattices. They had a speckled appearance (Waterman-Storer and Salmon 1997). Microtubules assemble at their ends from α/β-tubulin dimers into 25-nm-diameter cylindrical polymers with 1625 dimers per micron of microtubule (Desai and Mitchison 1997). To understand the origin of speckles on microtubules, it was necessary to consider how the images of fluorescent microtubules were formed by the microscope.

The resolution R, the smallest resolvable distance between two objects, in the epifluorescence microscope, can be estimated by the Rayleigh criterion, $R = 0.61 \, \lambda/NA_{obj}$, where λ is the wavelength of emitted light and NA_{obj} is the numerical aperture of the objective lens. For example, for X-rhodamine (620-nm emission)-labeled microtubules, imaged with a high-resolution 1.4 NA objective lens, R is 270 nm, a length that would contain approximately 440 tubulin dimers. Thus, no matter how many of these tubulin dimers are fluorescently labeled, this part of the microtubule will always appear in the image as a region of 270 nm in size. However, the intensity of this resolution-limited image region depends on the number of fluorophores it contains. Variations of the number of fluorescent tubulin dimers incorporated in such 270-nm regions result in intensity variations along the image of the microtubule (Fig. 1A). A pool of tubulin dimers containing a given fraction of fluorescent dimers, F, will produce a mean number of fluorescent dimers per resolution-limited region in the image. At very low values of F, the standard deviation of the number of fluorescent dimers, relative to the mean, is high. Thus, the contrast of adjacent resolution-limited image regions increases as F decreases, giving rise to the speckled appearance of microtubules assembled from a low fraction of fluorescent tubulin in high-resolution images (Fig. 1C). In contrast, microtubules assembled from higher fractions of fluorescent tubulin appear continuously labeled (Fig. 1B) (Waterman-Storer and Salmon 1998).

Since the original analysis of microtubules, the "speckle" method has been applied to other cytoskeletal proteins. When injected with low levels of fluorescently labeled actin, actin-rich structures, such as the lamellipodium of migrating cells, appear speckled in high-resolution fluorescence images (Fig. 1E) (Waterman-Storer et al. 1998; Verkhovsky et al. 1999; Waterman-Storer et al. 2000; Watanabe and Mitchison 2002). In addition, a green fluorescent protein (GFP) fusion of a microtubule-binding protein, ensconsin, when expressed in cells at very low levels, gave a speckled distribution along microtubules (Bulinski et al. 2001). These actin and ensconsin speckles formed differently from those in a single microtubule polymer. For example, for actin, the lamellipodium is filled with a cross-linked and dense three-dimensional network of 6-nm paired helical actin filaments (Small 1981; Svitkina et al. 1997; Pollard et al. 2000). Here, a fluorescent speckle may arise from fluorescently labeled actin monomers within multiple actin filaments, at various angles in the cross-linked network, within a resolution-limited image region (Fig. 1D). However, none of these filaments are detected individually. Thus, images of the entire meshwork show a relatively even distribution of fluorescent speckles (Fig. 1E). It should be noted that fine structural details may be lost in FSM images, as compared to images obtained at higher labeling ratios. However, what is gained is the ability to extract information on molecular dynamics (see Fig. 7 and below). Thus, a "speckle," in general terms, is a resolution-limited image region that is significantly higher in fluorophore concentration (i.e., fluorescence intensity) than its immediate neighbors.

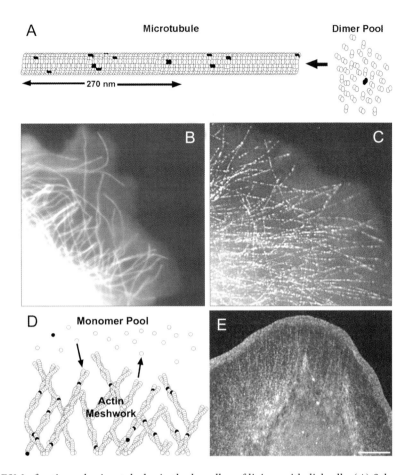

FIGURE 1. FSM of actin and microtubules in the lamellae of living epithelial cells. (*A*) Schematic diagram of the incorporation of a low amount of labeled tubulin dimers (black) together with unlabeled dimers (white) to form a microtubule with a sparse distribution of labeled subunits along its length; 270 nm is the limit of resolution of the light microscope for red fluorescence (620 nm, 1.4 NA). (*B*) Image of fluorescent microtubules in a living epithelial cell labeled with X-rhodamine tubulin in which labeled tubulin makes up ~10 % of the cellular tubulin pool. (*C*) Microtubule FSM image of a similar cell in which labeled tubulin makes up 0.5% of the tubulin pool. In the conventional image, high background fluorescence masks details in the central regions of the cell, and microtubules are continuously labeled along their lengths, devoid of fiduciary marks. In the FSM image, the background fluorescence is low. The full length of individual microtubules can be seen and they are covered with fiduciary marks. The images were taken on the same imaging system with the same objective lens. The exposure time was 0.2 seconds for the image on the left and 1.5 seconds for the image on the right. (*D*) Schematic diagram of how low levels of labeled (black) actin monomers coassemble with unlabeled monomers (white) into a meshwork of filaments as in the lamellipodium of migrating cells. (*E*) FSM image of an epithelial cell injected with a low level of X-rhodamine-labeled actin. Note the speckled appearance of the lamellipodium, which is made up of a very dense meshwork of cross-linked actin filaments. Bar, 10 μm.

For a fluorescent speckle to form, it is critical that the fluorescent molecules remain immobilized during the time required for image acquisition, for example, through incorporation into a macromolecular cytoskeletal structure (Watanabe and Mitchison 2002). The rapid movement of a small number of mobile fluorescent molecules does not give rise to speckles, but results in a low-level, evenly distributed background signal in the many pixels visited by the molecules during the 0.5–3-second exposure time required to acquire an image of the dim FSM specimen (e.g., in the authors' imaging system, unincorporated actin monomers diffuse at a rate of 63 pixels per second in the image space).

Speckle patterns on macromolecular assemblies act as a series of fiduciary marks on the structure that changes only with the binding and dissociation of fluorescent subunits to and from the structure (Waterman-Storer and Salmon 1998). These are much like the fiduciary marks on fluorescent structures produced by photomarking techniques (Waterman-Storer and Salmon 1997). In time-lapse movies, changes in the speckle pattern stand out to the eye and can be measured, with translation of speckles indicating movement of the structure and appearance (and disappearance) of speckles relating to the assembly (and disassembly) of the structure. Currently, major efforts are being made toward the automated computer-based analysis of speckle dynamics for extracting high spatiotemporal resolution molecular kinetic information from living cells containing fluorescent molecules.

DESIGNING A MICROSCOPE SYSTEM FOR TIME-LAPSE DIGITAL FSM

Time-lapse FSM requires the ability to image high-magnification, high-resolution diffraction-limited regions (~0.25 µm) containing few (1–10) fluorophores, the ability to detect very slight differences in image intensity between adjacent image regions, and the capacity to inhibit fluorescence photobleaching for long-term imaging (up to 1 hour). These requirements demand the use of a sensitive imaging system with little extraneous background fluorescence, efficient light collection, a camera with low noise, high quantum efficiency, high dynamic range, high resolution, and suppression of fluorescence photobleaching with illumination shutters and/or oxygen scavengers (Waterman-Storer et al. 1993; Mikhailov and Gundersen 1995; Waterman-Storer et al. 1998). In addition, all fluorescently labeled molecules must be functionally competent to assemble into the macromolecular structure of interest; otherwise, they contribute to diffusible background and reduce the speckle contrast. Recommendations for critical components of the imaging system for FSM imaging (Fig. 2) are discussed below.

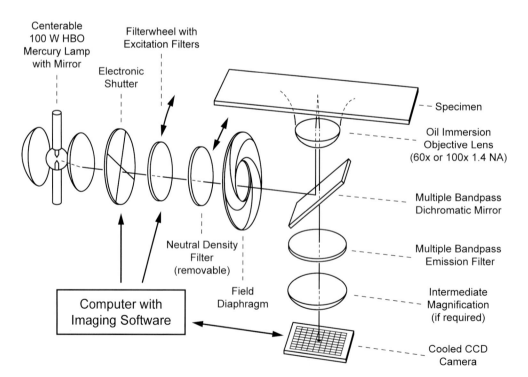

FIGURE 2. Schematic diagram of an inverted fluorescence imaging system suitable for multiple-wavelength FSM. See text for details.

Microscope and Objective Lens

FSM can be performed readily with either an upright or an inverted configuration microscope. However, for ease in handling open-topped chambers commonly used in live cell imaging, an inverted platform may be preferable. The microscope stand should be of biological research quality, with a substantial mass that resists vibration and minimizes temperature-induced expansion and contraction. If possible, the microscope should be mounted on a compressed-air vibration isolation table.

The microscope should be equipped with a high-quality epi-illuminator containing a 100-W HBO mercury arc lamp with a parabolic mirror that can be centered for achieving high intensity illumination. It is helpful to have manual control of the size of the field diaphragm in the epi-illuminator to reduce photodamage to the specimen and out-of-focus fluorescent "flare" in the image. The illuminator should be equipped with removable neutral density filters and an infrared blocking filter to minimize photo- and heat damage.

The objective lens (which also acts as the condenser in epi-illumination) must be of the highest numerical aperture available, i.e., 1.4 NA oil immersion. Magnification should be either 60×, 63×, or 100×, the choice being dependent on the spatial resolution of the camera detector (see below). If possible, the lens should not contain contrast-forming elements, such as phase rings, that partially block light transmission and should be corrected for chromatic aberration (apochromatic), particularly for multispectral FSM applications. When imaging a small specimen, it is not necessary for the lens to be flat-field-corrected ("Plan"). Indeed, exceptionally bright Super Fluor, Fluor, or Fluar lenses are often not Plan-corrected. However, for a wide-field view, a Plan Apochromat is recommended.

For imaging the weakly fluorescent FSM specimens, the light path in the microscope from the objective to the detector should be simple, containing as few intervening components (analyzers, wave plates, prisms, etc.) as possible. Optovars (magnification changers) should be removed except in cases where they are needed to match the microscope resolution to the detector resolution (see below). The camera port that utilizes the most direct path from the specimen should be used. In an upright microscope, this is directly over the objective lens. For an inverted configuration, this is the bottom ("Keller") port, underneath the microscope, which requires a hole in the table to accommodate the camera. If a hole is not possible, the side camera port is the next best choice.

Filters and Dichromatic Mirrors

The excitation and emission filters and the dichromatic mirror should be as efficient as possible at exciting the fluorophore, separating the excitation from emission, and collecting the emission of the fluorophore of choice. The use of long-pass instead of band-pass filters may maximize this efficiency, although long-pass filters can lead to high background. To be able to switch rapidly and automatically between wavelengths in multispectral FSM applications, excitation filters can be mounted separately in a filterwheel and a multi-band-pass dichroic mirror and emission filter mounted in the filter cube (Salmon et al. 1998).

Focus Control

It has been found that temperature-induced focal drift is a major problem in high-resolution live cell FSM imaging, since slight shifts in focus result in changes in speckle intensity that can be erroneously interpreted as cytoskeletal dynamics. Indeed, the aluminum parts making up stages, etc., on the microscope have a coefficient of expansion of about 220 nm per cm³ per degree temperature increase. In addition, mechanical drifts of a few microns per hour occur in the focus mechanisms of virtually all commercially available research microscopes. A focus lock helps to alleviate this problem. In any case, prior to purchasing a microscope, it is advisable to test the system for focus drift in a setting that is similar to the experimental environment (i.e., high-magnification oil immersion objective, air conditioner on, lasers or arc lamps running, etc.). If long-term time-lapse imaging (>1 hour) is required, it may be wise to invest in a closed-loop focus-control system with a linear encoder fitted to the microscope. This setup allows feedback on the distance between the objective lens mounting turret and the stage to drive a focus stepper or servo-motor. For this purpose, turnkey systems are available from several manufacturers (e.g., Applied Scientific Instruments; Ludl Electronic Products; Prior Scientific).

Illumination Control

Since fluorophores are subject to photobleaching, light exposure to the specimen should be kept to a minimum. This is accomplished using an electronically controlled shutter on the epi-illuminator, which opens only during image acquisition. The shutter should be mounted, with proper adapters, in the light path between the lamp-housing and the epi-illuminator, and should be mirrored on the surface facing the lamp to reflect heat away from the specimen. The shutter should operate quietly, quickly, reliably, and without excessive vibration. Shutters that are activated by a software-triggered pulse, via the serial or parallel port of the computer, can be obtained commercially (e.g., Uniblitz Shutters from Vincent Associates).

Cooled CCD Camera

Choice of the camera is one of the most critical make-or-break decisions in the design of an FSM imaging system. For imaging the low fluorescence of dim FSM specimens, the camera should be highly sensitive, extremely low-noise (remember that speckles look very much like noise!), and high spatial resolution, and, depending on the biological application, it may need high speed as well. The camera should be a scientific grade, slow-scan, cooled charge-coupled device (CCD) camera. To date, every intensified cooled CCD tested by the authors (both microchannel plate type or on-chip type) has had noise characteristics that obfuscate speckle detection.

Several critical properties of the CCD in the camera must be considered, including spatial resolution, well capacity, spectral sensitivity, illumination geometry, and readout geometry. These are detailed below.

The spatial resolution is determined by the number and physical size of the photosensitive "pixel" areas on the CCD chip. Pixels range from about 4×4 to 30×30 μm in size, and from approximately 200,000 to 2,000,000 or greater in number. The larger the pixel size, the more magnification is needed to resolve the higher frequency image details. Thus, smaller pixel size is better for FSM. The total number of pixels making up the CCD and the pixel size determine the imaging area, and the requirements for this will depend on the size of the sample. In addition, each pixel has a maximal number of photoelectrons that it can hold before it is saturated with charge, which will correspond to white saturation in the image. The greater this "full-well capacity," the greater the potential for a high dynamic range, after taking noise into consideration (see below).

The quantum efficiency of the CCD is defined as the probability that a photon hitting a pixel will be converted into a photoelectron that is detected by the camera, and this will vary depending on the wavelength of light. The chosen CCD should have the best quantum efficiency for emission wavelength of the chosen fluorophore.

Illumination geometry of a CCD refers to whether it is illuminated from the front or from the back. In general, back-illuminated CCDs have a much better quantum efficiency, but there are limits to the size of the pixels, the smallest available being 13×13 μm. For FSM applications, sensitive front-illuminated CCDs with smaller pixel sizes work quite well.

Once photons are converted to charge in the array of pixels, the charges must be read out to a computer and digitized into a gray scale for image viewing. Charges can be read out of the CCD in three basic geometries. Full-frame readout occurs as each row of pixel charges is transferred out of the CCD serially, one row after another. In frame transfer and interline transfer CCDs, either the entire pixel array or whole rows of pixels are rapidly transferred simultaneously to an array of pixels that are masked from light. The charges are then read out from the masked area while the imaging area is being exposed to light again. Frame and interline transfer CCDs are much faster and less noisy than the full-frame readout, although all three geometries are acceptable for FSM.

For any given CCD, camera manufacturers can use different electronic controls that effect the signal-to-noise properties of the final image. These include varying the cooling temperature, readout speed, dynamic range, and binning or subarraying. Since FSM imaging is really a battle for signal to noise at very low light levels, it is crucial to choose a manufacturer that offers the right CCD with the most noise-frugal electronics. Heat on the CCD can be erroneously interpreted as light during longer camera exposures, thereby contributing to noise. The coldest camera possible within a reasonable budget should be chosen.

Readout speed is a major factor in the overall signal-to-noise characteristics of the camera that is independent of the exposure time. The faster the readout speed (given in Hz, bits per second), the more error-prone the charge transfer, which translates to noise in the image. It is difficult to offer advice on this point because different biological applications may require faster image frame rates than others. The slowest readout speed for the process of interest should be chosen. For example, we use a 1.25-MHz readout for imaging actin dynamics, and a 10-MHz readout for studying endocytosis.

Dynamic range, or the number of distinguishable gray levels, is another critical factor. Although full well capacity is set for a given CCD, the number of gray levels into which this amount of charge is divided is not fixed. It can be encoded by 8, 10, 12, 14, or 16 bits of information per pixel, corresponding to 256; 1024; 4096; 16,384; or 65,536 (i.e., 2^8, 2^{10}, 2^{12}, 2^{14}, or 2^{16}) gray levels. However, it is not statistically possible to distinguish between two gray levels that differ by less than the noise level. Thus, the actual dynamic range is determined by the pixel full well capacity, divided by the readout noise. For example, if a camera advertises 12-bit dynamic range (4096 gray levels) and has a full well capacity of 23,000 photoelectrons, then it must have a noise level of (23,000/4096) 5.6 electrons per pixel or less to make use of the full 12-bit range. For FSM imaging, a broad dynamic range (at least 12-bit) is required so that differences between the intensity of 2 or 3 fluorophores can be detected quantitatively. Subarraying, or reading out only a specified portion of the CCD, can increase acquisition speed. Binning, in which the charges in a group of pixels are combined and read out as a single pixel to increase sensitivity, should not be used in FSM. This process effectively increases pixel size and decreases CCD resolution.

We currently use cameras from Hamamatsu Photonics, including the Orca II and the Orca II ER. Both have small pixel sizes (6.7 and 6.45 μm, respectively), low readout noise (~5 electrons in 14-bit mode), and two readout modes, a high precision 1.25-MHz 14-bit mode and a fast but noisier 10-MHz 12-bit mode. The Orca II ER is an improvement over the Orca II in that it has enhanced quantum efficiency at longer wavelengths.

Matching Microscope and Detector Resolution

A key factor in obtaining resolution-limited images of fluorescent speckles is matching microscope and camera resolution. The resolution-limited image region must be magnified

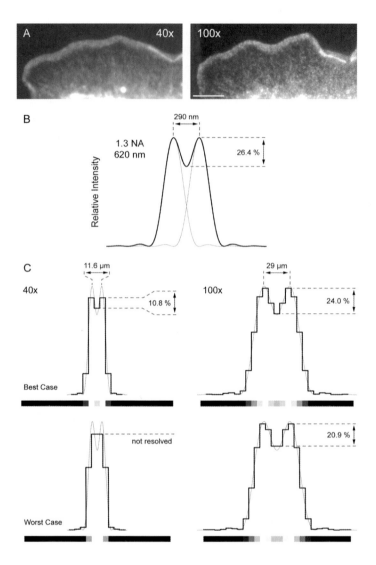

FIGURE 3. Effect of sampling frequency (magnification) on speckle contrast and resolution. (*A*) Images of the same edge region of a living newt lung epithelial cell injected with a low amount of X-rhodamine-labeled actin. The images were acquired with 1.3 NA 40× and 100× objective lenses, respectively, and an Orca II CCD camera (6.7 × 6.7-μm pixel size). The 40× panel was digitally zoomed to match the effective magnification of the 100× image. Speckles are clearly visible in the 100× image, but they are not resolved with the 40× lens. Bar, 10 μm. (*B*) Calculated intensity distribution of two point sources separated by 290 nm, the Rayleigh resolution limit for X-rhodamine fluorescence (620 nm) and 1.3 NA, imaged with an ideal lens system. The maximal theoretically achievable contrast is 26.4% of the peak intensity. (*C*) In the image plane, this distance is magnified to 11.6 μm with a 40× objective lens (*left panels*) and to 29 μm with a 100× lens (*right panels*). The graphs show the intensity received by 6.7-μm-wide pixel elements in the best case, when the central minimum of the theoretical distribution hits the center of a pixel, and in the worst case, when the distribution minimum falls between two pixels. The bars below the graphs show how this intensity distribution would appear on a row of pixels in an image acquired by a digital camera. With 100× magnification, 80–90% of the maximal theoretical contrast are sampled, resulting in good speckle contrast, whereas with 40×, even in the best case, only about 40% of the theoretical contrast can be achieved, resulting in very poor to no resolution of individual speckles. In practice, the presence of noise would further reduce contrast and resolution.

to an area on the CCD large enough to achieve a sufficiently high sampling frequency to be able to digitally resolve structures at the Rayleigh resolution limit (Fig. 3) (Stelzer 1998).

As a rule of thumb, magnifying the resolution-limited image region to 3×3 pixels on the CCD is sufficient to satisfy the Nyquist sampling criterion, i.e., to prevent the CCD from limiting the resolution of the imaging system and from producing aliasing between pixel rows. Any magnification over this value simply reduces the area of the specimen that is imaged, with no significant improvement in the information generated. The magnification (M) required to satisfy the Nyquist criterion is given by $M = 3 \, P_{width} / R$, where P_{width} is the width of a pixel and R is the size of the resolution-limited image region. Thus, for red fluorescence with a resolution limit of 0.27 µm and a camera with 6.7 µm pixels such as the Hamamatsu Orca II mentioned above and a 1.4 NA objective lens, the magnification required to satisfy the Nyquist criterion is 75×. Thus, either a 100× objective or a 60× with 1.25× intermediate magnification, whichever transmits more light, should be used.

Computer, Digital Image Acquisition Board, and Software for Control of Shutter and Image Acquisition

The most powerful computer, i.e., the fastest processor with the most random access memory (RAM), affordable should be used. Time-lapse FSM image series are large files, often on the order of 100 Mbytes or more, and serious computing power is necessary to view and manipulate them. A CD or DVD R/W device is the most economical way of archiving the large files, although a large networked file server is preferable. The image acquisition board should be chosen following the recommendations of the chosen camera and software manufacturer. Software should be capable of time-lapse digital image acquisition and triggering the shutter during camera exposure. It should allow easy viewing of time-lapse series as movies, with user-controlled playback rate and adjustment of brightness and contrast in the entire image series. The software should also provide basic image processing, including the ability to perform low-pass filtering and image arithmetic (subtraction, multiplication, etc.) and should be able to perform quantitative analysis of intensity, position, and distance. We currently use "Metamorph" software (Universal Imaging).

SPECIMEN PREPARATION AND FSM IMAGE ACQUISITION

We have used FSM extensively to study the dynamics of the microtubule and actin cytoskeletons in cells grown in tissue culture. The following sections briefly describe the setup used and the possible alternatives. Important aspects for successful FSM, and live cell imaging in general, are stressed. Note, however, that different cells and proteins may have their own specific requirements.

Microinjection

Although it is possible to use GFP-tagged proteins for FSM, we have found, for cytoskeletal components such as tubulin or actin, that microinjection of fluorescently labeled proteins gives better results. There are a variety of reasons for this:

- Modern chemical fluorophores are usually brighter than GFPs (higher extinction coefficient and better quantum yield) resulting in brighter speckles. This can be countered by attaching multiple GFPs to the protein of interest (Bulinski et al. 2001), but such a large tag might interfere with biological function and structural dynamics of the molecule.

- Dyes with longer excitation wavelengths (yellow or red) can be used, which are less damaging to cells and elicit less cellular autofluorescence than GFP.

- Photobleaching of synthetic fluorophores is largely oxygen-dependent and can be prevented using oxygen scavengers. It is more difficult to prevent bleaching of GFPs. We routinely label actin and tubulin with X-rhodamine succinimidyl ester, but other, even brighter, and more expensive dyes can be used. For protocols on how to prepare these fluorescently labeled proteins, see Waterman-Storer (2002). For FSM of X-rhodamine-labeled tubulin and actin dynamics, we typically use exposure times of 0.5–1 second and acquire images at a frame rate of 5–10 seconds.

Cells grown on coverslips, suitable for the intended observation chamber, can be microinjected directly in 35-mm tissue culture dishes. Coverslips should be cleaned thoroughly by sequential sonication in detergent, water, and ethanol (Waterman-Storer 1998) prior to use. It is usually not necessary to control the temperature during microinjection, but it may be advisable to add 10–20 mM K-HEPES to the cell culture medium. This will prevent the media from becoming too basic while the cells are away from the CO_2 incubator. A good 40× phase-contrast objective is recommended for use during microinjection. Ideally, the objective should be equipped with a correction collar to adjust for the optical properties of culture dish plastics and coverslip. After microinjection, the culture dishes are returned to the incubator for 2–4 hours to allow recovery from microinjection and incorporation of the injected protein into the structures to be observed.

We use microinjection needles home-pulled from borosilicate glass capillaries (1.0 mm outer diameter and 0.78 mm inner diameter) using a Sutter Instruments P-87 needle puller equipped with a 2.5 × 4.5-mm box filament (Sutter FB245B). The inner tip diameter of microinjection needles should be less than 1 μm, and the taper should be short to prevent clogging, breaking, and excess flexibility (Fig. 4). To pull good microinjection needles, some experimentation is required. The four-step program in Table 1 is useful as a starting point. Note that ramp is the heat-setting required to start melting the glass and must be determined empirically for different filament and glass types; the numbers in Table 1 are relative units specific for the Sutter needle puller.

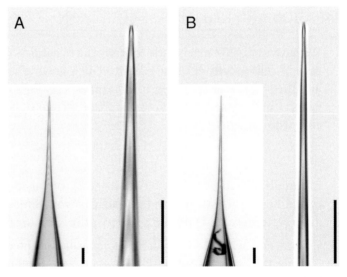

FIGURE 4. (*A*) A microinjection needle pulled from glass capillaries as described in the text; (*B*) a commercially available Eppendorf Femtotip II. Bars, 50 μm.

TABLE 1. Four-step program for pulling microinjection needles

	Heat	Pull	Velocity	Time	Pressure
1	Ramp + 10	100	10	250	500
2	Ramp	100	10	250	500
3	Ramp	100	10	250	500
4	Ramp + 10	100	15	250	500

Alternatively, commercially available microinjection needles, such as Eppendorf Femtotips, can be used. To prevent protein aggregation and needle clogging caused by protein absorption to the glass, microinjection needles can be coated by incubating them in a closed Petri dish with a few drops of hexamethyldisilazane (Pierce).

Immediately prior to injection, fluorescently labeled proteins should be centrifuged in a micro-ultracentrifuge, or an airfuge, at about 40,000g for 20 minutes at 4°C. This will remove any protein aggregates and other particles that might clog the needle. For cytoplasmic injection, a needle concentration of 0.5–1.0 mg/ml labeled tubulin or actin is generally recommended to obtain good FSM images. The final intracellular concentration of fluorescent protein will depend on the microinjection setup and the volume injected into a specific cell type. It will probably require some adjustment. We found that fluorescently labeled tubulin or actin can also be injected into the nucleus and will be exported into the cytoplasm within 2–3 hours. This is useful for studying cytoskeletal dynamics in cells that express additional proteins from co-injected plasmid DNA. For example, in PtK1 cells (Wittmann et al. 2003), nuclear injection of 100 μg/ml plasmid DNA will result in a high expression level within 2–3 hours from a cytomegalovirus (CMV) immediate-early promoter. However, for nuclear microinjection, the concentration of fluorescent protein should be increased by threefold to fourfold, since injection of large volumes into the nucleus can result in poor expression or cell death.

Live Cell Observation

Two major points must be considered for successful FSM of living cells: First, if using cells originating from a warm-blooded species, the observation chamber must be heated. Although there are numerous stage-heater designs available, a simple air stream incubator (Nevtek) that blows warm air across the microscope stage is the most flexible solution, providing relatively even heating to the sample. The objective lens, which is coupled to the specimen by immersion oil, will always be a major heat sink, and this can be problematic with heaters that heat only the cell chamber and not the rest of the microscope. In either case, heating will always create focus drift. To minimize this problem, the temperature of the microscope should be equilibrated for several hours prior to imaging.

Second, the observation chamber should be tightly sealed to prevent gas exchange so that oxygen depletion, which inhibits photobleaching of chemical fluorophores, can occur. Sealing the chamber minimizes evaporation of media, a major problem in open chambers, especially when using an air stream incubator. A sealed chamber also reduces problems associated with changes in pH caused by loss of CO_2 from the culture medium. If the chamber is sealed, it may not be necessary to add additional buffering agents.

In the authors' laboratory, PtK1 cells are routinely grown in modified Ham's F12 medium (Sigma), which contains 25 mM HEPES for pH buffering. The cells are observed in the same medium, supplemented with 0.5–1.0 unit/ml Oxyrase (Oxyrase Inc.), an oxygen-scavenging enzyme complex from *Escherichia coli*, to prevent photobleaching. Some cells, such as neurons, do not tolerate Oxyrase very well. In such cases, the Oxyrase concentration can be titrated back until a balance between cell survival and inhibition of photobleaching is reached. The use of high-glucose tissue culture medium to provide cells with a nutrient source for anaerobic metabolism might also help with cell viability. Note that direct observation in tissue culture medium is only possible with a good band-pass filter set. Long-pass filters allow too much background fluorescence from fluorescent media components, such as phenol red and vitamins (e.g., riboflavin), which will mask any specific fluorescent signal.

Although many different designs of live-cell observation chambers are available, a very simple aluminum slide chamber is a convenient solution for short observation times (up to 2 hours). Since these chambers contain only a small amount of medium (~80 μl), they might not sustain normal cell behavior for longer periods. The chambers consist of an aluminum slide (25 × 76 × 1.5 mm) with a hole in the middle and a counterbore on each face to fit a round coverslip (Fig. 5). These can be made *inexpensively* in quantity by any good machine shop. We routinely use 15-mm-round coverslips, but we also have chambers that hold 22-mm coverslips. For assembly of the slide chamber, silicon grease is applied to both counterbores. A clean coverslip is then inserted into one counterbore, and the chamber is filled with an excess of medium (~150 μl) and closed with a second coverslip on the underside of which the cells are growing. It is important to avoid air bubbles, and both coverslips should be tightly stuck to the grease. Excess medium can then be removed by aspiration, and the outside surface of the coverslip on which the cells are growing should be cleaned thoroughly with a cotton applicator and 70% ethanol to remove any grease that would otherwise contaminate the immersion oil and adversely affect image quality. After use, the aluminum slides can be cleaned by sonication, first in detergent, then water, and finally ethanol. Since the chamber is made to the same dimensions as a standard microscope slide, it can be used on any stage configuration fitted to either an inverted or upright microscope. Because of its thinness, it also allows the use of high-NA condensers for the acquisition of high-resolution DIC (differential interference contrast) or phase images. An obvious disadvantage is that, once it is sealed, nothing can be added to the cell culture. For experiments requiring the addition of drugs, more complicated perfusion chambers must be used.

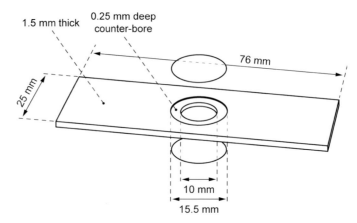

FIGURE 5. Diagram of a simple aluminum slide chamber used for live cell imaging. To assemble the observation chamber, a clean round coverslip is sealed into the counterbore on the underside of the slide with silicon grease. The chamber is filled with medium and closed on the other side of the slide with a second coverslip on which tissue culture cells are growing.

IMAGE PROCESSING FOR ENHANCING FSM IMAGES

Low-pass Filter to Reduce Noise

Since FSM operates at low fluorophore concentrations, and thus low light levels, images are inherently noisy. However, since camera noise has a higher spatial frequency (every pixel has an independent noise level) than fluorescent speckles (whose maxima should be separated by 3–4 pixels on an optimized imaging system; see Fig. 3), much of the high-frequency noise can be selectively suppressed with a low-pass filter. The simplest way to do this is by spatial convolution of the raw image with a 3 × 3 low-pass kernel (Fig. 6A). A kernel is a matrix of numbers that describe the relative contributions of the original pixel and its neighbors to the output pixel. In the simplest case of a low-pass kernel, as shown in Figure 6, this means that every pixel in the original image is replaced by the average of itself and its eight surrounding pixels,

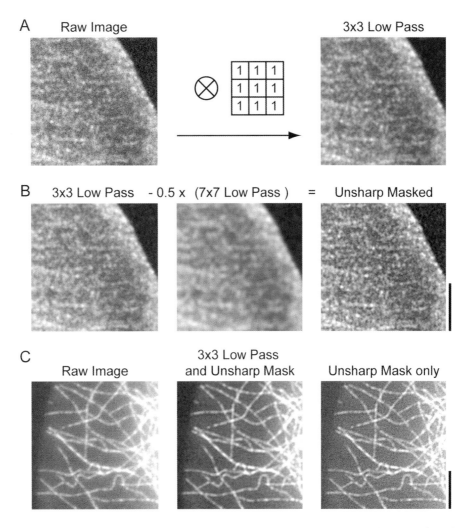

FIGURE 6. Basic image processing to enhance contrast in FSM images. (*A*) Reduction of random pixel noise in a dim actin FSM image by convolution with a 3 × 3 low-pass kernel. (*B*) Subtraction of an unsharp mask (in this case, a 7 × 7 low pass of the raw image multiplied by a scaling factor of 0.5) greatly enhances the speckle contrast as compared to the raw image in *A*. (*C*) In brighter, less noisy images such as this one of moderately speckled microtubules, low-pass filtering mainly leads to unwanted image blur and subtraction of the unsharp mask directly from the raw image gives a better result. Bars, 5 μm.

suppressing pixel values that are very different from their neighbors. A negative side effect of this filter is that it causes blurring of fine detail in the original image. Thus, at low noise levels, low-pass filtering might be counterproductive (Fig. 6C). Low-pass filters weighted with a Gaussian function introduce slightly less blur. Most imaging software packages, such as Metamorph, NIH image, or photo processing software like Adobe Photoshop, include these low-pass filtering functions.

Unsharp Mask to Increase Contrast

The contrast in FSM images can be enhanced by a sharpening procedure, known as "unsharp mask," performed with software mentioned above (Fig. 6B). First, the raw image is blurred through the application of a low-pass filter with a larger kernel size. A 7×7 low-pass kernel works well for actin and microtubule FSM images. This blurred image, the unsharp mask, is then multiplied by a scaling factor and subtracted from the raw image. The best kernel size and scaling factor must be determined empirically.

As a result, low-frequency image detail, such as out-of-focus haze, will be largely subtracted from the original image, enhancing high-frequency spatial detail and thus speckle contrast. In addition, this procedure equalizes the dynamic range of the entire image since more intense signals will be subtracted in bright image areas, therefore enhancing dim image features. However, unsharp masking will also enhance noise. The benefits and drawbacks of applying any of these filters must be weighed carefully.

RECENT DEVELOPMENTS AND FUTURE DIRECTIONS

Simultaneous Uniform and FSM Imaging of Filamentous Actin to Integrate Structural and Kinetic Information

As discussed above, in FSM images of the actin cytoskeleton, a speckle represents a locally significant increase in the number of fluorescently labeled actin monomers per resolution-limited image region, and its intensity may be contributed by fluorophores incorporated in many different actin filaments. In addition, since the speckle density is limited by the optical resolution of the microscope, speckle density is not directly correlated with the amount of actin polymer present, as demonstrated by computer simulations (Waterman-Storer and Danuser 2002). As a consequence, it is difficult to determine whether a speckle is in a region of relatively high or low actin density, and with which type of higher-order actin structure the speckle is associated (e.g., an actin filament bundle in a filopodium or a filament meshwork in a lamellipodium). This is particularly problematic at very low levels of labeled actin (when speckle contrast is highest),

It is often possible to infer differences in actin density from the speckle background intensity, or to associate a group of actin speckles with a filament bundle based on their co-transport at the same trajectory and velocity. However, we have recently devised a better method of relating actin speckle dynamics to their higher-order structural context. In this method, a mixture of differently labeled actins are coinjected into the cell. The mixture contains a low level of X-rhodamine-labeled actin (~0.4 mg/ml), for FSM imaging, and a high level (5–10 times more concentrated) of actin labeled with a spectrally distinct fluorophore, such as Oregon green or Alexa 488, for visualizing the higher-order structural dynamics of filamentous actin. Image pairs of X-rhodamine-labeled and green-labeled actin are then captured at intervals over time to near-simultaneously visualize the relationship of actin speckle dynamics with the overall distribution of actin filaments. This is accomplished using a multispectral imaging system in which the excitation wavelength is selected for each fluorophore by a filterwheel or similar

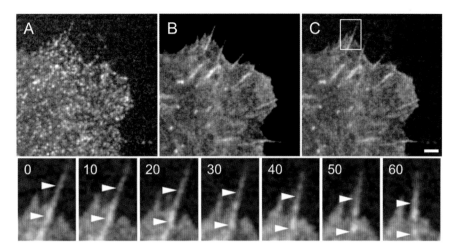

FIGURE 7. Dual-wavelength imaging of (*A*) X-rhodamine actin speckles at low fluorophore concentration and (*B*) the structural organization of filamentous actin labeled with a different fluorophore (Oregon green) coinjected at higher concentration in a chicken heart primary fibroblast. (*C*) Color overlay. Bar, 2 μm. This technique makes it possible to distinguish between speckles in different actin structures, as pointed out by arrowheads in the time-lapse series on the bottom of the figure. Time-series images were acquired every 10 seconds and are zoomed by a factor of 3 compared with *A*, *B*, and *C*.

device, and emission is selected with a multiple band-pass dichromatic mirror and emission filter. Images of the two fluorophores are captured in rapid succession with a single cooled CCD camera (Salmon et al. 1998; Adams et al. 2003). We use an exposure time of about 0.5 seconds for both images in each pair. However, because of the difference in brightness between the high fraction of green actin and low fraction of red actin, the intensity of the illumination for exciting the green fluorophore is reduced by about 50%. Following capture, images in each pair are color encoded to produce a single 24-bit RGB (red-green-blue) image for viewing.

This method of near-simultaneous uniform and FSM imaging of filamentous actin provides several advantages over actin FSM alone. For example, fluorescent speckles colocalized with filopodia at the leading edge can be distinguished from speckles in the lamellipodial meshwork (Fig. 7). By distinguishing these speckles, it becomes possible to compare their trajectories and lifetimes in specific structures, and it is possible to correlate actin dynamics and organization. Uniform and FSM imaging also allows the tracking of speckles into and out of particular actin structures. For example, speckles within the leading-edge lamellipodium can often be observed entering actin bundles, suggesting that lamellipodial filaments can flow and be incorporated into the ends of bundles. Furthermore, it will be possible to determine whether the rate of flow (from the speckles) correlates with the growth or shrinking of a specific bundle (from the uniform labeling). FSM provides information about the underlying dynamics within actin structures, whereas uniform labeling provides information about the organization and structure of the actin cytoskeleton, which can only be inferred from the speckles. These complementary techniques can be combined to provide both kinetic and structural information with high spatial and temporal resolution.

Computational Analysis of FSM Image Series

As noted, speckle dynamics in FSM image series encode information about the movement and assembly/disassembly of the macromolecular structure into which the fluorescent subunits are incorporated. Currently, kymographs of the FSM image series are used to provide rough estimates of speckle velocities (Waterman-Storer et al. 1998; Waterman-Storer 2002; Waterman-Storer and Danuser 2002). In kymographs, a line 1–5 pixels wide, whose axis is parallel to the direction of speckle movement, is extracted from each image of the series and pasted side by side in a time montage. As speckles move along the line, they appear as oblique streaks on the kymograph, the slope of which corresponds to the velocity of speckle movement. However, this analysis is limited to speckles moving along the same line; it averages speed variations over time and delivers only sparse data. As an alternative, speckles can be tracked manually using semiautomated tracking functions, such as the "track points" function on Universal imaging's MetaMorph software (Salmon et al. 2002; Watanabe and Mitchison 2002). However, this process is laborious, prone to error, and leaves the majority of information unexploited. The full potential of FSM will only be accessible if specialized computational tools are available for statistical analysis of speckle translocation and intensity changes. There is currently no commercial software available to track the ≤100,000 speckles per image that fluctuate just above background intensity (~5–30 gray levels on a 14-bit camera) in a typical movie made up of 1000×1000 pixel images.

The Danuser group has made efforts to develop a computational framework for specifically extracting the assembly/disassembly behavior of actin filaments within the cortical actin meshwork of epithelial cells from the time-lapse actin FSM image series (Waterman-Storer and Danuser 2002; Ponti et al. 2003). This turns out to be a computationally intensive problem for automated image analysis. The steps required to extract this information are outlined as follows. First, after noise filtering, speckles must be statistically selected as resolution-limited

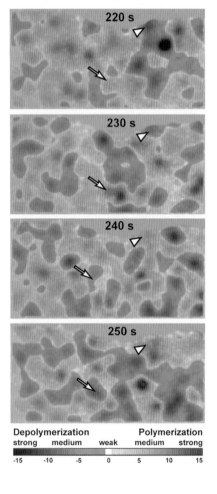

FIGURE 8. Computational FSM analysis of the turnover of filamentous actin in the cortex of a newt lung epithelial cell. Panels show kinetic maps over a period of 30 seconds. Two positions are selected to indicate the periodic cycling assembly and disassembly of the actin meshwork. The arrowhead points to a region cycling from depolymerization to polymerization and back to depolymerization. The arrow points to a region where the cycle is 180° phase-shifted. The 31 score classes in the kinetic maps are color-encoded as shown on the bottom of the figure.

image areas that are significantly brighter than their immediate surroundings and significantly different from image noise. The lifetimes of speckles can then be determined by measuring how long they remain higher in intensity than their local surroundings. However, since speckles in actin meshworks exhibit submicron positional fluctuations (probably due to Brownian motion), micro-motions of selected speckles must be tracked over time. In addition, local intensity maxima may arise (speckle births) in two ways: (1) by an increase in fluorescence intensity (addition of labeled monomer) within a resolution-limited image region leading to speckle birth by assembly or (2) by a decrease in fluorescence intensity (removal of labeled monomer) around a resolution-limited image region leading to speckle birth by disassembly. Thus, to relate the appearance and disappearance of individual speckles in an FSM image series to a local relative rate of actin assembly and disassembly, it is necessary to determine whether a speckle was born by actin assembly or local disassembly. The results of such an analysis are displayed as a pseudocolored map of polymerization and depolymerization, superimposed onto frames of the actin FSM movie (Fig. 8). Progress is also being made toward automated speckle trajectory and velocity analysis in actin FSM (Valloton et al. 2003), with the ultimate goal of establishing a generic tool for extracting assembly, disassembly, trajectory, and velocity values for speckles of any molecular origin. It is anticipated that these tools will be available for FSM analysis in the next few years to realize the full potential of FSM for measuring the kinetics of many macromolecular assemblies in vivo.

ACKNOWLEDGMENTS

The authors thank Gaudenz Danuser and Pascal Valloton (Swiss Federal Institute of Technology, Zurich, Switzerland) for sharing unpublished data. This work was supported by National Institutes of Health grants GM-61804-03 and GM-67230-01, a Young Investigators award from the Human Frontier Science Program (RGY5/2002), and the Jane Coffin Childs Memorial Fund for Cancer Research.

REFERENCES

Adams M.C., Salmon W.C., Gupton S.L., Cohan C.S., Wittmann T., Prigozhina N., and Waterman-Storer C.M. 2003. A high-speed multispectral spinning-disk confocal microscope system for fluorescent speckle microscopy of living cells. *Methods* **29:** 29–41.

Bulinski J.C., Odde D.J., Howell B.J., Salmon E.D., and Waterman-Storer C.M. 2001. Rapid dynamics of the microtubule binding of ensconsin in vivo. *J. Cell Sci.* **114:** 3885–3897.

Desai A. and Mitchison T.J. 1997. Microtubule polymerization dynamics. *Annu. Rev. Cell Dev. Biol.* **13:** 83–117.

Matz M.V., Lukyanov K.A., and Lukyanov S.A. 2002. Family of the green fluorescent protein: Journey to the end of the rainbow. *BioEssays* **24:** 953–959.

Mikhailov A.V. and Gundersen G.G. 1995. Centripetal transport of microtubules in motile cells. *Cell Motil. Cytoskel.* **32:** 173–186.

Mitchison T.J., Sawin K.E., Theriot J.A., Gee K., and Mallavarapu A. 1998. Caged fluorescent probes. *Methods Enzymol.* **291:** 63–78.

Pollard T.D., Blanchoin L., and Mullins R.D. 2000. Molecular mechanisms controlling actin filament dynamics in nonmuscle cells. *Annu. Rev. Biophys. Biomol. Struct.* **29:** 545–576.

Ponti A., Valloton P., Salmon W.C., Waterman-Storer C.M., and Danuser G. 2003. Computational analysis of F-actin turnover in cortical actin meshworks using fluorescent speckle microscopy. *Biophys. J.* **84:** 3336–3352.

Salmon E.D., Shaw S.L., Waters J., Waterman-Storer C.M., Maddox P.S., Yeh E., and Bloom K. 1998. A high resolution multi-mode digital microscope system. *Methods Cell Biol.* **56:** 185–214.

Salmon W.C., Adams M.C., and Waterman-Storer C.M. 2002. Dual-wavelength fluorescent speckle microscopy reveals coupling of microtubule and actin movements in migrating cells. *J. Cell Biol.* **158:** 31–37.

Small J.V. 1981. Organization of actin in the leading edge of cultured cells: Influence of osmium tetroxide and dehydration on the ultrastructure of actin meshworks. *J. Cell Biol.* **91:** 695–705.

Stelzer E.H.K. 1998. Contrast, resolution, pixelation, dynamic range and signal-to-noise ratio: Fundamental limits to resolution in fluorescence light microscopy. *J. Microsc.* **189:** 15–24.

Svitkina T.M., Verkhovsky A.B., McQuade K.M., and Borisy G.G. 1997. Analysis of the actin-myosin II system in fish epidermal keratocytes: Mechanism of cell body translocation. *J. Cell Biol.* **139:** 397–415.

Theriot J.A. and Mitchison T.J. 1991. Actin microfilament dynamics in locomoting cells. *Nature* **352:** 126–131.

Vallotton P., Ponti A., Waterman-Storer C.M., Salmon E.D., and Danuser G. 2003. Recovery, visualization, and analysis of actin and tubulin polymer flow in live cells: A fluorescent speckle microscopy study. *Biophys. J.* **85:** 1289–1306.

Verkhovsky A.B., Svitkina T.M., and Borisy G.G. 1999. Self-polarization and directional motility of cytoplasm. *Curr. Biol.* **9:** 11–20.

Wang Y.L. 1985. Exchange of actin subunits at the leading edge of living fibroblasts: Possible role of treadmilling. *J. Cell Biol.* **101:** 597–602.

———. 1989. Fluorescent analog cytochemistry: Tracing functional protein components in living cells. *Methods Cell Biol.* **29:** 1–12.

Watanabe N. and Mitchison T.J. 2002. Single-molecule speckle analysis of actin filament turnover in lamellipodia. *Science* **295:** 1083–1086.

Waterman-Storer C.M. 1998. Microtubule/organelle motility assays. In *Current protocols in cell biology* (ed. J.S. Bonifacio et al.), Unit 13.1. John Wiley, New York.

———. 2002. Fluorescent speckle microscopy (FSM) of microtubules and actin in living cells. In *Current protocols in cell biology* (ed. K.S. Morgan), Unit 4.10. John Wiley, New York.

Waterman-Storer C.M. and Danuser G. 2002. New directions for fluorescent speckle microscopy. *Curr. Biol.* **12:** R633–R640.

Waterman-Storer C.M. and Salmon E.D. 1997. Actomyosin-based retrograde flow of microtubules in the lamella of migrating epithelial cells influences microtubule dynamic instability and turnover and is associated with microtubule breakage and treadmilling. *J. Cell Biol.* **139:** 417–434.

———. 1998. How microtubules get fluorescent speckles. *Biophys. J.* **75:** 2059–2069.

———. 1999. Fluorescent speckle microscopy of microtubules: How low can you go? *FASEB J.* **13:** S225–S230.

Waterman-Storer C.M., Salmon W.C., and Salmon E.D. 2000. Feedback interactions between cell-cell adherens junctions and cytoskeletal dynamics in newt lung epithelial cells. *Mol. Biol. Cell* **11:** 2471–2483.

Waterman-Storer C.M., Sanger J.W., and Sanger J.M. 1993. Dynamics of organelles in the mitotic spindles of living cells: Membrane and microtubule interactions. *Cell Motil. Cytoskel.* **26:** 19–39.

Waterman-Storer C.M., Desai A., Bulinski J.C., and Salmon E.D. 1998. Fluorescent speckle microscopy, a method to visualize the dynamics of protein assemblies in living cells. *Curr. Biol.* **8:** 1227–1230.

Wittmann T., Bokoch G.M., and Waterman-Storer C.M. 2003. Regulation of leading edge microtubule and actin dynamics downstream of Rac1. *J. Cell Biol.* **161:** 845–851.

Wolf D.E. 1989. Designing, building, and using a fluorescence recovery after photobleaching instrument. *Methods Cell Biol.* **30:** 271–306.

Yvon A.M. and Wadsworth P. 2000. Region-specific microtubule transport in motile cells. *J. Cell Biol.* **151:** 1003–1012.

CHAPTER THIRTEEN

Polarization Microscopy with the LC-PolScope

Rudolf Oldenbourg

Marine Biological Laboratory, Woods Hole, Massachusetts 02543

INTRODUCTION

MOST OF THE IMAGING APPROACHES discussed in this volume are based on the use of fluorescent molecules and their detection inside living cells. Only a decade ago, fluorescence microscopy required that cells be killed, fixed, and then labeled in order to detect specific molecular species. At that time, live cell imaging was dominated by techniques such as phase-contrast, differential interference contrast, and polarized light microscopy, which do not require the introduction of exogenous dyes or markers. The rise of fluorescence microscopy in this field was made possible by spectacular advances in fluorescence imaging techniques and by the development of many vital molecular probes, in particular the green fluorescent protein (GFP). These newer techniques outshine the more traditional techniques which probe the subtle changes of phase and amplitude of light that has interacted with the specimen. Phase-sensitive techniques create images of living cells on the basis of their intrinsic optical properties, in particular their refractive index. While attention was focused on fluorescence imaging, the more traditional techniques for live cell imaging also made great advances, and they remain indispensable as complementary tools for imaging cell structure and function while avoiding the need for introducing fluorescent molecules into cells (Inoué and Spring 1997; Sluder and Wolf 1998).

Polarization microscopy uses polarized light to probe the local anisotropy of the specimen's optical properties, such as birefringence or dichroism. These optical properties, which will be discussed in more detail later, have their origin in molecular order, i.e., in the alignment of molecular bonds or fine structural form in the specimen. Hence, polarization microscopy provides structural information at a submicroscopic level, including the alignment of molecules in filament arrays and membrane stacks such as the mitotic spindle, f-actin arrays, and retinal rod outer segments. Therefore, the polarizing microscope provides information not only about where and when certain structures form and change inside the living cell, but also about some of the submicroscopic features of these structures. Most importantly, the polarizing microscope can perform live cell imaging dynamically, over both short and long time periods, at the highest resolution of the light microscope, and with minimal interference with the physiological conditions required for the maintenance of live and healthy cells.

The use of the polarizing microscope for studying living cells was pioneered more than 70 years ago by W.J. Schmidt, whose many original observations were published in a celebrated monograph (Schmidt 1937). Intrigued by Schmidt's observations and other early studies (Ambronn and Frey 1926; Frey-Wyssling 1953), other groups began to use and refine the technique for live cell imaging, in particular for studying the phenomenon of cell division (Mitchison 1950; Swann and Mitchison 1950; Inoué and Dan 1951; Inoué 1952, 1953; Inoué and Hyde 1957). For a recent review of polarized light microscopy of spindles see Oldenbourg (1999). For a discussion of the traditional polarized light microscope and its many uses in cell biology, see the recent article by Shinya Inoué (2002), who for more than 50 years has been the technique's strongest proponent. (See also Appendix III in the first edition of Inoué [1986]).

This chapter describes the use of a new type of polarized light microscope whose development began at the Marine Biological Laboratory about ten years ago. The new "LC-PolScope" is based on the traditional polarized light microscope, enhanced by the addition of liquid-crystal devices and special image-processing algorithms. The LC-PolScope measures polarization optical parameters at many specimen points simultaneously, at fast time intervals and at the highest resolution of the light microscope. It rapidly generates a birefringence map whose pixel brightness is directly proportional to the local optical anisotropy, unaffected by the specimen orientation in the plane of view. It also generates a map depicting the slow axis orientation of the birefringent regions. The basic LC-PolScope technology can be adapted to most research-grade microscopes and is available commercially from Cambridge Research and Instrumentation (CRI, http://www.cri-inc.com) under the trade name LC-PolScope.

The advantage of the LC-PolScope over the traditional polarizing microscope is its speed, sensitivity, and accuracy. As Shinya Inoué likes to point out, for a seminal study on DNA organization in sperm heads, it took him and two colleagues 3 years to analyze the polarized light images of three living cave cricket sperm (Inoué and Sato 1966). A similar analysis would take seconds using the LC-PolScope. It must be said that the LC-PolScope measures specimen properties that, in principle, can be measured with the traditional polarizing microscope. However, the vastly improved speed and completeness of measurement (birefringence parameters for all resolved specimen points simultaneously) make the LC-PolScope the instrument of choice for many studies that benefit from the analytic power and submicroscopic information gleaned from polarized light images. Nevertheless, in some cases, the traditional polarizing microscope might provide the required information, be it the organization of filaments, membranes, or chromophores in living cells (see the aforementioned article by Inoué 2002). The decision as to whether to use the LC-PolScope is sometimes based on observations with the traditional instrument. In some cases, relevant information can be found in the literature (e.g., see the extensive bibliography in Appendix III of Inoué 1986); in other cases, exploratory observations may be carried out using the traditional instrument itself. This chapter was written for experimental scientists who want to obtain a better understanding of the capabilities of the new instrument and to optimize its use.

In describing the use of the LC-PolScope for live cell imaging, the basic instrument will be discussed first, including the optical, electronic, and software components and the influence of their performance parameters on the sensitivity, resolution, and speed of LC-PolScope measurements. As with the traditional polarizing microscope, LC-PolScope technology can be implemented using many different design variations, either for adapting it to a special measurement setup, for improving the performance of the instrument, or for measuring different optical parameters in the specimen. Some of the simpler design variations of the system will also be described, followed by a discussion of the image acquisition and measurement procedures used in the LC-PolScope, its performance characteristics, and how to measure them. A good understanding of these issues allows the investigator to optimize the performance of the instrument for the specific study at hand. A number of LC-PolScope images of living cells are presented to give examples of the range of systems studied with this new incarnation of an old and powerful technique.

A glossary of polarization optical terms is presented at the end of the chapter, including a brief outline of the optical concepts of birefringence and retardance, and of their measurement using imaging optics.

LC-PolScope technology continues to advance with developments being made at the Marine Biological Laboratory (MBL) and CRI in the hope of extending its measurement capabilities beyond birefringence, to include dichroism and fluorescence polarization, and to increase the speed of measurement to reduce the effects of movement of live specimens. Methods are also being developed to measure polarization optical parameters in three dimensions, along with analytical tools for the interpretation of image content and for image restoration. All of these developments are aimed at improving our understanding of cell structure and function, as displayed in the architectural dynamics in living cells (for information on the current status of these advances, see http://www.mbl.edu/research/resident/lab_oldenbourg.html and http://www.cri-inc.com).

INSTRUMENTATION

The LC-PolScope instrumentation includes optical and electronic components and software for image processing and analysis. It is covered by a U.S. patent (Oldenbourg and Mei 1996), which is licensed to CRI (http://.cri-inc.com). The instrumentation guidelines presented here are based on the commercial version of the LC-PolScope. The optical components are inserted into a research-grade microscope stand. The stands produced by most major microscope manufacturers are suitable for accommodating the components, and their arrangement can be optimized using the guidelines discussed here.

Basic Optical Setup

The optical design of the LC-PolScope builds on the traditional polarized light microscope, introducing two essential modifications: The specimen is illuminated with nearly circularly polarized light and the traditional compensator is replaced by an electro-optic, universal compensator. The PolScope also requires the use of monochromatic light. In Figure 1, the schematic of the optical train shows the universal compensator located between the monochromatic light source (e.g., arc lamp with interference filter) and the condenser lens. The analyzer for circularly polarized light is placed after the objective lens. The universal compensator is built from two variable retarder plates and a linear polarizer (see text box The Universal Compensator). The variable retarder plates are implemented as electro-optical devices made from two liquid crystal plates. Each liquid crystal plate has a uniform retardance that depends on the voltage applied to the device. The voltage for each plate is supplied

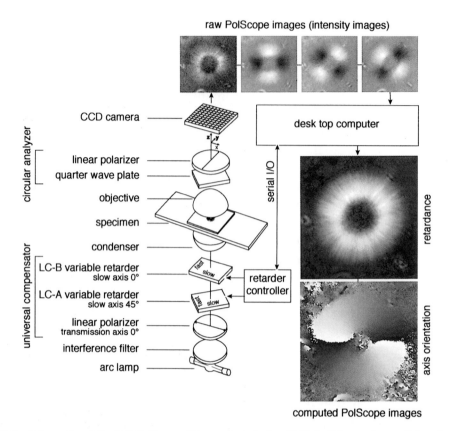

raw PolScope images (intensity images)

CCD camera

circular analyzer
linear polarizer
quarter wave plate

objective
specimen
condenser

universal compensator
LC-B variable retarder
slow axis 0°
LC-A variable retarder
slow axis 45°
linear polarizer
transmission axis 0°
interference filter
arc lamp

desk top computer

serial I/O

retarder
controller

retardance

axis orientation

computed PolScope images

FIGURE 1. Schematic of the LC-PolScope. The optical design (*left*) builds on the traditional polarized light microscope with the conventional compensator replaced by two variable retarders LC-A and LC-B. The polarization analyzer passes circularly polarized light and is typically built from a linear polarizer and a quarter-wave plate. Images of the specimen (*top row*, aster isolated from surf clam egg) are captured at four predetermined retarder settings that cause the specimen to be illuminated with circularly polarized light (*first, left-most image*) and with elliptically polarized light of different axis orientations (*second to fourth image*). Based on the raw PolScope images, the desk top computer calculates the retardance image and the axis orientation or azimuth image shown on the lower right.

by an electronic controller box connected to the retarders. The retarder controller box includes a microprocessor that stores liquid-crystal-related parameters for control of the variable retarder plates. The controller box is connected to a desktop computer using a serial interface. The desktop computer is also connected to the electronic camera, typically a CCD (charge-coupled device) camera, for recording the specimen images through the microscope optics. Specialized software synchronizes the image acquisition process with the liquid-crystal control and implements image-processing algorithms. These algorithms compute images that represent the retardance and principal axis orientation at each resolved image point.

It must be emphasized that during recordings of all four images in Figure 1, no part of the microscope setup or specimen was mechanically moved. To obtain the four images, only the voltages applied to the liquid-crystal devices were changed. Therefore, all four images recorded with the CCD camera are in perfect register, with corresponding picture elements representing intensities of the same object elements. Optimal image registration is particularly important for measurements requiring high spatial resolution.

The Universal Compensator

The universal compensator is built from two liquid-crystal devices **LC-A** and **LC-B**, each of which functions as a uniform linear retarder plate. The retardance of a single plate can be adjusted to between about 50 nm and 800 nm by applying a voltage between 10 V (800 nm) and 40 V (50 nm) to the liquid-crystal device (for a more detailed description of liquid-crystal variable retarders, see Khoo and Wu 1993). When changing the voltage, only the retardance changes, whereas the orientation of the slow (or fast) birefringence axis remains fixed.

As indicated in Figure 1, the slow axes of the linear retarders are mutually oriented at 45° to one another. In addition, the universal compensator includes a linear sheet polarizer (**polar**) whose transmission axis is oriented at 45° to the slow axis of **LC-A**. The polarizer and two liquid-crystal plates are bonded together to form a thin optical flat with antireflection coatings on each side. The clear aperture can be 25 mm or greater and the stack is about 6 mm thick.

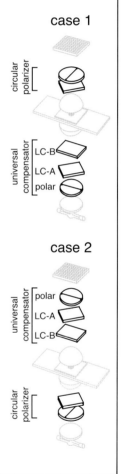

case 1

circular polarizer

universal compensator — LC-B, LC-A, polar

For measuring polarization optical properties of a specimen, the optical setup also includes a circular polarizer. As indicated in Figure 1 and to the right here, a circular polarizer is typically built as a sequence of a linear sheet polarizer and a fixed quarter-wave plate.

There are two basic choices for adding the universal compensator to the microscope optical train: either before the condenser lens (**case 1** on the right, also shown in Figure 1) or after the objective lens (**case 2**). In both cases, the specimen (and high numerical aperture optics) is sandwiched between the universal compensator and the circular polarizer. Furthermore, the wave plates (the quarter-wave plate of the circular polarizer and the LC wave plates of the universal compensator) are facing the specimen. In other words, in both cases the linear polarizers are the first and last elements in the polarization optical train.

Hence, in either case, the specimen, high NA optics, and fixed and variable retarder plates are forming a polarization optical train sandwiched between two linear polarizers. This polarization optical train can be analyzed using mathematical descriptions known as Müller matrices and Stokes vectors. Based on such an analysis, we have developed image acquisition procedures and algorithms that allow one to use the PolScope for measuring polarization optical properties of the specimen in every resolved image point simultaneously (see Retardance Measurement on p. 218).

Both arrangements are equivalent in their ability to measure polarization optical parameters of the specimen. For the optical purist, though, **case 1** might be preferable to **case 2**, which adds the universal compensator to the imaging path between objective and camera. Compared to the circular polarizer, the universal compensator represents a thicker optical flat and therefore might compromise the performance of the highly corrected objective lens. In practice, however, in most cases this difference is negligible.

Some additional optical properties, performance parameters, and possible modifications to the universal compensator and circular polarizer are discussed under the section below on Polarization Optical Components and throughout this chapter.

case 2

universal compensator — polar, LC-A, LC-B

circular polarizer

Design Specifics and Variations

Polarization Optical Components

Polarizers. The linear polarizers used in a PolScope should have high transmission and good extinction. Modern sheet polarizers are now available with transmission coefficients for unpolarized light of more than 40% and extinction coefficients of approximately 10^4. An ideal polarizer has a transmission of 50% and infinite extinction. The extinction is a measure of the purity of the polarization state of light, after it has passed through the polarizer. The extinction coefficient is the ratio of light transmitted through two polarizers in sequence, with their transmission axes aligned either parallel or perpendicular to each other (extinction = $I_\parallel / I_\perp$. Although Glan-Thompson and other prism polarizers have values that are close to the ideal, good sheet polarizers are typically sufficient for the LC-PolScope).

The purity of the polarization in a light microscope is usually limited by the polarization distortions introduced by high numerical aperture (NA) condenser and objective lenses. A pair of sheet polarizers might have an extinction ratio of 10^4 or higher, but if a condenser and objective lens are placed between them, the extinction usually drops to about 10^3 or below. (A pair of 1.4 NA oil immersion condenser and objective lenses has an extinction coefficient of about 200 if the condenser aperture diaphragm is fully open; partially closing the condenser aperture increases the extinction.) Owing to the development of calibration and correction procedures, extinction has a less critical role in the PolScope than in a traditional polarizing microscope. However, the sensitivity for measuring small retardance values, which might be below 0.01 nm, can be limited by the extinction of the setup (see the section on Performance Characteristics and How to Measure Them). The proper choice of lenses is discussed in the following section (Choice of Optics), which also refers to articles discussing the use of polarization rectifiers.

The circular polarizer used in the LC-PolScope is usually assembled from a linear polarizer and a quarter-wave plate (see text box The Universal Compensator). Both components are made of thin polymer sheets of high optical quality that are bonded together. When inserting the circular polarizer in the optical path, care must be taken to insert it right-side up. As illustrated in the text box The Universal Compensator, the correct orientation depends on the arrangements illustrated in cases 1 and 2. In either case, the quarter-wave plate of the circular polarizer faces the specimen. A separate linear polarizer can be used to test which side the quarter-wave plate is on. Looking through the circular and linear polarizer in tandem, against a white light source, rotate the two devices about their common normal axis and check for extinction. Good extinction can only be achieved if the quarter-wave plate is located outside the path between the two linear polarizers. If, however, the quarter-wave plate is between the two linear polarizers, then different rotation angles result in differently colored lights, but no extinction can be achieved.

Although it is important to insert the circular polarizer right-side up in the LC-PolScope optical train, it does not matter how it is rotated around the optical axis of the microscope. This is in stark contrast to the linear polarizers of a traditional polarizing microscope, which must be carefully rotated to their crossed position. This is a feature peculiar to the LC-PolScope, the explanation for which is left as an exercise for the interested reader.

In a LC-PolScope with a straight optical path, the circular polarizer usually transmits left circularly polarized light and absorbs right circularly polarized light. In some microscopes, a mirror is present in the optical path between universal compensator and circular polarizer. Upon reflection, circularly polarized light changes handedness. Therefore, in such a microscope, the circular polarizer usually used in the LC-PolScope must be exchanged for one that transmits circularly polarized light of the opposite handedness. An even better solution is to rearrange the polarization elements in such a way that no mirror is in the optical path between them. This is because mirrors, like lenses, usually introduce polarization aberrations and reduce the extinction of the setup.

Finally, note that the circular polarizer is typically made for a specific wavelength, unless the quarter-wave plate is specified for a range of wavelengths, a so-called achromatic quarter-wave plate. If frequent changes between two or more wavelengths are necessary, it may be worth using such an achromatic circular polarizer or obtaining a tunable circular polarizer made, for example, from a linear polarizer and a liquid-crystal variable retarder.

Universal compensator. The universal compensator is manufactured by Cambridge Research and Instrumentation and comprises two liquid-crystal variable retarders and a linear polarizer (see text box The Universal Compensator). These components are usually bonded together, and their polarization optical performance is assessed as a single unit. When considered as a single unit, the universal compensator can also be called a *variable polarizer* due to the following property. When monochromatic light passes through the device so that it first passes through the linear polarizer, and then through the two variable retarders, the light leaving the retarders can have any desired polarization state imposed upon it. This can be a linear polarization of any azimuth or elliptical polarization of any handedness, ellipticity, or azimuth (see Equations 2, 3, and 4 in Shribak and Oldenbourg 2003). When light shines through the device in the opposite direction, so that it first passes through the variable retarders and then through the linear polarizer, the device functions as a universal polarization analyzer. In this configuration, extinction can be achieved for any input polarization by setting the retardance of the liquid-crystal variable retarder plates to suitable values. On the basis of the retardance values needed for extinction, it is possible to calculate the polarization of the incoming light using equations equivalent to those mentioned earlier. This property can be used for some unconventional applications of the universal compensator. For example, in fluorescence imaging, when inserting the universal compensator in the illumination path or the imaging path (or both), the specimen can be illuminated with light of a specific polarization and/or the light from the specimen can be analyzed for any given polarization state.

With regard to the performance of the universal compensator, instructions on how to measure the extinction coefficient of the complete PolScope optical setup are provided in the section on Performance Characteristics and How to Measure Them.

In addition to extinction and transmission, another important performance parameter of the universal compensator is the time required to switch reliably between two polarization states, the so-called settling time. The settling time is determined by the liquid-crystal devices and is usually greater than 100 ms. However, faster settling times are possible and liquid-crystal devices can be constructed with settling times of 30 ms and less, usually at the expense of the range of retardance values to which the devices can be adjusted.

Choice of Optics

As indicated earlier, the polarization distortions introduced by the objective and condenser lenses limit the extinction that can be achieved in a polarizing microscope. But not all lenses are created equal. Most microscope manufacturers offer lenses that are designated "Pol" to indicate low polarization distortions which can arise from a number of factors, including stress or crystalline inclusions in the lens glass and the type of antireflection coatings used on lens surfaces. Some lens types are available only with differential interference contrast (DIC) designation. DIC lenses do not meet the more stringent Pol requirements but pass for use in differential interference contrast and can also be used with the LC-PolScope (without the Wollaston or Nomarski prisms specific to DIC, of course).

The polarization performance of the most highly corrected lenses, so-called Plan Apochromat objectives, is often compromised by the large number of lens elements, special antireflection coatings, and, in some cases, special types of glass used to construct the lenses (Inoué and Oldenbourg 1995). For some applications, these high quality lenses, which provide a large, highly corrected viewing field over a wide spectral range, is simply not required. If the objective lens is to be used only with the PolScope that requires monochromatic light and typically acquires images from a region near the center of the viewing field, a less stringent correction might well suffice. To find out what works best for a particular imaging situation, several lenses should be tested. It is also helpful to be able to select the best performing combination of condenser and objective lens from a batch of the same lens types.

Whenever possible or practical, oil immersion lenses should be used. This is because the transition of a light ray between two media of different refractive index ($n_{air} = 1.00$, $n_{glass} = 1.52$) introduces polarization distortions, especially for high numerical aperture (NA) lenses. The peripheral rays leaving the condenser front lens are highly tilted to the slide surface and their polarization is typically rotated when traversing the air-glass interface (Inoué 1952; Inoué and Hyde 1957). Oil, and to a lesser degree, water and other immersion liquids, greatly reduce polarization aberrations caused by air-glass interfaces between the specimen and the lenses. For a discussion of the origin of polarization distortions in lenses and optical systems, and of ways to reduce them, see Shribak et al. (2002). This publication also discusses the various conoscopic images that can be observed in a polarizing microscope equipped with crossed, or nearly crossed, polarizers and a Bertrand lens.

Lenses and other optical elements that are not placed between the universal compensator and the circular polarizer do not affect the extinction performance of the microscope setup. Therefore, in a dissecting microscope, for example, the polarizing elements and liquid-crystal devices might be better placed between the objective lens and the light source. The light source is typically an incandescent lamp with a diffuser plate in the base of the dissecting scope. An interference filter, followed by the fixed circular polarizer, can be placed on top of the diffuser plate, and liquid crystals can be mounted in a special holder in front of the objective lens. In a dissecting microscope, the rays captured by the objective to project the image have small tilt angles (low NA), as do the rays passing through the liquid crystals. These angles are too small to compromise the polarization performance of the liquid crystals.

Optical elements placed outside the polarization optical train do not affect the extinction, but they do contribute to the overall performance of the setup. Aside from considerations regarding image quality, the ability to measure small retardance features in the specimen also depends on the proper match between the resolution of the optical image and that of the electronic camera. The relationship between resolution, sensitivity, and speed of measurement will be discussed in the section on Performance Characteristics and How to Measure Them.

Inverted Microscope Stand

Inverted microscope stands, equipped with the required optical components, can be just as well suited to polarized light microscopy as upright stands. Discussed below are a few points to consider when using an inverted stand for the PolScope:

- Inverted stands frequently have a mirror in the optical path. This mirror has two effects:

  1. It causes circularly or elliptically polarized light to change handedness upon reflection (only relevant if the mirror is in polarization optical path, between the universal compensator and the analyzer). This means that the circular polarizer or analyzer might have to be exchanged for one with the opposite handedness.

  2. If the mirror is in the imaging path, it changes the symmetry of the raw LC-PolScope images, and the calculated azimuth angle must be inverted. A checkbox named "inverted microscope" is provided in the LC-PolScope software to accommodate this final inversion step. If this box was not checked at the beginning of the experiment and it is necessary to correct the angle later, the old azimuth value φ can simply be replaced by $(180° - \varphi)$.

- Plastic Petri dishes, which are often used with inverted scopes, cannot be used with the PolScope. However, plastic Petri dishes with a glass bottom are fine (see, e.g., http://www.glass-bottom-dishes.com/).

- If the chosen cell preparation is in a culture dish with a glass bottom, it might be worth considering transillumination using an immersion condenser that dips into the aqueous culture medium. Using a water immersable objective as a condenser lens will give the best results, but it requires modifications to the condenser mounting. As an alternative, a regular oil immersion condenser can be dipped into the medium. This setup will avoid the air/glass and air/water interface and the concomitant polarization distortions of a dry condenser lens. It also provides high-NA illumination for improved resolution. Of course, during the process, the sterility of the culture will be compromised. The condenser must also be carefully dried after contacting an aqueous medium.

Culture Chamber and Heating Stage

For a comprehensive discussion on culture chambers, heating stages, and other accessories for keeping live cells happy while they are mounted on the microscope, please refer to the special chapters in this volume. For observations using polarized light, the following issues should be noted:

- Absolutely avoid plastic components in the polarization optical path. Most plastic materials are highly birefringent and therefore greatly reduce the sensitivity and accuracy of the LC-PolScope measurement. In some cases, it is possible to replace the plastic material with suitable glass components, for example, the modified plastic Petri dish with glass bottom.

- Avoid equipment that introduces temperature gradients in the objective and condenser lenses, as well as in other optical components that are part of the polarization optical train of the microscope. Temperature gradients in glass cause stress and associated stress birefringence, which reduces the sensitivity of the PolScope. Stress birefringence also introduces a background retardance that can change over time, modulating the calibration of the instrument.

Choice of Light Source

The PolScope requires a stable, bright, monochromatic light source. Depending on the application, the requirements can be met by a halogen lamp, arc lamp (e.g., mercury or xenon), light-emitting diode (LED), or suitably scrambled laser light. For a more detailed discussion of light sources for live cell imaging, see Chapter 6. A brief discussion of the effect of different lamp characteristics on PolScope measurements is presented here.

Stability is paramount among the specifications for a light source suitable for the PolScope. The lamp intensity must be stable during the exposure time of a stack of raw PolScope images. For example, for measuring retardance values of approximately 0.1 nm using a standard LC-PolScope optimized for low-retardance measurements, the lamp intensity must be stable within 1%. An unstable light source is tantamount to varying the exposure level between raw PolScope images, which in turn has the same effect as introducing a uniformly birefringent plate in the polarization optical train. However, in some cases, PolScope images that were exposed using an unstable light source can be corrected after the fact using the background ratioing method discussed below in the section on Background Birefringence and How to Correct It.

Brightness is another important factor to consider in polarization microscopy. Obviously, the intensity of the light source is directly proportional to the time it takes to expose the camera detector to appropriate light levels. Furthermore, for measuring small retardance values, the PolScope is operated with universal compensator settings that are near extinction.

Hence, the light levels on the camera, and in the eye piece for direct observations, are low, and it can become difficult to align the instrument and to position and focus on the sample correctly in the absence of sufficient light. Therefore, a bright light source is needed for an efficient work flow, especially when working with live cells. Since cells are sensitive to light, the PolScope setup should include a shutter that is conveniently operated and blocks the light to the specimen whenever possible. Blocking the light when it is not needed is an efficient way of reducing light damage to the cells.

The PolScope requires monochromatic light for its operation. Light with a spectral bandwidth of 30 nm is an upper limit and should not be exceeded without risking loss of sensitivity (increased spectral width reduces the extinction of the setup). Therefore, most light sources, except for lasers, require an interference filter for selecting or narrowing the spectral range of the source. In contrast, a laser's spectral width is usually too narrow and its time coherence too long for wide-field imaging (Inoué and Spring 1997).

In our experience, a 100-W mercury arc lamp, as supplied by microscope manufacturers and equipped with a 546 nm interference filter (http://www.omegafilters.com or http://www.chroma.com), provides sufficiently bright, monochromatic light for measuring fast dynamic events, such as spindle dynamics during mitosis. Other strong emission lines in the spectrum of a mercury arc lamp have center wavelengths of 577 and 436 nm and can also be used after exchanging the interference filter and circular polarizer suitable for the selected wavelength. Typically, when an arc burner is used for more than 150 hours, the gap between the electrodes increases and the shape of the tip of the electrodes often becomes irregular, leading to an unstable arc geometry. As a result, the lamp will start to flicker. When lamp flickers become apparent on direct viewing, or the flickers can be recognized as elevated background retardance that changes erratically over time (see above), it is probably time to change the bulb. To improve resolution and minimize image artifacts, a fiber-based Ellis light scrambler, available from Technical Video, Ltd. (http://www.technicalvideo.com/), can be used. Finally, there are several ways to adjust the amount of light delivered to the specimen, including a diaphragm that varies the light level entering the fiber optic light scrambler and neutral density filters with calibrated attenuation factors. Avoid closing the condenser aperture to reduce the light level, because this will also decrease the resolution of the image.

In the future, LEDs are expected to replace traditional light sources for many applications, especially those requiring monochromatic light. Over the past few years, LEDs have improved significantly with respect to their brightness and they now provide higher intensity monochromatic light than halogen lamps. LEDs are highly efficient and stable light sources, whose brightness can be controlled by the source current. They can also be used to provide short light pulses with peak powers that are several times the direct current (DC) power limit (see, for example, http://www.lumileds.com).

Combining the PolScope with DIC and Fluorescence Imaging

For DIC, two matching prisms are added to the polarization optical path (for a discussion of the standard DIC arrangement, see Salmon and Tran 1998). The two prisms, called Wollaston or Nomarski prisms, are specially designed to fit in specific positions, one before the condenser and the other after the objective lens, and to match specific lens combinations. The prisms in their regular positions can be combined with the PolScope setup; in which case, the universal compensator and circular analyzer take the places of the linear polarizer and analyzer in a standard DIC arrangement. In fact, the same DIC prisms function equally well using either linearly or circularly polarized light.

For best results, the PolScope is first calibrated without prisms to find the extinction setting for the universal compensator. With the compensator set to extinction, the prisms are entered into the optical path. Typically, one of the two prisms has a mechanism for fine-adjusting its position, which in turn affects the brightness and contrast of the DIC image. In a PolScope setup, the mechanical fine adjustment or the universal compensator can be used to add a bias retardance to change the brightness and contrast of the DIC image. When using the universal compensator, retardance must be added or subtracted from the extinction setting of either the LC-A or LC-B retarder (see Fig. 1), depending on the orientation of the shear direction of the prisms.

For live cell imaging, it is very useful to be able to switch easily between PolScope and DIC imaging. DIC provides good contrast of many morphological features in cells using direct viewing through the eyepiece or by video imaging. When viewing specimens that have low polarization contrast, it is more effective to align the optics, including the visualization of the specimen, using DIC. After the optics have been aligned and the specimen is in focus, the DIC prisms are removed from the optical path and the PolScope-specific adjustments completed.

Fluorescence imaging can be combined with the PolScope in several ways. Fluorescence is commonly observed using epi-illumination, which requires a filter cube in the imaging path. The filter cube includes a dichromatic mirror and interference filters for separating the excitation and emission wavelengths. For best results, the filter cube should be removed for PolScope observations to avoid the polarization distortions caused by the dichromatic mirror. For observing fluorescence, on the other hand, the PolScope analyzer should be removed, because it attenuates the fluorescence emission by at least 50%. To meet both requirements, the PolScope analyzer can be mounted in an otherwise empty filter cube holder, which is moved into the optical path as the fluorescence cube is moved out, and vice versa.

If, however, the option of removing the fluorescence cube is unavailable, the cube can remain in the optical path for PolScope observations. In this case, the following points should be considered:

- For the light source of the PolScope, choose a wavelength that is compatible with the emission wavelength filter in the fluorescence cube. For example, fluorescein isothiocyanate (FITC) requires excitation with blue light (485 nm) and fluoresces in green. The FITC dichromatic beam splitter and a barrier filter passing green fluorescence light with a wavelength longer than 510 nm would be compatible with 546-nm light for PolScope observations using a mercury arc burner for the transmission light path.

- The PolScope must be callibrated with the fluorescence filter cube in place. The polarization distortions caused by the dichromatic mirror are partially counteracted by the calibrated settings of the universal compensator.

- The light source for one imaging mode must be blocked while observing with the other imaging mode.

- Removing the PolScope analyzer, while observing fluorescence, more than doubles the fluorescence intensity.

While beyond the scope of this article, it should also be noted that the universal compensator can be used to analyze fluorescence polarization. As described above in the section on Polarization Optical Components, the universal compensator can be regarded as a universal polarization analyzer or a universal polarizer. It can therefore have a valuable role in selectively exciting or detecting polarized fluorescence. Fluorescence polarization imaging has become a powerful tool for probing structure together with functional aspects in living cells and working model systems. Fluorescence polarization can be used as binding assay of small and medium-sized molecules to receptors and other binding sites (Peterman et al. 2001), and as a probe to differentiate between conformational states (Warshaw et al. 1998). The combination of molecular specificity and dynamic measurement of molecular orientation by fluorescence polarization analysis has revealed the rotational motion of actin filaments as they slide over myosin molecules fixed on a glass surface (Sase et al. 1997). Recently, Inoué et al. (2002) identified and analyzed the remarkable fluorescence polarization properties of single crystals of GFP. This work demonstrated that the highly polarized nature of GFP fluorescence provides a new method for dynamically observing, and quantifying, the changing orientation of fluorescent chromophores constituting (or attached to) functional molecular structures.

Specimen Preparation

Last, but certainly not least, the specimen and its mount must be considered as part of the polarization optical train. Several chapters in this volume discuss the preparation of viable cells suitable for observations in the light microscope. Other useful resources for polarized light microscopy include Inoué (2002a) and Oldenbourg (1999), which has a section on cell types suitable for live cell imaging of spindles and their preparation. The following advice is provided for the preparation of living cells for analysis using the PolScope.

- Prepare the specimen so that, when mounted in the microscope, at least one clear area without cells or birefringent material can be identified and moved into the viewing field when required. A clear area is needed for calibrating the PolScope and for recording a PolScope stack of background images. The background images provide a record of the combined effect of the polarization properties of all optical components, including the slide and coverslip, but not of the specimen itself. The background images are used to remove spurious background retardance from PolScope images, allowing precise measurement and imaging of the cells and structures of interest. Many cell cultures and cell-free systems can be prepared without special attention to this requirement. For example, a free area can often be found around sparsely plated cells. Cultures of free-swimming cells might have to be diluted before mounting a small drop of the suspension between slide and coverslip to observe free areas. Sometimes, it is helpful to add a tiny drop of oil or other nontoxic, immiscible liquid to the preparation. This clear drop can provide an area for calibrating the instrument and taking background images in a preparation that is otherwise dense with birefringent structures. For further discussions on collecting background images see the sections on Background Birefringence and How to Correct It and Performance Characteristics and How to Measure Them.

- Some cells tend to generate strong edge birefringence at their surface, which can obscure other details. To reduce this edge birefringence, the refractive index of the medium can be increased by adding polyethylene glycol, polyvinylpyrrolidone, or some other harmless polymer or protein substitute. The concentration of substitute required might vary from a few percent to 16% or more, depending on the average refractive index of the cytoplasm of the cells of interest. It is best to test various media containing different polymer concentrations by suspending cells in them and examining the birefringence in a slide and coverslip preparation with DIC or polarized light. At the optimal polymer concentration, no distinct cell boundary is visible, giving the impression that the organelles and cytoskeleton are suspended in space. After determining the optimal concentration for imaging, the medium should be tested for compatibility with growing cells. Cells should grow and develop normally in media with these types of additives. The molecular weight of the polymer should be approximately 40 kD or more to prevent its uptake into cells.

- Earlier in this chapter, the use of immersion optics whenever possible was recommended (see Choice of Optics). When using high-NA oil immersion optics, specimens should be prepared such that the cells or structures of interest are as close to the coverslip as possible. A layer of more than 10 μm of aqueous medium introduces enough spherical aberration to noticeably reduce the resolution of cell images, although the use of water immersion optics alleviates this problem (Inoué and Spring 1997).

- Finally, as discussed earlier, plastic Petri dishes must not be used to observe specimens, unless the dish has a glass bottom (see Inverted Microscope Stand).

IMAGE ACQUISITION AND MEASUREMENT

After the optical components have been aligned and the specimen has been correctly mounted, imaging can begin. The appearance of raw PolScope images depends on the settings of the universal compensator. To illustrate the workings of the universal compensator, four different PolScope images of the same specimen are shown at the top of Figure 1. The specimen is an aster, formed in lysate prepared from surf clam eggs (Palazzo et al. 1992). Asters are convenient test objects since the radial arrangement of their microtubules leads to a spherically symmetrical birefringence pattern. The centrosome in the middle of the picture shows little birefringence. Near the surface of the centrosome, the birefringence increases rapidly due to the increasing density of aligned microtubules. With increasing distance from the centrosome, the microtubule density and hence its birefringence decreases. At different points in the aster, the slow axis of birefringence is oriented parallel to the direction of the microtubule bundles at that point. The top left image of Figure 1 shows the specimen illuminated with circularly polarized light. The birefringent astral rays appear bright, regardless of their orientation, against the dark background. The analyzer blocks background light that has opposite circular handedness to the illuminating polarization. The following three images to the right show the specimen illuminated with elliptically polarized light. The ellipticity is the same for each of the three settings, but the orientation of the principal axes of the polarization ellipse is rotated by multiples of 45°. The intensity recorded in a specific image point of the aster depends on the density and orientation of the microtubule array at that point. The intensity is a function of the microtubule density and the mutual orientation between the polarization ellipse and the microtubule alignment. The algorithms that are implemented in the PolScope software evaluate the intensities and compute the corresponding birefringence and orientation values at each image point.

Calibration

Calibrating the LC-PolScope optimizes the liquid-crystal settings that are used to measure specimen retardance. During calibration, the specimen is usually moved aside, or, when necessary, removed from the optical path, to provide a clear view of an area that has no retardance (see Specimen Preparation). Calibration usually proceeds automatically, supported by the software that is part of the LC-PolScope instrument, as follows: With the clear area in view, the software (or user) selects a small region of interest (ROI), typically near the center of the image and measures the average intensity in that region. Subsequently, the retardance of one of the liquid crystal devices in the universal compensator is slightly changed and a new intensity reading is taken and compared to the previous reading. The sequence of changing retardance and reading the intensity in the ROI is repeated until the following settings are optimized:

- *Setting 1 (extinction):* Starting with nominal retardance values ($\lambda/4$, $\lambda/2$), the retardance of both LC-A and LC-B are changed until the intensity reading is at a minimum. The pair of optimized retardance values will be called ($R_{1,\text{LC-A}}$, $R_{1,\text{LC-B}}$). The optimized values are usually within 0.04 λ of the nominal values.

- *Setting 2:* Starting with Setting 1, a so-called "swing" retardance is added to LC-A. The swing retardance, or simply swing, is user-selected and depends on the expected sample retardance values. For imaging living cells, the swing retardance is typically set to 0.03 λ or 16 nm (λ = 546 nm). After the swing retardance is added to LC-A, the intensity is read. This intensity is used as reference I_{ref} for optimizing the remaining settings. Setting 2 requires no further adjustments. Hence, the retardance pair for Setting 2 is ($R_{1,\text{LC-A}}$ + swing, $R_{1,\text{LC-B}}$).

- *Setting 3:* Starting from Setting 1, the swing retardance is added to LC-B and the intensity is read. Subsequently, LC-B is varied, up or down, in small steps until the measured intensity is equal to I_{ref} (LC-A is kept constant at $R_{1,\text{LC-A}}$). The resultant retardance pair for Setting 3 is ($R_{1,\text{LC-A}}$, ~ $R_{1,\text{LC-B}}$ + swing), where the sign " ~ " indicates that the optimized value might vary from the start value.

- *Setting 4:* Starting from Setting 1, the swing retardance is subtracted from LC-B, and the retardance of LC-B is varied in small steps until the measured intensity is equal to I_{ref}. The resultant retardance pair for Setting 4 is ($R_{1,\text{LC-A}}$, ~ $R_{1,\text{LC-B}}$ − swing).

Recently, Shribak developed a series of algorithms that use from two to five settings of the universal compensator (Shribak and Oldenbourg 2003). Adding the following fifth setting improves the sensitivity and accuracy of retardance measurements:

- *Setting 5:* Starting from Setting 1, the swing retardance is subtracted from LC-A and the retardance of LC-A is varied in small steps until the measured intensity is equal to I_{ref}. The resultant retardance pair for Setting 5 is (~ $R_{1,\text{LC-A}}$ − swing, $R_{1,\text{LC-B}}$).

An abbreviated measurement process that uses only Settings 2 and 3 can be used. This speeds up the measurement process, but reduces image quality for preliminary examinations of birefringent specimens.

As discussed earlier, the universal compensator can be thought of as a variable polarizer. Setting 1 produces a circularly polarized light beam, whereas Setting 2 and higher produce elliptically polarized beams, each having equal ellipticity but different orientations of the principal axes.

Retardance Measurement

The images recorded of a birefringent specimen using the aforementioned Settings 1 through 5 of the universal compensator are evaluated for measuring the retardance at every resolved specimen point. The image algorithms used in the evaluation are integral to the software of the LC-PolScope system and are published elsewhere (Shribak and Oldenbourg 2003). The algorithms are briefly summarized here to give some insight into the relationship between the intensity values (I), measured by the camera in a given image point, the swing value, used in calibrating the liquid crystals, and the sought-after retardance and azimuth values of a specimen region. The summary is based on the original algorithm using the first four settings of the universal compensator.

Intermediate results A and B are determined first, based on the image intensities I_1 to I_4

and the swing value χ expressed as a phase angle (e.g., for a swing of $0.03\,\lambda$, $\chi = 0.03 \cdot 360° = 10.8°$):

$$A \equiv \frac{I_2 - I_3}{I_2 + I_3 - 2I_1} \cdot \tan \frac{\chi}{2}$$

$$B \equiv \frac{I_2 + I_3 - 2I_4}{I_2 + I_3 - 2I_1} \cdot \tan \frac{\chi}{2}$$

Using these intermediate results, expressions for the retardance R and azimuth φ are as follows:

$$R = \arctan\left[\sqrt{A^2 + B^2}\right] \text{ if } I_2 + I_3 - 2I_1 \geq 0$$

$$R = 180° - \arctan\left[\sqrt{A^2 + B^2}\right] \text{ if } I_2 + I_3 - 2I_1 < 0$$

$$\varphi = \frac{1}{2} \arctan\left(\frac{A}{B}\right)$$

The above formulas for the retardance and slow axis direction of a birefringent region apply to transparent specimens that possess linear birefringence. In the visible wavelength range, and for nonabsorbing biological specimens, linear birefringence is typically the dominant optical anisotropy. However, absorbing specimens, recognized by their colored appearance, might possess dichroism, which can affect retardance measurements. It is best to choose a wavelength for which the absorption is minimal. In addition, beware that the PolScope measures retardance values above half a wavelength ambiguously. For example, assume the specimen retardance $R_{specimen}$ is in the range $\lambda/2 < R_{specimen} < \lambda$. The measured retardance is then $R = \lambda - R_{specimen}$, and the azimuth angle is turned by 90°. For $\lambda < R_{specimen} < 3\lambda/2$, $R = R_{specimen} - \lambda$ and the measured azimuth angle is correct. However, a retardance in the range $0 < R_{specimen} < \lambda/2$ is measured correctly in both the retardance and azimuth value.

As discussed in the next section, it is necessary to distinguish between so-called instrument, or background, birefringence, and the specimen birefringence which is actually of interest.

Background Birefringence and How to Correct It

After calibrating the instrument, it is good practice to record a so-called background stack. Figure 2 shows images of a clear area of a thin, aqueous sample, sandwiched between a microscope slide and coverslip. The images were recorded after the calibration procedure was completed. The features seen in the images are not part of the specimen, but are contributed by other components of the optical setup. The retardance and azimuth image (panels 5 and 6 in Fig. 2) were computed using the algorithm of the LC-PolScope software. In the retardance image, a dark, almost vertical band crossing the center of the image can be recognized. The dark center is a result of the calibration procedure, which seeks settings for the liquid-crystal universal compensator that result in near-zero retardance in the center of the image. The exact shape of the dark region varies between different microscope setups and depends on the lenses and other optical components used. Outside the dark area, the image shows a smoothly varying retardance background that increases with distance from the central area. This smooth retardance background is due to the residual polarization distortions introduced by optical components, including the microscope objective and condenser lens. In addition to this background retardance, there are several localized retardance features, which are the result of contaminations (dirt), blemishes, and imperfections, that are sometimes located on surfaces near the camera face plate. All of these

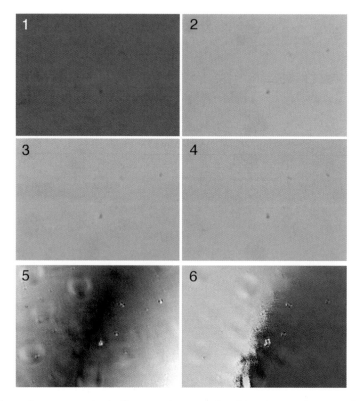

FIGURE 2. PolScope images acquired of an empty area of a thin aqueous layer sandwiched between a microscope slide and coverslip. The focus was set to the center of the aqueous layer which lacks any distinct feature. (*Panels 1–4*) Raw PolScope images representing the recorded intensities from the empty area using universal compensator Settings 1 through 4. (*Panel 5*) Computed retardance image, black representing zero retardance and white 1.5-nm retardance. (*Panel 6*) Azimuth image (black is 0° and white is 180°). These so-called background images illustrate the retardance and its azimuth contributed by components of the optical setup other than the focused layer in the specimen (see text). (Oil immersion Plan Apochromat objective 60×/1.4 NA, and oil immersion condenser.)

instrument or background retardance features are associated with slow axis orientations, as shown in panel 6 of Figure 2.

If left uncorrected, background retardance can reduce the quality of specimen images. Problems can occur when imaging specimens that contain structures with retardance values that are one to two orders of magnitude smaller than the average background retardance. Therefore, a background correction procedure has been developed, which removes the effect of instrument birefringence on the measured specimen retardance.

Figure 3 shows retardance images of an oral epithelial (cheek) cell, prepared by putting a drop of saliva between a slide and coverglass. Resolving the fine ridges on the surface of the cheek cell is a useful test of the achieved resolution and contrast in phase-based microscopy, such as phase contrast, differential interference contrast, and polarized light microscopy. Figure 3 shows two retardance images, one without background correction and one with correction. The difference is remarkable and illustrates the indispensability and efficacy of the procedure.

The theory of background correction is described in detail elsewhere (Shribak and Oldenbourg 2003). In practice, the correction requires a stack of raw PolScope images of a clear area in the specimen prep. The retardance image in Figure 3, for example, was corrected using a background stack that was recorded after the cheek cell was moved to the side using

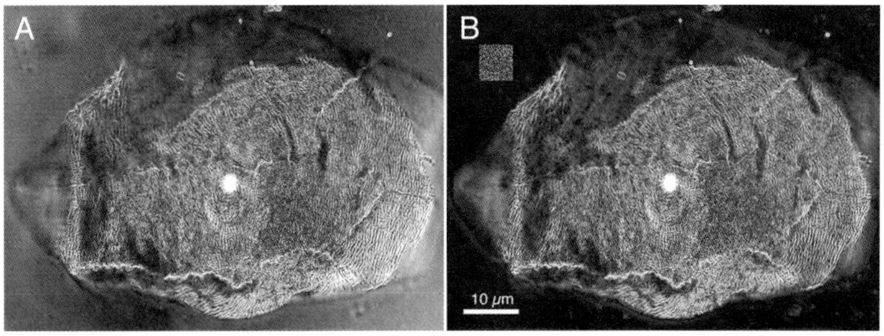

FIGURE 3. Retardance images of a buccal mucosal cell (cheek cell) illustrating the effect of background retardance and of its removal using background correction (same specimen preparation was used for Fig. 2). (*A*) Retardance image (white is 3-nm retardance) without background correction; (*B*) retardance image with background correction. In *B*, the birefringence of cell structures such as the fine ridges on the cell surface are faithfully reproduced. The clear area around the cell is dark, showing no retardance, as expected. In the top left corner of *B*, a square region of the background is contrast-enhanced (white is 0.3-nm retardance) to illustrate the measurement noise as discussed in Sensitivity and Noise Sources. In *A*, the retardance of the cell is contaminated by the background retardance. The background retardance is shown separately in Fig. 2.

the *x-y* translation stage of the microscope. For the typical experimental setup, the following steps are recommended:

1. Calibrate the microscope with the specimen in place while viewing an adjacent clear area that has no specimen retardance.

2. With the clear area in view, acquire a background stack that records the retardance due to instrument parts alone.

3. Move specimen under investigation into view.

4. Focus the object and acquire a sample stack that records the retardance of the object and of the instrument parts together.

5. When computing the object retardance, make sure that background correction is enabled and refers to the correct background stack.

The following points help to determine whether the background correction has worked:

- In the retardance image of the object, the birefringent parts appear bright against a dark, flat background.

- When looking at background areas with suitable contrast enhancement, a random noise level is observed. The expected average retardance value of the noise level is discussed in the section on Performance Characteristics and How to Measure Them.

- In the azimuth image, the background areas exhibit a random variation of the azimuth angle between 0° and 180°. This large range of random angles is very striking in azimuth images and serves as an easily recognizable indicator for the quality of the background correction.

The background correction procedure can use a single background stack repeatedly and can correct, for example, a time series of sample images measuring the subtle changes in birefringence in a living cell. The following is some further advice on background correction.

- If specimen parts that should have no retardance show some uniform retardance background, it may be time to record a new background stack. The uniform retardance background might be caused by a drift in the liquid-crystal settings due to changes in room temperature, for example (see below for background ratioing) .

- The background image should be slightly defocused before acquiring a background stack. Defocusing can avoid image features that are due to blemishes on the coverslip surface, for example, appearing in the background stack. If present in the background stack, such features will also show up in the background corrected sample image.

- Even though frame averaging may not be used for acquiring the sample stack, it should be used for acquiring the background stack. Frame averaging reduces the noise level in images. The noise level in the final, background-corrected sample image is the root mean square of the noise level in the background stack and the sample stack together. The expected noise level in acquired and computed images is discussed in more detail in the section on Performance Characteristics and How to Measure Them.

The final image correction procedure is known as background ratioing. This process can be helpful in cases where background-corrected sample images show an unwanted, uniform background retardance. This situation might arise while recording a time series of PolScope images, for example. Assume that, while the time series was being recorded, the liquid-crystal settings slowly drifted away from their calibrated values, due, for example, to a slow temperature increase in the room. The background-corrected sample images would then show a uniform background retardance that slowly increases from time point to time point. To avoid this situation, a new background stack could be recorded at appropriate time intervals. If, however, the recording of the time series prevents the specimen from being moved, or if the intruding background retardance was only noticed after the recording, then background ratioing can come to the rescue.

The process requires a background stack, a sample stack, and an area within the sample image that has zero retardance. The a priori knowledge that the identified sample area has zero retardance can be used to correct the entire sample stack. The correction modifies the intensity values in the raw PolScope images of the sample stack before calculating the retardance and azimuth images. Each of the raw PolScope images is modified separately as follows: (1) the average intensities in the identified area of the sample image and the background image are calculated; (2) the ratio (background/sample) of the average intensities is formed; and (3) all intensities in the sample image are multiplied with this ratio. After modification of the sample stack, the retardance and azimuth images are calculated using the usual background correction procedure. The procedure demonstrably corrects any small sample retardance that was measured with incorrectly calibrated liquid crystals. The procedure also corrects errors caused by an unstable light source (see Choice of Light Source). Liquid-crystal drifts and lamp flickers have similar effects on computed retardance and azimuth images. Both can be suppressed by the background ratioing procedure.

PERFORMANCE CHARACTERISTICS AND HOW TO MEASURE THEM

After aligning and optimizing the optical setup and carefully preparing the specimen, what performance can be expected from the LC-PolScope instrument? The performance characteristics of the PolScope can be divided into four categories: sensitivity, accuracy, speed, and resolution. Obviously, the four categories are interrelated, and for maximizing one category, some sacrifices will be made in another. The following section discusses these relationships and offers guidelines for optimizing the specifications that are typically required for live cell imaging.

Resolution

The spatial resolution of the PolScope image is mainly determined by the optical lenses and their proper use and alignment. The numerical aperture of the objective lens (NA_{obj}) and of the condenser lens (NA_{cond}) are equally important in determining the minimum distance (d_{min}) between two discrete object points that can still be separated in the image

$$d_{min} = \frac{\lambda}{NA_{obj} + NA_{cond}}$$

where λ is the wavelength of light used (Inoué and Oldenbourg 1995). To actually achieve this optical resolution in PolScope images, the following points must be considered:

- The specifications provided by the manufacturer of the lenses must be observed. This includes the coverslip thickness (usually Nr. 1.5 or 0.17 mm) required for the objective.

- Each additional optical component (filter, polarizer, compensator, etc.) in the imaging path should introduce less than a quarter-wave wavefront distortion.

- A high-NA oil immersion objective gradually loses its nominal resolution when focusing into an aqueous layer of more than 10-μm thick. Water immersion (or water-immersable) lenses should be used for focusing deeper into cells or tissues.

- When using high-NA lenses in the PolScope, sensitivity can be improved by partially closing the condenser aperture and increasing the extinction of the polarization optical train (see Choice of Optics). Using 1.4 NA oil immersion optics, for example, we usually close the condenser aperture down to 1.0 NA, choosing a compromise between reducing resolution and improving extinction.

- The resolution of the final PolScope image is also determined by the camera's resolution. A zoom lens before the camera provides some flexibility in adjusting the field size projected onto the camera face plate. The zoom factor is usually set so that the optical resolution matches the camera resolution, i.e., the distance d_{min} is spanned by 3 camera pixels (according to the Nyquist theorem, d_{min} should be spanned by at least 2 pixels). Under normal circumstances, there is no benefit in choosing a larger zoom factor. However, a smaller zoom factor increases the brightness of the image and allows the exposure time to be reduced enough to catch fast events in the cell. Thus, a zoom lens can provide the flexibility required to compromise between spatial and temporal resolution.

The resolution and distance calibration of a PolScope image can be measured using a number of different test objects:

1. The distance calibration is affected by the magnification of the objective and intermediate zoom lenses in the optical path. If all magnification factors are known, then the nominal distance calibration can be estimated by dividing the camera pixel size by the various magnification factors. For a reliable way of calibrating the image distance, a micrometer scale, available from microscope manufacturers and distributors, should be used.

2. To test the resolution, a single submicroscopic point or line object, such as a microtubule (Oldenbourg et al. 1998), small calcite crystal (Oldenbourg and Török 2000), and even a submicroscopic isotropic scatterer such as a latex bead, or a small organelle in a cell can be used. Even though the scatterer is isotropic, there is a residual edge effect that makes the retardance image of a submicroscopic bead look like a small doughnut (Török 2000). Single, submicroscopic objects tend to have very low retardance, usually below 1 nm. The distance d_{min} can be estimated from the half-width at half-maximum of the retardance image of the

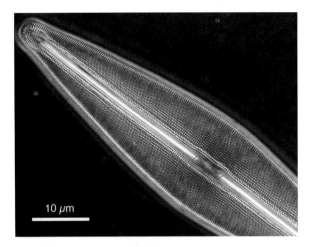

FIGURE 4. Retardance image of a *Frustulia rhomboides* diatom shell. Rows of pores are 0.29 μm apart. The diatom was imaged with a Plan Apochromat 60×1.4 NA oil immersion objective and Plan Achromat oil immersion condenser, with the condenser aperture set to about 1.0 NA. White corresponds to 4-nm retardance.

submicroscopic object. A single-point scatterer is also useful to test for the presence of spherical aberrations in the imaging optics. When spherical aberration is present, the defocused image looks different when changing the focus from above to below the point scatterer (e.g., focusing into a water layer with an oil immersion objective introduces spherical aberration). Spherical aberration is the main cause of an asymmetric point spread function.

3. The silica shell of diatoms is studded with pores and markings that have species-specific regular distances. Figure 4 shows the retardance image of a *Frustulia rhomboides* diatom with rows of pores that are 0.29 μm apart. The retardance of diatoms is due to edge birefringence induced by the mismatch between the mounting medium and the silica of the shell. When using a high-NA objective, the retardance measured around pores and markings is between 2 and 10 nm. The retardance pattern sensitively depends on the focus position and therefore is not suitable for calibrating the retardance and azimuth values of PolScope images. Test slides that include eight different species with spacings ranging from 0.27 μm to 1.25 μm are available from Carolina Biological Supply Company (http://www.carolina.com/).

4. Test slides with lithographed point and line objects, gratings of different periodicities, and other structures are an excellent way of testing the resolution and contrast transfer of microscope optics (Oldenbourg et al. 1996). Unfortunately, the availability of these test slides is limited. Magnification standards that include repeating parallel lines with pitches of 0.5 μm and longer are commercially available from Geller MicroÅnalytical Laboratory (http://www.gellermicro.com/).

Sensitivity and Noise Sources

The sensitivity of the PolScope is determined by the smallest retardance that can reliably be measured with the instrument. Sensitivity primarily depends on the noise level of PolScope images. It is useful to distinguish between the two categories of noise sources: optical noise and measurement noise. In a static sample, the two categories are easily distinguished by their spectral composition. Measurement noise is typically "white noise," changing rapidly at random in PolScope images, both from pixel to pixel and frame to frame. Optical noise, on the other hand, usually varies slowly from pixel to pixel and from frame to frame.

Optical noise originates from optical parts of the microscope that introduce unintended phase shifts in the polarized light beam, leading to spurious sample retardance in PolScope images. Sources of optical noise include miscalibrated liquid crystals of the universal compensator, differential transmission and phase shifts in microscope lenses, and optically anisotropic parts of the specimen itself. Miscalibrated liquid crystals and polarization distortions, introduced by optical components other than the specimen, are effectively compensated using the background-correction procedure described earlier. Unfortunately, the specimen itself is often a source of spurious birefringence. PolScope images of thick specimens typically show the retardance of the in-focus layer and of the out-of-focus layers together. Out-of-focus birefringence contributes to the optical noise and can reduce the sensitivity and accuracy of measuring the in-focus retardance in PolScope images. Suitable specimen preparations, avoiding the unnecessary stacking of specimen parts, for example, can reduce optical noise from out-of-focus parts. In addition, out-of-focus retardance can sometimes be estimated and its effect on in-focus retardance can be partially removed (J.R. LaFountain and R. Oldenbourg, in prep.).

Measurement noise originates almost exclusively from shot noise of the light detected by the camera. Shot noise, also called photon statistics noise, derives from the inherent ambiguity in counting a finite number of photons (Taylor 1997). Because raw PolScope images are typically exposed to near full capacity of the CCD sensor chip, other electronic noise arising from modern CCD cameras, such as read noise, dark current, and amplifier noise of the detector circuitry, is negligible compared with shot noise. Current CCD sensors in cameras suitable for use with the PolScope typically have a single well capacity of about 18,000 electrons, which corresponds to 18,000 detected photons per image pixel. The ambiguity associated with detecting 18,000 photons is $\sqrt{18,000}$ and the signal-to-noise ratio is $S/N = 18,000/\sqrt{18,000} = 134$. This is the maximum S/N possible for this type of sensor chip. Detecting fewer than 18,000 photons decreases the S/N. For acquiring raw PolScope images, the camera pixels are filled to around 10,000 electrons, on average, leading to $S/N = 10,000/\sqrt{10,000} = 100$. Hence, the shot noise of raw PolScope images recorded with this type of camera leads to a relative intensity error of around 1%.

To make use of the full well capacity, the camera gain should be set to a minimum. At this setting, the analog to digital conversion in the camera converts the charge in a full well to a number near the upper limit of the digital range (either 8- or 12-bit). The camera offset should be adjusted with the light blocked from the camera. Adjust the offset, so that when no light is falling on the camera face plate, the pixel reading is near but slightly larger than zero.

The noise in measured intensity values leads to ambiguities in the calculated retardance and azimuth values. A more detailed analysis shows that the ambiguity of the measured specimen retardance is the swing retardance multiplied by the relative intensity error. For example, an intensity shot noise of 1% and a Swing retardance of 16 nm leads to a retardance sensitivity of 0.16 nm.

It might be tempting to reduce the swing retardance in order to improve the retardance sensitivity of the instrument. This is indeed a possible solution but it comes with some pitfalls. These are discussed later in connection with the effect of the extinction of the polarization optical train on retardance sensitivity.

The ambiguity in the measured azimuth angle increases inversely with the measured retardance value. For retardance values near zero, below the instrument's sensitivity, the shot noise in raw PolScope images results in a random azimuth angle that is uniformly distributed between 0° and 180°. The random distribution of azimuth values measured in specimen areas with zero retardance is indeed a good indicator of successful background correction. Azimuth values that are associated with retardance values that are larger than twice the sensitivity typically have an ambiguity of 1° to 2°.

The retardance sensitivity of a PolScope setup can be estimated using the results of a calibration procedure. Figure 2 shows the retardance and azimuth images of a clear specimen area after calibrating a PolScope equipped with high-NA lenses and a digital CCD camera.

Most retardance features are due to optical noise, as discussed in the section on Background Birefringence and How to Correct It. The optical noise is removed by the background correction procedure, as illustrated in Figure 3. In panel B of Figure 3, the background area around the cheek cell is black and shows no retardance, except for the measurement noise. The measurement noise is made visible by contrast enhancement of a small square region in the top left corner of panel B. The mean retardance background due to measurement noise in panel B is 0.08 nm, using four frame averaging.

The shot noise of a single raw PolScope image can be improved by averaging several camera frames that are captured in quick succession while leaving the LC setting unchanged. Averaging N frames reduces the shot noise by a factor $1/\sqrt{N}$. The effectiveness of averaging though depends on the bit depth of a digital image. Gray scale images are typically pixel arrays, where each pixel is represented by either an 8- or 16-bit number. (8-bit integer values range between 0 and 255, 16-bit integers range between 0 and 65,536.) As discussed earlier, a well-exposed single frame has about 1% shot noise. Averaging eight such frames reduces the shot noise to a level below the digital resolution of 8 bits. Therefore, averaging of more than eight frames will improve image noise only if image data are stored using 16 bits per pixel.

Unfortunately, averaging does take time and therefore reduces the number of measurements that can be made within a certain time period. A faster alternative is to use a camera with a bigger well capacity so that more photons can be captured before the camera frame is read out.

The sensitivity of the PolScope also depends on the algorithm used to calculate the retardance and azimuth images. Recently, Michael Shribak developed a series of algorithms that use from two to five settings of the universal compensator (Shribak and Oldenbourg 2003). Increasing the number of settings used increases the sensitivity of the measurements.

The extinction of the polarization optical train in the PolScope has not yet been mentioned. It is ultimately the extinction factor that determines the lower limit of retardance that can reliably be measured with the PolScope. As discussed earlier, the extinction factor is the ratio between the intensity with parallel polars (i.e., the total intensity of polarized light) and the light level passing through the system when the polarizers are crossed (extinction $= I_{\parallel} / I_{\perp}$). The shot noise of the background light that passes, even at extinction, ultimately limits the sensitivity of the instrument. The following is a realistic numerical example: The ratio of the intensity with parallel polars to the intensity of one of the elliptical polarization settings (Settings 2 through 5, see above) is $1/\sin^2(\chi/2)$, where χ is the swing retardance in degrees. Assuming $\chi = 10.8°$ (16 nm retardance), this ratio becomes 113. With high-NA optics, on the other hand, an extinction factor of about 500 can be achieved. Hence, the average intensity at the elliptical polarization settings is only five times higher than the average intensity at extinction. Reducing the swing value by a factor of 2 decreases the intensity at the elliptical settings by a factor of 4 and the background light becomes dominant for all PolScope settings.

To measure the extinction of the PolScope, all optical components in the polarization optical train must be included and a clear area in the specimen viewed. The extinction setting must be optimized, either by manually changing the liquid-crystal settings in the software or by calibrating the instrument. The lamp intensity is then reduced and half a wavelength retardance is manually added to the extinction setting (e.g., changing Setting 1 from $[\lambda/4, \lambda/2]$ to $[3\lambda/4, \lambda/2]$). The light level on the camera will be very bright because the polarization optical train is now set for full transmission. If necessary, the lamp intensity can be further reduced. The light level near the center of the image is measured by choosing a short exposure time that results in medium gray pixels. Next, liquid crystals are adjusted back to the extinction setting and the exposure time is increased until the light level registered by the camera is equal to the light level at full transmission. The ratio of the two exposure times gives the extinction factor.

Accuracy

Accuracy is related to systematic errors caused by incorrect system parameters. The system parameters that influence the accuracy of the measured specimen retardance include the calibration of the swing retardance, the proper mutual orientation of the birefringence axes of the retarder and polarizer components, and the linearity of the camera response. A trivial but sometimes overlooked source of error is the azimuth reference angle entered into the PolScope software.

The optical components of the PolScope, including the camera, were manufactured and selected to satisfy a high standard of quality. Therefore, the accuracy of the instrument is expected to be of the same magnitude as the random errors introduced by the sources discussed earlier. If the accuracy of the complete setup must be determined, the instrument should be tested and calibrated by measuring a test sample with known retardance and/or azimuth values. There are several points to consider when selecting the test sample and test procedure:

- As a test sample, a variety of birefringent objects with known retardance can be used. Wave plates are available from microscope manufacturers and optical companies. Use wave plates with low retardance such as quarter-wave plates, or the $\lambda/10$ or $\lambda/30$ wave plate of a traditional Brace-Köhler compensator.

- The retardance of a quartz wedge increases linearly with distance in the direction of the taper. The continuous, linear increase with distance provides a sensitive test for the linearity of the retardance measurement. The retardance of quartz wedges usually varies from near zero at one end of the wedge to several thousand nanometers at the other end. Be aware that the PolScope measures retardance values above half a wavelength ambiguously, as discussed under Retardance Measurement.

- Use a setup that is as complete as possible, including the imaging optics. That said, for some large test specimens like a commercial wave plate or quartz wedge, it might be necessary to remove the objective and condenser lens and use the Bertrand lens to focus on the test specimen on the sample stage. The imaging optics should have only a minor influence on the accuracy of the instrument.

- Needle-shaped crystals, such as ammonium nitrate, can be used as azimuth standards, as described in Hartshorne and Stuart (1960). Although the retardance of the crystal depends on its size, its principal axes are exactly perpendicular and parallel to its long axis.

- When measuring a specimen, rotate it to several orientations and measure its retardance after each rotation. This will sensitively test the accuracy of the measured retardance and azimuth values versus known rotation angles (a makeshift rotation stage can be assembled from a Petri dish that is rotated in its stage holder).

Speed

The speed of the instrument is determined by the time taken to acquire the raw PolScope images. During live retardance imaging, the time period also includes the calculation of the computed retardance and/or azimuth images. The calculation usually takes a fraction of a second and cannot be influenced by the user. The time needed for acquiring the raw PolScope images, however, can vary substantially depending on a number of parameters, including the specifications of instrument parts (such as liquid crystal settling time, lamp intensity, camera sensitivity) and the required resolution and sensitivity in the computed PolScope images. These parameters have been discussed at length throughout the text and their influence on the speed of acquisition is summarized here.

- The liquid-crystal settling time can account for more than 50% of the acquisition time. Currently, the universal compensator offered by CRI can be built with liquid crystals that have settling times of either 150 ms or 30 ms.

- When using a small swing retardance and seeking low shot noise to achieve high sensitivity, the exposure time, including frame averaging, can dominate the total acquisition time. To keep exposure times short, use a bright light source, such as an arc lamp, and a camera with high quantum efficiency.

- Adjust the resolution of PolScope images to that required for the chosen application. Adjustments can be made by selecting the right objective lens, using a zoom lens before the camera and by either binning or selecting a subregion of the CCD sensor. Decreasing the zoom factor, for example, is an effective way of increasing image brightness and reducing exposure time. Binning or subregion read-out can substantially reduce the camera read-out time and subsequent processing time.

- Consider the sensitivity of PolScope images in the context of the chosen application. The sensitivity depends on the light level captured and the number of frames used for the calculation of retardance and azimuth images. A small reduction in sensitivity can substantially increase the speed of measurement.

Use of the Bertrand Lens

The use of the Bertrand lens or telescope ocular is highly recommended for inspecting the optical path outside the immediate vicinity of the specimen. Many problems, such as misaligned optical components, which cause shading, the presence of dust particles, which reduce the extinction, and air bubbles in immersion oil, which can wreak havoc on the image, are quickly recognized by looking down the optical path through a telescope ocular or with a Bertrand lens in place. The telescope ocular is exchanged for a regular eyepiece, whereas the Bertrand lens swings in the optical path, somewhere between the objective and the ocular, which remains in place. In fact, if neither telescope ocular nor Bertrand lens is available, by removing the regular ocular from the microscope and gazing down the empty tube (keep eye about a handwidth away from the tube end), it is possible to see the various optical components, albeit at low magnification. The Bertrand lens, which has a separate focusing mechanism and provides a magnified view of different optical planes, is generally the preferred device. In many microscope designs, the Bertrand lens can also be used with the camera, thus providing images, for example, of the objective back focal plane at different compensator settings (Fig. 5).

When properly aligned for Köhler illumination, the objective back focal plane and the condenser front focal plane are simultaneously in focus (Inoué and Oldenbourg 1995). Hence, using the Bertrand lens, the condenser aperture diaphragm can be adjusted to optimize the performance of the polarization optical train. For objectives with NAs of 0.7 or less, the condenser aperture should be set to limit the condenser NA to just under the objective NA. This setting reduces the chance of illuminating rays hitting the mounting of objective lens elements and thus contributing to stray light, which decreases image contrast and reduces extinction. For higher-NA objectives, the condenser NA is typically set somewhat smaller than the objective NA. This cuts out high-NA illumination rays, which carry the

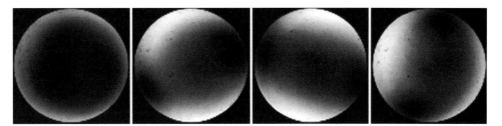

FIGURE 5. Objective lens back aperture of the PolScope equipped with high-NA oil immersion objective and condenser lenses. Images (*left to right*) were taken using Settings 1 through 4 of the universal compensator and with a Bertrand lens in the optical path. The images demonstrate the nature and appearance of polarization distortions introduced by high-NA lenses. The full diameter of the aperture corresponds to 1.4 NA of a Plan Apo 60×/1.4 NA oil immersion objective and apochromat condenser of same NA, both selected for low polarization distortions; homogeneous immersion of $n = 1.52$ between condenser and objective; no specimen in optical path. For a discussion of polarization aberrations in high-NA optical systems and ways to rectify them, see Shribak et al. (2002).

highest polarization distortions. Keeping the condenser NA at 1.0 or less considerably improves the extinction of the setup when using immersion lenses.

EXAMPLE IMAGES AND MOVIES

The following images of living cells and cell components illustrate the relationship between measured retardance and specific cell structures. An introductory text to this relationship is available in Appendix III of Inoué (1986). In addition to the references cited here, the reader is directed to the many references on polarized light imaging also listed in Appendix III of Inoué (1986) and to the extensive discussion of the origin of birefringence in mitotic spindles in Oldenbourg (1999).

The art and science of relating measured retardance and azimuth values to structural information on the molecular level of the specimen is still in its infancy. The potential information in PolScope images and time-lapse movies is enormous. They can reveal the assembly and disassembly of biological structures imaged in living cells that are functioning under physiological conditions. For some structures, such as the mitotic spindle, this promise has been partially realized and much was learned about the reality of spindle fibers and the reversible assembly and disassembly of spindle filaments, long before these notions were confirmed by other techniques (Inoué and Salmon 1995). Work continues on the development of advanced instrumentation and of analytical methods for recording the intrinsic optical properties of biological materials and for interpreting them in terms of their molecular organization. Figures 6 through 10 and Movies 13.1 and 13.2 on the accompanying DVD present examples of LC-PolScope images and time lapse records of living cells and cell components, illustrating the analytical strength and superb imaging that have been made available by this technique.

GLOSSARY OF POLARIZATION OPTICAL TERMS

The following is a brief introduction to terms that are relevant for observations with the LC-PolScope. For a more detailed explanation of these and other terms that describe physical phenomena or optical devices, see Hecht (1998), Born and Wolf (1980), Shurcliff (1962), and Chipman (1995). See also Appendix III of Inoué (1986).

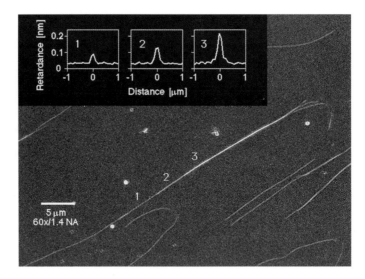

FIGURE 6. Spontaneously assembled microtubules, stabilized with taxol, adhere to the cover glass surface and were imaged with the LC-PolScope (retardance image). Most filaments are single microtubules (MTs), with the exception of one bundle containing one, two, and three microtubules. (*Inset*) Line scans across the filament axis at locations with one, two, and three microtubules. Note on the top right end, the bundle sprays into three individual microtubules. Reproduced, with permission, from Oldenbourg et al. (1998).

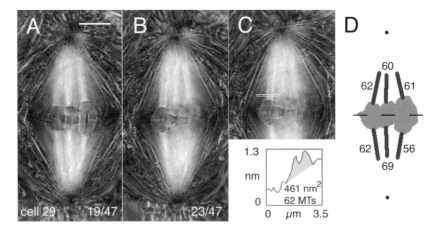

FIGURE 7. Birefringent kinetochore fibers in a spermatocyte of the crane fly *Nephrotoma suturalis* imaged with the LC-PolScope. *A* and *B* are two sections from a series of optical sections made through the cell at focus steps of 0.26 μm. (*Lower right*) Slice number/total slices in the focus series. In viewing these polarized light images, brightness represents the retardance (black = 0 nm and white = 2.5 nm retardance), irrespective of the orientation of the principal axis. Bar, 5 μm. (*C*) A duplicate image of *B* including the line from which retardance area data were obtained. The shaded area in the plot is the retardance area (461 nm²) of the fiber, which was evaluated for the number of kinetochore microtubules (62). (*D*) The metaphase positions of the three bivalent chromosomes in this cell upon projection of all images within the Z-focus series to make a two-dimensional profile. Two dots indicate the positions of the flagellar basal bodies within the centrosomes at the two spindle poles. The numbers on each kinetochore fiber indicate the number of kinetochore microtubules in that fiber, based on retardance area analysis. (A full account of this study is being prepared by J.R. LaFountain and R. Oldenbourg.)

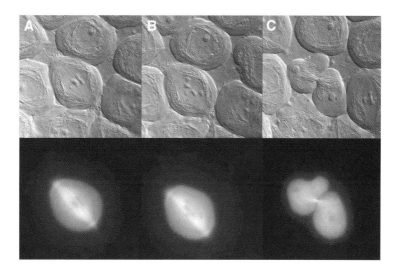

FIGURE 8. Crane-fly spermatocytes viewed with differential interference contrast and fluorescence microscopy, two imaging modes that are complementary to the PolScope (Fig. 7). Spindle microtubules are fluorescently labeled following injection of rhodamine-tubulin into the cytoplasm of one spermatocyte. A field of cells recorded with DIC optics is presented in *A*, *B*, and *C*. In the paired fluorescence images, the kinetochore fibers of the injected cell appear prominent on a less-fluorescent background, which contains unassembled rhodamine-tubulin subunits. Centrosomes are also quite fluorescent. Kinetochore microtubules shorten during anaphase *A*, as the injected cell progresses through anaphase and then through cytokinesis. The figure was kindly provided by Dr. James R. LaFountain, Jr. of the University at Buffalo.

FIGURE 9. Birefringent fine structure in the living growth cone of *Aplysia* bag cell neuron. This retardance image, recorded with the LC-PolScope, shows the peripheral lamellar domain containing radially aligned birefringent fibers which end in filopodia at the leading edge of the growth cone. The graph at the top (center) shows the retardance, in nanometers, measured along a 3-pixel-wide strip, indicated by a horizontal white bar. The measured peak retardance in radial actin fibers can be interpreted in terms of the number of actin filaments in a fiber. Near the bottom, the image shows the central domain, which is filled with vesicles and highly birefringent microtubule bundles. On the bottom left, a region is outlined and its contrast is enhanced to show clearly the diffuse birefringent patches of partially aligned actin networks located mainly in the transition region between the thin lamellar and thicker central domain. The arrow points to a spontaneously formed intrapodium with its strongly birefringent tail formed by parallel actin arrays. The image is one frame of a time-lapse record that reveals the architectural dynamics of the cytoskeleton including the formation of new filopodia and radial fibers and the continuous retrograde flow of birefringent elements in the peripheral domain. The top left corner of the image contains a gray wedge with retardance scale, the time in minutes and seconds elapsed since the first frame of the movie and a bar indicating the length scale. (Reproduced, with permission, from Katoh et al. 1999.)

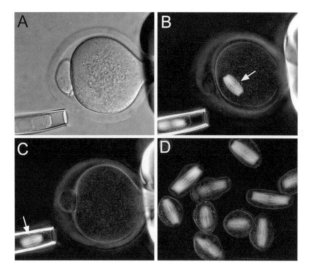

FIGURE 10. Mouse oocyte and enucleated spindles. (*A*) Mouse oocyte held in place by gentle suction of a holding pipette (oocyte diameter ~75 μm; DIC microscopy); (*B*) retardance image of the same oocyte, with birefringent spindle of meiosis II (white arrow); (*C*) the spindle (*arrow*) is aspirated into an enucleation pipette; (*D*) a batch of enucleated spindles and chromosomal karyoplasts. Chromosomes are aligned in the middle of spindles. The figure was kindly provided by Lin Liu of the Marine Biological Laboratory (for more detailed information, see Liu et al. 2000).

Analyzer

An analyzer is a polarizer that is used to analyze the polarization state of light (see Polarizer).

Azimuth

The azimuth is an angle that refers to the orientation of the slow axis of a uniformly birefringent region. The azimuth image refers to the array of azimuth values of a birefringent specimen imaged with the LC-PolScope. The azimuth is typically measured from the horizontal orientation with values increasing for counterclockwise rotation. Angles range between 0° and 180°, both of which indicate horizontal orientation.

Birefringence

Birefringence is a material property that can occur when there is molecular order, i.e., when the average molecular orientation is nonrandom, as in crystals or in aligned polymeric materials such as those comprising the mitotic spindle. Molecular order usually renders the material optically anisotropic, leading to a refractive index that, in general, changes with the polarization of the light. For example, assume that a beam of light passes through a mitotic spindle in a direction perpendicular to the spindle axis. The light that is polarized parallel to the spindle axis experiences a higher refractive index than the light that is polarized perpendicular to the spindle axis, i.e., the mitotic spindle is birefringent (due to the parallel alignment of microtubules; Sato et al. 1975). For all birefringent materials, the difference between the two indices of refraction is called birefringence

$$\text{birefringence} = \Delta n = n_{||} - n_{\perp}$$

where $n_{||}$ and $n_{\perp}$ are the refractive indices for light polarized parallel and perpendicular to one of the principal axes. The LC-PolScope measures the relative phase shift between two orthogonally polarized light beams, after they have traversed the birefringent material (see Retardance).

Compensator

A compensator is an optical device that includes one or more retarder plates and is commonly used to analyze the birefringence of a specimen. For a traditional polarizing microscope, several types of compensators exist that typically use a single fixed retarder plate mounted in a mechanical rotation stage. With the help of a compensator, it is possible to distinguish between the slow and fast axis direction and to measure the retardance of a birefringent object after orienting it with respect to the polarization axes of the microscope.

The LC-PolScope employs a universal compensator that includes two electro-optically controlled variable retarder plates. Using the universal compensator, it is possible to measure the retardance and slow axis orientation of birefringent objects that have any orientation in the plane of focus.

Dichroism

Dichroism is a material property that can occur in absorbing materials in which the light absorbing molecules are arranged in a nonrandom orientation. Dichroism refers to the difference in the absorption coefficients for light polarized parallel and perpendicular to the principal axis of alignment.

The measurement of optical anisotropy by the LC-PolScope is affected by the dichroism of absorbing materials. In nonabsorbing, clear specimens, however, dichroism vanishes and birefringence is the dominant optical anisotropy measured by the LC-PolScope. Like absorption, dichroism is strongly wavelength-dependent, whereas birefringence only weakly depends on wavelength.

Extinction

Extinction is defined as the ratio of maximum to minimum transmission of a beam of light that passes through a polarization optical train. Given a pair of linear polarizers, for example, the extinction is the ratio of intensities measured for parallel versus perpendicular orientation of the transmission axes of the polarizers (extinction = $I_{\parallel} / I_{\perp}$). In addition to the polarizers, the polarization optical train can also include other optical components, which usually affect the extinction of the complete train. In a polarizing microscope, the objective and condenser lens are located between the polarizers and significantly reduce the extinction of the whole setup (see the section on Polarization Optical Components).

Fast Axis

The fast axis describes an orientation in a birefringent material. For a given propagation direction, light that is polarized parallel to the fast axis experiences the lowest refractive index, and hence travels the fastest in the material (see also Slow Axis).

Optic Axis

The optic axis refers to a direction in a birefringent material. Light propagating along the optic axis does not change its polarization; hence, for light propagating along the optic axis, the birefringent material behaves as if it were optically isotropic.

Polarized Light

A beam of light is said to be polarized when its electric field is distributed nonrandomly in the plane perpendicular to the beam axis. In unpolarized light, the orientation of the electric field is random and unpredictable. In partially polarized light, some fraction of the light is polarized, whereas the remaining fraction is unpolarized. Most natural light is unpolarized (sun, incandescent light), but can become partially or fully polarized by scattering, reflection, or interaction with optically anisotropic materials. These phenomena are used to build devices to produce polarized light (see Polarizer).

Linearly Polarized Light

In a linearly polarized light beam, the electric field is oriented along a single axis in the plane perpendicular to the propagation direction.

Circularly Polarized Light

In circularly polarized light, the electric field direction rotates either clockwise (right circularly) or counterclockwise (left circularly) when looking toward the source. While the field direction rotates, the field strength remains constant. Hence, the endpoint of the field vector describes a circle.

Elliptically Polarized Light

In elliptically polarized light, as in circularly polarized light, the electric field direction rotates either clockwise or counterclockwise when looking toward the source. However, while the field direction rotates, the field strength varies in such a way that the endpoint of the field vector describes an ellipse. The ellipse has a long and a short principal axis that are orthogonal to each other and have fixed orientation. Any type of polarization (linear, circular, or elliptical) can be transformed into any other type of polarization by means of polarizers and retarders.

Polarizer

A polarizer, sometimes called a polar, is a device that produces polarized light of a certain kind. The most common polar is a linear polarizer made from dichroic material (e.g., a plastic film with small embedded iodine crystals that have been aligned by stretching the plastic), which transmits light of one electric field direction while absorbing the orthogonal field direction. Crystal polarizers are made of birefringent crystals that split the light beam into orthogonal linear polarization components. A polarizer that produces circularly polarized light, a circular polarizer, is typically built from a linear polarizer followed by a quarter wave plate.

The LC-PolScope employs a universal compensator that also serves as a universal polarizer in that it converts linear polarization into any other type of polarization by means of two variable retarders (see the section on Polarization Optical Components).

Retardance

Retardance is a measure of the relative optical path difference, or phase change, suffered by two orthogonal polarization components of light that has interacted with an optically anisotropic material. Retardance is the primary quantity measured by the LC-PolScope. Assume a nearly collimated beam of light traversing a birefringent material. The light component that is polarized parallel to the high refractive index axis travels at a slower speed through the birefringent material than the component polarized perpendicular to that axis. As a result, the two components, which were in phase when they entered the material, exit the material out of phase. The relative phase difference, expressed as the distance between the respective wave fronts, is called the retardance:

$$\text{Retardance} = R = (n_{||} - n_{\perp}) \cdot t = \Delta n \cdot t$$

where t is the physical path length or thickness of the birefringent material. Hence, retardance has the dimension of a distance and is often expressed in nanometers. Sometimes, it is convenient to express that distance as a fraction of the wavelength λ, such as $\lambda/4$ or $\lambda/2$. Retardance can also be defined as a differential phase angle, in which case $\lambda/4$ corresponds to 90° and $\lambda/2$ to 180° phase difference.

As a practical example, consider a mitotic spindle observed in a microscope that is equipped with low-NA lenses (NA $\leq$ 0.5). When the spindle axis is contained in the focal plane, the illuminating and imaging beams run nearly perpendicular to the spindle axis. Under these conditions, the retardance measured in the center of the spindle is proportional to the average birefringence induced by the dense array of aligned spindle microtubules. To determine Δn, it is possible to estimate the thickness, t, either by focusing on spindle fibers located on top and bottom of the spindle and noting the distance between the two focus positions or by measuring the lateral extend of the spindle when focusing through its center. The latter approach assumes a rotationally symmetric shape of the spindle. Typical values for the spindle retardance of crane fly spermatocytes (Fig. 7) and of other cells is 3–5 nm and the spindle diameter is about 30–40 µm, leading to an average birefringence of around 10^{-4}. It has been found that the retardance value of the spindle is largely independent of the NA for imaging systems using NA $\leq$ 0.5 (Sato et al. 1975).

On the other hand, when using an imaging setup that employs high-NA optics (NA > 0.5) for illuminating and imaging the sample, the measured retardance takes on a somewhat different context. For example, the retardance measured in the center of a microtubule image recorded with a LC-PolScope equipped with a high-NA objective and condenser lens is 0.07 nm (Fig. 6). A detailed study showed that the peak retardance decreased inversely with the NA of the lenses. However, the retardance integrated over the cross section of the microtubule image was independent of the NA (Oldenbourg et al. 1998). Although a conceptual understanding of the measured retardance of submicroscopic filaments has been worked out in the aforementioned publication, a detailed theory of these and other findings about the retardance measured with high-NA optics has yet to materialize.

Retarder

A retarder, or wave plate, is an optical device that is typically made of a birefringent plate. The retardance of the plate is the product of the birefringence of the material and the thickness of the plate. Fixed retarder plates are either cut from crystalline materials such as quartz, calcite, or mica, or they are made of aligned polymeric material. If the retardance of the plate is $\lambda/4$, for example, the retarder is called a quarter-wave plate.

A variable retarder can be made from a liquid crystal device. A thin layer of highly birefringent liquid crystal material is sandwiched between two glass windows, each bearing a transparent electrode. A voltage applied between the electrodes produces an electric field across the liquid-crystal layer that reorients the liquid-crystal molecules. This reorientation changes the birefringence of the layer without affecting its thickness or the direction of its slow axis.

Slow Axis

The slow axis describes an orientation in a birefringent material. For a given propagation direction, light polarized parallel to the slow axis experiences the highest refractive index and hence travels the slowest in the material (see also Fast Axis).

Wave Plate

See Retarder.

ACKNOWLEDGMENTS

The wisdom collected in this chapter is the result of more than 14 years of designing and building microscope instrumentation and working with microscopes to study living cells. During this time, I benefited from the encouragements and challenges of a large group of people, too numerous to name them all. I thank all of them including the following colleagues who provided critical support: Shinya Inoué, who set me on the path; Bob Knudson, who magically creates the parts; Ted Salmon and Phong Tran, who were early believers; Kaoru Katoh, David Keefe, and Lin Liu, who reaped the first fruits; Michael Shribak, who is unleashing a flurry of innovations; Grant Harris, who is bridging the digital divide; Jim LaFountain, who opened my eyes to the beauty of crane flies; Cliff Hoyt and Cathy Boutin, who bring this technology to the masses; and Nannette, Marie, and Helena, who give meaning and a place to call home. I also thank Michael Shribak, Jim Valles, and Shinya Inoué for comments on parts of this manuscript.

This work was supported by funding from the National Institute of General Medical Sciences and the National Institute of Biomedical Imaging and Bioengineering (grants GM-49210 and EB002045, respectively).

MOVIE LEGENDS

MOVIE 13.1. Meiosis I in spermatocyte of the crane fly *Nephrotoma suturalis* (first published as a Web supplement to Reider and Khodjakov 2003). This time-lapse movie was recorded with the LC-PolScope over a timespan of 4 hours at 30-second time intervals. The horizontal image width is 56 μm. The image brightness scales between black = 0 and white = 2 nm retardance.

MOVIE 13.2. Birefringent fine structure in the living growth cone of an *Aplysia* bag cell neuron. This time-lapse movie was recorded with the LC-PolScope over a time span of 15 minutes at 5-second time intervals. The horizontal image width is 150 μm. The image brightness scales between black = 0 and white = 0.8 nm retardance. A detailed analysis of the birefringent fine structure revealed in this movie was published in Katoh et al. 1999. (From Katoh et al. 1999 with permission of the American Society for Cell Biology.)

REFERENCES

Ambronn H. and Frey A. 1926. *Das Polarisationsmikroskop, seine Anwendung in der Kolloidforschung und in der Färberei*. Akademischer Verlag, Leipzig.

Born M. and Wolf E. 1980. *Principles of optics*. Pergamon Press, Elmsford, New York.

Chipman R.A. 1995. Polarimetry. In *Handbook of optics* (ed. M. Bass) vol. 2, pp. 22.1–22.37. McGraw-Hill, New York.

Frey-Wyssling A. 1953. *Submicroscopic morphology of protoplasm*. Elsevier, Amsterdam.

Hartshorne N.H. and Stuart A. 1960. *Crystals and the polarising microscope: A handbook for chemists and others*. Arnold, London.

Hecht E. 1998. *Optics*. Addison-Wesley, Reading, Massachusetts.

Inoué S. 1952. Studies on depolarization of light at microscope lens surfaces. I. The origin of stray light by rotation at the lens surfaces. *Exp. Cell Res.* **3:** 199–208.

———. 1953. Polarization optical studies of the mitotic spindle. I. The demonstration of spindle fibers in living cells. *Chromosoma* **5:** 487–500.

———. 1986. *Video microscopy*, 1st edition. Plenum Press, New York.

———. 2002. Polarization microscopy. In *Current protocols in cell biology* (suppl. 13) (ed. J.S. Bonifacino et al.), vol. 1, pp. 4.9.1–4.9.27. John Wiley & Sons, New York.

Inoué S. and Dan K. 1951. Birefringence of the dividing cell. *J. Morphol.* **89:** 423–456.

Inoué S. and Hyde W.L. 1957. Studies on depolarization of light at microscope lens surfaces II. The simultaneous realization of high resolution and high sensitivity with the polarizing microscope. *J. Biophys. Biochem. Cytol.* **3:** 831–838.

Inoué S. and Oldenbourg R. 1995. Microscopes. In *Handbook of optics* (ed. M. Bass) vol. 2, pp. 17.1–17.52. McGraw-Hill, New York.

Inoué S. and Salmon E.D. 1995. Force generation by microtubule assembly/disassembly in mitosis and related movements. *Mol. Biol. Cell* **6:** 1619–1640.

Inoué S. and Sato H. 1966. Deoxyribonucleic acid arrangement in living sperm. In *Molecular architecture in cell physiology* (ed. T. Hayashi and A.G. Szent-Gyorgyi), pp. 209–248. Prentice Hall, Englewood Cliffs, New Jersey.

Inoué S. and Spring K.R. 1997. *Video microscopy.* Plenum Press, New York.

Inoué S., Shimomura O., Goda M., Shribak M., and Tran P.T. 2002. Fluorescence polarization of green fluorescence protein. *Proc. Natl. Acad. Sci.* **99:** 4272–4277.

Katoh K., Hammar K., Smith P.J.S., and Oldenbourg R. 1999. Birefringence imaging directly reveals architectural dynamics of filamentous actin in living growth cones. *Mol. Biol. Cell* **10:** 197–210.

Liu L., Oldenbourg R., Trimarchi J.R., and Keefe D.L. 2000. A reliable, noninvasive technique for spindle imaging and enucleation of mammalian oocytes. *Nat. Biotechnol.* **18:** 223–225.

Mitchison J.M. 1950. Birefringence of Amœbæ. *Nature* **166:** 313–315.

Oldenbourg R. 1999. Polarized light microscopy of spindles. *Methods Cell Biol.* **61:** 175–208.

Oldenbourg R. and Mei G. 1996. Polarized light microscopy. *US Patent, Number 5,521,705.*

Oldenbourg R., Salmon E.D., and Tran P.T. 1998. Birefringence of single and bundled microtubules. *Biophys. J.* **74:** 645–654.

Oldenbourg R. and Török P. 2000. Point spread functions of a polarizing microscope equipped with high numerical aperture lenses. *Applied Optics* **39:** 6325–6331.

Oldenbourg R., Inoué S., Tiberio R., Stemmer A., Mei G., and Skvarla M. 1996. Standard test targets for high resolution light microscopy. In *Nanofabrication and biosystems: Integrating material science, engineering and biology* (ed. H.C. Hoch et al.), pp. 123–138. Cambridge University Press, England.

Palazzo R.E., Vaisberg E., Cole R.W., and Rieder C.L. 1992. Centriole duplication in lysates of *Spisula solidissima* oocytes. *Science* **256:** 219–221.

Peterman E.J.G., Sosa H., Goldstein L.S.B., and Moerner W.E. 2001. Polarized fluorescence microscopy of individual and many kinesin motors bound to axonemal microtubules. *Biophys. J.* **81:** 2851–2863.

Rieder C.L. and Khodjakov A. 2003. Mitosis through the microscope: Advances in seeing inside live dividing cells. *Science* **300:** 91–96.

Salmon E.D. and Tran P.T. 1998. High-resolution video-enhanced differential interference contrast (VE-DIC) light microscopy. *Methods Cell Biol.* **56:** 153–184.

Sase I., Miyata H., Ishiwata S.I., and Kinosita K., Jr. 1997. Axial rotation of sliding actin filaments revealed by single-fluorophore imaging. *Proc. Natl. Acad. Sci.* **94:** 5646–5650.

Sato H., Ellis G.W., and Inoué S. 1975. Microtubular origin of mitotic spindle form birefringence. Demonstration of the applicability of Wiener's equation. *J. Cell Biol.* **67:** 501–517.

Schmidt W.J. 1937. *Die Doppelbrechung von Karyoplasma, Zytoplasma und Metaplasma.* Bornträger, Berlin.

Shribak M. and Oldenbourg R. 2003. Techniques for fast and sensitive measurements of two-dimensional birefringence distributions. *Applied Optics* **42:** 3009–3017.

Shribak M., Inoué S., and Oldenbourg R. 2002. Polarization aberrations caused by differential transmission and phase shift in high NA lenses: Theory, measurement and rectification. *Optical Eng.* **41:** 943–954.

Shurcliff W.A. 1962. *Polarized light, production and use.* Harvard University Press, Cambridge, Massachusetts.

Sluder G. and Wolf D.E. 1998. Video microscopy. In *Methods in cell biology* (ed. L. Wilson and P. Matsudaira), vol. 56, pp. 1–315. Academic Press, San Diego.

Swann M.M. and Mitchison J.M. 1950. Refinements in polarized light microscopy. *J. Exp. Biol.* **27:** 226–237.

Taylor J.R. 1997. *An introduction to error analysis.* University Science Books, Sausalito, California.

Török P. 2000. Imaging of small birefringent objects by polarised light conventional and confocal microscopes. *Optics Commun.* **181:** 7–18.

Warshaw D.M., Hayes E., Gaffney D., Lauzon A.-M., Wu J., Kennedy G., Trybus K., Lowey S., and Berger C. 1998. Myosin conformational states determined by single fluorophore polarization. *Proc. Natl. Acad. Sci.* **95:** 8034–8039.

Confocal Microscopy, Deconvolution, and Structured Illumination Methods

John M. Murray

University of Pennsylvania, Department of Cell and Developmental Biology,
Philadelphia, Pennsylvania 19104-6058

INTRODUCTION

WHEN A THICK SPECIMEN IS VIEWED through a conventional microscope, the sum of a sharp image of an in-focus region is seen, plus blurred images of all of the out-of-focus regions. The depth of field (i.e., the distance between the top and bottom of the in-focus region at a fixed setting of the focus knob) is less than 1 μm for the high numerical aperture (NA) objective lenses that are used for fluorescence microscopy. Thus, even when viewing a specimen as thin as 5 μm, 80% of the light may be coming from out-of-focus regions. The result is a low-contrast image composed of an intensely bright, but very blurred background on which is superimposed the much dimmer in-focus information.

In this chapter, "thick" and "thin" refer to the thickness of the fluorescent sample; overall specimen thickness per se does not increase the background. As the overall specimen thickness increases beyond 5–10 μm, however, two other factors begin to degrade the image quality. When the illumination or imaging path intersects regions of widely different refractive index such as small granules or organelles, their curved surfaces act as microlenses to deflect the light in random directions. The consequence of multiple deflections may be to distort the light path enough to introduce aberrations into the image, or even to scatter the light completely out of the field of view.

One way to eliminate the high background, scattering, and aberrations is to slice the thick specimen into many thin sections. This approach requires fixation, dehydration, and embedding, so it has limited application to the microscopy of living cells, but several other methods do work well with living samples. These methods can be grouped into three classes: primarily "optical" (e.g., confocal microscopy, multiphoton microscopy), primarily "computational" (e.g., deconvolution techniques), and mixed (e.g., structured illumination) approaches. These techniques make it possible to see details within thick specimens (e.g., the interiors of cells within living tissue) by optical sectioning, without the artifacts associated with physically sectioning the specimen.

Which Method to Use?

All of these methods address problems encountered in imaging thick specimens. For routine qualitative observation of relatively thin specimens (<3 μm), it is probably much quicker and less frustrating to use conventional ("wide-field") microscopy. There are a few situations, however, in which the benefits of these more complex methods are important enough, even for a thin specimen, to warrant the extra cost, inconvenience, and time.

The most common application to thin specimens is when intrinsic contrast is very low, so that any loss of contrast, even the minimal decrease due to a small amount of out-of-focus light, complicates interpretation of the data. In this situation, all of these methods can usually improve contrast for any sample thicker than about 1 μm. Another common application is to enable accurate measurement of the amount of a fluorescent component present in a cell, a task in which deconvolution methods excel. Finally, in the relatively rare case in which even a modest enhancement of resolution would change the interpretation of the data, then confocal, deconvolution, and some of the structured illumination methods are capable of delivering a small improvement over a conventional microscope.

For thicker objects that produce a moderate amount of out-of-focus light (typically 5–30 μm), any of the methods discussed here (and multiphoton microscopy, discussed in Chapter 15) should give a dramatically better result than a conventional microscope. When the sample is living (i.e., photobleaching and phototoxicity constrain exposure), and the signal is weak or the contrast is low, methods that must use photomultipliers for detection (e.g., point-scanning methods, confocal or multiphoton) have a severe handicap compared to methods that can use charge-coupled device (CCD) cameras (e.g., deconvolution, disk-scanning confocal microscopy, and structured illumination), because of the much higher quantum efficiency of CCDs. With very thick specimens that produce an overwhelming amount of out-of-focus light, only point-scanning (confocal or multiphoton) microscopy will give a satisfactory result.

How much is a "moderate" amount of out-of-focus light? Typically, the image seen through a conventional microscope will be too blurred to discern details, but one will be able to locate the region of interest and at least roughly set the focus level. If the view through a conventional microscope is virtually featureless and gives no landmarks for choosing the appropriate area or for setting the focus, two methods—point-scanning confocal or multiphoton microscopy—can be used to produce extremely useful images from terrible specimens. From very thick specimens, however, it is not realistic to expect a final image quality comparable to the best that a conventional microscope produces with a thin specimen, for reasons that will be considered in this chapter.

DECONVOLUTION METHODS

The goal of deconvolution is to improve the images of thick objects by computationally removing the out-of-focus blur. The strategy is to calculate the structure of a hypothetical object that could have produced the observed, partially focused image. The calculation is based on fundamental optical principles, in particular, a quantitative understanding of the effects of defocus, and may also take into account prior information or guesses about the specimen. The method commonly employed is to refine iteratively an initial guess about the true object until the estimated image (i.e., the estimated object appropriately blurred by the effects of defocus) corresponds to the actual observed image.

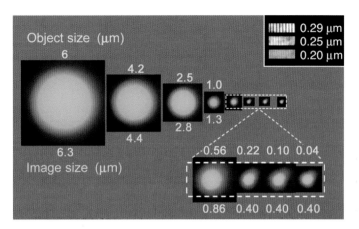

FIGURE 1. Epifluorescence microscope images of 8 beads of known, decreasing size. The images of the fluorescent beads are shown in green; all are adjusted to have the same maximum brightness. The actual brightness varies by ~1000-fold. The true diameter of each bead is given above its image, and the diameter measured from the image is listed below it. The four smallest beads are shown at two different magnifications. The apparent diameters measured in the image are slightly larger than the true diameters, and the apparent diameters are the same for the three smallest beads, even though their true diameters differ by more than fivefold. These three images reveal the PSF of this microscope. The images in the upper right corner, obtained with the same microscope setup, show the appearance of three gratings with spacings of 0.29, 0.25, and 0.20 μm between the white bars. Note that the contrast between the black and white bars decreases as the spacing gets smaller. In the actual object, the grating contrast is the same for all three scales. The bead images in the upper row are displayed at the same magnification as the gratings.

Optical Principles

Successful application of these techniques requires an appreciation of how an image is formed by a microscope and what happens to an image when the lens is defocused. For this purpose, it is helpful to introduce the twin concepts of *point spread function* (PSF) and *contrast transfer function* (CTF). Both of these concepts describe the relationship between a real object and the image that is formed of it by an optical system. The PSF describes this relationship in terms of the image of a very small object, effectively a single point. Although the microscope can "see" objects that are vanishingly small, the limited resolution of a microscope prevents the image from accurately portraying their size, no matter what magnification is applied.

An illustration of this phenomenon is given in Figure 1. Note that below a certain object size, images of every object appear the same. Increasing the magnification does not help; the image can be made larger, but not sharper. This limiting image is called an "Airy disk," after the British astronomer G.B. Airy who first recognized its significance in 1834. Notice that for the microscope used for this figure, the Airy disk is not quite a perfect circle. The three smallest beads all appear to be slightly elliptical, with a weaker tail extending toward the upper right, although electron microscopy shows that the beads are nearly perfect spheres. As Airy was the first to point out, the shape of this limiting image provides no information about the shape of the object—the Airy disk is an intrinsic property of the optical system itself. The Airy disk shows that the image of a point-like object is not a single point, but is spread into a fuzzy disk, hence the name "point spread function." The PSF is often used as a means of quantitatively characterizing the performance of an imaging system.

The grating images in Figure 1 show that the size of the PSF sets the resolving power of an optical system. An optical system forms an image by substituting its PSF for every point in the object, and then sums all of these PSFs to make the image. The width of the PSF determines how far apart two points in the object must be to avoid being smeared together in the image. If the PSF of the optical system is broad, two points must be rather well separated to prevent overlap of the two corresponding PSFs in the image. If their PSFs overlap extensively, the two points will appear to be just a single point, smeared into an indistinct average like the lines in the image of the 0.2-μm grating (Fig. 1, upper right box).

Suppose that the experiment in Figure 1 included not an individual 0.04-μm bead, but rather a pair of 0.04-μm beads separated by 0.08 μm, twice their own diameter. This pair would still be a smaller "object" than the single 0.22-μm bead, and their joint image would have the same shape as any of the three limiting images in Figure 1. In other words, 0.08 μm is well below the resolution of this microscope. It is important to realize, however, that the Airy disk for the pair of beads would be twice as bright as the Airy disk from a single bead, i.e., the imaging process is a *linear* operation. The total intensity in an image of A and B is exactly the sum of the total intensity in an image of A plus the total intensity in an image of B.

A single number, such as the Rayleigh criterion, is often quoted for the resolving power of a microscope. The Rayleigh criterion is the radius of the Airy disk, given numerically by 0.6 λ/NA, where NA is the numerical aperture of the objective lens and λ is the wavelength of light that forms the image. However, using a single number for the resolving power is somewhat misleading, because there is no sharp cutoff. As the size of small details approaches the resolution of the imaging system, these details do not suddenly disappear. Instead, their contrast in the image becomes a smaller and smaller fraction of their true contrast in the object, until finally the image contrast approaches the size of the random fluctuations due to noise, and they then become "invisible." The images of the gratings in Figure 1 show this progressive loss of contrast with decreasing size.

A complete description of the resolving power of an optical system thus requires information about the variation in the ratio of image contrast to object contrast as a function of size. The PSF contains this information, but it is revealed more clearly in its twin, the contrast transfer function (CTF). This function describes the extent to which contrast variations in the object are faithfully replicated in the image. Perfect contrast transfer means that image contrast equals object contrast. The CTF is usually expressed as a ratio, so that perfect contrast transfer means that the CTF has a value of 1.0 (in reality, the CTF is always less than 1.0).

It is reasonable to expect that information about some features of the object might be transferred into the image more faithfully than others. For instance, the image may be a nearly perfect representation of the large-scale features of the object, but contain much less information about the very smallest details. This will always be true for images obtained from an optical microscope, because one cannot see clearly those details of the object that are small compared to the wavelength of light (i.e., details on the scale of the PSF). Thus, the CTF is a function of the size of the feature being observed (Fig. 2, left). Normally, the CTF is shown in graphical form, plotting the ratio of image contrast to object contrast (vertical axis) against the reciprocal of size (i.e., spatial frequency). The CTF is simply a different representation of the same information that is given by the PSF of an optical system. Mathematically, the CTF is the Fourier transform of the PSF, and vice versa.

When we speak of image resolution, we are therefore making a statement about the *ratio* of image-to-object contrast at small spacings. There is always an interaction between image contrast, image signal-to-noise ratio, and image resolution, even though it is sometimes convenient to think of these as independent properties. What is normally referred to as *visibility* is determined by all three of these parameters, as well as properties of the display system and the observer. Keep in mind that much useful information can often be extracted concerning features that, by eye, are "invisible."

The example of a typical, good microscope CTF and PSF shown on the left side of Figure 2 represents the case in which the specimen is thin and lies exactly in the focal plane of the

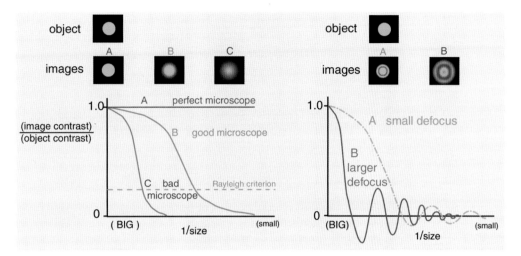

FIGURE 2. *(Left)* Schematic representation of some contrast transfer functions, and the corresponding PSFs (object-image pairs for a very small object). *(A)* A perfect (impossible!) microscope; *(B)* a typical, good microscope; *(C)* a poor or improperly used microscope. The dashed line lies at a relative contrast of 25%, corresponding to the Rayleigh criterion for the resolving power of a microscope. *(Right)* Images of a tiny object from a microscope at two different values of defocus reveal the 3D nature of the PSF and CTF. Objects in some size ranges appear to change from black to white or vice versa as the focus level changes between the indicated values. An example of this contrast reversal in a transmitted light image is shown in Figure 4.

objective lens. In fact, the CTF and PSF are three-dimensional (3D) functions. Their third dimension is revealed by comparing image to object when the object is displaced vertically from the focal plane of the lens. As the focus changes, concentric rings appear in the PSF, and the CTF develops ripples and in some regions becomes negative. For features in the size range corresponding to these negative oscillations of the CTF, dark parts of the object will appear bright in the image and vice versa (Fig. 2, right). As the degree of defocus increases, the CTF becomes increasingly oscillatory, with the contrast reversals affecting ever larger features in the image.

The image of a small fluorescent bead (i.e., the PSF) develops concentric rings as the lens moves away from focus (Fig. 2, top right). Changing the focus of the lens means that the 3D PSF is viewed at different levels along the optical axis. A vertical slice of the complete 3D PSF, viewed from the side, is shown in Figure 3 (Born and Wolf 1999).

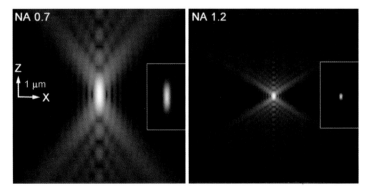

FIGURE 3. Vertical sections through the 3D PSFs calculated for lenses of two different numerical apertures (NAs). The contrast has been greatly enhanced to show the weak side lobes (which are the rings of the Airy disk, viewed edge-on). The small insets show the PSFs at their true contrast level. Note that vertical smearing (proportional to NA²) is much more strongly affected by the lens NA than is lateral smearing (proportional to NA).

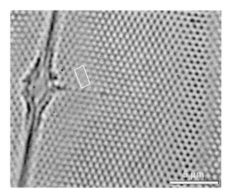

FIGURE 4. Contrast reversal because of defocus. (*Left*) A bright-field image of a diatom taken through a conventional microscope using a 60 × NA 1.4 lens. (*Right*) An enlargement of a small portion of the image on the left, showing contrast reversals because of changing amounts of defocus. The diatom shell is curved, being thinner at the edges. As a result, the distance from the lens to the surface of the diatom varies, i.e., the view includes a range of defocus values. Over this range, the CTF changes sign several times: From left to right, the holes change from black to white, to black again, and finally to white on the right-hand edge. The white rectangle indicates a narrow band midway between a black and a white hole region where the contrast of the holes is low, i.e., the CTF is nearly zero for structures of this size at this value of defocus.

It surprises most people that microscopes can produce such wildly "incorrect" images as depicted in Figures 2 (right) and 3. A striking example is shown in Figure 4, but the same effect can easily be observed on almost any specimen. In bright-field, a small high-contrast feature such as a small dust particle or a scratch demonstrates the contrast reversal phenomenon clearly. If a good dry or oil-immersion lens is used to carefully focus up and down by a small distance on either side of the correct focus, the particle will oscillate from bright to dark and back again. If the focus can be controlled carefully enough, the position can be found, halfway between a bright and a dark oscillation, where the particle becomes practically invisible (i.e., CTF ~0 for details of this size).

To reiterate, a microscope substitutes its 3D PSF for each point in the 3D object, and then sums all of those innumerable PSFs to give the final 3D image. The mathematical operation called *convolution* precisely describes this "substitute and sum" process. The distribution of intensities in the 3D image is the result of *convolving* the object intensity distribution with a 3D PSF. The 3D PSF is an intrinsic property of the optical system and does not depend on the object.

Deconvolving Wide-field Microscope Images

The 3D PSF (or equivalently, the 3D CTF) contains all of the information needed to predict the appearance of a known object when viewed through the corresponding optical system, for any choice of focus. For the microscopist, however, the appearance (i.e., the image) of the object is known, and the real structure is unknown. It is in principle possible to go "backward" in a one-step calculation from the observed appearance to the actual structure using the mathematical procedure called *deconvolution*. In practice, for realistic signal-to-noise ratios, a much better approach is to perform this calculation in a multistep, iterative process.

To illustrate this procedure, imagine that a 3D object consists of a stack of discrete, two-dimensional (2D) planes. Normally, the imaging data will also be a stack of 2D image planes, collected by changing the fine focus of the microscope by a small increment between

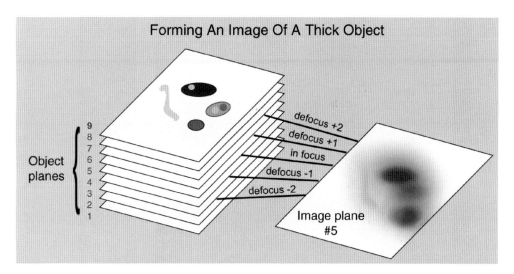

FIGURE 5. A simplified conception of how one plane in the microscopic image of a thick object is formed. The process can be thought of as adding the in-focus image of one object plane to the images of neighboring object planes viewed at different amounts of defocus. For clarity, the process is illustrated for only two neighbors on each side of the in-focus plane. In reality, each image plane receives contributions from all object planes.

successive images. Consider plane number 5 of the observed stack of images in Figure 5. When this image was recorded, what the detector "saw" was the sum of an in-focus view of object plane 5, plus a view of object plane 6 blurred by one increment of defocus, plus object plane 4 blurred by one increment of defocus in the opposite direction, plus planes 7 and 3 blurred by plus and minus two increments of defocus respectively, and so forth. To deconvolve the image data, an initial guess is made at the real structure of the object in planes 1–4 and 6–9. Using the known CTF appropriate for each plane's defocus, these initial estimates are blurred and added together to estimate the contribution from out-of-focus blur to the *observed* plane 5. The sum of blurred object planes 1–4 and 6–9 is subtracted from the image data for plane 5, and the result is an estimate for the in-focus appearance of the object in plane number 5. Repeating these steps for all nine planes generates an improved estimate of the object. The entire sequence of operations on all nine planes is repeated in a loop until the object estimate no longer changes significantly (Fig. 6).

This description of iterative deconvolution is a simplified rendering of one of the early methods for deconvolving light microscope (LM) images (Castleman 1979; Agard 1984; Agard et al. 1989). Several other computational approaches have been reported (Erhardt et al. 1985; Fay et al. 1989; Holmes 1992; Carrington et al. 1995), and each of the commercially available imaging packages incorporates its own additional proprietary features for improving the speed and accuracy of the convergence. These features include the application of various constraints at each cycle of the iteration. Smoothness constraints can be used to enforce the physical impossibility of seeing intensity fluctuations on a scale much smaller than the known resolution of the microscope. Another possible constraint is the requirement that all values of intensity in the object must be greater than zero. One commercial package employs the so-called "blind-deconvolution" approach (Holmes 1992), in which the 3D PSF of the optical system is also estimated from the data to be deconvolved, instead of being experimentally measured from a separate 3D image of a small bead or computed from a theoretical model (Hopkins 1955; Stokseth 1969; Born and Wolf 1999).

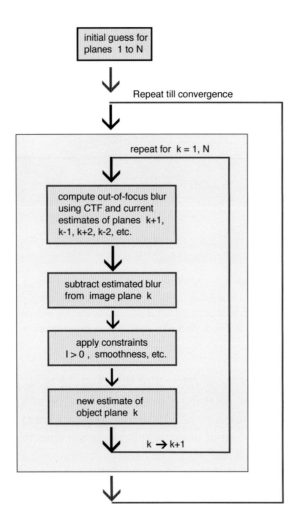

FIGURE 6. Flow chart for the constrained iterative deconvolution algorithm.

Constrained iterative deconvolution is a computationally intensive task. It demands significant computer power and requires 10–20 minutes for a typical 3D data set even on today's fast processors. Why would one choose this approach rather than the quicker and, at least on the surface, simpler approach of confocal microscopy? The resolution achieved by the two methods is comparable, and with samples that are not excessively thick, the two methods are approximately equal in their ability to remove the out-of-focus background light that degrades contrast. Where deconvolution plus wide-field microscopy is clearly superior to confocal microscopy is in the quality of the image data, measured as signal-to-noise ratio.

The reasons for the difference are discussed later in the section on Interpreting the Results, but the point to note here is that with living samples, attaining a higher signal-to-noise ratio becomes a top priority. If examining only fixed cells by, for example, labeling with antibodies, then quantitative measurements of the fluorescence intensity are not usually worthwhile due to the huge uncertainties inherent in immunocytochemical detection. However, it is now possible to see protein molecules in living cells. Suddenly, the types of questions asked have changed beyond recognition because the molecules can be directly counted (Femino et al. 1998; see Chapters 34 and 35). The yield of information from live cell experiments can be enormously increased if the signal-to-noise ratio in the images is high enough to extract reliable quantitative estimates of fluorophore distribution (Swedlow et al. 2002).

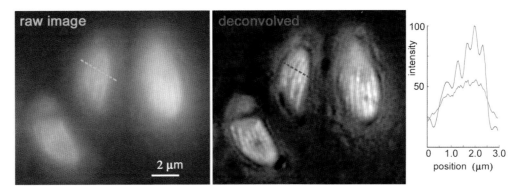

FIGURE 7. Deconvolution of wide-field microscope images of the parasite *Toxoplasma gondii* expressing YFP-tubulin (Swedlow et al. 2002). (*Left*) A focal plane near the plasma membrane from a 3D stack of raw images of YFP fluorescence in several living parasites. (*Middle*) The same focal plane after processing of the 3D stack by constrained iterative deconvolution. Microtubules are clearly visible as bright striations. (*Right*) A plot of the change in intensity along the dashed lines in the raw data and in the data after deconvolution. Deconvolution greatly improves the signal-to-noise ratio by restoring the out-of-focus light to its proper location. (Specimen kindly provided by Dr. Ke Hu, Univ. of Pennsylvania. Images and deconvolution courtesy of Paul Goodwin, Applied Precision Inc.)

There are occasions, particularly if real-time evaluation is needed, when a rough contrast enhancement procedure is useful. For this purpose, there are several less-rigorous computational methods for enhancing the contrast in images that have been degraded by high background. It is very important to distinguish between the mathematically linear 3D deconvolution as described above and the several varieties of fast simple "deblurring" algorithms (nearest neighbor, multi-neighbor, unsharp masking), which are fundamentally 2D, nonlinear operations. *Only the linear operation of 3D deconvolution restores image intensities so that they correspond quantitatively to the intensity distribution in the object.* In doing so, this operation increases the signal-to-noise ratio of the data (Fig. 7), and the output images are appropriate for all forms of quantitative measurement. Deblurring algorithms can make the image look better, but the signal-to-noise ratio is often degraded. The output data from these nonlinear procedures may be acceptable for distance measurements on high contrast objects, but they are *not* useable for quantitative measurements of fluorophore distribution.

Practical Aspects and Tips for Generating Reliable Images

Modern objective lenses are nearly perfect when used under the conditions for which they were designed. Living cells, however, with their structural variation and optical inhomogeneities (e.g., the refractile blobs in the phase-contrast images of Fig. 8) are far from the optical engineer's ideal object. Furthermore, when the goal is live cell imaging, a beautiful image of a dead cell loses to a mediocre image of a happy cell; when choices have to be made, optics is compromised to improve the cell's environment, rather than the other way around. As a consequence, it is necessary to be able to recognize the more common forms of optical aberration (Cagnet 1962; Agard et al. 1989; McNally et al. 1994; Keller 1995) that afflict live cell imaging and to do what one can to compensate (Hell and Stelzer 1995; see the sections on Confocal Microscopy and Interpreting the Results).

Spherical aberration is the manifestation of a difference in focal position of paraxial rays compared to peripheral rays. It is usually induced by the presence of material with the "wrong" (i.e., not anticipated by the optical engineer) refractive index between the lens surface and the focal plane. For instance, using an oil immersion lens to image a sample immersed in water will cause serious spherical aberration unless the sample is within a few micrometers of the coverslip. Use of a water immersion lens avoids this particular problem, but currently, these lenses are often afflicted with other more complex aberrations that interfere with deconvolution. To some extent, proper choice of immersion oil can compensate for the induced spherical aberration (Hiraoka et al. 1990). This compensation, however, is adequate only over a limited range of depths within the water and at the expense of introducing some *chromatic aberration* (Scalettar et al. 1996), because the dispersion of water is quite different from that of glass or immersion oil. Chromatic aberration is a manifestation of a difference in focal position and in magnification for light of different wavelengths. It causes a lateral and axial shift in apparent position of objects of different color, obviously a significant problem when trying to determine colocalization of different fluorophores. The apparent shift becomes worse with increased distance from the optic axis.

The success of the iterative deconvolution procedure depends critically on the accuracy of the image data and of the PSF. Computer software and hardware are the easy inexpensive components of a quantitative 3D imaging system. The difficult expensive parts are the optical, mechanical, and electronic components that are required to collect image data that are precise and artifact-free and have a high signal-to-noise ratio. To ensure the reliability of the raw data, the imaging conditions must be painstakingly optimized (Wallace et al. 2001), and there are several important preprocessing steps that must be performed to correct various artifacts typically present in 3D epifluorescence image data (Hiraoka et al. 1990; Scalettar et al. 1996; McNally et al. 1999; Markham and Conchello 2001). For a superb guide to the practical aspects of deconvolving light microscope images, see Wallace et al. (2001).

Limitations

As with any other technique, limitations to the use of deconvolution exist: Some specimens are unsuitable, and some suitable specimens challenge the currently available computational methods. One straightforward limitation, not unique to deconvolution methods, is the need for a certain minimal signal-to-noise ratio in the input data. All of the algorithms have the potential for amplifying noise. If the input signal is too noisy (i.e., the noise is large compared to the contrast between signal and background), the algorithms will fail and the output will be meaningless. Effectively, this limits deconvolution methods to specimens of "moderate" thickness (Swedlow et al. 2002; see Introduction).

Other limitations are not intrinsic to the method itself, but are imposed by limited computational resources. For instance, the algorithms assume that the PSF of the optical system is the same for all points in the field of view ("shift-invariance"), because the computations required to take account of a spatially variant PSF are not feasible for most applications (McNally et al. 1994). It is easy to show experimentally that this assumption is routinely violated in images of typical large (i.e., non-yeast) eukaryotic cells. Incorporating a measurement of local optical inhomogeneities based on differential interference contrast (DIC) imaging into an algorithm that allows for space-variant deconvolution is one promising approach to this problem (Kam et al. 2001). Another assumption that is incorporated into most algorithms (e.g., in smoothing filters that are used to constrain the intermediate calculations) is that the data are "stationary" in a statistical sense, which would require that the power spectrum be the same for every small region of the image. This is far from true for any biological sample, particularly when the image records fluorophore distribution. Again, this assumption is not a necessary feature of the restoration algorithms (Castleman 1996), but a simplification to reduce the computational load. The practical consequence of violating this assumption is that small features can become unstable after a number of iterations and suddenly disappear from the calculated result even though they may be present in the raw data. To avoid being misled by this artifact, sensible controls must be designed and the experiments repeated using independent methods and different conditions.

Notwithstanding these minor difficulties, deconvolution of images from a wide-field microscope is a tremendously powerful technique that can yield information unobtainable in any other way (Swedlow et al. 2002). It is an invaluable tool that is becoming increasingly important with the rapid progress in methods for visualizing gene products in living cells.

CONFOCAL MICROSCOPY

The goal of this method is to improve imaging of thick objects by physically removing the out-of-focus light before the final image is formed (Minsky 1961; Petráň et al. 1968; Brakenhoff et al. 1979; Carlsson et al. 1985; Amos et al. 1987). The method takes advantage of differences in the optical paths followed by in-focus and out-of-focus light, selectively blocking the latter while allowing the former to pass to the detector.

Optical Principles

Confocal microscopes differ from conventional (wide-field) microscopes because they do not "see" out-of-focus objects. In a confocal microscope, most of the out-of-focus light is excluded from the final image, greatly increasing the contrast and hence the visibility of fine details in the specimen. Figure 8 shows a comparison of images of a thick specimen viewed by both wide-field and confocal microscopes. Figures 9 and 10 schematically illustrate how a confocal microscope works. On the left of Figure 9 is a wide-field microscope. A light source, in conjunction with a condenser, distributes light uniformly across the area of the specimen under observation. The diagram illustrates the paths followed by light arising from the specimen, passing up through the objective lens and eventually reaching a detector (e.g., film, video camera, or retina). Three paths are shown, corresponding to light arising from three locations in the sample. The first location is in the center of the field of view and in the focal plane of the objective lens. The heavy dashed lines in Figure 9 are the limits of the bundle of light rays that contribute to the image from this point. Similarly, the lighter dashed lines mark the rays from a second point in the same plane but displaced horizontally from the first point. Finally, light represented by the dotted lines is coming from a third point located below the first point (i.e., from an out-of-focus plane). This light contributes to the blurred background.

The right side of Figure 9 shows how the background is eliminated simply by adding a pinhole aperture to the wide-field microscope. Note that behind the objective lens, all of the light rays are brought together at a crossover point, the location of the intermediate image plane of the microscope. Normally, the microscope oculars are focused on this plane to form the final, fully magnified image. The location of this crossover plane along the vertical axis of the microscope is different for different light rays, depending on the distance of the corresponding point in the specimen from the front of the objective lens. The crossover point for light rays from the illustrated out-of-focus plane (dotted lines) is below that for rays from the in-focus plane (dashed lines). As illustrated, a pinhole aperture at the correct height will pass the converged rays from the in-focus point, but block nearly all of the dispersed rays from points higher or lower than the focal plane. (The geometry is slightly

FIGURE 8. Images of a thick fluorescent specimen from a confocal and a conventional microscope. The sample is a chick embryo labeled with propidium iodide and antibody against the carboxy-terminal glutamic acid form of α-tubulin (FITC-label). (*Top left*) Low-magnification, wide-field, phase-contrast image of the entire embryo. The sample is ~0.5 mm thick and contains a high density of refractile globules that scatter light efficiently. (*Top right*) Phase-contrast image at the same magnification as the fluorescence images. (*Middle row*) Conventional epifluorescence images showing (*left*) propidium iodide and (*right*) glu-tubulin distribution. The large amount of out-of-focus light severely reduces contrast. (*Bottom left*) Optical section obtained by confocal microscopy of exactly the same field and focal plane as the middle row. (*Bottom right*) Higher-magnification confocal view of a portion of the same field. Mitotic nuclei with condensed chromatin can be readily identified. (*Dotted white ellipse*) Bundles of tubulin are also seen. The mitotic spindle in these cells is formed predominantly of the tyrosinylated form of α-tubulin, and hence is not seen. (Sample kindly provided by Dr. Camille DiLullo, Philadelphia College of Osteopathic Medicine, Philadelphia.)

different in so-called "parallel beam" confocal systems, but the principle is identical [Amos et al. 1987; Shao et al. 1991].) Out-of-focus points therefore contribute little to the final image and are essentially invisible. A side effect of the pinhole aperture is that most of the in-focus points also become invisible; only the rays from the central spot are allowed to pass through the aperture.

Because all of the specimen will be invisible except for the tiny spot imaged through the pinhole aperture, there is no need to illuminate an extended area. There are three good reasons for restricting the incoming light to the minimum necessary area. First, light going to other parts of the specimen will be scattered, and inevitably some of it will leak through the pinhole aperture, degrading the contrast in the image. Second, all of the illuminated area will

Wide Field **Confocal**

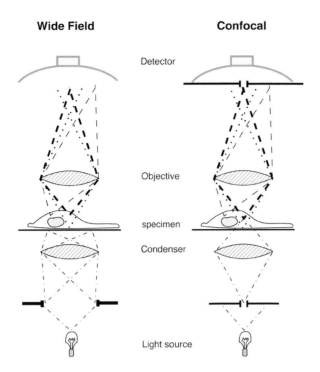

Detector

Objective

specimen

Condenser

Light source

FIGURE 9. Schematic illustration of the operating principle of the confocal microscope. (*Left*) A conventional, or *wide-field*, microscope. The specimen is illuminated over an extended region by a light source and condenser. Light rays arising from three points in the specimen are shown. The dashed lines emanate from two points in the focal plane, one centrally located (*darker dashed lines*), the other off-axis (*lighter dashed lines*). The third point is on-axis but located below the plane of focus (*dotted lines*); it gives a blurred image at the detector. The detector forms an image from the sum of all the simultaneously arriving light rays. (*Right*) A *confocal* microscope. Two pinhole apertures have been introduced. The upper aperture allows only the focused light rays from the on-axis, in-focus point of the specimen to pass to the detector. The lower aperture restricts the illumination so that it is focused on the point seen by the upper pinhole aperture.

be subject to photobleaching. Third, restricting the illumination to a single, focused point gives a dramatic improvement in the discrimination against points above and below focus; i.e., it enhances the vertical resolution.

The reason for this enhancement is as follows. If the incoming illumination is focused sharply to a point in the focal plane, then regions above or below this focal point will receive dispersed, much less intense, illumination. In fact, the intensity of illumination falls off as the square of the axial distance from the focal plane (i.e., the intensity within the cone of illumination is inversely proportional to the cross-sectional *area* of the cone). Thus, when using this type of focused spot illumination in combination with the pinhole-blocked detector not only will the pinhole aperture reject most of the light from out-of-focus planes, but the light emitted from those planes will be less than it would have been with wide-field illumination. By exactly the same reasoning, the lateral resolution of the microscope will also be enhanced if a focused spot of illumination is used. These two modifications, limiting the area "seen" by the detector and the area illuminated by the light source, are the key ingredients of a confocal microscope. A confocal microscope is simply a light microscope in which both the field of view of the objective lens and the region of illumination have been restricted to a single point in the same focal (*con*focal) plane (Wilson and Sheppard 1984).

To gain the optical sectioning capability of the confocal microscope, other aspects of the microscope's performance have been sacrificed. Field of view has been traded for increased axial resolution. The pinhole aperture effectively excludes light from out-of-focus planes, but it also restricts the field of view laterally to a spot the size of the de-magnified pinhole. Thus, to gain the advantages conferred by the confocal pinhole, one must give up the convenience of acquiring an image from an extended area in parallel. The confocal image must be built sequentially by scanning one or more spots over the specimen until the region of interest has been covered (Fig. 10).

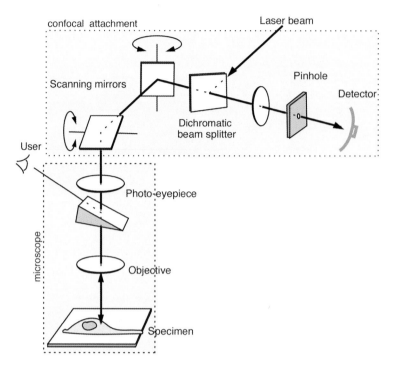

FIGURE 10. A typical laser-scanning confocal microscope. The instrument consists of a conventional fluorescence microscope (enclosed in the *lower shaded rectangle*) attached to a confocal scanning unit (*upper shaded rectangle*) comprising a pair of scanning mirrors, a laser, some wavelength-selective filters, a pinhole aperture, and a photomultiplier detector. The laser illumination is directed down the photo-tube of the microscope, having been deflected by the rapidly oscillating scanning mirrors so that it sweeps across the specimen in a raster pattern. Fluorescent light emitted by the sample passes back up through the phototube, is "descanned" by the scanning mirrors, and passes through the dichromatic beam splitter (which removes any reflected laser light) to the pinhole aperture. Light originating from the focal plane passes through the pinhole to the detector, but all other light is blocked. For reflectance imaging, the dichromatic beam splitter is replaced by a half-silvered mirror. A sliding prism allows visual (nonconfocal) observation through the usual binocular eyepieces, using the normal microscope lamps for illumination.

Instruments

To make a useful image, one needs to see much more than a single tiny spot of the sample. In principle, a complete image could be built by scanning the specimen to and fro under a fixed spot of illumination or by scanning the objective lens, the illumination, or the pinhole itself. In practice, because the scanning must be very fast to generate an image in an acceptable time, some types of scanning are much easier than others. There are three major types of confocal microscopes currently available. In the simplest type, a single diffraction-limited spot is held stationary on the optical axis of the microscope while the specimen is moved (*specimen scanning*). In the *beam-scanning* instruments, a laser beam is focused to a single diffraction-limited spot or line that is deflected over the stationary specimen using oscillating mirrors or acousto-optical deflectors. This type of microscope employs a single, fixed pinhole or slit in front of the detector. In the *pinhole scanning* instruments, a disk (called a Nipkow disk) containing an array of more than 10^4 pinholes rotates in the illumination path, sweeping approximately 1000 spots over the specimen simultaneously.

Specimen Scanning

There are important optical advantages associated with the stationary beam of the specimen-scanning type of instruments (Brakenhoff et al. 1979). All of the imaging takes place exactly on the optical axis, which minimizes many of the lens aberrations that plague the beam-scanning instruments. Alignment of the fixed optical path is also greatly simplified compared to systems with moving optical components. The primary disadvantage is that the specimen must be moved, along with specimen holder, chambers, and for living cells, liquid culture medium. The total mass of moving material is much larger than in a beam-scanning instrument, creating many opportunities for vibration and loss of positional accuracy. For the scan to be completed within a reasonable time, the mechanical accelerations must be large, and only certain specimens are suitable. Even with the lightest specimens, the time resolution of specimen-scanning instruments is much worse than for other types of confocal microscopes and can often be problematic for living samples. Sweeping a beam of light over a stationary specimen can be done much more rapidly than moving a specimen under a stationary light beam. On the other hand, when the illumination and imaging light travel separate paths, as in all forms of transmitted light imaging (e.g., bright-field, phase-contrast, DIC), only the stationary-beam, specimen-scanning instruments are truly confocal. The beam-scanning instruments are confocal only when the objective lens also serves as the condenser (e.g., epi-illumination fluorescence or reflection modes), for reasons that will be explained below.

Beam Scanning: Single-Spot Mode

In beam-scanning confocal microscopes (Fig. 11), the illumination is scanned while the specimen is held stationary (Carlsson et al. 1985). In the single-spot mode, a small (diffraction-limited) spot is swept over the specimen by means of a rapidly oscillating mirror interposed between the light source and the condenser lens (which is also the objective lens in epifluorescence mode). Because a useful image often consists of 10^5 to 10^6 pixels, the dwell time for each pixel must be kept very short to accumulate a useful image in a reasonable length of time. The need for fast scanning places stringent demands on the source of illumination, because the number of photons collected per pixel also decreases as the dwell time is shortened. To collect a 512×512 pixel image in one second, the scanning spot of light can dwell on each point for 4 μsec at most. During this time, as many photons as possible must be collected so

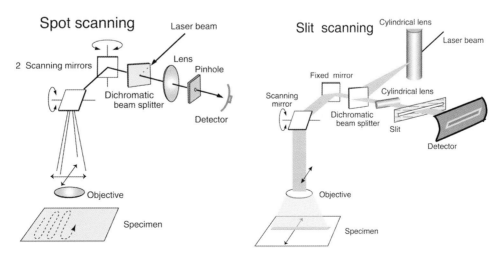

FIGURE 11. Comparison of spot-scanning and line (slit)-scanning modes of confocal microscopy.

that the statistical noise in the image is minimized. For this reason, a very intense source of light is needed, which in most instances means a small laser.

Beam Scanning: Line Mode

A second form of beam-scanning microscope (Fig. 11, right) has been developed that is, strictly speaking, only partially confocal. "Slit-scanning" instruments replace the round pinhole aperture behind the objective lens with a long, very narrow slit (Lichtman et al. 1989; Amos and White 1995). The illumination is shaped into a single narrow line, focused to the same line on the specimen that is seen by the slit aperture. Scanning is necessary in only one direction, because the slit is long enough to admit light from points all the way across the field of view. The resolution and contrast in the image are no longer completely isotropic, but in practice, the image quality is almost equal to that achieved with spot-scanning instruments for some types of sample, and the images are collected in a fraction of the time.

Pinhole (Tandem) Scanning

The renaissance in optical microscopy that has been under way for the last two decades can in some respects be traced to the stir caused in the biological research community by the appearance of a remarkable new type of microscope, a homemade tandem scanning confocal, and its equally remarkable Czech inventor, Mojmír Petráň (Egger and Petráň 1967; Petráň et al. 1968; Egger et al. 1969; an earlier, undeveloped microscope was reported by Minsky 1961, 1988). This small device, carried by Dr. Petráň in his coat pocket, provided the first glimpse of the promise of confocal microscopy. It was a nightmare to align, but the images were astonishing and inspired the development of more user-friendly instruments (Boyde et al. 1983; Petráň et al. 1986; Xiao and Kino 1987).

In tandem-scanning microscopes, the spot of illumination and the detector pinhole move over the field of view in tandem. The same pinhole is used for forming the spot of illumination on the input light path and blocking out-of-focus light on the return path. Multiple spots are formed and imaged in parallel by means of an array of pinholes arranged in an Archimedean spiral on a rotating disk (Fig. 12).

Modern tandem-scanning microscopes are enormously improved from early days, and they offer two important advantages compared to spot-scanning methods: First, the detector is a

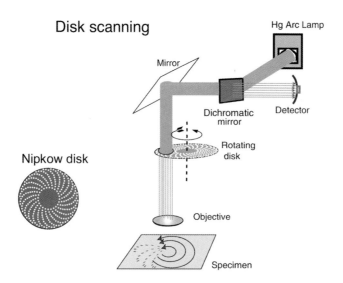

FIGURE 12. Tandem (disk)-scanning mode of confocal microscopy. These instruments make use of an array of pinhole apertures on a rotating disk.

CCD (60–70% quantum efficiency) instead of a photomultiplier (10–20% quantum efficiency), and second, for some bright specimens, direct visual observation in real time is possible. At any one instant, only a few percent of the field of view is illuminated (1000 pinholes of the ~25,000 on the disk), but the disk rotates fast enough that the moving spots fuse into a seemingly uniform image. Tandem-scanning instruments can in principle use broad-spectrum light sources such as high-pressure arc lamps for illumination. This would allow a wider selection of fluorescent probes and would eliminate one source of noise that commonly contaminates images from laser-based instruments. In practice, severe loss of illumination intensity at the disk with conventional light sources limits their use to reflected light imaging rather than fluorescence.

To avoid interference and maintain confocality, adjacent pinholes on the disk must be separated from each other by a distance that is large compared to their diameter. For example, if they are positioned 10 diameters apart, only 1% of the incoming illumination will pass through the disk. To circumvent this inefficiency, one commercially available design uses a second disk, consisting of an array of tens of thousands of microlenses, positioned above the "pinhole" disk. Each lens collects light and focuses it on one pinhole (which is actually a small transparent area etched in the opaque coating of a clear plate), thus increasing the effective aperture of each pinhole. With this design, approximately 25% of the illumination light actually passes through the disk. Nevertheless, coherent laser illumination is still necessary, because with conventional illumination, the focused spots from the lenslet array are much larger than the pinholes and light throughput is severely reduced (Watson et al. 2002). A completely different design has been described that has even better efficiency than the dual-disk lenslet array, in which a random mask of opaque and transparent patches, with 50% overall transmittance, replaces the array of pinholes (Juskaitis et al. 1996). Another promising approach that has been demonstrated (Hanley et al. 1999, 2000; Heintzmann et al. 2001), but not yet commercially available, is the replacement of the spinning disk with a stationary array of approximately 10^6 individually addressable micro-mirrors on a chip (digital micromirror device [DMD], Texas Instruments).

Imaging Modes

Confocal microscopes can form an image using several different sources of optical information from the specimen. A "reflectance" image can be formed by using the light that is scattered (backscatter) from the specimen in a backward direction, i.e., back along the path of the incoming epi-illumination. This light has the same wavelength as the original illumination. Colloidal gold labels are easily visualized in reflectance mode. The insoluble precipitates formed by the Golgi stain procedure for neurons (Fig. 13) or horseradish peroxidase (HRP) oxidation of substrates such as diaminobenzidine also give bright backscatter images.

The confocal microscope also can easily be configured to detect the interference pattern between light reflected from cell membranes and that reflected from the underlying substrate (Sato et al. 1990). This technique, known as interference reflection contrast microscopy (IRM) (Izzard and Lochner 1976; DePasquale and Izzard 1987, 1991), works well with intact living cells (Fig. 14).

Fluorescence emitted by the specimen is the most common source of optical information used to generate confocal microscope images. In this case, the image-forming light has a different wavelength from the illumination, so the fluorescence and reflectance signals can be separated using dichromatic beam splitters as in conventional epifluorescence microscopes. A particularly valuable feature of commercial instruments is the ability to acquire simultaneous, perfectly registered images from multiple different fluorescent labels with (in some instruments) independent control of the trade-off between sensitivity and resolution in each channel.

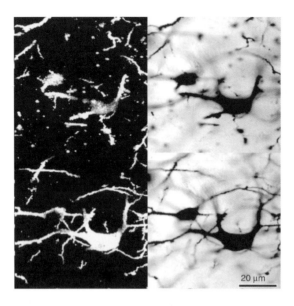

FIGURE 13. Confocal reflectance and nonconfocal transmitted light images of Golgi-stained neurons. *(Left)* Two optical sections made by using reflectance mode imaging in a laser-scanning confocal microscope. The silver precipitate gives a very bright backscattered image. *(Right)* The corresponding nonconfocal transmitted light (bright-field) images.

As well as being able to form an image from the emitted fluorescent light that is collected by the objective lens, most beam-scanning confocals have the useful ability to simultaneously collect illuminating light that passes through the specimen and thus acquire a transmitted (e.g., bright-field, phase-contrast, or DIC) image in parallel with the backscatter or epifluorescence image. The quality of these scanned, transmitted images is usually higher than could be obtained using a conventional wide-field microscope, and they should be in perfect register with the simultaneously acquired fluorescence or reflectance confocal image. It is important to realize, however, that the transmitted light image is *not* confocal, because there is no pinhole between the specimen and the transmitted light detector. The crucial difference between the transmitted and epifluorescent light is that only the latter encounters the scanning

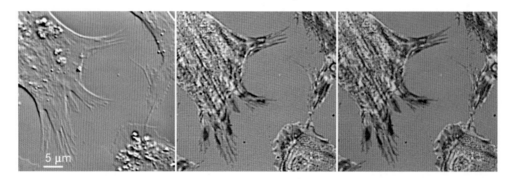

FIGURE 14. Confocal interference reflectance contrast imaging (IRM). Transmitted light DIC *(left)*, single wavelength (488 nm) IRM *(middle)*, and dual-wavelength (488, 633 nm) IRM *(right)* images of living fibroblasts from dissociated embryonic quail heart tissue growing on a glass substrate. In the IRM images, the contrast arises from interference between the laser light reflected from two surfaces (e.g., basal cell membrane and glass substrate). The image was acquired at full aperture with a 63 × NA 1.2 water immersion lens, but the high degree of coherence of the illuminating laser light nevertheless causes the higher-order fringes (e.g., from interference between reflections off the apical and basal cell membranes) to have much higher contrast than with conventional illumination (Izzard and Lochner 1976; Sato et al. 1990). The dual-wavelength IRM image is quite useful for discriminating between the zero-order *(black)* fringes, which occur at the same location for both wavelengths, and first- or higher-order *(green, yellow, red)* interference bands, which occur at different positions for the two laser lines. (Specimen kindly provided by Dr. J. Sanger, University of Pennsylvania.)

mirrors en route to the detector. Epifluorescent light from every point in the scanned field of view is reflected from the mirrors back along exactly the same path that the incoming laser illumination traversed: In other words, the epifluorescent light is *descanned* by the scanning mirrors, and thus forms a stationary beam that can pass through a fixed pinhole. The transmitted light is not descanned, and thus is *not* stationary anywhere along its path. In principle, a second set of mirrors could be introduced below the condenser, synchronized with the mirrors on the input side, to descan the exiting transmitted light. In practice, technical difficulties of this and other approaches to descanning (Goldstein et al. 1990; Art et al. 1991; Dixon et al. 1991; Dixon and Cogswell 1995) have blocked commercialization of a spot- or tandem-scanning instrument that is truly confocal in transmitted light.

Lasers and Fluorescent Labels

The total power required for imaging typical specimens is quite modest (~0.1 mW) when compared to the power of commonly available lasers, but the intensity at the focal spot can be enormous (MW/cm^2). It is important to use the minimum power necessary to acquire each image, which usually means reducing the beam intensity 10–100-fold by using neutral density filters or an acousto-optic modulator in the illumination path. At very low laser power, the strength of the emitted fluorescence will increase directly in proportion to increases in the intensity of the illumination. However, as the power is increased, the emitted light will reach a constant value, which occurs when the strength of illumination is high enough to excite every fluorescent molecule in the illuminated spot. Increases of illumination power beyond this point result in no further increase in emission from the fluorescent molecules in the focal plane. Away from the plane of focus, where the illumination is less intense, increases in laser power will continue to excite more and more fluorescent molecules. This is an undesirable effect, because the out-of-focus light does not contribute to the image, and the excited molecules are subject to photobleaching. In this respect, multiphoton imaging (see Chapter 15) can have an advantage over confocal imaging, because the light intensity is not high enough to generate multiphoton absorption events outside the focal spot. Some fluorophores, however, photobleach faster with multiphoton than with single-photon excitation (Patterson and Piston 2000).

Each type of laser emits light at a set of characteristic wavelengths, so the type of laser available determines the fluorophores that can be imaged. Table 1 shows the wavelengths available from some of the more common lasers and the major peaks in the spectrum from a mercury arc lamp. It is important to remember that the peaks in the arc lamp spectrum are much broader than the spectral lines from the lasers and, in addition, significant emission

TABLE 1. Visible laser and mercury arc emission wavelengths

Source	Emission Wavelengths (nm)											
Argon	{351}	{364}	458	466	477	488	496	502	514			
Krypton	{337}	{356}	468	476	482	521	531	568	647			
Argon/krypton	488	568	647									
Helium/neon	543	594	604	612	629	633	1152					
Helium/cadmium	325	442	534	539	636							
Solid state	???	405	457	532	635	650	670	685	694	750	780	810	???
Mercury arc	313	334	365	405	436	546	577					

Unbracketed values are the lines available from the commonly used low-power (5–50 mW), air-cooled lasers. New, solid-state lasers are introduced frequently; other wavelengths will probably be available soon. Many more lines, including those in the UV ranges that are listed in brackets, are available from the large, high-power (1–5 W), water-cooled versions of the gas lasers. In contrast to the lasers, the mercury arc lamp emits at all wavelengths between its major peaks, at a level of 5–10% of the peak intensity.

(5–10% of peak intensity) occurs at all wavelengths between the peaks. Thus, the range of fluorophores that can be excited by mercury arc illumination is much broader than that for any single laser. Table 2 shows the wavelengths for peak excitation and the range over which the excitation efficiency is at least 25% of the peak for some commonly used fluorophores.

TABLE 2. Excitation and emission ranges for common fluorophores

Fluorophore	Ex_{max}	Em_{max}	25% Ex	5% Em
For labeling nucleic acids				
Acridine orange	500	526	450–520	496–631
7-Amino-actinomycin D	540	655	460–600	570 to >780
Chromomycin A3	430	550	370–470	450–700
DAPI	359	461	310–395	385–600
Ethidium homodimer	528	617	450–570	550–750
Hoechst 33258	346	460	315–438	390–610
Propidium iodide	536	620	470–580	560 to >750
Quinacrine	420	490	360–460	435–600
TOTO-1	514	533	470–535	515–650
YOYO-1	491	509	450–515	475–615
For labeling membranes/organelles				
di-I	540	565	495–570	540–680
di-OC6 (3)	478	496	445–505	480–610
Rhodamine 123	505	534	470–530	460–610
For labeling ions				
BCECF	505	530	425–550	495–605
fluo-3	506	526	455–540	480–680
indo-1	339	405,490	260–370	370–580
For labeling protein (intrinsic fluorescence)				
ECFP	426	476	350–475	440–600
EGFP	491	512	430–514	470–600
EYFP	515	532	478–540	507–630
Tetrameric RFP (DsRed)	558	583	473–578	550–700
Monomeric RFP (mRFP1)	584	607	516–604	556–732
For labeling protein (covalent labels)				
Alexa Fluor 488	495	519	450–520	460–610
Alexa Fluor 546	556	573	510–580	540–680
Alexa Fluor 568	578	603	520–610	560–720
Alexa Fluor 594	590	617	530–625	570 to >750
Alexa Fluor 633	632	647	560–660	610–750
AMCA	343	442	< 250–376	390–546
Bodipy Fl	503	512	450–515	485–580
Cascade Blue	395	420	340–420	400–510
Cy-3	555	570	495–575	540–660
Cy-5	650	670	585–680	635–750
Fluorescein (FITC), Cy2	490	525	450–520	460–610
Lucifer yellow	430	540	380–470	470–660
NBD	460	534	420–510	480–630
Rhodamine	555	570	495–580	540–680
Texas Red	590	615	540–610	570–720

The wavelengths for maximum excitation (ex) and emission (em) are listed. The last two columns give the range of wavelengths over which excitation efficiency is at least 25% and emission intensity is greater than 5% of the maximum values. For more information, spectra, and an excellent discussion of applications of these and other fluorophores, the Molecular Probes CD-ROM catalog or web site (http://www.probes.com/) is invaluable. mRFP1 is from Campbell et al. (2002).

Simultaneous Imaging of Multiple Labels

The distribution of wavelengths available from the light source becomes especially important when two or more fluorophores must be imaged in the same specimen. In general, one can expect problems with cross-talk between two channels ("bleed-through") when the emission ranges of two fluorophores overlap significantly and one of them is much more strongly excited than the other. Table 2 lists the range of wavelengths over which emission intensity is at least 5% of the maximum for each fluorophore. This range can be used together with the range of excitation wavelengths to predict when problems are likely to arise. For example, a pair of "rhodamine-class" and "fluorescein-class" labels (with spectra similar to rhodamine and fluorescein) is one popular combination for double-labeling experiments in conventional epifluorescence microscopy, using mercury arc illumination at 495 nm and 546 nm. However, pairs of these fluorophores often give unsatisfactory results in confocal microscopes that use only an argon or argon/krypton laser, because these lasers do not emit appropriate wavelengths for efficient excitation of the rhodamine-class dye. Some instruments attempt to use the 488-nm and 514-nm lines of the argon laser to excite the fluorescein class (typical peak excitation at 490 nm) and rhodamine class (excitation optimum ~ 550 nm). Only the first fluorophore is efficiently excited at 488 nm, but it has an extended long wavelength "tail" of emission that completely overlaps the emission spectra of the second dye. At 514 nm, both dyes are excited equally (~20% of maximum excitation). This combination of spectral properties and laser excitation wavelengths thus leads to severe problems with bleed-through.

The problem is solved by using different fluorophores and/or different lasers. For example, the 488-nm argon and 543-nm green helium/neon laser lines work well with these two dyes, or the fluorescein-class dye can be used, plus a longer-wavelength ("Texas Red class") dye using the 488-nm argon and the 567-nm line of the argon/krypton laser. As one increases the number of fluorophores used simultaneously, the probability of significant cross-talk approaches 100%. There are several approaches this problem, and new methods are introduced regularly. One approach is to give up simultaneous data acquisition and instead employ sequential scans with single laser lines. This works for combinations of fluorophores that have overlapping emission spectra but can be individually excited by different laser lines. It does not solve the problem when both emission and excitation spectra overlap extensively. An alternative is to greatly increase the spectral resolution of the detection apparatus (using a dispersive element such as a prism, grating, or acousto-optic deflector) so that arbitrary wavelength bands of emission can be collected (see Chapter 15). Although it is sometimes possible to choose a narrow range of wavelengths that gives acceptable discrimination between fluorophores with partially overlapping excitation/emission spectra, the detected signal becomes weaker as the detection window is narrowed. Additional criteria can also be introduced for "gating" the fluorescence output. For instance, fluorophores that have similar emission and excitation spectra may nevertheless have quite different fluorescent lifetimes. Very short pulsed excitation and time-gated detection synchronized to the laser pulses then allow discrimination based on fluorescent lifetime (Gadella et al. 1993; Cole et al. 2001). Fluorescence polarization (or "hole-burning" plus time-gated polarization-sensitive detection) offers additional possibilities for discrimination (Massoumian et al. 2003).

Inevitably, the number of signals that need to be detected separately will exceed the capability of the detection system to discriminate, so ultimately, the problem of fluorophore cross-talk will have to be addressed by postacquisition processing. In this approach, multiple images contaminated with cross-talk among different fluorophores are collected, each with a different combination of excitation/emission wavelengths. Reference "images" are also collected from pure samples of each fluorophore individually, using the same set of excitation/emission combinations. The contributions of each fluorophore are then determined by solving the appropriate system of simultaneous linear equations for each pixel of the image. Applications of this computational method to karyotyping routinely discriminate among more than 20 different "colors" of fluorescent label (e.g., the SKY system from Applied Spectral Imaging) (Schrock et al. 1996; Ried et al. 1997).

Specimen Preparation

For a comprehensive discussion of preparing specimens for live cell imaging, see Chapter 17. A few points that are particularly important for confocal microscopy are mentioned here. Confocal microscopy is compatible with any of the conventional specimen preparation methods, including imaging of unprepared living tissue. Modest-resolution images of material down to a depth of about 0.2 mm below the surface can be obtained from many tissues if the working distance of the objective is large enough. Thicker slices can often be examined completely if they are mounted between two thin coverslips and imaged from both sides. For the highest-resolution work, spherical aberration introduced by the sample limits the maximum depth to about 0.2 mm. However, severe attenuation of both the incoming laser illumination and exiting fluorescence emission, because of scattering by local inhomogeneities in the refractive index of the sample, often limits the quality of confocal images deeper than 0.05 mm. Attenuation is sometimes less for multiphoton microscopy, because of the longer wavelength used for illumination and the fact that a detector pinhole is unnecessary; thus, multiphoton microscopy has an advantage over confocal for deep imaging in some tissues (see Chapter 15). When mounting thick specimens, care must be taken to avoid compressing them while at the same time minimizing the distance between the coverslip and the specimen. It is also important to use small coverslips when possible. Large coverslips flex with each motion of the objective lens, causing fluid displacements and specimen motion.

Time-lapse imaging of living samples for extended periods at 37°C poses several problems for confocal microscopy. The cycling of the heater for the sample chamber results in vertical movements that shift the plane of focus. Because immersion lenses are usually required, the sample chamber will be strongly thermally coupled to the objective lens and will require a separate objective lens heater, whose cycling also changes the focal plane. In a typical microscope room, the cycling of the laboratory air conditioner/heater system causes additional focal shifts. With well-designed chambers and lens heaters, and a modern laboratory building with good environmental stability, these thermally induced shifts are small enough to be barely noticeable under visual observation. Even under these ideal conditions, however, the shifts are still about tenfold larger than the vertical resolution of the confocal microscope. To compensate for these shifts in unattended time-lapse imaging, the vertical extent of each 3D stack must be increased both above and below the specimen by the size of the thermal shift. For instance, to be certain of completely capturing a 5-μm-thick cell, images might have to be taken over a 15-μm span, tripling the total exposure and making it very difficult to do extended time-lapse imaging. A third problem arises when using water immersion lenses. At 37°C, the water between the sample chamber and the lens quickly evaporates, and it is impossible to replace it without interrupting the time-lapse study and moving the specimen.

Specifications provided by a microscope manufacturer reveal the fundamental problem. According to Zeiss, the overall thermal response for their Axiovert inverted microscope system, including lenses, stage, and focus drive, is 10 μm of focal plane shift per 1°C temperature

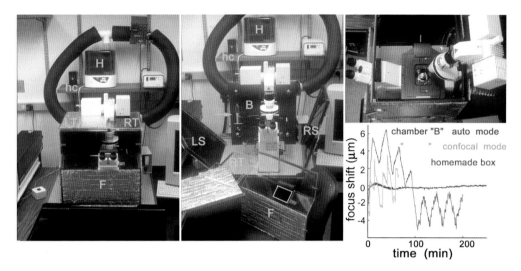

FIGURE 15. "Homemade" confocal microscope environmental chamber for live cell imaging. *(Left)* Front view of the temperature- and humidity-controlled box that encloses the entire inverted microscope except for the mercury arc and tungsten lamps. Orange letters indicate the separate pieces of the box: (H) Ultrasonic humidifier (Vicks); *(hc)* relative humidity controller (RHCN-3A, Omega Engineering); (T) heater-air circulator (Air-Therm, World Precision Instruments). *(Middle)* The disassembled pieces of the box, which are made of a 6-mm acrylic sheet, that fit snugly together and are locked into place with clasps. The front piece (F) has a rectangular aperture through which the eyepieces protrude *(orange rectangle)*. The front piece and right half of the top (RT) are clear acrylic. The left and right sides (LS, RS), left half of top (LT), and back panel (B) are opaque, black acrylic covered with reflective thermal insulating "bubble wrap." The back panel and floor plate remain permanently mounted on the microscope. The remainder of the box disassembles for use at room temperature. *(Upper right)* Top view from the right side, with the top of the box removed. The condenser lamp housing post of the microscope tilts backward for access to the stage, pushing a swiveling panel that is set into the back panel of the box. *(Lower right)* Graphs showing focal-plane shifts with different sample chambers. Blue and green lines represent shifts with a commercial sample chamber and objective lens heater in "auto" and "confocal" modes. The red line represents the shifts with the homemade box. After an initial equilibration period, the focal plane with the homemade box is stable to within ~0.2 μm.

change. Therefore, to reduce the focal shifts to below the vertical resolution of the confocal microscope, the temperature of the microscope must be stabilized to better than ± 0.04°C. That is not a realistic goal when the room temperature fluctuates by 1 or 2°C and the difference between the room temperature and the sample/lens temperature is 15–20°C. Moving the microscope to a "warmroom" held at 37°C is one solution, but the humidity must be very high to reduce the evaporation rate of the immersion water. A more user-friendly solution is to enclose the entire microscope except for the mercury arc lamp in a box held at 37°C and 80–90% humidity (Fig. 15).

Photobleaching and Phototoxicity

These two phenomena are in most cases actually the same process viewed from two different perspectives. When the emphasis is on accurate measurement of the 3D distribution of a fluorophore, then the primary concern is with photobleaching. When the emphasis is on observing the fluorophore distribution in a physiologically "normal" state, then phototoxicity will be the foremost concern. As a general rule, for quick observations of living cells at a single time point, photobleaching is the relevant phenomenon, and it is usually a surmountable problem. For repeated observations of the same cell, phototoxicity is *always* a major problem and usually necessitates accepting compromises that limit image quality.

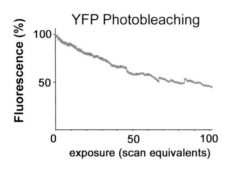

FIGURE 16. Loss of fluorescence of yellow fluorescent protein (YFP) because of photobleaching. Fluorescence from a YFP fusion protein expressed in a living cell (a non-diffusible cytoskeletal component) was recorded using 514-nm light in a laser-scanning confocal microscope (Zeiss LSM 510). The illumination intensity, dwell time per pixel, and photomultiplier gain were adjusted to give a high-quality image using a pinhole diameter equivalent to ~1 Airy disk. The average fluorescence intensity in a small region was then recorded over the course of repeated scans. The fluorescence decreased by ~1/2 after 100 scans.

Modern confocal microscopes can acquire a high-quality digital image with much lower illumination than is necessary for visual observation of the same sample. For example, Figure 16 shows the photobleaching of yellow fluorescent protein (YFP) in a living cell observed by confocal microscopy. In this specimen, as is often the case, the error caused by photobleaching is small for a single image, or even a moderately large 3D stack of images. Long before photobleaching makes the intensity measurement inaccurate, however, photo-toxicity will have made the experiment irrelevant (see Chapter 17). To keep cell damage to a minimum, careful attention must be paid to optimizing the microscopy. The goal is to extract as much information as possible from the limited number of photons that the cell will tolerate before phototoxicity becomes unacceptable.

When the sample is not too thick, the much larger quantum efficiency of CCD cameras compared to photomultipliers gives disk-scanning confocals, wide-field deconvolution, and structured illumination methods an important advantage over point-scanning confocal microscopes. For very thick, living specimens for which photobleaching is a serious problem, multiphoton illumination may provide some sample-dependent improvement (Patterson and Piston 2000).

Deconvolution of Confocal Images

Although the confocal pinhole is said to remove out-of-focus light, this removal is never perfect, and confocal images always show residual effects of the ripples in the 3D PSF caused by defocus. An example is shown in Figure 17, a confocal reflectance image of the same diatom imaged by wide-field microscopy in Figure 4. Notice that the phase reversals are seen in the

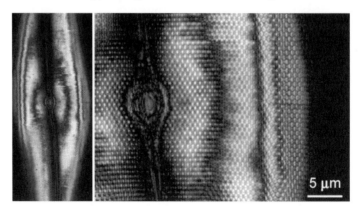

FIGURE 17. Contrast reversal because of defocus in a confocal image. *(Left)* A reflectance image of a diatom taken with a confocal microscope (Olympus Fluoview 300) using a 60 × NA 1.4 lens. *(Right)* An enlargement of a small portion of the image on the left, showing contrast reversals because of changing amounts of defocus. Compare with Figure 4.

confocal image; the holes change from black to white across the curved diatom surface. This result suggests that confocal images might benefit from deconvolving the 3D PSF, and indeed such a benefit has been reported (Shaw and Rawlins 1991; Cox and Sheppard 1995; van der Voort and Strasters 1995; Verveer et al. 1999; Boutet de Monvel et al. 2001). However, there are several considerations that might make investigators wary of the results of deconvolution, and several more reasons why they might conclude it is not worth the effort.

The first challenge is to determine the correct 3D PSF for a confocal microscope. In theory, it can be computed (Wilson and Sheppard 1984), but in practice, the computation is inaccurate, as it is for conventional wide-field microscopy with high-NA objectives. Can one then experimentally determine a 3D PSF, as is routinely done for wide-field microscopes? Unfortunately, the images from confocal microscopes often suffer from "patterned noise" artifacts (Fig. 18). Some of the patterns repeat with a periodicity that is a significant fraction of the entire field of view, which means that they are not represented in the images

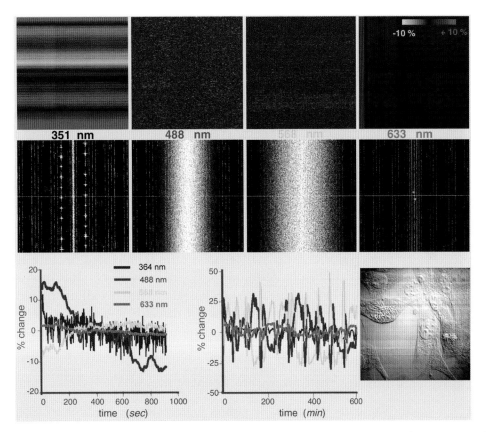

FIGURE 18. Fluctuations in illumination intensity and polarization in a confocal microscope (Zeiss LSM510). Mean-corrected images of a uniform specimen *(top row)* and their power spectra *(middle row)* acquired with four different laser lines. The images would ideally be of uniform intensity, thus entirely black after subtraction of the mean value, but instead show artifactual intensity fluctuations. Deviations from the mean value of up to ±10% are color-coded according to the scale in the top right corner. The power spectra show that each laser contributes patterned noise with a complex mixture of periodicities. Contrast of the power spectra has been enhanced to make the weaker features visible in the print. Some of the patterns have repeat lengths that are much longer than a single scan line. *(Bottom row)* The average intensity in a 100 × 100-pixel image of a uniform, stable fluorescent sample is plotted for a series of 1000 images acquired at 1-second intervals *(left)* or 600 images at 1-minute intervals *(middle)*. The effect of these artifacts is to severely degrade the image signal-to-noise ratio (see Fig. 19). The horizontal stripes in the DIC image at the lower right are caused by random changes in plane of polarization that accompany the fluctuations in intensity of the illumination.

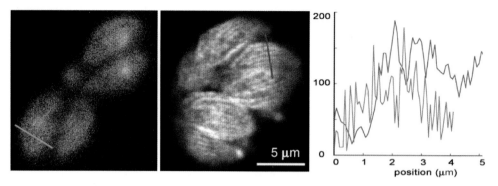

FIGURE 19. Confocal images of transgenic *Toxoplasma gondii* expressing a YFP–α-tubulin fusion protein. Microtubules near the cell surface are included in these single optical sections. *(Left)* A typical confocal image; *(middle)* a superior image (Olympus confocal, much less intensity fluctuation), the best ever recorded from this specimen. Compare the signal-to-noise ratio in this image with the wide-field image of Figure 7. *(Right)* The intensity profile along the red and blue lines in the images.

of tiny beads that are used for PSF measurements. Effectively, the PSF varies across the field of view, contrary to the assumptions of typical deconvolution algorithms. A second concern is the poor signal-to-noise ratio of confocal images (Figs. 18 and 19). This makes it quite difficult to measure the (local) 3D PSF to the accuracy required for reliable deconvolution and greatly exacerbates the tendency of deconvolution to amplify noise in the raw data. These concerns apply even to images of thin specimens. If a specimen is thick enough that a confocal microscope must be used (instead of using wide-field plus deconvolution), then the 3D PSF is certain to be seriously degraded by spherical and chromatic aberration (see Fig. 23), and this distortion will change dramatically between the top and bottom of the 3D data stack. In this situation, deconvolution is unlikely to give a correct result.

Finally, it could reasonably be argued that deconvolution comes too late to correct the most important defects of confocal images. The great benefit of deconvolving wide-field images is that the signal-to-noise ratio is enhanced because at least some of the out-of-focus light can be restored to the in-focus plane, thus increasing its total information content. In a confocal image, however, virtually all of the out-of-focus light is blocked by the pinhole before reaching the detector, and thus cannot be retrieved by deconvolution. This removes much of the motivation for using deconvolution methods with confocal images.

Practical Aspects and Tips for Generating Reliable Images

The currently available confocal microscopes are rather delicate, unstable instruments. Typically, they are controlled by complex computer programs that are prone to unexplained crashes and failed operations or missing functions. For these reasons and more, acquiring high-quality confocal images that are a faithful representation of the sample is a slow often frustrating process. Even with a complete novice at the controls, an image of some kind will usually appear on the screen, but distinguishing image from garbage takes time and considerable care regardless of how much experience the operator has.

Below are listed some guidelines that may help in adjusting the microscope parameters to obtain interpretable confocal images (see also the Troubleshooting Guide). For the preliminary adjustment of imaging parameters, choose an area of the specimen that is roughly equivalent to the area that will be recorded but is not the best area. The chosen area will be rendered unusable during the setup phase.

1. Choose the appropriate combination of laser, dichromatic mirror (beam splitter), and emission filter (see the section on Simultaneous Imaging of Multiple Labels).

2. Decide what pixel spacing is appropriate for collecting the information needed from this particular sample, and set the magnification or electronic zoom factor accordingly.

> Do not over-sample (the pixel spacing should be only slightly smaller than the Nyquist criterion; i.e., slightly less than one half of the spatial resolution required for the experiment). *Use the highest NA objective available.* For thick samples mounted in aqueous media, a water immersion objective with correction collar set to minimize spherical aberration is the best choice (see the section on Interpreting the Results). For thinner samples, an oil immersion lens may be acceptable, but the refractive index of the immersion oil must be carefully selected to minimize the spherical aberration for each specimen.

3. Estimate the imaging parameters.

    a. Set the pinhole initially to ~1 Airy disk diameter.

    b. Set the laser to the minimum power that gives a decent signal at maximum gain on the detector.

4. Find the linear range of the detector system. Use a pseudocolor lookup table (LUT) that highlights underflow (intensity = 0) and overflow (intensity = 255 for 8-bit or 4096 for 12-bit detection) in color, but is gray scale in between the two.

    a. With the laser off or set to zero power, scan at the speed to be used for the specimen, and adjust the "offset" (dark current compensator) so that the recorded image intensity is minimized but there are no pixels at zero intensity.

    b. Find a region of the specimen that is likely to be the brightest. With the laser on, decrease the detector gain until the recorded intensity in the brightest region of the image is safely below the saturation value (e.g., 200 of a maximum 255 for an 8-bit system).

5. Find the linear range for the specimen. Check that the recorded fluorescence emission increases linearly with an increase in laser power up to at least twice the laser power that will be used for imaging.

> If the emission does not increase in proportion to laser power (i.e., ground state depletion is occurring), temporal resolution (work with lower laser power and longer scan times), spatial resolution (work with lower laser power and increased pinhole diameter or pixel size), or both must be sacrificed.

6. Verify that settings are below the instantaneous damage threshold. With the scan speed and laser power set at the preliminary values determined in Steps 2–5, monitor the image intensity in a small area of the specimen over the course of numerous repeated scans.

> One would like to be able to scan dozens of times before the cumulative photobleaching reaches 50%. If the fluorescence is bleaching too much, sacrifice either temporal resolution (use longer intervals between scans in a time-lapse series), spatial resolution (increase the pinhole diameter and spacing between optical sections; increase the *xy* pixel size), or both. If the fluorescence is not bleaching measurably, the experiment is going to be quick, or photodamage is not a concern, then decrease the photomultiplier gain (which will decrease the noise), decrease the pinhole size (which may improve contrast and - resolution), and increase the laser power to maintain maximum intensities just below saturation.

7. Iteratively readjust the parameters according to Steps 4–6 until the image signal-to-noise ratio is optimized.

TROUBLESHOOTING GUIDE

This guide is primarily for confocal microscopes, but the same general principles apply to all methods. For troubleshooting deconvolution, see McNally et al. (1999), McNally et al. (1994), and particularly Wallace et al. (2001).

When poor images are obtained, the first question to be answered is whether the problem is with the specimen or with the equipment. An enormous amount of frustration and wasted time can be avoided if standard samples are available that can be used to compare the system performance at the moment with its performance in the past (i.e., on a day when good images were obtained). Four simple specimens are useful for this purpose: a resolution test target such as used for Figure 20; an optically flat mirror or bare glass slide; small beads, 0.2–0.5 μm in diameter and labeled with multiple fluorophores (e.g., Molecular Probes TetraSpeck) that are excited by all of the laser lines and detected through all of the filter sets on the instrument; and a solution of fluorescent dyes covering similarly broad excitation and emission spectra (e.g., a mixture of DAPI, Alexa Fluor 488, Alexa Fluor 594, and Alexa Fluor 633). The beads should be spread into a film on a coverslip, allowed to dry, and mounted on a thin layer of anti-fade solution or optical cement (e.g., Epo-Tek #301). To make the fourth standard specimen, a generous layer of the solution of fluorescent dyes should be sealed under a coverslip.

On a day when the equipment seems to be in good working order, collect and store a 2D image of the resolution test target at optimum focus, a 3D stack of images of the fluorescent beads, and an image of the fluorescent dye solution a few micrometers underneath the coverslip. Use the smallest pinhole. Collect similar images for all laser lines and all filter/detector channel combinations. Also collect an "X–Z" scan of the beads as in Figure 22. For these measurements, place any gain, sensitivity, background, dark level, or other adjustments in manual mode. Experiment to find settings of these parameters that give a zero intensity reading in the absence of illumination and a peak intensity reading that is just below saturation for the in-focus illuminated beads (see above section on Practical Aspects and Tips for Generating Reliable Images). Carefully record these settings along with the objective lens used and the parameters relevant to illumination and signal intensity (laser tube current, neutral density filters, acousto-optic tunable filter [AOTF] settings, beam splitters, pixel spacing, pixel integration time, etc.). These measurements serve as a calibration that can be repeated later when the performance of the system becomes questionable. Some common problems affecting confocal microscope systems are listed below.

Symptom: Image intensity is decreased over the whole field of view (at all magnifications).

Likely causes: If the image is at first bright but then gets dimmer, the problem is either photobleaching or drift of the focus level. Check for focus drift by collecting a reflectance image of a mirror or bare glass surface using the smallest possible pinhole. (This image is very sensitive to focus level.) If the image is always dim, then there is probably a misalignment of the confocal pinhole or of the internal mirrors of the laser. If the mirrors of the laser are misaligned, some wavelengths will be affected more strongly than others. For instance, the argon/krypton and krypton lasers are very prone to loss of their yellow (568 nm) and red (647 nm) lines while retaining the blue (488 or 476 nm).

Symptom: Image intensity is decreased over the periphery of the field of view (more pronounced at lower magnification).

Likely causes: If the effect is seen in both fluorescence and reflectance images, then it is caused by either a misalignment (most systems) or an intrinsic design problem (older Bio-Rad systems). If the effect is much more pronounced in fluorescent images than in reflectance images, chromatic aberration is indicated.

Symptom: Resolution is poor.

Likely causes: If the problem is apparent in both thin (e.g., fluorescent beads) and thick samples, then the fault is probably in the alignment of the confocal optics (but first make sure that the objective lens is clean!). Verify that the laser beam is correctly centered on the axis of the objective lens and that the entire back aperture of the lens is filled with incoming light. If the problem is restricted to thick samples, then spherical aberration is probably the culprit. The newer water immersion objectives with a correction collar greatly ameliorate this problem for thick samples in aqueous media, but with the drawback of slightly decreased resolution for optically ideal specimens (i.e., very thin samples with a refractive index the same as glass and positioned immediately adjacent to the coverslip).

Symptom: Focusing for maximum brightness does not give the sharpest image. The image can be made bright or sharp, but not both.

Likely causes: The system is not *con*focal. The focal plane for the illumination system does not coincide with the focal plane for the imaging system. For visible wavelength illumination and imaging, the pinhole or an intermediate lens that focuses the light on the pinhole is probably misaligned. If the illumination or imaging wavelengths are ultraviolet or far red, then a misaligned collimator lens is the likely culprit.

Symptom: Alternating stripes of higher and lower intensities appear in the image.

Likely causes: Mechanical vibration, defects on the scanning mirrors, or an electronic oscillation in the laser or the detector circuits are the likely causes. To decide among these possibilities, collect an image of the fluorescent dye test sample with the largest available pinhole aperture. This image is very insensitive to vibration but is still sensitive to electronic oscillations and mirror defects. Mirror defects cause a fixed pattern of bright-dark stripes that does not change between images. Most electronic oscillations (and mechanical vibrations) give a different pattern with each image. Confocal systems from some manufacturers use single-mode polarization-preserving fiber-optic coupling of the laser to the scan head. Some of these systems are very prone to fluctuating illumination intensity and polarization angle. Painstaking rotational alignment of the fiber polarization axis with the laser polarization axis mitigates the effect, but within a few days, the fiber drifts out of alignment and the stripes reappear.

Symptom: A circular bright spot or a set of rings appears at a fixed point in every image.

Likely cause: A reflection of the laser beam off an internal glass surface is being detected. If fluorescence images are being collected, then an inappropriate set of filters is being used (reflected laser light is getting through). Most systems now include a quarter-wave plate and polarizer combination to minimize the problem in reflectance mode imaging. One of these elements probably has been rotated.

Symptom: Images from different fluorophores are misregistered.

Likely cause: Displacement of a very short wavelength (e.g., DAPI channel) or very long wavelength (e.g., CY5 or Alexa 633 channel) from the middle-wavelength channels is usually due to misalignment of a ultraviolet/visible or infrared/visible collimator lens (a necessary component of the illumination path that corrects for the small residual chromatic aberration present in all currently available objective lenses). Be aware that some manufacturers' service personnel will not check collimator lens alignment unless specifically asked, so a recent inspection by unsupervised service personnel provides no assurance that the system is correctly aligned.

STRUCTURED ILLUMINATION METHODS

The goal of these techniques is to improve the images of thick objects by a combination of optical and computational manipulations. By a wonderfully simple manipulation, the blurring caused by defocus can be turned into an effective tool for separating in-focus from out-of-focus light, when the light is right (i.e., structured).

Investigators normally strive to achieve completely uniform illumination (i.e., completely "unstructured") across the entire field of view, so that variations in intensity across the image arise solely from variations in the structure of the object. Contrary to what one might expect, superimposing artifactual intensity fluctuations across the image by using carefully patterned nonuniform illumination can actually *increase* the amount of information about the object that is stored in the image.

Development of new techniques for using structured illumination to enhance microscope performance is proceeding rapidly and in many different directions (Bailey et al. 1993; Neil et al. 1997, 1998, 2000; Wilson et al. 1998; Gustafsson et al. 1999; Hanley et al. 1999, 2000; Gustafsson 2000; Cole et al. 2001; Heintzmann et al. 2001; Dubois et al. 2002. For an excellent survey, see Gustafsson 1999). At this time, most of the techniques have been demonstrated only on very thin specimens and are available only in a few specialized laboratories. In this chapter, we describe one method (Neil et al. 1997, 2000) that does work quite well with thick specimens and is now commercially available from at least two vendors.

Optical Principles

The basic principle is shown in Figure 20, which illustrates in a different way the same information contained in Figures 1 through 3. Figure 20 shows an image of a resolution test target, a series of gratings of different spacing, which was tilted by about 15° before

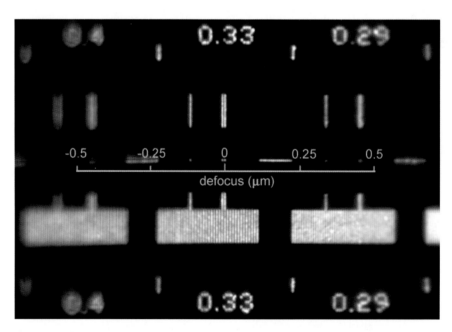

FIGURE 20. Image of a resolution test target that was tilted by ~1.5° on the microscope stage. The spacing of the gratings in micrometers is marked on the specimen. Due to the tilt, the defocus varies across the image as indicated by the orange scale. The specimen-to-lens distance increases from right to left. Note that the black and white bars of the 0.29-μm grating are fairly distinct at the left end but smeared to an average gray on the right. 60 × NA 1.4 objective, 546 nm illumination.

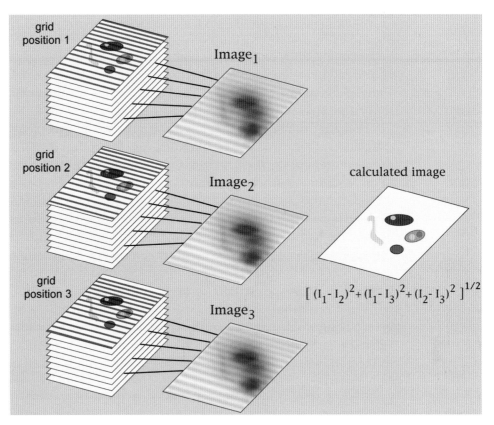

FIGURE 21. Structured illumination employed for optical sectioning. The light source is masked by a regular grid, which casts a pattern of stripes on the in-focus object plane. Three images of the thick object are acquired, shifting the grid by 1/3 repeat between each acquisition. Each image is the sum of contributions from the in-focus plane, which is shadowed by sharply defined stripes, plus blurred out-of-focus planes without distinct stripes. The square root of the sum of the squared difference images removes the out-of-focus blurred light, giving the in-focus plane.

being photographed. The tilt, which is from right to left in the image, displaces the sample from the focal plane by an amount that increases, in opposite directions, away from the center of the field. Notice that the 0.33-μm grating is sharp and well resolved in the center, but becomes quite blurred toward the left. The left side of the 0.29-μm grating is just resolved at a defocus of 0.25 μm, but the right side is completely smeared out when the defocus increases to 0.5 μm. How can this "problem" be turned into a "solution?"

Imagine the use of a grid of evenly spaced lines as a mask through which the specimen is illuminated (Fig. 21). The grid is placed in the light path of the microscope, with appropriate lenses to bring its shadow on the specimen into sharp focus in exactly the focal plane of the objective lens. If the specimen is thin, the result is simply an image of the in-focus specimen crossed by a set of sharply demarcated shadows where the illumination has been interrupted by the mask. If the specimen is thick, then superimposed on this in-focus image will be a blurred image of the other planes of the specimen (see Figs. 5 and 21). However, the contribution from the out-of-focus planes of the specimen will *not* be modulated by the lines of the grating; away from the in-focus plane, the shadow of the grating quickly becomes smeared and the illumination is uniform, at a level that is the average of the light and dark bars. Note in Figure 20 that a defocus of 0.5 μm is sufficient to completely eliminate the 0.29-μm modulation.

A simple algebraic combination of three images of the object illuminated through a grating shifted by exactly one third of its period between each image yields an in-focus image uncontaminated by out-of-focus blur (Neil et al. 1997): $I_{\text{in-focus}} = [(I_1 - I_2)^2 + (I_1 - I_3)^2 + (I_2 - I_3)^2]^{1/2}$. The simple sum of the three raw images, $(I_1 + I_2 + I_3)$, is identical to the normal wide-field image with no grating. The best optical sectioning is obtained when the period of the grating is slightly larger than the diameter of the Airy disk (Neil et al. 1997).

The grating can be shifted rapidly and precisely (e.g., by a piezo-electric element), and the calculation is very fast. The rate-limiting step for optical sectioning by this method is thus the acquisition of the three images. Compared to wide-field plus deconvolution, the maximum acquisition rate would be threefold slower, but the finished product, the in-focus optical sections, is available immediately instead of after 15–20 minutes as for the deconvolution computation. The total exposure is slightly greater than for wide-field plus deconvolution, because the bars of the grating are typically not completely opaque. The z-axis resolution is comparable to what is achieved by either confocal or deconvolution methods.

The first commercial versions of this methodology have only recently been introduced (http://www.thales-optem.com/optigrid, http://www.zeiss.de/us/micro/home.nsf), so very few investigators have experience with it and consequently its limitations are not yet fully known. One would expect that the computation would fail for extremely thick samples, when the background out-of-focus light becomes overwhelming. In that case, the contrast of the gird lines in the raw data will be very low, and the difference between I_1, I_2, and I_3 would become comparable to the noise level. For a discussion of some of the technical problems that need to be addressed to make the technique useful in practice, see Cole et al. (2001). The simplicity of the hardware component means that it is inexpensive and can be user-installed as an add-on to virtually any modern microscope. It is safe to predict that this and other variants of structured illumination microscopy will quickly supplant expensive confocal microscopes for many applications on modestly thick specimens, particularly living samples.

INTERPRETING THE RESULTS

The Meaning of "Optical Section"

Although the images produced by confocal microscopy, deconvolution, and structured illumination methods are referred to as optical "sections," they differ from true sections in that their top and bottom edges are not sharply defined. In a physical section that has been cut by a knife in a microtome, there is no ambiguity about which section contains each point of the original object, at least not at the resolution of the light microscope. A specified point in the cell was either included in one particular microtome section or it was not, but there is no intermediate state. An optical section, however, includes some locations fully (i.e., present at their true intensity), and other locations above and below at less than their true intensity. There is no sharp cutoff that demarcates what is included in the optical section and what is excluded. Instead, there is a continuous decrease in the ratio of image intensity to object intensity for locations further and further away from the mid-point (i.e., the CTF falls off steadily with distance from the focal plane; see Figs. 2 and 3).

A measurement that is commonly used as the analog of section "thickness" for optical sections is the width at half-amplitude (full-width at half-maximum, FWHM) of the curve that describes the relative intensity of points at different distances from the midpoint of the section. This curve can be measured by collecting a series of closely spaced optical sections

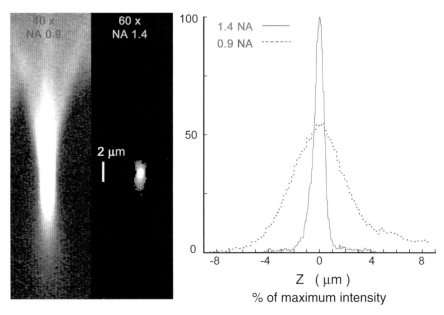

FIGURE 22. *X–Z* scans of 0.9-μm-diameter fluorescent beads and measured axial intensity profiles. The left side of the image shows a scan using a 40 × NA 0.9 objective lens. The right half is a scan of a different sample using a 60 × NA 1.4 lens. Intensity for the two images has been adjusted to the same maximum value; in reality, the 40 × NA 0.9 image is much dimmer. The round bead appears much more elongated with the lower-NA objective for two reasons. Decreasing the NA affects vertical resolution much more severely than lateral resolution. The left scan is also markedly asymmetrical above and below focus, indicative of spherical aberration, which causes further elongation. The right scan (of a different sample) also shows a small amount of spherical aberration. Bar, 2 μm. The graph shows the measured axial intensity profiles for these two situations. In these curves, spherical aberration is manifest as asymmetry of the profile to the right and left of the peak (e.g., 40 × curve beyond 5 μm).

(0.1 μm for the highest NA lenses) of a small, bright object. A small fluorescent bead is a good specimen for this measurement, but any bright object that is small compared to the expected FWHM can be used. A plot of the total intensity of the image of the object in each optical section should give a curve similar to those in Figure 22, the axial intensity profile of the 3D PSF for the optical system (equivalent to a vertical line through the center of Fig. 3). Confocal microscopy, deconvolved wide-field images, or structured illumination microscopy should give a PSF with an axial FWHM ≤0.6 μm for the highest NA objective lenses. Much smaller values of FWHM have been obtained, but not with samples and equipment that are realistic for use in live cell imaging experiments (Bailey et al. 1993; Hell et al. 1997; Gustafsson et al. 1999).

Spherical Aberration

The FWHM of the vertical PSF decreases with the square of the NA of the objective lens. For confocal systems, the width also decreases with decreasing pinhole size. For all microscopes, the vertical PSF is very sensitive to the presence of spherical aberration (Fig. 22). Unfortunately, a certain amount of spherical aberration must often be accepted when examining living specimens (Fig. 23). The highest-NA oil immersion objectives are designed for work with a specimen that is located immediately beneath a coverslip connected to the lens by immersion oil. Images of thick specimens, for which these conditions are not uniformly possible, are increasingly degraded by spherical aberration as the focal plane is lowered. The problem is

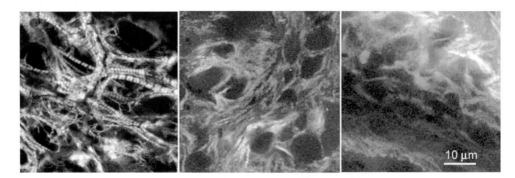

FIGURE 23. Confocal optical sections of thick tissue. The developing heart in a chick embryo, labeled with fluorescent antibody against cardiac myosin, was imaged with a 60 × NA 1.4 oil immersion lens. *(Left)* The first optical section of a 3D stack of 160 sections spaced at 0.5-μm increments. The first section was ~20 μm below the coverslip. *(Middle)* Section 30; *(right)* Section 130. Spherical aberration increasingly degrades the resolution, so that the 2.2-μm myofibrillar striations, clearly visible in section 1, are barely resolved in Section 130. (Sample kindly provided by Howard Holtzer, University of Pennsylvania.)

greatly ameliorated by using the newer, long-working-distance water immersion objectives. These objectives are designed to be used *with* a coverslip, in contrast to older designs (e.g., those used by electrophysiologists for patch-clamp studies). They are very expensive, but with thick specimens immersed in water, they perform better than standard, very short-working-distance high-NA oil immersion objectives. The PSF of these new water immersion lenses is not yet as good as with the best high-NA oil objectives, so their performance in deconvolution is limited.

As illustrated in Figure 22, spherical aberration leads to an asymmetrical response to defocus; the image looks different when defocused by the same amount in opposite directions from the in-focus plane. When present, significant spherical aberration is readily visible. A convenient way to check for it is to find a very small, very bright "dot" of fluorescence and observe its appearance as the lens is defocused by a small amount (a few micrometers) in either direction. A typical observation is the appearance of bright rings on one side of focus and general fuzziness without rings on the other side. With the water-immersion lenses, this asymmetry can often be eliminated over an extended range of focal planes in a thick specimen by careful adjustment of the correction collar. With oil immersion lenses, one can choose an immersion oil with a refractive index that minimizes the spherical aberration at the depth in the specimen where the major interesting features lie, but optical performance in other focal planes will be degraded (Fig. 23).

Chromatic Aberration

With modern, highly corrected objective lenses and an ideal sample, all wavelengths of light in the visible range should be focused to the same point with an accuracy of better than 0.3 μm (Keller 1995). However, if the mixture of refractive indices in the sample deviates from the design parameters of the objective lens enough to cause noticeable spherical aberration, then chromatic aberration is also likely to be induced. The lens design is specific not only for a particular arrangement of refractive indices, but also for the way in which those refractive indices vary with wavelength (dispersion). Sample dispersion often does not match design specifications, and the result of this mismatch is a shift in focal position according to wavelength. Figure 24 shows this behavior in a series of optical sections of a single bead. In this case, the chromatic aberration is caused not by the sample, but by misalignment of collimator lenses that

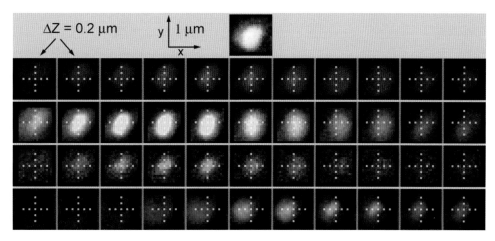

FIGURE 24. Chromatic aberration in confocal images. Eleven optical sections, spaced at 0.2-μm intervals, of a single, 0.2-μm-diameter bead labeled with four different fluorophores. The dotted cross in each section marks the *x-y* location of the center of the bead in the green image at best focus. Note that the other three wavelength bands are shifted in *x*, *y*, and *z* with respect to the green. The yellow and blue images also reveal some spherical aberration. A composite color image for the sixth optical section is shown at the top. *(Red)* Excitation (ex) 633 nm, emission (em) >650 nm; *(yellow)* ex 568 nm, em 585–615 nm; *(green)* ex 488 nm, em 505–600 nm; *(blue)* ex 364 nm, em 385–470 nm. 63 × NA 1.2 water immersion lens.

are supposed to correct the confocal system for residual chromatic aberration in the objective. This correction allows the system to be used with UV, infrared, and visible light.

As the composite image at the top of Figure 24 shows, chromatic aberration is a serious problem in determining colocalization of different fluorescent molecules. In an extended sample, complicated artifactual shifts are observed because the direction and amount of apparent shift between different colors varies with position in the field of view. Interpretation of slight differences in localization demands careful control experiments to rule out chromatic aberration.

Chromatic aberration is a particular concern with confocal microscopes (Hell and Stelzer 1995; Keller 1995). If the focal spot for the illumination wavelength is not in the same position as the focal spot seen through the pinhole aperture, then the system is not *con*focal. In this situation, image intensity is severely decreased.

Signal-to-Noise Ratio

Live cell imaging is considerably more demanding than observation of fixed samples. The motivation to invest the extra time and effort is often the need to observe *changes* in the distribution or amount of some fluorescent molecule. To be confident that an observed change is real, rather than the result of random fluctuations in the image intensity, the investigator must consider the signal-to-noise ratio (S/N) in the image.

Suppose that an experiment requires detecting changes of 25% or more in the local concentration of a fluorescent molecule. As an example, consider the cells shown in Figure 7. These cells will be used to follow the incorporation or removal of YFP-tubulin at the growing and shrinking ends of microtubules. Because this specimen is thin enough that the background from out-of-focus light is not overwhelming, it is appropriate for any of the microscopic techniques described in this chapter.

The S/N must be large enough to be confident that a 25% change in fluorescence at the end of a microtubule is not caused by random noise. To draw this conclusion with 95%

confidence, the expected fractional change in signal (i.e., 0.25 S) must be greater than twice the standard deviation (i.e., greater than twice the noise, 2N). Thus, S/N must be at least 8 for this measurement to succeed. If the required precision is expressed as a fractional change P (i.e., 25% precision in the measurement of S means P = 0.25), then for 95% confidence, S/N must be greater than 2/P.

To decide if this experiment is feasible, a rough estimate of the signal and its standard deviation (s) is needed. First, an area of a typical sample that is similar to the region of interest must be located. Two images must be collected in rapid succession from this area without changing anything between exposures. Photobleaching must be avoided. In the absence of noise, these two images should be identical, so the difference between them can be used to estimate the noise level in a typical measurement. In fact, the standard deviation of this difference image is $\sqrt{2}$ times the standard deviation of a single pixel in the original images. The noise in the measurement of microtubule fluorescence will be larger than s because the background must be subtracted. If the ratio of background to total intensity is b, then the standard deviation in one pixel of the background corrected microtubule fluorescence image will be a factor $\sqrt{(1 + b)}$ larger than the standard deviation in the raw intensity measurement. In Figure 7, $b \approx 0.95$ before deconvolution and 0.8 after deconvolution. Finally, the size of the target area must be taken into account. If the target includes n pixels and the average fluorescence per pixel of microtubule after background correction is denoted by F, then

$$S/N \cong \frac{F}{s\sqrt{(1 + b)/2n}}$$

For the experiment in Figure 7, F was ~50 (of 4096 maximum for the 12-bit CCD) averaged over a 5 × 5 pixel box, s was ~12, and b was 0.95, so S/N ≈ 21. It can be concluded that the images collected under these conditions are indeed good enough to detect the hoped-for change in YFP-tubulin incorporation.

Now suppose one contemplated doing this experiment with a laser-scanning confocal microscope instead of a wide-field plus deconvolution. Doing this simple calculation beforehand would save a lot of time and frustration, because one would find that the experiment cannot be done with an ordinary confocal! Typical numbers (Zeiss LSM510) are roughly F = 40 (of 255 maximum for 8-bit detection), s = 30, and b = 0.5, which for the same size target give an S/N ≈ 4 (compare Figs. 7 and 19).

Point-scanning confocal microscope images are much noisier than wide-field images for several reasons. Photomultipliers, the usual detectors on point-scanning confocal systems, are about fourfold less efficient than a good CCD camera (quantum efficiency [QE] ~15% vs. ~60%). Thus, for the same exposure (photobleaching, phototoxicity), the wide-field image would be formed from four times as many photons and have twice the S/N as the confocal image. In addition, many confocal microscopes add a large amount of unnecessary noise to the image, generated by electronic artifacts in the detector circuitry (Fig. 18) and random fluctuations in the illumination intensity. Figure 25 shows a direct comparison of the noise in the illumination of a laser-scanning confocal and a wide-field microscope. It should be emphasized that this extra noise in the confocal is entirely unnecessary and is simply a matter of poor design in most of the commercially available instruments.

Disk-scanning confocals, which use CCD detectors, should have an S/N at least twofold better than point-scanning instruments for the same exposure levels. The noise level in images from the structured illumination method described above will be increased by the processing steps needed to calculate the in-focus image, but model calculations suggest that this effect will be less than a factor of 2. Thus, the structured illumination technique also has

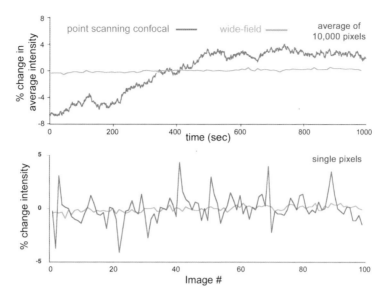

FIGURE 25. Fluctuations in illumination intensity in confocal versus wide-field microscope. *(Top)* The average intensity in a 100 × 100-pixel image of a uniform, stable, fluorescent sample is plotted for a series of 1000 images acquired at 1-second intervals, for a point-scanning confocal and a wide-field microscope with CCD detector. *(Bottom)* The intensity from a single pixel in a sequence of 100 successive images of the same specimen is plotted.

the potential for giving an image S/N ratio substantially higher than the commercially available point-scanning confocals.

Why Confocal?

After this comparison of signal-to-noise ratios with a thin sample, it is perhaps worth restating the message of the Introduction to this chapter: Each of the microscopic techniques described here has its own realm within which it reigns unchallenged, and outside of which it must take second place to other approaches. When specimen thickness is modest, the S/N of the raw data acquired with a wide-field microscope is high enough to allow excellent contrast enhancement by restoration methods such as deconvolution or structured illumination. However, when the sample is very thick, out-of-focus background reduces the contrast, and only point-scanning confocal or multiphoton techniques are useful. As a counterpoint to the S/N comparison for a thin sample, Figure 26 shows a comparison between point-scanning confocal microscopy and wide-field microscopy plus deconvolution with a sample that is definitely in the confocal realm.

Three-dimensional Reconstruction

The output from confocal microscopy, deconvolution, or structured illumination methods is typically a stack of optical sections, collected by changing the focus by a constant amount between each image acquisition. If the goal is to collect enough information to permit a 3D reconstruction of a specimen at the highest resolution possible, then adjacent optical sections must be spaced at increments of roughly half the FWHM of the vertical PSF. The resolution in the vertical direction will always be worse than in the horizontal direction, by ~3 × at high NA. Thus, it is common in 3D reconstructions, particularly from confocal microscopes, for objects to appear elongated in the vertical direction (see Fig. 22). The software supplied for

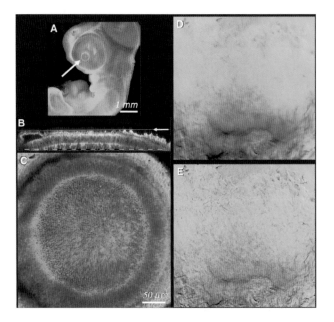

FIGURE 26. Comparison of laser-scanning confocal and wide-field microscope performance in imaging a thick tissue (Swedlow et al. 2002). A 5-day-old quail embryo stained with Alexa488-phalloidin *(green)* and DAPI *(blue)* was imaged by laser-scanning confocal *(A–C)* and wide-field microscopy *(D)*. *(A)* A low-magnification survey image of the entire embryo. The head is at the top of the figure. The arrow points to the developing eye, the region shown at higher magnification in B–E. *(B)* "X–Z" section (parallel to optical axis) showing position of coverslip (arrow) and location of the focal plane shown in C *(dashed line)*. *(C)* A single optical section ~50 μm below the exterior surface of the embryonic eye. Concentrations of actin at cell cortices are visible. *(D)* The same embryo imaged by wide-field microscopy and recorded with a CCD camera (Alexa488 only). The image shown is one of 60 recorded from a series of optical sections. *(E)* The same optical section as in D after restoration by deconvolution. No cellular details are visible in wide-field images of this thick tissue. Bars: *(A)* 1 mm; *(C)* 50 μm; *(B,D,E)* same scale as C. (Images D and E, courtesy of Jason Swedlow, University of Dundee.)

deconvolution often compensates for this effect, and a similar compensation (essentially a one-dimensional edge sharpening filter) is helpful on confocal images.

Display of 3D image stacks in a form that efficiently and faithfully transmits the data to the viewer is a challenging task. Simplified representations of 3D image data are essential for display purposes. Software abounds for "3D-rendering" or various other forms of displaying 3D intensity distributions, and the output can be visually striking and quite persuasive. However, injudicious use of 3D rendering software can also produce misleading representations that exaggerate the contrast (hence the reliability of the segmentation between different regions of the image) and the resolution (rendering fuzzy intensity gradients as sharp boundaries). As always, *Caveat emptor!*

REFERENCES

Agard D.A. 1984. Optical sectioning microscopy: Cellular architecture in three dimensions. *Annu. Rev. Biophys. Bioeng.* **13:** 191–219.

Agard D.A., Hiraoka Y., Shaw P., and Sedat J.W. 1989. Fluorescence microscopy in three dimensions. *Methods Cell Biol.* **30:** 353–377.

Amos W.B. and White J.G. 1995. Direct view confocal imaging systems using a slit aperture. In *Handbook of biological confocal microscopy* (ed. J.B. Pawley), pp. 403–415. Plenum Press, New York.

Amos W.B., White J.G., and Fordham M. 1987. Use of confocal imaging in the study of biological structures. *Appl. Optics* **26:** 3239–3243.

Art J.J., Goodman M.B., and Schwartz E.A. 1991. Simultaneous fluorescent and transmission laser scanning confocal microscopy. *Biophys. J.* **59:** 155a.

Bailey B., Farkas D.L., Taylor D.L., and Lanni F. 1993. Enhancement of axial resolution in fluorescence microscopy by standing-wave excitation. *Nature* **366:** 44–48.

Born M. and Wolf E. 1999. *Principles of optics: Electromagnetic theory of propagation, interference and diffraction of light.* Cambridge University Press, England.

Boutet de Monvel J., Le Calvez S., and Ulfendahl M. 2001. Image restoration for confocal microscopy: Improving the limits of deconvolution, with application to the visualization of the mammalian hearing organ. *Biophys. J.* **80:** 2455–2470.

Boyde A., Petráň M., and Hadravsky M. 1983. Tandem scanning reflected light microscopy of internal features in whole bone and tooth samples. *J. Microsc.* **132:** 1–7.

Brakenhoff G.J., Blom P., and Barends P. 1979. Confocal scanning light-microscopy with high aperture immersion lenses. *J. Microsc.* **117:** 219–232.

Cagnet M. 1962. *Atlas optischer Erscheinungen (Atlas of optical phenomena).* Springer, Berlin.

Campbell R.E., Tour O., Palmer A.E., Steinbach P.A., Baird G.S., Zacharias D.A., and Tsien R.Y. 2002. A monomeric red fluorescent protein. *Proc. Natl. Acad. Sci.* **99:** 7877–7882.

Carlsson K., Danielsson P.E., Lenz R., Liljeborg A., Majlof L., and Aslund N. 1985. Three-dimensional microscopy using a confocal scanning laser microscope. *Optics Lett.* **10:** 53–55.

Carrington W.A., Lynch R.M., Moore E.D., Isenberg G., Fogarty K.E., and Fay F.S. 1995. Superresolution three-dimensional images of fluorescence in cells with minimal light exposure. *Science* **268:** 1483–1487.

Castleman K.R. 1979. *Digital image processing.* Prentice-Hall, Englewood Cliffs, New Jersey.

———. 1996. *Digital image processing.* Prentice Hall, Englewood Cliffs, New Jersey.

Cole M.J., Siegel J., Webb S.E., Jones R., Dowling K., Dayel M.J., Parsons-Karavassilis D., French P.M., Lever M.J., Sucharov L.O., et al. 2001. Time-domain whole-field fluorescence lifetime imaging with optical sectioning. *J. Microsc.* **203:** 246–257.

Cox G. and Sheppard C.J.R. 1995. Effects of image deconvolution on optical sectioning in conventional and confocal microscopes. *Bioimaging* **1:** 82–95.

DePasquale J.A. and Izzard C.S. 1987. Evidence for an actin-containing cytoplasmic precursor of the focal contact and the timing of incorporation of vinculin at the focal contact. *J. Cell Biol.* **105:** 2803–2809.

———. 1991. Accumulation of talin in nodes at the edge of the lamellipodium and separate incorporation into adhesion plaques at focal contacts in fibroblasts. *J. Cell Biol.* **113:** 1351–1359.

Dixon A.E. and Cogswell C. 1995. Confocal microscopy with transmitted light. In *Handbook of biological confocal microscopy* (ed. J.B. Pawley), pp. 479–490. Plenum Press, New York.

Dixon A.E., Damaskinos S., and Atkinson M.R. 1991. A scanning confocal microscope for transmission and reflection imaging. *Nature* **351:** 551–553.

Dubois A., Vabre L., Boccara A.C., and Beaurepaire E. 2002. High-resolution full-field optical coherence tomography with a Linnik microscope. *Appl. Optics* **41:** 805–812.

Egger M.D. and Petráň M. 1967. New reflected light microscope for viewing unstained brain and ganglion cells. *Science.* **157:** 305–307.

Egger M.D., Gezari W., Davidovits P., Hadravsky M., and Petráň M. 1969. Observation of nerve fibers in incident light. *Experientia* **25:** 1225–1226.

Erhardt A., Zinser G., Komitowski D., and Bille J. 1985. Reconstructing 3-D light-microscopic images by digital image-processing. *Appl. Optics* **24:** 194–200.

Fay F.S., Carrington W., and Fogarty K.E. 1989. Three-dimensional molecular distribution in single cells analysed using the digital imaging microscope. *J. Microsc.* **153:** 133–149.

Femino A.M., Fay F.S., Fogarty K., and Singer R.H. 1998. Visualization of single RNA transcripts in situ. *Science* **280:** 585–590.

Gadella T., Jovin T.M., and Clegg R.M. 1993. Fluorescence lifetime imaging microscopy (FLIM): Spatial resolution of microstructures on the nanosecond time scale. *Biophys. Chem.* **48:** 221–239.

Goldstein S.R., Hubin T., Rosenthal S., and Washburn C. 1990. A confocal video-rate laser-beam scanning reflected-light microscope with no moving parts. *J. Microsc.* **157:** 29–38.

Gustafsson M.G. 1999. Extended resolution fluorescence microscopy. *Curr. Opin. Struct. Biol.* **9:** 627–634.

———. 2000. Surpassing the lateral resolution limit by a factor of two using structured illumination microscopy. *J. Microsc.* **198:** 82–87.

Gustafsson M.G., Agard D.A., and Sedat J.W. 1999. I5M: 3D widefield light microscopy with better than 100 nm axial resolution. *J. Microsc.* **195:** 10–16.

Hanley Q.S., Verveer P.J., Arndt-Jovin D.J., and Jovin T.M. 2000. Three-dimensional spectral imaging by hadamard transform spectroscopy in a programmable array microscope. *J. Microsc.* **197:** 5–14.

Hanley Q.S., Verveer P.J., Gemkow M.J., Arndt-Jovin D., and Jovin T.M. 1999. An optical sectioning programmable array microscope implemented with a digital micromirror device. *J. Microsc.* **196:** 317–331.

Heintzmann R., Hanley Q.S., Arndt-Jovin D., and Jovin T.M. 2001. A dual path programmable array microscope (PAM): Simultaneous acquisition of conjugate and non-conjugate images. *J. Microsc.* **204:** 119–135.

Hell S.W. and Stelzer E.H. 1995. Lens aberrations in confocal fluorescence microscopy. In *Handbook of biological confocal microscopy* (ed. J. Pawley), pp. 347–354. Plenum, New York.

Hell S.W., Schrader M., and van der Voort H.T. 1997. Far-field fluorescence microscopy with three-dimensional resolution in the 100-nm range. *J. Microsc.* **187:** 1–7.

Hiraoka Y., Sedat J.W., and Agard D.A. 1990. Determination of three-dimensional imaging properties of a light microscope system. Partial confocal behavior in epifluorescence microscopy. *Biophys. J.* **57:** 325–333.

Holmes T.J. 1992. Blind deconvolution of quantum-limited incoherent imagery: Maximum-likelihood approach. *J. Opt. Soc. Am. A* **9:** 1052–1061.

Hopkins H.H. 1955. The frequency response of a defocused optical system. *Proc. R. Soc. Lond. Ser. A* **231:** 91–103.

Izzard C.S. and Lochner L.R. 1976. Cell-to-substrate contacts in living fibroblasts: An interference reflexion study with an evaluation of the technique. *J. Cell Sci.* **21:** 129–159.

Juskaitis R., Wilson T., Neil M.A., and Kozubek M. 1996. Efficient real-time confocal microscopy with white light sources. *Nature* **383:** 804–806.

Kam Z., Hanser B., Gustafsson M.G., Agard D.A., and Sedat J.W. 2001. Computational adaptive optics for live three-dimensional biological imaging. *Proc. Natl. Acad. Sci.* **98:** 3790–3795.

Keller H.E. 1995. Objective lenses for confocal microscopy. In *Handbook of biological confocal microscopy* (ed. J.B. Pawley), pp. 111–126. Plenum Press, New York.

Lichtman J.W., Sunderland W.J., and Wilkinson R.S. 1989. High-resolution imaging of synaptic structure with a simple confocal microscope. *New Biol.* **1:** 75–82.

Markham J. and Conchello J.A. 2001. Artefacts in restored images due to intensity loss in three-dimensional fluorescence microscopy. *J. Microsc.* **204:** 93–98.

Massoumian F., Juskaitis R., Neil M.A., and Wilson T. 2003. Quantitative polarized light microscopy. *J. Microsc.* **209:** 13–22.

McNally J.G., Karpova T., Cooper J., and Conchello J.A. 1999. Three-dimensional imaging by deconvolution microscopy. *Methods* **19:** 373–385.

McNally J.G., Preza C., Conchello J.A., and Thomas L.J. 1994. Artifacts in computational optical-sectioning microscopy. *J. Optic. Soc. Am. A* **11:** 1056–1067.

Minsky M. 1961. "Microscopy apparatus." U.S. Patent #3,013,467.

———. 1988. Memoir on inventing the confocal scanning microscope. *Scanning* **10:** 128–138.

Neil M.A., Juskaitis R., and Wilson T. 1997. Method of obtaining optical sectioning by using structured light in a conventional microscope. *Optics Lett.* **22:** 1905–1907.

———. 1998. Real time 3D fluorescence microscopy by two beam interference illumination. *Optics Commun.* **153:** 1–4.

Neil M.A., Squire A., Juskaitis R., Bastiaens P.I., and Wilson T. 2000. Wide-field optically sectioning fluorescence microscopy with laser illumination. *J. Microsc.* **197:** 1–4.

Patterson G.H. and Piston D.W. 2000. Photobleaching in two-photon excitation microscopy. *Biophys. J.* **78:** 2159–2162.

Petráň M., Hadravsky M., Benes J., and Boyde A. 1986. In vivo microscopy using the tandem scanning microscope. *Ann. N.Y. Acad. Sci.* **483:** 440–447.

Petráň M., Hadravsky M., Egger M.D., and Galambos R. 1968. Tandem scanning reflected light microscope. *J. Opt. Soc. Am. A.* **58:** 661–664.

Ried T., Koehler M., Padilla-Nash H., and Schrock E. 1997. Chromosome analysis by spectral karyotyping. In *Cells: A laboratory manual. Subcellular localization of genes and their products* (ed. D.L. Spector et al.), vol. 3, pp. 113.1–113.9. Cold Spring Harbor Laboratory Press, Cold Spring Harbor, New York.

Sato M., Sardana M.K., Grasser W.A., Garsky V.M., Murray J.M., and Gould R.J. 1990. Echistatin is a potent inhibitor of bone resorption in culture. *J. Cell Biol.* **111:** 1713–1723.

Scalettar B.A., Swedlow J.R., Sedat J.W., and Agard D.A. 1996. Dispersion, aberration and deconvolution in multi-wavelength fluorescence images. *J. Microsc.* **182:** 50–60.

Schrock E., du Manoir S., Veldman T., Schoell B., Wienberg J., Ferguson-Smith M.A., Ning Y., Ledbetter D.H., Bar-Am I., Soenksen D., et al. 1996. Multicolor spectral karyotyping of human chromosomes. *Science* **273:** 494–497.

Shao Z.F., Baumann O., and Somlyo A.P. 1991. Axial resolution of confocal microscopes with parallel-beam detection. *J. Microsc.* **164:** 13–19.

Shaw P.J. and Rawlins D.J. 1991. The point-spread function of a confocal microscope—Its measurement and use in deconvolution of 3-D data. *J. Microsc.* **163:** 151–165.

Stokseth P.A. 1969. Properties of a defocused optical system. *J. Optic. Soc. Am.* **59:** 1314.

Swedlow J.R., Hu K., Andrews P.D., Roos D.S., and Murray J.M. 2002. Measuring tubulin content in *Toxoplasma gondii:* A comparison of laser-scanning confocal and wide-field fluorescence microscopy. *Proc. Natl. Acad. Sci.* **99:** 2014–2019.

van der Voort H.T.M. and Strasters K.C. 1995. Restoration of confocal images for quantitative image analysis. *J. Microsc.* **178:** 165–181.

Verveer P.J., Gemkow M.J., and Jovin T.M. 1999. A comparison of image restoration approaches applied to three-dimensional confocal and wide-field fluorescence microscopy. *J. Microsc.* **193:** 50–61.

Wallace W., Schaefer L.H., and Swedlow J.R. 2001. A workingperson's guide to deconvolution in light microscopy. *BioTechniques* **31:** 1076–1082.

Watson T.F., Juskaitis R., and Wilson T. 2002. New imaging modes for lenslet-array tandem scanning microscopes. *J. Microsc.* **205:** 209–212.

Wilson T., Neil M.A., and Juskaitis R. 1998. Real-time three-dimensional imaging of macroscopic structures. *J. Microsc.* **191:** 116–118.

Wilson T. and Sheppard C.J.R. 1984. *Theory and practice of scanning optical microscopy.* Academic Press, London.

Xiao G.O. and Kino G.S. 1987. A real-time scanning optical microscope. *SPIE Scan. Imag. Tech.* **809:** 107–113.

ADDITIONAL READING

World Wide Web Virtual Library: Microscopy http://www.ou.edu/research/electron/www-vl

Davidson M. 2003. Molecular expressions: Images from the microscope http://micro.magnet.fsu.edu/index.html

Inoué S. and Spring K.R. 1997. *Video microscopy: The fundamentals.* Plenum Press, New York.

Taylor C.A. 1978. *Images: A unified view of diffraction and image formation with all kinds of radiation.* Wykeham Publications, New York.

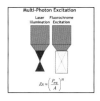

Multiphoton and Multispectral Laser-scanning Microscopy

Mary E. Dickinson

*Biological Imaging Center, Beckman Institute, California Institute of Technology,
Pasadena, California 91125*

INTRODUCTION

FLUORESCENCE MICROSCOPY HAS BECOME AN INVALUABLE and common tool for research scientists in a wide variety of fields. Consistent developments in fluorescent probes and proteins, better imaging technology, and more robust protocols for live-cell imaging have made fluorescence imaging one of the most versatile tools available to researchers. This chapter focuses on two advanced forms of fluorescence imaging that increase the information that can be obtained from biological samples: multiphoton laser-scanning microscopy (MPLSM) and multispectral laser scanning microscopy (MSLSM). In a broad sense, these techniques have advantages over standard confocal laser-scanning microscopy by enhancing the ability to collect data in multiple dimensions. MPLSM improves the ability both to collect data from deep in the sample (along the *z* axis) and to obtain more data over time with less lethality, whereas MSLSM reveals the spectra of markers in biological samples and enhances the color separation that is possible, enabling more markers to be viewed simultaneously. Live imaging of whole, intact animals is possible using imaging techniques such as magnetic resonance imaging (MRI), but the lack of resolution and the availability of contrast agents make it difficult to study molecular and cellular events. Both of the techniques featured in this chapter offer submicron resolution for depths of less than a millimeter and benefit greatly from the availability of a large number of fluorescent probes to examine specific cellular and biochemical events. This chapter is devoted to the practical considerations that affect the success of imaging experiments and is designed as a starting point for scientists who are considering the use of these techniques in their own research.

BASIC PRINCIPLES OF MULTIPHOTON LASER-SCANNING MICROSCOPY

MPLSM emerged as a technique in the early 1990s (Denk et al. 1990), although multiphoton excitation has been predicted since the 1930s (Goppert-Mayer 1931). In MPLSM, fluorescent molecules are excited by the quasi-simultaneous absorption of two or more near-infrared photons. Multiphoton, also referred to as nonlinear excitation, has a quadratic dependence, producing excitation only in a small focal volume; thus, out-of-focus fluorescence does not contribute to the acquired image, reducing background, and photodamage outside the plane of focus is greatly reduced. In practical terms, MPLSM makes it possible to acquire images with a high signal-to-noise ratio by using a wavelength that is less harmful to live cells. The use of near-infrared light makes it possible to image deeper in the specimen, due to less scatter and absorption of the incident light. However, the efficiency of multiphoton excitation depends on criteria that differ from those favoring single-photon excitation events.

In single-photon excitation, a fluorescent molecule or fluorochrome (also called a chromophore) absorbs a high-energy photon within a certain wavelength range and then, within nanoseconds, releases a photon of longer wavelength (lower energy). The absorption of a photon results in the excitation of the molecule, by displacing an electron from the ground state to an excited state. For single-photon excitation, the excitation is directly proportional to the incident photon flux of the source, since each photon has an equal probability of exciting a molecule in the ground state. As the molecule relaxes back to the ground state, some energy is lost through nonradiative exchange (heat or vibration within the molecule), but the rest is shed as a photon of light. The energy loss accounts for the Stokes' shift (or red-shift) seen between the excitation and the emission wavelength. Multiphoton excitation of the fluorochrome is induced by the combined effect of two or more lower-energy near-infrared photons and can be achieved by two photons of the same or different wavelengths, but with a single laser source, two photons of approximately the same wavelength are used. The probability of two-photon excitation is proportional to the intensity squared (I^2), because a quasisimultaneous absorption of two photons is necessary. It follows that for three-photon excitation, the probability of three-photon absorption is the intensity cubed (I^3). The emission characteristics of the excited fluorochrome are unaffected by the different absorption processes (Figure 1).

To improve the efficiency of multiphoton excitation, ultrafast lasers emitting short pulses of light at a rapid frequency are used (for a review, see Wise 2000). Using these lasers, very large peak intensities can be achieved in a repeated pulse train to sustain fluorophore excitation exclusively at the focal plane where the density of the photon flux is sufficiently high. Although several laser sources have been utilized, Titanium:Sapphire (Ti:Sapphire) lasers have become the most common light source for this application. Further developments in laser technology now include full automation and software control of Ti:Sapphire lasers, resulting in turnkey laser sources for multiphoton microscopy that are convenient to use and provide enough power over a large enough wavelength range to be useful for many applications (Coherent-Inc.; Spectra-Physics).

If we examine the parameters that affect the probability of two-photon excitation, we find that excitation probability relies on (1) how efficiently a given molecule is excited via a nonlinear event, which is a property of the fluorochrome itself, and (2) the density of the photon flux or, in other words, how well the photons at the plane of focus are concentrated, which depends on the output of the laser, in addition to how well the incident beam is focused. Two-photon excitation probability can be expressed as

$$n_a \propto \delta \left(\frac{P_{avg}^{2}}{\tau f^{2}} \right) \left(\pi \frac{NA^2}{hc\lambda} \right)^2$$

where the n_a is the probability of excitation, δ is the excitation cross-section of a dye, P_{avg} is the average power of the incident beam at the sample, τ is the duration of the pulse or pulse-

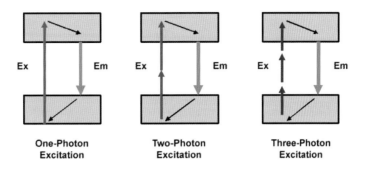

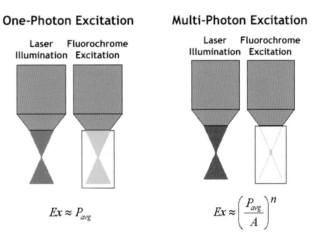

$$Ex \approx P_{avg}$$

$$Ex \approx \left(\frac{P_{avg}}{A} \right)^n$$

FIGURE 1. Principles of multiphoton excitation. (*Top*) Three diagrams showing theoretical differences between single-, two-, and three-photon excitation with respect to the energy levels of the fluorochrome. (*Bottom*) Diagram showing the difference between one- and multiphoton excitation in the sample. For single-photon excitation, fluorochrome excitation is directly proportional to the photon flux of the incident light, whereas for two-photon excitation, excitation depends on the square of the intensity of the incident light and on the intensity cubed for three-photon excitation. Thus, multiphoton excitation is limited to the focal plane.

width, f is the repetition frequency of the pulse train from the laser, NA is the numerical aperture of the lens, h is Planck's constant, c is the speed of light, and λ is the wavelength (from Denk et al. 1990). Given this, multiphoton excitation is favored when using molecules with large cross-sections, high peak power excitation sources, and high-NA objective lenses. Discussed below are how these parameters can be optimized to produce the best quality images revealing the most information.

Optimizing Multiphoton Excitation in Biological Samples

Many factors contribute to the success of biological imaging experiments, as evidenced by the many chapters in this book. Summarized below are some of the common areas that have the greatest effect on success for multiphoton microscopy.

Choosing a Dye

Although multiphoton excitation appears to be conceptually simple, two factors make choosing the best dye for the experiment a complicated choice. First, it is difficult to predict whether a molecule will efficiently absorb two lower-energy photons simultaneously (Xu et al. 1987, 1996; Xu 2000). Drastic differences in multiphoton absorption between different molecules have been identified, and it is difficult to predict, either by the structure of a molecule or by its single photon properties, how efficiently a dye will absorb simultaneous low-energy photons, although some theories are emerging (Albota et al. 1998; Rumi et al. 2000). Second, the wavelengths for maximum multiphoton excitation are difficult to predict (Xu 2000). Deriving the multiphoton excitation wavelength maximum is not as simple as doubling the single-photon excitation wavelength maximum, although in some cases, this can be a good place to start. The absorption efficiency at a given wavelength must be measured and is reflected in the multiphoton cross section (usually referred to as δ) for a given fluorochrome (for a review, see Xu 2000). In most cases, this value is determined for a given dye using a specially constructed multiphoton fluorimetry system. Cross sections for some dyes have been characterized (Xu et al. 1987, 1996; Xu 2000; Bestvater et al. 2002; W. Zipfel, unpubl. Pic. http://cellscience.bio-rad.com/products/multiphoton/Radiance2100MP/mpspectra.htm.). These data are generally gathered from solutions of dyes in a cuvette and quantified against a known standard. However, the biological environment may affect the absorbance spectrum of some dyes. Recently, the multiphoton excitation peak has been determined directly in biological samples, providing the ability to optimize multiphoton excitation on the microscope under the same conditions in which the experiment will be performed (Dickinson et al. 2003).

The Laser Source

Since the density of the photon flux has an important role in excitation efficiency, both the performance of the laser and the optics of the system are important considerations in optimizing efficiency of excitation. As mentioned above, ultrafast Ti:Sapphire lasers have become the most popular laser source for MPLSM, and several commercial sources for Ti:Sapphire lasers are available. For most applications, a broadly tunable source is desired because different fluorochromes with different excitation peaks may be used for different applications. The laser systems currently available to microscope users can be tuned either manually across a broad range from 690 to approximately 1000 nm (such as with the Coherent MIRA 900 or the Spectra-Physics Tsunami) or be tuned via software-controlled motorized mirrors (such as with the Coherent Chameleon or the Spectra-Physics Mai Tai), with only small sacrifices to the tuning range and power output. Most dyes can be excited by two-photon absorption between 710 and 950 nm. The maximum power output over the tuning range is a function of the fluorescence spectra of the Ti:Sapphire crystal and the optics within the cavity. For the Ti:Sapphire cavity to lase, the crystal is pumped by a 532-nm doubled vanadate pump source. Typically, 5–10-W pump sources are used, and as more pump power is added, more Ti:Sapphire output is seen. The Ti:Sapphire lasers currently marketed can produce a large amount (well over 2 W) of average power at the peak of the tuning curve. Typically, the peak of the Ti:Sapphire output is near 780–800 nm (a wavelength range that might be used to optimally excite fluorescein or rhodamine using two-photon absorbance), but overall power output of these lasers usually falls to lower values at the wings of the tuning curve, toward 700 nm and past 950 nm. For the manually tunable lasers, it is possible to improve the power output of the laser at the wings by installing optics in the cavity that are optimized for specific wavelengths; however, above 930 nm, significant water absorption bands exist, thus limiting power output. Purging the laser cavity with nitrogen or by recirculating dry air will provide the best performance and mode-locking stability. In addition to their other advantages, software tunable lasers are constructed with sealed and dry cavities for maxium performance throughout the tuning range.

In addition to the power output and the wavelength, two other parameters also affect the efficiency of fluorochrome excitation: the frequency of the pulse train (abbreviated as f above) and the duration of the pulse (often referred to as the pulsewidth or τ). The frequency of the pulse train is a fixed value and depends on the geometry of the laser cavity. For most Ti:Sapphire lasers, this range is between 70 and 100 MHz. Thus, the duration between pulses is in the range of 10 ns and fits well with the lifetime of most fluorochromes (Denk et al. 1995).

The pulsewidth is an important and often elusive parameter. According to the equation given above, in the discussion on Basic Principles of MSLSM, the shorter the pulsewidth, the greater the peak power and the greater the likelihood of an excitation event. That said, it is also true that the shorter the pulsewidth exiting the laser, the more the pulse will be broadened as it encounters glass on the way to the specimen. When the pulsewidth becomes broader, more average power is needed to achieve the same excitation efficiency. Thus, if power is limiting or if the specimen is sensitive to the level of average power, optimizing pulsewidth can be important.

Pulse broadening is due to an effect termed group velocity dispersion (GVD). GVD is encountered when pulses of light encounter normal dispersive media such as glass (for a review, see Wolleschensky et al. 2002). Pulses, centered at a particular wavelength, are produced as the result of constructive and destructive interference of several wavelengths lasing at once. The more frequencies or colors that are combined, the larger the bandwidth and the shorter the pulsewidth. Photons of different wavelengths in the pulse do not pass through glass at equal rates, lengthening the pulse duration, so less GVD is seen when there is less bandwidth to the pulse (less difference in the wavelengths). For instance, consider the difference in pulse broadening between an 80- and 180-fs pulse traveling through a single objective lens (assume the objective lens has a dispersion parameter of 2000 fs^2). The 80-fs pulse will stretch to approximately 105 fs, whereas the 180-fs pulse will not be stretched at all (181 fs) as it passes through the same lens. Considering all the glass that is encountered in an entire laser-scanning microscope system, it can be expected that an 80-fs pulse will be somewhere between 290 and 450 fs at the sample, whereas a 180 fs pulse will be between 220 and 270 fs at the sample. It is possible to compensate for the dispersive effects of the optics in the microscope system, or even compensate for fiber delivery (Helmchen et al. 2002; Wolleschensky et al. 2002); however, precompensation systems add to alignment complexities and can often result in limiting the amount of available power at the sample.

Measuring Bandwidth and Pulsewidth

Manually tunable ultrafast lasers allow the user to optimize the length of the pulse exiting the cavity by adjusting the prism pair that compensates for GVD within the laser cavity. Information about the duration of the pulse exiting the laser head can be inferred by measuring the bandwidth. This can be done quite easily by using a spectrometer (IST-Rees; Ocean-Optics). A pulse duration of 150–200 fs generally corresponds to an approximately 6–8-nm bandwidth, depending on the particular laser. Although this provides a useful benchmark, spectrometers do not provide a direct measure of pulsewidth and are not designed to take measurements at the sample plane of the microscope where multiphoton excitation efficiency is most critical.

Directly measuring the pulsewidth of ultrafast laser output requires a device called an autocorrelator. Most autocorrelators take advantage of a nonlinear effect called second harmonic generation (SHG) in which light of twice the input energy (half the wavelength) is generated when light of sufficiently high intensity (large peak power) is focused in a crystal (termed a doubling crystal) or other such media. Autocorrelators measure pulsewidth by splitting the input light into two paths of equal intensity, such as in a Michelson interferometer. A variation in the two paths is introduced, thereby producing a slight delay between pulses. The two paths are focused onto a doubling crystal, and a detector is used to measure the intensity of the SHG signal generated by the input. The intensity of the SHG signal as a function of the delay in the pulse gives the pulsewidth.

In the past, it has been difficult to measure the pulsewidth at the sample on an MPLSM system. Although it is possible to measure the pulsewidth produced by the laser, as described above, a measure of the pulsewidth at the sample is needed because of the effects of GVD. Recently, an autocorrelator has been developed for this purpose that is quite easy to use (APE, Berlin, Germany). The Carpe autocorrelator makes two measurements, one from the laser path outside the microscope and then another using an external sensor that is placed on the microscope stage. Thus, it is possible to determine the amount of pulse broadening due to dispersion in the microscope and to optimize the excitation efficiency related to pulsewidth. With such an autocorrelator, the laser output can be optimized to provide the shortest pulsewidth at the sample, producing the best excitation efficiency for a given fluorochrome.

Choosing Optics and Objective Lenses for MPLSM

For nonlinear microscopy, objective lenses should be optimized for the following parameters:

- *Long working distance with a high NA:* Excitation efficiency increases as the NA of the objective lens increases. However, many long-working-distance objective lenses have a low NA. Therefore, it is best to choose an objective lens with the highest NA for the working distance needed. NA has an effect not only on excitation efficiency, but also on collection efficiency, so NA is extremely important. Low-magnification high-NA lenses can be better for collecting an emission signal from deep within the sample (Oheim et al. 2001), and often offer more working distance than higher-magnification lenses with a similar NA.

- *High transmission in the NIR and the visible-wavelength range:* Many objective lenses used for biomedical microscopy are corrected for the UV/Visible range, and the efficiency of transmission of these lenses often decreases significantly in the near-infrared (NIR) range. Thus, both the laser output and the transmission of the objective can combine to lower the power available to the sample at longer wavelengths. Further on in this chapter, we will see that this effect is significant when characterizing the multiphoton excitation spectra of fluorochromes using the microscope. If excitation energy at longer wavelengths is limiting, lenses corrected for better transmission in the infrared may be worth considering. For instance, a comparison of the transmission of the Zeiss 40× Achroplan lens with that of the Zeiss 40× IR-Achroplan lens at 900 nm demonstrates, almost 25% higher transmission using the infrared-corrected version of this lens.

- *Limited pulse broadening:* Objective lenses should be minimally dispersive to reduce the chance that short pulses will be lengthened en route to the sample, which will again reduce the peak intensity. In addition to the group velocity dispersion, chromatic aberration of lenses leads to pulse distortions. Specifically, a radius-dependent group delay is introduced

(Kempe and Rudolph 1993; Netz et al. 2000). Therefore, different radial portions of the beam across the pupil of the objective lens arrive at different times at the focal region and cause a temporal broadening of the pulse, resulting in lower peak intensity in the focal region. This effect is also referred to as propagation time difference (PTD). Known dispersion and PTD values for some lenses can be found in (Wolleschensky et al. 2002).

- *Limiting or correcting for focal plane mismatch:* Chromatic aberrations can also affect the ability to both excite and collect photons at the same focal plane, due to the mismatch in focal distance between two such different wavelengths (see Wokosin and Girkin 2002 and references therein). Correcting for this mismatch properly can ensure not only that visible and NIR beams are focused at the same plane, but also that the greatest collection efficiency for the visible emission signal will be realized. Corrections for chromatic aberrations can be made either by using a collimator to correct for the mismatch by focusing the NIR beam at the visible plane or by using a wider collection lens or nondescanned detection to recover photons that are not focused back through to the detector. Although correcting for the collection can mitigate the loss of collection signal, proper collimation of the incident beams will ensure that images produced by the simultaneous use of visible and NIR beams can be overlayed and that photochemistry performed by the NIR beam can be imaged accurately using excitation by the visible lasers.

Optimizing Laser Alignment into the Microscope

Poor alignment of the laser into the microscope can have a very profound effect on excitation efficiency. The following protocol describes a sequence for optimizing alignment of the laser into the scanhead and for ensuring the best overlay between the visible and NIR excitation beams.

Aligning the NIR Laser to the Optical Path of the Microscope Using a VIS Laser

1. Place a mirror slide (part 453001-9062, Carl Zeiss, Inc.) on the stage and focus on the slide using a low-magnification (10×) lens.

2. Configure the system to reflect one of the visible laser lines down to the sample by choosing the appropriate primary beam-splitter selection.

3. While scanning, make sure that the visible laser line is reflected back out of the scan head. Use the focus knob if necessary to change the spot size.

4. The visible and NIR beams can be walked together using the routing mirrors (see Fig. 2). Place a piece of lens paper in the path close to where the NIR beam enters the scan head. Adjust the mirror farthest from the scan head using both the *x* and *y* knobs or adjustment screws to overlay the visible and NIR beams.

5. Hold the lens paper in the path closest to the laser. Align the overlay of the two beams by adjusting the routing mirror closest to the scan head.

6. Repeat Steps 4 and 5 until no further improvements can be made.

7. Switch to an open position in the objective turret. Place a piece of lens paper on the stage to determine whether the NIR spot is centered. If it is not centered, adjust the mirror closest to the scan head to center the beam.

   Many lasers have an elliptical beam shape and may not fill the entire aperture.

External alignment of the Titanium:Sapphire laser into the microscope scan head

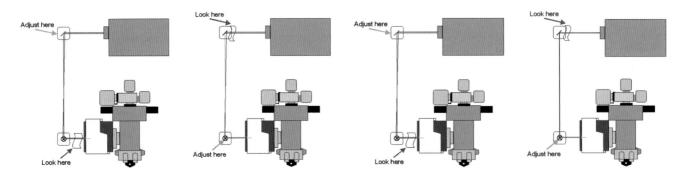

Alignment of the Titanium:Sapphire laser to the optical path of the microscope using a reflective grid slide

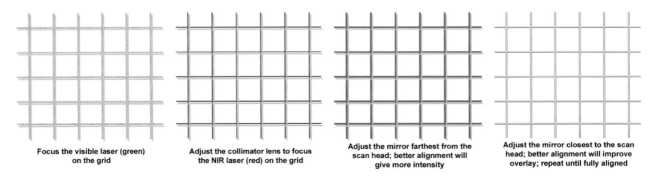

Focus the visible laser (green) on the grid

Adjust the collimator lens to focus the NIR laser (red) on the grid

Adjust the mirror farthest from the scan head; better alignment will give more intensity

Adjust the mirror closest to the scan head; better alignment will improve overlay; repeat until fully aligned

FIGURE 2. Alignment of an external Ti:Sapphire laser to the microscope using visible laser alignment as a guide (see text).

8. Use a slide with a thin mirrored grid (part 474028-0001, Test Grid Specimen for LSM, Carl Zeiss, Inc.) to ensure that the NIR and visible lasers are aligned well to one another. Place the grid slide on the stage and focus on the grid pattern using the objective lens of choice.

9. Set up an imaging configuration that allows separate channels for the reflected laser light from a visible line and the NIR line. Use a very low amount of power from the Ti:Sapphire laser—too much power will burn the slide. Confirm that these channels are correct by switching off one line at a time, the corresponding image should disappear.

10. While imaging, adjust the focus of the visible reflection and then adjust the collimation lens if there is one available. Optimizing this lens ensures that the visible and NIR excitation planes are parfocal (Fig. 2).

11. To begin, peak the mirror farthest from the scan head. Better alignment is indicated by an increase in intensity (Fig. 2). Next adjust the alignment of the two beams on the grid using the mirror closest to the scan head (Fig. 2).

12. Repeat the sequence until no further improvements can be made (Fig. 2).

Detecting Photons: Descanned vs. Nondescanned

Nondescanned detectors (NDDs) are used to improve the collection of emission photons, especially from deep within the sample. In a confocal microscope, pinholes and detectors are aligned with respect to the incident light so that the position of the pinhole accurately rejects light that is out of the plane of focus. For multiphoton microscopy, excitation is limited to the focal plane so a pinhole is not required. In fact, since all fluorescence that is produced is a result of excitation at the focal plane, collecting scattered photons improves the signal in the image. NDDs, placed very near to the emission signal at the sample, can be used to collect scattered photons that might be lost before reaching the detectors within the scan head. For highly scattering samples, this approach can greatly improve light collection, improving the signal to noise within the image and increasing the detection efficiency when deep imaging is needed. Because of the increased ability to collect scattered light, many users immediately notice that stray light from fluorescent bulbs or monitors in the room can contribute a significant amount of noise. Therefore, very dark conditions are recommended when using NDDs.

Although a single NDD can offer significant improvement, signal from multiple NDDs can be added. Because fluorescence emission occurs in a radial fashion from the focal point, NDDs can be placed above and below the sample; the signal from the detectors in the epi (through the objective lens) path and the forward (toward the condenser) path can be summed to offer more sensitivity if proper collection optics are used. High-NA condensers can often be useful for collecting the light scattered in the forward direction. In addition to MPLSM, other nonlinear imaging modalities, such as second harmonic generation (SHG) imaging (Campagnola et al. 2001; Millard et al. 2003) or coherent anti-Stokes raman scatter (CARS) imaging (Muller et al. 2000; Cheng et al. 2002; Nan et al. 2003) often make use of nondescanned detection. For these methods, forward detection can be more sensitive than epi detection.

Live Cell Imaging

MPLSM is currently being applied to a large variety of experiments and many imaging protocols have evolved to look at different samples. The popularity of multiphoton imaging, in part, is due to the availability of robust culture methods and the creativity of many microscopists who have devised clever ways of preparing living samples for light microscopy. Some of these protocols are reviewed in other chapters in this volume, or in recent reviews (Helmchen and Denk 2002; Hadjantonakis et al. 2003; Ragan et al. 2003). One area in which MPLSM has been well-utilized is intravital imaging (see Chapters 22 and 23). MPLSM has been used to follow cellular processes, often for days, weeks, or months, in whole rodents. Neurons in the brain (Helmchen et al. 1999; Svoboda et al. 1999; Trachtenberg et al. 2002; Chaigneau et al. 2003; D'Amore et al. 2003; Stosiek et al. 2003) and neuromuscular junctions (Keller-Peck et al. 2001; Walsh and Lichtman 2003) have been studied, as well as cells in the immune system (Miller et al. 2002, 2003; Padera et al. 2002), in implanted tumors (Brown et al. 2001; Wang et al. 2002), and in the kidney (Dunn et al. 2002, 2003). These methods utilize minimally invasive procedures and provide evidence about cell behaviors that can only be gleaned by watching cells in their native environment.

Intravital imaging methods are increasing as technology becomes available. Many advances have been made toward engineering fiber optic multiphoton microscopes (Helmchen et al. 2001; Helmchen 2002; Helmchen and Denk 2002; Jung and Schnitzer 2003). Recently, a fiber optic, head-mounted MPLSM system has been used to image from the brain as the rodent subject is free to move about in its environment (Helmchen et al. 2001). Such methods make long-term observations possible and provide direct evidence of the cellular and molecular consequences of animals interacting with their environment. In addition to having an impact on basic research, fiber optic methods may enable the use of MPLSM in surgical and clinical procedures.

MULTISPECTRAL LASER-SCANNING MICROSCOPY

During the last several years, there has been an increased interest in multicolor fluorescence experiments. The development of new dyes and fluorescent proteins has made it possible to label a wide range of cell types and to generate specific protein tags in live animals (for a review, see Hadjantonakis et al. 2003). This has increased the need for more advanced methods to discriminate between different dyes and fluorescent markers simultaneously. In response to this need, multispectral imaging has become an important tool for laser-scanning microscopists. Although the first multispectral imaging systems were developed for wide-field microscopy, this chapter focuses on the methods used in laser-scanning microscopy, as this is a rapidly growing field on its own. For further information about wide-field multispectral imaging, see the following references (Eng et al. 1989; Wiegmann et al. 1993; Morris et al. 1994; Schrock et al. 1996; Zangaro et al. 1996; Mooney et al. 1997; Richmond et al. 1997; Wachman et al. 1997; Garini et al. 1999; Ornberg et al. 1999; Zeng et al. 1999; Levenson and Hoyt 2000; Tsurui et al. 2000; Farkas 2001; Ford et al. 2001; Yang et al. 2003).

Traditional Approaches for Fluorochrome Separation

When more than one fluorochrome is used, fluorescence cross-talk can make it difficult to interpret the results of an experiment. This is particularly true if colocalization experiments are performed or if quantitative measurements are made, e.g., for fluorescence resonance energy transfer (FRET) (see Chapter 8) and for fluorescence recovery after photobleaching (FRAP) (see Chapter 7). Several strategies can be used to eliminate or reduce cross-talk.

Traditional Filter Methods

In laser-scanning microscopy (see Chapter 14), laser illumination is used as the excitation source, and coated glass is commonly used to separate emission signals. Typically, a beam splitter splits light into two paths, one transmitted, one reflected, and then a band-pass filter further refines the spectral composition of the light collected by the detector, such as a photomultiplier tube (PMT). The dichroic and band-pass filter combination perform two functions: To limit excitation light from reaching the detector and to collect as much emission signal as possible for one fluorochrome, without allowing the emission signal from the other fluorochrome(s) contaminate the signal. Thus, separation of fluorescent signals depends on the characteristics of the filters that are used. For closely overlapping spectra, the narrower the band-pass filter, the better the spectral separation; in this case, however, fewer photons are collected.

Multitracking

In cases where there is overlap in the excitation spectra of two fluorochromes, laser illumination can be carefully controlled during the scan cycle to reduce or avoid cross-talk. This approach, called multitracking, allows excitation and emission collection of each fluorochrome separately and is accomplished by the fast switching of laser lines, either by line or by frame. For instance, for two dyes such as fluorescein and rhodamine, the 488-nm line used to excite fluorescein would be turned on as the laser scans across the line, collecting fluorescein emission using a fluorescein filter in front of a detector; then, on the return, the 488-nm would be turned off and the 543-nm line used to excite rhodamine would be turned on. As the 543-nm line scans across the sample, rhodamine signal is acquired using a rhodamine band-pass filter. This avoids exciting both dyes at once and can significantly reduce spectral cross-talk, particularly when the choice of emission filters is limited. The fast switching of the laser lines is conferred through an Acousto-Optic Tunable Filter (AOTF) or similar device that has a very rapid response time.

Quantitative Cross-talk Correction Using Ratiometric Analysis

For quantitative experiments, fluorescence cross-talk and background signal can be particularly problematic. In these instances, correction methods based on the ratio of intensities can be used to quantify the fluorescence from each of the probes in a sample or to estimate changes in a ratiometric indicator (for a review, see Helmchen 2000). By measuring the amount of cross-talk present when using single-labeled samples, it is possible to estimate the amount of signal bleed-through that is likely to occur when fluorochrome pairs are imaged. Often, two different excitation wavelengths and/or two different channels for emission collection are used to determine the amount of signal obtained when each fluorochrome is excited. Once the ratio of signal is determined for each dye, or for a single dye with and without an indicator present, the pattern can be applied to the experimental situation.

Correction methods based on a ratio of fluorescence intensities have been used to determine FRET efficiencies (for a review, see Berney and Danuser 2003). FRET occurs via a dipole-dipole interaction between fluorochromes such that excitation of the donor fluorochrome results in emission of an acceptor fluorochrome through a nonradiative exchange of energy from one molecule to the other (see Chapter 8). Because FRET efficiency falls off as the distance increases between fluorochromes, FRET is a valid way to examine protein-protein interactions (Herman et al. 2001; Periasamy 2001; Periasamy et al. 2001). For FRET to occur, there must be overlap between the donor emission and acceptor excitation spectra. Often this means that there is cross-talk between acceptor and donor emission in addition to coexcitation of both the donor and the acceptor. In many cases, it is not possible to separate dyes using narrow-band-pass filters or multitracking, and correction methods are needed to determine donor and acceptor fluorescence. This method has been used successfully in many cases for cross-talk elimination and FRET measurements, but methods such as these are complicated by the concentration of different components, background fluorescence, the spectral response of the detector, and differences between detectors and noise, making it cumbersome to acquire and process multiple images and difficult to control for all variables (for a review, see Berney and Danuser 2003). In addition, this method may require optimized filters for each set of fluorochromes, and, although these correction schemes are generally successful for two fluorochromes, they are not very effective for more than two labels.

Multispectral Laser-scanning Microscopy: Image Acquisition

Careful choices of fluorescent labels, narrow filters, multitracking and ratio imaging can reduce cross-talk in a large number of cases; however, there are some instances when spectral overlap cannot be eliminated by these strategies. Despite the efforts of dye and filter manufacturers, there are often cases when it is not possible or feasible to choose nonoverlapping probes for certain experiments. This is particularly true when multiple fluorescent protein variants are used in the same experiment (see Chapter 1). In addition, naturally occurring fluorescence or fluorescence produced by fixatives or other treatments may mask the signal of some fluorochromes and can bleed through to many channels, making these difficult to eliminate. In addition, when a large number of probes are used, spectral cross-talk is inevitable. In these instances, multispectral imaging can be used to achieve cross-talk elimination. Multispectral imaging has been used extensively in remote sensing analysis which involves studying how light waves are emitted and scattered off the surface of the earth and within the atmosphere. Characterizing the spectral frequencies that are present allows the different structures and terrains to be identified. Similar approaches are now being used to determine fluorochrome identity.

Multispectral imaging is a method used to collect spectral information from samples with a mixed set of fluorochromes by collecting a series of images in which spectral information is known for each pixel (Levenson and Hoyt 2000; Dickinson et al. 2001, 2003; Farkas 2001; Lansford et al. 2001; Hiraoka et al. 2002; Zimmermann et al. 2003). Spectral information in an image can be used to determine the identity of the dye and shows how dye fluorescence is affected by environmental changes. To determine the spectral content of each pixel within a field of view, images are collected at a series of wavelength bands (x, y, λ). Images acquired in a wavelength series, similar to a time series or a z series, are called a Lambda stack, image cube, or spectral cube (Fig. 3). Figure 3 shows an example of a lambda stack of a group of 3T3 cells that express either cyan fluorescent protein (CFP) or green fluorescent protein (GFP) in the nucleus. The spectrum of any field of view can be revealed by graphing how the average intensity changes with respect to wavelength. In this example, simply graphing the spectra of individual nuclei makes it obvious which nuclei have which dye, but for more complex samples with spatial and spectral overlap of signals, dye identity can be determined using mathematical approaches (discussed later in the chapter).

To acquire a Lambda stack, the microscope system must be capable of collecting images at different emission or excitation wavelengths. In contrast to the push broom approach employed in wide-field microscopy and remote sensing (for a review, see Farkas 2001),

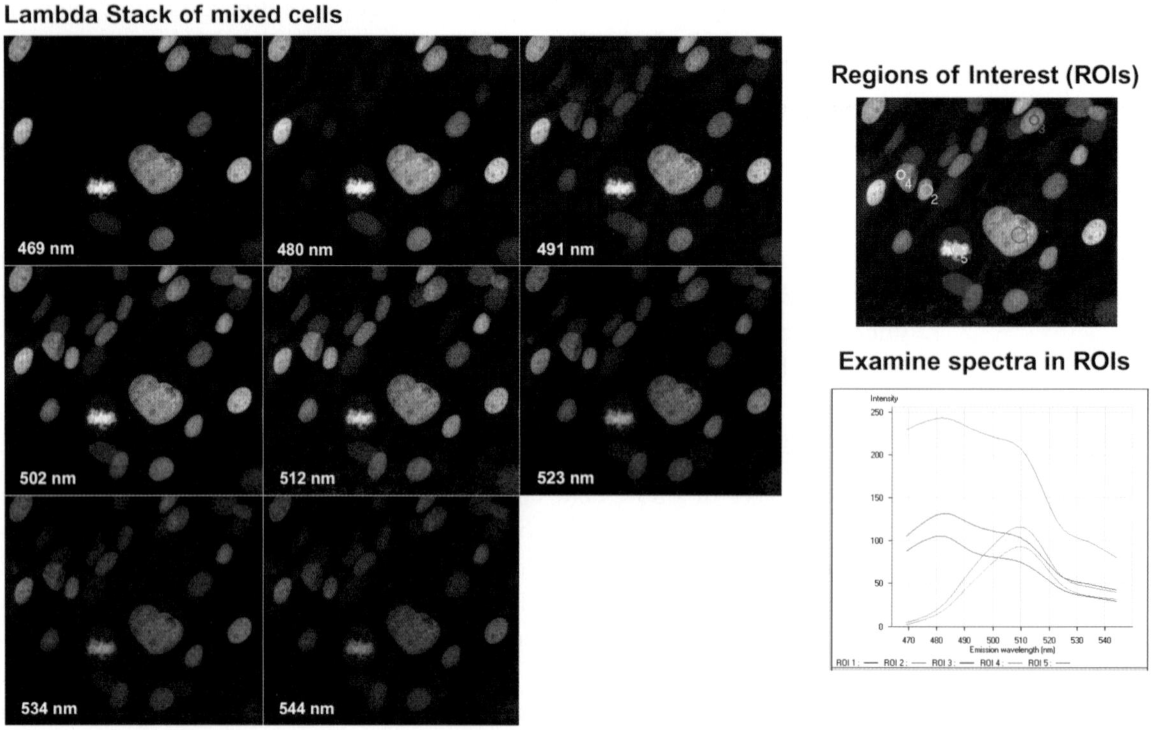

FIGURE 3. Spectral analysis of a mixed group of cells. Lambda stack images of individual cells that express either CFP or GFP targeted to the nucleus. Selecting a region of interest for each nucleus reveals the spectrum of fluorescent protein emission.

where spectral information is collected a line at a time, in laser-scanning microscopy, acquiring a Lambda stack is more like sweeping each floor tile, separating the yield into different bins for each tile. Different methods of acquiring data at wavelength bands can be used to generate a Lambda stack. These are summarized below.

Acquiring Image Stacks at Variable Emission Wavelengths

Initial attempts to add spectra detection to laser-scanning microscopes involved coupling a spectrometer to the microscope to characterize the wavelength content of the emission signal (Hanley et al. 1998; Favard et al. 1999; Haralampus-Grynaviski et al. 2000). In most instances, a CCD (charge-coupled device) camera was used as a detector, limiting the sensitivity and resolution of the collected signal. Nevertheless, these successful implementations led the way to the development of PMT-based spectral detectors that can be built in to the descanned path of the LSM.

Multiple filters placed in front of a PMT offer a simple approach for building an imaging spectrometer. The same detector is used, and different band-pass filters or different long-pass and short-pass filter pairs are rotated in front of it to acquire bands of equal bandwidth successively. This approach is now being used on some confocal and multiphoton laser-scanning systems (Bio-Rad; Konig et al. 2000). The larger the number of filters, the greater the flexibility of the band pass collected and the spectral range. This method benefits from having defined, reproducible bands, but a major drawback to the filter-based approach is that repetitive scanning is necessary as an image is collected at each wavelength increment. This process can be slow and increases the chance that photobleaching will occur during acquisition. Tunable filter devices (Liquid Crystal Tunable Filters, Acousto-Optic Tunable Filters, etc.) can be used for greater flexibility, but the poor transmission efficiency of some filters, due in part to polarization, can reduce the signal to noise (Farkas 2001; Lansford et al. 2001).

Greater demand for multispectral strategies for live-cell imaging has pushed for the development of rapid acquisition schemes that limit the amount of times the field of view is exposed to excitation light. A new device, called the META detector (Carl-Zeiss), uses a multichannel detector to collect bands of emission signals that have been separated using a diffraction grating (Dickinson et al. 2001, 2002; Haraguchi et al. 2002). Using this device, multiple bands of data can be acquired in parallel instead of in series, greatly reducing the amount of time needed to collect a Lambda stack. The grating allows for the precise register of 10.7-nm bands onto each channel of a 32-channel multi-anode detector, and a separate image can be made from each channel that is activated. In a single scan, eight images, corresponding to eight different wavelength bands, can be acquired simultaneously to produce a Lambda stack that covers 85.6 nm of the visible spectrum. If a larger part of the spectrum must be sampled, additional scans can be made or adjacent channels can be binned to double, triple, or quadruple the width of the band detected. The limitation to the number of images collected in a Lambda stack relates only to the rate of signal processing. Future improvements in this area may make it possible to image the entire spectrum at 10-nm resolution in a single scan. In addition, with the use of a multichannel detector and a grating, the spectral registration to different bands is highly reproducible. Because of this, acquired spectra are consistent between experiments and resemble spectral information provided by other sources (Beechem et al. 2003). A similar approach utilizing a 16-channel PMT detector and a grating has shown similar capabilities (Buehler et al. 2003).

Another approach to acquiring multiple bands in parallel utilizes a prism to separate emission wavelengths and movable gates that allow a specific band of light to pass to a detector

(Leica-Microsystems-Inc.), much like a prism monchrometer. Mirrored gates allow what is not detected along the primary path to be reflected to additional detectors to acquire multiple image bands at once. In this case, the number of scans in the Lambda stack is limited by the number of detectors available on the system, usually three to four, regardless of the bandwidth of the signal in each image, and care must be taken to calibrate the input from multiple detectors.

In addition to Lambda stack acquisition, with the implementation of methods for variable band-pass detection, LSMs now have a greater capacity than ever before for flexible emission collection. In some cases, fluorochrome separation fails because the best band-pass filter for separation is not present on the system. Some approaches outlined above offer the ability to freely configure the wavelength range of detection, making it possible to design custom band-pass settings for the fluorochromes in use. With the META detector, for example, this is accomplished by activating a series of channels in the multichannel array, and the signal collected by those channels is binned to make an image. The wavelength depends on the position of the channel in the detector, and the width of the band pass depends on the number of detector channels that are active (can be as small as 10.7 nm). Eight images, all with different band-pass configurations, can be collected. For prism-gate systems, the width of the gate specifies the width of the band collected, and the position with respect to the prism determines the wavelength content. The number of different images depends on the number of detectors. For systems using filter pairs, choices are limited to the filters that are present, but this certainly extends the range of possibilities beyond single band-pass filters. All of these approaches offer a tremendous advantage for avoiding cross-talk and provide the flexibility necessary to keep up with the growing number of available fluorescent conjugates and proteins.

Multiphoton, Multispectral Analysis

Multiphoton excitation produces fluorescence with the same spectral emission properties as single photon excitation (for reviews, see Denk et al.1995; Xu 2000). Any of the methods discussed above can be used to acquire MP-MSLSM Lambda stacks. However, as discussed earlier in the chapter, NDDs can be used to improve the detection efficiency, especially during deep tissue imaging. Thus, it would be advantageous to develop methods for spectral separation that can be used to acquire nondescanned photons.

In addition to characterizing fluorchrome emission, spectral information can also be collected by relating fluorescence intensity to varying excitation wavelength. MPLSM utilizes a continuously tunable Ti:Sapphire laser for excitation, making it possible to alter the excitation wavelength to collect an excitation Lambda stack. Recently, the author's laboratory has utilized the software-controlled, tunable Chameleon laser to acquire Lambda stacks with variable multiphoton excitation (Dickinson et al. 2003). Because this approach requires that only the excitation wavelength be varied, signal can be collected with any PMT including an NDD. By varying the excitation signal, multispectral nondescanned detection is possible, providing a powerful tool for eliminating autofluorescence and separating overlapping fluorochromes deep within tissues. This method can be used for eliminating spectral cross-talk and also enables the user to determine how the tuning curve of the laser, objective transmission, and environment of the fluorochrome effects dye excitation. By determining excitation curves directly on the microscope, users can quickly determine the ideal wavelength for the most efficient excitation of a single fluorochrome or labels used in combination. Interestingly, some fluorochromes that have very closely overlapping excitation spectra with single-photon excitation, such as CFP and YFP, or Alexa 488 and Sytox Green, display significantly less overlap in their multiphoton excitation curves (Dickinson et al. 2003; W. Zipfel, unpubl.), and for triple fluorochrome applications, this may result in more robust signal separation (Dickinson et al. 2003). Collecting excitation Lambda stacks can be slow since it requires sequential scans, but for some fluorochromes, only two or three scans may be necessary for cross-talk elimination.

Mathematical Approaches to Fluorochrome Separation

As mentioned above, fluorochromes have unique spectral signatures. Once a Lambda stack or spectral cube is collected, these signatures can be used to assign the proper dye identities to individual pixels. This process has been called emission or excitation fingerprinting, depending on how the Lambda stack is collected. Two methods for determining the contribution of different fluorochromes to the image are discussed here. The first, called linear unmixing, is a very reliable method for determining dye identity using known reference spectra for the fluorochromes present in the sample, and the second, principle component analysis, can help to identify unknown fluorochromes that contribute to the signals in the image.

Linear Unmixing

Simply put, linear unmixing is a way of matching the spectral variations in the Lambda stack with the known spectral variations for the fluorochrome labels that are used. The fluorescence intensity in a pixel in which fluorochromes are colocalized and have overlapping emission spectra will be the sum of the intensity of each label (Lakowicz 1999). Thus, the pixel is said to be linearly mixed where the sum spectrum, $S(\lambda)$ equals the proportion or weight, A, of each individual spectrum $R(\lambda)$: $S(\lambda) = A_1 R_1(\lambda) + A_2 R_2(\lambda) + A_3 R_3(\lambda)...$, etc.

Consider the example in Figure 4. The orange spectrum is the sum of the two-component spectra shown in pink and blue according to the proportions listed on the graph. In fact, for every possible combination of dye concentrations, the sum spectrum can be predicted. Therefore, if we wanted to determine the spectral content of each pixel in an image, we could match the shape of the sum spectrum from each pixel to a known sum spectrum in a library, much like matching a fingerprint in a database. For instance, if the shape matched the sum spectrum in the first panel of Figure 4, then we would conclude that the two dyes are evenly mixed in that pixel, whereas if it matched the orange spectrum in the second pixel, we would conclude that the pixel contained 75% blue and 25% pink, and so on.

Linear unmixing algorithms are used to very quickly solve the problem described above. A matrix of values representing the sum spectra is matched against a library of predicted

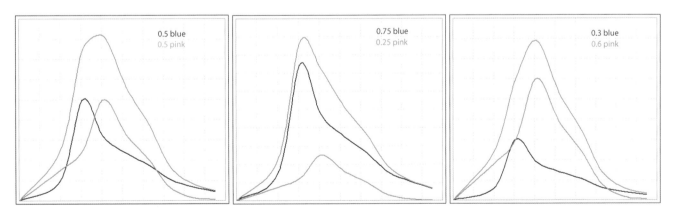

FIGURE 4. Theoretical linear mixing. Emission signal from two different components yield different sum spectra when signals overlap. With known components, it is possible to predict the sum spectra of any concentration of components.

spectra according to whatever best-fit parameters are imposed. Once the weight of each spectral component is determined, the Lambda stack can be separated into individual images for each fluorescent label. The intensity of each pixel in the unmixed images is the total collected intensity of the pixel multiplied by the proportion of spectra in the pixel.

Practical Considerations for Linear Unmixing

The success of linear unmixing depends on using reference spectra that faithfully represent the spectra in the sample to ensure the best fit. Thus, controls must be imaged under the same conditions as the mixed sample, pixel saturation should be avoided, and Lambda stack acquisition must be reliable. The optical components of each microscope system will impose some bias, making it impossible to use spectra from other sources as reference spectra. In fact, the best reference spectra are those that are acquired in exactly the same way the Lambda stack is acquired (same objective, offset, amplifier gain, wavelength range, dichroics, etc.) or those chosen from nonoverlapping regions within the lambda stack itself. In addition, it is often helpful to consider any background signal from mounting media or laser light reflections as a known reference spectra in order to properly separate background from specific fluorochrome signals. Spectral signals that do not match those in the spectral library are considered residual and are not assigned. If a residual image is part of the output, it is possible to see how well the reference spectra fit the acquired spectra. It is often the case that the residual signal comes from pixels that are saturated or from background in the sample.

Linear unmixing can be used to separate fluorescent signals that have subtle differences in emission spectra (Dickinson et al. 2001). To determine the limits of spectral separation using linear unmixing and the META detector, we performed a pairwise comparison of seven commonly used dyes with peak emissions from 509–531 nm (Dickinson et al. 2002). The results of these comparisons showed that many of these dye combinations could be separated using linear unmixing, despite the similarity in the spectra. In fact, dyes with as little as 4–5 nm of separation in the spectra could be unmixed if the fluorescence signals were balanced. Dyes such as enhanced GFP and FITC (fluorescein isothiocyanate) or Alexa 488, which have 7 nm of difference, could be unmixed reliably, even if very different levels of signal were present (Fig. 5). These experiments illustrate the power of multispectral imaging for eliminating

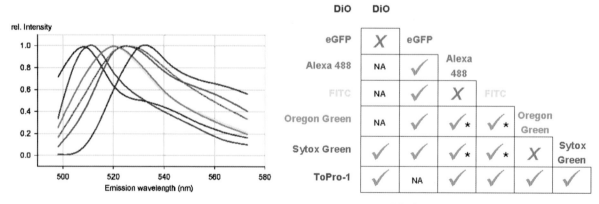

FIGURE 5. Summary of data from a pairwise comparison of the linear unmixing of green dyes. Spectra are shown to the left of individual fluorochromes listed on the right. Signals that could be unmixed are indicated by a check mark, whereas those that could not be unmixed are indicated by an X.

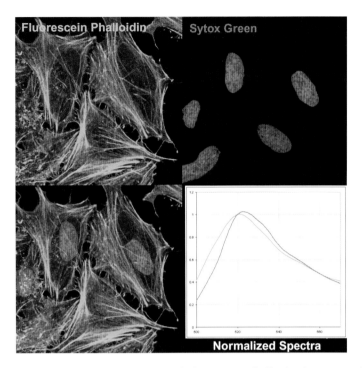

FIGURE 6. Linear unmixing of HeLa cells labeled with fluorescein phalloidin (actin; *green*) and Sytox green (nuclei; *red*). Linear unmixing produces two individual images for each dye shown at the top. The overlay of the individual images is shown at the bottom with the emission spectra of the two dyes. This figure was acquired first as both a *z* stack and a Lambda stack. A maximum intensity projection of the *z* stack was used for linear unmixing.

cross-talk. The result of linear unmixing for a sample labeled with Sytox Green and fluorescein is shown in Figure 6.

It is often asked how many bands of data or how many images in a Lambda stack are needed for accurate spectral separation. In some cases, particularly in the case of two components, a Lambda stack with images at only two wavelengths may be needed; however, the two wavelength bands that are collected must be representative of the spectral differences between the dyes. Although they may provide accurate separation, acquiring data from throughout the emission spectra will increase the signal that is collected and may result in more sensitive detection. If samples contain more than one dye, more than two channels are needed to resolve all the components properly.

Principle Component Analysis

Although the best way to properly assign the identity of pixels in a Lambda stack is to use linear unmixing based on known components, it is also possible to identify individual component spectra in a Lambda stack or spectral cube using principle component analysis (PCA). This type of analysis is performed routinely in a large number of fields from astronomy to music to find patterns in data sets. The main goal of PCA is to reduce a complicated series of overlapping spectra into a series of individual peaks. This is done by mapping spectral intensities onto a set of orthogonal axes in order to reveal variances. PCA works best when spectra are spatially distinct in the sample so that individual spectral components are abundant. Once individual components are detected, they can be used for linear unmixing in order to produce images containing separate components.

Applications of MSLSM

MSLSM has only recently been introduced as an option for commercial systems for laser-scanning microscopy, yet several groups have already applied these methods to FRET analysis (Haraguchi et al. 2002; Zimmermann et al. 2003; Nashmi et al. 2004), live-cell imaging of fluorescent protein variants (Haraguchi et al. 2002; Bertera et al. 2003), and elimination of autofluorescence (Dickinson et al. 2003). All of these instances represent key advantages of multispectral imaging over other methods, as cross-talk elimination without linear unmixing is extremely difficult if not impossible. In addition, the ease and speed of Lambda stack acquisition and linear unmixing make it an attractive option for many researchers.

CONCLUSIONS

Improvements in imaging technology have made it possible to use multidimensional methods for fluorescence imaging and quantitative analysis in biological samples. The methods described here add to the flexible strategies that are available to researchers interested in answering complex questions at the level of cells and molecules. The more tools available to isolate individual components and then study them in the context of other related components and processes, the greater the chance that our interpretations will reflect what really happens in biological systems.

ACKNOWLEDGMENTS

I thank (alphabetically) Greg Bearman, Scott Fraser, Zbi Iwinski, Richard Levenson, Eva Simbuerger, Sebastian Tille, and Ralf Wolleschensky for discussions and contributions to this chapter.

REFERENCES

Albota M., Beljonne D., Bredas J.L., Ehrlich J.E., Fu J.Y., Heikal A.A., Hess S.E., Kogej T., Levin M.D., Marder S.R., et al. 1998. Design of organic molecules with large two-photon absorption cross sections. *Science* **281:** 1653–1656.

APE (Berlin, Germany). *http://www.ape-berlin.com.*

Beechem J.M., Haugland R., Janes M., Clements I., Kilgore J., Salisbury J., and Ignatius M. 2003. Advanced spectroscopy of fluorescent indicators/dyes imaged in-situ with wavelength-resolved confocal laser-scanning microscopy (CLSM). *Biophys. J.* **84:** 587a–587a.

Berney C. and Danuser G. 2003. FRET or No FRET: A quantitative comparison. *Biophys. J.* **84:** 3992–4010.

Bertera S., Geng X., Tawadrous Z., Bottino R., Balamurugan A.N., Rudert W.A., Drain P., Watkins S.C., and Trucco M. 2003. Body window-enabled in vivo multicolor imaging of transplanted mouse islets expressing and insulin-Timer fusion protein. *BioTechniques* **35:** 718–722.

Bestvater F., Spiess E., Stobrawa G., Hacker M., Feurer T., Porwol T., Berchner-Pfannschmidt U., Wotzlaw C., and Acker H. 2002. Two-photon fluorescence absorption and emission spectra of dyes relevant for cell imaging. *J. Microsc.* **208:** 108–115.

Bio-Rad, Inc. (Hercules, California). *http:// www.biorad.com.*

Brown E.B., Campbell R.B., Tsuzuki Y., Xu L., Carmeliet P., Fukumura D., and Jain R. K. 2001. In vivo measurement of gene expression, angiogenesis and physiological function in tumors using multiphoton laser scanning microscopy. *Nat. Med.* **7:** 864–868.

Buehler C., Kim K.H., Greuter U., Schlumpf N., and So P.T.C. 2003. Multi-color two-photon scanning microscopy using a 16-channel photomultiplier. In *Progress in biomedical optics and imaging. Multiphoton microscopy in the biomedical sciences* (ed. A. Periasamy and P.T.C. So), vol. 426, pp. 217–230. Proceedings of the International Society for Optical Engineering (SPIE), Bellingham, Washington.

Campagnola P.J., Clark H.A., Mohler W.A., Lewis A., and Loew L.M. 2001. Second-harmonic imaging microscopy of living cells. *J. Biomed. Opt.* **6:** 277–286.

Carl Zeiss (Jena, Germany). *http:// www.zeiss.com.*

Chaigneau E., Oheim M., Audinat E., and Charpak S. 2003. Two-photon imaging of capillary blood flow in olfactory bulb glomeruli. *Proc. Natl. Acad. Sci.* **100:** 13081–13086.

Cheng J.X., Jia Y.K., Zheng G., and Xie X.S. 2002. Laser-scanning coherent anti-Stokes Raman scattering microscopy and applications to cell biology. *Biophys. J.* **83:** 502–509.

Coherent Inc. (Santa Clara, California). *http://www.coherentinc.com.*

D'Amore J.D., Kajdasz S.T., McLellan M.E., Bacskai B.J., Stern E.A., and Hyman B.T. 2003. In vivo multiphoton imaging of a transgenic mouse model of Alzheimer disease reveals marked thioflavine-S-associated alterations in neurite trajectories. *J. Neuropathol. Exp. Neurol.* **62:** 137–145.

Denk W., Piston D.W., and Webb W.W. 1995. Two-photon molecular excitation in laser-scanning microscopy. In *Handbook of biological confocal microscopy* (ed. J.B. Pawley), pp. 445–458. Plenum Press, New York.

Denk W., Strickler J.H., and Webb W.W. 1990. Two-photon laser scanning fluorescence microscopy. *Science* **248:** 73–76.

Dickinson M.E., Bearman G., Tilie S., Lansford R., and Fraser S.E. 2001. Multi-spectral imaging and linear unmixing add a whole new dimension to laser scanning fluorescence microscopy. *BioTechniques* **31:** 1272, 1274–1276, 1278.

Dickinson M.E., Simbuerger E., Zimmermann B., Waters C.W., and Fraser S.E. 2003. Multiphoton excitation spectra in biological samples. *J. Biomed. Opt.* **8:** 329–338.

Dickinson M.E., Waters C.W., Bearman G., Wolleschensky R., Tilie S., and Fraser S.E. 2002. Sensitive imaging of spectrally overlapping fluorochromes using the LSM 510 META. In *Progress in biomedical optics and imaging. Multiphoton microscopy in the biomedical sciences* (ed. A. Periasamy and P.T.C. So), vol. 426, pp. 123–136. Proceedings of the International Society for Optical Engineering (SPIE), Bellingham, Washington.

Dunn K.W., Sandoval R.M., and Molitoris B.A. 2003. Intravital imaging of the kidney using multiparameter multiphoton microscopy. *Nephron Exp. Nephrol.* **94:** e7–11.

Dunn K.W., Sandoval R.M., Kelly K.J., Dagher P.C., Tanner G.A., Atkinson S.J., Bacallao R.L., and Molitoris B.A. 2002. Functional studies of the kidney of living animals using multicolor two-photon microscopy. *Am. J. Physiol. Cell. Physiol.* **283:** C905–C916.

Eng J., Lynch R.M., and Balaban R.S. 1989. Nicotinamide adenine dinucleotide fluorescence spectroscopy and imaging of isolated cardiac myocytes. *Biophys. J.* **55:** 621–630.

Farkas D.L. 2001. Spectral microscopy for quantitative cell and tissue imaging. In *Methods in cellular imaging* (ed. A. Periasamy), pp. 345–361. Published for the American Physiological Society by Oxford University Press, New York.

Favard C., Valisa P., Egret-Charlier M., Sharonov S., Herben C., Manfait M., Da Silva E., and Vigny P. 1999. A new UV-visible confocal laser scanning microspectrofluorometer designed for spectral cellular imaging. *Biospectroscopy* **5:** 101–115.

Ford B.K., Volin C.E., Murphy S.M., Lynch R.M., and Descour M.R. 2001. Computed tomography-based spectral imaging for fluorescence microscopy. *Biophys. J.* **80:** 986–993.

Garini Y., Gil A., Bar-Am I., Cabib D., and Katzir N. 1999. Signal to noise analysis of multiple color fluorescence imaging microscopy. *Cytometry* **35:** 214–226.

Goppert-Mayer M. 1931. Ueber Elementarakte mit zwei Quantenspruengen. *Ann. Phys. (Leipzig)* **9:** 273–295.

Hadjantonakis A.K., Dickinson M.E., Fraser S.E., and Papaioannou V.E. 2003. Technicolour transgenics: Imaging tools for functional genomics in the mouse. *Nat. Rev. Genet.* **4:** 613–625.

Hanley Q.S., Verveer P.J., and Jovin T.M. 1998. Optical sectioning fluorescence spectroscopy in a programmable array microscope. *Appl. Spectrosc.* **52:** 783–789.

Haraguchi T., Shimi T., Koujin T., Hashiguchi N., and Hiraoka Y. 2002. Spectral imaging fluorescence microscopy. *Genes Cells* **7:** 881–887.

Haralampus-Grynaviski N.M., Stimson M.J., and Simon J.D. 2000. Design and applications of a rapid-scan spectrally resolved fluorescence microscopy. *Appl. Spectrosc.* **54:** 1727–1733.

Helmchen F. 2000. Calibration of fluorescent calcium indicators. In *Imaging neurons: A laboratory manual* (ed. R. Yuste et al.), pp. 32.1–32.9. Cold Spring Harbor Laboratory Press, Cold Spring Harbor, New York.

———. 2002. Miniaturization of fluorescence microscopes using fibre optics. *Exp. Physiol.* **87:** 737–745.

Helmchen F. and Denk W. 2002. New developments in multiphoton microscopy. *Curr. Opin. Neurobiol.* **12:** 593–601.

Helmchen F., Tank D.W., and Denk W. 2002. Enhanced two-photon excitation through optical fiber by single-mode propagation in a large core. *Appl. Opt.* **41:** 2930–2934.

Helmchen F., Fee M.S., Tank D.W., and Denk W. 2001. A miniature head-mounted two-photon microscope. High-resolution brain imaging in freely moving animals. *Neuron* **31:** 903–912.

Helmchen F., Svoboda K., Denk W., and Tank D.W. 1999. In vivo dendritic calcium dynamics in deep-layer cortical pyramidal neurons. *Nat. Neurosci.* **2:** 989–996.

Herman B., Gordon G., Mahajan N., and Centonze V. 2001. Measurement of fluorescence resonance energy transfer in the optical microscope. In *Methods in cellular imaging* (ed. A. Periasamy), pp. 257–272. Published for the American Physiological Society by Oxford University Press, New York.

Hiraoka Y., Shimi T., and Haraguchi T. 2002. Multispectral imaging fluorescence microscopy for living cells. *Cell. Struct. Funct.* **27:** 367–374.

IST Rees. (Horseheads, New York). *http://www.istspectech.com.*

Jung J.C. and Schnitzer M.J. 2003. Multiphoton endoscopy. *Opt. Lett.* **28:** 902–904.

Keller-Peck C.R., Walsh M.K., Gan W.B., Feng G., Sanes J.R., and Lichtman J.W. 2001. Asynchronous synapse elimination in neonatal motor units: studies using GFP transgenic mice. *Neuron* **31:** 381–394.

Kempe M. and Rudolph W. 1993. Femtosecond pulses in the focal region of lenses. *Phys. Rev. A* **48:** 4721–4729.

Konig K., Riemann I., Fischer P., and Halbhuber K.J. 2000. Multiplex FISH and three-dimensional DNA imaging with near infrared femtosecond laser pulses. *Histochem.Cell Biol.* **114:** 337–345.

Lakowicz J.R. 1999. *Principles of fluorescence spectroscopy.* Kluwer Academic/Plenum, New York.

Lansford R., Bearman G., and Fraser S.E. 2001. Resolution of multiple green fluorescent protein color variants and dyes using two-photon microscopy and imaging spectroscopy. *J. Biomed. Opt.* **6:** 311–318.

Leica Microsystems Inc. (Bannookburn, Illinois) *http:// www.leica.com.*

Levenson R.M. and Hoyt C.C. 2000. Spectral imaging and microscopy. *Am. Lab.* **32:** 26–34.

Millard A.C., Campagnola P.J., Mohler W., Lewis A., and Loew L.M. 2003. Second harmonic imaging microscopy. *Methods Enzymol.* **361:** 47–69.

Miller M.J., Wei S.H., Cahalan M.D., and Parker I. 2003. Autonomous T cell trafficking examined in vivo with intravital two-photon microscopy. *Proc. Natl. Acad. Sci.* **100:** 2604–2609.

Miller M.J., Wei S.H., Parker I., and Cahalan M.D. 2002. Two-photon imaging of lymphocyte motility and antigen response in intact lymph node. *Science* **296:** 1869–1873.

Mooney J.M., Vickers V.E., An M., and Brodzik A.K. 1997. High-throughput hyperspectral infrared camera. *J. Opt. Soc. Am.* **14:** 2951–2961.

Morris H.R., Hoyt C.C., and Treado P.J. 1994. Imaging spectrometers for fluorescence and raman microscopy — Acoustooptic and liquid-crystal tunable filters. *Appl. Spectrosc.* **48:** 857–866.

Muller M., Squier J., De Lange C.A., and Brakenhoff G.J. 2000. CARS microscopy with folded BoxCARS phasematching. *J. Microsc.* **197:** 150–158.

Nan X., Cheng J.X., and Xie X.S. 2003. Vibrational imaging of lipid droplets in live fibroblast cells with coherent anti-stokes raman scattering microscopy. *J. Lipid Res.* **44:** 2202–2208.

Nashmi R., Dickinson M.E., McKinney S., Jareb M., Labarca C., Fraser S.E., and Lester H.A. 2004. Effects of localization, trafficking, and nicotine-induced upregulation in clonal mammalian cells and in cultured midbrain neurons. *J. Neurosci.* **23:** 11554–11567.

Netz R., Feurer T., Wolleschensky R., and Sauerbray R. 2000. Measurement of the pulse-front distortion in high numerical aperture lenses. *Appl. Phys.* **B70.**

Ocean Optics (Dunedin, Florida). *http:// www.oceanoptics.com.*

Oheim M., Beaurepaire E., Chaigneau E., Mertz J., and Charpak S. 2001. Two-photon microscopy in brain tissue: Parameters influencing the imaging depth. *J. Neurosci. Methods* **111:** 29–37.

Ornberg R.L., Woerner B.M., and Edwards D.A. 1999. Analysis of stained objects in histological sections by spectral imaging and differential absorption. *J. Histochem. Cytochem.* **47:** 1307–1314.

Padera T.P., Stoll B.R., So P.T., and Jain R.K. 2002. Conventional and high-speed intravital multiphoton laser scanning microscopy of microvasculature, lymphatics, and leukocyte-endothelial interactions. *Mol. Imaging* **1:** 9–15.

Periasamy A. 2001. Fluorescence resonance energy transfer microscopy: A mini review. *J. Biomed. Opt.* **6:** 287–291.

Periasamy A., Elangovan M., Wallrabe H., Barroso M., Demas J.N., Brautigan D.L., and Day R.N. 2001. Wide-field, confocal, two-photon, and lifetime resonance energy transfer imaging microscopy. In *Methods in cellular imaging* (ed. A. Periasamy), pp. 295–308. Published for the American Physiological Society by Oxford University Press, New York.

Ragan T.M., Huang H., and So P.T. 2003. In vivo and ex vivo tissue applications of two-photon microscopy. *Methods Enzymol.* **361:** 481–505.

Richmond K.N., Burnite S., and Lynch R.M. 1997. Oxygen sensitivity of mitochondrial metabolic state in isolated skeletal and cardiac myocytes. *Am. J. Physiol.* **273:** C1613–C1622.

Rumi M., Ehrlich J.E., Heikal A.A., Perry J.W., Barlow S., Hu Z.Y., McCord-Maughon D., Parker T.C., Rockel H., Thayumanavan S., Marder S.R., Beljonne D., and Bredas J.-L. 2000. Structure-property relationships for two-photon absorbing chromophores: Bis-donor diphenylpolyene and bis(styryl)benzene derivatives. *J. Am. Chem. Soc.* **122:** 9500–9510.

Schrock E., du Manoir S., Veldman T., Schoell B., Wienberg J., Ferguson-Smith M.A., Ning Y., Ledbetter D.H., Bar-Am I., Soenksen D., Garini Y., and Ried T. 1996. Multicolor spectral karyotyping of human chromosomes. *Science* **273:** 494–497.

Spectra Physics. (Mountain View, California). *http://www.spectraphysics.com/.*

Stosiek C., Garaschuk O., Holthoff K., and Konnerth A. 2003. In vivo two-photon calcium imaging of neuronal networks. *Proc. Natl. Acad. Sci.* **100:** 7319–7324.

Svoboda K., Helmchen F., Denk W., and Tank D.W. 1999. Spread of dendritic excitation in layer 2/3 pyramidal neurons in rat barrel cortex in vivo. *Nat. Neurosci.* **2:** 65–73.

Trachtenberg J.T., Chen B.E., Knott G.W., Feng G., Sanes J.R., Welker E., and Svoboda K. 2002. Long-term in vivo imaging of experience-dependent synaptic plasticity in adult cortex. *Nature* **420:** 788–794.

Tsurui H., Nishimura H., Hattori S., Hirose S., Okumura K., and Shirai T. 2000. Seven-color fluorescence imaging of tissue samples based on Fourier spectroscopy and singular value decomposition. *J. Histochem. Cytochem.* **48:** 653–662.

Wachman E.S., Niu W., and Farkas D.L. 1997. AOTF microscope for imaging with increased speed and spectral versatility. *Biophys. J.* **73:** 1215–1222.

Walsh M.K. and Lichtman J.W. 2003. In vivo time-lapse imaging of synaptic takeover associated with naturally occurring synapse elimination. *Neuron* **37:** 67–73.

Wang W., Wyckoff J.B., Frohlich V.C., Oleynikov Y., Huttelmaier S., Zavadil J., Cermak L., Bottinger E.P., Singer R.H., White J.G., et al. 2002. Single cell behavior in metastatic primary mammary tumors correlated with gene expression patterns revealed by molecular profiling. *Cancer Res.* **62:** 6278–6288.

Wiegmann T.B., Welling L.W., Beatty D.M., Howard D.E., Vamos S., and Morris S.J. 1993. Simultaneous imaging of intracellular $[Ca^{2+}]$ and pH in single MDCK and glomerular epithelial cells. *Am. J. Physiol.* **265:** C1184–C1190.

Wise F.W. 2000. Lasers for multiphoton microscopy. In *Imaging neurons: A laboratory manual* (ed. R. Yuste et al.), pp. 18.1–18.9. Cold Spring Harbor Laboratory Press, Cold Spring Harbor, New York.

Wokosin D.L. and Girkin J.M. 2002. Practical multiphoton microscopy. In *Confocal and two-photon microscopy: Foundations, applications, and advances* (ed. A. Diaspro), pp. 207–235. Wiley-Liss, New York.

Wolleschensky R., Dickinson M.E., and Fraser S.E. 2002. Group-velocity dispersion and fiber delivery in multiphoton laser scanning microscopy. In *Confocal and two-photon microscopy: Foundations, applications, and advances* (ed. A. Diaspro), pp. 171–189. Wiley-Liss, New York.

Xu C. 2000. Two-photon cross sections of indicators. In *Imaging neurons: A laboratory manual,* (ed. R. Yuste et al.), pp. 19.1–19.9. Cold Spring Harbor Laboratory Press, Cold Spring Harbor, New York.

Xu C., Zipfel W., Shear J.B., Williams R.M., and Webb W.W. 1996. Multiphoton fluorescence excitation: New spectral windows for biological nonlinear microscopy. *Proc. Natl. Acad. Sci.* **93:** 10763–10768.

Xu Y.W., Zhang J.R., Deng Y.M., Hui L.K., Jiang S.P., and Lian S.H. 1987. Fluorescence of proteins induced by two-photon absorption. *J. Photochem. Photobiol. B* **1:** 223–227.

Yang V.X., Muller P.J., Herman P., and Wilson B.C. 2003. A multispectral fluorescence imaging system: Design and initial clinical tests in intra-operative Photofrin-photodynamic therapy of brain tumors. *Lasers. Surg. Med.* **32:** 224–232.

Zangaro R.A., Silveira L., Manoharan R., Zonios G., Itzkan I., Dasari R.R., VanDam J., and Feld M.S. 1996. Rapid multiexcitation fluorescence spectroscopy system for in vivo tissue diagnosis. *Appl. Opt.* **35:** 5211–5219.

Zeng H.S., Weiss A., MacAulay C., and Cline R.W. 1999. System for fast measurements of in vivo fluorescence spectra of the gastrointestinal tract at multiple excitation wavelengths. *Appl. Opt.* **38:** 7157–7158.

Zimmermann T., Rietdorf J., and Pepperkok R. 2003. Spectral imaging and its applications in live cell microscopy. *FEBS Lett.* **546:** 87–92.

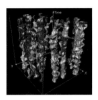

CHAPTER SIXTEEN

Analyzing Live Cell Data and Tracking Dynamic Movements

Wolfgang Tvaruskó, Julian Mattes, and Roland Eils

Intelligent Bioinformatics Systems, DKFZ, Im Neuenheimer Feld 580, D-69120 Heidelberg, Germany

INTRODUCTION

ALTHOUGH BIOLOGICAL PROCESSES AND STRUCTURES are dynamic in nature, most investigations into their mechanisms have been carried out in fixed specimens using, for example, immunocytochemistry and fluorescence in situ hybridization. This incongruity has been imposed in large part by a lack of efficient vital markers. Recent developments in in vivo microscopy techniques, including that of green fluorescent protein (GFP) and its spectral variants (Chapter 1) (see Chalfie et al. 1994), have allowed the nondistructive investigation of a wide range of dynamic processes in living cells (e.g., see Lippincott-Schwartz and Smith 1997; Misteli and Spector 1997; Heun et al. 2001a). Such investigations have revealed unexpectedly high levels of activity in many cellular structures, including nuclear subcompartments, that were previously thought to be rather sedentary (see Marshall et al. 1997; Misteli et al. 1997; Platani et al. 2000; Misteli 2001; Muratani et al. 2002).

The examples presented here focus on cells expressing GFP-tagged proteins, visualized by excitation with an appropriate wavelength of light. Images are captured digitally, making them instantly available for image processing. Unfortunately, the processes and structures under investigation can be damaged by irradiation at certain wavelengths, and prolonged exposure can lead to cell death (see White and Stelzer 1999; for an in-depth discussion, see Chapter 17). This is a particularly important consideration when constructing time-lapse image series. Minimizing the total light exposure during image acquisition (by reducing the light intensity, shortening the exposure time, and/or reducing the number of images taken in a particular series) leads to a reduction in the signal-to-noise ratio (SNR) and a loss of spatial and temporal resolution. A new method has recently been devised that is a compromise between capture time and absolute intensity, which results in an enhanced SNR (see Tvaruskó 2000).

Early live cell studies characterized organelle dynamics by simply analyzing time-lapse movies manually, by visual inspection. Although this can provide an overall qualitative impression of movement patterns, it cannot be used to address the underlying mechanisms or to provide quantitative data. The visual approach also fails to improve poor temporal resolution caused by minimal light exposure. The manual approach also suffers because of the sheer volume of data that must be analyzed. A typical three-dimensional image, with spatial dimensions of $512 \times 512 \times 15$ ($X \times Y \times Z$), representing a single time step, is approximately 4 MB. Simultaneous measurement in three color channels, for 100 time steps, results in a data set of more than 1 GB. This makes manual evaluation very time-consuming, and the results tend to be user-biased. For an unbiased, quantitative, and reproducible study of subcellular dynamics, automated image-processing techniques (e.g., object detection, motion estimation, and visualization) are required.

This chapter begins with a discussion of concepts for the automated analysis of multidimensional image data acquired from live cell microscopy and their application to the dynamics of subcellular compartments. The steps involved in quantitative analysis and visualization of live cell imaging data are considered in the following section, along with some of the available software packages. The section on Applications provides a number of recent examples in which such analyses have yielded new insights into dynamic cellular processes. Finally, the section on Computational Methods for Qualitative Analysis and Visualization provides a theoretical explanation of the applied techniques and how they are interrelated.

WORKFLOW FOLLOWED IN QUANTITATIVE ANALYSIS

The typical workflow in computational imaging is presented in Figure 1. Once the microscope images have been acquired and preprocessed to improve the SNR, they can be directly visualized by volume rendering and further processed to obtain quantitative data. After volume rendering, an automated estimation of the optical flow of the images' gray values in continuous space is carried out so that the mobility of structures can be quantified (see Section 2). Flow estimation yields the normal component of motion of spatial contours, based on pixel gray values, and image registration is used to measure elastic or rigid changes of form (see Section 3). The model deformation is derived from two- or three-dimensional data sets or image regions. It is often used to correct for global movement prior to further quantitative analysis. Segmentation can be thought of as the basis for registration, particle tracking, and surface rendering (see Section 1). Various approaches for segmentation exist, from simple thresholding to the detection of meaningful boundaries using level lines. For multiple objects in motion, single-particle tracking is the most direct method used, in which a particle is tracked over different time steps (Section 2). It provides access to parameters such as velocity, acceleration, and the diffusion coefficient. Surface reconstruction in time-space is obtained after segmentation of contours in each individual section and gives rise to volumetric measurements such as volume and surface area over time (Section 3). All of these processes lead to accurate estimates of quantitative parameters (Section 4).

More theoretically, the values of a two-dimensional image can be thought of as a two-dimensional surface, or as a manifold in three-dimensional space. Extending this metaphor to higher dimensional spaces, such as four-dimensional time spaces, image processing and computer graphics can be used to analyze biological structures as deformable surfaces in time-space (Fig. 2).

Software Packages

The following software packages can accommodate the most commonly used procedures for image processing and the production of computer graphics. No one package covers all

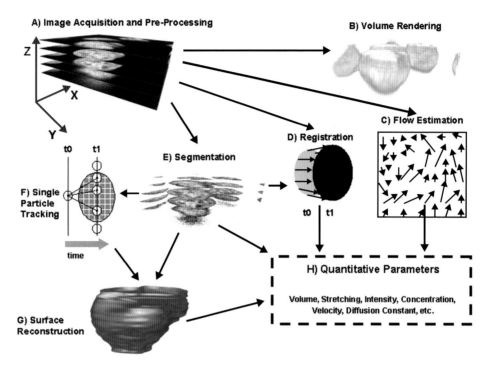

FIGURE 1. The typical workflow in computational imaging is presented. (*A*) Images are acquired by microscopy and preprocessed to improve the SNR. (*B*) A better interpretation of three-dimensional imaging data is achieved with rendering tools. (*C*) Optical flow fields are computed directly from time-space data. (*D*) Image registration is used to measure elastic or rigid changes of form. It is also often used to correct for global movement before further quantitative analysis. (*E*) To identify objects, images are segmented into their constituent parts. (*F*) For image sequences containing small objects, single-particle tracking is the most direct method used, in which correspondent objects must be found. (*G*) Based on extracted objects, time-space is reconstructed. Methods for transition of smooth surfaces are referred to as morphing. (*H*) All of these processes lead to accurate estimates of quantitative parameters, such as velocity, acceleration, and diffusion coefficient.

procedures, and more advanced methods, such as level-set approaches, must be incorporated by the user in the form of macros or program modules.

TILL Photonics visTRAC

The TILL Photonics visTRAC software package (available from Olympus) is suitable for the acquisition and quantitative analysis of data relating to time-resolved cellular processes using fluorescence microscopes. Features of the package include automated object identification, with anisotropic diffusion filtering and edge-based segmentation; single-particle tracking, based on fuzzy-logic algorithms; graphical visualization in an interactive virtual reality viewer; and quantitative analysis of dynamic parameters. For more information, see http://www.till-photonics.de.

Amira

Amira (from TGS Indeed Visual Concepts) is a powerful three-dimensional visualization toolbox, which includes volume rendering and graphical surface rendering techniques. Basic tools for quantitative analysis are also available. For more information, see http://www.tgs.com.

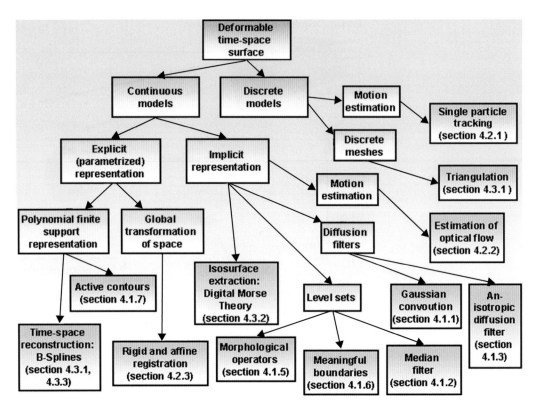

FIGURE 2. The taxonomy shows the interrelation between the different fields of preprocessing, filtering, segmentation, iso-surface extraction, geometric modeling, motion analysis, and graphical reconstruction.

ImageJ

ImageJ is a free software package available from the National Institutes of Health. It is suitable for quantitative image analysis, and many specialized plug-ins are available. Additional modules can be programmed by the user via the Java interface. For more information and software download, see http://rsb.info.nih.gov/ij/.

LSM 5

The LSM 5 Software package from Zeiss is used to control the Zeiss confocal microscope and for quantitative image analysis. A three-dimensional module is available for graphical display of four-dimensional imaging data. For more information, see http://www.zeiss.com.

Voxelshop pro, Imaris, Surface

These software packages are available from Bitplane AG. Voxelshop pro is a module to automatically separate and quantify objects characterized by gray levels or texture. Imaris is a high-quality software package to process and visualize three-dimensional images. It has been designed to accept most microscopic image formats and offers a range of functions. Surface automatically converts a volume image into a geometric object. For more details, see http://www.bitplane.com.

AnalySIS

AnalySIS, available from SIS, is compatible with most cameras and microscopes. It is used for image acquisition and import, image display and editing, archiving, documentation, image processing, measuring, and analysis, including a macroscripting language. For more details, see http://www.soft-imaging.net.

AnalyzeDirect: Analyze

Analyze, from AnalyzeDirect, includes volume rendering, virtual endoscopy, segmentation, image registration, surface rendering, measurements, and many other imaging functions. For more details, see http://www.analyzedirect.com.

APPLICATIONS

In vivo and computational images of GFP-tagged proteins have revealed the dynamic organization of various subcompartments in the interphase nucleus (for review, see Janicki and Spector 2003), including nuclear speckles (for review, see Lamond and Spector 2003). Labeled pre-mRNA splicing factor assemblies were investigated by computational imaging of live cell microscopy images. Images were analyzed for regulated dynamics by segmentation and tracking (Figs. 3 and 4; Movies 16.1 and 16.2) (see Eils et al. 2000). It was shown that the velocity and morphology of speckles, as well as budding events, were related to transcriptional activity.

Studies of nuclear architecture have revealed that chromatin (see Marshall et al. 1997) undergoes slow diffusional motion and that this movement is confined to relatively small regions in the nucleus. Importantly, the constraint on this motion is regulated throughout the cell cycle (see Heun et al. 2001b; Vazquez et al. 2001). A long-standing question has been whether nuclear compartments can also undergo directed, energy-dependent movements,

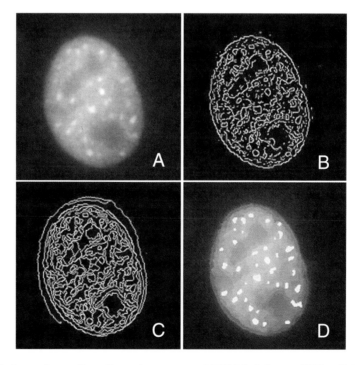

FIGURE 3. Single frame from a live cell movie sequence of GFP-labeled pre-mRNA splicing factor SF2/ASF. (*A*) Unprocessed image of initial time step. (*B*) After a standard nonmaximum suppression algorithm, including a weak hysteresis formulation, candidate edge pixels are identified. (*C*) Edges were traced based on two parameters, namely, local orientation and equal probability of belonging to adjacent regions. (*D*) Based on the induced region neigborhood, graph speckles are detected as regions with locally maximal intensity. For comparison, the unprocessed image (*A*) is overlaid with the detected vesicles in white.

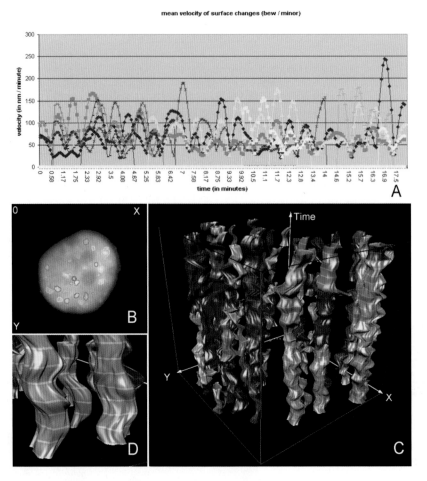

FIGURE 4. Quantification and visualization of surface dynamics for nuclear speckles enriched in pre-mRNA splicing factors in transcriptionally active cells (see Eils et al. 2000). (*A*) Quantification of surface velocities. For display reasons, velocities for 12 representative of a total of 74 speckle trajectories for this cell are shown. For each speckle surface, velocities are calculated of corresponding boundary points in adjacent time sections. (*B*) A single frame from the movie sequence shows highly variable morphology of nuclear speckles. Detected speckles are outlined in colors corresponding to the trajectory color in *C* (Movie 1). (*C*) Spatiotemporal reconstruction of speckles. The surface reconstruction can be visualized in a multidimensional virtual reality viewer that allows real time user interaction (Movie 2, OpenInventor SceneViewer, Template Graphics Software Inc., San Diego, California). (*D*) Detail from *C*. Note that B-spline interpolation, which does not tend to oscillations even for a large number of sample points, was chosen for time-space reconstruction of speckles.

thereby providing a potential mechanism of regulated gene expression. Computational imaging revealed that several nuclear subcompartments do undergo transport that is dependent on metabolic energy (see Calapez et al. 2002; Muratani et al. 2002; Platani et al. 2002).

The investigation of the role of dynamic tension in cells has involved, for example, the analysis of polarized light images of the flow of the actin network and motion of actin bundles and filopodia in crawling neurons. These investigations were used to test and verify a model of actin polymerization in motile cells (Chapter 13) (see Oldenbourg et al. 2000). A study that tracked nuclear envelope breakdown (NEBD), quantification of volumes, and visualization of four-channel images with labeled chromatin, lamin-B receptor, nucleoporin, and tubulin (see Beaudouin et al. 2002), revealed that the initial piercing of the nuclear envelope during NEBD is achieved by spindle microtubules. To further investigate stresses on the nuclear envelope during breakdown, a crosswise grid pattern was bleached onto the nuclear envelope before breakdown. Stresses detected during hole formation were compared, with respect to the positions of the hole, for the different grid vertices, thus providing information about localized stresses during the tearing process (see Mattes et al. 2001). Conversely, the effect of stress on the morphology of cells has been measured using an experimental approach of imposing known stresses on cells in solid-state culture. Changes in height, width, volume, and surface area of the cell are measured in three-dimensional confocal microscopy images, helping to understand the mechanotransduction response (see Guilak 1995).

The positioning of chromosomes during the cell cycle has been investigated in live mammalian cells with a combined experimental and computational approach. In contrast to the random behavior predicted by a computer model of chromosome dynamics, a striking order of chromosomes was observed throughout mitosis (Gerlich et al. 2003). Furthermore, strong similarities between daughter and mother cells were found for mitotic single-chromosome positioning. These results support the existence of an active mechanism that transmits chromosomal positions from one cell generation to the next.

COMPUTATIONAL METHODS FOR QUANTITATIVE ANALYSIS AND VISUALIZATION

The analysis of live cell imaging data usually involves the isolation (segmentation) and tracking of fluorescent structures over time. This leads to the extraction of space- and time-related data and further computation. For practical applications, these image analysis modules can be combined in different ways (see Section 2 and Fig. 1). For instance, to characterize the movement of a cell (see Germain et al. 1999), the image gray values are used directly to estimate the parameters of an affine transformation, thus obviating the need for segmentation (Section 2). For interpretation, a mathematical analysis of the parameters of the affine transformation can allow visualization of a velocity vector field. Other algorithms can track multiple presegmented objects, taking into account only the center of gravity and some attributes of each object (see Tvaruskó et al. 1999). Still other procedures can extract the object surface or contour and quantify the surface area and volume on the basis of the extracted data (see Gerlich et al. 2001). This allows corresponding surface points, in different time steps, to be assigned to each other, and the whole image volume to be mapped through time. This process is referred to as point set registration (see Brown 1992; Lavalée 1996).

In the following discussion, various image-processing techniques are presented in the sequence in which they can be applied: pre-processing (Section 1), segmentation (Section 1), motion estimation (Section 2), visualization (Section 3), and evaluation (Section 4). Starting with the preprocessing in Section 1, the different filters are associated with their corresponding scale-space concept, formulated as a partial differential equation (PDE). Although the concept is reduced to its pure mathematical form, the PDE, a more intuitive description of each filter is also provided. Central to a deeper understanding and evaluation of the various segmentation approaches is the concept of a level set (see Section 1). Demanding contrast invariant image analysis is reduced to set analysis in Mathematical Morphology. This is directly connected with the introduced scale-space in the nonlinear filter in Section 1. From the very beginning, investigators in the field of biological imaging recognized the importance of the space-time domain for motion estimation. Section 2 discusses a selected set of techniques that are particularly suitable for motion analysis of cellular structures and compartments. Visualization of motion also implies the selective reconstruction of the continuity of space-time (Section 3). When information is extracted from space-time in the form of isosurfaces, again the concept of level sets comes to the fore, although in higher dimensions.

The topological organization of the data requires advanced techniques such as Digital Morse Theory (Section 3). Finally, the data in Section 4 must be related to the biological experiment to produce new biological insights.

SECTION 1: IMAGE SEGMENTATION

"Low-level" image analysis aims to extract reliable, local geometric information from a digital image, often referred to as features. Features are used to extract the shape of an object from a segmented image. Segmentation subdivides an image into its constituent parts or objects, which are defined as homogeneous and disjointed regions (image segments), separated by boundaries (closed edges). In principle, two opposite approaches to segmentation can be taken: region-oriented segmentation or contour (boundary)-oriented segmentation. Region-oriented segmentation focuses on the detection of similar features between neighboring pixels, which are then merged and assigned to image segments. The homogeneity criterion must be set empirically for each application (see Zhu and Yuille 1996). Contour-oriented segmentation detects differences between neighboring image points and is therefore based on the image gradient. Contours are drawn between homogeneous image segments at locations where the gradient is strong. Because light intensity across contours behaves, at least in a first approximation, as a discontinuous step function, one of the goals in image analysis has been to find edges in an image (see Canny 1986; Desolneux et al. 2002a).

Once a segmented image has been resolved to visualize its regions and contours, the theory of Mathematical Morphology proposes a list of contrast-invariant operators, such as dilations, erosions, openings, and closings to manipulate the regions and contours (see Sear 1982). This has important consequences: Several geometric PDEs, namely, the curvature motions (see Equation 3), can be considered as the common asymptotic denominators to many "morphomath" operators (see Guichard et al. 2002). These PDEs allow the association of linear scale-space doctrine (Section 2) with Mathematical Morphology, related to level sets (see below), which can lead to deep insights into image processing. Depending on the initialization of the contour and on the choice of the respective parameters, region-oriented or contour-oriented approaches can lead to the same segmentation result (see Desolneux et al. 2002b). However, due to the different scale spaces involved, the performance of both approaches is often very different in practice and strongly depends on the specific application.

Image Preprocessing by Linear Convolution Filters

The quality of images acquired by live cell microscopy is reduced by interference from a variety of noise sources. White noise, which interferes at both low and high frequencies (see Mac Donald 1962), is often introduced by the recording device (e.g., charge-coupled device [CCD] camera or photomultiplier tube).

Noise remaining after digitization leads to the widespread appearance of contours, anomalous or otherwise (see Marr and Hildreth 1980). The image analysis process is a smoothing process that allows true contours to be distinguished from noise. Linear filtering in the form of the heat equation has been proposed for this initial smoothing process:

$$\frac{\partial I(x, y, t)}{\partial t} = div \cdot \nabla I(x, y, t), I(0) = I \tag{1}$$

Equation (1) is an ordinary differential equation applied to the image $I = I(0)$. Higher values of the abstract time scale, t, imply stronger filtering. The heat equation is the asymptotic state of any iterative linear smoothing and is equivalent to the convolution by a Gauss kernel at different scales (see Morel and Caselles 1999). The basic idea is that a window of some finite size (scale) and shape is scanned across the image. The output pixel value is the weighted sum of the input pixels, within the window where the weights are the values of a Gaussian function assigned to every pixel of the window itself. The window with its weights is called the convolution kernel (see Jähne 1997).

The solutions of the heat equation define the linear causal and isotropic scale-space (see Lindeberg 1994). Scale-space means that instead of describing features of an image at a given location, they are described at a given location and at a given scale. The scale quantifies the amount of the smoothing performed on the image in terms of the abstract time-scale (Equation [1]) before computing a feature, such as the contours.

Nonlinear Filters

Many noise-reducing filters are based on low-pass filtering and are therefore suitable for reducing shot noise, which affects the high-spatial-frequency domain. However, because sharp contours and edges are, in the main, part of the high-frequency spectrum, important features can be lost when using nonspecific low-pass filters: Irrespective of the gray value structures present in the image, a Gaussian convolution kernel assigns the same weight to all pixels equidistant from the output pixel. Two images I_1 and I_2 are said to be perceptually equivalent if there is a continuous nondecreasing function G such that

$$I_1 = G(I_2) \tag{2}$$

Simply cutting off high spatial frequencies results in a nonuniform reduction in the contrast of the image (blurring effect), leading to a substantial loss of information. According to the Wertheimer principle, visual perception should be independent of the image contrast (see Wertheimer 1923). As a consequence, demanding contrast invariance rules out the linear scale-space defined by the heat equation. Instead, it leads to a scale-space of the kind

$$\frac{\partial I(x, y, t)}{\partial t} = |\nabla I| G\left(div \cdot \frac{\nabla I}{|\nabla I|}, t\right) \tag{3}$$

where G is as defined in Equation (2), and

$$div \cdot \frac{\nabla I}{|\nabla I|} = curv\,(I)$$

has a differential geometry interpretation as the curvature of the gradient lines.

At first sight, the partial differential in Equation (3) looks rather complicated. However, with $G = const$, the median filter obeys the motion equation in Equation (3) asymptotically (see Morel and Caselles 1999). The median filter is the most common nonlinear filter, and it replaces every pixel of an image with the median of the pixel intensities in a neighborhood. For a pixel near an edge, pixels on the same side of the edge will be in a majority in a square or circular neighborhood, and so the median will be within the intensity range of pixels on that side. Thus edges remain sharp, unlike in linear filtered images, and contrast invariance is not violated. This method is particularly effective when the noise pattern consists of strong, spike-like components and when preservation of edge sharpness is important.

Anisotropic Diffusion Filters

Although the heat equation is, under sound invariance requirements such as causality and isotropy, the only good linear smoother, there are many nonlinear ways to smooth an image (see above). In principle, with nonlinear filters, any function of the neighborhood pixels can be defined.

A new generation of more sophisticated filters for image restoration was proposed with the rough idea of smoothing out irrelevant, homogeneous regions and, to instead, enhance the boundaries (see Perona and Malik 1987). Thus, the diffusion should look like the heat equation (Equation [1]) when the image gradient magnitude is small, whereas the inverse heat equation is applied when the image gradient magnitude is large. Anisotropic diffusion can be thus seen as a robust estimation procedure that estimates a piecewise, smooth image from a noisy input image (see Black et al. 1998) leading to the following partial differential equation:

$$\frac{\partial I(x, y, t)}{\partial t} = div \cdot \left(g_\lambda(|\nabla I|)\nabla I \right) \qquad (4)$$

where $I(x,y,0) = I(x,y)$. The approach of this type of filter is based on the statistical properties of the noise: Image areas containing structure, and strong contrast between edges, will have a higher variance than areas containing noise only. In the robust statistical filtering process, the variance value of each image segment will determine whether noise reduction is to be applied in that area. Thereby, the diffusion process is governed by an "edge-stopping function" g_λ (with λ being a contrast threshold) that is closely related to the error norm and influences function in the robust estimation framework (see Black et al. 1998, Tvaruskó et al. 1999). As a result of this process, noise is reduced and sharp edges are maintained in the image (Fig. 5A–F). However, besides the smoothing scale t, there is the contrast threshold λ, motivated by the noise model and restoration task. As opposed to shape analysis approaches (see below), this approach does not demand compliance with contrast invariance for an image restoration task (Equation [4]).

Segmentation by Thresholding

In general, an image segmentation should allow the isolation of interesting objects based on a threshold applied to the transformed image or region characteristics. Thresholding, as a region-oriented segmentation technique, classifies image points as object or background according to their different intensity values. The threshold l assigns any image point (x, y) at $I(x, y) > l$ to objects, whereas points $I(x, y) < l$ are regarded as background. Unless the object in the image has extremely steep edges, the exact value of l can have considerable effect on the boundary position and thus the apparent size of the extracted object. Subsequent size measurements are thus rather sensitive to the threshold gray value. For this reason, investigators need a consistent method to establish the threshold. An image containing an object on a contrasting background has a bimodal gray-level histogram. The two peaks correspond to the relatively large number of points inside and outside of the object. The dip between the peaks corresponds to the relatively few points around the edge of the object. This dip is commonly used to determine the threshold (see Jähne 1997).

Level Sets and Morphological Operators

For shape analysis, the contrast invariance requirement rules out the heat equation and all models stated before, except the curvature motion of the median filter (see above). Under the usual invariance requirements for image processing, including the contrast invariance by Wertheimer (1923), all image multiscale analyses should have the form of a curvature motion as in Equation (3) (see Chen et al. 1991).

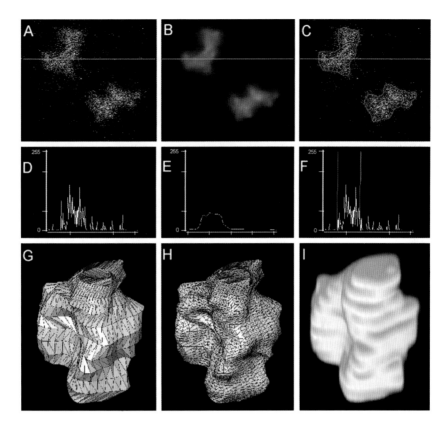

FIGURE 5. Detailed description of three-dimensional reconstruction technique (reprinted, with permission, from Gerlich et al. 2001). (*A–F*) Detection of fluorescent structures after applying anisotropic diffusion and thresholding. (*A*) Single optical slice from a four-dimensional imaging experiment. Chromatin labeled by H2B-YFP. (*B*) Same slice processed by anisotropic diffusion filtering. (*C*) Segmentation of two daughter nuclei by thresholding. (*D*) Line profile of gray values as indicated by green line in *A*. Note that noise strongly disturbs the signal. (*E*) Line profile from *D*. After the preprocessing noise is reduced without perturbing essential edge information. (*F*) Line profile from *C* with red vertical lines indicating the edge position. (*G–I*) Comparison of visualization techniques. The chromatin signal from one daughter nucleus is visualized using surface or volume rendering. (*G*) Surface reconstruction using linear interpolation. After segmentation, the object outlines in individual optical slices are represented by discrete pixels. The reconstruction of a continuous surface requires a parameterized contour running through all pixels of each outline. Equally parameterized contour points from adjacent slices were connected by straight lines. Thereby, a continuous surface is obtained. The triangulated mesh is overlaid on the surface model. (*H*) Surface reconstruction with B-spline interpolation. Cubic B-splines were used to generate smooth curves connecting all outline pixels. Then, equally parameterized contour points on the neighboring outlines were connected, again using B-splines. From these curves, five intermediate contours per optical slice were obtained to increase spatial resolution of the surface model and avoid sharp edges as visible in *A*. (*I*) Conventional volume rendering of the chromatin from one daughter nucleus. Although the overall morphology appears to be similar with all three methods, only the reconstruction with B-spline interpolation achieves a smooth and realistic visualization of the data, with a distinct display of small-scale features.

Contrast invariance led Matheron (1975) to reduce image analysis to a set analysis, namely, the analysis of level sets. For a given level l, thresholding provides a binary image and assumes that each object is formed by a white region (with $I(x,y) \geq l$) entirely separated from other white pixels by black background pixels. Such regions are called upper level sets or confiners. The boundaries of level sets define the level lines. Calculated for different levels l, the level sets define a tree structure, called the topographic map (see Caselles

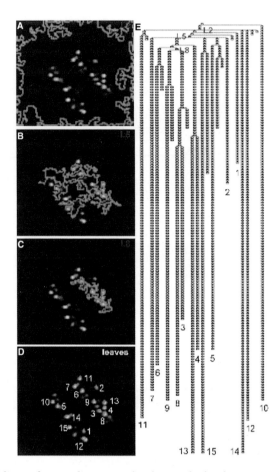

FIGURE 6. (*A–D*) Level lines of succeeding upper level sets. The levels L2, L5, and L8 in *A* through *C* correspond to bifurcation nodes of the confinement tree in *E*, whereas the 15 detected telomeres in *D* correspond to the numbered leaves of the tree in *E*. (*D*) Identification of telomeres after filtering has been applied to the tree. (*E*) The topographic map represents a multiscale analysis of the gray value image. Note that the bifurcation nodes of the levels L2, L5, and L8 in *A–C* are marked in red. At a bifurcation node, two confiners of the higher level fuse.

et. al 1999) or confinement tree, as illustrated in Figure 6E. According to the theory of Mathematical Morphology, all of the image shape information is contained in the level sets (see Sear 1982) or level lines, respectively. In addition, the topology of level lines changes effortlessly at saddle points and permits the level lines to merge or to split simply by evolving their level.

The identification of each region (confiner) and the decision of whether it is related to a region in the previous or succeeding level in the confinement tree is computationally more demanding than the calculation of the binary image. However, a novel algorithm was proposed that is capable of calculating the confinement tree for all available gray levels of the image but requiring only slightly more computation time than the calculation of the binary image obtained by thresholding for one given threshold *l* (see Mattes and Demongeot 2000).

Connected operators are closely related to confinement trees and can be obtained by filtering the tree according to some criteria (see above), followed by a reconstruction of the image based on the filtered tree. In this reconstruction process, the gray value of each pixel is defined by the levels for which the pixel is in a confiner that has been deleted or not. This yields powerful and simple denoising operators, the so-called "Extrema killers" (see Vincent

1993), that delete confiners from the confinement tree according to a certain filtering crite- rion (see Salembier et al. 1998; Jones 1999). Other more basic operators are "opening," "clos- ing," or "thinning," etc. The iteration of these contrast invariant operators results asymptot- ically in the motion equation (Equation [3] above).

Edge-based Segmentation and Meaningful Boundaries

The only drawback of level lines is that there are so many of them. Before describing how to reduce the number of level lines drastically without changing the image aspect, some addi- tional information on edge-based segmentation will be presented.

According to a basic principle of perception, attributed to Helmholtz, an observed geo- metric structure is perceptually "meaningful" if its number of occurrences would be very small in a random deviation: In this context, edges can be considered as large deviations from randomness (see Desolneux et al. 2002a). Therefore, edge detection algorithms are used to establish the boundaries (closed edges) of objects within an image. First, each pixel and its immediate neighborhood are examined in order to determine whether it is on the boundary of an object. Pixels that exhibit the required "boundary" characteristics are labeled as edge points. An image with labeled edge points normally shows each object outlined in edge points, but the boundaries are generally not closed (see Figure 3B). Thus, as a second step, edge-point linking is required to create closed connected boundaries.

If a pixel lies on the boundary of an object, its neighborhood is a zone of gray-level tran- sition. The two characteristics of principal interest are the slope and direction of that transi- tion, i.e., the magnitude and direction of the gradient vector. Edge detection operators are often convolution kernels that implement directional derivatives, examining each pixel neighborhood and quantifying the slope and direction of the local gray-level transition. The Canny-Deriche filter is roughly understood as the detection of maxima of the norm of the gradient in the direction of the gradient (see Canny 1986; Deriche 1987). The nonmaximum suppression algorithm (see Nevatia and Babu 1980; Tvaruskó et al. 1999) determines candi- date edge pixels, if the gradient is maximal compared with the two neighbors in the direc- tion of the gradient, and if it has a potential predecessor and successor (see Fig. 3B). Thereby, multiple responses to a single edge are suppressed and the formation of closed contours is assisted. To obtain closed borderlines, boundaries can be traced by taking into account local orientation (approximated by the direction of the gradient) and equal probability of pixels belonging to adjacent regions (see Fig. 3C). This assumes a motion equation for boundary pixels including a variance force term (see Zhu and Yuille 1996; Tvaruskó 2000). As a result of the tracing algorithm, closed borderlines enclosing homogeneous regions are obtained, which can be used to build a region neighborhood graph.

Each node of the graph is associated with morphological parameters such as mean inten- sity, shape, and size, according to the assigned region within the image. Objects can finally be detected by application of a selection criterion for these nodes, e.g., local maximal inten- sity (see Fig. 3D).

If all possible boundaries in an image are considered as level lines, an opposite approach can be taken. Level lines are curves provided directly by the image itself (see above). They are a way of defining global contrast-insensitive candidates as edges. Edges are orthogonal to the gradient. As a consequence, level lines are adequate indicators of local edges. The meaning- ful geometric event will be a strong contrast along a level line of an image. If a length condi- tion is applied, a contrasted boundary is any closed curve with sufficiently strong contrast that is orthogonal to the gradient of the image at each of its points. This allows defining and computing edges and boundaries (closed edges) by a practically parameter-free method (see Desolneux et al. 2002a). The assumed statistical a priori model for the image is a uniform random noise model. Only geometric structures like edges are defined as the best counter examples, i.e., the least expected in a random model. A major advantage of this Minimum a Posteriori is its reduction to a single robust log-dependent rate: The difference between the false alarm rate and the logarithm of the image size (see Desolneux et al. 2000).

Active Contours

If the initial segmentation is suboptimal, it can often be improved by methods that refine the boundary initially assigned to an object by integrating global image features. The active contour, or snake, is a common tool for such nonlocal refinement. Snakes are particularly well suited when the object is to be tracked through an image sequence and is moving both locally and globally. This is because the contour determined at the current time step can be used as initialization for the next step. Active contour techniques have also been used for the precise detection of object surfaces in three-dimensional recordings, by minimizing appropriate energy terms and expressing the quality of the fit of the deformable surface to the image (see Xu et al. 2000; Gebhard et al. 2001).

The active contour, or snake, was originally introduced by Kass et al. (1988). An initial curve is put into the image according to the segmentation result of a previous time step, or as derived from an initial, simpler segmentation method. Alternatively, it is placed interactively, or automatically, roughly around the object's borders. An energy function is associated with the curve, which has an *internal* term, penalizing stretching or bending, and an *external* term, assessing the fit into the image in each point of the curve. According to this, a snake must be as smooth as possible and have as strong a contrast as possible. The external term can be calculated using local properties of the image (e.g., gray level and gradient) or using previously extracted edges. During the minimization of the energy function, the snake reacts to the image and moves in a smooth continuous manner toward the desired object boundary.

Increasingly simple models for the energy function have been devised more recently. These are no longer a hybrid combination of contrast and smoothness. A recently proposed class of models that reduces the number of model parameters to just one is actually optimal. The snake energy in this class of models is simply an average contrast across the curve, where not only the magnitude of the gradient is taken into account, but also that the gradient is, as much as possible, normal to the curve (see Kimmel and Bruckstein 2001). If the contrast is not reversed along the sought objects, evolving snakes simply coincide with the well-contrasted level lines of the image (see above). Conversely, the meaningful level lines of the image remain almost unchanged by the snake evolution equation (see Desolneux et al. 2002b). Thus, two image-analysis structures, with very different origins, namely, the variational snakes and the maximal meaningful level lines, arrive at almost exactly the same numerical results.

All snake models seek local minima, the global minima being irrelevant, because the best snake, in terms of energy, is reduced to a single point. Although snakes can accurately detect object boundaries in principle, a general limitation is their strong dependence on precise parameter settings that must be determined empirically for specific applications, especially when the three-dimensional topology is complex and noise levels are high. Recent efforts to automate parameter adjustments have led to snakes being more generally applicable (see Gebhard et al. 2001, 2002). However, there are so many ways of changing the large set of local minima of the snake energy terms that an appropriate user interaction is needed. In comparison, to maximize meaningful boundaries (see above), edge detection with the snake model is much more difficult and an automatic algorithm with no parameters seems unreachable.

SECTION 2: MOTION ESTIMATION

Motion analyses of biological images have always been performed in the space-time domain (see Adelson and Bergen 1986). Motion in two-dimensional image sequences is understood as orientation in three-dimensional space-time (see Jähne 1993). One classification of schemes for motion computation distinguishes between similarity matching and optical flow methods. Conceptually, correspondence or "feature-matching" methods are discrete in time, whereas optical flow methods are formulated in a time-space continuum. Note, however, that feature-based methods can be used to reconstruct the continuity of time-space (see Section 3 and Fig. 4C,D).

Object-based feature matching is essentially a two-step approach: First, robust features of single objects are extracted through segmentation, and then objects are matched in consecutive images (see Fig. 1E,F). This approach is most appropriate for tracking a large number of small particles that move individually and independently from each other. In this circumstance, the movement of the gravity center of a particle is often the only movement considered. The movement of the different points within the particle are not deemed significant. Conversely, optical flow techniques determine the local motion directly, using the pixel gray values. Velocity is estimated in each image point as orientation in continuous space-time and leads to complete motion fields.

For the determination of a more complex movement, for each detected object or in the whole-image scene, a third technique called *image registration* has been developed (see Fig. 1D). A computer "registers" (apprehends and allocates) certain objects in the real world, as they appear in the computer's internal model. To avoid the possibility that registration is ill-posed, knowledge of underlying processes is incorporated into regularization schemes. If the model does not adequately reflect reality, the results will be dismissed as incorrect.

Single-particle Tracking

For image sequences containing small objects with large displacements (with respect to the object's diameter), dynamic analysis can be achieved by single-particle tracking in time-space (see Adamczyk and Rimai 1988; Grant 1994). For each object in a given time frame, corresponding objects in a previous time frame must be found. Importantly, for an object in the current frame, the corresponding object in a previous frame is not necessarily the nearest object ("correspondence problem"). Therefore, the correspondence is established, or not, depending either on object features or on inter-object relationships of the objects inside a frame. Object features can be dynamic criteria, such as displacement and acceleration of an object, as well as area/volume, mean gray value, or shape of the object. Assuming that optical flow is continuous (see below), essential information on corresponding objects in subsequent images should be similar. However, distortions in the imaging process, such as noise, bleaching, and illumination differences, as well as changes in focal position, might considerably distort the assumption of time-space continuity. Standard region-based matching techniques do not, in general, give satisfactory results (see Anadan 1989). A more reliable tracking approach involves fuzzy-logic-based analysis of the tracking parameters (see Hering et al. 1997; Tvaruskó et al. 1999).

Fuzzy theory assumes that all things are a matter of degree (see Kosko 1992). Fuzzy systems behave as associative memories mapping close inputs to close outputs without requiring a mathematical description of how the output functionally depends on the input. A fuzzy system relies on linguistic "rules," encoded in a numerical, fuzzy associative memory mapping the FAM rules. According to a dynamic particle model, the velocity of an object is assumed to remain relatively constant. To compare two objects in consecutive images, differences in velocity and deviation of the expected extrapolated position from the potential new position are measured. In addition, differences in total intensity and area are computed and translated into fuzzy rules.

Estimation of the Local Optical Flow

Optical flow or image flow means that the apparent motion at the image plane is based on visual perception and has the dimension of velocity. When the optical flow is determined from two consecutive images, it appears as a *displacement vector (DV)* from the features of the first to the second image, i.e.,

$$I([x,y], t) = ([x - u, y - v], 0) \tag{5}$$

where DV is equal to $\vec{v} = (u, v)^T$ (see Jähne 1997). It is not equivalent to the real motion occurring in the image scene as, for instance, movements inside a region of homogeneous intensity cannot be detected (see Schunk 1986; Fleet 1990; Haussecker and Fleet 2001). A dense representation of displacement vectors is known as a *displacement vector field (DVF)*. However, it is important to note that the optical flow is a term inherent to continuous space. Classical approaches to estimate the optical flow (see Horn and Schunk 1981) are based on the motion constraint equation (MCE), which is derived as a first-order Taylor expansion of Equation (5), from an assumption that intensity is conserved (called continuity of the optical flow), $dI(x,y,t)/dt = 0$:

$$\nabla I(x, y, t) \cdot \vec{v} + I_t(x, y, t) = 0 \tag{6}$$

where $I_t(x,y,t)$ denotes the partial time derivative of $I(x,y,t)$, $\nabla I(x,y,t) = (I_x[x,y,t], I_y[x,y,t])^T$ the image gradient, and $\nabla I \cdot \vec{v}$ denotes the usual dot product.

The MCE relates the spatial and temporal derivatives of I linearly with $\vec{v}$. In effect, Equation (6) yields the normal component of motion of spatial contours of constant intensity. Therefore, MCE is an underdetermined equation (with respect to $\vec{v}$) and either a so-called regularization must be applied in order to relate the velocities in different points (see Haussecker and Fleet 2001) or a parametric approach as described below must be used. A synopsis of methods based on the hypothesis of intensity conservation is provided elsewhere (see Barron et al. 1994; Mitiche and Bouthemy 1996).

Parametric Image Registration

A model deformation may be driven iteratively by applying global transformations of the space embedding to the model, instead of applying local deformation of the model inside space. This approach is referred to as registration, as opposed to local deformations using active contour models (see Section 1) (see Montagnat et al. 2001). Registration (often called model-based matching) of two two- or three-dimensional data sets can be most concisely described as finding a spatial deformation that relates the data sets, i.e., that transfers the first data set into the second. Characteristic features of one shape should thereby be mapped to the corresponding points of the other shape.

In the context discussed here, the considered transformations are given in a closed and parametric form; i.e., for a given parameter value, a transformation is specified, which allows calculation, for each point in space, of the transformed point by simple algebraic operations. An example is a rotation where, after specifying the rotation angles, the transformation of each point can be calculated by a matrix multiplication. A parametric image registration algorithm specifies the parameters of a given parameterized transformation in a way that physically corresponding points, at two different time steps, are brought as close together as possible. Such algorithms have been formulated in various ways: Some have been formulated on previously extracted surface points (see Besl and McKay 1992; Lavalée and Szeliski 1995), whereas others register the images directly using value differences.

In cell biology, parametric registration has been used mainly for automated correction of rotational and translational movements in a time series. This allows the visual interpretation of continuous time-space reconstructions to be enhanced by revealing only local dynamics, especially if the movement is a result of the superposition of two or more independent motions. It also makes the estimation of the local dynamics more robust in this case. For example, when tracking particles inside the cell nucleus, the global movement of the nucleus as a whole must be compensated for. The inverse case can be true as well: Only the global movements are of interest; local movements are considered artifacts. In both cases, *rigid* transformations must be compensated for. When dealing with deformations, *nonrigid* registration algorithms, which differ with respect to the so-called "motion model" (see Fieres et al. 2001; Mattes et al. 2002), must be applied, and the strategy to find the desired parameters varies. Most commonly, a cost or error function is defined and an optimization method is chosen that iteratively adjusts the parameters until an optimum has been reached (minimum of the cost function). A classical similarity criterion, based on extracted point features, measures the Euclidean distance of nearest points in the two data sets. The squared sum of these distances can be minimized using the Levenberg-Marquardt optimization algorithm (see Press et al. 1992). More recently, several intensity-based similarity measures, e.g., the sum of squared intensity differences, or the mutual information, have been suggested (Roche et al. 1999).

SECTION 3: VISUALIZATION

Two-dimensional time-lapse live cell recordings are widely used for the investigation of dynamic fluorescently labeled structures with high spatial resolution (see Belmont et al. 1999; Lippincott-Schwartz et al. 1999; Eils et al. 2000; Tsukamoto et al. 2000; Muratani et al. 2002). An important step for the interpretation of dynamic imaging data has been the development of computational methods for time-space visualization (see Fig. 4) (see Tvaruskó et al. 1999; Rustom et al. 2000). In contrast to the display of time series as movies, additional information concerning, for example, colocalization events, is gained from the continuous time-space display.

A detailed analysis of complex dynamic processes in cells would ideally be perfomed in three spatial dimensions over time. Such four-dimensional imaging experiments can be performed, for example, on confocal or epifluorescence microscopes. Large and complex data sets, typically 5,000–10,000 single images, are thereby generated, but they cannot be analyzed appropriately without computational tools for the visual and quantitative inspection in space and time (see Gerlich et al. 2001). Early studies have explored four-dimensional data sets by simply browsing through an image gallery and by highlighting interactively selected structures (see Thomas et al. 1996). To facilitate interpretation, many four-dimensional experiments are preprocessed by a projection step that reduces dimensionality (e.g., projection of image stacks to the *x-y* plane). A better interpretation of four-dimensional imaging data can often be achieved with specialized visualization tools. The data are visualized using either volume rendering (see Chen et al. 1995; Max 1995) or isosurface extraction (see Cline et al. 1988; Karron and Cox 1995a,b). Volume rendering techniques achieve a satisfactory display of biological structures, but this method is limited to pure visualization and does not deliver quantitative information. In addition, the high anisotropy, typical for live cell imaging with low *z* resolution, limits the quality of this visualization technique (see Fig. 5G–I).

Three-dimensional Time-space Reconstruction

The trajectories generated in the single-particle tracking process can be reconstructed and visualized in three-dimensional time-space. The dynamics of objects with constant shape over time are well described by the dynamic repositioning of their gravity centers. Interpolation techniques are used to generate continuous curves from gravity centers. Classical interpolation techniques, such as linear interpolation, are not appropriate for the reconstruction of trajectories. In contrast, cubic splines allow the generation of smooth curves with second-order continuity. Therefore, continuous derivatives exist up to the order of two, with the first and second derivatives corresponding to velocity and acceleration, respectively (see Fig. 4A) (see Tvaruskó et al. 1999; Rustom et al. 2000).

Splines are functions that are defined piecewise on intervals, in particular as cubic polynomials, and fulfill smoothness criteria at the knots between individual segments. Cubic basic spline curves, often referred to as cubic B-splines, are a generalization of interpolating cubic polynomial splines and are more flexible with respect to changes of the curve (see Fig. 5H). For detailed mathematical description of cubic B-splines see Watt and Watt (1992). B-splines were chosen for interpolation between assigned gravity centers of discrete time steps because they are stable in a geometric sense; e.g., they tend not to oscillate even for a large number of sample points. The idea of splines is to interpolate a smooth curve between points with minimal bending, which is built up from segments connected at the given points.

A second approach was designed for visualization of two-dimensional objects that change their shape dynamically over time. This problem accounts for a continuous shape reconstruction from a series of two-dimensional images in time-space. Cubic B-Splines were used to generate smooth curves approximating the contour pixels in individual time sections. In the next step, contour points from corresponding contours at different time steps are connected, again using cubic B-splines (see Fig. 5C,D). Under the assumption that the transformation of one contour in its adjacent contour is sufficiently smooth and leaves the order of contour points unchanged, the optimal transformation is found by minimizing the integral over all Euclidean distances between corresponding contour points. For minimization of this energy term, a recursive contour splitting approach can be chosen (see Tvaruskó et al. 1999).

Isosurface Extraction

Surface reconstructions approximate a selected structure by a list of polygons. The structure is displayed by projecting all of the polygons onto a plane that is perpendicular to a selected viewing direction. The user can examine the displayed structure by interactively changing the viewing direction. Although the rendering algorithms are well-developed, the generation of a polygon list that represents the surface appropriately can be difficult. As a first step in isosurface extraction, data are segmented by thresholding, i.e., identifying objects defined by the sets of points in three-dimensional space where a suitable interpolating function I is larger than or equal to a real value l. The topological boundary of these objects will generally correspond to the level surfaces of *I*. In the degenerative case, multiple surface components are created. When the volume data are given on cubes, a classic algorithm, called Marching Cube, is used to construct a polygonal mesh of the three-dimensional surface (see Cline et al. 1988). More advanced techniques use Digital Morse Theory (J. Cox et al., unpubl.) for the topological organization of the data and extraction of isosurfaces. The drawback of isosurface extraction methods is that the surface of many biological structures cannot be defined using a single intensity value, resulting in loss of relevant information. In addition, the anisotropic resolution characteristic of most optical microscopic data sets becomes very obvious at viewing angles that are perpendicular to the imaging axis. It is a major goal to enhance image quality by regaining spatial resolution in four-dimensional live cell imaging experiments.

To take into account the high degree of anisotropy typical of four-dimensional live cell recordings, B-spline reconstruction can be applied to *z* slices at one time step (see Gerlich et al. 2001). In contrast to the approach described above, time sections are replaced by *z* sections. For the reconstruction of a continuous three-dimensional surface from optical slices, equally parameterized contour points from adjacent *z* slices are connected, again by using cubic B-splines. From these curves, intermediate contours between optical slices can be obtained to increase spatial resolution of the surface model and to avoid sharp edges (see Fig. 5G,H). The specific advantage of an interpolated surface rendering is a very distinct display of small-scale features, as compared to volume rendering (Fig. 5I). Moreover, surface reconstruction allows direct access to quantitative data (Section 4).

Visualization of Four-dimensional Data

A sequence of three-dimensional reconstructions can be used to visualize a four-dimensional data set. However, besides the number of *z* sections per stack, the temporal resolution is limited in live cell experiments. Thus, with the surface reconstructions of cellular structures at distinct time points in hand, it can be of great interest to infer continuous motion from these data (see Gerlich et al. 2001). Interpolation over time is especially useful when the experiment is carried out over longer time courses, thus enlarging the time-lapse interval between consecutive image stacks.

To achieve temporal interpolation, an algorithm must be developed for the smooth transition from one surface reconstruction to the next. Methods for this task are referred to as morphing. The morphing problem consists of two major tasks. For objects that are represented as point meshes, such as are generated by the three-dimensional surface reconstruction, corresponding points must be identified in subsequent time frames. Such correspondence can be obtained as a result of motion estimation (Section 2). In the absence of a motion estimation step, correspondence can be established using identical parameterization in subsequent frames (see Xu et al. 2000). In the next step, interpolation over time between these corresponding surface points is carried out. Once again, using B-splines, a continuous reconstruction of entire four-dimensional data sets is obtained (see Movies 16.3 and 16.4) (see Gerlich et al. 2001).

Visualization of the Computed Motion-Model

Several techniques have been used to depict the motion occurring in the image sequence after quantitative evaluation. Classically, motion vectors regularly placed on the image, trajectories, or deformed grids give the user an impression of the underlying motion. Whereas motion vectors provide more details (see Haussecker and Fleet 2001), deformed grids allow comprehension of the global movement and the bending of the space (see Fieres et al. 2001). Trajectories are useful for providing continuous visualization of the motion at specific points (Fig. 4C,D) (see Tvaruskó et al. 1999; Eils et al. 2000; Rustom et al. 2000; Muratani et al. 2002). Techniques to visualize scalar quantitative values associated with each point on a surface or in space use color, gray value intensities, or patterns (see Ferrant et al. 2001), and these features can be combined with the techniques mentioned above. The statistical analysis of the extracted quantitative values provides further possibilities for evaluating these values.

SECTION 4: QUANTITATIVE ANALYSIS

A great advantage of combined segmentation and surface reconstruction is the immediate access to quantitative information that corresponds to visual data (see Tvaruskó et al. 1999; Gerlich et al. 2001). The binarized object representation can be used directly to measure volume over time. Moreover, the gray values in the segmented area of corresponding original images can be measured to determine the amount and concentration of fluorescently labeled protein in the segmented cellular compartments. After motion estimation, the velocity of the detected object mass center or, alternatively, of each point on the object surface, or even of each point in space, is available. With this, parameters such as acceleration (see Eils et al. 2000), tension, or bending (see Bookstein 1989), and diffusion coefficients (see Marshall et al. 1997; Tvaruskó et al. 1999; Eils et al. 2000; Heun et al. 2001b; Vazquez et al. 2001; Chubb et al. 2002) can be determined. Statistical analysis provides further possibilities for motion characterization. For instance, the peaks of velocity histograms report which velocity values occur most frequently (see Uttenweiler et al. 2000). In addition, the movements of different objects (or of the same object at different time steps) can be compared by their velocity histograms.

A challenge for future work is to extract more specific parameters by fitting a biophysical model to the data. This has been achieved in medical image analysis (see Ferrant et al. 2001), where the human brain was modeled using Finite Element Methods, giving insight into forces occurring during brain deformations.

CONCLUSIONS

In this chapter, concepts of quantitative image analysis of multidimensional microscopy data have been summarized. In particular, the chapter focused on recently developed approaches for image segmentation, motion estimation, and visualization that were designed to suit the specific demands of live cell microscopy (e.g., low signal-to-noise, high anisotropy). The applicability of the described methods is demonstrated by a number of applications.

Preprocessing and segmentation techniques were embedded in a more general theory of image processing reduced to set analysis. In a shortcut, well known to physics, invariance requirements lead to different concepts of scale-space and its implication for motion equations of level sets. On the other hand, central to motion analysis and visualization is the fragmented reverse and explicit reconstruction of continuity in time-space. Reconstruction is influenced through assumed models of motion implicit to registration techniques. Dynamic processes are not only perceived in higher dimensions, but also understood in terms of their models, and finally evaluated.

Most importantly, this chapter points out that objectivity in the interpretation of dynamic imaging data can be achieved only through automated computational tools, such as those described here. Quantitative tools for measuring velocity, acceleration, diffusion coefficients, volume changes, concentrations, and distances in four-dimensional data sets are now available. A more consequent realization of the mathematical concepts underlying image processing in the form of practical algorithms and tools will further improve the situation for the end user.

ACKNOWLEDGMENTS

We thank Chaitanya Athale and Daniel Gerlich for conceptual input to this chapter and Professor Jäger for continuous support.

MOVIE LEGENDS

MOVIE 16.1. Live cell movie sequence of GFP-labeled SF2/ASF. Speckles detected by contour-based segmentation algorithms are displayed by color outlines. During the imaging experiments, small globular structures were frequently seen to bud off from larger speckles. In the time step before these buds start, new trajectories of speckles are shown with two contours, one contour indicating the beginning of a new trajectory.

MOVIE 16.2. Spatiotemporal reconstruction of speckles. Based on contours from segmented individual speckles at distinct time sections as shown in Movie 16.1, a continuous time-space reconstruction was computed. Note that the colors of the trajectories correspond to the colors of the speckle contours in Movie 16.1.

MOVIE 16.3. Assembly of image stack from optical slices and reconstruction of a three-dimensional surface model of LBR-EYFP patches (*green*) on chromatin surface (*red*) (from Gerlich et al. 2001 with permission of the Nature Publishing Group). First, the preprocessed optical image slices of a single image stack are displayed sequentially. Then, they are removed stepwise to reveal the three-dimensional reconstructed surface model. For further details, see Figure 5.

MOVIE 16.4. Morphed reconstruction of nuclear envelope (NE) and chromatin expansion (from Gerlich et al. 2001 with permission of the Nature Publishing Group). Four intermediate states of chromatin and NE surfaces per time frame were interpolated, thereby increasing temporal resolution.

REFERENCES

Adamczyk L. and Rimai A.A. 1988. 2D particle tracking velocimetry (PTV): Technique and image processing algorithms. *Exp. Fluids* **6:** 373–380.

Adelson E.H. and Bergen J.R. 1986. The extraction of spatio-temporal energy in human and machine vision. In *Proceedings on motion: Representation and analysis*, pp. 151–155. IEEE Computer Society Press, Washington.

Anandan P. 1989. A computational framework and an algorithm for the measurement of visual motion. *Int. J. Comp. Vision* **2:** 283–310.

Barron J.L., Fleet D.J., and Beauchemin S.S. 1994. Systems and experiment—Performance of optical flow techniques. *Int. J. Comp. Vision* **12:** 43–77.

Beaudouin J., Gerlich D., Daigle N., Eils R., and Ellenberg J. 2002. Nuclear envelope breakdown proceeds by microtubule-induced tearing of the lamina. *Cell* **108:** 83–96.

Belmont A.S., Dietzel S., Nye A.C., Strukov Y.G., and Tumbar T. 1999. Large scale chromatin structure and function. *Curr. Opin. Cell Biol.* **11:** 307–311.

Besl J.P. and McKay N.D. 1992. A method for registration of 3D shapes. *IEEE Trans. PAMI* **14:** 239–256.

Black M.J., Sapiro G., Marimont D., and Heeger D. 1998. Robust anisotropic diffusion. *IEEE Trans. Image Process.* **7:** 421–432.

Bookstein F.L. 1989. Principal warps: Thin-plate splines and the decomposition of deformations. *IEEE Trans. PAMI* **11:** 567–585.

Brown L.G. 1992. A survey of image registration techniques. *ACM Comput. Surveys* **24:** 325–376.

Calapez A., Pereira H.M., Calado A., Braga J., Rino J., Carvalho C., Tavanez J.P., Wahle E., Rosa A.C., and Carmo-Fonseca M. 2002. The intranuclear mobility of messenger RNA binding proteins is ATP dependent and temperature sensitive. *J. Cell Biol.* **159:** 795–805.

Canny J.F. 1986. A computational approach to edge detection. *IEEE Trans. PAMI* **8:** 679–698.

Caselles, V., Coll B., and Morel J.M. 1999. Topographic maps and local contrast changes in natural images. *Int. J. Comp. Vision* **33:** 5–27.

Chalfie M., Tu Y., Euskirchen G., Ward W.W., and Prasher D.C. 1994. Green fluorescent protein as a marker for gene expression. *Science* **263:** 802–805.

Chen H., Swedlow J.R., Grote M., Sedat J.W., and Agard D.A. 1995. The collection, processing, and display of digital three-dimensional images of biological specimens. In *Handbook of biological confocal microscopy* (ed. J.B. Pawley), chap. 13, pp. 197–210. Plenum Press, New York.

Chen Y.-G., Giga Y., and Goto S. 1991. Uniqueness and existence of viscosity solutions of generalized mean curvature flow equations. *J. Differ. Geom.* **33:** 749–786.

Chubb J.R., Boyle S., Perry P., and Bickmore W.A. 2002. Chromatin motion is constrained by association with nuclear compartments in human cells. *Curr. Biol.* **12:** 439–445.

Cline H.E., Lorensen W.E., and Ludke S. 1988. Two algorithms for the three-dimensional reconstruction of tomograms. *Med. Phys.* **15:** 320–327.

Deriche R. 1987. Using Canny's criteria to derive a recursively implemented optimal edge detector. *Int. J. Comp. Vision* **1:** 167–187.

Desolneux A, Moisan L., and Morel J.-M. 2000. Maximal meaningful events and applications to image analysis. *CMLA, ENS Cachan.* At http://www.cmla.ens-cachan.fc.

———. 2002a. Edge detection by helmholtz principle. *J. Math. Imaging Vision* **14:** 271–284.

Desolneux A., Moisan L., and Morel J.-M. 2002b. Variational snake theory. In *Geometric level set methods in imaging, vision and graphics* (ed. S. Osher and N. Paragios), pp. 79–97. Springer-Verlag, Berlin.

Eils R., Gerlich D., Tvarusko W., Spector D.L., and Misteli T. 2000. Quantitative imaging of pre-mRNA splicing factors in living cells. *Mol. Biol. Cell* **11:** 413–418.

Ferrant M., Nabavi A., Benoit M., Kikinis R., and Warfield S.K. 2001. Real-time simulation and visualization of volumetric brain deformation for image guided neurosurgery. In Proceedings of *SPIE Medical Imaging 2001* (vol. 4319) (ed. S.K. Mun), pp. 366–373. SPIE, San Diego.

Fieres J., Mattes J., and Eils R. 2001. A point set registration algorithm using a motion model based on thin-plate splines and point clustering. In *Pattern recognition DAGM 2001, vol.* 2191, pp. 76–83. Springer Verlag, Berlin.

Fleet D.J. 1990. "Measurement of image velocity." Ph. D. thesis, University of Toronto, Canada.

Gebhard M., Eils R., and Mattes J. 2002. Segmentation of 3D objects using NURBS surfaces for quantification of surface and volume dynamics. In *International Conference on Diagnostic Imaging and Analysis (ICDIA)*, vol. 3, pp. 125–130. Shanghai, China.

Gebhard M., Mattes J., and Eils R. 2001. An active contour model for segmentation based on cubic B-splines and gradient vector flow. In *Lecture notes in computer science (MICCAI 2001)* (ed. W.J. Nressen and M.A. Viergever), vol. 2208, pp. 1373–1375. Springer-Verlag, Berlin.

Gerlich D., Beaudouin J., Gebhard M., Ellenberg J., and Eils R. 2001. Four-dimensional imaging and quantitative reconstruction to analyse complex spatiotemporal processes in live cells. *Nat. Cell Biol.* **3:** 852–855.

Gerlich D., Beaudouin J., Kalbfuss B., Daigle N., Eils R., and Ellenberg J. 2003. Global chromosome positions are transmitted through mitosis in mammalian cells. *Cell* **112:** 751–764.

Germain F., Doisy A., Ronot X., and Tracqui P. 1999. Characterization of cell deformation and migration using a parametric estimation of image motion. *IEEE Trans. Biomed. Eng.* **46:** 584–600.

Grant I., ed. 1994. *Selected papers on particle image velocimetry.* SPIE Optical Engineering Press, Bellingham, Washington.

Guichard F., Moisan L., and Morel J.-M. 2002. A review of P.D.E. models in image processing and image analysis. *J. Physique IV* **12:** 137–154.

Guilak F. 1995. Compression-induced changes in the shape and volume of the chondrocyte nucleus. *J. Biomech.* **28:** 1529–1541.

Haussecker H.W. and Fleet D.J. 2001. Computing optical flow with physical models of brightness variation. *IEEE Trans. PAMI* **23:** 661–673.

Hering F., Leue C., Wierzimok D., and Jähne B. 1997. Particle tracking velocimetry beneath water waves. Part I: Visualization and tracking algorithms. *Exp. Fluids* **23:** 472–482.

Heun P., Taddei A., and Gasser S.M. 2001a. From snapshots to moving pictures: New perspectives on nuclear organization. *Trends Cell Biol.* **11:** 519–525.

Heun P., Laroche T., Shimada K., Furrer P., and Gasser S. 2001b. Chromosome dynamics in the yeast interphase nucleus. *Science* **294:** 2181–2186.

Horn B.K.P. and Schunck, B.G. 1981. Determining optical flow. *Artif. Intell.* **17:** 185-204.

———, ed. 1993. *Lecture notes in computer science: Spatio-temporal image processing.* Springer-Verlag, Berlin.

Jähne B., ed. 1997. *Digital image processing—Concepts, algorithms, and scientific applications.* Springer-Verlag, Berlin.

Janicki S.M. and Spector D.L. 2003. Nuclear choreography: Interpretations from living cells. *Curr. Opin. Cell Biol.* **15:** 149–157.

Jones R. 1999. Connected filtering and segmentation using component trees. *Comput. Vision Image Understand.* **75:** 215–228.

Karron D. and Cox J. 1995a. The spiderweb algorithm for extracting 3D objects from volume data. In *Lecture notes in computer science: Proceedings of the First International Conference, CVRMed'95* (ed. N. Ayache), vol. 905, pp. 481–486. Springer-Verlag, Berlin.

———. 1995b. Extracting 3D objects from volume data using digital morse theory. In *Lecture notes in computer science: Proceedings of the First International Conference, CVRMed'95* (ed. N. Ayache), vol. 905, pp. 481–486. Springer-Verlag, Berlin.

Kass M., Witkin A., and Terzopoulos D. 1988. Snakes: Active contour models. *Int. J. Comp. Vision* **1:** 321–331.

Kimmel R. and Bruckstein A. 2001. Regularized laplacian zero crossings as optimal edge integrators. In *Proceedings of image and vision computing, IVCNZ01.* University of Otago, Danedin, New Zealand.

Kosko B., ed. 1992. *Neural networks and fuzzy systems.* Prentice Hall, New Jersey.

Lamond A.I. and Spector D.L. 2003. Nuclear speckles: A model for nuclear organelles. *Nat. Rev. Mol. Cell. Biol.* **4:** 605–612.

Lavallée S. 1996. Registration for computer-integrated Surgery: Methodology, state of the art. In *Computer integrated surgery* (ed. R. Taylor et al.), pp. 77–97. MIT Press, Cambridge, Massachusetts.

Lavallee S. and Szeliski R. 1995. Recovering the position and orientation of free-form objects from image contours using 3D distance maps. *IEEE Trans. PAMI* **17:** 378–390.

Lindeberg T., ed. 1994. *Scale-space theory in computer vision.* Kluwer Academic Publishers, Netherlands.

Lippincott-Schwartz J.L. and Smith C.L. 1997. Insights into secretory and endocytic membrane traffic using green fluorescent protein chimeras. *Curr. Opin. Neurobiol.* **7:** 631–639.

Lippincott-Schwartz J., Presley J.F., Zaal K.J., Hirschberg K., Miller C.D., and Ellenberg J. 1999. Monitoring the dynamics and mobility of membrane proteins tagged with green fluorescent protein. *Methods Cell Biol.* **58:** 261–281.

Mac Donald D.K.C., ed. 1962. *Noise and fluctuations.* John Wiley, New York.

Marr D. and Hildreth E. 1980. Theory of edge detection. *Proc. Royal Soc. Lond. B* **207:** 187–217.

Marshall W.F., Straight A., Marko J.F., Swedlow J., Dernburg A., Belmont A.S., Murray A.W., Agard D.A., and Sedat J.W. 1997. Interphase chromosomes undergo constrained diffusional motion in living cells. *Curr. Biol.* **7:** 930–939.

Matheron G., ed. 1975. *Random sets and integral geometry.* John Wiley, New York.

Mattes J. and Demongeot J. 2000. Efficient algorithms to implement the confinement tree. In *Lecture notes in computer science* (ed. G. Borgefors et al.), vol. 1953, pp. 392–405. Springer-Verlag, Berlin.

Mattes J., Fieres J., and Eils R. 2002. A shape adapted motion model for non-rigid registration. In *Proceedings of SPIE Medical Imaging 2002: Image Processing* (ed. M. Sonka and M.J. Fitzpatrick), vol. 4684, pp. 518–527. San Diego, California.

Mattes J., Fieres J., Beaudouin J., Gerlich D., Ellenberg J., and Eils R. 2001. New tools for visualization and quantification in dynamic processes: Application to the nuclear envelope dynamics during mitosis. In *MICCAI 2001 Lecture notes in computer science,* vol. 2208, pp. 1323–1325. Springer-Verlag, Berlin.

Max N. 1995. Optical models for direct volume rendering. *IEEE Trans. Visual. Comput. Graphics* **1:** 99–108.

Misteli T. 2001. Protein dynamics: Implications for nuclear architecture and gene expression. *Science* **291:** 843–847.

Misteli T. and Spector D.L. 1997. Applications of the green fluorescent protein in cell biology and biotechnology. *Nat. Biotechnol.* **15:** 961–964.

Misteli T., Caceres J.F., and Spector D.L. 1997. The dynamics of a pre-mRNA splicing factor in living cells. *Nature* **387:** 523–527.

Mitiche A., and Bouthemy P. 1996. Computation and analysis of image motion: A synopsis of current problems and methods. *Int. J. Comp. Vision* **19:** 29–55.

Montagnat J., Delingette H., and Ayache N. 2001. A review of deformable surfaces: Topology, geometry and deformation. *Image Vision Comput.* **19:** 815–828.

Morel J.-M. and Caselles V. 1999. The geometry of digital images. In *Mathematisches Forschungsinstitut Oberwolfach, Seminar (03.10.99-09.10.99)*, pp. 17–19. At http://www.info.de/seminars/seminars_1999.html.

Muratani M., Gerlich D., Janicki S.M., Gebhard M., Eils R., and Spector D.L. 2002. Metabolic-energy-dependent movement of PML bodies within the mammalian cell nucleus. *Nat. Cell Biol.* **4:** 106–110.

Nevatia R. and Babu K.R. 1980. Linear feature extraction and description. *Comput. Graph. Image Process.* **13:** 257–269.

Oldenbourg R., Katoh K., and Danuser G. 2000. Mechanism of lateral movement of filopodia and radial actin bundles across neuronal growth cones. *Biophys. J.* **78:** 1176–1182.

Perona P. and Malik J. 1987. Scale space and edge detection using anisotropic diffusion. *CVWS87* **12:** 16–22.

Platani M., Goldberg I., Lamond A.I., and Swedlow J.R. 2002. Cajal body dynamics and association with chromatin are ATP-dependent. *Nat. Cell Biol.* **4:** 502–508.

Platani M., Goldberg I., Swedlow J.R., and Lamond A.I. 2000. In vivo analysis of cajal body movement, separation, and joining in live human cells. *J. Cell Biol.* **151:** 1561–1574.

Press W.H., Teukolsky S.A., and Vetterling W.T., eds. 1992. *Numerical recipes in C—The art of scientific computing.* Cambridge University Press, England.

Roche A., Malandain G., Ayache N., and Prima S. 1999. Towards a better comprehension of similarity measures used in medical image registration. In *Proceedings of 2nd International Conference on Medical Image Computing and Computer-Assisted Intervention (MICCAI'99)* (ed. C. Taylor, and A. Colchester), vol. 1679, pp. 555–566. Springer-Verlag, Cambridge.

Rustom A., Gerlich D., Rudolf R., Heinemann C., Eils R., and Gerdes H.H. 2000. Analysis of fast dynamic processes in living cells: High-resolution and high-speed dual-color imaging combined with automated image analysis. *BioTechniques* **28:** 722–728, 730.

Salembier P., Oliveras A., and Garrido L. 1998. Antiextensive connected operators for image and sequence processing. *IEEE Trans. Image Process.* **7:** 555–570.

Schunk B.G. 1986. Image flow continuity equation for motion and density. In *Proceedings of the Workshop on Motion: Representation and Analysis*, pp. 89–94. IEEE Comp. Soc., Washington.

Sear J., ed. 1982. *Image analysis and mathematical morphology.* Academic Press, New York.

Thomas C., DeVries P., Hardin J., and White J. 1996, Four-dimensional imaging: Computer visualization of 3D movements in living specimens. *Science* **273:** 603–607.

Tsukamoto T., Hashiguchi N., Janicki S.M., Tumbar T., Belmont A.S., and Spector D.L. 2000. Visualization of gene activity in living cells. *Nat. Cell Biol.* **2:** 871–878.

Tvaruskó W. 2000. "Zeitaufgelöste Analyse und Visualisierung von dynamischen Prozessen in lebenden Zellen." Ph.D. thesis, Universität Heidelberg, Germany.

Tvaruskó W., Bentele M., Misteli T., Rudolf R., Kaether C., Spector D.L., Gerdes H.H., and Eils R. 1999. Time-resolved analysis and visualization of dynamic processes in living cells. *Proc. Natl. Acad. Sci.* **96:** 7950–7955.

Uttenweiler D., Veigel C., Steubing R., Götz C., Mann S., Haussecker H., Jähne B., and Fink R. 2000. Motion determination in actin filament fluorescence images with a spatio-temporal orientation analysis method. *Biophys. J.* **78:** 2709–2715.

Vazquez J., Belmont A.S., and Sedat J.W. 2001. Multiple regimes of constrained chromosome motion are regulated in the interphase *Drosophila* nucleus. *Curr. Biol.* **11:** 1227–1239.

Vincent L. 1993. Grayscale area openings and closings, their efficient implementation and applications. In *Proceedings of 1st Workshop on Mathematical Morphology and Its Applications to Signal Processing* (ed. J. Serra and Ph. Salembier), pp. 22–27. Barcelona, Spain.

Watt A. and Watt M., eds. 1992. *Advanced animation and rendering techniques*, Addison-Wesley, New York.

Wertheimer M. 1923. Untersuchungen zur lehre der gestalt, II. *Psychol. Forsch.* **4:** 301–350.

White J. and Stelzer E. 1999. Photobleaching GFP reveals protein dynamics inside live cells. *Trends Cell Biol.* **9:** 61–66.

Xu C., Pham D.L., and Prince J.L. 2000. Medical image segmentation using deformable models. In *SPIE Handbook of medical imaging: Medical image processing and analysis* (ed. J.M. Fitzpatrick and M. Sonka) vol. 2, pp. 129–174. SPIE, Wellingham, Washington.

Zhu S.C. and Yuille A.L. 1996. Region competition: Unifying snakes, region growing, and bayes/mdl for multiband image segmentation. *IEEE Trans. PAMI* **18:** 884–900.

SECTION 2

Imaging of Live Cells and Organisms

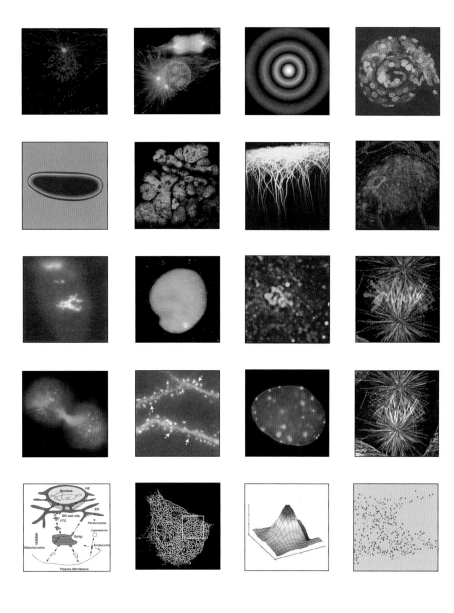

CHAPTER SEVENTEEN

In Vivo Imaging of Mammalian Cells

Jason R. Swedlow and Paul D. Andrews

Division of Gene Regulation and Expression, MSI/WTB Complex,
University of Dundee, Dundee DD1 5EH, Scotland

Melpomeni Platani

Gene Expression Programme, European Molecular Biology Laboratory,
Meyerhofstrasse 1, D-69117 Heidelberg, Germany

INTRODUCTION

THERE ARE A NUMBER OF INSTANCES IN WHICH THE USE OF LIVE CELL IMAGING has provided critical insight into cellular and tissue function. As such, it has become a requisite analytical tool for use in cell biology, neurobiology, and developmental biology as well as a routine method practiced in many biomedical research laboratories.

The technical requirements for performing live cell imaging include the appropriate cells for the imaging experiment as well as the necessary digital image acquisition system. The availability of fluorescent protein (FP) technology allows the molecular specificity of fluorescent markers to be used in a genetically encoded manner (Tsien 1998). Moreover, it is now possible to obtain commercial turnkey systems for digital imaging using a number of different imaging modalities.

The theoretical barriers to performing routine live cell imaging have largely been removed, but there remains a large number of technical considerations for carrying out these experiments. The goal of this chapter is to describe many of the methods and considerations for performing a successful imaging experiment in living cells. This chapter does not, however, provide a single recipe for success: The approach is much too empirical and depends on careful observation of the cells under study. Instead, we describe the methods and considerations we use to perform these experiments. We focus our discussion on the use of fluorescence microscopy in live cell imaging, although most of the points covered are relevant to any type of imaging.

CELL ENGINEERING FOR LIVE CELL IMAGING

A wide range of fluorescent reporters is now available for use in live cell microscopy. These reporters can be fused to the gene encoding the protein of interest (or fluorescent probes may be used to label the protein directly). The constructs are then introduced directly into the cell and used to track expression of the protein and its localization.

Functionality of Fluorescent Proteins

The introduction of green fluorescent protein (GFP) has driven the revolution in modern cell biology (Tsien 1998). GFP fusions can be easily generated by standard recombinant techniques and these fusions act as faithful reporters of the fusion's expression and localization (Chalfie et al. 1994). Nonetheless, as with all tags and fusions, it is critical to ascertain whether the particular FP fusion is functional within the living cell (see Chapters 1 and 2 for details).

We also note the continued use of small molecule fluorophores for live cell imaging. In some cases, these fluorophores have specific molecular characteristics that make them specific reporters for cellular function. Membrane-specific probes (such as $DiOC_6$) and ion-sensitive probes (i.e., Fura, Indo, etc.) provide methods for labeling single cells and thick tissues.

Transfection Strategies

A wide variety of schemes are available for introducing FP fusions into cells in preparation for live cell imaging. In general, cells or tissues can be transiently transfected with plasmids bearing the FP fusions of interest and then examined at a later time. The inherent convenience and flexibility of transient transfections, especially for examining the behavior of various domains or mutants, often make them the appropriate method. In addition, it is possible to approximate a pulse-chase experiment using transient transfection, especially when the plasmid is introduced by microinjection (Sleeman and Lamond 1999). It is critical to note, however, that transient expression of FP fusions is necessarily an overexpression technique. For cases in which the localization of the FP fusion depends on other cellular factors or is a component of a multiprotein complex, the observed localization may differ from that of the endogenous protein. It is therefore advantageous to consider generating a cell line in which the FP fusion is stably expressed at relatively low, defined levels. Selection for the expression plasmid is achieved by coexpression of a resistance marker (e.g., G418, puromycin, etc.). Once a stable cell clone is obtained, the expression of the FP fusion can be biochemically analyzed and characterized. This strategy can help demonstrate the functional properties of the FP fusion. Where possible, one should confirm the ability of the FP fusion to rescue a cell bearing a mutation or deletion in the endogenous gene (see Chapter 19). An alternative strategy involves designing an expression system that employs a regulatable expression cassette, so that the levels of expression of the fusion can be altered during an experiment. A number of these cassettes are commercially available (e.g., Tetracycline System, BD Biosciences Clontech, Palo Alto, CA) and can be of use when expression of the fusion must be manipulated during an experiment or when high levels of expression are not tolerated for long periods.

STRATEGIES FOR MAINTAINING CELL VIABILITY AND HEALTH DURING IMAGING

Environmental control is critical for a successful live cell imaging experiment. Aspects of the environment that may be manipulated include the selection of the chamber in which the cells are grown; the temperature within the chamber; and the growth conditions, including the nature of the media, the osmolality, etc. These are discussed in detail in the sections below.

Chambers for Live Cell Imaging

Environmental chambers provide a mechanism for viewing the cells on the microscope stage and keeping them close to optimal growth conditions. In general, chambers include a glass window, usually the thickness of a coverslip, through which the cells are imaged. Temperature control can be achieved using an external source of heated air, a metal heat sink under thermistor control coupled directly to the chamber, or thin coatings of metal oxides applied to a coverslip to directly heat the cells. Chambers are usually either "open" to the atmosphere and therefore generally more accessible or "closed" and therefore often sealed to protect cells from evaporation, etc. (see below). An open chamber will usually allow access to the cells within it, thus permitting microinjection or other manipulations to the cells to be easily carried out before the experiment. Closed chambers usually make microinjection difficult, but most closed chambers include ports that permit the addition of a fresh medium or drugs, etc., during the imaging experiment, usually through a gravity-driven manifold or a pump. When new solutions are added to a chamber, it is critical that before addition they be equilibrated to the same temperature as the cells. Furthermore, many cells are sensitive to shear, so perfusion of cells should be performed at low flow rates.

There are a number of chambers that may be used for different applications. The basic types of chambers include the following.

1. *Dish-based chambers.* This class of chambers essentially consists of a coverslip mounted on the bottom of a standard tissue culture dish (e.g., MatTek, Ashland, MA). These chambers are simple to use, but they are not tightly sealed, so it is important to consider the amount of medium evaporation occurring during an experiment. Finally, these do not include any heating system, so they are normally mounted on a microscope equipped with its own heating system. This type of chamber is usually easy to use and best suited for simple short-term visualization experiments.

2. *Simple sealed chambers.* A large number of designed and published small sealed chambers are suitable for live cell visualization. A complete list of these designs is beyond the scope of this chapter, however, some simple but very effective chambers are available (see Inoué and Spring 1997; EMBL 1998; Reider and Cole 1998; and those listed in Chapter 18 of this volume). All of these are sealed systems: When placed at the appropriate temperature, they can maintain cells in a healthy state for many hours.

3. *Integrated, commercial systems.* Many commercial systems are available that integrate a chamber to hold cells and allow control of temperature, gas exchange, and media perfusion (e.g., Bioptechs, Inc., Harvard Apparatus, Inc.). Versions for both open and closed configurations are available; many of these are designed for quite specific applications and it is therefore worthwhile to consider the various options available.

In our experience, most laboratories have favored chamber designs that reflect their experience using a given system and these preferences differ. The key point is that, in a given experiment, the chamber must maintain an optimal environment for cell function. Because the requirements of experiments differ, design preferences differ. As discussed below, the best approach is to test a number of different systems and identify the one that is most suited to the cells and the experiment.

Temperature Control

Cellular function is exquisitely sensitive to temperature—changes of even 2°C can have profound effects on cell physiology. A variety of methods are available for controlling the temperature of cells on the microscope stage. Many of the commercial systems described above include heating elements directly coupled to the chamber. This strategy provides a simple integrated solution, but limits the temperature control to the chamber itself. The whole microscope, especially the stage and objective lens, will act as a large heat sink, so it is often necessary to consider heating this as well; systems specifically designed for objective lens heating are commercially available (e.g., Bioptechs, Inc.). Alternatively, a large box can be built around the microscope and heated with warm air (e.g., Buck Scientific, Norwalk, CT). In this case, much of the microscope can be equilibrated to a single temperature. The advantage is the elimination of any movement resulting from thermal expansion of the microscope components. However, any air currents moving around the sample must be minimized. Access to the microscope and its components may be limited, so it may be worthwhile to construct the box from relatively inexpensive components if many modifications are to be made to the system.

A final consideration is the temperature control of the whole microscope. Even small changes of 1°C can cause elements in the optical train to move, resulting in focus or alignment shifts. Placing air conditioning ducts near the microscope will often cause local cooling and changes in focal position. For long-term observation, many investigators build a large thermostatically controlled box around the stage or even place the whole microscope into a room maintained at 37°C. The exact strategy used depends on the specific application, but it is absolutely critical to consider these issues when designing a live cell imaging system and the room in which it will be housed.

Cell and Media Conditions

Mammalian cells must be grown and maintained in specified conditions, usually in a defined medium supplemented with growth factors and/or animal serum, containing a buffering system to maintain the cells at an optimum pH. When cells are to be imaged for even short periods of time, these same medium conditions must be reproduced in the live cell chamber. It is critical to consider the following issues when choosing the medium for live cell imaging.

1. *pH.* All cell media use buffers to maintain the media at the optimal pH for cell growth. Many media depend on an atmosphere of 5% CO_2 (e.g., DMEM, etc.) for their pH buffering. If such media must be used during imaging, then an atmosphere of 5% CO_2 must be maintained. This can be achieved either by sealing the cell imaging chamber after purging it with a 5% CO_2 supply or by maintaining the chamber (and often the microscope body itself) in a 5% CO_2 atmosphere. Alternatively, it is possible to use media that do not depend on a controlled atmosphere. In the authors' laboratory, either a Liebovitz medium or a commercially supplied CO_2-independent medium (GIBCO-BRL) has been used. In all cases, the viability and growth rates of cells in these modified media are confirmed before use in imaging experiments.

An alternative strategy is to add HEPES to the medium (10–20 mM, pH 7.2). This is most commonly done to control the pH in CO_2-dependent media in the absence of a 5% CO_2 atmosphere. Although this is convenient, HEPES has been reported to be cytotoxic when exposed to light. In our experience, some HeLa cell lines are driven into apoptosis 8–12 hours after illumination in tissue culture grade HEPES. At the moment, the mechanistic basis for this is unknown but may in fact be caused by contaminants within commercial preparations of HEPES. Regardless of the method chosen, its effect on cell viability should be evaluated.

2. *pH indicators.* For convenience, most cell culture media are supplemented with a pH indicator such as phenolphthalein. These compounds are weakly fluorescent, but given the low levels of signal often available in living cells, they present significant sources of noise in live cell experiments. Therefore, media that are not supplemented with pH indicators should be used for fluorescence live cell imaging.

3. *Free radical scavengers.* Live cells are very sensitive to the production of free radicals during fluorescence excitation. Some investigators have supplemented their media for cell imaging with free radical scavengers such as ascorbate (vitamin C) or TroloX (a derivative of vitamin E). We have preferred to limit fluorescence excitation levels (e.g., short exposure times), but in cases where this strategy does not sufficiently protect cells, such compounds can be tried.

4. *Oxygen depletion systems.* Photodamage in live cell imaging often occurs through reactions with an oxygen free radical. It therefore follows that depleting oxygen should decrease photodamage. This strategy has been used by some researchers, but it is worth noting that lowering oxygen tension can be deleterious to cells: Hypoxic stress signaling pathways have been characterized in mammalian cells (Harris 2002). The most common strategy involves the use of a commercial oxygen depletion system (Oxyrase; http://www.oxyrase.com/). This product is a preparation of *E. coli* membranes and thus contains endotoxin, so it should only be used in situations in which responses to this component will not affect the assay. In cases in which the sample is not live cells but an in vitro reaction, a very common and effective approach is to use an enzymatic system based on a combination of catalase and glucose oxidase (e.g., Hyman and Mitchison 1993) that ensures depletion of soluble O_2 from the system.

5. *Osmolality.* The concentration of ions and nutrients in the live cell experiment will initially be set by the medium chosen for the experiment. However, live cell imaging chambers often have small volumes and even at room temperature will rapidly suffer evaporation, thus changing the osmolality of the medium (the problem is much worse when the medium is heated to 37°C). Therefore, special care must therefore be taken when assembling cells into chambers. In addition, during the imaging experiment evaporation must be minimized either by using a sealed system, by covering the medium in an open chamber with oil, or by humidifying the chamber during assembly.

Examination of Cells Before and During the Imaging Experiment

Once the cells are assembled in the chamber and the chamber is mounted on the microscope stage, it is usually necessary to visualize the cells to check their overall health and to identify appropriate cells for imaging, set the imaging parameters, etc. For example, cells that are highly vacuolated and/or exhibit swollen mitochondria and cytoplasmic blebs or are not well attached to the coverslip should not be pursued for data collection since they may not be healthy. Regardless of the imaging mode, cell exposure to light must be minimized. The sensitivity of cells to UV light is well known, but cells are at least equally sensitive to blue light around the wavelengths used for GFP and cyan fluorescent protein (Gorgidze et al. 1998; Ohara et al. 2002). For a fluorescence imaging experiment, all visualization of cells during the setup of the experiment must be performed under as low a light level as possible and as quickly as possible. This can be achieved by inserting neutral density filters in the excitation light path. Ideally, the digital imaging system should be used for all fluorescence images during setup. Visualization by eye usually takes at least 1–2 seconds, which is at least tenfold longer than is often required to get an image of sufficient quality for cell selection, focusing, etc. Alternatively, cells can be found using either bright field or differential interference contrast (DIC) imaging modes (it is wise to use a green or interference filter to improve contrast and minimize blue light). In general, at all points of the experiment, it is recommended that only enough light be used to just see the cells (see further discussion below). This minimizes any chance of inducing photodamage in the cells.

DESIGN OF THE ACQUISITION SYSTEM: MAXIMIZING SIGNAL TO NOISE

Live cell imaging usually requires a consideration of the relative requirements for resolution and the need to maximize the signal-to-noise ratio (S:N) in the recorded data. Considered here are sources of noise and how they may be minimized. Next, the various elements in the optical path are discussed, with suggestions for how they may be optimally configured to minimize noise and thereby maximize signal.

Sources of Noise

In imaging there are four major sources of noise: detector noise, illumination noise, poisson or shot noise, and stray light, each considered in detail below. In general, reducing systematic sources of noise by carefully choosing detectors and optimizing illumination systems can improve the S:N.

1. *Detector noise.* Every photon detector introduces noise into every measurement it makes. The exact details of detector noise are specific to each type of detector. In scanning microscopes (including laser scanning confocal microscopes [LSCMs] and multiphoton microscopes [MPMs]), the photomultiplier tube (PMT) can produce spurious electrons within the signal amplification system (Art 1995). In wide-field microscopes (WFMs) and spinning disk confocal microscopes, the charge-coupled device (CCD) has a background dark current associated with each pixel. Cooling the CCD and other improvements in CCD design have reduced the contribution of dark current to very low levels (Inoué and Spring 1997). However, reading each pixel from the CCD and converting the analog signal to a digital value contribute an additional noise component—the major noise component in the current generation of cooled cameras. This "read noise" is available on the data sheet for any camera (also usually available online, e.g., http://www.roperscientific.com/life_sciences.html and http://www.hamamatsu. com). For live cell imaging using fluorescence, we usually use cooled CCDs with read noise values below 6 e^- RMS.

2. *Illumination noise.* Any digital imaging experiment assumes that the illumination is constant across the sample field and between different images. In most imaging systems, the optical properties across the sample field are usually fairly constant (spatial invariance).

However, in scanning microscopes, there may be differences in the illumination dose delivered between different pixels (Zucker and Price 2001; Swedlow et al. 2002). In WFMs, the illumination field can be made fairly uniform by placing an optical fiber between the light source and the microscope illumination path (Kam et al. 1993), and any remaining illumination gradient can be corrected computationally by using a flat field correction based on a uniform fluorescent sample (Chen et al. 1995). Nonetheless, an optical fiber does not correct any temporal variations in lamp output. Fluctuations of up to 10% from the mean are common, but they can be corrected using stabilized power supplies and/or by measuring power deviations and correcting them (Chen et al. 1995).

3. *Poisson or shot noise.* Photon measurement is inherently a statistical process, with an uncertainty associated with signal detection (Art 1995). For a measurement of N photons, the uncertainty is $(N)^{1/2}$, so the S:N is given by $(N)^{1/2}/N$, or $(N)^{1/2}$. Maximizing N leads to an improvement in the S:N. A simple way to improve the S:N then is to collect more signal, usually by taking longer exposures or by increasing the amount of excitation light. Unfortunately, this often is not possible, since these measures usually increase cell photo damage.

CCDs can be used to circumvent this problem using a technique known as "binning." Normally, the pixels in a CCD camera are read by transferring a row of pixels into a read register. Each pixel in the read register is then individually read (Abramowitz et al. 2000). In a binned image, each signal from a square group of pixels (2×2, 3×3, etc.) on the CCD is read out together and assigned to a single pixel (Abramowitz and Davidson 2000). This is achieved by transferring multiple rows into the read register (two rows in 2×2 binning, three rows in 3×3 binning, etc.) and then reading groups of 2, 3, etc. pixels together. For 2×2 binning, there is a twofold loss in resolution, a fourfold increase in signal, and a twofold improvement in the S:N. This improvement can be significant, especially in signal-limited applications. A key point in binning is that the addition of read noise occurs after the binning process, so even though 2×2, 3×3, etc. pixels are being combined, there is only one read event. If a 2×2 box of pixels were computationally averaged after data collection, the result would be inferior to binning because of the noise contribution to each of the four pixels. In summary, binning may sacrifice some spatial resolution, but it provides a significant increase in the S:N, especially when performing experiments on photosensitive cells that require low light levels and short exposures.

4. *Stray light or spurious photons.* Recent interest has developed in very high-resolution, high-sensitivity fluorescence imaging in which single fluorescent molecules are detected and tracked over time (Chapter 34; Vale et al. 1996). This technique requires sensitive detection systems, usually a CCD camera fitted with an intensifier (Inoué and Spring 1997) or more recently, an electron multiplying CCD camera. In fact, the key to making these approaches work has been reducing out-of-focus light using evanescent wave fluorescence excitation and reducing any sources of stray light in the microscope, including those caused by dirty optical elements (Funatsu et al. 1997). An important lesson using this technique for all signal-limited imaging is the importance of removing sources of noise in the optical path.

Microscope Optics

Another aspect of optimizing the S:N is the collection and transfer of photons by the microscope optics. Any of the elements in the optical path can degrade the S:N. It is therefore critical to consider the following elements in a microscope designed for live cell imaging.

1. *Objective lens numerical aperture.* For fluorescence imaging, it is common to use objective lenses with very high apertures to maximize the amount of light that is collected from the sample. The aperture width is given by the numerical aperture (NA) and is written on the side of all objective lenses. The equation $NA = \eta\sin(\alpha/2)$ defines the NA in terms of η, the refractive index of the medium in front of the lens, and α, the acceptance angle of the lens aperture (for further details, see Inoué and Spring 1997). The resolution of an image recorded with a given objective lens is a function of the NA (Inoué and Spring 1997; Wallace et al. 2001). For fluorescence imaging, however, the light gathering power, and therefore the signal recorded by an objective lens, is proportional to NA^4, so a small increase in NA can yield a significant improvement in signal. It is for this reason that live cell fluorescence imaging often requires high NA lenses and high refractive index immersion media (e.g., oil, $\eta = 1.51$ or water, $\eta = 1.33$).

2. *Objective lens magnification.* The brightness of the signal is inversely proportional to the square of the magnification, so it is sometimes worthwhile to consider using a lower magnification objective when imaging dim samples. For example, if the S:N were limiting in a sample imaged with a 100x 1.4NA lens, changing to a 60x 1.4NA, 63x 1.4NA, or even a 40x 1.3NA objective lens would likely provide a significant improvement in the S:N. The loss of resolution will be small compared to the improvement in the S:N. It is always worthwhile to consider magnification and binning together to assess the final magnification necessary for a given experiment.

3. *Objective lens correction.* Objective lens manufacturers offer a number of different types of lenses. In general, each lens is designed for a limited series of applications; it is advisable to try matching the design principles to the correct application. For example, Olympus, Inc. manufactures a 100x plan-apochromat (PlanApo) 1.4NA objective, a good high-resolution lens. The specification for this lens, however, includes a large amount of chromatic correction so that it can be used for transmitted light techniques including DIC. The result is that light throughput is sacrificed for chromatic correction. This lens has not worked well for us for live cell imaging—its resolution is exceptional, but it does not perform well in signal-limited applications. In contrast, Olympus also makes a 100x UPlanApo 1.35NA lens that is not as well corrected (we have detected a small amount of axial chromatic aberration in this lens) but it has significantly higher light throughput, even with a lower NA, because of fewer correcting elements within the lens. Because we rarely attempt high-resolution colocalization in live cell experiments, we sacrifice a small amount of color correction for a significantly improved S:N. Finally, some objective lenses include specific optical elements (such as phase retardation plates for phase contrast optics) that significantly reduce light throughput. This reduction is rarely a problem in phase contrast, but it can significantly reduce the signal in fluorescence. For this reason, we rarely use phase objectives for live cell imaging.

4. *Fluorescence filter sets.* For fluorescence imaging, interference filters and chromatic mirrors are used to select appropriate wavelengths of light for the chosen fluorophores (Flynn et al. 1998). Because successful live cell imaging depends on the S:N, the choice of filters and chromatic mirrors is absolutely critical for a successful experiment. It is now common to use band-pass filters (Abramowitz and Davidson 1998), rather than simple long-pass filters, to select the fluorophore emission. Band-pass filters are available from Chroma Technology, Inc. and Omega Optical, Inc. for almost any fluorophore or combination of fluorophores. Where possible, choose these for the specific fluorophore(s) being used. For example, using an FITC filter set designed for use in a multifluorophore

experiment for imaging a living cell labeled with only GFP will needlessly sacrifice some GFP fluorescence that would help improve the S:N. This is because the filter set is more restrictive than the experiment warrants. Simply choosing the correct filter set can make a significant difference in signal detection.

5. *Aberrations that reduce the S:N.* Any defects in the optical path will affect the S:N in the final image (Inoué and Spring 1997). Spherical aberration is the most common aberration in imaging of living cells with high NA lenses. This arises from a mismatch between the cells' medium (usually an aqueous medium, $v = 1.33$) and the refractive index of the immersion medium. Spherical aberration manifests itself as a misfocusing of the signal, such that objects are significantly elongated along the optical axis (Wallace et al. 2001). Since the signal is spread over a larger volume, the S:N is reduced. This problem can be largely eliminated by using a water immersion lens. Alternatively, the refractive index of the immersion medium or the thickness of the coverslip can be adjusted (Hiraoka et al. 1990). The refractive index changes with temperature, so the optimal combination of immersion medium and coverslip will differ between room temperature and 37ºC.

IMAGE ACQUISITION: KEEPING THE CELLS ALIVE AND GETTING THE DATA

Once the cells are appropriately engineered and assembled into an environmental chamber, and the microscope is appropriately configured, one can record image data from the living cells. Deciding on the correct imaging parameters will determine the success of the experiment.

2D Versus 3D Imaging

All cells are three dimensional (3D). Imaging with high NA lenses produces images with a relatively narrow depth of field (~700 nm for a 1.4NA lens), so any single image may only sample a small section of the cell. For some samples (e.g., the leading edge of a migrating cell), imaging at a single focal plane ("optical section") may adequately cover the sample. In other cases (e.g., in the cell nucleus), the object is much thicker and only a portion of an object will be imaged in a given focal plane. Moreover, postprocessing methods such as deconvolution usually require sampling of out-of-focus information to generate an accurate reconstruction of the object (Chapter 15; Swedlow et al. 1997; McNally et al. 1999; Wallace et al. 2001). These methods can significantly improve contrast, even in images from live cells with a poor S:N (Swedlow et al. 1993). However, recording a series of optical sections subjects cells to significantly more light at each time point. In addition, recording extra images takes time and may significantly slow data collection. This may mean that time points cannot be taken rapidly enough or even that cellular components move during optical sectioning.

Photodamage

During the imaging experiment it is critical to keep light exposure as low as possible. In general, this means reducing illumination dose using neutral density filters, simply reducing the power of the illumination source, or keeping the time of exposure to fluorescence excitation light as short as possible using shutters or other devices.

Cells are intrinsically photosensitive, and adding fluorophores only exacerbates this sensitivity. On excitation of a fluorophore by the absorption of a photon (Flynn et al. 2003), the fluorophore remains in an excited state ("S_1"), usually for a few nanoseconds. "Good" fluorophores will decay back to the ground state by releasing a photon—the fluorescence emission. An S_1 fluorophore, however, can convert to an alternative triplet state ("T_1") that is highly reactive and can cause the formation of free radicals, especially O_2^-. These reactive free radicals can chemically modify cellular substituents, thus damaging the cell. The generation of free radicals and subsequent "photodamage" always occurs during fluorescence excitation and can only be limited, not prevented. Cells have intrinsic enzymatic mechanisms for converting free radicals to less harmful compounds, so as long as these systems are not saturated, cells tolerate fluorescence excitation. In practice, this means that low levels of excitation must always be used during live cell imaging experiments.

Choosing the exact light attenuation and exposure time is almost always an empirical art. The best strategy for a new experiment for which the correct parameters are unknown is to attenuate the light as much as possible (e.g., use a neutral density [ND] filter with an optical density of 1.0) and to use short exposure times (<100 msec) such that the structures are visible in the acquired image, but only barely so. If the cells tolerate this light level through a time-lapse experiment, then the light can be increased in subsequent experiments until a workable compromise is achieved between the S:N and cell viability. It is important to note that there is often a nonlinear relationship between the amount of total light exposure a cell can tolerate and the length of individual exposure times. In our experience, we have found that cells are healthiest when exposed to very brief pulses of light, since extended exposures (>0.5 s) are often lethal to the cell. In one cell line, for example, attenuating the fluorescence excitation with an ND 1.0 filter and using 100-ms exposures was well tolerated as long as there was ~0.5 s between individual exposures. Increasing the rate of exposure, using otherwise identical imaging parameters to 7/second, caused vesiculation and blebbing. Although the cause of this behavior is unknown, we assume that after each exposure to fluorescence excitation, cellular antioxidant enzymes must reduce the level of free radicals and this process takes time. Rapid or long exposures exceeded this limit and thus produced a significant, and in this case unacceptable, dose of free radicals inside the cell.

Variability in Behavior Between Different Cells

Many live cell experiments are performed on one or just a few cells. It is critical to note that cells in culture can be quite heterogeneous and present a wide variety of phenotypes. This heterogeneity can be the result of cells in different positions in the cell cycle or possibly intrinsic differences among cells. A recent study of transcription activation in a serum starvation assay highlighted the wide variety of responses among individual cells, even though large differences between control and treatments were scored in a population assay such as an immunoblot (Levsky et al. 2002). For this reason, it is often necessary to record data from many individual cells to gain a statistically significant sampling of cellular behavior and dynamics. Collecting such large data sets can be laborious. In our laboratory, we use a DeltaVision restoration microscope (Applied Precision, LLC) fitted with a 3D motorized stage to record time-lapse images of multiple cells in a single data collection run. This approach has provided enough data to perform statistically valid quantitative analyses of intranuclear dynamics in vivo (Platani et al. 2002). In total, we collected some 350 time-lapse images of GFP labeled cells under a variety of conditions. Note that this approach generates large amounts of data that challenge standard storage and analysis methods. As a result, we recently developed informatics tools designed to store and analyze these vast assemblies of image data (Swedlow et al. 2003).

Choice of Acquisition System

A wide variety of digital imaging systems is now available for live cell imaging. The type of system used depends largely on the type of sample and the kind of experiment to be performed (see Chapter 15). Most systems will behave as reasonably linear signal detection systems, so they can be used for quantitative imaging. WFMs tend to have relatively low intrinsic noise levels and work well with weakly fluorescent samples (Swedlow et al. 2002). For thick samples (i.e., ≥ 30 μm), however, the amount of out-of-focus light recorded in each focal plane dominates any signal, and better performance is achieved with an LSCM (Inoué 1995; Parry-Hill et al. 2003). An alternative approach is to use a spinning disk confocal microscope (SDCM), where an array of illumination beams rapidly scans the sample and produces a confocal image (Maddox et al. 2003). For very thick samples (i.e., ≥ 100 μm), the absorption and scatter from the samples become limiting and multiphoton microscopes are the most appropriate system (Helmchen and Denk 2002; Piston and Davidson 2003). The different illumination patterns in these systems are illustrated in Figure 1. In general, it is necessary to try a few of these systems to ascertain the correct approach for a given sample. With the proliferation of these systems in the last few years, it is often possible to visit a nearby laboratory to test whether a specific system works well for a given sample.

Rate of Acquisition

The biological process under study determines how quickly data must be acquired. Ca^{2+} sparks occur over milliseconds, vesicle movement over seconds, mitosis over minutes, and apoptosis over hours. Sampling these different timescales appropriately will require different hardware and software. Before purchasing an imaging system, it is critical to match experimental requirements to the capabilities of the system and to demonstrate that the system can collect meaningful data on an appropriate timescale.

EVALUATING THE RESULTS

Establishing Criteria for Cell Viability

In any live cell imaging experiment the imaging conditions must be minimally perturbing to the cell and must not significantly affect the experimental result. Most critically, criteria for the health of the cells under study must be developed so that the success of the experiment can be evaluated not only on whether a given result was achieved, but also on whether the imaging process itself did not damage the cells. The exact criteria used will vary depending on the experiment. Most obviously, the process under study should occur to the same degree and with the same kinetics as that observed in fixed cells. It is, however, also important that the conditions used for the experiment (medium, buffering strategy, and chamber) do not substantially change growth rates or mitotic or apoptotic indices. The effects of each of these components should be separately assessed. For example, cells can be grown in the medium used for the imaging experiment in a standard incubator and assessed for viability, mitotic index, etc. In addition, cells can be installed in the environmental chamber for a number of hours and then similarly assessed, in the absence of imaging. This strategy isolates the various contributions to cell health and identifies any that should be modified.

During an imaging experiment, cells are exposed to high doses of illumination and, in fluorescence, the generation of reactive species (see above). It is therefore wise to assume that at least some damage has occurred during experiment and to assess whether it is sig-

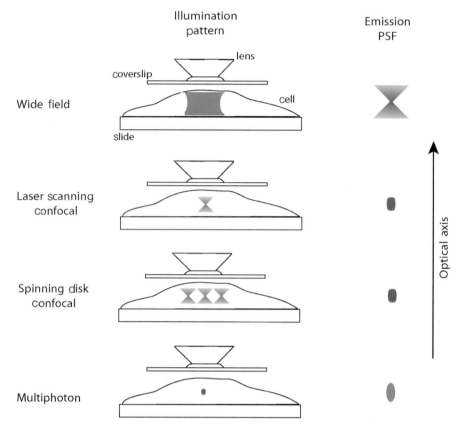

FIGURE 1. Illumination and detection in scanning microscopy and the WFM. The WFM, LSCM, SDCM, and MPM use different illumination and detection strategies to form an image. The diagram shows a schematic of an objective lens, coverslip, cell, and slide in a typical layout (objects are not to scale). In the center of the figure, the volume of fluorescence excitation for each method is shown. The right side of the figure shows the distribution of fluorescence emission from each point source in the sample, observed with the optical axis running vertically on the page. Note that for clarity, the right side of the figure has been magnified. The WFM illuminates throughout the field as well as above and below the focal plane. Each point source is spread into a shape resembling a double-inverted cone known as the point-spread function (Inoué and Spring 1997). Only the central part of this shape is in focus; the rest is in blurred out of focus light. In a real sample filled with point sources, blurred light degrades the image. In contrast, the LSCM, SDCM and MPM scan the image with a focused laser (Inoué 1995; Helmchen and Denk 2002; Parry-Hill et al. 2003; Piston and Davidson 2003). The pattern of excitation is a PSF, but a pinhole in the optical path in the LSCM and SDCM prevents any out-of-focus fluorescence from hitting the photomultiplier tube detector; therefore, only in-focus light is recorded. The LSCM has a single pinhole and a single focused laser spot that is scanned across the specimen. In the SDCM, an array of holes fitted with microlenses are placed on a spinning disk such that the holes rapidly sweep over the specimen and create an image recorded with an area camera (Maddox et al. 2003). In the MPM, the place at which photon flux is high enough to excite fluorophores with more than one photon is at the in-focus position of the PSF. Thus, fluorophore excitation only occurs in focus. Because all fluorescence emanates from in-focus fluorophores, no pinhole is needed and the emitted fluorescence generates an in-focus image. (Adapted, with permission, from Andrews et al. 2002.)

nificant enough to affect basic cellular processes. To do this, after the completion of the imaging experiment, leave the cells in the chamber on the microscope stage and examine them over a number of hours to determine if they initiate apoptosis or enter and complete mitosis. This is most easily done by taking time-lapse images at widely spaced time intervals (10–20 mins) over a number of hours. These criteria depend on the appropriate pathways present in the cells used; some mammalian cells have defunct cell cycle checkpoints, so one should consider the appropriateness of mitotic and apoptotic progression as cell health criteria in a given cell type.

Finally, after an imaging experiment, a very easy assay of cell health is to compare the cells that were imaged with a neighboring field that was not imaged. Do these two populations look comparable by phase or DIC? In the case of fluorescence imaging, is the final localization the same in both populations? If the cells are followed for many hours after the experiment, do the imaged cells behave similarly to the nonimaged ones? These comparisons often reveal whether significant damage has occurred to the cell(s) under study.

Examining Cells in Long-term Time Lapse

It is often the case that live cell experiments, especially in fluorescence, are performed over relatively short time periods, i.e., 30–60 min. While these studies will reveal exciting results, many of the cells' responses to damage, especially apoptosis, require many hours to develop a visible phenotype. Therefore, to ensure that cells are truly healthy, it is often advisable to image cells to perform an experiment and then simply follow the cells over the next 24 hours, preferably by phase or DIC (see Fig. 1). This time-lapse analysis can have relatively crude temporal resolution (e.g., one image every 15 minutes), but it will reveal whether the cells that have been imaged are reacting to the fluorescence experiment over the long term.

Monitoring Events in Fixed Cell Time Point Assays

Live cell imaging experiments can be extraordinarily powerful, but it is often very helpful to complement them, where possible, with fixed cell assays to help validate the live cell experiment. This may seem counterintuitive—live cell imaging assays are often held as a "gold standard" for cell and molecular dynamics. However, because of their technical difficulty and the risk of damaging cells during the experiment, demonstrating at least crudely similar kinetics and events in time points from a fixed cell assay can help validate the live cell assay. In some cases, this is simply not possible, either because the kinetics of an event are too fast or because the nature of an experiment (e.g., fluorescent speckle microscopy as a measurement of cytoskeletal dynamics [Chapter 12; Waterman-Storer and Danuser 2002]) cannot be meaningfully performed within the fixed cell. Moreover, the point of the live cell experiment is to reveal events or properties that are not observed or easily interpreted in fixed cells. Nonetheless, it is worthwhile to consider the use of this approach as a method of confirming events seen in longer time-lapse experiments to confirm the absence of effects of extended illumination.

CONCLUSIONS

Live cell imaging stands as a powerful technique for the analysis of molecular dynamics within cells. Continued advances in imaging techniques and probe design, many discussed in this volume, demonstrate the power of this approach and ensure its future as an important tool in modern biology. Nonetheless, the technical care and expertise required for a successful experiment is considerable. Taking care to maximize the S:N while minimizing damage to the cells under study will almost always result in a successful experiment.

ACKNOWLEDGMENTS

We thank members of Swedlow laboratory for helpful discussions. The authors are especially grateful to the academic and commercial faculty of the Analytical and Quantitative Light Microscopy and Optical Microscopy in the Biological Sciences Courses (Marine Biological Laboratory) and the In Situ Hybridization, Immunocytochemistry, and Live Cell Imaging Course (Cold Spring Harbor Laboratory) for helpful and stimulating discussions. Work in our laboratory is supported by grants from the Wellcome Trust and Cancer Research, UK. J.R.S is a Wellcome Trust Senior Research Fellow.

REFERENCES

Abramowitz M. and Davidson M.J. 1998. Optical Microscopy Primer, Olympus America and Florida State University. http://www.micro.magnet.fsu.edu/primer/techniques/fluorescence/filters.html

Abramowitz M. and Davidson M.W. 2000. Optical Microscopy Primer, Olympus America and Florida State University. http://www.micro.magnet.fsu.edu/primer/digitalimaging/concepts/binning.html

Abramowitz M., Tchourioukanov K.I., and Davidson M.J. 2000. Optical Microscopy Primer, Olympus America and Florida State University. http://www.micro.magnet.fsu.edu/primer/java/digitalimaging/ccd/shiftregister/index.html

Andrews P.D., Harper I.S., and Swedlow J.R. 2002. To 5D and beyond: Quantitative fluorescence microscopy in the postgenomic era. *Traffic* **3:** 29–36.

Art J. 1995. Photon detectors for confocal microscopy. In *Handbook of biological confocal microscopy* (ed. J.B. Pawley), pp. 183–196. Plenum Press, New York.

Chalfie M., Tu Y., Euskirchen G., Ward W.W., and Prasher D.C. 1994. Green fluorescent protein as a marker for gene expression. *Science* **263:** 802–805.

Chen H., Swedlow J.R., Grote M., Sedat J.W., and Agard D.A. 1995. The collection, processing, and display of digital three-dimensional images of biological specimens. In *Handbook of biological confocal microscopy* (ed. J. Pawley), pp. 197–210. Plenum Press, New York.

EMBL, Heidelberg. 1998. http://www.embl-heidelberg.de/~jwhite/microscopy/slide_chamber/

Flynn B.O., Tchourioukanov K.I., and Davidson M.J. 2003. Optical Microscopy Primer, Olympus America and Florida State University. http://www.micro.magnet.fsu.edu/primer/java/jablonski/lightandcolor/index.html

Flynn B.O., Long J.C., Parry-Hill M.J., and Davidson M.W. 1998. Optical Microscopy Primer, Olympus America and Florida State University. http://www.micro.magnet.fsu.edu/primer/techniques/fluorescence/fluorhome.html

Funatsu T., Harada Y., Higuchi H., Tokunaga M., Saito K., Ishii Y., Vale R.D., and Yanagida T. 1997. Imaging and nano-manipulation of single biomolecules. *Biophys. Chem.* **68:** 63–72.

Gorgidze L.A., Oshemkova S.A., and Vorobjev I.A. 1998. Blue light inhibits mitosis in tissue culture cells. *Biosci. Rep.* **18:** 215–224.

Harris A.L. 2002. Hypoxia—A key regulatory factor in tumour growth. *Nat. Rev. Cancer* **2:** 38–47.

Helmchen F. and Denk W. 2002. New developments in multiphoton microscopy. *Curr. Opin. Neurobiol.* **12:** 593–601.

Hiraoka Y., Sedat J.W., and Agard D.A. 1990. Determination of three-dimensional imaging properties of a light microscope system. *Biophys. J.* **57:** 325–333.

Hyman A.A. and Mitchison T.J. 1993. An assay for the activity of microtubule-based motors on the kinetochores of isolated Chinese hamster ovary chromosomes. *Methods Cell Biol.* **39:** 267–276.

Inoué S. 1995. Foundations of confocal scanned imaging in light microscopy. In *Handbook of biological confocal microscopy* (ed. J. Pawley), pp. 1–17. Plenum Press, New York.

Inoué S. and Spring K.R. 1997. *Video microscopy*. Plenum Press, New York.

Kam Z., Jones M.O., Chen H., Agard D.A., and Sedat J.W. 1993. Design and construction of an optimal illumination system for quantitative wide-field multi-dimensional microscopy. *Bioimaging* **1:** 71–81.

Levsky J.M., Shenoy S.M., Pezo R.C., and Singer R.H. 2002. Single-cell gene expression profiling. *Science* **297:** 836–840.

Maddox P.S., Moree B., Canman J.C., and Salmon E.D. 2003. Spinning disk confocal microscope system for rapid high-resolution, multimode, fluorescence speckle microscopy and green fluorescent protein imaging in living cells. *Methods Enzymol.* **360:** 597–617.

McNally J.G., Karpova T., Cooper J., and Conchello J.A. 1999. Three-dimensional imaging by deconvolution microscopy. *Methods* **19:** 373–385.

Ohara M., Kawashima Y., Katoh O., and Watanabe H. 2002. Blue light inhibits the growth of b16 melanoma cells. *Jpn. J. Cancer Res.* **93:** 551–558.

Parry-Hill M.J., Fellers T.J., and Davidson M.J. 2003. Optical Microscopy Primer, Olympus America and Florida State University. http://www.micro.magnet.fsu.edu/primer/techniques/confocal/index.html

Piston D.W. and Davidson M.J. 2003. Optical Microscopy Primer, Olympus America and Florida State University. http://www.micro.magnet.fsu.edu/primer/techniques/fluorescence/multiphoton/multiphotonintro.html

Platani M., Goldberg I., Lamond A.I., and Swedlow J.R. 2002. Cajal body dynamics and association with chromatin are ATP-dependent. *Nat. Cell Biol.* **4:** 502–508.

Rieder C.L. and Cole R.W. 1998. Perfusion chambers for high-resolution video light microscopic studies of vertebrate cell monolayers: Some considerations and a design. *Methods Cell Biol.* **56:** 253–275.

Sleeman J.E. and Lamond A.I. 1999. Newly assembled snRNPs associate with coiled bodies before speckles, suggesting a nuclear snRNP maturation pathway. *Curr. Biol.* **9:** 1065–1074.

Swedlow J.R., Sedat J.W., and Agard D.A. 1993. Multiple chromosomal populations of topoisomerase II detected in vivo by time-lapse, three-dimensional wide field microscopy. *Cell* **73:** 97–108.

———. 1997. Deconvolution in optical microscopy. In *Deconvolution of images and spectra* (ed. P.A. Jansson), pp. 284–309. Academic Press, New York.

Swedlow J.R., Goldberg I., Brauner E., and Sorger P.K. 2003. Informatics and quantitative analysis in biological imaging. *Science* **300:** 100–102.

Swedlow J.R., Hu K., Andrews P.D., Roos D.S., and Murray J.M. 2002. Measuring tubulin content in Toxoplasma gondii: A comparison of laser-scanning confocal and wide-field fluorescence microscopy. *Proc. Natl. Acad. Sci.* **99:** 2014–2019.

Tsien R.Y. 1998. The green fluorescent protein. *Annu. Rev. Biochem.* **67:** 509–544.

Vale R.D., Funatsu T., Pierce D.W., Romberg L., Harada Y., and Yanagida T. 1996. Direct observation of single kinesin molecules moving along microtubules. *Nature* **380:** 451–453.

Wallace W., Schaefer L.H., and Swedlow J.R. 2001. A workingperson's guide to deconvolution in light microscopy. *BioTechniques* **31:** 1076–1097.

Waterman-Storer C.M. and Danuser G. 2002. New directions for fluorescent speckle microscopy. *Curr. Biol.* **12:** R633–R640.

Zucker R.M. and Price O. 2001. Evaluation of confocal microscopy system performance. *Cytometry* **44:** 273–294.

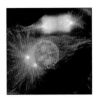

A Sealed Preparation for Long-term Observations of Cultured Cells

Greenfield Sluder, Joshua J. Nordberg, and Frederick J. Miller

Department of Cell Biology, University of Massachusetts Medical School, Worcester, Massachusetts 01605

Edward H. Hinchcliffe

Department of Biological Sciences, University of Notre Dame, Notre Dame, Indiana 46556

INTRODUCTION

THE CONTINUOUS LONG-TERM OBSERVATION OF CULTURED CELLS on the microscope has always been a technically demanding undertaking. Traditionally, this has involved the use of a bulky or inconvenient apparatus, such as open-dish preparations requiring control of environmental CO_2/humidity levels and perfusion chambers (for review, see Focht 1998; Rieder and Cole 1999). Even traditional sealed chambers, such as Rose chambers, require periodic changes of the culture medium using syringes. This chapter describes a sealed preparation that allows the continuous long-term observation of cultured mammalian cells on upright or inverted microscopes without environmental CO_2 control (see Movie 18.1 on the accompanying DVD). The preparation allows for optical conditions consistent with high-quality imaging and good cell viability for at least 100 hours (Hinchcliffe et al. 2001).

The preparation is an aluminum support slide with a square aperture cut in its center. The coverslip bearing the cells is attached to the top of the slide with a thin layer of silicone grease, and the bottom of the slide is similarly covered with a clean coverslip of the same size. The thickness of the slide is intended to coordinately maximize the volume of the medium while maintaining optical properties that allow Koehler illumination with standard condensers. The chamber is filled in equal parts with HEPES-buffered media containing fetal calf serum and a low-viscosity fluorocarbon oil. These oils have a high solubility for atmospheric gasses. For FC43, the solubility of air is 26 ml/100 ml, and for FC40 it is 27 ml/100 ml, compared to the solubility of air in distilled H_2O, which is 1.9 ml/100 ml (see the 3M Fluorinert product guide, 1997). The inclusion of the oil in the preparations is intended to provide a source of oxygen and perhaps a sink for some of the CO_2 produced by the cells. Although the inclusion of fluorocarbon oil in the preparation may not be necessary for short-term (~24 hours) observations, particularly with cells that are sparsely plated, long-term cell viability is assured when the oil is present.

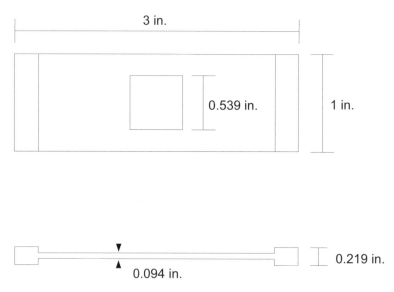

FIGURE 1. Dimensions of the aluminum support slide.

DETAILS OF SUPPORT SLIDE CONSTRUCTION

The aluminum support slide can be fabricated by a professional machine shop. The dimensions of the slide are shown in Figure 1; it consists of a 1 × 3-inch aluminum slide 3-mm thick, with a square hole cut in its center and sized to allow a 22 × 22-mm coverslip to slightly overlap the margins of the hole. The ridges going across the ends of the slide on both the top and bottom are necessary to provide clearance for the top and bottom coverslips so that they do not drag on the stage. These ridges are put on both top and bottom surfaces so that a preparation can be moved between an upright and an inverted microscope without having to be rebuilt. However, there is one compromise with this chamber design. When used with upright microscopes for transillumination observations (e.g., phase contrast or differential interference contrast), the cells sit slightly higher above the stage than they would on a conventional glass slide. For upright microscopes, this may make it difficult to achieve perfect Koehler illumination for transmitted light with high-numerical-aperture (NA) condensers, because the condenser travel is often limited by the microscope manufacturer to the level of the stage. To avoid this difficulty, we use one of two strategies. First, we use long-working-length condensers with NAs of 0.6 or less. Most of our work with live cells utilizes objectives with magnifications of 40× or less, so the condenser numerical aperture has not been limiting. Second, when we want to use a condenser of high NA, we remove the setscrew above the condenser focus dovetail to allow the condenser to come slightly above the stage.

DETAILS OF PREPARATION, CONSTRUCTION, AND USE OF THE SEALED CULTURE CHAMBER

1. Clean the coverslips used in the construction of this preparation prior to use by sonicating them in distilled H_2O containing a small amount of detergent and then rinsing the coverslips several times in distilled H_2O (Sluder et al. 1999). Store cleaned coverslips in jars containing 95% ethanol.

Although the utility of these preparations is not sensitive to coverslip thickness, use number 1.5 (i.e., 0.17-mm thick) coverslips, because microscope objectives (without coverslip correction collars) are designed for this thickness of glass. Use of number 1 or number 2 coverslips introduces spherical aberration that degrades image quality.

2. To prepare the coverslips for use in growing cells, pass each coverslip through the flame of an alcohol burner to burn off excess ethanol.

3. To ensure a uniform and lasting seal, *do not* apply silicone grease to the margins of wet coverslips coming out of a tissue culture dish. Instead, grease the margins of the coverslips before the cells are grown on them by applying a thin and uniform coating of silicone vacuum grease ("High Vacuum Grease" from Dow Corning Corp.) to the margins of the coverslip with a small spatula.

4. Place the coverslips in a 100 × 20-mm tissue culture dish. To ensure sterility, place the dish with coverslips in a tissue culture hood and expose it to UV light for 10 minutes.

5. Following sterilization, plate the cells onto coverslips and culture them in ~10 ml of media. For our applications, we use the media appropriate for each cell type supplemented with 12.5 mM HEPES, 10% fetal calf serum, and a 1:100 dilution of the antibiotic/antifungal reagent sold by GIBCO (catalog number 15240-062).

6. Wipe the aluminum support slide with a tissue soaked with 70% ethanol and then briefly pass the slide through a flame to remove residual alcohol.

7. Use a small spatula to apply a thin rim of silicone grease around the top and bottom margins of the opening.

Use a thin even layer of silicone grease. Thicker layers, although they may work well, can lead to dimensional instability as the preparation "settles" at 37°C. This can lead to a constantly changing focus for the first few hours even though the microscope may have a stable focus mechanism.

8. Flame a cleaned blank coverslip (from Step 1) to remove the alcohol and attach the coverslip to the bottom of the support slide. Use the back of a pair of curved forceps to gently tamp the coverslip to ensure a good seal.

9. Place the prepared slide in a plastic culture dish and expose it to UV light for 10 minutes in a tissue culture hood.

10. Warm the media and fluorocarbon oil to 37°C in a water bath.

11. Use a sterile 1-ml pipette to fill the chamber halfway with oil (~350 μl).

12. Use a fresh sterile 1-ml pipette to add medium until the oil at the margins of the opening just barely overflows from the chamber (again, ~350 μl).

13. Remove a coverslip with cells from the culture dish and aspirate off excess medium. Quickly place the coverslip, silicone grease side down, on the chamber. Tamp the coverslip with forceps to ensure a good seal. Aspirate off any excess media and oil that may have flowed over onto the top of the coverslip.

14. Wash the top of the preparation prior to use to prevent salts from the medium forming crystals on the coverslips once they air dry.

15. Fill a 1-liter beaker with 37°C water and place a squirt bottle of distilled H_2O in it to warm.

16. When the water is warm, take the assembled chamber out of the incubator. Use a small spatula to apply a small amount of melted 1:1:1 Vaseline:lanolin:paraffin (VALAP) to the edges of the top coverslip to provide an extra seal. Use the squirt bottle to gently wash the top coverslip and aspirate off any excess water.

Keep the amount of VALAP used to a minimum to reduce the chance that any will catch on the objective when the preparation is in use. This soft waxy material is difficult to remove from optical surfaces.

17. Proceed with microscopy—the culture preparation is now ready for observation.

NOTES ON CELLS GROWN IN THESE PREPARATIONS

Although a survey of the suitability of this preparation for a wide variety of cultured cells has not been conducted, it has been used extensively in our laboratories for BSC-1, CV-1, COS-7, CHO, mouse embryo fibroblasts, and hTERT RPE1 cells. The FC47, FC40, FC43, and FC77 oils all work well for these cells, and no significant differences in the performance of the different oils for these cell types have been observed. Although we have not systematically explored the length of time cells can be grown in this preparation, most cells appear viable and have normal interphase morphology at 250 hours. Normally, observations of cells stop after 70–120 hours because by that time the cells have become confluent. Over the course of 100 hours, all of the cells types show constant motility, and mitoses continue for the duration of the observations with no noticeable prolongation of the cell cycle at later times. Note, however, that at later times, cells often develop small spherical inclusions that are phase-bright. The identity of these inclusions remains obscure, but they may be large pinocytotic vesicles. In any case, they do not appear to have an adverse impact on cell motility, mitosis, or gross morphology.

ENVIRONMENTAL CONDITIONS

Since this preparation is sealed, control of environmental CO_2 is not needed; the preparations are designed to work under normal atmospheric conditions. However, the cells under observation need to be maintained at 37°C. In our laboratories, the entire microscope is enclosed in a box, and a proportional feedback control apparatus is used to blow warmed air into the enclosure. Cardboard boxes work well when configured so that the video camera and the mercury arc lamp (when present) are external to the enclosure. The oculars should project from the top edge of the box. A more elegant and user-friendly setup is a Plexiglas enclosure with sliding doors custom fabricated for the particular microscope. Again, the oculars, camera, and arc lamp housing should be located outside of the box. Three alternative heating strategies include placing the microscope in a 37°C room, enclosing the volume around the stage with a custom-built Plexiglas box, and warming the preparation alone with a temperature-controlled support apparatus on the stage. This last strategy suffers because there will be a temperature gradient from the margins of the preparation to the cells under observation. The gradient is particularly severe when a water or oil immersion objective is used, which acts as an efficient heat sink, unless the objective is equipped with a heated collar or other heating device.

SOURCE OF FLUOROCARBON OILS

The oils we use are manufactured by the 3M Corporation and are part of their Fluorinert series of performance liquids. These are short (primarily 8 carbon) hydrocarbons that are fully substituted with fluorine, and they are extremely inert, immiscible with water, and do not present any recognized significant health hazards. However, it is recommended that all users consult the Material Safety Data Sheets provided by the 3M Corporation for potential health hazards before using these oils. Parenthetically, we suggest that users do not carelessly dispose of any significant quantity of these oils, because they evaporate and are a "greenhouse gas." Some, such as FC77, have an extremely long life in the atmosphere (2300 years).

These oils are available from three manufacturer's representatives in three-quarter gallon (or 11 pounds) amounts for $30–55 per pound depending on the oil. The 3M Corporation does not directly sell anything but truckload quantities of these oils. Small sample quantities are available from Acuity Technical Sales (New Hampshire) 800-554-4905; AMS Materials (Florida) 904-230-0536; and Semitorr (Oregon) 503-682-7052.

MOVIE LEGEND

MOVIE 18.1. BSC-1 (monkey kidney epithelial) cells imaged by time-lapse video microscopy. The sequence shows an individual cell and its subsequent daughter cells undergoing three rounds of cell division (mitosis). The cells are imaged by phase contrast microscopy. Frames are captured every three minutes using a CCD camera coupled to a personal computer.

REFERENCES

3M Corporation. 1997. *Fluorinert liquids, product and contact guide.* Engineering Fluids and Systems, 3M Specialty Chemicals Division. St. Paul, Minnesota.

Focht D.C. 1998. Observation of live cells in the light microsocope. In *Cells: A laboratory manual.* Vol. II. *Light microscopy and cell structure* (ed. D.L. Spector et al.), pp. 75.1–75.15, Cold Spring Harbor Laboratory Press, Cold Spring Harbor, New York.

Hinchcliffe E.H., Miller F.J., Cham M., Khodjakov A., and Sluder G. 2001. Requirement of a centrosomal activity for cell cycle progression through G1 into S phase. *Science* **291:** 1547–1550.

Rieder C.L. and Cole R. 1999. Perfusion chambers for high-resolution light microscopic studies of vertebrate cell monolayers: Some considerations and design. *Methods Cell Biol.* **56:** 253–275.

Sluder G., Miller F.J., and Hinchcliffe E.H. 1999. Using sea urchin gametes for the study of mitosis. *Methods Cell Biol.* **61:** 439–472.

CHAPTER NINETEEN

Live Cell Imaging of Yeast

Daniel R. Rines,[a] Dominik Thomann,[b] Jonas F. Dorn,[b]
Paul Goodwin,[c] and Peter K. Sorger[a]

[a]Massachusetts Institute of Technology, Cambridge, Massachusetts 02139; [b]Bio Micro Metrics Group,
Laboratory for Biomechanics, Swiss Federal Institute of Technology, CH-8952 Schlieren,
Switzerland; [c]Applied Precision, LLC, Issaquah, Washington 98027

INTRODUCTION

THE DEVELOPMENT OF CLONING VECTORS for green fluorescent protein (GFP) and the simplicity of yeast reverse genetics allow straightforward labeling of yeast proteins in living cells. Budding and fission yeast are therefore attractive organisms in which to study dynamic cellular processes such as growth, cell division, and morphogenesis using live cell fluorescence microscopy. This chapter focuses on methods to culture, mount, and observe budding yeast cells using three-dimensional (3D) microscopy, but the methods are broadly applicable to other types of cells and other imaging techniques. The emphasis is on 3D imaging, because yeast cells are roughly spherical, and most organelles in yeast move in three dimensions. Three-dimensional imaging also makes it possible to apply image restoration methods (e.g., deconvolution) to obtain sharper images with better definition. This is important, because yeast cells are small (haploid *Saccharomyces cerevisiae* cells have a diameter of approximately 4–5 μm) relative to the resolution of even the best optical microscope (~0.25 μm).

Making a live cell movie involves trade-offs between the number of optical slices collected per time point, the brightness of each slice, the sampling frequency, and the movie's duration. Properties of the sample and limitations in the imaging system link these variables together in a complex fashion. One unavoidable limitation is the total amount of light to which a live yeast cell can be exposed without causing photobleaching or disrupting critical cellular processes. Upon extended illumination, chemical and GFP-based fluorophores become sufficiently bleached that they can no longer be detected above background. Moreover, cells are sensitive to light and will die, or arrest cell division if overexposed. Like other cell types, *S. cerevisiae* appears to be more sensitive to blue light (~420–480 nm) than to near-UV or infrared irradiation. This sensitivity creates a serious problem, because it overlaps the excitation wavelength for most variants of GFP. The practical impact of photobleaching and phototoxicity is that the investigator must limit the frequency of sampling, the number of slices per time point, and the duration of a movie so that the cells remain below their exposure limit.

351

Microscopes are subject to fundamental limitations in optical resolution, acquisition speed, and the signal-to-noise response, all of which are of practical interest in live cell experiments and must be balanced to achieve the highest-quality results. For instance, short exposures increase the sampling rate but decrease the signal-to-noise ratio (SNR) and effective image resolution. Long exposures improve SNR and resolution but increase photodamage and reduce the frequency of sampling. All microscopy, whether fixed or live cell, must begin with high-quality samples, accurately aligned optics, and mounting techniques that minimize aberrations in image formation. However, there is no single best way to set the exposure time, section thickness, duration, or other critical features of a live cell movie. Instead, the limitations of the sample and the instrument must be evaluated with respect to the goals of an individual experiment. For example, if a dynamic process is monitored with bright, 3D stacks collected every 5 seconds over a period of 5 minutes, the acquisition of such a movie would cause significant photobleaching. This would be a valid way to examine rapid processes in mitosis but would not allow the overall timing of cell division to be elucidated. For cell cycle studies, phototoxicity negatively affects cell growth. The intensity of illumination and frequency of sampling must therefore be minimized, with a consequent reduction in temporal resolution. This chapter briefly discusses the physics that underlie the limitations inherent in optical microscopy and the ways live cell imaging can be optimized within these limits.

SPATIAL AND TEMPORAL RESOLUTION

The two most frequent questions in microscopy concern resolution: What is the smallest object that can be detected? How close can two objects be and remain distinct in the image? The resolution of optical microscopes is limited by the diffraction of light as it passes through circular apertures in the objective lens and other optical elements. All microscopes are subject to this effect, which is typically referred to as the "diffraction limit." Light from a point source object, which is necessarily below the diffraction limit, is spread out to generate a blurred distribution known as the point spread function (PSF) (see Chapter 15). This distribution is narrower in the *x*-*y* plane than along the *z*-axis and its central portion resembles an elongated ellipse standing on end (Fig. 1A,B). Occasionally, the PSF will be referred to as the "impulse response" of the microscope

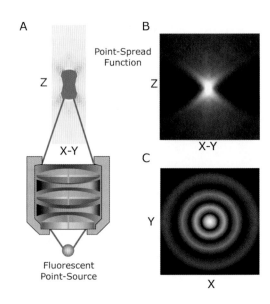

FIGURE 1. Point spread function (PSF). (*A*) Light collected from a fluorescent point source (e.g., a small bead) is smeared. The smearing effect is a well-understood phenomenon and can be modeled computationally (Thomann et al. 2002). (*B*) Complex pattern of smearing is commonly referred to as the PSF and is a characteristic of lens magnification, construction, immersion medium, coverglass, and any cellular components between the focal plane and lens. A perfect PSF will have symmetric cones of light above and below the point source. This effect is best observed by optical *z* sectioning through a single point source. Light from the *z* stack, when viewed directly from the side, generates the pattern or distribution shown here. (*C*) The large circular pattern represents a view of the PSF from above and is commonly referred to as an Airy disc. The signal intensity is represented by the density shading.

because it represents the output in response to a point source impulse or input. It turns out to be remarkably useful to examine the PSF of an objective lens under real imaging conditions (see below). Imaging small point sources in a real biological sample (e.g., fluorescently tagged spindle pole bodies) can be used to confirm the quality of images by checking that the blurred image resembles a symmetric PSF. Asymmetry in the PSF along the z-axis is known as "spherical aberration" and is only corrected by selecting an immersion oil with a different refractive index (see the section on Refractive Index Matching Using Immersion Media). Normally, the elliptical shape would be seen along the z axis (Fig. 1B) as well as a bright circle with multiple concentric rings of rapidly decreasing amplitude in the x-y plane (Fig. 1C). The central part of this pattern is referred to as the Airy disc and contains about 85% of the total light; the concentric rings contain the remaining 15%. Restoration, or deconvolution microscopy, uses information about the PSF to increase the sharpness of an image by attempting to recreate the original object from its blurred image (see Chapter 15).

Spatial resolution is typically defined using the Rayleigh criterion, in which two point sources are considered to be distinctly resolvable when they are separated by no less than a distance d_{xy}^R from each other in the *x-y plane* (perpendicular to the optical axis) and d_z^R in the *z plane* (parallel to the optical axis) (Inoue 1995):

$$d_{xy}^R = 0.61\frac{\lambda}{NA} \quad d_z^R = 2\frac{\lambda n}{NA^2}$$

The Rayleigh distance is the point at which the first minimum in the diffraction-induced pattern (Fig. 2A) of one point source overlaps the maximum in the second (Fig. 2B,C). Resolution increases with higher numerical aperture (*NA*) in the objective lens and decreases with higher-wavelength (λ) or refractive index (*n*) values. Thus, with a 100 × 1.4 NA objective, the Rayleigh resolution limit with green light is approximately 250 nm in the x-y plane and 700 nm in the z plane. It is important to note that this resolution can only be achieved under optically ideal circumstances in which spherical aberration has been minimized (see the section on Refractive Index Matching Using Immersion Media). Image restoration methods are designed to compensate for optical blurring, but not for sample

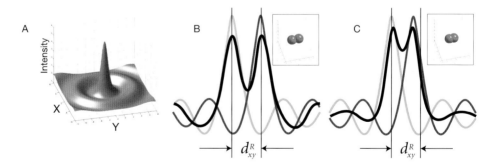

FIGURE 2. Rayleigh resolution criteria. Green spheres in insets represent two identical point sources in close proximity. Green curves show fluorescent intensity of one point source; red curves represent intensity of other source; black curves show the cumulative signal from both point sources. (A) A three-dimensional representation of an Airy distribution. Zeroth order is represented by the large peak in the center and successive orders occur in the circular radial bands. (B) When two objects are in close proximity to one another where the Airy patterns overlap by less than 42% of the maximal intensities, the objects are individually resolvable (equal intensity objects). Rayleigh resolution limit (d^R) represents the minimum distance the two objects can be to each other. (C) If the objects move together with more than 42% of their maximal intensity overlapping, the ability to resolve the objects as distinct is no longer possible. Notice that the two maxima are closer than the Rayleigh resolution limit.

preparation problems or poor instrument alignment. Fortunately, yeast cells are sufficiently small that we need not concern ourselves with sample-induced light scattering and depth-dependent aberrations that normally complicate image collection when working with tissues and other thick specimens.

In live cell imaging, we are also concerned with temporal resolution. Sampling must be sufficiently frequent to capture critical dynamics without loss of information. For instance, a dynamic process described by a periodic waveform (e.g., the oscillation of chromosomes) must be sampled at no less than the Nyquist frequency, which is defined as twice the highest frequency in the original waveform. Discrete sampling below the Nyquist frequency results in misestimation of the original waveform's frequency. Temporal aliasing is a serious concern in live cell imaging of yeast because wide-field 3D microscopes are currently unable to capture an image stack in less than 1–2 seconds. Thus, we cannot hope to follow a process with a frequency above 0.25–0.5 Hz.

IMAGE INFORMATION CAPACITY

The Rayleigh formulation for resolution assumes noise-free images. In reality, noise is a major determinant of both the detectability and resolvability of objects. For example, even if two point sources are resolvable by Raleigh criteria, it may be impossible to distinguish them if the amplitude of the noise exceeds the depth of the minimum in the intensity profile. When considering the detectability of small faint objects, it is also important to consider both the level of the noise and the blurring effect of the optics. We can consider the impact of noise on images in a general sense by thinking about a microscope simply as an information transfer system with three spatial (x, y, z) and one temporal channel. As first shown by Shannon, the information capacity, C, of such a system is $C = \beta \log(1 + SNG)$, where β is the continuous bandwidth of the microscope in all four channels. As the SNR falls, the information capacity (the information about the image that can be conveyed by the microscope) falls and is completely lost when the SNR reaches zero. The bandwidth is itself a function of the cutoff frequencies of the objective lens. To understand why this cutoff exists, recall that a lens decomposes an image into its frequency components such that the lowest-frequency components (those containing the least detail) are toward the center of the optical axis and the higher-frequency components (those containing more image detail) are toward the edge (Fig. 3). Thus, as the half-angle ($<\theta>$) of collected light increases, so does the objective's ability to collect higher frequencies and better resolve small objects. Viewed in another way, the fixed diameter of the objective limits the maximum frequency that can pass, the cutoff frequency, ν_c, to an extent that varies with $<\theta>$:

$$\nu_c = \frac{2NA}{\lambda} \quad NA = n(\sin \theta)$$

All frequencies larger than ν_c, which would define the fine structures of the object, are not conveyed by the objective lens and therefore cannot appear in the image (n = refractive index of medium). The precise relationship between resolution and SNR depends on specific properties of the image and the way it is processed, but a typical result from point object tracking is shown in Figure 4. The important point here is that it is ultimately both the cutoff frequency of the objective lens and the SNR that limit resolution and detectability under low-light conditions. Factors in the imaging system that affect SNR are discussed in detail below.

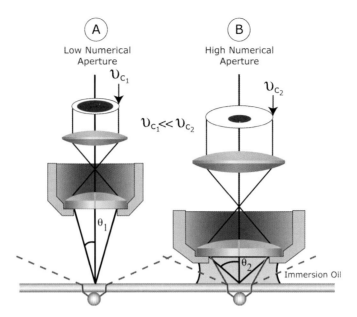

FIGURE 3. Comparing light collected from different NA objectives. Shown are two lenses with different NAs (low NA on the left, high NA on the right). Black lines represent half-angle cone of light collected by the objective as determined by the NA of the lens; the greater the half-angle ($<\theta>$), the larger the NA value. The lens on the left (*A*) has a smaller NA than the one on the right (*B*); thus, θ_1 is less than θ_2. (*Red dashed lines*) Normal dispersion of emitted light from a single fluorescent point source after passing through a coverglass. Note that higher-NA objectives collect more light emitted from sample. In addition, use of refractive matched immersion oil limits excess refraction and enhances the number of photons collected by an objective (*solid red line*). The effects on image detail are illustrated about each objective with respect to the cutoff frequency. The higher-NA lens produces a more defined image, because it can collect the higher frequencies not available in the low-NA lens.

BRIGHTNESS

An additional question that is frequently asked, and one of particular relevance for live cell imaging, is how bright an object must be to be detectable. Detectability increases with the intrinsic brightness (*B*) of the object being imaged and the SNR of the image. Increasing the SNR is a significant problem in live cell imaging, because photobleaching and photodamage prevent us from increasing the intensity or duration of the illumination indefinitely. Instead, we must ensure that incident (excitation) photons are efficiently converted into fluorophore (emitted) photons. In other words, as many emitted photons as possible need to be collected by the objective and passed through the microscope to the camera, and the camera must be optimally configured to convert these emitted photons into an electronic signal. We consider each of these issues in turn, and in all cases concentrate on epifluorescence imaging.

The intrinsic brightness of a fluorescently tagged object depends on the number of fluorophores per unit volume (density), the probability that an excitation photon will be absorbed (the extinction coefficient of the fluorophore), and the probability that a photon will be emitted by the fluorophore in response to an absorbed photon (the quantum yield). It is also important to consider the difference between the excitation and emission wavelengths (the Stokes' shift). When the excitation and emission wavelengths are close, it is difficult to block all of the excitatory photons and collect only the much less numerous ($\sim 10^{-6}$-fold) emitted photons. GFP has a relatively small extinction coefficient, quantum yield, and

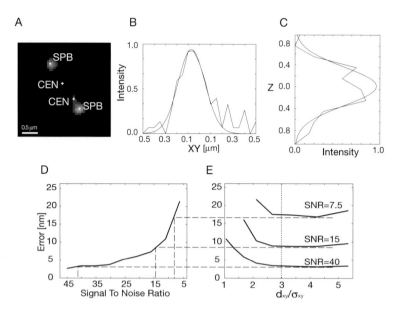

FIGURE 4. Dependence of localization error and resolution on SNR in a typical application. (*A*) Maximum intensity projection of an *S. cerevisiae* mitotic spindle. Spindle pole bodies (SPB) are marked with GFP-tagged Spc42p, and the centromere proximal DNA is marked on chromosome III (CEN) with a TetO/TetR GFP-tagging system (Straight et al. 1996; Ciosk et al. 1998). (*B*) Gaussian approximation of the PSF for a fluorescent point source in *xy* plane. (*C*) Same as *B* with respect to *z* plane. Note that distribution of Gaussian is greater in *z* plane compared to the *xy* plane. The combination of the two approximations describes the shape of a fluorescent point source in 3D. (*D*) The error in localizing a single-point source object as a function of the SNR. (*E*) The localization error as a function of the point-to-point distance calculated for three different SNR values. Note that the horizontal axis points from high to low SNR. The graph confirms the well-known fact that center positions of features with a known intensity distribution, in our case the distribution of the PSF, can be determined with sub-20-nm precision (Bobroff 1986). For SNR > 15, the precision even reaches the single-nanometer range. With shorter separation distances, the accuracy decreases; the Rayleigh limit is denoted by a dashed vertical line. d_{xy} is the separation distance between two spots and σ_{xy} is the width of the Gaussian distribution. (*D* and *E*, reprinted with permission, from Thomann et al. 2002.)

Stokes' shift when compared to chemical fluorophores, but some GFP variants are better than others (Tsien 1998). Moreover, enhanced GFP has good photostability, making it the best fluorophore for live cell imaging. Preliminary evidence also suggests that the intrinsic brightness of a protein can be increased considerably by fusing it to two tandem GFP tags. When deciding what proteins in a multiprotein complex to label, it is also helpful to choose those that are concentrated in a small area (for simplicity, we are ignoring possible complications from proximity-induced fluorescence quenching). Thus, to track the growth and movement of mitotic spindles in living cells, we decided to GFP-tag Spc42p, a protein that is present in many copies at the spindle pole bodies, rather than tubulin, which forms extended microtubule-based structures (Fig. 4A). The spindle pole bodies are smaller than the diffraction limit of the microscope (~0.24 μm) (Bullitt et al. 1997) and therefore appear as bright spots, independent of magnification. It is relatively simple to determine the center of these fluorescent spots and thereby measure the length and orientation of the spindle (Fig. 4B,C). In contrast, it is difficult to measure with precision the extent of tubulin in the spindle.

Choosing an Objective

Although many aspects of a microscope's optics affect the brightness of a feature, none are as important for fluorescent imaging as the NA of an objective lens (Figure 19.3) (for a review, see Spector and Goldman 2003). The observed brightness of a fluorescent image (*b*) is:

$$b \approx B \left(\frac{NA^4}{M^2} \right) E$$

where *B* is the fluorescent brightness of the object, *M* is the magnification, *NA* is the objective lenses numerical aperture, and *E* is the transmission efficiency of the optics. All other factors being equal, it is important to use the highest-NA lowest-magnification lens available. In the case of diffraction-limited point-source objects, magnification does not affect brightness. As a practical matter, all high-NA microscope objectives are also high magnification and are typically available with 60–100× magnifications. For yeast, a 100× 1.4 NA lens is usually optimal. With regard to the transmission efficiency of the optics (*E*), it is important to avoid objectives designed for phase contrast, because the phase ring causes approximately 30% reduction in transmission. Similarly, it is important to remove differential interference contrast (DIC)-phase plates and prisms from the optical path, since they also reduce light transmission by as much as 70%. Fully color-corrected flat-field (Plan Apochromat) objectives typically exhibit 75–85% transmission for 400–700-nm light. It is possible to increase the transmission by a small but significant extent by using less than fully color-corrected objectives (for more details, see Chapter 17). Overall, the rule in live cell imaging is to use the highest-NA objective that is designed for epifluorescent imaging, rather than for phase contrast.

Filters and Dichroic Mirrors

To select the desired excitation and emission wavelengths, epifluorescence microscopes make use of three optical elements: an excitation filter, a dichroic mirror (beam splitter), and an emission filter. A typical arrangement is detailed in Figure 5, and a full range of absorption spectra can be downloaded from Chroma Technology Corp., Omega Optical, and other vendors. Broad-spectrum light from the mercury or xenon burner first passes through a series of infrared and UV-blocking filters near the rear of the microscope (Fig. 5A) and then through an interference filter that allows only excitation wavelengths to pass (Fig. 5B). Excitatory illumination is then reflected by the dichroic mirror toward the objective lens and the sample (Fig. 5C). Emitted photons from the sample are collected by the objective and pass through the dichroic and then through an emission filter on their way to the camera or occulars (Fig. 5D). Filter manufacturers such as Omega Optical (www.omegafilters.com) and Chroma Technology Corp. (www.chroma.com) offer filters in sets suitable for various applications. In many cases, all of the optical elements are mounted in a single holder known as a filter cube. In microscopes with automatic wavelength selection, the dichroic mirror is mounted in the cube, whereas the excitation and emission filters are mounted in separate motorized wheels.

We have encountered a considerable degree of confusion surrounding the selection and assembly of filter sets. The issue is easy to understand, however, if one makes reference to spectral plots and the geometry of an epifluorescence microscope (Fig. 5). Wavelength filtering in epifluorescence microscopes relies on a special property of dichroics: They reflect light shorter than a characteristic wavelength but are transparent to longer wavelengths (polychroic beam splitters have several transitions between reflection and transmission). Short-pass and long-pass emission and excitation filters are described by a wavelength that denotes the point of inflection between transmission and absorption (Fig. 6A,B). Alternatively, band-pass filters are described by the wavelength of peak emission (the center wavelength) and the width of the allowable band (the full width at half-maximum transmission; e.g., 520/20). By examining the characteristics of each element in a set, it is possible to choose sets for a par-

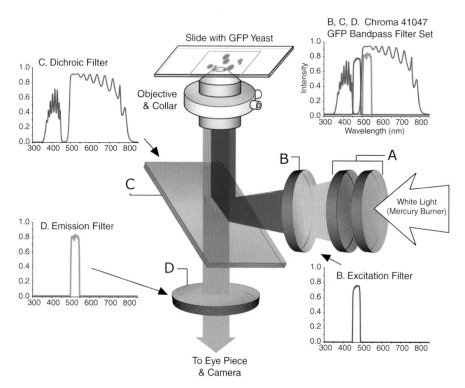

FIGURE 5. Filtering elements in a typical epifluorescent microscope system. (*A*) Light consisting of all wavelengths from the visible spectrum enters a set of neutral-density, UV, and infrared filters to block excessively damaging light rays. (*B*) Wavelengths are then filtered by passing through the excitation filter. Only those photons between 450 and 500 nm pass through this filter before being reflected by the angled beam splitter (dichroic) to the objective lens and to the specimen slide. (*C*) The beam splitter not only selectively reflects photons from the excitation wavelength, but also allows photons emitted from the specimen to pass directly through. In this example, only those photons between 450 and 500 nm are reflected. The photons emitted from the GFP molecules (510–530 nm) and any background fluorescence pass down through the splitter toward the emission filter. (*D*) Light slipping through the beam splitter is selectively blocked so that only those photons between 500 and 550 nm are passed through to the eyepiece or CCD camera. Thus, the emission filter helps to reduce autofluorescent signals and excitation wavelengths that slip through the beam splitter. The graph of Chroma 41047 shows the overlapping image spectra from each of the individual elements for the GFP band-pass filter set.

ticular application, but some experimentation with individual elements is advisable to fully optimize the optical train.

Live cell imaging of GFP-labeled yeast cells presents some special problems in filter set design and a trade-off must be made between sensitivity and selectivity. Because the Stokes' shift of GFP is small, it is common to pass all of the emitted photons from the sample while sacrificing somewhat on the excitation side. In some yeast strains, autofluorescence is a significant problem and filter sets must be assembled by the user to minimize the background. We have found that a long-pass GFP (Chroma, HQ500LP) emission filter provides good sensitivity for single-color recording of yeast cells expressing enhanced GFP. When autofluorescence is a problem, particularly with *ade*– strains (see below), use of the band-pass GFP (Chroma, HQ525/50m) emission filter is preferred. In dual-color imaging, filter and fluorophore combinations must be examined closely for overlapping wavelengths. However, combinations with cyan fluorescent protein (CFP)/GFP or CFP/YFP (yellow fluorescent protein) are usually quite successful. It is best if the less-abundant protein is tagged with GFP, since GFP is significantly brighter than CFP and is less likely to photobleach. It should also be noted that CFP and YFP can be used as a FRET (fluorescence resonance energy transfer) pair. For detailed discussions on FRET, see Chapters 8 and 9 in this volume.

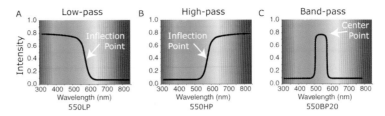

FIGURE 6. Filter types. (*A*) Low-pass filter selectively allows only wavelengths below 550–600 nm to pass in this example. Longer-wavelength photons are blocked by the element. (*B*) High-pass filter designed to allow wavelengths above 550–600 nm to pass through. (*C*) Band-pass filter blocks both short and long wavelengths while allowing only light from 525–625 nm to pass through.

The shortest exposures and lowest possible illumination should be used for live-sample excitation. With our GFP-labeled spindles and tagged chromosomes, we typically collect 50-ms exposures per *z* section and use neutral density (ND) filters to reduce the incident illumination fivefold to tenfold. A broad range of neutral density filters are available and should be kept on hand for this purpose.

RESOURCES

Omega Optical: www.omegafilters.com. Curv-o-matic viewing of filter spectra on line.
Chroma Technology: www.chroma.com. Excellent on-line handbook of filter basics, GFP filter brochure, and absorption spectra.

Setting up the Camera

Camera design and configuration are critical variables in determining the brightness of images and the extent to which features can be discriminated. Because they are invariably more efficient than color cameras, we consider here only the monochromatic charge-coupled device (CCD) cameras. In a microscope system connected to a monochromatic camera, exposures are taken with different excitation and emission filters and color is introduced after the fact by assigning a hue to each exposure based on the properties of the filter. The construction of CCD cameras is a relatively complex topic that we will not cover in any detail (see Chapter 6), but to optimize camera performance, it is helpful to have a basic understanding of their operation (Fig. 7). At the core of a CCD camera is a chip containing a rectangular area of photosensitive detectors—pixels—that convert photons into electrons with a particular quantum efficiency (QE varies with wavelength and is typically 60–80% for blue-green light). After an exposure is taken, the number of electrons generated at each pixel is determined by an analog-to-digital converter (ADC). Cameras typically have only one ADC and the electrons in each pixel of the CCD must therefore be moved to the ADC via a series of positional shifts, until the entire chip has been read (Fig. 7). Thus, even though a CCD camera collects photoelectrons at each pixel in a parallel process, the pixels are read serially. Serial readout significantly limits the rate at which CCD cameras operate, and thus, the fewer the number of pixels read, the greater the rate at which successive frames can be recorded (commonly referred to as frame rate). One simple way to increase frame rate is to limit the area of the CCD being read by the ADC. Yeast cells are sufficiently small, even with a 100× objective, where a 64 × 64 or 128 × 128-pixel area of the CCD is usually sufficient. A second method to increase frame rate is to add the contents of adjacent pixels together

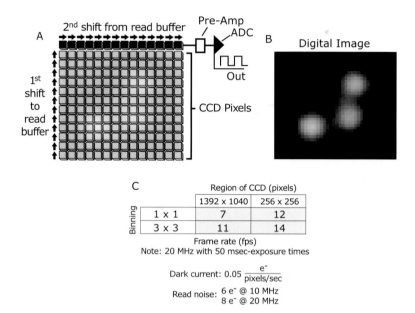

FIGURE 7. CCD chip architecture. (*A*) Photons coming from the microscope hit the surface of the CCD photosensitive detectors. These photons are registered and incrementally counted as electrons for the desired exposure duration. The electron count is converted into signal intensity by a single read buffer (ADC) at the top of the array. To read each line from the CCD array, the information must be shifted up one line at a time until the entire image is transferred into the read buffer and read out sequentially. (*B*) Once complete, the raw digital image can be reconstructed from the individual intensity values. (*C*) Properties of a typical high-performance interline CCD camera.

prior to their arrival at the ADC in a process known as on-chip binning. Besides faster read-out, binning increases the SNR at the cost of lower image resolution. To understand the impact of binning on an image, note that the pixels in a CCD array perform a discrete sampling of the image, which is a continuously varying two-dimensional waveform. The pixel array must therefore, by Nyquist criteria, sample the image at twice the frequency of the highest spatial frequencies in the image. The highest frequency in the image is determined by the cutoff frequency ν_c of the objective. We can therefore determine the pixel spacing, p_x, as follows:

$$p_x \leq \frac{\lambda M}{4NA}$$

With a 100 × 1.4 NA objective, the largest acceptable pixel spacing is 8.9 μm. For instance, a Roper CoolSNAP HQ interline CCD camera has a pixel size of 6.25 × 6.25 μm without binning, and with a 100 × 1.4 NA objective lens, oversamples the image 1.4-fold (relative to the Nyquist criteria). This is reasonable for small diffraction-limited objects. In contrast, 2 × 2 binning (to create 12.5 × 12.5 μm superpixels) would be detrimental.

CCD cameras are subject to three sources of noise: dark noise, photon noise, and read noise. Dark noise refers to the random generation of electrons, within the photosensitive elements of the CCD array, from heat rather than by photon absorption. Modern microscope cameras cool the CCD to between −30°C and −50°C, effectively making dark noise negligible. Photon noise, also known as shot-noise, arises from variations in the flux of photons in a beam of constant intensity. Photon noise exhibits Poisson statistics, varying with the square root of the signal, and is an intrinsic feature of photons that cannot be eliminated by changes in camera design and setup. Read noise refers to uncertainty in the measurement of photoelectron number introduced by limitations in the preamplifier, ADC, and other electronics in the camera (www.roperscientific.com). Read noise is a function of the quality of the camera and the care that has been taken in designing the electronics, but for a given camera, the faster the read rate of the camera, the greater the noise. Ideally, imaging should be performed under conditions in which the physics of photon counting and not the camera design is limiting. Under these conditions, commonly referred to as photon-limited imaging, photon noise is the greatest contributor to overall noise. However, it is not always possible to work in this range with live samples, and we typically find ourselves recording in the instrument-limited range. Under these circumstances, it is important to find the lowest-noise cameras and lowest-noise settings. For example, with a CoolSNAP camera recording 50-ms exposures on a 128×128 area of the CCD, we have observed a dramatic increase in SNR but little decrease in frame rate by dropping from a 20-MHz read rate to 10 MHz.

In summary, for live cell imaging of yeast, it is usually best to use only a fraction of the area of a megapixel CCD camera. This makes it possible to increase the frame rate while keeping the ADC read rate as low as possible to increase SNR. In general, avoid binning the image, but before accepting this nostrum, it is best to perform the simple calculation mentioned above to see if this is also true for the microscope being used. If it allows bining then the image SNR will increase, because the signal becomes photon-noise-limited at a lower overall intensity. Before leaving the topic of cameras, however, it is worth mentioning interline CCD architectures, which have a significant impact on frame rate. In an interline CCD, a set of masked (nonphotosensitive) pixels are interleaved between the primary photosensitive detector lines. Such interline CCDs have a primary recording array and a second masked array. An entire frame can be shifted in parallel from the recording to the masked array, and the masked array can then be digitized while the primary array is recording a subsequent exposure. This makes it possible to record successive frames without a shutter. Historically, interline CCDs have had the drawback that the interline masks reduce the fraction of the chip that responds to light and therefore reduce camera sensitivity. Recent cameras circumvent this problem by including a small lens for each pixel that focuses the light from masked elements onto photosensitive elements, thereby achieving both increased speed and increased sensitivity. Interline CCD cameras currently represent the best option for high-speed live cell imaging (Fig. 7C).

RESOURCES

Roper Scientific: www.roperscientific.com. Good technical library from a leading camera manufacturer.
Hamamatsu Photonics: www.hamamatsu.com. Leading camera manufacturer.
Molecular Expressions Primer: www.microscopy.fsu.edu. Excellent source of microscopy information; many interactive demos for understanding important concepts.
Essentials from Cells: A Laboratory Manual. Edited by D.L. Spector and R.D. Goldman (2003), Cold Spring Harbor Laboratory Press, New York.

PREPARING AND MOUNTING SAMPLES FOR LIVE CELL IMAGING

Quite often the most tedious parts of live cell microscopy are keeping cells growing proficiently and preventing them from floating around during an imaging session. The following sections describe procedures for selecting appropriate strains, preparing the culture, and maintaining cells in a suitable environment.

Strains and Growth

Autofluorescence can be a significant problem in yeast if cultures are grown under poor conditions, especially when working with *ade–* strains. In *ade–* strains, a colored intermediate in adenine biosynthesis, phosphoribosylaminoimidazole, accumulates to high levels in the absence of exogenous adenine and is highly fluorescent when excited with blue light (Stotz and Linder 1990). In fact, it is the phosphoribosylaminoimidazole that gives *ade–* yeast cells the distinctive red color used in genetic sectoring assays (Ishiguro 1989). Autofluorescence can be minimized by using ADE+ genetic backgrounds and growing cells in synthetic complete (SC) medium supplemented with essential amino acids and 20 μg/ml extra adenine. Additionally, cultures should be maintained below 5×10^6 cells/ml for 4–10 generations, and the medium should be refreshed prior to mounting and imaging.

Slide Preparation

Although it is possible to mount live yeast cultures directly under a coverglass, we have found it quite difficult to avoid physical damage and to hold cells in a fixed position using direct mounting. Thus, we recommend mounting cells on a pad of 1.2% agarose formed in a slide that contains a shallow depression. This system not only holds the cells in place better and prevents accidental shearing, but also provides cells with a significant volume of nutrients to help maintain cell viability. The use of agarose pads is particularly important for time-lapse experiments, and we have successfully used them to maintain cells at wild-type growth rates for 8 hours or more (Fig. 8).

Preparing Slides with an Agarose Cushion

1. To prepare the agarose solution, combine 10 μl of SC medium supplemented with a Complete Supplement Mixture of amino acids (Bio101, www.Bio101.com), 20 μg/ml of additional adenine, and a carbon source (e.g., 2% w/v glucose) with 1.2% (w/v) agarose. Heat the solution in a microwave to completely melt the agarose.

2. Add ~200 μl of melted agarose solution to a 60ºC prewarmed slide fabricated with a shallow 18-mm hemispherical depression (VWR, www.vwr.com) (Fig. 9A). Quickly

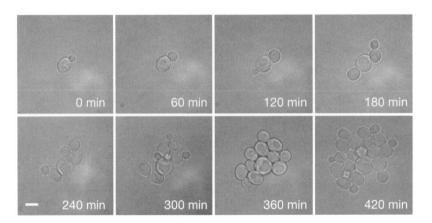

FIGURE 8. *S. cerevisiae* cells growing on an agarose specimen slide. Wild-type cell culture grown in medium to log phase and mounted at $\sim 1 \times 10^7$ cells/ml using the techniques described in this chapter. The slide was maintained at 30ºC for just over 8 hours. Series of phase-contrast images collected at regular intervals (every 60 minutes) to monitor growth rates. Culture doubles every 90–100 minutes based on cell counts. Bar, 4 μm.

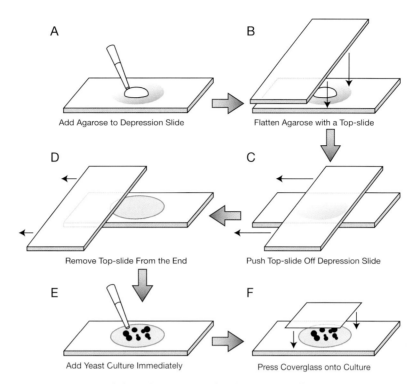

FIGURE 9. Preparing specimen slide with agarose pad and mounting of *S. cerevisiae* cells. (*A*) 200 μl of hot agarose/SC-Complete Medium Mixture is added to a 60°C prewarmed depression slide using a pipette. (*B*) A prewarmed top slide is placed directly on the agarose-filled depression and pressure is held evenly across the top slide for 60 seconds while the agarose solution hardens. (*C*) The top slide is gently removed by sliding it toward one end of the depression slide. (*D*) After removal of the top slide, excess agarose is removed from the depression slide to create a flat and even surface. (*E*) 2.2 μl of yeast culture is immediately added to the center of the agarose pad using a small pipette tip. (*F*) A dust-free petroleum-edged coverglass is gently placed on top of the yeast sample and sealed by applying light amounts of pressure to the four sides of the coverglass. Large air bubbles can be removed by gently applying pressure to the coverglass in a radial direction.

cover the agarose-filled depression with a regular microscope slide (the top slide) by placing it directly onto the agarose. Using the thumbs, press evenly against the top slide for ~1 minute to allow the agarose to harden (Fig. 9B).

> Excess agarose will squeeze out from the edges. It is essential that no air bubbles be trapped in the agarose during this process and that the agarose forms a smooth flat surface with the slide. To prevent dehydration, leave the top slide in place until cells are ready for mounting.

3. Prepare a 22 × 22-mm, no. 1.5 (0.16–0.19-mm), coverglass (VWR) for mounting by first physically removing dust particles from the cover glass with a dry artist's paint brush and hand blower (e.g., Bergeon Blower 3B-750 from www.watchmakertools.com). Make sure to remove any particles from the coverglass since they will prevent a tight seal with the agarose pad.

> Avoid using precision wipes (e.g., Kimwipes) as they can leave scratches and fluorescent strands on the glass. In addition, avoid Dustoff or similar compressed gas products because they usually contain small, highly fluorescent particles.

> For tips on choosing the best cover glasses, see the shaded box on Coverglass Selection at the end of this protocol.

4. Apply a very fine band of 100% pure petroleum jelly (e.g., Vaseline) to the extreme edge of the coverglass. Do this to all sides by adding a very small amount of jelly with a toothpick

to one corner and running a finger gently along the extreme edge to distribute the petroleum jelly evenly.

> The petroleum jelly prevents rapid evaporation of the cell culture medium and creates a better seal when the coverglass is applied to the agarose bed.

Mounting Cells

5. To pellet the cells, transfer 1 ml of log-phase yeast culture to a microfuge tube and microfuge at low speed (4000 rpm) for 1 minute. Avoid overspinning.

6. Remove the supernatant from the microfuge tube and resuspend the cell pellet in 0.3–0.5 ml of fresh SC medium (as in Step 1 but without agarose).

> The volume used to resuspend the cells depends on the size of the cell pellet in Step 5 and the desired final cell density on the microscope slide.

7. Remove the top slide to expose the agarose pad by gently pushing the top slide along the length of the depression slide (Fig. 9C,D). This step should release the seal between the two slides and leave a very smooth surface on top of the depression slide. If any large bits of agarose remain, carefully remove them.

8. Transfer 2.2 μl of resuspended cell culture to the smooth agarose and gently place the coverglass with petroleum jelly over the cell culture (Fig. 9E,F). To create a tight seal, apply a very slight amount of pressure at the extreme edges of the coverglass. Do not press too hard because the cells could be damaged or crushed between the coverglass and microscope slide.

9. Spot a small amount of nail polish at the four corners of the coverglass to prevent it from sliding or pulling away from the agarose bed during the imaging session.

COVERGLASS SELECTION

The coverglass is an important and often overlooked component of the optical path. Objective lenses are typically designed to work with a particular thickness of coverglass, usually no. 1.5 (0.16–0.19-mm thickness), and proper selection can enhance image quality. For the most precise work, it is good practice to measure the thickness of the coverglass and to use only those within ±0.02 mm of nominal dimension. A digital micrometer suitable for measuring thickness can be obtained from any tool supply house (e.g. www.mscdirect.com).

INSTRUMENTATION: ENVIRONMENTAL CONTROL AND MICROSCOPE OPTICS

Environmental Control Devices

Temperature-sensitive yeast strains are well suited to live cell microscopy and present an opportunity to determine the role of a single gene on a dynamic process in a living cell. However, special consideration must be given to control the temperature of the yeasts' environment, because the objective and slide are coupled thermally via immersion oil, and thus the objective acts as a heat sink. Live cell yeast microscopy can be performed successfully on either upright or inverted microscopes, and commercial environmental control systems are available for both. With yeast cells, it is not necessary to use the elaborate chambers developed for animal cells, since the regulation of oxygen and CO_2 levels is not critical. When choosing or building an environmental control system, it is important to keep in mind the geometric constraints imposed by high-NA objectives. These include a relatively large-diameter lens, direct contact with the sample (via immersion oil), and short working distance. In our experience, by using a stage heater, an objective heater, and the depression slide system discussed in the previous section, it is possible to maintain live yeast cells at temperatures from 15°C to 40°C for many hours. The best way to perform rapid shifts from permissive to nonpermissive temperatures is to use preheated objectives, rather than to change the temperature of a single objective. Note that such a shift entails a compromise in the choice of mounting oil (see below).

Building a Thermally Regulated Slide Holder

A simple stage heater can be fabricated from a brass block measuring $4 \times 2 \times 0.5$ inches. One advantage of this simple water-based thermal system is that it appears to be more stable thermally than traditional resistive systems, which cycle on and off. In addition, a refrigerated circulator allows cold-sensitive mutants to be chilled below room temperature.

A completed thermal unit should make clean contact with the microscope objective and look similar to the detailed arrangement shown in Figure 10D (e.g., inverted microscope configuration).

1. Drill a 1-inch hole through the block to permit bright-field illumination. Drill intersecting holes through the length of the block, cap them with seals, and connect the block via tubing to a circulating water bath to create a continuous channel for heated or cooled water (Fig. 10A).

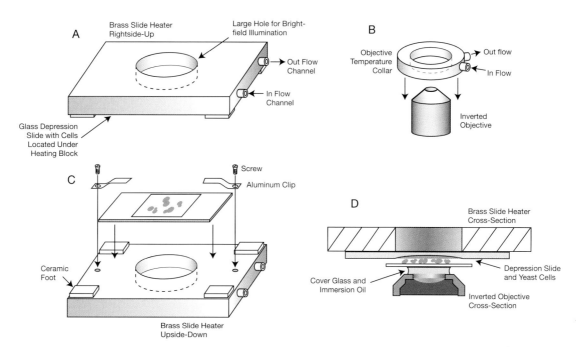

FIGURE 10. Temperature controller block and objective lens assembly. (A) Liquid-controlled thermal block for maintaining specimen slides at desired temperature. Channels through the block allow liquid to flow in through one port and out the other. The block can be fabricated with a large hole bored directly through the center to allow bright-field illumination on the specimen slide (normally attached to the bottom of the block for inverted microscope configurations). Flat ceramic blocks can be added to the bottom of the block to insulate the block and specimen slide from the microscope stage. (B) Commercial liquid-regulated thermal collar for an objective lens. The ring is incorporated into the liquid circulating system, typically provided from a temperature-controlled circulating water bath. (C) Thermal block in upside-down configuration. Small aluminum clips and screws are used to hold the specimen slide to the underside of the block for inverted microscope configurations. Screws and clips can be added to the top of the block for upright configurations. (D) Fully assembled thermal block with specimen, slide, and mounting oil placed over inverted microscope objective.

2. Fit a temperature collar (available from Bioptechs Inc., www.bioptechs.com) over the objective to regulate its temperature (Fig. 10B) and to prevent heat from being absorbed into the microscope base. Exercise caution in using the collar since it is typically wide enough that excessive stage movement can cause the objective to crash into the stage.

3. Insert small spring-steel clips to hold the slide firmly to the brass block and ceramic foot pads to prevent heat loss through contact with the microscope stage (Fig. 10C).

Commercial Environmental Systems

Commercial temperature-control systems are available in a wide range of configurations, styles, options, and price ranges. Some systems are similar to the basic system described above and contain nothing more than an aluminum slide holder connected to a resistive heating coil (www.instec.com). Others, typically designed for mammalian cell microscopy, are elaborate temperature-regulated circulating dishes (www.bioptechs.com). These systems are not well suited to yeast work, because most mammalian studies use adherent cells that bind tightly to the coverglass, whereas yeast cells float or spin around continuously. Finally, acrylic chambers that enclose most of the microscope are also available and can be used to control the air temperature as well as that of the microscope (www.lis.ch/thebox.html). This type of system has the advantage of keeping the entire microscope, stage, and objective at the desired temperature, but the cost is significantly greater than a stage-mounted controller. Regardless of what system is used, the temperature of the slide must be monitored as close to the sample as possible using a small thin-film RTD temperature sensor and a digital thermometer (www.omega.com). We find that the actual temperature of the sample typically differs by 1–3°C from the temperature of the circulating bath.

Limiting Geometric Aberrations

Imperfections in lens geometry or sample mounting directly affect the shape of the PSF (point spread function) in distinct ways (Fig. 11). A tilted or misaligned lens skews the PSF so that it is

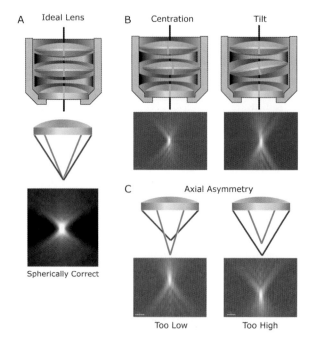

FIGURE 11. Lens aberrations and spherical aberration. (*A*) An ideally shaped lens should collect photons symmetrically from a single focused point shown by the red and green lines. With the correct immersion medium, the light from a single fluorescent point source results in a spherically correct or symmetric PSF. (*B*) Improperly aligned elements within the objective lens results in PSF deformation. A lens can be misaligned laterally (e.g., centration) or in a slightly tilted configuration. Lens construction problems cause the PSF to be altered perpendicular to the *x-z* plane or bent asymmetrically about the point source origin. (*C*) Aberrations in the PSF can also occur when imperfections in the curvature of the glass cause light to be collected from different axial-shifted positions in space and result in axial asymmetry (also known as spherical aberration). Spherical aberrations can additionally result from using an immersion oil with the wrong refractive index. All aberrations lead to a reduction in the signal intensity strength and distortions in the final image.

no longer perpendicular to the optical stack or is bent in half (Fig. 11B). A PSF of this type suggests a significant problem with tilt and centration in the objective and cannot be corrected by the user. When the PSF symmetry is disrupted in the longitudinal or axial dimension (Fig. 11C), spherical aberration is the problem. Spherical aberration can arise from problems with the objective or the use of immersion media with the wrong refractive index. The small size of a yeast cell relative to the resolution limit of an optical microscope requires that spherical aberration be minimized, since even small axial aberrations lead to relatively large errors. Spherical aberration can be reduced dramatically by selecting an immersion medium with the proper refractive index.

Refractive Index Matching Using Immersion Media

Immersion lenses are designed to work with either oil- or water-based media but not both, as specified on the side of the lens housing. The higher the refractive index of the immersion oil, the greater the extent to which light is bent into the objective, and the highest-NA lenses are therefore oil immersion (see Fig. 3). Environmental factors, particularly temperature, affect the refractive index of the mounting and immersion medium and mismatches in either cause spherical aberration. By changing the refractive index of the immersion medium within a narrow range, it is possible to correct for differences in the culture medium, coverglass thickness, and temperature. Kits with immersion oils having refractive indexes between 1.500 and 1.534 in increments of 0.002 can be obtained from Applied Precision (www.api.com) or Cargille Laboratories (www.cargille.com). Choosing which oil to use for a given temperature is determined empirically, based on a visual analysis of the PSF with different oils. When the angle of light dispersion is greater above the point source than below it, the refractive index of the mounting medium is too high and vice versa (Fig. 11C). When the PSF is nearly symmetric in z, the correct refractive index must be determined. We find that when imaging live yeast at 30°C with no. 1.5 coverglass, a refractive index of 1.518 is typically appropriate; at 37°C, the correct refractive index can be as high as 1.526 or 1.528. Although oil matching can be tedious at first, the process becomes routine over time and the results are worth the effort.

Experimental Determination of PSF

Determining the PSF of an objective under typical imaging conditions can help to identify flaws in the objective and problems with sample preparation. The following protocol describes an approach for the experimental determination of PSF.

1. Dilute a set of small fluorescein (FITC)-conjugated beads in the same medium as is used for mounting cells SC medium supplemented with a Complete Supplement Mixture of amino acids (Bio101, www.Bio101.com), 20 μg/ml of additional adenine, and a carbon source).

2. Add 2.2 μl of the bead mixture to a coverglass coated with poly-L-lysine, and seal it with nail polish onto a standard microscope slide.

   > An initial immersion oil with a refractive index of 1.516 is a good starting point for imaging at room temperature.

3. While focusing the lens above, through, and beyond the point source (using the motorized system), collect sections at regular intervals (e.g., every 0.020 μm).

4. Examine by eye the symmetry and extent of blurring once the stack of optical z-sections are rotated up by 90° and viewed from the side (Fig. 11A).

   > The spherically correct PSF is completely symmetric both above and below the point source as well as circularly symmetric.

5. Change the immersion oil refractive index until a symmetric PSF is obtained.

PHOTODAMAGE

As discussed above, phototoxicity and photobleaching are significant problems in live cell analysis. The exact mechanism of GFP photobleaching has not been elucidated. What is known is that the protein is neither destroyed nor degraded; rather, continuous or long-term repeated exposure to excitation wavelengths irreversibly alters the GFP molecule so that it can no longer fluoresce. This property can be utilized in fluorescence recovery after photobleaching (FRAP) experiments (see Chapter 7) and is a common approach to studying cytoskeletal dynamics.

Phototoxicity indirectly describes a general class of harmful effects on live cells based on either long-term or short but extreme exposures to a light source. Phototoxicity is much more difficult to discern than photobleaching, since it is harder to measure directly and can be a consequence of DNA damage (UV light) or protein damage (infrared light). In addition, excessive excitation of a fluorophore can lead to oxide radical formation and may also negatively impact cell growth. To avoid unpredicted effects during live cell acquisition, it is important to incorporate interference filters into the microscope system. Quite often, multiple filters can be piggybacked on top of one another. Alternatively, special filters can be ordered that have the appropriate filters fused together (www.chroma.com and www.omegafilters.com). Performing 3D microscopy on cells exaggerates the negative effects of photobleaching and phototoxicity. The acquisition of multiple optical sections requires continuous or repeated exposure by a magnitude based on the number of sections taken. Consequently, performing live cell analysis with fluorescent light mandates attention to detail and proper microscope configurations.

To minimize photobleaching, a band-pass excitation filter (e.g., Chroma, HQ470/40x) and an infrared-blocking filter can be combined in series with the emission filter to block unwanted infrared and UV irradiation. In addition, neutral-density filters can be used to reduce the intensity of excitatory illumination across all wavelengths. These filters are described by a number that increases in value as the percent transmission falls. Such filters are available from 0.100 (80% transmission) up to 3.0 (0.10% transmission). With all filters, maintenance is also required over time. This is especially true for those filters immediately exposed to an intense light source, such as excitation and neutral-density filters. We have found that these elements will fade, develop small imperfections, and experience pinhole defects over time and should be replaced after about 6 months of heavy use.

The best exposure time and neutral-density filter combination must be determined empirically and depends on how frequently repeated exposure is required and on the dynamics of the process being imaged. We have found that 50 ms and a 0.500 neutral-density filter works best for rapid acquisition movies where 14–20 optical *z* sections are taken every 5 seconds. This combination allows us to acquire a movie with 120 time points up to 10 minutes long using enhanced GFP-labeled spindle pole bodies and single chromosome tags.

FUTURE DEVELOPMENTS

During the next several years, expect important developments to occur in live cell microscopy and automated image analysis. Machine vision methods have the potential to greatly accelerate the rate at which image information can be converted into quantitative information for mechanistic and screening studies. In addition, machine vision methods can extend the resolution limit of microscopes beyond the Rayleigh limit and can process images that are too indistinct for human interpretation. Machine vision therefore has the potential to overcome many of the practical problems associated with live cell imaging, including phototoxicity and photobleaching, large data streams, and subjective data analysis.

REFERENCES

Bobroff N. 1986. Position measurement with a resolution and noise-limited instrument. *Rev. Sci. Instrum.* **57:** 1152–1157.

Bullitt E., Rout M.P., Kilmartin J.V., and Akey C.W. 1997. The yeast spindle pole body is assembled around a central crystal of Spc42p. *Cell* **89:** 1077–1086.

Ciosk R., Zachariae W., Michaelis C., Shevchenko A., Mann M., and Nasmyth K. 1998. An ESP1/PDS1 complex regulates loss of sister chromatid cohesion at the metaphase to anaphase transition in yeast. *Cell* **93:** 1067–1076.

Inoue S. 1995. Microscopes. In *Handbook of optics*, vol. 2, 2nd edition (ed. M. Bass et al.), pp. 17.1–17.50. McGraw-Hill, New York.

Ishiguro J. 1989. An abnormal cell division cycle in an AIR carboxylase-deficient mutant of the fission yeast *Schizosaccharomyces pombe. Curr. Genet.* **15:** 71–74.

Spector D.L. and Goldman R.D. 2003. *Essentials from cells: A laboratory manual* (Laboratory Essentials on CD). Cold Spring Harbor Laboratory Press. Cold Spring Harbor, New York.

Stotz A., and Linder P. 1990. The *ADE2* gene from *Saccharomyces cerevisiae*: Sequence and new vectors. *Gene* **95:** 91–98.

Straight A.F., Belmont A.S., Robinett C.C., and Murray A.W. 1996. GFP tagging of budding yeast chromosomes reveals that protein-protein interactions can mediate sister chromatid cohesion. *Curr. Biol.* **6:** 1599–1608.

Thomann D., Rines D.R., Sorger P.K., and Danuser G. 2002. Automatic fluorescent tag detection in 3D with super-resolution: Application to the analysis of chromosome movement. *J. Microsc.* **208:** 49–64.

Tsien R.Y. 1998. The green fluorescent protein. *Annu. Rev. Biochem.* **67:** 509–544.

CHAPTER TWENTY

Live Imaging of
Caenorhabditis elegans

Benjamin Podbilewicz

Department of Biology, Technion-Israel Institute of Technology, Haifa, 32000, Israel

Yosef Gruenbaum

Department of Genetics, The Institute of Life Sciences, The Hebrew University of Jerusalem, Givat Ram, Jerusalem, 91904 Israel

INTRODUCTION

CAENORHABDITIS ELEGANS IS AN IDEAL ORGANISM for applying various live microscopy techniques. It is a small (adults are ~1 mm long) free-living soil nematode that feeds on bacteria. The two sexes are males, which produce sperm, and hermaphrodites, which produce both oocytes and sperm and can reproduce by self-fertilization. The hermaphrodite has a short life cycle of approximately 3.5 days and a simple body containing only 959 somatic nuclei. However, it contains all of the major cell types, including muscle, neurons, intestine, hypodermis, and germ cells, among others. Fertilization in *C. elegans* occurs inside the hermaphrodite, where development of the embryo starts. Eggs are laid and the embryo continues developing until it hatches 800 minutes after the first cleavage. The larva develops through four stages (L1–L4) and emerges to become a mature adult. The adult hermaphrodite lays about 300 eggs during the first 4 days of reproductive life.

There are several major advantages to the use of *C. elegans* for live microscopy. The organism is transparent, so it is possible to microscopically analyze the whole animal throughout its entire life. Its complete cell lineage is known, making it possible to follow developmental and differentiation processes in real time. In addition, its entire genome has been sequenced. Furthermore, the development of transgenic techniques, as well as RNA interference (RNAi) methods and sophisticated genetic analyses, and the availability of a large collection of mutant lines all make *C. elegans* especially attractive for live microscopy.

This chapter summarizes useful techniques for preparing *C. elegans* for live microscopic analysis, provides several examples using these techniques illustrated with figures and movies (on the accompanying DVD), and includes a brief troubleshooting guide at the end of the chapter (Table 1). The chapter begins with a short history of *C. elegans* as seen through the microscope.

C. ELEGANS AND MICROSCOPY: A BRIEF HISTORY

Nematodes in general and C. elegans in particular are elegant organisms for curious investigators wishing to study cellular and subcellular processes in living systems and organisms. Fundamental cell theories were proposed and studied using light microscopy of living and fixed nematodes beginning in the nineteenth century. For example, in 1875, Bütschli reported the first observation of polar bodies during oogenesis (Chitwood and Chitwood 1974). In 1883, E. van Beneden discovered the cytological basis of heredity by watching how homologous chromosomes separate (Chitwood and Chitwood 1974). Toward the end of the nineteenth and the beginning of the twentieth centuries, Theodor Boveri established the concept of differentiation of nematode cells and confirmed the cellular basis of development by observing the early separation between the "soma" (the cells that ultimately form the body, differentiating into the various tissues) and the "germ line" (embryonic blastomeres containing the complete genetic information leading to immortal germ cells that form gametes) (Schierenberg 1997). In 1963, Sidney Brenner wrote a prophetic letter to Max Perutz that a mere 40 years later is of significant historical importance: "We propose to identify every cell in the worm and trace lineages. We shall also investigate the constancy of development and study its genetic control" (Wood 1988).

To identify every neuron, muscle, epithelial cell, and gonadal cell in any organism and to trace the lineages, it is necessary, but not sufficient, to have a good light microscope. John Sulston discovered that by just looking at nematodes, all embryonic and postembryonic divisions could be traced by direct microscopic observation of living worms over three and a half days (Sulston and Horvitz 1977). Since the 1970s, John Sulston, John White, Robert Horvitz, and their followers have used and continue using Nomarski differential interference optics (DIC) to discover numerous fascinating inter- and intracellular phenomena (Sulston and Horvitz 1977; Singh and Sulston 1978; Horvitz and Sulston 1980; Sulston and White 1980; Sulston et al. 1980; White 1980). Among the many dynamic processes that could be studied in living nematodes are meiosis, oogenesis, spermatogenesis, fertilization, embryonic and postembryonic development, neurogenesis, organogenesis, behavior, aging, and programmed cell death. The late 1970s and early 1980s also saw the important technical development of laser microsurgery, used in combination with genetic methods, for studying cell-fate determination (Sulston and Horvitz 1977). Important scientific breakthroughs have resulted from combining DIC and fluorescence microscopy to study cellular processes such as dye filling in neuronal cells in living animals (Hedgecock et al. 1985), generation of polyploidy (Hedgecock and White 1985), programmed cell death (Hedgecock et al. 1983; Ellis and Horvitz 1986), and the generation of cell polarity (Strome and Wood 1983; Hill and Strome 1988; Kirby et al. 1990; Mains et al. 1990; Guo and Kemphues 1995; Hutter and Schnabel 1995).

The application of transgenic techniques coupled with the use of DIC optics has enabled transformation-rescue experiments to be performed in C. elegans. In C. elegans, such technically challenging experiments have been mostly dependent on microinjection of DNA constructs directly into the syncytial gonad of partially dehydrated hermaphrodites while observing the living animals with DIC optics. Thanks to the leadership and vision of Andrew Fire and Craig Mello, C. elegans researchers are now able to construct transgenic worms and to perform transformation-rescue experiments to clone genes (Mello and Fire 1995). In 1994, Martin Chalfie and collaborators initiated a new revolution in live microscopy by using green fluorescent protein (GFP) reporters in C. elegans (Chalfie et al. 1994). The development of confocal microscopy (White and Amos 1987; White et al. 1987) permitted the tracking of GFP and its derivatives in neuronal development (Bargmann and Horvitz 1991; Bargmann et al. 1993; Sengupta et al. 1994; Zipkin et al. 1997; Kerr et al. 2000). The combination of confocal microscopy and deconvolution together with microinjection of fluorescent molecules into living embryos, larvae, and adults is now widely used by *C. elegans* researchers (Hird and White 1993; Hird 1996; Hirose et al. 2003). Microinjection technology is also used for expression of RNAs, as well as for antisense and RNAi studies (Fire et al. 1991, 1998; Goodwin et al. 1993; Crittenden et al. 1994; Guo and Kemphues 1995; Seydoux et al. 1996).

In summary, throughout the history of this small biologically attractive nematode, direct observation of developmental and behavioral processes in live cells has proceeded a remarkable array of important scientific contributions and breakthroughs. Currently, *C. elegans* researchers use live microscopic techniques involving multidimensional, multiphoton, and multicolor applications developed by and for the research community (Hird and White 1993; Mohler and White 1998; Skop and White 1998; Miller et al. 1999; Squirrell et al. 1999; Rabin and Podbilewicz 2000; Zipperlen et al. 2001; Eliceiri et al. 2002; Inoue et al. 2002).

PREPARATION OF SAMPLES FOR LIVE IMAGING

NOTE: <!> indicates hazardous material. See Cautions Appendix.

Live imaging requires sample preparation that preserves, as much as possible, optimal growth conditions. It also requires that the specimen not move during the imaging period and that the microscope illumination causes minimal photodamage to the worms. The details of sample preparation depend on the developmental stage of the worm to be studied.

Mounting of Early Embryos

Early-stage embryos are collected and transferred onto an agar pad or coated coverslip (prepared as described in the following protocols). There is little movement in early *C. elegans* embryos until about the 1.5-fold stage.

1. Collect early-embryonic-stage embryos by dissecting adults in depression slides under a dissecting microscope using two sharp needles as scissors. Perform the dissection in egg salts solution.

> EGG SALTS SOLUTION (Edgar 1995; Keating and White 1998)
>
> 118 mM NaCl
>
> 40 mM KCl <!>
>
> 3.4 mM $MgCl_2$ <!>
>
> 3.4 mM $CaCl_2$ <!>
>
> 5 mM HEPES (pH 7.4)

2. Transfer dissected eggs either to an agar pad or to a poly-L-lysine-coated coverslip (for preparation of agar pads and coated coverslips, see protocols below). To avoid sample drying, make sure that there is enough egg salts solution on top of the embryos.

3. Spread vacuum grease around the edges of the coverslip and place a slide on top of the coverslip. Make sure that no bubbles are trapped beneath the coverslip.

Later-stage Embryos, Larva, or Adults: Mounting the Organism and Slowing of Movement

When early embryos are examined in uterus or at later embryonic, larval, or adult stages, the movement of the animal may be slowed down or eliminated as follows.

1. Place the animals on an agar pad or a poly-L-lysine-covered coverslip in a drop of M9 medium (for preparation of agar pads and coated coverslips, see protocols below).

M9 MEDIUM (Sulston and Hodgkin 1988)

3 g of KH_2PO_4 <!>

6 g of Na_2HPO_4 <!>

5 g of NaCl

1 ml of 1 M $MgSO_4$ <!>

per liter (water or phosphate-buffered saline)

---

2. Mount the samples in a manner similar to that described above for early embryos (Steps 2–3 of preceding protocol).

3. To slow down or immobilize movement of the animal, use one of the following approaches:

   a. Temperature can be used to inhibit worm movement. Keeping the animals at 4°C prior to placing them on the microscope stage is the least harmful. Cooling an animal is, of course, reversible, allowing recovery of the worm after imaging is completed. However, if cooling is not maintained throughout the imaging procedure, the worm will warm up and start to move, making further live imaging difficult. In an apparatus for temperature-controlled microscopy (Rieder and Cole 1998), worms may be anesthetized using temperatures between 4°C and 8°C (Rabin and Podbilewicz 2000). A simple device consisting of a standard temperature-controlled water bath, a temperature-controlled microscope stage, a microscope lens heat exchanger (a water jacket that fits the oil immersion 60× objective), and a temperature measurement device has been used successfully not only to anesthetize worms, but also to study the temperature dependence of developmental processes (Rabin and Podbilewicz 2000).

   b. Chemicals can be used as an alternative to temperature for inhibiting worm movement. The addition of sodium azide (3–10 mM) to M9 medium stops all movement but at the expense of slowly killing the animal. The most common chemical for paralyzing worms is 1 mM levamisole, added to M9 medium. Levamisole anesthetizes the worm, and washing the animal with fresh M9 medium reverses its anesthetic effect.

4. In all cases, place the sample as quickly as possible onto the microscope stage and begin to collect the data (see the sections below on Observation of Nematodes and Data Collection).

Preparation and Use of Agar Pads for Live Microscopy

This method was modified from Sulston and Horvitz (1977). See WORMATLAS for a figure at http://www.wormatlas.org/anatmeth/agarpad.htm.

1. Melt 1–2 ml of 3–5% agar in a test tube. Maintain the melted agar at 60°C in a dry heat block. Make sure that no air bubbles form in the agar.

2. Prepare a short Pasteur pipette to transfer the agar by cutting it at the neck and leaving it immersed in distilled H_2O in a test tube kept at 60°C.

3. Prepare two "spacer" slides raised by one or two layers of adhesive tape.

4. Place a clean slide between the two spacer slides in a parallel orientation.

5. Use a Pasteur pipette to add a drop of agar and flatten it with a second slide (this slide may be siliconized) in a perpendicular orientation such that it is supported by the spacer slides. Make sure that no bubbles are formed.

6. After the agar is hard, carefully remove the perpendicular top slide. The agar slab should be 0.2–0.5 mm thick.

7. Add a small drop (~2 μl) of egg salts solution for early embryos or M9 buffer for larvae and adults.

> These solutions may contain 3–10 mM azide <!> or 1 mM levamisole to stop movement of the larvae and adults (see preceding protocol).

8. Quickly transfer nematodes using either a human eyelash taped to a toothpick, a platinum wire attached to a Pasteur pipette, or a mouth pipette apparatus with drawn out capillaries of proper size. Elute the worms into the small drop of buffer with gentle shaking.

9. Gently place a 12 × 12-mm number 1 coverslip onto the agar, taking care not to break the agar surface.

> The small volume of liquid and the pressure of the coverslip prevents the worms from thrashing, although they can still move.

10. Remove excess agar protruding beyond the coverslip with a razor blade. Use a fine paintbrush to seal the remaining agar pad with liquefied Vaseline (kept in a test tube at 60°C).

> For long-term observations of living nematodes, sealing is critical to prevent evaporation. Small air bubbles formed after placing the coverslip will be absorbed by the agar.

11. For extensive periods of observation of living larval specimens, coat the center of the coverslip with a very thin layer of *Escherichia coli* OP50 transferred from a Petri dish with a platinum wire. Worms stay in the area of the bacterial lawn where they will feed.

> Nematodes mounted as described above can be observed live using Nomarski optics and then removed for other studies such as immunolabeling or electron microscopy.

Preparation and Use of Agarose Pads for Microinjection

1. Wash 25 × 75-mm glass slides with distilled H_2O and soap. Rinse them twice with double-distilled H_2O. Allow the slides to air-dry and then store them in a clean box until ready for use.

2. Melt 2% agarose in double-distilled H_2O. Keep the melted agarose at 60°C in a water bath or in a dry heat block.

3. Place one drop (~0.2 ml) of the melted agarose on a clean slide and immediately cover with a 22 × 40-mm number 1 coverslip placed perpendicular (at 90°) to the slide.

4. Allow the agarose to solidify and separate the coverslip from the slide. Mark the surface of the coverslip containing the agarose.

5. Dry the coverslip overnight with the agarose side up at room temperature on a clean surface or bake them for several hours at 80°C.

6. Store the coverslips in a clean dry place to avoid dust and humidity.

7. Place a drop of liquid paraffin (BDH Chemicals 140172W) on the agarose pad. Use an eyelash taped to a toothpick to carefully transfer adult hermaphrodites containing a few eggs in the uterus to the agarose pads. Allow them to adhere.

8. Allow the worms to sit for few minutes on the slide, which cause partial dehydration. They are now ready for microinjection.

9. Microinject the worms using DIC optics (Nomarski) with a 40× dry objective and an inverted microscope as described by Mello and Fire (1995).

Preparation and Use of Poly-L-Lysine Coverslips for Observation of Embryos

1. Prepare a solution of 0.1% (1 mg/ml) high-molecular-weight poly-L-lysine (Sigma P 1524) in double-distilled H_2O.

2. Clean 22 × 22-mm number 1 coverslips with distilled H_2O and soap. Rinse them twice with double-distilled H_2O. Dry the coverslips in a 65°C oven, and allow them to cool to room temperature.

3. Place an ~150-μl drop of poly-L-lysine in the middle of each coverslip and incubate them for 30 minutes at room temperature.

4. Use a micropipette to remove excess poly-L-lysine. Mark the side of the coverslip containing the poly-L-lysine.

5. Dry the coverslips for 3 hours in a 65°C oven. Alternatively, place them on a hot plate for a few minutes until they are dried. Store the coverslips in a clean slide box until ready for use.

6. Add a drop of egg salts solution. Transfer embryos to the coverslip using either a mouth pipette with a capillary of proper diameter, an eyelash taped to a toothpick, a platinum pick, or a micropipette.

> Use glass pipettes or drawn out capillaries to transfer nematodes. For very precious experimental samples, use siliconized glass transfer pipettes. *Do not* use plastic tips to transfer *C. elegans* samples because nematodes stick to plastic.

7. Use a 1-ml syringe with a micropipette tip attached to its end to deliver petroleum jelly or vacuum grease onto the periphery of the coverslip ~2 mm from its edge.

8. Use fine tweezers to gently cover the coverslip containing the embryos with a second coverslip or a slide. Avoid trapping air bubbles.

9. Press the coverslips gently to remove excess liquid.

10. Observe the embryos using Nomarski (DIC) optics.

OBSERVATION OF NEMATODES

Live nematodes can be studied at all levels of microscopic resolution. At low magnification, dissection microscopes are useful, whereas at higher magnification and resolution, a wide range of optical methods can be employed, including DIC optics, phase contrast, fluorescence, polarized light, confocal, and dark field. However, DIC optics are used for most applications. Some other helpful recommendations for observation include the following.

- To localize worms under the microscope, use 4×, 10×, or 20× objectives.

- For microinjection, use 40× Plan Fluor or Plan Apochromat (numerical aperture [NA] 0.9 and higher) dry objectives.

- Maximum resolution and magnification are obtained with 100×, 1.4 NA Plan Apochromat oil immersion objectives used for lineaging, observation of cell fate, and to follow intracellular and intercellular events.

- Simpler lenses than Plan Apochromats are usually better for Nomarski, such as Plan or Plan Fluor (selected for less strain birefringence).

- For fluorescence microscopy, it is better to use 60× or 63× 1.4 NA Planachromat or Plan Fluor objectives, which transmit the appropriate wavelengths of light with higher efficiency.

- Leica, Zeiss, Olympus, and Nikon research microscopes all have equally fine optics for *C. elegans* live microscopy. Nikon oil immersion objectives (NA = 1.4) have the longest working distance, permitting deeper optical sectioning using confocal microscopy.

DATA COLLECTION

Embryonic and postembryonic cell lineages can be followed by direct observation with excellent spatial and temporal resolution. Using the lineage maps described by Sulston and Horvitz (1977) and Sulston et al. (1983) as guides, it is possible by microscopy to identify all nuclei in a worm and to follow all embryonic and postembryonic cell lineages with excellent spatial and temporal resolution. Drawings of the positions of nuclei are still in use today as standards for following the pathways of nematode cells. This method allows only a relatively small number of cells to be followed in a single worm, the only limitation being the short-term memory of the microscopist.

Nuclei, nucleoli, and intracellular granules are clearly resolved by Nomarski optics, due to differences in their refractive indices and the transparency of *C. elegans* (Sulston and Horvitz 1977). Although it is possible to follow by direct microscopic observation of living specimens all of the nuclei of different cell types as they are born, migrate, divide, and die, cell boundaries are very often not visible. It is thus necessary to stain specific cell types in the worms using antibodies and in situ hybridization and by tagging the products of reporter genes to follow cells, not only their nuclei (Sulston and Horvitz 1977; Podbilewicz and White 1994).

Photographic Records

Photographic records of living specimens require the use of low levels of illumination. Prolonged exposure to light necessary for photographic recordings can cause the worms to warm up and result in heat shock and photo-oxidative damage. It is therefore important to keep the temperature between 20°C and 22°C.

Video Microscopy

Video microscopy has been used to monitor *C. elegans* during early embryonic development (Deppe et al. 1978). These videos allow cell lineages or other events such as neuronal migration, cell death, or cell behavior to be followed in a reproducible quantitative fashion; however, it is difficult to follow later embryonic divisions occurring inside the embryo.

Multiple Focal Plane Time-lapse Recording Systems

Most cell lineages can be obtained by direct microscopic observation. Many cells, however, change their position during development, which requires live imaging of optical sections along the *z* axis. When required, the digital images obtained by optical, fluorescence, or confocal microscopy can be deconvoluted or used to make movies or stereo pairs. For Nomarski imaging, for example, we use a microscope equipped with DIC optics, a

Hamamatsu C2400 video camera with analog black-level subtraction, and a computer-controlled motorized focus drive unit. The original system used for this purpose was described by Mike Thomson in the Laboratory of Molecular Biology Electronics Workshop of the MRC in Cambridge U.K. (Hird and White 1993). The current system was developed by Charles Thomas at the University of Wisconsin in Madison and is available by downloading (http://www.loci.wisc.edu/4d/native/4d.html). The later version of software allows investigators to specify the number of focal planes per time point, the distance between each focal plane, the time interval between each three-dimensional set of recordings, and the total number of time points. A single video frame is recorded at each focal level in a scan, and at specified time intervals, scans are repeated. The images can be played either forward or backward in time and up or down in different focal levels at a given time point, allowing the tracking of small nuclei, granules, or other objects in three dimensions over time in a living specimen (Hird and White 1993; Thomas et al. 1996).

Digital Imaging

Digital imaging coupled to a confocal laser-scanning system or a wide-field microscope equipped with deconvolution software allows the recording of fluorescently labeled specimens using time-lapse imaging with the efficient removal of out-of-focus information (for overviews, see Chapters 14 and 15). Sequential images collected along the z axis and projections of whole animals can be obtained using commercially available software packages (e.g., www.scioncorp.com) or free software available from the National Institutes of Health (http://rsb.info.nih.gov/). Four-dimensional microscopic observations can be obtained using many commercially available systems including those manufactured by Zeiss, Nikon, Leica, Olympus, PerkinElmer, Deltavision, Bio-Rad, and others. These systems can be used to simultaneously collect both fluorescence and DIC images. Because *C. elegans* specimens are relatively thick (20–50 μm), it is convenient to collect 15–150 optical sections that are 0.1–1 μm apart, depending on the specific application. Consecutive time points are usually taken seconds to minutes apart depending on the process under study, taking care to turn off the illumination between scans and always taking care not to induce photodamage.

EXAMPLES OF LIVE IMAGING OF *C. ELEGANS*

First Divisions in Early Embryos and Chromatin Dynamics in Lamin Mutants

To image early cleavages and chromatin dynamics, it is convenient to use histone H2B fused to GFP (Praitis et al. 2001) or lamin::GFP (Liu et al. 2000). Time-lapse movies can be obtained using conventional confocal microscope systems and their included software. Early embryos dissected from transgenic hermaphrodites are placed with egg salts on agar pads. Chromatin dynamics can be followed easily, and wild-type embryonic cells can be compared with mutants or RNAi-treated embryos (e.g., lamin; Liu et al. 2000) (for normal lamin staining, see Fig. 1 and for chromosomal defects in *lamin(RNAi)* embryos, see Movie 20.1 on the accompanying DVD).

Simultaneous Nomarski and GFP Imaging of Elongating Epidermal Cells in Embryos

To image embryos using time-lapse movies in different focal planes, it is convenient to use the four-dimensional imaging system developed by John White and collaborators (Hird

TABLE 1. Troubleshooting guide for *C. elegans* live imaging

Problem	Solution
Animals move too much during imaging	Use less buffer or remove some buffer by capillary action with a piece of paper.
Larvae move and leave field of view	Place a thin coat of OP50 *E. coli* in the center of the coverslip. The worms will stay there and feed.
Embryos twitch or move too much	Use a temperature-controlled stage and a jacket surrounding the immersion objective to lower the temperature to 4–8°C.
Animals move	Use azide (3–10 mM) or levamisole (1 mM).
Focus drift during time-lapse microscopy	Keep the temperature in the room uniform using a good AC system (20–22°C).
Small focus drift during time-lapse microscopy	Adjust manually by moving the stage 1–2 clicks up or down until focus stabilizes (trial and error). Babysitting the embryos during imaging is sometimes necessary.
Out-of-focus fluorescence	Use confocal or deconvolution microscopy.
Animals do not stick to poly-L-lysine-coated coverslips	Use water instead of buffer. High salts may prevent binding. Alternatively, make fresh poly-L-lysine-coated coverslips.
Bubbles in the closed observation chamber	Press the coverslip gently with tweezers to displace the bubbles, or prepare a new chamber and gently place the coverslip using fine tweezers.

and White 1993; Skop and White 1998; Eliceiri et al. 2002). The software for this imaging system is freely available from John White at the University of Wisconsin in Madison. Suitable commercial software is also available with different confocal or deconvolution microscopy systems (see the section on Data Collection). Embryos are placed on poly-L-lysine-coated coverslips in a small drop of egg salts solution (Edgar 1995). Using a 1-ml syringe, a thin layer of high-vacuum grease (or petroleum jelly) is applied around the edge of the coverslip, and a second coverslip or a slide with an agar pad is placed on top. The embryos attach to the poly-L-lysine-coated coverslip, and the vacuum grease or petroleum

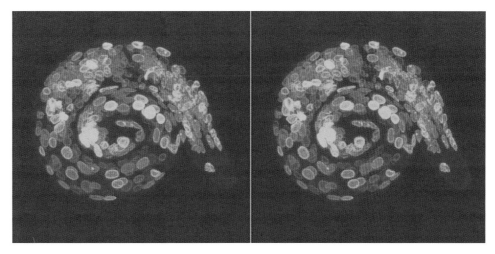

FIGURE 1. A stereo pair showing three-dimensional views of transgenic L1 larva expressing Ce-lamin fused to GFP (Liu et al. 2000). The animal was viewed with a Bio-Rad MRC-1024 confocal scanhead coupled to a Zeiss Axiovert 135M inverted microscope. The VoxBlast 3-Dimensional Measurement and Volume Visualization Software (VayTek, Fairfield, Iowa) were used for the three-dimensional reconstruction.

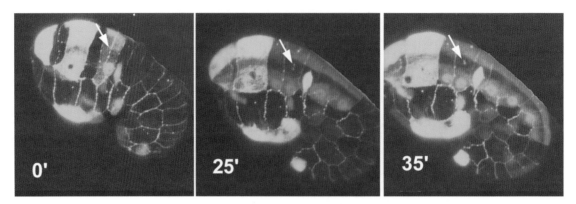

FIGURE 2. Projections of embryo expressing *AJM-1::*GFP in the zonulae adherens (ZA) and *eff-1p::*GFP in the cytoplasm of epidermal cells during skin syncytiogenesis (Rabin and Podbilewicz 2000; Mohler et al. 2002; Shemer and Podbilewicz 2002). The animal was viewed with a Bio-Rad MRC-1024 confocal scanhead coupled to a Nikon E800 microscope. The Lasersharp software was used for the three-dimensional reconstruction of the three *z*-series images shown that represent three time points in minutes. Arrows point to the ZA between two dorsal epithelial cells before (0 minute) and after (25 and 35 minutes) cell fusion. Strong cytoplasmic and nuclear staining is from the transcriptional reporter gene *eff-1promoter::*GFP. Cytoplasmic mixing confirms cell fusion, and ZA disappearance shows plasma membrane fusion.

jelly seals the chamber and protects the embryos from being squashed. This closed chamber allows extended observation of embryos expressing multiple fluorescence signals (GFP) and DIC (Rabin and Podbilewicz 2000) (Fig. 2 and Movies 20.2 and 20.3 on the accompanying DVD).

Analysis of Individual Larvae or Adults Using Confocal Microscopy and Three-dimensional Reconstructions of Vulval Rings

To analyze organogenesis of the vulva, including the stages of cell generation, cell migration, ring formation, and intra-ring cell fusion, a laser-scanning confocal microscope is used to follow different GFP reporters expressed in the vulval cells (Sharma-Kishore et al. 1999; Shemer et al. 2000). It is possible to follow individual worms throughout development, using successive recovery of the worms as described in the section on Preparation of Samples for Live Imaging, even after exposure to laser confocal microscopy. We anesthetize larvae with 0.01% levamisole to completely block movement in an agar pad. After rapid analysis using the 488-nm line of the 100-mW argon laser at 0.3% intensity (Rabin and Podbilewicz 2000), the coverslip is slowly removed, and the worms are gently transferred from the slide into a fresh plate and recovered with a drop of M9 at 20°C. After 2 hours or more, the worms are returned to a new agar pad with anesthesia, imaged, and recovered again (Shemer and Podbilewicz 2002) (Movies 20.4 and 20.5 on the accompanying DVD).

ACKNOWLEDGMENTS

We thank John White and members of our laboratories for suggestions for this manuscript. This work was funded by grants from the Israel Science Foundation (ISF), the National Institutes of Health (GM-64535), the German Cancer Research Center (DKFZ) to Y.G., and the Israel Science Foundation, the Charles H. Revson Foundation (203/00-

2), and the Fund for the Promotion of Research at the Technion and Human Frontier Science Program (HFSP) to B.P.

MOVIE LEGENDS

MOVIE 20.1. Arrested embryos injected with *lmn-1* dsRNA. The expression of the *lmn-1* gene (Liu et al. 2000) was down-regulated in *C. elegans* hermaphrodites expressing histone H2B-GFP (AZ212 line, kindly provided by Dr. J. Austin) (Praitis et al. 2001). The F_1 progeny of hermaphrodites injected with double-stranded RNA of *lmn-1* were mounted on an agar pad with egg salts solution, sealed with vacuum grease, and viewed with a Zeiss Axioplan II microscope equipped with an epifluorescence illuminator. An Axiocam CCD camera and the AxioVision Image Analysis package were used to collect the time-lapse data every 30 seconds. The video shows that most nuclei have disorganized chromatin. One nucleus shows the typical anaphase bridges during mitosis.

MOVIE 20.2. Simultaneous imaging of Nomarski (*blue*) and *eff-1promoter*::GFP (*green*) in developing embryos reveals timing of cell fusion. Confocal microscopy was performed as in Figure 2 (Rabin and Podbilewicz 2000). The initial expression of the reporter gene was observed in the precursors of the epidermal cells. Cytoplasmic mixing of the GFP signal in the dorsal and ventral hypodermis occurs during elongation (morphogenesis) of the embryo, reflecting cell fusion between epidermal cells (Mohler et al. 2002).

MOVIE 20.3. Expression of *AJM-1*::GFP in epithelial cells during embryonic morphogenesis. Animation of confocal projections collected every 5 minutes shows changes in epidermal cell morphology and cell fusion during elongation.

MOVIE 20.4. *z* series of two vulval toroids expressing *egl-17/FGFpromoter*::GFP (Burdine et al. 1998). In the L4 stage, the vulva is composed of a stack of seven rings or toroids (Sharma-Kishore et al. 1999). Only two vulval syncytial cells express fibroblast growth factor (FGF/egl-17). The vulD ring on top (dorsal) contains two nuclei and the vulC toroid contains four nuclei. Two additional rings above vulD and three ventral rings do not express this reporter gene at this stage.

MOVIE 20.5. Projection of the *z* series and rotation of two vulval rings expressing fibroblast growth factor (egl-17) from Movie 20.4. Using software from Lasersharp (Bio-Rad), it is possible to animate the vulval rings. The center (hole) allows the passage of eggs to the outside and sperm in the opposite direction during mating.

REFERENCES

Bargmann C.I. and Horvitz R.H. 1991. Chemosensory neurons with overlapping functions direct chemotaxis to multiple chemicals in *C. elegans*. *Neuron* **7:** 729–742.

Bargmann C.I., Hartwieg E., and Horvitz R.H. 1993. Odorant-selective genes and neurons mediate olfaction in *C. elegans*. *Cell* **74:** 515–527.

Burdine R.D., Branda C.S., and Stern M.J. 1998. Egl-17(fgf) expression coordinates the attraction of the migrating sex myoblasts with vulval induction in *C. elegans*. *Development* **125:** 1083–1093.

Chalfie M., Tu Y., Euskirchen G., Ward W.W., and Prasher D.C. 1994. Green fluorescent protein as a marker for gene expression. *Science* **263:** 802–805.

Chitwood B.G. and Chitwood M.B. 1974. *Introduction to nematology*. University Park Press, Baltimore.

Crittenden S.L., Troemel E.R., Evans T.C., and Kimble J. 1994. GLP-1 is localized to the mitotic region of the *C. elegans* germ line. *Development* **120:** 2901–2911.

Deppe U., Schierenberg E., Cole T., Krieg C., Schmitt D., Yoder B., and von Ehrenstein G. 1978. Cell lineages of the embryo of the nematode *Caenorhabditis elegans*. *Proc. Natl. Acad. Sci.* **75:** 376–380.

Edgar L.G. 1995. Blastomere culture and analysis. In *C. elegans: Modern biological analysis of an organism* (ed. H.F. Epstein and D.C. Shakes), vol. 48: pp. 303–320 . Academic Press, San Diego.

Eliceiri K.W., Rueden C., Mohler W.A., Hibbard W.L., and White J.G. 2002. Analysis of multidimensional biological image data. *BioTechniques* **33:** 1268–1273.

Ellis H.M., and Horvitz H.R. 1986. Genetic control of programmed cell death in the nematode *C. elegans*. *Cell* **44:** 817–829.

Fire A. 1986. Integrative transformation of *Caenorhabditis elegans*. *EMBO J.* **5:** 2673–2680.

Fire A., Albertson D., and Harrison S.W. 1991. Production of antisense RNA leads to effective and specific inhibition of gene expression in *C. elegans* muscle. *Development* **113:** 503–514.

Fire A., Harrison S.W., and Dixon D. 1990. A modular set of *lacZ* fusion vectors for studying gene expression in *Caenorhabditis elegans*. *Gene* **93:** 189–198.

Fire A., Xu S., Montgomery M.K., Kostas S.A., Driver S.E., and Mello C.C. 1998. Potent and specific genetic interference by double-stranded RNA in *Caenorhabditis elegans*. *Nature* **391:** 806–811.

Goodwin E.B., Okkema P.G., Evans T.C., and Kimble J. 1993. Translational regulation of *tra-2* by its 3′ untranslated region controls sexual identity in *C. elegans*. *Cell* **75:** 329–339.

Guo S. and Kemphues K.J. 1995. *par-1*, a gene require for establishing polarity in *C. elegans* embryos, encodes a putative ser/thr kinase that is asymmetrically distributed. *Cell* **81:** 611–620.

Hedgecock E.M. and White J.G. 1985. Polyploid tissues in the nematode *Caenorhabditis elegans*. *Dev. Biol.* **107:** 128–138.

Hedgecock E.M., Sulston J.E., and Thomson J.N., 1983. Mutations affecting programmed cell deaths in the nematode *Caenorhabditis elegans*. *Science* **220:** 1277–1279.

Hedgecock E.M., Culotti J.G., Thomson J.N., and Perkins L.A. 1985. Axonal guidance mutants of *Caenorhabditis elegans* identified by filling sensory neurons with fluorescein dyes. *Dev. Biol.* **111:** 158–170.

Hill D.P. and Strome S. 1988. An analysis of the role of microfilaments in the establishment and maintenance of asymmetry in *Caenorhabditis elegans* zygotes. *Dev. Biol.* **125:** 75–84.

Hird S. 1996. Cortical actin movements during the first cell cycle of the *Caenorhabditis elegans* embryo. *J. Cell Sci.* **109:** 525–533.

Hird S.N. and White J.G. 1993. Cortical and cytoplasmic flow polarity in early embryonic cells of *Caenorhabditis elegans*. *J. Cell Biol.* **121:** 1343–1355.

Hirose T., Nakano Y., Nagamatsu Y., Misumi Y., Ohta H., and Ohshima Y. 2003. Cyclic GMP-dependent protein kinase EGL-4 controls body size and lifespan in *C. elegans*. *Development* **130:** 1089–1099.

Horvitz H.R. and Sulston J.E. 1980. Isolation and genetic characterization of cell-lineage mutants of the nematode *Caenorhabditis elegans*. *Genetics* **96:** 435–454.

Hutter H. and Schnabel R. 1995. Specification of anterior-posterior differences within the AB lineage in the *C. elegans* embryo: Polarising induction. *Development* **121:** 1559–1568.

Inoue T., Sherwood D.R., Aspoeck G., Butler J.A., Gupta B.P., Kirouac M., Wang M., Lee P.Y., Kramer J.M., Hope I., Burglin T.R., and Sternberg P.W. 2002. Gene expression markers for *Caenorhabditis elegans* vulval cells. *Mech. Dev.* **119:** S203–S209.

Keating H.H. and White J.G. 1998. Centrosome dynamics in early embryos of *Caenorhabditis elegans*. *J. Cell. Sci.* **111:** 3027–3033.

Kerr R., Lev-Ram V., Baird G.V.P., Tsien R.Y., and Schafer W.R. 2000. Optical imaging of calcium transients in neurons and pharyngeal muscle of *C. elegans*. *Neuron* **26:** 583–594.

Kirby C., Kusch M., and Kemphues K., 1990. Mutations in the *par* genes of *Caenorhabditis elegans* affect cytoplasmic reorganization during the first cell cycle. *Dev. Biol.* **142:** 203-215.

Liu J., Rolef Ben-Shahar T., Riemer D., Treinin M., Spann P., Weber K., Fire A., and Gruenbaum Y. 2000. The *Caenorhabditis elegans* lamin gene is essential and is required for nuclear organization, mitotic progression, chromosome segregation and spatial organization of nuclear pore complexes. *Mol. Biol. Cell* **11:** 3937–3947.

Mains P.E., Kemphues K.J., Sprunger S.A., Sulston I.A., and Wood W.B. 1990. Mutations affecting the meiotic and mitotic divisions of the early *Caenorhabditis elegans* embryo. *Genetics* **126:** 593–605.

Mello C. and Fire A. 1995. DNA transformation. *Methods Cell Biol.* **48:** 451–482.

Mello C.C., Kramer J.M., Stinchcomb D., and Ambros V. 1991. Efficient gene transfer in *C. elegans*: Extrachromosomal maintenance and integration of transforming sequences. *EMBO J.* **10:** 3959–3970.

Miller D.M., Desai N.S., Hardin D.C., Piston D.W., Patterson G.H., Fleenor J., Xu S., and Fire A. 1999. Two-color GFP expression system for *C. elegans*. *BioTechniques* **26:** 914–921.

Mohler W.A. and White J.G. 1998. Stereo-4-D reconstruction and animation from living fluorescent specimens. *BioTechniques* **24:** 1006–1010.

Mohler W.A., Shemer G., del Campo J., Valansi C., Opoku-Serebuoh E., Scranton V., Assaf N., White J.G., and Podbilewicz B. 2002. The type I membrane protein EFF-1 is essential for developmental cell fusion in *C. elegans*. *Dev. Cell* **2:** 355–362.

Podbilewicz B. and White J.G. 1994. Cell fusions in the developing epithelia of *C. elegans*. *Dev. Biol.* **161:** 408–424.

Praitis V., Casey E., Collar D., and Austin J. 2001. Creation of low-copy integrated transgenic lines in *Caenorhabditis elegans*. *Genetics* **157:** 1217–1226.

Rabin Y. and Podbilewicz B. 2000. Temperature-controlled microscopy for imaging living cells: Apparatus, thermal analysis, and temperature dependency of embryonic elongation in *C. elegans*. *J. Microsc.* **199:** 214–223.

Rieder C. and Cole R. 1998. Perfusion chambers for high-resolution video light microscopic studies of vertebrate cell monolayers: Some considerations and design. *Methods Cell Biol.* **56:** 253–275.

Schierenberg E. 1997. Nematodes, the roundworms. In *Embryology: Constructing the organism* (ed. S.F. Gilbert and A. M. Raunio), pp. 131–148. Sinauer Associates, Sunderland, Massachusetts.

Sengupta P., Colbert H.A., and Bargmann C.I. 1994. The *C. elegans* gene *odr-7* encodes an olfactory-specific member of the nuclear receptor superfamily. *Cell* **79:** 971–980.

Seydoux G., Mello C.C., Pettit J., Wood W.B., Priess J.R., and Fire A. 1996. Repression of gene expression in the embryonic germ lineage of *C. elegans*. *Nature* **382:** 713–716.

Sharma-Kishore R., White J.G., Southgate E., and Podbilewicz B. 1999. Formation of the vulva in *C. elegans*: A paradigm for organogenesis. *Development* **126:** 691–699.

Shemer G. and Podbilewicz B. 2002. LIN-39/Hox triggers cell division and represses EFF-1/Fusogen-dependent vulval cell fusion. *Genes Dev.* **16:** 3136–3141.

Shemer G., Kishore R., and Podbilewicz B. 2000. Ring formation drives invagination of the vulva in *C. elegans*: Ras, cell fusion and cell migration determine structural fates. *Dev. Biol.* **221:** 233–248.

Singh R.N. and Sulston J.E. 1978. Some observations on moulting in *Caenorhabditis elegans*. *Nematologica* **24:** 63–71.

Skop A.R. and White J.G. 1998. The dynactin complex is required for cleavage plane specification in early *Caenorhabditis elegans* embryos. *Curr. Biol.* **8:** 1110–1116.

Squirrell J.M., Wokosin D.L., White J.G., and Bavister B.D. 1999. Long-term two-photon fluorescence imaging of mammalian embryos without compromising viability. *Nat. Biotechnol.* **17:** 763–767.

Strome S. and Wood W.B. 1983. Generation of asymmetry and segregation of germ-line granules in early *C. elegans* embryos. *Cell* **35:** 15–25.

Sulston J. and Hodgkin J. 1988. Methods. In *The nematode* Caenorhabditis elegans (ed. W.B. Wood), pp. 587–606. Cold Spring Harbor Laboratory, Cold Spring Harbor, New York.

Sulston J.E. and White J.G. 1980. Regulation and cell autonomy during postembryonic development of *Caenorhabditis elegans*. *Dev. Biol.* **78:** 577–597.

Sulston J.E. and Horvitz H.R. 1977. Postembryonic cell lineages of the nematode *Caenorhabditis elegans*. *Dev. Biol.* **56:** 110–156.

Sulston J.E., Albertson D.G., and Thomson J.N. 1980. The *Caenorhabditis elegans* male: Postembryonic development of nongonadal structure. *Dev. Biol.* **78:** 542–576.

Sulston J.E., Schierenberg E., White J.G., and Thomson J.N. 1983. The embryonic cell lineage of the nematode *Caenorhabditis elegans*. *Dev. Biol.* **100:** 64–119.

Thomas C., DeVries P., Hardin J., and White J. 1996. Four-dimensional imaging: Computer visualization of 3D movements in living specimens. *Science* **273:** 603–607.

White J.G. 1980. The astral relaxation theory of cytokinesis revisited. *BioEssays* **2:** 267–272.

White J.G. and Amos W.B. 1987. Confocal microscopy comes of age. *Nature* **328:** 183–184.

White J.G., Amos W.B., and Fordham M. 1987. An evaluation of confocal versus conventional imaging of biological structures by fluorescence light microscopy. *J. Cell Biol.* **105:** 41–48.

Wood W.B. 1988. *The nematode* Caenorhabditis elegans. Cold Spring Harbor Laboratory Press, Cold Spring Harbor, New York.

Zipkin I.D., Kindt R.M., and Kenyon C.J. 1997. Role of a new rho family member in cell migration and axon guidance in *C. elegans*. *Cell* **90:** 883–894.

Zipperlen P., Fraser A.G., Kamath R.S., Martinez-Campos M., and Ahringer J. 2001. Roles for 147 embryonic lethal genes on *C. elegans* chromosome I identified by RNA interference and video microscopy. *EMBO J.* **20:** 3984–3992.

CHAPTER TWENTY ONE

Time-lapse Cinematography in Living *Drosophila* Tissues

Ilan Davis and Richard M. Parton

Wellcome Trust Centre for Cell Biology, University of Edinburgh, Edinburgh EH9 3JR, Scotland, United Kingdom

INTRODUCTION

THE FRUIT FLY, *DROSOPHILA MELANOGASTER*, has been an extraordinarily successful model organism for studying the genetic basis of development and evolution. It is arguably the best-understood complex multicellular model system, owing its success to many factors (Table 1). Perhaps its greatest strength is the diversity of easily studied tissue types, and a vast array of genetic and molecular tools (St Johnston 2002), which have been developed over many years by the highly cooperative community of fly researchers.

Recent developments in imaging techniques, in particular, sophisticated fluorescence microscopy methods and equipment, now allow cellular events to be studied at high resolution in living material. This ability has enabled the study of features that tend to be lost or damaged by fixation, such as transient or dynamic events. Coupling live cell imaging with the experimental advantages of *Drosophila* has led to great advances in cell and developmental biology.

Although many of the techniques of live cell imaging in *Drosophila* are shared with the greater community of cell biologists working on other model systems, studying living fly tissues presents unique difficulties in keeping the cells alive, introducing fluorescent probes, and imaging through thick hazy cytoplasm. This chapter outlines the major tissue types amenable to study by time-lapse cinematography and different methods for keeping them alive. It then describes various imaging and associated techniques best suited to following changes in the distribution of fluorescently labeled molecules in real-time in these tissues. Finally, likely future developments in imaging methods and their possible application to *Drosophila* are described.

PREPARATION OF MATERIAL FOR LIVE CELL IMAGING

NOTE: <!> indicates hazardous material. See Cautions Appendix.

A crucial first step prior to imaging is the preparation of the experimental material (see Chapter 17). The main aims are to ensure physiological relevance and to achieve the best con-

TABLE 1. *Drosophila* as a model organism

Size and Diversity of Tissue Types

• Small and easy to culture in useful quantities in the lab.
• Prolific breeders, short life cycle (~10 days; 25°C).
• Contains the majority of complex tissue types found in mammals, but more accessible to manipulation.
• Good for imaging; small enough to examine whole organism under microscope and large enough to isolate individual tissues (e.g., embryos ~150 × 150 × 400 µm).
• Giant salivary glands with polytene chromosomes (easily identified bands) simple karyotype (4 chromosomes).

Genetically Tractable

• Many simple genetic screens over many years have identified, and continue to identify, new mutations.
• Sequenced genome (*D. melanogaster* and *D. pseudobscura*).
• Many genetic tricks, such as P-element transformation, germ line clones, somatic clones, many tissue-specific expression lines, very easy RNAi on tissue culture cells.
• A variety of existing GFP-expressing lines available upon request from individual research groups or stock centers.

Background

• Large community of highly cooperative investigators and communal genome efforts easily accessible via Flybase Web site.
• Ready availability of mutant lines.
• History of the application of imaging techniques to fixed and live material.

Drawbacks

• Cannot store lines frozen very easily.
• Transgenic line development takes a few months.
• Homologous recombination still difficult.
• Do not "self" as with hermaphrodite nematodes.

ditions for good image quality. However, in the case of live cell imaging, it is often necessary to balance image quality against optimizing tissue viability. The optimal conditions for preparing samples for imaging while maintaining cell viability vary with the tissue type.

Drosophila Tissue Types Are Amenable to Time-lapse Imaging

A variety of different tissue types are routinely used to investigate different kinds of questions in axis specification, cell differentiation, and organogenesis. The different tissues are derived from different stages of the life cycle of the flies: from fertilized eggs through larvae to adults. An exhaustive list of examples of work on living tissue in flies is beyond the scope of this chapter, but a few key examples are presented (Table 2).

Collection and Mounting of *Drosophila* Embryos

The following description refers to the steps of the protocol as outlined in Figure 1. For culturing *Drosophila,* see Sullivan et al. (2000).

1. Place flies (~5–12 days after hatching) in a "cage" (a plastic cylinder with a nylon mesh top). Tape a "wine agar" plate (Davis 2000) sprinkled with yeast powder or smeared with yeast paste to the bottom.

2. After the desired time (e.g., a 2-hour collection, aged for 2 hours at 25°C for blastoderm embryos), exchange the plate, and then age it as required.

3. Add distilled H_2O to the removed plate. Use a soft brush to loosen the embryos from the agar surface.

4. Pour suspended embryos into a "filter tube" (a plastic tube, of diameter 2–5 cm, with nylon mesh fixed to the bottom). Using a plastic weigh boat as a washing basin, rinse the embryos twice with room-temperature distilled H_2O to clean off yeast and debris.

TABLE 2. Examples of live cell imaging of *Drosophila* tissues

Tissue Type	Subject	Approach	Reference
Egg chambers	Border-cell migration	GFP fluorescence	Rorth (2002)
	Microtubules	*Tau*-GFP, tubulin-GFP, Exu-GFP	Grieder et al. (2000); Micklem et al. (1997); Theurkauf and Hazelrigg (1998)
	mRNA transport	AlexaFluor546-RNA, FITC-RNA	Cha et al. (2001); Glotzer et al. (1997); MacDougall et al. (2003)
Spermatogenesis	Cell division	GFP, FRAP	Noguchi and Miller (2003)
Embryos	Lipid droplet motility	DIC	Welte et al. (1998)
	Asymmetric cell division	*Tau*-GFP, *Src*-GFP	Kaltschmidt et al. (2000)
	Asymmetric cell division	FRAP GFP-Aurora	Berdnik and Knoblich (2002)
	Neuronal cell ablation	GFP	Hidalgo and Brand (1997)
	Marking clones	Caged-FITC-Dextran	Vincent and O'Farrell (1992)
	mRNA transport	AlexaFluor546-RNA	Bullock and Ish-Horowicz (2001); Wilkie and Davis (2001)
Larvae (dissected)	Salivary gland chromosomes	DIC, GFP	Vazquez et al. (2001); Vazquez et al. (2002)
	Imaginal discs	GFP-Hh	Entchev et al. (2000)
	Imaginal discs	Argos-GFP	Greco et al. (2001)
Pupae	Asymmetric cell division	GFP	Gho et al. (1999)
Adults	Eye in whole adult	Rhodapsin	Mollereau et al. (2000); Pichaud and Desplan (2001)
	Brain, calcium	Camgaroo	Yu et al. (2003)

5. "Dechorionate" the embryos (i.e., remove the egg shell or chorion) in ~7% fresh hypochlorite solution (diluted bleach) <!> for 1–2 minutes with gentle agitation. Monitor the extent of dechorionation under a dissecting microscope.

6. Wash the embryos twice more with room-temperature distilled H₂O. Dechorionated embryos tend to float.

7. Blot excess water from the filter tube, and use a fine brush to transfer the dechorionated embryos to a water agar plate for arrangement.

8. View the embryos under a dissecting microscope, space them at least one embryo-width apart to avoid problems of hypoxia. Arrangement in rows facilitates injection.

9. Cut out an agar block around the aligned embryos and use it to convey them to a slide.

10. Coat coverslips with "heptane glue." Application of ~20 μl can be made over a 1.5 × 3-cm area by applying the glue at an angle and "streaking" left to right. The dried glue should only just be visible on the coverslip.

   The glue stock solution is prepared from Scotch double-sided cellotape (3 meters of "scrunched up" tape in 200 ml of heptane <!>, shaken for 3 hours, and then decanted). The glue should be clear and very thin.

11. Press the coverslip firmly against the agar block to transfer the embryos. The agar block prevents crushing and desiccation.

12. At this stage, gently dry the embryos in a chamber with a small amount of silica gel for 4–10 minutes. Slight dehydration removes surface water and also allows the embryos to be flattened or injected more easily. Avoid excessive dehydration as it leads to flaccid defective embryos.

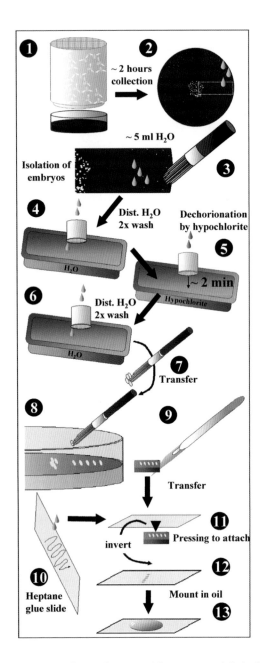

FIGURE 1. Collection and mounting of *Drosophila* embryos (for details, see the steps in the section on Collection and Mounting of *Drosophila* Embryos).

13. Cover the embryos with oxygen-rich halocarbon oil (Halocarbon Products Corporation, Series 700).

> If access to the embryos is not required, a Teflon membrane (such as may be obtained from YSI Incorporated; standard membrane kit, 5793) helps to maintain viability for prolonged observations and improves optical clarity (Fig. 2). The preparation is now ready to view under the microscope.

Isolation of *Drosophila* Egg Chambers

The following description refers to the steps of the protocol as outlined in Figure 3.

1. Prepare well-fed female flies (up to 5 days after hatching) and anesthetize them with CO_2.

2. Place individual females (distinguished from males by the clearer dark banding of the abdomen, less obvious black posterior, and lack of dark sex combs on the front legs) ventral side up in halocarbon oil (Halocarbon Products Corporation, Series 95) on a glass slide.

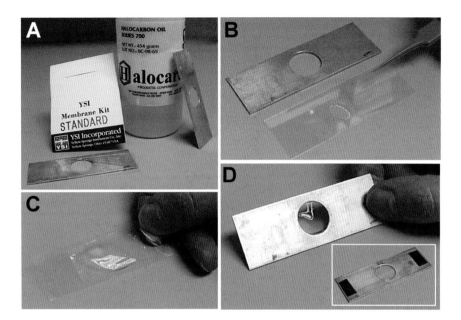

FIGURE 2. Mounting material in halocarbon oil. (*A*) Halocarbon oil, Teflon membranes, and aluminium coverslip mounts. (*B*) Laying a piece of membrane over the mounting medium. (*C*) Embryos on a coverslip, held by heptane glue, mounted in series 700 halocarbon oil and covered with a Teflon membrane. (*D*) Mounting the coverslip to an aluminium coverslip mount (an aluminium "slide") to fit certain stage plates or to be used on an upright microscope. (*Inset*) Reverse side view.

3. View the slide under a dissecting microscope using a dark base plate.

4. Use two pairs of fine-nosed tweezers to sacrifice each fly by crushing its head. Carefully remove the abdomen. Pulling the gut away from the abdomen makes things easier later on.

5. Open the exoskeleton of the abdomen by pinching individual plates at the center with both tweezers and drawing apart.

6. Identify the ovaries, which should almost fill the posterior and appear as two roughly tear-shaped structures.

7. Handling the ovaries only at their posterior end (the milky white end that contains the very old egg chambers), draw the ovaries out of the exoskeleton.

8. Transfer the ovaries to a drop of halocarbon oil on a fresh slide.

9. Holding an ovary at its posterior end with tweezers, use a fine (1-μm tip) tungsten needle to slowly draw out individual ovarioles ("strings" of egg chambers). For descriptions of ovary organization, see Spradling (1993).

> Individual egg chambers can be isolated from ovarioles. The relative positions of egg chambers in an ovariole reveal a serial sequence of developmental stages. The whole operation must be performed within a few minutes to maintain viability. As egg chambers isolated under oil tend to stick firmly to the glass coverslip, the preparation is immediately ready to be examined under the microscope.

Methods for Keeping Various Fly Tissues Alive on the Microscope Stage

Drosophila tissues commonly used for experimental purposes are relatively simple to maintain in a physiological state on the microscope stage. Although there are many common considerations

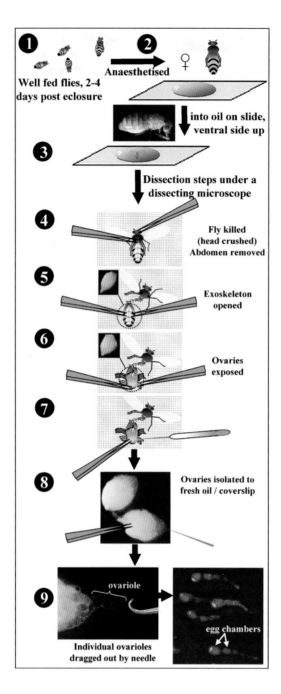

FIGURE 3. Isolation of *Drosophila* egg chambers (for details, see the steps in the section on Isolation of *Drosophila* Egg Chambers).

for the different kinds of tissues, such as temperature (generally, 18–37°C), sensitivity to hypoxia, and dehydration, the optimal conditions for viability vary under different experimental circumstances as outlined in Table 3.

An aqueous-based medium is not suitable for mounting early-to-midstage egg chambers, because it causes a change in microtubule organization and follicle cell morphology. Such egg chambers survive better in halocarbon oil (Halocarbon Products Corporation, Series 95), which affords good oxygenation, has low viscosity, and thus prevents dehydration and hypoxia. Halocarbon oil also has good optical properties for imaging, as it has a refractive index (RI) similar to that of glycerol. The disadvantage of halocarbon oil is that changes in the composition of the medium are not possible, as they are in the case of aqueous solutions. However, halocarbon oil is more convenient than aqueous media for injection of material into the eggs, which offers a means of adding but not removing compounds. In contrast, late-staged egg

TABLE 3. Optimal conditions for culturing *Drosophila* tissues during imaging

Tissue	Optimized Culture Conditions	Imaging Conditions
Early-to mid-stage egg chambers dissected from well-fed females (Fig. 3)	Halocarbon oil (Series 95) Avoids dehydration/hypoxia	Early stages: Problem of activated and loose MT organization in aqueous media. Oil has higher refractive index than water. Good environment for injection.
Late-stage egg chambers (as above)	Grace's medium (Sigma) Provides ionic, osmotic balance and nutrients	Late stages not susceptible to problems of early stages.
Embryos, dechorionated and dehydrated (Fig. 1)	Halocarbon oil (Series 700, RI similar to glycerol)	Prevents excess dehydration while avoiding hypoxia (which causes changes to the cell cycle). Good environment for injection.
	Halocarbon oil with Teflon breathable membrane	Better dehydration prevention for long-term development studies. Better bright-field imaging. Can help squash specimen for greater optical clarity (reduced spherical aberration).
	Aqueous medium	Water immersion objectives, which have longer working distance with high NA.

chambers develop better in aqueous-based culture media, such as Grace's medium, than in halocarbon oil, although the basis of this difference is unclear.

To optimize imaging quality, it is important to choose microscope objectives that are appropriate for the refractive index of the tissue or mounting medium used (see also the section below on Overcoming the Special Problems of Imaging Thick *Drosophila* Specimens). For aqueous media, water immersion objectives (e.g., the 20× 0.75 numerical aperature [NA] and 60 × 1.2 NA WI) are useful, or with upright microscopes, dipping objectives (e.g., 20× to 60× W Plan Fluor long working distance) are useful. Such objectives are available from all of the leading manufacturers. For higher-refractive-index media and tissues, glycerol and multi-immersion objectives are available (e.g., the Nikon 20 × 0.75 NA multi-immersion and Leica HCX CS PL APO 63 × 1.3 NA glycerol immersion).

In addition to choosing the appropriate medium, it is also useful to consider culture chambers and methods of securing the specimens to the stage. The choice of culture chamber is dependent on whether an inverted or upright microscope is used, although most labs favor the former. A common method of imaging living embryos is to literally glue them to a coverslip, cover them with halocarbon oil, and secure the coverslip to the stage of an inverted microscope with slide clips (see Fig. 1 and 2, and Table 3) (Davis 2000). Egg chambers can be similarly mounted, except that glue is not required, as they naturally stick to glass under halocarbon oil. When using oil immersion objectives on such coverslips, an aluminium slide with a small hole (see Fig. 2) can be used as a support, thus avoiding unwanted focus changes due to bending of the coverslip.

Maintaining the specimen and the stage at a constant temperature is not only important for avoiding focal shift problems, but also very important physiologically, for example, when imaging temperature-sensitive mutations at the restrictive temperature (Wilkie and Davis 2001). For controlling the temperature of the specimen, we use the Bioptechs stage (http://www.bioptechs.com/), but many other stages are also available.

FLUORESCENT REAGENTS FOR LIVE CELL IMAGING AND THEIR INTRODUCTION INTO CELLS

Most biological specimens, including *Drosophila* tissues, are relatively transparent, so that details of internal and intracellular morphology are difficult to image in untreated living

TABLE 4. Examples of fluorescent reagents used for live cell imaging in *Drosophila*

Marker	Use	Source, Reference
Rhodamine B	General staining of fly eggs	Sigma; I. Davis (unpubl.)
nlsGFP	General marker for all nuclei in all tissues	Davis et al. (1995)
Numerous stains and dyes	For numerous cell components	Molecular Probes
Caged-FITC–Dextran-nls	Marking clones of cells by photoactivation	Molecular Probes; Vincent and O'Farrell (1992)
Actin-GFP	Marking actin in sperm cells	Noguchi and Miller (2003)
Rhodamine-histone	Tracing cell lineage	Molecular Probes; Minden et al. (1989)
Rhodamine-tubulin	Microtubule cytoskeleton dynamics	Molecular Probes; Robinson et al. (1999)
Histone-GFP	Marker for chromosomes and cell division	Yucel et al. (2000)
Tubulin-GFP	Microtubule cytoskeleton dynamics, axon marker	Grieder et al. (2000)
Tau-GFP	Microtubule cytoskeleton dynamics	Micklem et al. (1997)
Ncd-GFP	MT motor protein decorates spindles in mitosis but not interphase MTs	Endow and Komma (1997)
Wg-GFP	Movement of Wg signal	Pfeiffer et al. (2002)
Hh-GFP	Movement of Hh signal	Entchev et al. (2000)
LacI-LacO-GFP	Following the movement of individual chromosome sites	Marshall et al. (1997)
Centrosomin-GFP	Following dynamics of centrosome movement	Megraw et al. (2002)

specimens using simple bright-field techniques. Although there are methods for improving contrast in bright-field imaging (see the section on Contrast-enhancing Bright-field Methods), fluorescence microscopy offers greater advantages and possibilities for increasing contrast and determining the specific localization of molecules in cells. Fluorescence microscopy has proven to be particularly well suited to the study of live material, now that an ever-increasing and bewildering array of different "vital" probes are available for tracking cellular components and activities (Haugland 2003), including some that have been applied to *Drosophila* (Table 4). The remainder of this section outlines the three methods most commonly used to introduce an appropriate label into *Drosophila* tissue without unduly perturbing the process.

External Application of Dyes

In principle, the simplest method of introducing membrane-permeable reagents into *Drosophila* embryos or other tissues is to soak the tissue in the reagent. However, in practice, this is only easily achieved in imaginal discs, late-stage egg chambers, and other tissues that can be dissected into aqueous-based media. The waxy vitelline membrane surrounding the embryo prevents membrane-permeable reagents from accessing the plasma membrane, and early- and mid-stage egg chambers must be dissected directly into halocarbon oil (see Table 3), making it difficult to add aqueous reagents after dissection. There are solutions to these problems, but they come at a cost. Embryonic vitelline membranes can be permeabilized with heptane or octane before adding the membrane-permeable solutions, but this treatment yields embryos with very fragile membranes that exhibit nonspecific lethality and sickness. Thus, results using this procedure must be interpreted with care. Membrane-permeable reagents can be introduced into early- to mid-stage egg chambers either by feeding adult female flies on a sugar and dye mixture on filter paper or by injecting the dyes directly into the abdomen of the mothers.

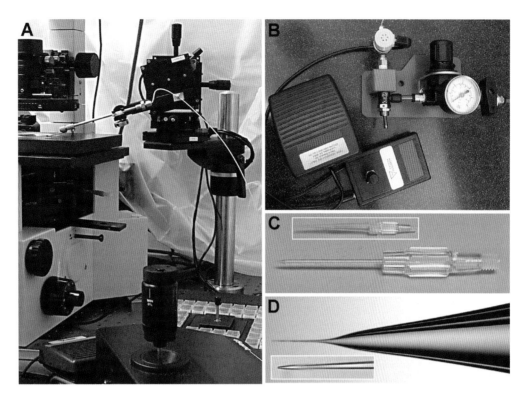

FIGURE 4. Injection apparatus. (*A*) Vibration-isolated table-mounted Burleigh micromanipulator system with joystick fine control and ability to rotate/lift away the injection needle and bring it back to virtually the same spot. Other manipulators allow direct mounting on the microscope body. (*B*) Components of our simple gas-powered injection system. *Clockwise from top:* simple desktop regular assembly allowing pressure adjustment and venting, the solenoid that controls pulsed release of gas is colored yellow; pulse-length regulator; footswitch to trigger pulses (components from Tritech Research, Los Angeles). Pressure is provided from a standard nitrogen gas cylinder and regulator (not shown). (*C*) Commercially available prepulled sterile glass needles (Eppendorf) offer convenience at a cost. (*Inset*) The needle with protective sheath. (*D*) The same needle as above at increased magnification showing the distinctive taper, useful for entering deep into embryos without excess damage. With any needle puller (e.g., the Sutter P97), custom shapes and tip sizes may be produced.

Microinjection

The most direct method of introducing any fluorescent reagent into certain living *Drosophila* tissues, whether or not the reagent is membrane permeable, is microinjection (Figure 4; Movie 21.1). Microinjection affords the advantage of delivering a high concentration of material at a very precisely controlled time. For embryos and egg chambers, with practice and suitable equipment, injection is the most reliable and quantitatively reproducible technique. However, there are several disadvantages to microinjection.

- Requires complex apparatus.
- Limit on the numbers of specimens that can be processed.
- Limit on the final concentration of reagent once injected into the cell (injected volume is usually no more than 5% of the total cell volume).
- Danger of physically damaging the cells, especially when they are small in size.

  For best results, it is worth taking several precautions.

1. Briefly centrifuge the material to be injected to avoid aggregates that can block the injection needles.

    Needles can often be unblocked by breaking the needle tip on a broken edge of a piece of coverslip under halocarbon oil, allowing the needle to be used for multiple injections.

2. Use only newly pulled or commercially supplied prepulled needles (such as Eppendorf femto tips).

3. If possible, make sure the needle has a slight positive "back pressure" from the injection system to reduce the chance of blocking it with cytoplasmic debris.

4. Anchor the cells or hold them down in some way during microinjection.

5. Embryos should be neither too dehydrated nor too "turgid" (see the earlier section on Collection and Mounting of *Drosophila* Embryos). This is best judged from experience.

Fluorescent Proteins

Arguably, the most important and versatile method of introducing fluorescent tagging into living *Drosophila* cells is the use of endogenously expressed recombinant fluorescent proteins (Chapter 1 and Table 4). There is an ever-increasing number of different colors, flavors, combinations, and permutations of fluorescent proteins to choose from (Zhang et al. 2002), all of which have the advantage of interfering less with cell function than is the case with either microinjection or permeabilization. In addition, virtually limitless amounts of labeled material can be produced. However, a major disadvantage of fluorescent proteins is the lead-up time for making the constructs and transgenic flies, if they are not already available (Roberts and Standen 1998; Davis 2000; see also Chapter 2). Although transient expression of constructs is possible (instead of making transgenic flies), this method has not been used extensively in *Drosophila* (Bossing et al. 2002).

With injection of a fluorescent dye into a cell or tissues, the exact time of introduction is known, thus time-lapse experiments that follow the fate, translocation, or diffusion rates of the probe are possible. Such studies are much more difficult with fluorescent protein expression, even with inducible promoters, because there is a lag of several hours before the protein exhibits fluorescence (Davis et al. 1995). Recent developments in the field, however, provide a solution to this problem. The use of photoactivatable forms of green fluorescent protein (GFP) (see Chapter 32; Patterson and Lippincott-Schwartz 2002) or "Kaede" (Ando et al. 2002), both of which change their spectral properties after irradiation with short-wavelength blue light (~405 nm), allows the kinetics of the "activated" population of protein molecules to be followed.

The versatility of fluorescent proteins is increasing dramatically as the technique is freed from the need to directly tag the proteins of interest. There is also the promise of a new generation of fluorescent-protein-based "reporter" molecules that may be used as indirect probes for nonprotein cellular components (Zhang et al. 2002), such as the recent calcium-ion-sensing camgaroo (Yu et al. 2003; see also Chapter 1). Indirectly tagging of DNA and RNA with GFP-tagged nucleic-acid-binding proteins is also possible, but it has a long lead-up and development time compared to the direct tagging of proteins.

HOW TO SELECT APPROPRIATE IMAGING EQUIPMENT AND METHODOLOGY

The wide range of microscopy equipment and techniques are truly bewildering and ever-expanding. In the past two decades, there have been many revolutions in light microscopy techniques made possible by improvements in optics, detector technology, and computers.

Furthermore, there is no indication that the rate of development of new equipment is slowing down. A description of the full range of available imaging possibilities is beyond the scope of this chapter; however, we have attempted to provide an overview of available options and important considerations applicable to imaging *Drosophila* cells and tissues (Table 5). It is assumed that the reader is familiar with the basics of bright-field and fluorescence microscopy. Numerous excellent comprehensive references and Web resources are available on the subject (Murphy 2001; also see Web Resources at the end of this chapter).

Evaluating a Microscope System

When deciding which kind of microscope system to use or purchase, the first consideration is what is most important for the imaging: speed, sensitivity, resolution, multi-wavelength discrimination, or cell viability. Many of these requirements are mutually exclusive, so compromises must usually be made. The second consideration is whether the system should be tailored to a specialized application or be a general-purpose instrument. In general, systems with more flexibility in the choice of hardware may be better customized for specialized tasks, but are likely to have more "technical" problems. Choosing "off the shelf solutions" with software that has been through its first teething problems can be more time-efficient. Whatever type of system is being considered, the need to test the instruments using a particular experimental material cannot be overemphasized. Company representatives demonstrating equipment should be advised about the choice of fluorochromes and magnifications, so that they can bring the appropriate filter sets, excitation sources, and objective lenses. In addition to the biological material of interest, it is worth using sets of slides of fluorescent beads such as Inspec and Tetraspec (from Molecular Probes) to provide quantitative data for support (Davis 2000; Swedlow et al. 2002).

FIGURE 5. An Olympus IX70 inverted microscope/Applied-Precision DeltaVision motorized stage. Note the condenser and the micromanipulator, which can be rotated out of the way to accommodate a stage cover (*inset*).

The Microscope

The first decision is whether to use an upright ("view from above") or an inverted ("view from below") microscope. In general, for live cell work, an inverted microscope (Fig. 5) offers more advantages if high resolution and multidimensional imaging are required. On an inverted microscope, the specimen is generally more accessible, simplifying the microinjection process and the use of growth chambers and environmental control chambers. Inverted microscopes also offer better mechanical stability for mounting CCD cameras.

Contrast-enhancing Bright-field Methods

Bright-field imaging is desirable in addition to fluorescence as it provides a reference for the location of the fluorescent signal. Contrast-enhancing bright-field methods such as Nomarski, also known as differential interference contrast (DIC), and phase contrast are very useful in specific cases. For example, DIC can be used to observe lipid droplet motility (Welte et al. 1998) or patterning in living embryos (I. Davis, unpubl.), and phase contrast works well to image mitosis in the developing testis (Fuller 1993). There are also several ways of "combining" bright-field and fluorescent imaging. If the fluorescence signal is very bright, then one can keep the DIC polarizer and wollaston prism in place while imaging fluorescence and accept a large loss in fluorescence intensity. However, there are specialized filter cubes that allow simultaneous acquisition of DIC images and fluorescent images without loss of fluorescence intensity (e.g., custom cube U-MWIB/DIC-SP from Olympus, made by Chroma).

Choosing a Fluorescence Imaging System to Use with *Drosophila* Tissues

The two predominant types of epifluorescence imaging used in biological research—wide-field systems and "optical sectioning" systems, in their varying forms—have all been used to image *Drosophila* tissues (Table 5). A detailed treatment is not appropriate here as many excellent descriptions of these different technologies may be found (examples in Table 5; also see Chapters 15 and 16).

In general, optical sectioning techniques (including multiphoton and spinning-disc confocals) perform better than wide-field systems on bright signals with a lot of blur in thick specimens, such as embryos (Fig. 6–8; Movie 21.2) and tissues dissected from larvae and adults. Wide-field deconvolution tends to outperform confocal microscopy on fainter signals and more sensitive material (such as following mRNA dynamics; Wilkie et al. 2001; MacDougall et al. 2003) where an image, however hazy, is still visible on the original unprocessed data. In practice, it has been observed that wide-field combined with deconvolution can be more sensitive and quantitative and achieve higher resolution than confocal imaging (Swedlow et al. 2002). Confocals are "more convenient" because the results are ready without lengthy processing and a need to image many *z* sections. However, for rapid image capture of sensitive processes, for example, cytoskeleton dynamics, or following fast RNA particle movements in *Drosophila* embryos and egg chambers, only spinning-disc confocals can really compete with wide field. Figure 7 shows a comparison of wide-field and spinning-disc confocal microscopy. Using *Tau*-GFP associated with the microtubule cytoskeleton in egg chambers as the test specimen, we found that it was possible to image for longer with a PerkinElmer UltraView spinning disc and an Orca-ER-cooled CCD (Hamamatsu) than with PMT-based scanning confocals, or using a wide-field system with a similar CCD camera (Coolsnap-HQ, Roper). However, it was necessary to use maximum laser power and bin 2×2 pixels to obtain good image quality. After a couple of minutes of continual imaging, photobleaching and cytoskeletal disruption were evident.

TABLE 5. Summary of major imaging techniques

Imaging Technique	Features	Notes	General References
Wide-field	Limited image quality due to contribution of out-of-focus blur from above and below the focal plane.	Most useful with thin tissues or culture cells or with low-NA, low-magnification dry objectives.	Murphy (2001); Davis (2000); Wallace et al. (2001)
Deconvolution wide-field (3D imaging)	Postimage acquisition processing delivers increased signal to noise, resolution, and contrast. Requires multiple z sections.	The most commonly applied imaging technique for high-resolution analysis of structure and dynamic processes (Davis 2000).	Wallace et al. (2001)
Deconvolution wide-field (2D imaging)	Limited deblurring or deconvolution approaches applied to 2D xy image data to sharpen image detail.	Useful increase in signal to noise and contrast when speed requirements/tissue sensitivity preclude z stack collection (Figs. 6 and 7; MacDougall et al. [2003]).	Parton and Davis (2004)
Point scanning confocal (or LSCM)	An optical technique that eliminates the contribution of out-of-focus light to the image to produce sharp "optical sections." Alternative to deconvolution for elimination of out-of-focus blur.	Useful in brightly labeled thick, hazy, or scattering material. Very good for high-resolution structural studies but suffers from speed limitations and its potential to affect biological processes (Figs. 6 and 8).	Pawley (1995); Swedlow et al. (2002)
Spinning disc or multifocal confocal	Similar to above but by using simultaneous multiple excitation beams has advantages in speed of image capture and less problems of dye photobleaching and phototoxicity.	Increasing in popularity as an alternative to wide-field deconvolution for rapid dynamic processes (Fig. 7).	Diaspro (2001); Pawley (1995)
Multiphoton	Optical sectioning by the principal of multiple low-energy photon (700–1100 nm) absorption that occurs at extremely high illumination intensity, limiting dye excitation to an ~1-µm-thick focal plane.	Promises the ability to image deeper in thick, hazy, or scattering material with improved cell viability. Limited application to *Drosophila* so far (Fig. 8; Movie 21.2).	Diaspro (2001); Amos (2000)

Optimizing Excitation and Emission for Imaging *Drosophila* Tissues

Optimizing the collection of photons for a given dose of illumination light is largely a matter of selecting appropriate combinations of dye, excitation source, filter sets, detector, and objective. Failure to do this impacts image quality, dye bleaching, and cell viability. *Drosophila* cells and tissues are no exception, so the rules of good imaging practices apply (for an overview, see Pawley 1995).

In general, longer excitation wavelengths cause less damage to biological specimens and induce less autofluorescence, but the exact characteristics depend on the *Drosophila* tissue being imaged. For example, yolk is particularly fluorescent in UV and blue. It is better to replace UV excitation with excitation at 405–440 nm, use alternative dyes, or use multiphoton excitation. We tend to use green-excited red-emitting dyes (Rhodamine-like spectral characteristics) when imaging in the presence of yolk.

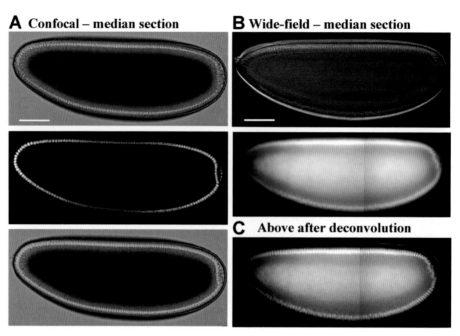

A Confocal – median section

B Wide-field – median section

C Above after deconvolution

FIGURE 6. Comparison of confocal and wide-field images of stage-4 syncytial blastoderm embryos (approximately cycles 14–15) expressing GFP in the nuclei. (*A*) Confocal images (20× water immersion objective 0.7 NA): bright-field, GFP fluorescence, merged. (*B*) Undeconvolved wide-field image (20× dry objective 0.75 NA): bright-field, GFP fluorescence. (*C*) Image in *B* after deconvolution: bright-field, GFP fluorescence. Bar, 50 μm.

If the cellular process being imaged is particularly susceptible to damage from imaging, then try to determine the cause of this perturbation.

- *Photodamage* is caused by the excitation light itself, in the absence of the fluorescent molecule. It is generally not significant, except with UV light or the high-level infrared irradiation used in multiphoton imaging.

- *Phototoxicity* is caused when the fluorescent molecule itself (upon irradiation) causes the damage, by generating free radicals or localized heating. This can be particularly significant and is a major reason for keeping the dye concentration and irradiation to a minimum.

- *Photobleaching* is the process by which the excited light damages the fluorescent molecule, so that it is no longer able to emit light. Photobleaching is dependent on the exact fluorescent molecule used and the conditions within the cells. All dyes possess a limit to the number of times they can be excited, known as the quantum yield, which is believed to be independent of whether excitation is continuous or intermittent. However, nonlinear photobleaching and phototoxic effects may also occur and could account for some of the benefits experienced with spinning-disc confocals.

To avoid the effects of photodamage, phototoxicity, and photobleaching when imaging live cells, it is best to minimize exposure times and attenuate the excitation power whenever possible, for example, by using neutral-density filters. It is important to maximize the efficiency of the imaging by matching the choice of fluorescent molecule with the excitation and emission filter sets, the bright peaks of the illumination source (http://www.olympusmicro.com/primer), and the quantum efficiency of the detector. Protective infrared and UV filters can also be useful to reduce the damaging effects of excitation light. Filter sets, particularly excitation filters (UV to blue) should be treated as consumables and checked annually for damage. It is important to note that the excitation and emission curves of fluorescent molecules reported in a company's product information may be generated under very different conditions compared to in vivo

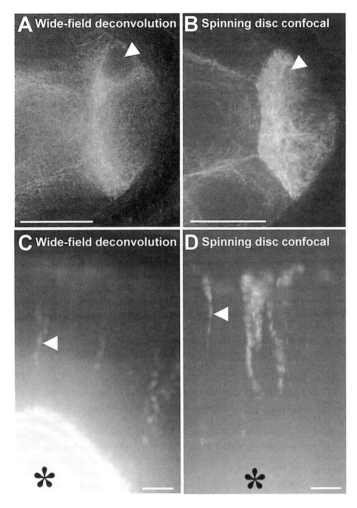

FIGURE 7. Comparison of wide-field deconvolution and spinning-disc confocal images. An Olympus 60× NA 1.2 water-immersion objective was used throughout. (*A, B*) Stage-8 egg chambers expressing *Tau*-GFP associated with the microtubule cystoskeleton. The location of the oocyte nucleus is indicated by a white arrowhead. Egg chambers are oriented with the oocyte posterior to the right and dorsal side up. (*C, D*) Particle tracks of fluorescently labeled *ftz* mRNA (see Movie 21.1; Wilkie and Davis 2001) after injection of labeled mRNA into stage-4 syncytial blastoderm embryos (approximately cycles 14–15). Particle movement is represented by the projected time-series images (image/2 seconds). The paths of two individual particles are indicated by white arrowheads. Particles move from the site of injection, near the center of the embryo (bottom of the image, indicated by an asterix), toward the peripherally located nuclei (top of the image, not shown). RNA eventually accumulates on the side of the nuclei toward the outer membrane. Bars: (*A, B*) 50 μm; (*C, D*) 10 μm.

imaging conditions. This is particularly important with multiple dyes when trying to ensure that their spectra are sufficiently separated to allow them to be covisualized. The spectral detection options now available on several confocal systems (such as the Leica SP2 AOBS) offer increased light efficiency and much greater freedom to optimize excitation and emission.

The choice of objective and its use is paramount for image quality and is considered again in more detail below with reference to specific *Drosophila* tissues. The next most important hardware choice that determines image quality is the detector. Point-scanning confocal systems generally come with optimized photomultipliers so the choice is already made. Spinning-disc confocals and wide-field systems which rely on CCD cameras offer more freedom of choice but may be limited by the availability of software drivers and capture cards for the operating software. A detailed comparison of different CCD cameras is beyond the scope of this chapter and has been covered elsewhere (Amos 2000). Where a choice is available, do not skimp on the cost of the detector! For live cell work, where speed (for following developments) and sensitivity (for reducing the necessity for damaging excitation exposure) are generally paramount, a high-quality cooled CCD (~£20,000) is the usual choice for wide-field fluorescence systems.

OVERCOMING THE SPECIAL PROBLEMS OF IMAGING THICK *DROSOPHILA* SPECIMENS

Imaging large multicellular structures presents challenges that are not encountered when imaging single-celled organisms. In this respect, *Drosophila* tissue tends to be easier to work with than many larger model systems, such as *Xenopus* and mouse, but harder to image than *Caenorhabditis elegans*. Most objective lenses, with some notable exceptions, are designed by the microscope manufacturers for optimum performance at the inner surface of a #1.5 coverslip (0.15–0.17 mm thick), often used with an immersion medium of refractive index matching that of the glass (usually 1.515–1.518). The optical quality of the image is compromised as soon as structures that are within the specimen are imaged (with its uncharacterized optical properties). The deeper the imaging, the greater the aberration. The most important of these aberrations is spherical aberration, which is caused by different angles of cones of light originating from a particular point of the specimen being collected by the objective and focused to different focal points (Davis 2000). In addition to imaging into biological material, spherical aberration can also result from poorly corrected objectives, refractive index mismatch between the immersion and mounting medium, or incorrect setting of coverslip thickness correction collars. The consequence of spherical aberration is a loss of resolution of detail or blurring of the image (Davis 2000). With confocals, spherical aberration contributes even more significantly to signal attenuation with imaging depth. Refractive index mismatch between immersion medium, mounting medium, and specimen is also the cause of the attenuation of working-distance, experienced usually with high-NA oil immersion objectives. Lack of working-distance causes the objective to hit the coverslip before it has reached the desired imaging depth. Chromatic aberration is also greater with depth, causing different wavelengths of light from the same focal depth in the specimen to be focused at different imaging planes on the detector (a problem for colocalization studies) (Davis 2000).

In our experience, spherical aberrations can be reduced using a number of combined approaches, which must be determined empirically. Refractive index inhomogeneities and scattering properties of each kind of *Drosophila* tissue are different.

- Image as near as possible to the surface of the coverslip. Partly squashing living specimens, for example, with a Teflon-breathable membrane (described in the section on Collection and Mounting of *Drosophila* Embryos), can improve the imaging quality considerably and reduce the depth required for imaging the structures of interest (Davis 2000).

- Choose the objective lens carefully. We find that a dry 20×/NA 0.75 lens is excellent for imaging deep, if high spatial resolution is not required. Higher numerical aperture lenses suffer from the problems of imaging deep to a far greater extent. Try to use oil immersion lenses (e.g., 40×/1.35) that have an aperture adjustment or water immersion lenses (e.g., 60×/1.2W) with a coverslip correction collar that can be used empirically to correct spherical aberration. Spherical aberration can be judged by eye on a wide-field microscope. When viewing a very small bright feature, the so-called "Airy rings" above and below the focal plane should look symmetric. If they are asymmetric, then spherical aberration is present.

- When using immersion lenses, match the refractive index of the immersion media as far as possible to the mounting media and cell content (water, glycerol, or immersion oil).

- Immersion oils of higher refractive indices can be used to "correct" spherical aberration (Hiraoka et al. 1990). Standard oils have a refractive index that matches the glass in the coverslip (1.515–1.518). Cargille oils are sold in a range of refractive indices (1.512–1.534). Oils with even greater refractive indices exist (e.g., 1.65), but they are expensive, toxic, and inconvenient to use.

- Thinner coverslips (#1.0 or 0) can in some cases be used to "correct" spherical aberration.

- Motorized spherical aberration correction lenses hold promise for correcting spherical aberration, but they are just becoming available (Intelligent Imaging Innovations).

Unfortunately, the problems of imaging at greater depth are not confined to spherical aberration. *Drosophila* tissues reflect, absorb, and disperse light as it passes through the specimen, mostly caused by variations in refractive indices across the tissue in three dimensions. Embryos, late egg chambers, and salivary glands are among the worst tissues to image from this point of view. A very laborious solution to this problem has been developed. It involves mapping the refractive index of living tissues in three dimensions using DIC and applying these variations to deconvolution algorithms (Kam et al. 2001), but its routine application will probably require more than a tenfold increase in computer processing speeds. Imaging with longer-wavelength visible dyes (Kam et al. 2001) has been used to improve deep imaging by reducing spherical aberration and scattering. Multiphoton imaging using infrared irradiation has been shown to reduce the problems of spherical aberration and scattering and has improved depth penetration, but it is expensive and tricky to implement. In our hands, multiphoton imaging of nuclear localization signal (nls)-GFP in late-stage embryos has proven to be successful (Fig. 8; Movie 2A,B). However, our attempts at imaging

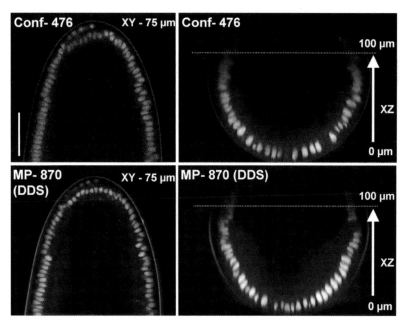

FIGURE 8. Comparison of confocal and multiphoton imaging of the same stage-4 syncytial blastoderm embryo (cycles 14–15) expressing GFP in the nuclei. Single median *xy* views 75-nm deep and *xz* line scan views up to 100-μm deep are compared for confocal (excitation 476 nm) and multiphoton (excitation 870 nm) imaging using a Bio-Rad Radiance 2000 confocal/multiphoton. The same 60 × 1.2 NA water immersion objective was used throughout. Multiphoton emission was detected using the direct detector system (DDS), avoiding the need for a confocal pinhole. Note the faster rate of signal fall off with depth using confocal compared to multiphoton. The vitteline membrane appears to be more autofluorescent at the multiphoton excitation wavelengths. Bar, 50 μm.

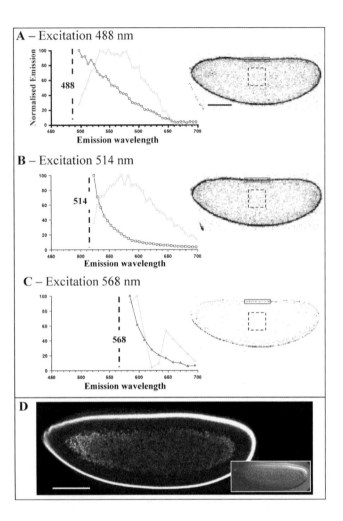

FIGURE 9. Qualitative examination of autofluorescence in a single wild-type (Oregon R) syncytial blastoderm stage embryo. Data were collected on a Leica Confocal with a spectral analysis scanhead (TCS-SP). For three different excitation wavelengths, the autofluorescence emission is examined; emission intensities are normalized. (*A*) Excitation at 488 nm, emission 500–700 nm; (*B*) excitation 514 nm, emission 520–700 nm; (*C*) excitation 568 nm, emission 580–700 nm. Two regions were sampled in each case corresponding to the vitteline membrane (*gray box; gray trace* on plots) and yolk (*black box, black trace* on plots). It is clear that emission from the yolk overlaps with that expected for GFP (main emission ~500–550). What is not shown here is that as excitation wavelength increases, the relative intensity of autofluorescence emission falls; see the corresponding images (shown "in negative" for clarity). (*D*) A similar embryo imaged by multiphoton at 780-nm excitation; the signal has been accumulated to exaggerate autofluorescence of the yolk and vitteline membrane. Bars, 50 μm.

the fainter microtubule cytoskeleton, by detecting *Tau*-GFP in egg chambers, led to rapid cytoskeletal and cellular disruption unless considerable care was taken to optimize the excitation pulse duration.

Another major problem with imaging thick *Drosophila* tissues is autofluorescence. Embryos (Figs. 6B and 9) and late-stage egg chambers suffer from this problem greatly, due to the presence of yolk granules. The chorion and vitteline membranes also contribute to autofluorescence and a lack of transparency, although the chorion can be removed using bleach. Autofluorescence can also be reduced by imaging at longer wavelengths and by choosing the most appropriate filter sets. For example, the fluorescein isothiocyante (FITC) HQ filter set from Chroma removes at least some yellow autofluorescence when imaging GFP.

LIVE CELL IMAGING AS AN EXPERIMENTAL APPROACH IN *DROSOPHILA*

At its simplest, imaging can be used for visual genetic screens for analysis of morphological characteristics (Schüpbach and Wieschaus 1986) or fluorescence of the distribution of proteins in living embryos (Merabet et al. 2002). The ability to examine the distribution of proteins in living cells can save considerable labor compared with similar kinds of screens using antibody staining (Seeger et al. 1993). Such simple observations of GFP fusion proteins can also provide very valuable complementary information to the use of antibody staining, which can suffer from penetration or inaccessibility problems (Micklem et al. 1997; Newmark et al. 1997). However, live cell imaging becomes particularly powerful when used to study dynamic biological processes that cannot be analyzed in fixed material. This is most important in the study of very rapidly evolving processes. Examples (see also Table 2) include lipid droplet motility in embryos (Welte et al. 1998), the movement of chromosomal loci within the nucleus (Vazquez et al. 2001, 2002), diffusion of protein signals in imaginal discs (Greco et al. 2001), rapidly moving protein particles in the oocyte (Theurkauf and Hazelrigg 1998), and RNA particles (Wilkie and Davis 2001; MacDougall et al. 2003).

Live cell imaging is particularly powerful in *Drosophila* when used in combination with a number of well-established genetic tricks. Simply studying a process in mutant tissue can be extremely revealing, especially when studying dynamic processes in hypomorphic viable alleles, which appear normal in fixed material. For example, the speed of motility of RNA particles is lower in flies expressing mutant motor protein, despite the finding that their localization appears normal by in situ hybridization. In such cases, strong alleles cannot be studied, as they are lethal. Lethal alleles can be studied under the correct circumstances if germ-line clones (Chou and Perrimon 1992) can survive until the stage required for analysis (Brendza et al. 2000). The UAS/Gal4 activation system (Brand et al. 1994) can also be used to activate a dominant-negative protein and study its effects at a particular time and location, even though it may be lethal (Duncan and Warrior 2002). In *Drosophila*, the array of genetic tricks is large and is described in an excellent review (St Johnston 2002).

Improvements in imaging techniques have facilitated the design of very sophisticated experiments to answer difficult biological questions. For example, caged-FITC has been used to mark cells in embryos and follow their fate in real time (Vincent and O'Farrell 1992). Careful quantification of the signal and analysis of its movement over time has led to conclusions regarding the mechanism of movement of chromosomes in the nucleus (Vazquez et al. 2001). There are many new and emerging fluorescence techniques, described in the next section, whose importance will certainly grow in the next decade.

NEW AND EMERGING FLUORESCENCE MICROSCOPY TECHNIQUES AND THEIR APPLICATION TO *DROSOPHILA*

Many recent developments in microscopy hardware hold a lot of promise for imaging living *Drosophila* tissues. For example, "filter-free" fluorescence imaging options (Leica) and spectral detection options (Zeiss) have started to emerge on confocal systems. These free the user from the constraints of filter blocks and dichroic mirrors. Rapid automated mechanical devices for rapid focusing or correction of spherical aberration can be used for very fast multidimensional imaging. There are continuous advances in CCD technology, the most recent of which are new less noisy methods of amplifying the signal (e.g., 512f from Roper and iXON from Andor both use a new Electron Multiplying CCD chip from Marconi).

There has also been great excitement regarding the use of new imaging methods that often exploit new developments in hardware or software. Examples include FRAP (fluorescence recovery after photobleaching; see Chapter 7) (Entchev et al. 2000; Berdnik and Knoblich 2002; Noguchi and Miller 2003), available now for wide-field as well as point-scanning confocal systems and used to study the movement of otherwise undetectable weak fluorescence in and out of biological structures with time. FRET (fluorescence resonance energy transfer; see Chapters 8 and 9) is used to establish close molecular interactions. FLIM (fluorescence lifetime imaging; see Chapter 8) is used to study how long fluorescence persists after excitation has stopped, which is affected strongly by the individual nature and local environmental conditions of a fluorescent molecule.

Fluorescent dye technology is also evolving rapidly. The new Alexafluor-dye series (Molecular Probes) have increased brightness and photostability. "Quantum dot" technology is also available commercially for the first time. These are highly photostable tiny crystals with fluorescence emission peaks defined by their size and promise the next generation of photostable labels. Many new fluorescent proteins from new sources with improved characteristics and red-shifted emissions are likely to be commonly used. In addition, split GFP and "circular permutated GFPs," which activate florescence upon ligand binding, are likely to provide useful alternatives to FRET (van Roessel and Brand 2002). Another likely important advance will be the further development of photoactivatable GFPs (Ando et al. 2002; Patterson and Lippincott-Schwartz 2002; see also Chapter 32). Yet another novel development is that of nuclease-resistant RNA- and DNA-based fluorescent probes that have the potential to report on mRNA molecules in vivo (Bratu et al. 2003). These "molecular Beacons," based on specifically targeted base sequences, were able to specifically localize endogenous RNA transcripts in living *Drosophila* oocytes.

Microscopy has always been a useful addition to the diverse range of techniques available to the *Drosophila* researcher. The application of new microscopy methods to live cell imaging is providing innovative ways to answer questions that were previously inaccessible. They are likely to increase in prominence in the *Drosophila* community in the coming decade and beyond.

ACKNOWLEDGMENTS

We thank members of the Davis Group past and present who have contributed to the current knowledge of the lab. We also thank Jochen Arlt and the COSMIC imaging facility (Physics Dept, University of Edinburgh) for helpful discussions and access to advanced imaging equipment, and Simon Bullock and David Ish-Horowicz (CR-UK, London) for discussions and use of their spinning-disc confocal.

MOVIE LEGENDS

MOVIE 21.1. Injection of fluorescently labeled mRNA into a stage-4 syncytial blastoderm *Drosophila* embryo (approximately cycle 14–15). RNA localizes to the periphery above the nuclei over ~3 minutes. Imaging by DIC and wide-field fluorescence microscopy, 20× NA 0.75 objective.

MOVIE 21.2. Multiphoton and confocal imaging deep into a late stage (12–18 hours after laying) *Drosophila* histone-2A GFP embryo. Confocal excitation of the GFP was at 475 nm and multiphoton excitation was at 870 nm. Emission was collected above 520 nm. (*Top*) Imaging using a 20× 0.75 NA multi-immersion objective with water. (*Bottom*) Imaging of the same embryo with a 60× 1.2 NA water immersion objective. Note the improved imaging depth possible with multiphoton but the poorer lateral resolution and optical sectioning capabilities.

REFERENCES

Amos W.B. 2000. Instruments for fluorescence imaging. In *Protein localization by fluorescence microscopy* (ed. V.J. Allan), pp. 67–108. Oxford University Press, England.

Ando R., Hama H., Yamamoto-Hino M., Mizunu H., and Miyawaki A. 2002. An optical marker based on the UV0 induced green-to-red photoconversion of a fluorescence protein. *Proc. Natl. Acad. Sci.* **99:** 12651–12656.

Berdnik D. and Knoblich J.A. 2002. *Drosophila* Aurora-A is required for centrosome maturation and actin-dependent asymmetric protein localization during mitosis. *Curr. Biol.* **12:** 640–647.

Bossing T., Barros C.S., and Brand A.H. 2002. Rapid tissue-specific expression assay in living embryos. *Genesis* **34:** 123–126.

Brand A.H., Manoukian A.S., and Perrimon N. 1994. Ectopic expression in *Drosophila*. *Methods Cell Biol.* **44:** 635–654.

Brendza R.P., Serbus L.R., Duffy J.B., and Saxton W.M. 2000. A function for kinesin I in the posterior transport of *oskar* mRNA and Staufen protein. *Science* **289:** 2120–2122.

Bullock S.L. and Ish-Horowicz D. 2001. Conserved signals and machinery for RNA transport in *Drosophila* oogenesis and embryogenesis. *Nature* **414:** 611–616.

Cha B., Koppetsch B.S., and Theurkauf W.E. 2001. In vivo analysis of *Drosophila* bicoid mRNA localization reveals a novel microtubule-dependent axis specification pathway. *Cell* **106:** 35–46.

Chou T.B. and Perrimon N. 1992. Use of a yeast site-specific recombinase to produce female germline chimeras in *Drosophila*. *Genetics* **131:** 643–653.

Davis I. 2000. Visualising fluorescence in *Drosophila*—Optimal detection in thick specimens. In *Protein localisation by fluorescence microscopy: A practical approach* (ed. V.J. Allan), pp. 131–162. Oxford University Press, England.

Davis I., Girdham C.H., and O'Farrell P.H. 1995. A nuclear GFP that marks nuclei in living *Drosophila* embryos—Maternal supply overcomes a delay in the appearance of zygotic fluorescence. *Dev. Biol.* **170:** 726–729.

Diaspro A. 2001. *Confocal and two-photon microscopy: Application and advances.* Wiley-Liss, New York.

Duncan J.E. and Warrior R. 2002. The cytoplasmic Dynein and Kinesin motors have interdependent roles in patterning the *Drosophila* oocyte. *Curr. Biol.* **12:** 1982–1991.

Endow S.A. and Komma D.J. 1997. Spindle dynamics during meiosis in *Drosophila* oocytes. *J. Cell Biol.* **137:** 1321–1336.

Entchev E.V., Schwabedissen A., and Gonzalez-Gaitan M. 2000. Gradient formation of the TGF-beta homolog Dpp. *Cell* **103:** 981–991.

Fuller M.T. 1993. Spermatogenesis. In *The development of* Drosophila melanogaster (ed. M. Bate and A.M. Arias), vol. 1, pp. 71–147. Cold Spring Harbor Laboratory Press, Cold Spring Harbor, New York.

Gho M., Bellaiche Y., and Schweisguth F. 1999. Revisiting the *Drosophila* microchaete lineage: A novel intrinsically asymmetric cell division generates a glial cell. *Development* **126:** 3573–3584.

Glotzer J.B., Saffrich R., Glotzer M., and Ephrussi A. 1997. Cytoplasmic flows localize injected *oskar* RNA in *Drosophila* oocytes. *Curr. Biol.* **7:** 326–337.

Greco V., Hannus M., and Eaton S. 2001. Argosomes: A potential vehicle for the spread of morphogens through epithelia. *Cell* **106:** 633–645.

Grieder N.C. de Cuevas M., and Spradling A.C. 2000. The fusome organizes the microtubule network during oocyte differentiation in *Drosophila*. *Development* **127:** 4253–4264.

Haugland R.P. 2003. *Handbook of fluorescent probes and research products*, 9th edition. Molecular Probes, Eugene, Oregon.

Hidalgo A. and Brand A.H. 1997. Targeted neuronal ablation: The role of pioneer neurons in guidance and fasciculation in the CNS of *Drosophila*. *Development* **124:** 3253–3262.

Hiraoka Y., Sedat J., and Agard D. 1990. Determination of three-dimensional imaging properties of a light microscope system. *Biophys. J.* **57:** 325.

Kaltschmidt J.A., Davidson C.M., Brown N.H., and Brand A.H. 2000. Rotation and asymmetry of the mitotic spindle direct asymmetric cell division in the developing central nervous system. *Nat. Cell Biol.* **2:** 7–12.

Kam Z., Hanser B., Gustafsson M.G., Agard D.A., and Sedat J.W. 2001. Computational adaptive optics for live three-dimensional biological imaging. *Proc. Natl. Acad. Sci.* **98:** 3790–3795.

MacDougall N., Clark A., MacDougall E., and Davis I. 2003. *Drosophila* gurken (TGFalpha) mRNA localizes as particles that move within the oocyte in two dynein-dependent steps. *Dev. Cell.* **4:** 307–319.

Marshall W.F., Straight A., Marko J.F., Swedlow J., Dernburg A., Belmont A., Murray A.W., Agard D.A., and Sedat J.W. 1997. Interphase chromosomes undergo constrained diffusional motion in living cells. *Curr. Biol.* **7:** 930–939.

Megraw T.L., Kilaru S., Turner F.R., and Kaufman T.C. 2002. The centrosome is a dynamic structure that ejects PCM flares. *J. Cell. Sci.* **115:** 4707–4718.

Merabet S., Catala F., Pradel J., and Graba Y. 2002. A green fluorescent protein reporter genetic screen that identifies modifiers of *Hox* gene function in the *Drosophila* embryo. *Genetics* **162:** 189–202.

Micklem D.R., Dasgupta R., Elliott H., Gergely F., Davidson C., Brand A., Gonzalez-Reyes A., and St Johnston D. 1997. The *mago nashi* gene is required for the polarisation of the oocyte and the formation of perpendicular axes in *Drosophila*. *Curr. Biol.* **7:** 468–478.

Minden J.S., Agard D.A., Sedat J.W., and Alberts B.M. 1989. Direct cell lineage analysis in *Drosophila melanogaster* by time-lapse, three-dimensional optical microscopy of living embryos. *J. Cell. Biol.* **109:** 505–516.

Mollereau B., Wernet M.F., Beaufils P., Killian D., Pichaud F., Kuhnlein R., and Desplan C. 2000. A green fluorescent protein enhancer trap screen in *Drosophila* photoreceptor cells. *Mech. Dev.* **93:** 151–160.

Murphy D.B. 2001. *Fundamentals of light microscopy and electronic imaging*, Wiley-Liss, New York.

Newmark P.A., Mohr S.E., Gong L., and Boswell R.E. 1997. Mago nashi mediates the posterior follicle cell-to-oocyte signal to organize axis formation in *Drosophila*. *Development* **124:** 3197–3207.

Noguchi T., and Miller K.G. 2003. A role for actin dynamics in individualization during spermatogenesis in *Drosophila* melanogaster. *Development* **130:** 1805–1816.

Parton R.M. and Davis I. 2004. Lifting the fog: Image restoration by deconvolution. In *Cell biology: A laboratory handbook*, 3rd edition (ed. J.E. Celis). Academic Press, San Diego. (In press.)

Patterson G.H. and Lippincott-Schwartz J. 2002. A photoactivatable GFP for selective photolabeling of proteins and cells. *Science* **297:** 1873–1877.

Pawley J.B. 1995. *Handbook of biological confocal microscopy*, 2nd edition. Plenum Press, New York.

Pfeiffer S., Ricardo S., Manneville J.B., Alexandre C., and Vincent J.P. 2002. Producing cells retain and recycle Wingless in *Drosophila* embryos. *Curr. Biol.* **12:** 957–962.

Pichaud F. and Desplan C. 2001. A new visualization approach for identifying mutations that affect differentiation and organization of the *Drosophila* ommatidia. *Development* **128:** 815–826.

Roberts D. and Standen G. 1998. The elements of *Drosophila* biology and genetics. In Drosophila: *A practical approach* (ed. D. Roberts), pp. 1–54. Oxford University Press, England.

Robinson J.T., Wojcik E.J., Sanders M.A., McGrail M., and Hays T.S. 1999. Cytoplasmic dynein is required for the nuclear attachment and migration of centrosomes during mitosis in *Drosophila*. *J. Cell Biol.* **146:** 597–608.

Rorth P. 2002. Initiating and guiding migration: Lessons from border cells. *Trends Cell.Biol.* **12:** 325–331.

Schüpbach T. and Wieschaus E. 1986. Maternal-effect mutations altering the anterior-posterior pattern of the *Drosophila* embryo. *Roux's Arch. Dev. Biol.* **195:** 302–317.

Seeger M., Tear G., Ferres-Marco D., and Goodman C.S. 1993. Mutations affecting growth cone guidance in *Drosophila:* Genes necessary for guidance toward or away from the midline. *Neuron* **10:** 409–426.

St Johnston D. 2002. The art and design of genetic screens: *Drosophila melanogaster*. *Nat. Rev. Genet.* **3:** 176–188.

Spradling A.C. 1993. Development genetics of oogenesis. In *The development of* Drosophila melanogaster (ed. M. Bate and A.M. Arias), vol. 1: pp. 1–70. Cold Spring Harbor Laboratory Press, Cold Spring Harbor, New York.

Sullivan W., Ashburner M., and Hawley R.S, eds. 2000. Drosophila *protocols.* Cold Spring Harbor Laboratory Press, Cold Spring Harbor, New York.

Swedlow J.R., Hu K., Andrews, P.D., Roos D.S., and Murray J.M. 2002. Measuring tubulin content in *Toxoplasma gondii*: A comparison of laser-scanning confocal and wide-field fluorescence microscopy. *Proc. Natl. Acad. Sci.* **99:** 2014–2019.

Theurkauf W.E., and Hazelrigg T.I. 1998. In vivo analyses of cytoplasmic transport and cytoskeletal organization during *Drosophila* oogenesis: Characterization of a multi-step anterior localization pathway. *Development* **125:** 3655–3666.

van Roessel P. and Brand A.H. 2002. Imaging into the future: Visualizing gene expression and protein interactions with fluorescent proteins. *Nat. Cell. Biol.* **4:** E15–20.

Vazquez J., Belmont A.S., and Sedat J.W. 2001. Multiple regimes of constrained chromosome motion are regulated in the interphase *Drosophila* nucleus. *Curr. Biol.* **11:** 1227–1239.

Vazquez J., Belmont A.S., and Sedat J.W. 2002. The dynamics of homologous chromosome pairing during male *Drosophila* meiosis. *Curr. Biol.* **12:** 1473–1483.

Vincent J.P., and O'Farrell P.H. 1992. The state of engrailed expression is not clonally transmitted during early *Drosophila* development. *Cell* **68:** 923–931.

Wallace W., Schaefer L.H., and Swedlow J.R. 2001. A workingperson's guide to deconvolution in light microscopy. *BioTechniques* **31:** 1076–1078, 1080, 1082.

Welte M.A., Gross S.P., Postner M., Block S.M., and Wieschaus E.F. 1998. Developmental regulation of vesicle transport in *Drosophila* embryos: Forces and kinetics. *Cell* **92:** 547–557.

Wilkie G.S., and Davis I. 2001. *Drosophila* wingless and pair-rule transcripts localize apically by dynein-mediated transport of RNA particles. *Cell* **105:** 209–219.

Yu D., Baird G.S., Tsien R.Y., and Davis R.L. 2003. Detection of calcium transients in *Drosophila* mushroom body neurons with camgaroo reporters. *J. Neurosci.* **23:** 64–72.

Yucel J.K., Marszalek J.D., McIntosh J.R., Goldstein L.S., Cleveland D.W., and Philp A.V. 2000. CENP-meta, an essential kinetochore kinesin required for the maintenance of metaphase chromosome alignment in *Drosophila*. *J. Cell Biol.* **150:** 1–11.

Zhang J., Campbell R.E., Ting A.Y., and Tsien R.Y. 2002. Creating new fluorescent probes for cell biology. *Nat. Rev.* **3:** 916–918.

WEB RESOURCES

Microscopy and Imaging

http://www.olympusmicro.com/primer/
http://www.zeiss.de/de/micro/home_e.nsf
http://www.light-microscopy.com/website/sc_mqm.nsf
http://www.api.com/lifescience/spectris.html
http://www.cairnweb.com/menus/menustub_main.html

Microscopy and Imaging-associated Products

http://www.universal-imaging.com/products/metamorph/
http://www.roperscientific.com/
http://www.bioptechs.com/
http://www.intelligent-imaging.com/
http://www.prior.com/uk/products/product.html
http://www.sutter.com/
http://www.chroma.com/
http://www.omegafilters.com/
http://www.probes.com/
http://www.tritechresearch.com/cgi-bin/shop/minj.html?id=mJjP9jiC

General Information on Microscopy, Imaging, and Image Processing

http://www.wisc.edu/zoology/faculty/fac/Paw/pawpapers.html
http://listserv.acsu.buffalo.edu/cgi-bin/wa?S1=confocal
http://rsb.info.nih.gov/nih-image/
http://www.openmicroscopy.org/
http://www.microscopy.fsu.edu/
http://www.microscopyu.com/
http://msg.ucsf.edu/IVE/
http://www.yale.edu/rosenbaum/gfp_gateway.html

Outstanding Resource for *Drosophila*

http://flybase.bio.indiana.edu/

CHAPTER TWENTY TWO

Single-cell Imaging in Animal Tumors In Vivo

Jeffrey Wyckoff, Jeffrey Segall, and John Condeelis

*Albert Einstein College of Medicine, Bronx,
New York*

INTRODUCTION

ANALYSIS OF THE INDIVIDUAL STEPS IN METASTASIS is crucial if insights at the molecular level are to be linked to the cell biology of cancer. However, the discovery of specific genes that correlate with metastatic potential cannot be related to a mechanism unless the metastatic step at the cellular level in which the gene is involved is identified.

A technical hurdle to achieving the analysis of the individual steps of metastasis is the fact that, at the gross level, whole tumors are heterogeneous in both animal models and patients, being composed of many different cell types with variable compositions. Human primary tumors show extensive variation in all properties ranging from growth and morphology of the tumor through tumor cell density in the blood and formation and growth of metastases. An additional problem is that the metastatic cascade has been studied almost exclusively at the level of extravasation and beyond. Paracrine interactions between cells and the entry of tumor cells into the circulation in the primary tumor, believed to be important factors in tumor metastasis (Wyckoff et al. 2000), have not been observed directly. For example, intravasation, an early and potentially rate-limiting step in the metastatic cascade, has been assayed in various ways (Liotta et al. 1974; Butler and Gullino 1975; Glaves 1986), but it has never been observed to be directly correlated with metastatic potential nor connected to the expression of genes thought to be essential for metastasis. Methods that directly image and measure intravasation at the cellular level within the primary tumor in the context of metastatic potential will be extremely important in determining the mechanism of action of metastasis genes. The recent application of multiphoton-based intravital imaging (MPIVI) to animal models of cancer in which the tumor cells express green fluorescent protein (GFP) has made the observation of cell behavior and tissue structure possible at the early steps of invasion and metastasis in the primary tumor (Ahmed et al. 2002; Wang et al. 2002; also see Chapter 24). MPIVI also presents an opportunity for students of cell motility to study the ways in which cells move in vivo, allowing, for the first time, direct comparisons of cell motility mechanisms in vivo and in vitro. For more information about MPIVI and its applications to the study of cell behavior see our Web sites: www.aecom.yu.edu/aif/welcome.htm and www.aecom.yu.edu/aif/intravital_imaging/introduction.htm.

We have developed a model that directly examines the behavior of carcinoma cells in live nonmetastatic and metastatic primary tumors in situ. Nonmetastatic and metastatic rat breast cancer cell lines were each prepared to constitutively express GFP. Upon subcutaneous injection of these cells into the mammary fat pad of female Fischer 344 rats, fluorescent primary tumors form whose tumorigenesis and metastasis are unaffected by expression of GFP (Farina et al. 1998). This model allows the behavioral phenotype of cells within metastatic and nonmetastatic tumors to be described and determined (Neri et al. 1982; Farina et al. 1998; Wyckoff et al. 2000b). In principle, this model can be extended to any type of tumor cell that grows as a primary tumor subcutaneously. It is the first model that allowed direct observations of invasive and metastatic behaviors in an intact orthotopically grown primary tumor while in a live animal.

An unexpected dividend of GFP-based intravital imaging is the ability to image motile nonfluorescent host cells against the background of fluorescent tumor cells. We have observed small, rapidly moving cells that are likely to be host immune cells. Sections of the primary tumor stained with hematoxylin and eosin (H&E sections) confirm that host immune cells are present in the primary tumor, supporting this interpretation. Increased numbers of immune system cells could contribute to increased metastasis by producing chemotactic factors or degrading extracellular matrix barriers, particularly around blood vessels (Christensen et al. 1996; Nielsen et al. 1996; Zeng and Guillem 1996; Dong et al. 1997; Ohtani 1998).

By extending this technology to transgenic mice, we are able to explore by the same means tumors that develop directly from mammary ductal epithelium in a fashion analogous to human breast tumor formation (Ahmed et al. 2002). With transgenic mouse systems, different cell types, such as macrophages, can be visualized, which aids in the interpretation of host involvement in the metastatic cascade. Ultimately, it should be possible to combine different fluorochromes and different genetic alterations to explore different cancers and diseases at the cellular level in complex living systems, with the goal of obtaining a better understanding of cell behaviors and interactions in vivo.

ANIMAL MODELS: ORTHOTOPIC RAT MODEL AND TRANSGENIC MICE

With advances in fluorescent protein and transgenic mouse technologies, in vivo imaging has become an important tool in studies of development, cancer, and disease. Many cell types can be transfected with GFP or similar fluorescent proteins and introduced into a live animal. Alternatively, transgenic mice can be made that express these proteins in specific cells or tissues. However, to perform MPIVI in an animal and make meaningful observations, it is essential that the target organ be accessible and proper respiratory activity be maintained without the disruption of blood flow or target organ function. To this end, we have developed mammary tumor models that permit the direct observation of cell behavior in live animals.

The initial model used for direct intravital imaging at the single-cell level in living primary tumors utilized the injection of GFP-labeled adenocarcinoma cells from a well-defined parental cell line (Neri et al. 1982) into the mammary fat pad of female Fischer 344 rats (Farina et al. 1998; Wyckoff et al. 2000; Wang et al. 2002). The technique has been developed further to include the use of transgenic mice that have been bred to incorporate several differing fluorescent and oncogenic transgenes (Faust et al. 2000; Ahmed et al. 2002).

Cell Lines

MTLn3-GFP cells were created by Lipofectamine transfection of parental MTLn3 cells with pEGFPN1 (Clontech) (Farina et al. 1998). MTC-GFP cells were created using retroviral transfection of parental MTC cells with a retroviral GFP expression construct. The GFP sequence was excised from pEGFPN1 using *Nsi*I and *Eco*RI and subcloned into the *Bam*HI/*Eco*RI site of pLXSN. This construct was then transfected into Phoenix cells (Gary Nolan, Stanford University, California) using standard methods (Miller and Rosman 1989; Kinsella and Nolan 1996) and allowed to grow for 24–48 hours. The supernatant was collected, filtered, and spun at 1000 rpm for 5 minutes, and 1 ml was overlaid on a confluent plate of MTC cells. Positive clones were selected by neomycin selection and GFP fluorescence. Stable cells were cultured as parental cells, as described previously (Segall et al. 1996). Cell growth rate and morphology of transfected cells were determined to be the same as those of parental cells, and fluorescence was shown to remain constant for 30 passages. Extensive histopathology studies for the MTLn3-GFP cells (Farina et al. 1998) and the MTC-GFP cells were carried out to confirm that their metastatic potentials were similar to those of the parental cell lines. To create tumors, 1×10^6 cells were injected under the second nipple anterior from the tail of a Fischer 344 rat and allowed to grow for 2.5 weeks.

Transgenic Mouse Models

To relate the overexpression of key signaling molecules to the onset of primary breast tumors, the transgenic mouse has served as an important research tool to assess the tissue-specific action of oncogenes in vivo. Of particular importance are transgenic mouse models of breast cancer that resemble the human disease in both etiology and histology. These mouse models are derived from the expression of either the polyomavirus middle T antigen (PyMT) or Neu/ErbB-2 under the control of the mammary epithelium-specific mouse mammary tumor virus (MMTV) long-terminal-repeat promoter. The MMTV promoter has been used in a number of studies to generate expression of proteins in the mammary gland (Gunzburg et al. 1991; Webster and Muller 1994; Bottinger et al. 1997). Expression studies using this promoter to drive β-galactosidase expression indicate that the major site of expression in female mice is in the epithelial cells of the mammary gland. We have crossed these tumor-bearing mice with GFP transgenics developed to be able to image single-cell behavior intravitally (Ahmed et al. 2002).

Transgenic mice expressing either normal or activated forms of Neu have been generated by Muller and colleagues (Muller et al. 1998; Siegel et al. 1999). Mice harboring the activated transgene (NDL1) develop mammary adenocarcinomas with an average onset of 135 days (± 37), and about 57% of those animals develop lung metastases within an average of 45 days after appearance of the primary tumor. ErbB-2 family members directly interact with a diverse set of signaling proteins such as Src PTKs, Shc, PLC-δ, and PI3K. The *PyMT* is a related oncogene that exerts its oncogenic effects through its association with Src PTKs. The signaling path from Src onward in *PyMT* mice is believed to be very similar to that in activated Neu mice (Muller et al. 1998). These models are particularly relevant to the study of human breast cancer because ErbB-2 is overexpressed in a large percentage of primary breast cancers, and the degree of its overexpression predicts a poor clinical prognosis in both lymph-node-positive and lymph-node-negative patients (Price et al. 1997). Metastasis of adenocarcinomas involves the escape of cells from the primary tumor via either lymphatics or blood vessels, transport to and arrest in a target organ, and growth of metastases in the target organ (Muller et al. 1998). Each of these steps is a multicomponent process, with potentially different tumor cell properties and molecules playing critical roles in different steps (Price et al. 1997). Ideally, high-resolution methods for the analysis of metastasis at the cellular level, such as imaging of cells within tumors, when combined with genomic approaches, could be used to accurately evaluate the roles of specific gene products in the individual steps of metastasis at the cellular level. Similarly, as new therapies are developed, imaging could be used to precisely evaluate the effects of specific treatments on the individual steps in the metastatic cascade at the cellular level.

GENERATION OF MMTV-GFP TRANSGENIC MICE

To generate MMTV-GFP transgenic mice, we used a P206 expression vector. It contains the long version of the MMTV promoter, which shows expression mainly in mammary glands and, rarely, slight leaky expression in salivary glands. Subcloning of enhanced GFP was done between *Hind*III and *Eco*RI sites. The plasmid was prepared with the Endofree Plasmid Kit (Qiagen). DNA was prepared for microinjection by digestion of P206-GFP plasmid with *Sal*I and *Spe*I. Six-week-old FVB/N female mice (Taconic Farms, Germantown, New York) were superovulated by an intraperitoneal (i.p.) injection with pregnant mare serum (PMS; 5 IU in 0.1 ml of 1× phosphate-buffered saline, followed 46–48 hours later by an i.p. injection of human chorionic gonadotrophin (hCG, 5 IU in 0.1 ml of distilled H_2O). Superovulated FVB/N females were mated with FVB/N males the night before injection, and CD1 females, to be used as foster mothers, were mated with vasectomized CD1 males. The next day, FVB/N females possessing a copulation plug were sacrificed, and the fertilized one-cell mouse embryos were isolated from the oviduct. The pronuclei of these zygotes were injected with prepared injection fragment (5 μg/ml DNA solution). Following microinjection, viable eggs were transferred into the oviduct of pseudopregnant CD1 mice.

Southern Blots

To identify transgenic progeny, genomic DNA was extracted from 1-cm tail fragments using the Promega DNA extraction kit (Promega). The nucleic acid pellet was resuspended in 100 μl of DNA Rehydration Solution (TE; 10 mM Tris, 1 mM EDTA). Genomic DNA was quantitated using DNA standards. Genomic DNA (5 μg) was cut with *Bam*H1 overnight at 37°C. After gel electrophoresis and Southern transfer, the nylon filters were hybridized with a GFP probe labeled with the AlkPhos DNA-labeling system (Amersham).

Genotyping by PCR

Tail tips (~0.5–1 cm) from mice 2–3 weeks old were digested in tail lysis buffer together with 10 mg/ml proteinase K. The DNA was precipitated with isopropanol.

- *For GFP genotyping:* Primers oIMR0872 (5′-AAG TTC ATC TGC ACC ACC G-3′) and oIMR0873 (5′-TGC TCA GGT AGT GGT TGT CG-3′), designed by Jackson Labs, produce a 475-bp PCR (polymerase chain reaction) product.

- *For MT genotyping:* The forward primer 5′-GGA AGC AAG TAC TTC ACA AGG G-3′ and the reverse primer 5′-GGA AAG TCA CTA GGA GCA GGG-3′ produce a 475-bp PCR product.

- *For Neu genotyping:* The forward primer 5′-CGG AAC CCA CAT CAG GCC-3′ and the reverse primer 5′-TTT CCT GCA GCC TAC GC-3′ produce a 600-bp PCR product.

TABLE 1. Other transgenics in use for MPIVI

Transgenic Mice Genotype	Comments	References
MMTV-PyMT × MMTV-GFP	Relatively weak GFP labeling of tumor cells.	Ahmed et al. (2002)
MMTV-HER2/neu × MMTV-GFP	Relatively weak GFP labeling of tumor cells.	Ahmed et al. (2002)
MMTV-PyMT × lys-GFP[Ki]	*lys-GFP*[Ki] is a GFP knock-in at the lysozyme gene in macrophages and granulocytes that allows for the direct visualization of tumor- and tissue-associated macrophages in living mice.	Faust et al. (2000)
MMTV-HER2/neu × lys-GFP[Ki]	*lys-GFP*[Ki] is a GFP knock-in at the lysozyme gene in macrophages and granulocytes that allows for the direct visualization of tumor- and tissue-associated macrophages in living mice.	Faust et al. (2000)
MMTV-PyMT × MMTV-CRE/ CAG-CAT-EGFP	Mice contain a Cre-activatable GFP transgene driven by the β-actin promoter to label epithelial cells in the mammary of female mice. Strong GFP labeling of tumor cells.	Kawamoto et al. (2000); Sato et al. (2001); Ahmed et al. (2002)
MMTV-HER2/neu × Wap-CRE/ CAG-CAT-EGFP	Mice contain a Cre-activatable GFP transgene driven by the β-actin promoter to label epithelial cells in the mammary of female mice. Strong GFP labeling of tumor cells.	Kawamoto et al. (2000); Sato et al. (2001); Ahmed et al. (2002)
MMTV-PyMT × Tie 2-GFP	Labels endothelial cells, which allows visualization of the vasculature within the tumor.	Motoike et al. (2000)
MMTV-HER2/neu × Tie 2-GFP	Labels endothelial cells, which allows visualization of the vasculature within the tumor.	Motoike et al. (2000)

IMAGING OF LIVING TUMORS AND THEIR VASCULATURE

The Rat Model

For imaging orthotopic tumors in our rat model, 1×10^6 cells are injected under the second nipple anterior from the tail of a Fischer 344 rat and allowed to grow for 2.5 weeks. After 2.5 weeks, the rat was placed under anesthesia with 5% isoflurane and maintained for the course of the imaging session between 2.5% and 0.5% isoflurane to control the rate of breathing. The tumor is exposed by creating a skin flap. This is done by inserting a scalpel between the epidermis and dermis to peel back the epidermis to expose and optically couple the dermis to the objective. The mammary tumor lies below the dermis and is readily imaged as are cells within the extracellular matrix of the dermis. If done properly, the extent of tissue damage is equivalent to that caused by a tattoo and produces minimal inflammation or disruption of the blood supply and microvasculature. In imaging the tumors, we have found that fat and dense extracellular matrix can be a detriment to a quality image. The surgery performed on a tumor must then be done carefully to remove even the most miniscule amount of fat layers covering the tumor. This part of the surgery must be carried out carefully and precisely

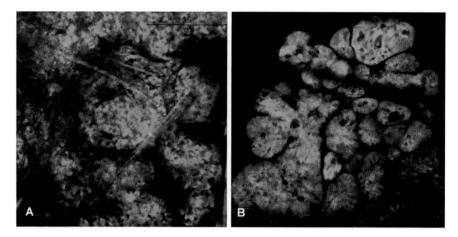

FIGURE 1. Multiphoton microscopy and GFP technology allows for the direct imaging of tumors in the whole animal. (*A*) Tumor from orthotopically injected MTLn3 cells labeled with GFP (*green*) is seen as a single optical section. Collagen signal is generated by second harmonic generation (*purple*) and cell-to-matrix interactions can be visualized. (*B*) An MMTV-PyMT tumor generated in a mouse with the *MMTV-CRE/ CAG-CAT-EGFP* transgene was imaged as a single optical section using multiphoton microscopy. The tumor can be seen as morphologically closer to human disease than the orthotopic model.

so as not to cut through the vasculature across the top of the tumor or to disrupt the cells in the tumor; destroying the microenvironment. This is particularly important in highly necrotic tumors, which can burst during surgery.

The animal is placed onto a Bio-Rad MRC-600 confocal microscope or Bio-Rad Radiance2000 Multiphoton microscope, using a 20× objective (Fig. 1A) and imaged in time lapse, with a single image being taken every minute. On average, each animal is imaged for 20–30 minutes and can be maintained on the microscope for 3–4 hours without special heating and ventilation equipment. Maintaining breathing without drift or movement of the image is key to successful MPIVI. This is achieved by maintaining the level of isoflurane administered to keep breathing even and steady while keeping the animal alive. Because areas near the chest cavity are more difficult to image, the animal must be kept more sedate, which can affect the length of an imaging session. Most injectible anesthetics, such as ketamine, do not suppress breathing and cannot be monitored or controlled during a long session.

The Transgenic Mouse Model

Transgenic tumor-bearing mice are allowed to develop tumors for 10–16 weeks prior to imaging. For imaging, mice are placed under anesthesia with 5% isoflurane and then maintained for the course of the imaging session at 3.5–1.5% isoflurane to control breathing. Exposure of the dermis for the mouse is the same as that described for the rat. The animal is placed on either the inverted or upright microscope depending on the application and imaged at 960 nm for GFP fluorescence with the Bio-Rad Radiance2000 Multiphoton microscope, using a 20× objective (Fig. 1B).

Vasculature Visualization

For visualizing vasculature, 200 µl of rhodamine-dextran (molecular mass: 2 million daltons; Sigma Chemical Co.) at 20 mg/ml in Dulbecco's PBS is injected into the tail vein of the rat or mouse after anesthesia, but before surgery. The vasculature in the tumor is then visualized using the rhodamine channel of the Bio-Rad confocal or by a comparable filter in the PMT of the Bio-Rad Radiance2000 Multiphoton. To image macrophages with rhodamine-dextran, the animals are left to sit for 2 hours postinjection to allow the macrophages to take up the dextran by phagocytosis.

Multiphoton Microscopy

A 10-W Millenium Xs laser (Spectra Physics) is used to run the Radiance2000 Multiphoton system (Bio-Rad), which provides output of 850 mW at 960 nm. The laser must be manually adjusted by turning the prism and slit width to reach mode lock at the appropriate wavelength. For GFP fluorescence, 960 nm is the optimal imaging wavelength, but if more power is needed, GFP can still be detected at 890 nm in which the power achieved is about 1 W. This is also closer to the optimal wavelength for rhodamine.

The Radiance2000 system has been set up to allow for conversion between an Olympus BX 50wi upright scope and an Olympus IX70 inverted scope. Images are obtained from the orthotopic rat model and from differing transgenic models on the inverted microscope using a 20×, 0.75, numerical aperature (NA) objective for time-lapse images. Time-lapse images are taken at 60-second intervals for 30 minutes. The images are collected using Bio-Rad's Lasersharp 2000 software at 50 lines per second. Images are processed using NIH Image 1.61/ppc, ImageJ, and Adobe Photoshop. For experiments involving the use of collection needles for imaging the cells (described below), an upright microscope is used for imaging to allow close control of insertion of the needle relative to the imaging field.

Advantages of Multiphoton Microscopy Over Confocal Microscopy

We have shown that multiphoton excitation fluorescence microscopy has important advantages over other imaging techniques such as confocal microscopy, particularly for the study of live cells and/or for thick tissues (Farina et al. 1998; Wang et al. 2002). The following are the advantages of the multiphoton instruments.

- The lack of fluorophore excitation in regions away from the focal plane minimizes fluorophore bleaching and the generation of toxic by-products during imaging.

- Multiphoton images are less prone to degradation by light scattering. This is because the longer wavelengths used for excitation suffer less scattering from microscopic refractive index differences within the sample, allowing much greater penetration of the tissues (Centonze and White 1998). In addition, as all the resolution is defined by the geometry of the excitation beam, the fluorescence emission is unaffected by light scattering. The insensitivity of multiphoton imaging to light scattering is particularly advantageous for the study of living specimens due to the presence of many refractive index interfaces in living tissues.

- The longer wavelengths used in multiphoton microscopy have the added benefit of second harmonic generation to image collagen fibers (Campagnola et al. 2001; Williams et al. 2001). This phenomenon is not found in conventional microscopy and allows for imaging of cell-to-matrix interactions such as adhesion and degradation.

A valuable benefit resulting from the second harmonic-generated image is the ability to directly estimate the amount and condition of the extracellular matrix (ECM) adjacent to carcinoma cells in tumors. The amount, integrity, and size of ECM fibers can be determined by measuring the signal intensity and morphology of fibers in images obtained from the

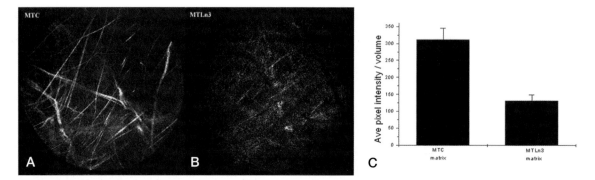

FIGURE 2. MTC tumors have more collagen-containing ECM than MTLn3 tumors. By imaging the second harmonics of ECM using multiphoton microscopy, the difference in collagen content between the two tumor types was determined. (*A*) MTC tumors at 35 μm depth into the tumor can be seen to have more matrix than the MTLn3 at tumor (*B*) at the same depth of focus. (*C*) By calculating the pixel intensity for the captured volume of a *z* series of each tumor, MTC tumors are seen to have on average 2.4 times more matrix than MTLn3-generated tumors (*n* = 4). (Reprinted, with permission, from Wang et al. 2002.)

second harmonic-generated light emitted from the ECM (Wang et al. 2002). An important insight resulting from this analysis is the demonstration that adjacent to carcinoma cells in the primary tumor, metastatic MTLn3-derived tumors contain twofold to threefold fewer ECM fibers than nonmetastatic MTC-derived tumors (Fig. 2C). This is determined by calculating the pixel intensity of a volume using NIH Image software. Furthermore, the morphology of the fibers shows discontinuities in second harmonic emissions along the fibers, suggesting that proteolysis has occurred at intervals along the ECM fibers (compare panels A and B in Fig. 2). Finally, these fibers are observed to support the rapid and directional linear migration of carcinoma cells in metastatic tumors (Wang et al. 2002). This is an exciting observation because chemotaxis of carcinoma cells in response to epidermal growth factor (EGF)-like ligands resulting from proteolysis of ECM could be an important property of invasive tumors (Liotta and Kohn 2001). However, patterned proteolysis of extracellular fibers associated with migrating carcinoma cells has not been observed directly in tumors until the advent of second harmonic intravital imaging.

The second harmonic signal is emitted as polarized light when the electrons in the π orbitals of α-helix chains, such as found in collagens, are excited by a long wavelength of light (Campagnola et al. 2001; Williams et al. 2001). We have found that we are able to excite the second harmonic signal with wavelengths of 760–960 nm and image them through a filter with a 450–480 nm cutoff.

Figure 3A shows two series of optical sections obtained by stepping at 12-μm intervals through a primary tumor in a live animal. The left series in Figure 3 was obtained with a conventional confocal microscope, and the series on the right was obtained with the multiphoton microscope using an infrared laser and an external detector. Comparison of these two *z* series demonstrates the superiority of the multiphoton microscope in penetrating deep within the live tissue to generate high-resolution confocal images. Figure 3B illustrates that the relative amount of bleaching is greatly reduced in the multiphoton microscope as demonstrated by reexposure of the same optical planes to laser light. Collagen fibers visualized by second harmonic generation can be seen in the images generated by the multiphoton microscope.

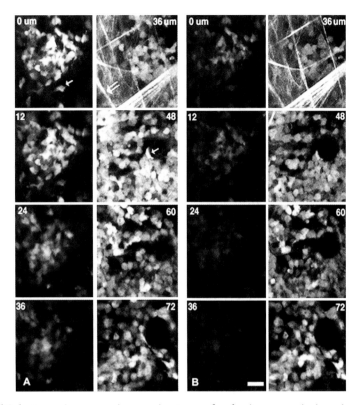

FIGURE 3. Multiphoton microscopy is superior to confocal microscopy in imaging primary tumors more deeply and with less photodamage. (*A*) Initial *z* series of images taken with the confocal (*left*) and with the multiphoton (*right*) microscope showing the greater depth of imaging possible with multiphoton excitation. (*B*) A second *z* series of the same focal planes illustrates dramatic improvement in GFP stability in the multiphoton microscope (*right*) compared to the confocal microscope (*left*). Collagen imaged by second harmonic generation can be seen in all multiphoton images (right in *A* and *B*). (Reprinted, with permission, from Wang et al. 2002.)

DIRECT VISUALIZATION OF THE INVASIVE POPULATION OF CELLS DERIVED FROM A LIVING TUMOR

We have developed a method to selectively collect invasive cells from live primary tumors in intact rats using microneedles filled with matrigel and containing chemoattractants to mimic chemotactic signals from blood vessels and associated cell layers. This assay models invasion of tumor cells into neighboring connective tissue and blood vessels (Wyckoff et al. 2000a). The model in Figure 4 shows a diagrammatic description of the experiment. Furthermore, a collection of these cells can be visualized using an upright microscope, with an attached micromanipulator, on the multiphoton system.

Preparation and Handling of Collection Needles

Needles (33 gauge) are prepared on the day of the experiment by filling them with 1:10 matrigel and L15-BSA (the isotonic equivalent of 5% FBS) or L15-BSA with a final concentration of 25 nM. All needles contained 0.01 mM EDTA (pH 7.4) to sequester heavy metals that might be released from the needle. A rat or mouse is anesthetized using 5% isoflurane and laid on its back on the microscope stand. The isoflurane is reduced to 2%, and a small patch of skin is removed to expose the tumor. The 33-gauge needles are held

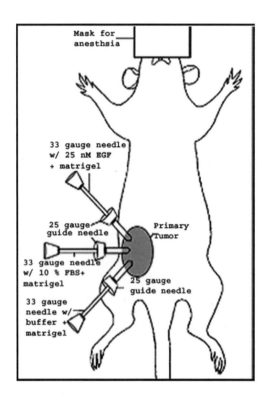

FIGURE 4. Method for using needles for in vivo cell collection. Needles (inner diameter. 102 μm) filled with matrigel and buffer, 25 nM EGF, or 10% FBS are placed in 25-gauge guide needles that are inserted into the primary tumor of an anesthetized rat. (Reprinted, with permission, from Wyckoff et al. 2000a.)

in place inside of 25-gauge needles, which are inserted first into the tumor in a specially designed holder which allows the simultaneous insertion of three separate 25-gauge needles into the tumor. The three 25-gauge needles are first inserted into the tumor with blocking wires inside to keep the interior of the needles from filling with tumor. They are positioned with a specially designed needle holder held in a micromanipulator for stability into the tumor. The guide wires are removed and the 33-gauge needles containing matrigel with or without chemoattractant (such as EGF) are inserted. The animal is kept under monitored anesthesia for 4 hours. Time-lapse observations are made during the collection of cells using a 20×, 0.75 NA objective. Afterward, the needle contents are expelled onto a coverslip and mixed 1:1 with DAPI, and the cells are counted immediately.

For EGF dose-response curves, the needles are filled as above with final concentrations of EGF of 5, 10, 25, and 50 nM. The experiment is then performed as described above. Three different concentrations of EGF are used in each experiment to control for differences between tumors. Figure 5 shows a dose-response curve for EGF in our rat model.

GENE DISCOVERY BASED ON BEHAVIORAL ANALYSIS

Combining MPIVI with cDNA microarrays has permitted the identification of genes and pathways that correlate with cell behavior in vivo (Wang et al. 2002). Differences in cell behavior between nonmetastatic and metastatic cells in culture, and within live primary tumors, can be correlated with results from cDNA microarray analyses to identify potentially important genetic determinants for breast cancer invasion and metastasis. Using multiphoton microscopy, five major differences in carcinoma cell behavior have been found between the nonmetastatic and metastatic primary breast tumors involving ECM, cell motility, and chemotaxis. Behavioral dif-

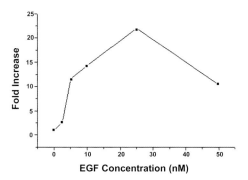

FIGURE 5. A dose-response curve for EGF-dependent collection of carcinoma cells from a primary tumor of the rat breast. The collection of the motile population of cells in the tumor is concentration-dependent and exhibits a dose response similar to that observed in vitro with cultured cells. By using microneedles filled with matrigel and varying concentrations of EGF, a 20-fold increase of motile cells is observed at 25 nM over matrigel alone.

ferences were correlated with seven categories of molecules that were differentially expressed and related to these behaviors. We found that ECM composition, actin nucleation factors, molecules involved in mechanical stability and survival, and cell polarity and chemotaxis showed large and consistent differences in gene expression (Wang et al. 2002).

Measurement of Cell Behavior

Cell motility, polarity, intravasation, and adhesion are visualized by time-lapse multiphoton microscopy by taking an image at 1-minute intervals for at least 30 minutes. Currently, each image requires 10 seconds for collection to provide good spatial resolution for cells moving 10 μm/min or less. The images are then assembled into time-lapse movies in NIH Image or Image J. Animation of the movies allows for the detection of cell motility, intravasation, and cell protrusion. From these observations, DIAS analysis (Soll 1995) can be used to determine such variables as speed, persistence, distance, and direction (Farina et al. 1998; Wang et al. 2002; Condeelis and Segall 2003).

Cell polarization, especially orientation toward blood vessels, is determined by observing the shape of the cells within the field. The polarization of cells around blood vessels is used to determine whether the cells are oriented toward the vessels due to a cytokinetic response or are just randomly oriented. If cells within two cell diameters of the vessel are polarized—defined as a cell with a distinct leading edge or, when the edge is indistinguishable within the image, as a cell whose length is at least twice its width—toward the vessel and cells more than two cell diameters away from the vessel are not, a cell is counted as being oriented toward the vessel (Farina et al. 1998).

Host cells can be imaged as shadows crawling on top of fluorescent carcinoma cells (Farina et al. 1998) (see Movie 22.1 on the accompanying DVD), and macrophages can be imaged by loading through intravenous injection with rhodamine-labeled dextran (see Movie 22.2), which they then phagocytose. Finally, specific cell types can be imaged in transgenic mice by expression of GFP by cell-type-specific promoters (see Movie 22.3), for example, using the lysine or CSF-1 promoter to drive GFP expression in macrophages (Faust et al. 2000).

Cell-matrix interactions can also be inferred due to the ability to image ECM fibers. These interactions include cell motility along matrix fibers and adhesion (Wang et al. 2002; Condeelis and Segall 2003). To image adhesion, a movie showing just the matrix channel is compared with a movie of the combined cell and matrix channels. Sites of cell colocalization with the matrix which correlate with shadows that appear on the matrix are counted as adhesion sites. Although adhesion cannot be directly quantitated, differences in the ability of cells to adhere can be determined by comparing their differences in cell/matrix colocalization or by differences in cell/matrix locomotion (Wang et al. 2002). Comparative matrix density between differing tumor volumes is determined by calculating the pixel intensity from a reconstructed *z* series of just the matrix channel (Wang et al. 2002).

Microarray Procedures

Total RNA was isolated from MTLn3 and MTC cells or tumors using Trizol reagent (Life Technologies). The quality of the RNA was verified by electrophoresis on formaldehyde 1.2% agarose gels.

Microarray analysis was performed by using cDNA microarrays made at the cDNA Microarray Facility Albert Einstein College of Medicine (Cheung et al. 1999). About 9700 mouse genes (Incyte Genomics) were precisely spotted onto a single glass slide. Detailed descriptions of microarray hardware and procedures are available from http://www.aecom.yu.edu/home/molgen/facilities.html. The use of these mouse arrays with rat RNA has been validated and shown to have more than 90% correlation between these two species (M. Dabeva, pers. comm.).

Microarray analysis was performed in triplicate for each experiment. For each hybridization, cDNA targets were prepared from the RNA samples obtained from MTLn3 (Cy5-labeled) and the MTC (Cy3-labeled) cells or tumors. Labeling and hybridization were performed as follows:

1. First-strand cDNA probes were generated by incorporation of either Cy3-dUTP or Cy5-dUTP (Amersham Pharmacia) during reverse transcription of 100 µg of total RNA.

2. The resulting cDNA probes were purified and concentrated, denatured at 94°C, and hybridized to an arrayed slide overnight at 50°C.

3. Slides were then washed in 1× SSC/0.1% SDS for 10 minutes and subsequently in 0.2× SSC/0.1% SDS for 20 minutes.

4. Slides were then rinsed and dried for scanning. Data from the hybridization reactions were collected using a two-color laser-scanning confocal microscope that has been custom-designed and built in our laboratory for the maximum sensitivity required to measure low-abundance mRNAs.

5. GenePix Pro 3.0 (Axon Instruments, Inc) was used to generate raw data files containing measurements of signal and background fluorescence emissions of Cy3 and Cy5, respectively, for each element.

Quality Control, Data Analysis, and Statistics

The primary data were analyzed and those spots not passing the initial quality-control test using default parameters set in the Genepix 3.0 image analysis program were flagged and not analyzed further. Normalization and data processing were carried out on data passing the quality-control test as described elsewhere (Zavadil et al. 2001; Chauhan et al. 2002). Briefly, the net signal intensity (NSI) of each spot in both channels (Cy5 as channel 1 and Cy3 as channel 2) was determined by subtracting the local background from the signal intensity values and then subjected to log transformation. For normalization of the two channels, the first calculation of the overall intensity for each channel (I_{ch1} and I_{ch2}) was performed by calculating the natural logarithms of the intensity values of all genes, determining the average of these values, and then taking the exponential of the average. The ratio of the overall intensity in channel 1 over that in channel 2 was then calculated as $R = I_{ch1}/I_{ch2}$. The intensities for both channels were then balanced by multiplying the NSI of each spot in channel 2 by the factor *R*. After this normalization procedure, the normalized intensity ratio was then calculated for each gene. In our experimental conditions, ratios of 1.6 and larger (up-regulated) or 0.6 and smaller (down-regulated) were chosen as significantly different gene expression levels between two samples hybridized to the same array spot (Chauhan et al. 2002). Genes showing consistent differential expression across replica arrays were extracted for further analysis. These genes were grouped on the basis of their function in cancer invasion and metastasis by searching PubMed, Swissprot (http://ca.expasy.org/sprot/), and Online Mendelian Inheritance in Man (OMIM) databases. Functional categories were correlated with cell behavior observed in the primary tumor and in culture (Wang et al. 2002).

MOVIE LEGENDS

MOVIE 22.1. In vivo motility of a carcinoma cell in an orthotopically injected tumor. GFP-labeled MTLn3 cell (*green*) is seen crawling along matrix (*purple*) imaged by second harmonic generation.

MOVIE 22.2. In vivo motility of macrophages in a transgenic mouse. Macrophages (*green*) are seen crawling in a PyMT-generated tumor among matrix (*blue*) imaged by second harmonic generation.

MOVIE 22.3. In vivo motility of carcinoma cells crawling in a PyMT-generated tumor in a transgenic mouse. Carcinoma cells (*green*) contain a CRE-activated EGFP transgene driven by the β-actin promoter.

REFERENCES

Ahmed F., Wyckoff J., Lin E.Y., Wang W., Wang Y., Hennighausen L., Miyazaki J., Jones J., Pollard J.W., Condeelis J.S., and Segall J.E. 2002. GFP expression in the mammary gland for imaging of mammary tumor cells in transgenic mice. *Cancer Res.* **62:** 7166–7169.

Bottinger E.P., Jakubczak J.L., Haines D.C., Bagnall K., and Wakefield L.M. 1997. Transgenic mice overexpressing a dominant-negative mutant type II transforming growth factor beta receptor show enhanced tumorigenesis in the mammary gland and lung in response to the carcinogen 7,12-dimethylbenz-[a]-anthracene. *Cancer Res.* **57:** 5564–5570.

Butler T.P. and Gullino P.M. 1975. Quantitation of cell shedding into efferent blood of mammary adenocarcinoma. *Cancer Res.* **35:** 512–516.

Campagnola P.J., Clark H.A., Mohler W.A., Lewis A., and Loew L.M. 2001. Second-harmonic imaging microscopy of living cells. *J. Biomed. Opt.* **6:** 277–286.

Centonze V.E. and White J.G. 1998. Multiphoton excitation provides optical sections from deeper within scattering specimens than confocal imaging. *Biophys. J.* **75:** 2015–2024.

Chauhan B.K., Reed N.A., Zhang W., Duncan M.K., Kilimann M.W., and Cvekl A. 2002. Identification of genes downstream of Pax6 in the mouse lens using cDNA microarrays. *J. Biol. Chem.* **277:** 11539–11548.

Cheung V.G., Morley M., Aguilar F., Massimi A., Kucherlapati R., and Childs G. 1999. Making and reading microarrays. *Nat. Genet.* (suppl.) **21:** 15–19.

Christensen L., Wiborg Simonsen A.C., Heegaard C.W., Moestrup S.K., Andreasen J.A., and Andreasen P.A. 1996. Immunohistochemical localization of urokinase-type plasminogen activator, type-1 plasminogen-activator inhibitor, urokinase receptor and alpha(2)-macroglobulin receptor in human breast carcinomas. *Int. J. Cancer* **66:** 441–452.

Condeelis J. and Segall J.E. 2003 Intravital imaging of cell movement in tumors. *Nat. Rev. Cancer* **3:** 921–930.

Dong Z., Kumar R., Yang X., and Fidler I.J. 1997. Macrophage-derived metalloelastase is responsible for the generation of angiostatin in Lewis lung carcinoma. *Cell* **88:** 801–810.

Farina K.L., Wyckoff J.B., Rivera J., Lee H., Segall J.E., Condeelis J.S., and Jones J.G. 1998. Cell motility of tumor cells visualized in living intact primary tumors using green fluorescent protein. *Cancer Res.* **58:** 2528–2532.

Faust N., Varas F., Kelly L.M., Heck S., and Graf T. 2000. Insertion of enhanced green fluorescent protein into the lysozyme gene creates mice with green fluorescent granulocytes and macrophages. *Blood.* **96:** 719–726.

Glaves D. 1986. Detection of circulating metastatic cells. *Prog. Clin. Biol. Res.* **212:** 151–167.

Gunzburg W.H., Salmons B., Zimmermann B., Muller M., Erfle V., and Brem G. 1991. A mammary-specific promoter directs expression of growth hormone not only to the mammary gland, but also to Bergman glia cells in transgenic mice. *Mol. Endocrinol.* **5:** 123–133.

Kawamoto S., Niwa H., Tashiro F., Sano S., Kondoh G., Takeda J., Tabayashi K., and Miyazaki J. 2000. A novel reporter mouse strain that expresses enhanced green fluorescent protein upon Cre-mediated recombination. *FEBS Lett.* **470:** 263–268.

Kinsella T.M. and Nolan G.P. 1996. Episomal vectors rapidly and stably produce high-titer recombinant retrovirus. *Human Gene Ther.* **7:** 1405–1413.

Liotta L.A. and Kohn E.C. 2001. The microenvironment of the tumour-host interface. *Nature* **411:** 375–379.

Liotta L.A., Kleinerman J., and Saidel G.M. 1974. Quantitative relationships of intravascular tumor cells, tumor vessels, and pulmonary metastases following tumor implantation. *Cancer Res.* **34:** 997–1004.

Miller A.D. and Rosman G.J. 1989. Improved retroviral vectors for gene transfer and expression. *BioTechniques* **7:** 980–982, 984–986, 989–990.

Motoike T., Loughna S., Perens E., Roman B.L., Liao W., Chau T.C., Richardson C.D., Kawate T., Kuno J., Weinstein B.M., Stainier D.Y., and Sato T.N. 2000. Universal GFP reporter for the study of vascular development. *Genesis* **28:** 75–81.

Muller W.J., Ho J., and Siegel P.M. 1998. Oncogenic activation of Neu/ErbB-2 in a transgenic mouse model for breast cancer. *Biochem. Soc. Symp.* **63:** 149–157.

Neri A., Welch D., Kawaguchi T., and Nicolson G.L. 1982. Development and biologic properties of malignant cell sublines and clones of a spontaneously metastasizing rat mammary adenocarcinoma. *J. Natl. Cancer Inst.* **68:** 507–517.

Nielsen B.S., Timshel S., Kjeldsen L., Sehested M., Pyke C., Borregaard N., and Dano K. 1996. 92 kDa type IV collagenase (MMP-9) is expressed in neutrophils and macrophages but not in malignant epithelial cells in human colon cancer. *Int. J. Cancer.* **65:** 57–62.

Ohtani H. 1998. Stromal reaction in cancer tissue: Pathophysiologic significance of the expression of matrix-degrading enzymes in relation to matrix turnover and immune/inflammatory reactions. *Pathol. Int.* **48:** 1–9.

Price J.T., Bonovich M.T., and Kohn E.C. 1997. The biochemistry of cancer dissemination. *Crit. Rev. Biochem. Mol. Biol.* **32:** 175–253.

Sato M., Watanabe T., Oshida A., Nagashima A., Miyazaki J.I., and Kimura M. 2001. Usefulness of double gene construct for rapid identification of transgenic mice exhibiting tissue-specific gene expression. *Mol. Reprod. Dev.* **60:** 446–456.

Segall J.E., Tyerech S., Boselli L., Masseling S., Helft J., Chan A., Jones J., and Condeelis J. 1996. EGF stimulates lamellipod extension in metastatic mammary adenocarcinoma cells by an actin-dependent mechanism. *Clin. Exp. Metastasis.* **14:** 61–72.

Siegel P.M., Ryan E.D., Cardiff R.D., and Muller W.J. 1999. Elevated expression of activated forms of Neu/ErbB-2 and ErbB-3 are involved in the induction of mammary tumors in transgenic mice: Implications for human breast cancer. *EMBO J.* **18:** 2149–2164.

Soll D.R. 1995. The use of computers in understanding how animal cells crawl. *Int. Rev. Cytol.* **163:** 43–104.

Wang W., Wyckoff J.B., Frohlich V.C., Oleynikov Y., Huttelmaier S., Zavadil J., Cermak L., Bottinger E.P., Singer R.H., White J.G., Segall J.E., and Condeelis J.S. 2002. Single cell behavior in metastatic primary mammary tumors correlated with gene expression patterns revealed by molecular profiling. *Cancer Res.* **62:** 6278–6288.

Webster M.A. and Muller W.J. 1994. Mammary tumorigenesis and metastasis in transgenic mice. *Semin. Cancer Biol.* **5:** 69–76.

Williams R.M., Zipfel W.R., and Webb W.W. 2001. Multiphoton microscopy in biological research. *Curr. Opin. Chem. Biol.* **5:** 603–608.

Wyckoff J.B., Segall J.E., and Condeelis J.S. 2000a. The collection of the motile population of cells from a living tumor. *Cancer Res.* **60:** 5401–5404.

Wyckoff J.B., Jones J.G., Condeelis J.S., and Segall J.E. 2000b. A critical step in metastasis: In vivo analysis of intravasation at the primary tumor. *Cancer Res.* **60:** 2504–2511.

Zavadil J., Bitzer M., Liang D., Yang Y.C., Massimi A., Kneitz S., Piek E., and Bottinger E.P. 2001. Genetic programs of epithelial cell plasticity directed by transforming growth factor-beta. *Proc. Natl .Acad. Sci.* **98:** 6686–6691.

Zeng Z.S. and Guillem J.G. 1996. Colocalisation of matrix metalloproteinase-9-mRNA and protein in human colorectal cancer stromal cells. *Br. J. Cancer* **74:** 1161–1167.

CHAPTER TWENTY THREE

Long-term, High-resolution Imaging in the Neocortex In Vivo

Brian E. Chen, Joshua T. Trachtenberg, Anthony J.G.D. Holtmaat, and Karel Svoboda

Howard Hughes Medical Institute, Cold Spring Harbor Laboratory, Cold Spring Harbor, New York 11724

INTRODUCTION

THE PHYSIOLOGICAL OUTPUT OF A NEURON is determined to a large degree by the nature and organization of the synaptic connections it receives. Governing the organization of dendrites, axons, and the synaptic connections that link them are factors both intrinsic and extrinsic to the organism. During postnatal life, sensorimotor experience is the dominant extrinsic factor influencing the activity of neuronal networks. Experience-dependent activity sculpts the structure of axons and dendrites as well as the efficacy and/or organization of impinging synapses. In the neocortex, elucidating the mechanisms of structural plasticity is essential to an understanding of the emergent network properties and fundamental cognitive phenomena, such as memory formation. The predominant method of characterizing structural plasticity relies on postmortem examinations of neuronal morphology in fixed tissue. Only a single time point is collected, which requires population comparisons to reveal significant differences in control and experimental conditions (Volkmar and Greenough 1972; Zito and Svoboda 2002). Steady-state dynamics is not detectable by such approaches. Time-lapse imaging microscopy has revealed a remarkable array of dynamic activities in dendritic structures in developing cortical tissue in vitro (Dailey and Smith 1996; Engert and Bonhoeffer 1999; Maletic-Savatic et al. 1999), the developing cortex in vivo (Lendvai et al. 2000), and even the adult neocortex (Grutzendler et al. 2002; Trachtenberg et al. 2002). This chapter briefly reviews the elements relevant to chronic in vivo imaging and then presents experimental procedures.

Two-Photon Microscopy in the Neocortex

Chronic high-resolution in vivo imaging of the structure of neurons in the cortex became possible with the invention of 2-photon laser scanning microscopy (2PLSM) (Fig. 1) (Denk et al. 1990). This technique has key advantages over conventional, single-photon excitation techniques, such as confocal microscopy (Denk and Svoboda 1997). Two-photon excitation is the near-simultaneous (within femtoseconds) absorption of two photons coinciding on a

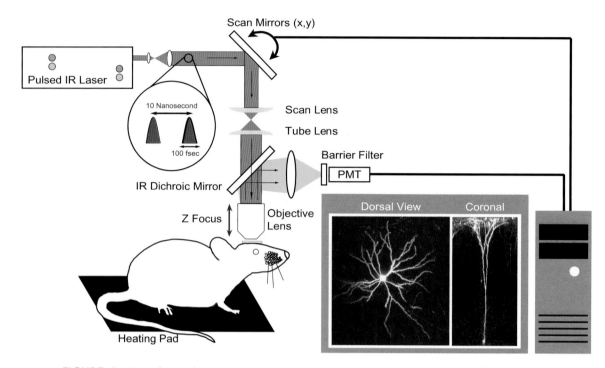

FIGURE 1. Two-photon laser-scanning microscopy in vivo. Images were acquired with a custom-built 2PLSM. The light source was a pulsed, mode-locked pulsed Titanium:Sapphire laser running at λ ~910 nm. The repetition rate was ~100 MHz and the pulse duration ~100 fsec. The power measured at the back-focal plane of the objective was <200 mW. The laser beam is expanded and coupled into a pair of gal-vanometer scan mirrors. The mirrors are imaged onto the back-focal plane of the objective using a scan lens and a microscope tubelens. The objective (typically Zeiss, 40×, 0.8 NA) focuses the excitation beam to a diffraction-limited spot. Fluorescence photons collected by the objective are reflected by a dichroic mirror and imaged onto a PMT detector. The image is constructed in a computer.

fluorophore. The absorption rate depends quadratically on the illumination intensity and is therefore confined to a small volume around the focal point. Scattered excitation light is too weak to generate fluorescence. Thus, the signal is generated exclusively in a tiny focal volume, and all emitted fluorescence photons constitute useful signals. A related advantage is that the longer wavelengths used to generate 2-photon excitation penetrate scattering tissue more efficiently than the shorter wavelengths used to generate single-photon excitation of the same fluorophores. The scattering length increases approximately linearly with increasing wavelengths (Oheim et al. 2001). In the cortex the scattering length at 800-nm wavelength, $l_s^{(800)}$, measured in vivo is approximately 100 μm (Kleinfeld et al. 1998) and appears to decrease with increasing age (Oheim et al. 2001). Typically, imaging is limited to the most superficial 500 μm of the tissue. The upper layers (1–4) of neocortex are ideal for high-resolution 2PLSM imaging.

The 2PLSM is basically a laser-scanning microscope (Fig. 1) (Pawley 1995). The light source is typically a mode-locked Titanium:Sapphire laser producing excitation light with wavelengths in the range of 780–980 nm. Fluorescence is collected in an epifluorescence mode and the image is constructed in a computer.

Fluorescent Molecules

Imaging the morphology of neurons requires fluorescent molecules in the cell of interest. Traditionally, synthetic small-molecule fluorophores have been delivered using bulk-labeling techniques or single-cell injection techniques. Bulk labeling gives weak and nonspecific labeling, and single-cell injections are extremely difficult (Svoboda et al. 1997).

Currently, the fluorophore of choice for in vivo imaging is the green fluorescent protein (GFP) and related fluorescent proteins (XFPs). The power of GFP as a protein tag is well-established (see, e.g., discussions of fluorescent proteins in Chapters 1 and 2). It is often not appreciated, however, that modern XFPs are excellent fluorophores: They have large extinction ratios and quantum efficiencies (comparable to some of the best synthetic fluorophores) over a large spectral range, and they are quite resistant to photobleaching (Tsien 1998; Campbell et al. 2002; Nagai et al. 2002). GFP has also been mutated to actively indicate subcellular activity (e.g., pH changes and Ca^{2+} influx) (Miyawaki et al. 1997; Miesenbock et al. 1998; Nagai et al. 2001a,b).

A general concern when using GFP to visualize small cellular compartments, such as dendritic spines, is the level of expression needed. Using GFP strictly as a cytoplasmic marker, we estimate that >1 μM of GFP is required to reliably detect dendritic spines, the smallest neuronal compartments. The protein concentrations needed for optimal imaging are higher than the vast majority of endogenous proteins expressed by neurons.

Fluorescent Protein Expression In Vivo

Delivery of foreign gene(s) to neurons in vivo remains a difficult problem, in part because these are postmitotic cells and not easily accessible. Ideally, one would like to label neurons with cell-type specificity, noninvasively, without cytopathic side effects, and with controllable efficiency. Standard transfection methods, such as chemical transfection (calcium phosphate precipitation, liposomes, lipids, cationic polymers) and biollistic gene transfer, have not been successful in vivo (Washbourne and McAllister 2002).

The use of recombinant viral vectors and germ-line gene transfer has been increasingly useful for expressing foreign genes in the adult brain. Recombinant viral vector technology exploits the ability of particular viruses to introduce transgenes into postmitotic neurons. Some recombinant viruses such as Sindbis-, herpes-simplex-, adeno-, adeno-associated- and lenti-based viral vectors can transduce brain cells with a remarkable, but as yet poorly understood, efficiency for neurons (Hermens and Verhaagen 1998; Kay et al. 2001; Washbourne and McAllister 2002).

Recombinant viruses can be produced relatively quickly and ultimately allow expression of different genes of interest in combinations under cell-type-specific promoters in a variety of species. The disadvantage of using viruses is their potential toxicity, although recent progress in this area, due to efforts to improve the safety of vectors for gene therapy, has greatly diminished many of the adverse effects of some viral vectors on their hosts (Kay et al. 2001). Another disadvantage of viral vectors in the intact brain is the fact that they usually need to be pressure-injected, which requires surgery and potentially induces an injury. Furthermore, the expression of recombinant proteins driven by some viral vectors (adenovirus, adeno-associated virus, and lentivirus) can be slow, with delays ranging from days to weeks postinfection. Germ-line transgenesis provides another powerful method for labeling neurons in vivo. In transgenic mice, the gene of interest is integrated into the genome of all body cells, but the temporal and spatial patterns of expression are in large part determined by cell-type-specific regulatory elements that are included in the transgene construct. The great advantages of using transgenic mice over transfection techniques include the nonimmunogenic (noncytopathic) expression of the recombinant protein and the stability and reproducibility of expression. Expression can be controlled spatially and temporally using tissue-specific and inducible promoters. Spatial control can be achieved by choosing from a limited variety of minimal promoters (Holtmaat et al. 1998; Wells and Carter 2001) or bacterial artificial chromosomes (Heintz 2001) for transgenesis. Other advantages include the fact that targeted cell types of a transgenic mouse can be imaged after only a single surgery, the mice can be crossed with other transgenic mice for differential labeling of neurons, and fluorescent expression can persist from development to adulthood. The main practical disadvantages of using transgenic mice are the long turnaround time needed from DNA construct to obtaining transgenic mouse lines and the often high labeling density within a particular neuronal population, which makes high-resolution optimal microscopy difficult. In addition, in many cases, expression levels are too low for in vivo imaging.

CHRONIC IN VIVO IMAGING

Our first attempts at imaging dendritic structures over time scales of days were in the developing rat barrel cortex, using Sindbis-virus-mediated expression of enhanced GFP in neurons (Chen et al. 2000). This approach has fundamental limitations. First, Sindbis virus is cytotoxic after about 4 days of expression, principally due to high levels of foreign gene expression (Agapov et al. 1998). Second, the rat dura mater degrades image quality. Nevertheless, the same neuron and dendrites could be imaged for a few days.

These disadvantages were overcome by the development of transgenic mice that express spectral variants of GFP at high levels in cortical pyramidal neurons (Feng et al. 2000). We have used two lines with different expression patterns (YFP-H and GFP-M) (Fig. 2). In both lines, XFP expression is sufficiently high to image individual dendritic spines in vivo. In the GFP-M mouse, neurons are labeled in a sparse manner, but the labeled cells express GFP at high concentrations, giving the appearance of Golgi stains. In addition, the mouse dura is sufficiently thin so that it can be left intact without degradation of imaging. In these mice, neuronal structure was imaged chronically through an optical chamber (Levasseur et al. 1975; Brown et al. 2001; Trachtenberg et al. 2002). These optical chambers are hermetically sealed and maintain the cortical environment intact with respect to, for example, intracranial pressure, cerebral fluid composition, and prevailing gas tensions (Fig. 3).

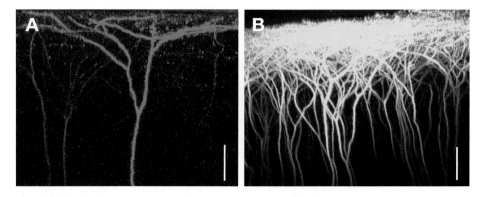

FIGURE 2. In vivo images of the upper layers of cortex in transgenic mice expressing YFP and GFP. Both mouse lines express fluorescence mostly in Layer V pyramidal neurons in the cortex. (A) Coronal view in a GFP-M mouse. (B) Coronal view in a YFP-H mouse. The dense expression pattern, compared to the GFP-M line, obscures individual Layer I dendrites. Side views are maximal projections in the x or y dimension (with z distance along the ordinate axis). Bars, 50 μm.

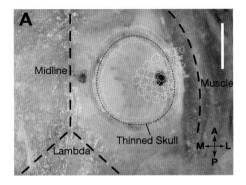

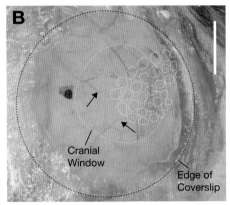

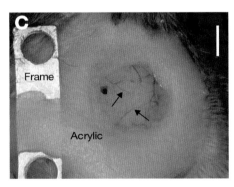

FIGURE 3. Experimental preparation for in vivo imaging. (*A*) The circumference of the region of skull to be excised is thinned. Black pen marks are used to align barrel cortex. The diameter of the craniotomy is ~4 mm. (*B*) The intact, exposed dura is covered with agar and a circular coverglass. (*C*) A small frame is embedded in dental cement to stabilize imaging. Arrows point to surface blood vessels. Bars, 2 mm.

Neuronal Health Under the Cranial Window

The health of the cortex under the cranial window was verified electrophysiologically, anatomically, and ultrastructurally. Whisker stimulation evoked normal local field potentials and intracellularly recorded postsynaptic potentials in neurons in the barrel cortex underlying the optical window even after the optical window had been left in place for a number of weeks. Additionally, membrane potential measurements from cells immediately below the imaging window were not significantly different from those recorded from neurons in normal age-matched controls. Cortical circuitry was similarly unaffected by the imaging window. Spontaneous synaptic activity and spontaneous fluctuations of membrane potentials, which reflect the strengths of synaptic inputs, were identical in implanted animals and controls (Trachtenberg et al. 2002).

Anatomically, densities of spines measured in fixed brains were not different from densities immediately below the imaging window. In addition, the spine density in our in vivo images (.392 ± .01 spines/μm) did not vary with experimental duration and is identical to measurements in fixed tissue from the same animals. Nor was there any indication of tissue damage at the ultrastructural level. Hallmarks of cortical damage, such as reactive microglia and increased extracellular space, were not observed in tissue that had been chronically imaged in vivo and subsequently examined under an electron microscope (Trachtenberg et al. 2002).

Image Acquisition

In each animal, the dendrites of a single Layer II/III or V pyramidal neuron and their spines can be imaged daily for several weeks. The cell and the respective regions of interest can be located each day by using the unique vascular pattern in the cortex as a reference. By using a picture of the surface vasculature taken on the first day of imaging with a digital camera (Fig. 3), the blood vessels can be aligned each day and the objective can be positioned within a distance of about 100 μm of the cell. The unique branching patterns of the fluorescence dendrites are then used to identify those specific regions of the cell that were imaged previously (Fig. 4). Fortunately, in the adult cortex, dendrites do not grow or retract appreciably in the adult animal (Trachtenberg et al. 2002) and are easily identifiable each day. In the example in Figure 4, all dendrites that were imaged on the first day are re-identified in subsequent imaging sessions. For each cell, typically up to 10 fields of view of 45 × 45 μm can be selected for imaging. At this magnification,

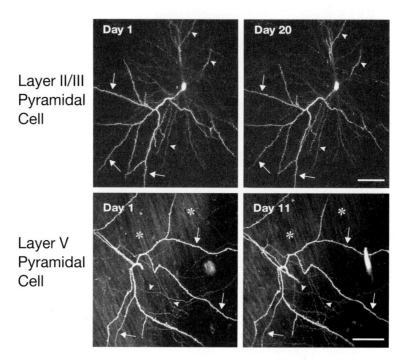

FIGURE 4. Long-term chronic imaging of GFP-expressing pyramidal neurons in adult mice. Typical examples of dendrites imaged over long times in vivo, shown as a projection of a stack of optical sections. (*Top*) Layer II/III pyramidal neuron was imaged over its full extent for 20 days. Both apical (*arrows*) and basal (*arrowheads*) dendrites are shown. Note that the basal dendrites appear dim due to their deep (~300 μm) location. (*Bottom*) Apical dendritic tufts of a Layer V pyramid extending into Layer I. Both dendritic branches (*arrows*) and axons (*arrowheads*) were stable. Vertical streaks (*asterisks*) are caused by autofluorescence in the dura, in the most superficial optical sections. Bars, 50 μm.

FIGURE 5. Long-term chronic in vivo imaging of dendritic spines. (*A*) Low magnification of a Layer V neuron apical dendritic tuft, shown as an *x-y* projection. Arrows point to locations where dendrites were hidden by surface blood vessels. Bar, 50 μm. (*Inset*) Region *a′* was imaged over a period of 25 days. Examples of the images of this region taken at days 1, 19, and 24 are given in *B, C,* and *D*, respectively. Bar, 10 μm.

dendritic spines are clearly resolvable, and images are acquired at 512 × 512 pixels, or about 0.0879 μm/pixel, with 1 μm *z*-steps (Fig. 5).

With the use of 2PLSM, it is straightforward to simultaneously excite two fluorophores with the same excitation wavelength. The emitted fluorescence can be separated spectrally using a dichroic mirror and detected using a pair of photo detectors. For example, we have injected Alexa Red–Dextran into the tail vein of a transgenic mouse and imaged the neocortical vasculature with respect to fluorescent neurons (Fig. 6).

Detection Issues

Imaging tiny structures, such as dendritic spines, has certain limitations. The axial resolution of our three-dimensional images is limited to >1.5 μm (Pawley 1995). We therefore cannot reliably resolve spines that project mainly along the optical axis, below or above the dendrite. In addition, because imaging conditions can change over the long experimental time scales, we adjust the excitation intensity with each imaging session to keep the brightness of the parent dendrite constant, thus achieving similar fluorescence levels for every timepoint.

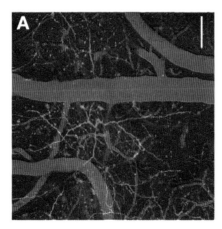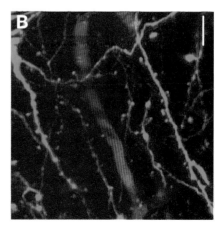

FIGURE 6. Dual-color in vivo imaging of blood flow and neurons in an adult mouse. A red dye (Alexa Red–Dextran) was injected into the tail vein of a GFP transgenic mouse, labeling the vasculature in the brain (*red*). (*A*) Top view of apical dendrites expressing GFP (*green*) and cortical blood vessels filled with Alexa Red–Dextran (*red*). Bar, 50 μm. (*B*) Higher-magnification view of axons and dendrites intermingled with capillaries. The dark stripes in the blood vessels indicate the excluded volumes of the nonfluorescent red blood cells. Bar, 10 μm.

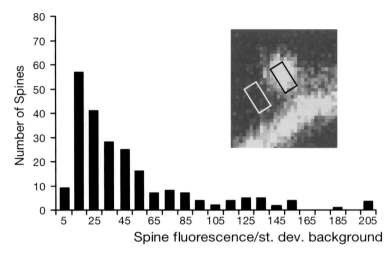

FIGURE 7. Spine fluorescence is larger than background fluorescence. The graph shows the distribution of the spine fluorescence-to-noise ratio of 227 randomly sampled ($n = 4$ mice) spines. Spine fluorescence was measured within a 4×8-pixel area (*black rectangle*) in the spine, and the intensity of the background (BG) was measured in a region of the same size next to the spine (*white rectangle*). The noise (S.D. of BG) was computed over ten independent local measurements of background. On average, spines have an intensity of ~42 standard deviations higher than the background. Even the smallest spines have a signal-to-noise ratio of 5 and can therefore easily be detected. The lengths of the rectangles are 0.70 μm.

The excitation intensity for imaging spines should be sufficiently high to detect all, including the dimmest, spines. An attempt to quantify signal-to-noise ratio is shown in Figure 7.

Postmortem Location of the Imaged Cell in Relation to Cortical Areas

NOTE: <!> Indicates hazardous material; see Cautions Appendix.

To identify the position of an imaged cell, for example, in relation to the barrel cortex map, red fluorescent latex microspheres are pressure-injected into the cortex using a glass micropipette at the end of the final imaging experiment. Several injections are made as close to the cell of interest as possible using the digital pictures of the unique vascular profile around the cell to position the micropipette tip. Mice are then transcardially perfused with 4% paraformaldehyde in phosphate buffer. The hemisphere containing the imaged cell is removed, flattened, and postfixed for 24–48 hours. Vibratome sections (100 μm) are then cut tangential to the cortical surface, and the sections containing the Layer IV barrels (usually sections 2–5) are stained for cytochrome oxidase. The first section, which usually contains the imaged apical dendrites, is used to determine the position of the injected beads relative to the position of the cell. Once this is known, the position of the cell relative to the barrel map is then determined by finding the position of the red fluorescent beads in the cytochrome-oxidase-stained sections containing the barrel map.

4% PARAFORMALDEHYDE

4% paraformaldehyde (10 g of paraformaldehyde) in 0.1 M NaPhosphate buffer (250 ml of 0.1 M NaPhosphate buffer)
Heat and stir at 60°C until clear. Adjust pH to 7.4.

NAPHOSPHATE BUFFER (0.1 M)

0.017 M NaPhosphate monobasic (3.0 g of NaPhosphate monobasic)
0.083 M NaPhosphate dibasic (14.2 g of NaPhosphate dibasic)
Dissolve in distilled H_2O (1250 ml of distilled H_2O)

Preparation of Mice for Imaging

Surgical Methods

1. Deeply anesthetize transgenic mice with an intraperitoneal injection of ketamine (0.13 mg g^{-1} body weight) and xylazine (0.01mg g^{-1}) mixture, or with an intraperitoneal injection of urethane (1.5 mg g^{-1} body weight) for acute experiments. Monitor the depth of anaesthesia periodically by lack of a toe-pinch reflex.

ANESTHESIAS:
KETAMINE/XYLAZINE MIXTURE

12.5 mg/ml ketamine (2.6 ml [100 mg/ml bottle])
0.65 mg/ml xylazine (0.2 ml [100 mg/ml bottle])
in distilled H$_2$O (20 ml of H$_2$O)

URETHANE MIXTURE

20% urethane <!> (1 g of urethane)
in a 0.9% NaCl solution (5 ml of saline)

> Urethane can only be used for nonrecovery studies. Humane treatment of animals must be observed at all times and should follow the institutional care and use guidelines for animals.

2. Administer dexamethasone (0.02 ml at 4 mg ml^{-1}) subcutaneously to minimize the cortical stress response during chronic experiments.

3. Sterilize all surgical tools and remove the flap of skin (~1 cm^2) covering the skull over the barrel cortex.

4. Remove all underlying fascia and push the lateral muscle (temporalis) ventrally, avoiding bleeding.

5. Apply a thin layer of cyanoacrylate to the skull and wound margins to stop bleeding and prevent the seepage of serosanguinous fluid.

   > *Optional:* A small amount of dental acrylic can be applied over the muscle and wound margins to seal the circumference.

6. Mark the cortical area of interest (barrel cortex at P35 is −2-mm Bregma, 4.25-mm lateral).

7. Use a dental drill (1/4 bit) to slowly thin the circumference of a 5 × 5-mm^2 region of the skull. Avoid overheating, which would cause bruising of the dura. Take care not to puncture the thinned skull while drilling.

   > The cortical surface vasculature should be visible where the skull is thinned and there should be minimal bleeding from the skull and none from the brain (Fig. 3A).

8. Use a sharp syringe to gently and superficially perforate the thinned skull. Lift up the skull, exposing the dura. Use Gelfoam to control any bleeding from the dura and skull. Prior to use, cut the Gelfoam (Pharmacia-Upjohn 09-0353-01 or Davol Avitene Ultrafoam 1050050) into ~7 mm^2 blocks and soak in cortex buffer.

STERILE CORTEX BUFFER

125 mM NaCl	(7.21 g of NaCl)
5 mM KCl	(0.372 g of KCl)
10 mM glucose	(1.802 g of glucose)
10 mM HEPES	(2.38 g of HEPES)
2 mM CaCl$_2$	(2 ml of 1M CaCl$_2$)
2 mM MgSO$_4$	(2 ml of 1M MgSO$_4$)
in distilled H$_2$O	(1 l of H$_2$O)
Adjust pH to 7.4.	

---

9. Construct an optical chamber by covering the clear, unblemished dura with a thin layer of 1.5% agarose and a custom-made circular coverglass (7-mm diameter, no. 1 thickness), flush with the skull. Remove the extraneous agar and dry the remaining liquids on the skull (Fig. 3B).

1.5% AGAROSE

Warm 0.375 g of low-melting-point agarose (Sigma A 9793) in 25 ml of cortex buffer (Step 8) until fully dissolved and clear of precipitate. This solution must be made fresh and sterile for each experiment and kept hot with mixing during the procedure. Before applying the hot agarose to the cortex, allow it to cool briefly, and check that the temperature is not above ~37°C, by squirting a drop onto the back of the hand.

---

10. Seal the optical window with dental acrylic, covering the exposed skull, wound margins, and coverglass edges. Embed a clean, small titanium bar with tapped screw holes into the acrylic to stabilize the animal for subsequent imaging sessions, keeping the screw threads clear of acrylic (Fig. 3C). Allow acrylic to thoroughly dry before manipulating mouse or frame.

Imaging

If chronic experiments are to be performed, proceed to Step 12. For acute experiments, proceed to Step 11.

11. Perform acute experiments for the duration of the anesthesia (typically up to 1 hour after surgery for ketamine/xylazine or for 5 hours under urethane anesthesia). Proceed to Step 13.

12. For chronic experiments, return the mouse to its cage after surgery and allow 7–10 days for recovery before imaging. For chronic imaging of ~45 minutes daily, use a dose of anesthesia that is one-half the surgical dose, or 0.075 mg g^{-1} ketamine and 0.005 mg g^{-1} xylazine mixture. The cranial window can remain clear for months, without the use of antibiotics. To reduce stress for the mouse, minimize the size of the metal frame used to stabilize the head so that the mouse can freely move in its cage.

13. Proceed with imaging experiments.

> In all experiments involving in vivo imaging of GFP/YFP expressing neurons in the cortex, we use a custom-built 2PLSM. As a light source, we use a Titanium:Sapphire laser (Tsunami, Spectra-Physics), running at λ ~910 nm, pumped by a 10-W solid-state laser (Millenia X, Spectra-Physics). The objective (40×, 0.8 NA) and scan lens are from Zeiss, the trinoc from Olympus, and the photomultiplier tube from Hamamatsu. Images are acquired with custom software (MatLab).

TROUBLESHOOTING

If the transgenic mouse line chosen for in vivo imaging expresses the fluorescent protein at high enough concentrations in the cells of interest, the limiting steps to high-resolution chronic imaging are in the protocol itself. For example,

- Perform sterile surgeries must be performed consistently and quickly, and make sure that the mouse's head and body is stabilized.

- Make sure that the imaging window is optically clear throughout the long-term imaging period, aside from eventual regrowth of the skull after a few months. This can be accomplished by keeping the surgeries as sterile as possible, and avoiding bleeding from the dura. During the surgery, absorb the blood with moist Gelfoam, allowing it to clot after a few minutes.

- Take care not to damage the dura by overheating during excessive skull thinning: keep the skull moist and cool during drilling.

- In the properly stabilized anesthetized mouse, movement artifacts should not occur during imaging if the precise amount of agarose is applied. This can be achieved with practice by varying the viscosity, concentration, and amount of agar applied after skull removal. As long as the regions of interest remain unobscured, high-resolution in vivo imaging of the fine structure of neurons can be performed daily for time scales of months.

REFERENCES

Agapov E.V., Frolov I., Lindenbach B.D., Pragai B.M., Schlesinger S., and Rice C.M. 1998. Noncytopathic Sindbis virus RNA vectors for heterologous gene expression [see comments]. *Proc. Natl. Acad. Sci.* **95:** 12989–12994.

Brown E.B., Campbell R.B., Tsuzuki Y., Xu L., Carmeliet P., Fukumura D., and Jain R.K. 2001. In vivo measurement of gene expression, angiogenesis and physiological function in tumors using multiphoton laser scanning microscopy. *Nat. Med.* **7:** 864–868.

Campbell R.E., Tour O., Palmer A.E., Steinbach P.A., Baird G.S., Zacharias D.A., and Tsien R.Y. 2002. A monomeric red fluorescent protein. *Proc. Natl. Acad. Sci.* **99:** 7877–7882.

Chen B.E., Lendvai B., Nimchinsky E.A., Burbach B., Fox K., and Svoboda K. 2000. Imaging high-resolution structure of GFP-expressing neurons in neocortex in vivo. *Learn. Mem.* **7:** 433–441.

Dailey M.E. and Smith S.J. 1996. The dynamics of dendritic structure in developing hippocampal slices. *J. Neurosci.* **16:** 2983–2994.

Denk W. and Svoboda K. 1997. Photon upmanship: Why multiphoton imaging is more than a gimmick. *Neuron* **18:** 351–357.

Denk W., Strickler J.H., and Webb W.W. 1990. Two-photon laser scanning microscopy. *Science* **248:** 73–76.

Engert F. and Bonhoeffer T. 1999. Dendritic spine changes associated with hippocampal long-term synaptic plasticity. *Nature* **399:** 66–70.

Feng G., Mellor R.H., Bernstein M., Keller-Peck C., Nguyen Q.T., Wallace M., Nerbonne J.M., Lichtman J.W., and Sanes J.R. 2000. Imaging neuronal subsets in transgenic mice expressing multiple spectral variants of GFP. *Neuron* **28:** 41–51.

Grutzendler J., Kasthuri N., and Gan W.B. 2002. Long-term dendritic spine stability in the adult cortex. *Nature* **420:** 812–816.

Heintz N. 2001. Bac to the future: The use of bac transgenic mice for neuroscience research. *Nat. Rev. Neurosci.* **2:** 861–870.

Hermens W.T. and Verhaagen J. 1998. Viral vectors, tools for gene transfer in the nervous system. *Prog. Neurobiol.* **55:** 399–432.

Holtmaat A.J., Oestreicher A.B., Gispen W.H., and Verhaagen J. 1998. Manipulation of gene expression in the mammalian nervous system: Application in the study of neurite outgrowth and neuroregeneration-related proteins. *Brain Res. Rev.* **26:** 43–71.

Kay M.A., Glorioso J.C., and Naldini L. 2001. Viral vectors for gene therapy: The art of turning infectious agents into vehicles of therapeutics. *Nat. Med.* **7:** 33–40.

Kleinfeld D., Mitra P.P., Helmchen F., and Denk W. 1998. Fluctuations and stimulus-induced changes in blood flow observed in individual capillaries in layers 2 through 4 of rat neocortex. *Proc. Natl. Acad. Sci.* **95:** 15741–15746.

Lendvai B., Stern E., Chen B., and Svoboda K. 2000. Experience-dependent plasticity of dendritic spines in the developing rat barrel cortex *in vivo. Nature* **404:** 876–881.

Levasseur J.E., Wei E.P., Raper A.J., Kontos A.A., and Patterson J.L. 1975. Detailed description of a cranial window technique for acute and chronic experiments. *Stroke* **6:** 308–317.

Maletic-Savatic M., Malinow R., and Svoboda K. 1999. Rapid dendritic morphogenesis in CA1 hippocampal dendrites induced by synaptic activity. *Science* **283:** 1923–1927.

Miesenbock G., Angelis D.A.D., and Rothman J.E. 1998. Visualizing secretion and synaptic transmission with pH-sensitive green fluorescent proteins. *Nature* **394:** 192–195.

Miyawaki A., Llopis J., Heim R., McCaffery J.M., Adams J.A., Ikura M., and Tsien R.Y. 1997. Fluorescence indicators for Ca^{2+} based on green fluorescent proteins and calmodulin. *Nature* **388:** 882–887.

Nakai J., Ohkura M., and Imoto K. 2001a. A high signal-to-noise Ca(2+) probe composed of a single green fluorescent protein. *Nat. Biotechnol.* **19:** 137–141.

Nagai T., Sawano A., Park E.S., and Miyawaki A. 2001b. Circularly permuted green fluorescent proteins engineered to sense Ca^{2+}. *Proc. Natl. Acad. Sci.* **98:** 3197–3202.

Nagai T., Ibata K., Park E.S., Kubota M., Mikoshiba K., and Miyawaki A. 2002. A variant of yellow fluorescent protein with fast and efficient maturation for cell-biological applications. *Nat. Biotechnol.* **20:** 87–90.

Oheim M., Beaurepaire E., Chaigneau E., Mertz J., and Charpak S. 2001. Two-photon microscopy in brain tissue: Parameters influencing the imaging depth. *J. Neurosci. Methods.* **111:** 29–37.

Pawley J.B., ed. 1995. *Handbook of biological confocal microscopy.* Plenum Press, New York.

Svoboda K., Denk W., Kleinfeld D., and Tank D.W. 1997. In vivo dendritic calcium dynamics in neocortical pyramidal neurons. *Nature* **385:** 161–165.

Trachtenberg J.T., Chen B.E., Knott G.W., Feng G., Sanes J.R., Welker E., and Svoboda K. 2002. Long-term in vivo imaging of experience-dependent synaptic plasticity in adult cortex. *Nature* **420:** 788–794.

Tsien R.Y. 1998. The green fluorescent protein. *Annu. Rev. Biochem.* **67:** 509–544.

Volkmar F.R. and Greenough W.T. 1972. Differential rearing effects on rat visual cortical plasticity. *Science* **176:** 1445–1447.

Washbourne P. and McAllister A. K. 2002. Techniques for gene transfer into neurons. *Curr. Opin. Neurobiol.* **12:** 566–573.

Wells T. and Carter D.A. 2001. Genetic engineering of neural function in transgenic rodents: Towards a comprehensive strategy? *J. Neurosci. Methods.* **108:** 111–130.

Zito K. and Svoboda K. 2002. Activity-dependent synaptogenesis in the adult mammalian cortex. *Neuron* **35:** 1015–1017.

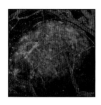

CHAPTER TWENTY FOUR

Intravital Microscopy of Normal and Diseased Tissues in the Mouse

Rakesh K. Jain, Edward B. Brown, Lance L. Munn, and Dai Fukumura

Edwin L. Steele Laboratory, Department of Radiation Oncology, Massachusetts General Hospital and Harvard Medical School, Boston, Massachusetts 02114

INTRODUCTION

THE PAST 30 YEARS HAVE WITNESSED SPECTACULAR ADVANCES in our understanding of the molecular origins of cancer and other diseases. These advances have led to the identification of various genes associated with angiogenesis, oncogenesis, and other pathological processes, as well as to the development of a vast array of therapeutic agents. The grand challenges now are (1) to relate the expression of these genes to their function in an intact organism and (2) to deliver these novel therapeutics to their targets in vivo (Jain 1998). Currently, gene expression, physiological function, and drug delivery are typically measured with techniques that either have poor spatial resolution (millimeter to centimeter) or are destructive to the tissue under study. Techniques with poor spatial resolution preclude the study of biological events occurring at cellular and subcellular levels (1–10 μm). Destructive techniques limit our ability to provide insight into the dynamics of biological processes.

Intravital microscopy combined with chronic window preparations overcomes these limitations (Jain et al. 2001, 2002). Furthermore, the recent availability of in vivo reporters such as green fluorescent protein (GFP), as well as transgenic mice and/or cell lines containing these reporter gene constructs, is presenting new opportunities for functional genomics. The animal models and imaging techniques described here are within an overall framework of tumor pathophysiology, but they are easily applicable to other biological questions and can be performed in healthy tissues, as well as for various diseases.

Conventional epifluorescence and bright-field microscopy have traditionally been the tools of choice for intravital imaging and are still the most common technologies available. More recently, multiphoton microscopy has been adapted to animal studies, and it provides advantages over these conventional, single-photon technologies (see Chapter 22). The multiphoton laser-scanning microscope (MPLSM) can image up to 700 μm inside living tissue with three-dimensional resolution of less than 1 μm (Brown et al. 2001). Although this depth penetration is a vast improvement over epifluorescence and confocal microscopy, it is still

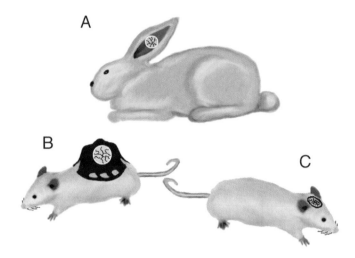

FIGURE 1. Chronic window preparations. Rabbit ear chamber (*A*), mouse dorsal skin chamber (*B*), and cranial window models (*C*) are used for high-resolution longitudinal observation of tumor growth, angiogenesis, physiological processes, and gene expression. (Adapted, with permission, from Jain et al. 2001.)

insufficient to image most internal tissues. Surgical tissue preparations are necessary to image deeper tissues with high resolution. Based on the type of surgical method used, the tissue preparation can be divided into three broad categories:

- *Chronic-transparent windows* (e.g., dorsal skinfold chamber, cranial window; Fig. 1).
- *Acute (exteriorized) tissue preparations* (e.g., mesentery; Fig. 2).
- *In situ preparations* (e.g., tail lymphatics; Fig. 3).

Each of these preparations can be used to study normal tissue or an implanted tumor. Each preparation has advantages and disadvantages. Consequently, a combination of several methodologies is normally required to examine the effect of tissue microenvironment on gene expression, physiology, and drug delivery.

This chapter begins with a brief historical perspective followed by descriptions of the surgical procedures for making various tissue preparations. Next, imaging techniques and the accompanying analysis used to extract parameters of interest are outlined. Finally, important insights obtained with these approaches are summarized and future possibilities are discussed.

> Perform all surgical procedures described in this chapter with the animal under appropriate anesthesia (refer to individual references for details) and with full approval by the institutional animal care and use committee. During a surgical procedure or intravital microscopy, maintain animal body core temperature at 36–37°C using a heating pad or similar device.

CHRONIC WINDOW PREPARATIONS

In 1924, Sandison developed the first transparent window chamber, implanted in the ear of a rabbit (Sandison 1924). This preparation allowed continuous, noninvasive, long-term monitoring of angiogenesis during wound healing (Clark et al. 1930; Clark and Clark 1932). Tumor angiogenesis was first studied in this window by Ide et al. (1939), using a Brown-Pearce carcinoma.

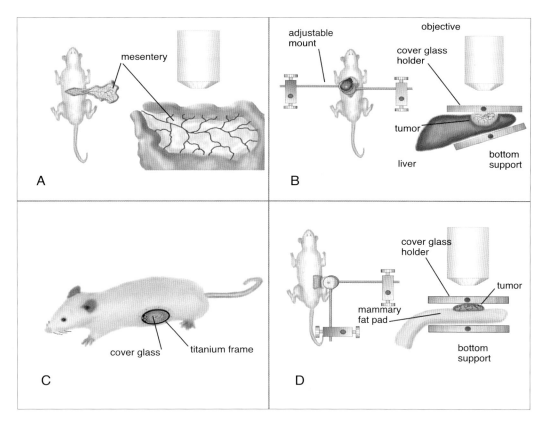

FIGURE 2. Acute tissue preparations. Mesentery (*A*), mouse liver (*B*), pancreas (*C*), and mammary fat pad models (*D*) are used for acute observations and/or organ-specific tumor microcirculation. (Adapted, with permission, from Jain et al. 2001.)

In the 1940s, Algire adapted the Sandison chamber to the dorsal skin in mice and carried out pioneering studies of angiogenesis during wound healing and tumor growth (Algire 1943a, b; Algire and Chalkley 1945; Algire and Legallis 1948). Similar chronic windows have been developed for the dorsal skin of other rodents (e.g., rats and hamsters), the hamster cheek pouch, and the cranium of the mouse and rat (for references, see Table 1).

The dorsal skin chamber in rodents is the most widely used window preparation. It is relatively easy to establish using standard equipment compared to some of the other preparations. The cranial window may last up to 1 year compared to 30–40 days for the dorsal window, and it is an immunoprivileged site. The main disadvantage of the cranial window is that the visualization of microvessels requires, in most cases, epi-illumination and the injection of a contrast agent such as a fluorescent marker. The protocols for preparation of each of these window chambers are presented below.

Rabbit Ear Chamber

Transparent chambers are surgically implanted in the ears of male New Zealand white rabbits (2–3 kg body weight) using the following procedure (Zawicki et al. 1981; Nugent and Jain 1984).

1. Punch four holes in the ear, being careful to avoid large blood vessels. Make three outer perforations for positioning the chamber and a central puncture (5.4 mm diameter) for housing the window.

TABLE 1. Examples of intravital preparations for tumor studies

Models	Species	Tumor	Year	References
Chronic window preparations				
Ear chamber	rabbit	Brown-Pearce CA[a]	1939	Ide et al. (1939)
		VX2 CA: intra-arterial injection	1958	Wood (1958)
		VX2 CA: multifocal growth	1984	Dudar and Jain (1984); Nugent and Jain (1984)
Dorsal skin chamber	mouse	various CAs, SAs[a], melanomas	1943	Algire (1943a,b)
		hepatoma 134	1961	Kligerman and Henel (1961)
		mammary CA	1971	Reinhold (1971)
	nude mouse	human amelanotic melanoma	1984	Falkvoll et al. (1984)
	SCID mouse	human tumor xenograft	1992	Leunig et al. (1992)
	hamster	amelanotic melanoma	1981	Asaishi et al. (1981)
	rat	ascites hepatoma	1971	Yamaura et al. (1971)
		rhabdomyosarcoma	1977	Reinhold et al. (1977)
		rat SA	1979	Endrich et al. (1979a,b)
		rat mammary CA	1989	Dewhirst et al. (1989)
Cranial window	rat and mouse	various rodent and human tumors	1994	Yuan et al. (1994)
Pancreas window	mouse	human pancreatic CA	1999	Tsuzuki et al. (2001)
Lung window	mouse	Lewis lung CA	2000	Al-Mehdi et al. (2000); Funakoshi et al. (2000)
Acute (exteriorized) preparations				
Cheek pouch	hamster	chemically induced SAs	1950	Lutz et al. (1950)
		human tumors	1952	Chute et al. (1952); Toolan (1954)
		melanomas, CA, human angiopericytoma	1965	Goodall et al. (1965)
		malignant neurilemmoma	1973	Eddy and Casarett (1973)
Mesentery	rabbit	VX2 CA: intra-arterial injection	1961	Zeidman (1961)
	rat	murine colon CA	1990	Norrby et al. (1990); Yanagi and Ohsima (1996)
Cremaster muscle	rat	Warker 256 CA, chondrosarcoma	1986	Heuser and Miller (1986)
Liver	mouse	human adenocarcinoma	1997	Fukumura et al. (1997)
Mammary gland	mouse	human mammary CA	1998	Monsky et al. (1998–2002)
Pancreas	mouse	human pancreatic CA	1999	Schmidt et al. (1999); Tsuzuki et al. (2001)
Lung	mouse	Lewis lung CA	2000	Funakoshi et al. (2000)
In situ preparations				
Eye anterior chamber	frog	renal CA	1939	Luckè and Schlumberger (1939)
	guinea pig	human tumor	1952	Greene (1952)

TABLE 1. *(Continued)*

Models	Species	Tumor	Year	References
Anterior chamber/ iris assays	rabbit	mouse mammary tumor	1976	Gimbrone and Gullino (1976)
		mouse mammary papilloma	1977	Brem et al. (1977)
		hyperplastic rat mammary gland	1977	Maiorana and Gullino (1978)
Corneal micropocket assay	rabbit	Brown-Pearce CA, VX2 CA	1974	Gimbrone et al. (1974); Ausprunk and Folkman (1977)
	mouse	murine mammary CA and SA	1979	Muthukkaruppan and Auerbach (1979)
Tail lymphatics	mouse	murine fibrosarcoma	2000	Leu et al. (2000)
Specialized models				
Individual microvessel perfusion	mouse	human adenocarcinoma	1996	Lichtenbeld et al. (1996)
Angiogenesis gel assay	mouse	various angiogenesis factors	1996	Dellian et al. (1996)
GFP used to track cancer cells	mouse	CHO cells, murine mammary CA, human adenocarcinoma	1997	Chishima et al. (1997); Naumov et al. (1999); Chang et al. (2000); Li et al. (2000); Brown et al. (2001)
GFP used as an intravital gene reporter in transparent windows	transgenic mouse	murine mammary and liver CA	1998	Fukumura et al. (1998)
	mouse	engineered human glioma	2001	Fukumura et al. (2001)

[a] CA indicates carcinoma and SA indicates sarcoma.

2. Carefully retract the epidermis on both sides of the ear around the puncture. Place a molded plate on the inside of the ear and align it with the existing holes, while positioning a thin (~200 µm) coverglass on the outside of the cartilage. Fasten the molded plate and the mica glass, which sandwich the central puncture to form the chamber, to each other by three threaded rods and six hex nuts.

3. Pull the retracted skin taut over the edges of the coverglass and molded plate to protect the exposed area. Administer a light covering of antiseptic, and mount two plastic covers to enclose and protect the chamber.

   Granulation tissue grows in the chamber (thickness, 40 ± 5 µm; diameter, 5.4 mm) over ~8 days and reaches maturity at ~40 days postoperation. At this time, the chamber is ready for normal (granulation) tissue study or tumor implantation.

4. For tumor implantation (Nugent and Jain 1984), carefully remove the coverglass which forms the top plate of the transparent chamber. Mince a tumor, excised from the flank of a tumor-bearing host, and place it in 0.9% NaCl solution. Spread the tumor-containing solution uniformly over the coverglass.

5. Replace the coverglass so that it is flush against the intact normal tissue.

   If this procedure causes tissue damage, exclude the damaged tissue from the study. An angiogenic response is observed 3–4 days postimplant and the tumor-bearing chamber is ready for intravital microscopy ~10 days postimplant.

440 ■ CHAPTER 24

6. For intravital microscopy, place the animal in a dorsal recumbent position in a cradle that restricts head movement while still maintaining proper circulation to the chamber. Extend the ear containing the chamber horizontally to the specimen plane of an intravital microscope. Secure the chamber to the microscope stage with an aluminum adapter (Nugent and Jain 1984).

Dorsal Skin Chamber Preparation

Dorsal skin chambers are implanted in mice using the following procedure (Leunig et al. 1992). For other species, see references in Table 1.

1. Depiliate the back of the animal. Use two titanium frames (weight 3.2 g), mirror images of each other, to sandwich a double layer of skin, which is gently stretched from the back of the mouse and held in place with sutures.

2. Remove one layer of skin in a 15-mm-diameter circle, and cover the remaining layer, consisting of epidermis, subcutaneous tissue, and striated muscle, with a glass coverslip laid in a depression in one of the frames. Hold it in place with a retaining ring.

3. Allow the animal to recover from surgery for 48 hours before tumor implantation or in vivo microscopy studies.

4. For implantation of tumor cells, position the animal in a transparent polycarbonate tube (25 mm inner diameter [ID]) with the dorsal skinfold chamber extending from a slot running down one side of the tube.

5. Carefully remove the coverslip of the chamber, and implant either 2 µl of tumor cell suspension ($\sim 10^5$ cells) or an $\sim$1-mm piece of tumor tissue at the center of the dorsal chamber. Seal the chamber with a new coverslip.

6. To image the chamber, anesthetize the mouse (Leunig et al. 1992a), lay it on its side on a heated stage, and gently clamp the chamber to the microscope stage with a custom-built metal frame.

Cranial Window Preparation

Cranial windows are implanted in mice and rats using the following procedure (Yuan et al. 1994).

1. Fix the head of the animal within a stereotactic apparatus.

2. Make a longitudinal incision of the skin between the occipet and the forehead. Remove a circular region of skin on top of the skull, and scrape off the periosteum underneath to the temporal crests.

3. Draw a 6-mm circle over the frontal and parietal regions of the skull. Use a high-speed drill with a burr tip to make a groove (0.5-mm diameter) on the margin of the drawn circle. Make this groove thinner by cautious drilling of the groove until the bone flap becomes loose.

4. Use a malis dissector to separate the bone flap from the dura mater underneath. Remove the bone flap, place Gelfoam on the cut edge, and rinse the dura mater continuously with saline.

5. Make a small incision close to the sagittal sinus. Insert iris microscissors into the incision. Cut the dura and arachnoid membranes completely from the surface of both hemispheres, avoiding any damage to the sagittal sinus.

6. Place a 1-mm piece of the tumor tissue in the center of the window. Seal the window with a 7-mm coverglass glued to the bone with histocompatible cyanoacrylate glue.

7. The measurements of functional properties are made when tumors have reached the desired size. Anesthetize the animal and lay it on a simple stereotactic apparatus, with the head fastened to minimize breathing motion artifacts. Position the surface of the cranial window perpendicular to the objective lens.

Pancreatic Tumor Preparation

Pancreatic cancer has a poor prognosis, and treatment strategies conducted from preclinical research have not succeeded in extending patient's survival appreciably. This newly developed abdominal window allows both direct intravital microscopy and chronic observation during pancreas tumor growth and the response to treatment (Tsuzuki et al. 2001).

1. For tumor implantation, perform a small left lateral laparotomy. Gently exteriorize the pancreas and fix a small piece of tumor to the serosal side of the pancreas with a 5-0 prolene suture. Suture and close, first the abdominal wall, and then the skin.

2. When tumors become 6–8 mm in diameter (~4 weeks after tumor implantation), implant an abdominal wall window to observe tumor microcirculation.

3. Reopen the skin and abdominal wall, and gently exteriorize the pancreas, which contains a growing tumor. Suture a portion of normal pancreas to the outer side of the abdominal wall with 5-0 prolene suture to keep the pancreas and the tumor outside the abdominal cavity.

4. Attach a titanium ring with 8 holes around the edge with sutures to the abdominal wall. This holds the pancreas and tumor inside the window. Place a circular glass coverslip (11 mm diameter) on top to seal the window (Fig. 2C).

5. For the subsequent intravital microscopy, Anesthetize the mice and place them inside a plastic tube (25 mm ID) that contains a slit of 14 × 37 mm width. The abdominal window fits in the slit of the plastic tube and is fixed by adhesive tape.

Lung Window Preparation

Observation of tumors growing in the lung microenvironment by intravital microscopy has been limited to a few acute preparations (Al-Mehdi et al. 2000; Funakoshi et al. 2000). We have recently developed a chronic thoracic window model in mice that allows temporal observation of the microcirculation of tumors implanted on the pleural surface without surgical manipulation at the time of observation. This window model allows the study of pleural metastases of various origins, as well as primary lung cancer in an orthotopic setting. The thoracic window is prepared as follows.

1. Remove the hair on the left chest area. Place the animal in the right lateral position. Make a transverse skin incision from the ventral to the dorsal side of the left chest, caudal to the scapula. Bluntly dissect subcutaneous fat and connective tissue. Sharply resect the chest muscle exposing the left chest wall.

2. Place the animal in a supine position. Perform a tracheostomy and tracheal intubation with PE-10 polyethylene tubes (ID 0.28 mm, OD 0.61 mm; Becton Dickinson) through a median cervical incision. Return the animal to the right lateral position and connect it to a mechanical ventilator (Rodent Ventilator, Harvard, Holliston, Massachusetts). Ventilate the animal at a tidal volume of 15 ml/kg and a respiratory rate of 50 per minute (Kleinman and Radford 1964).

3. Resect together the ventral 6-mm portions of the fifth and sixth ribs with the chest wall, leaving a hole in the chest wall of ~7 × 6 mm. Cauterize thoroughly the resected margins of the ribs and chest wall to avoid latent bleeding. Cover this hole with a circular glass coverslip (8 mm diameter). Apply biocompatible cyanoacrylate adhesive and cement mixture around the coverslip to secure it and to seal the chest cavity.

4. Perform intrathoracic puncture using a 30-gauge needle with a 3-ml syringe to withdraw the remaining air in the chest cavity, allowing the lung to inflate and attach to the coverslip. Suture the skin and muscles with 5-0 Ethibond (Ethicon, Somerville, New Jersey). Sequentially terminate the mechanical ventilation as the animal recovers adequate, spontaneous breathing. Remove the tracheal tube, suture the hole in the trachea with 8-0 Ethibond, and close the cervical incision with 5-0 Ethibond.

5. One week later, remove the coverslip for implantation of a 1-mm tumor fragment, which is placed on the pleural surface. After tumor implantation, replace the coverslip using cyanoacrylate adhesive and cement.

6. For intravital microscopy, position the animal in the right lateral position on the stage (see liver tumor preparation) (Fukumura et al. 1997b). To minimize chest wall movement due to respiration, suture two strings to the skin, one cranial and one caudal to the coverslip. Using these strings attached to the adjustable arms on the stage, slightly lift the chest wall.

ACUTE (EXTERIORIZED) PREPARATIONS

To image tumors grown in internal organs with high resolution, it is sometimes necessary to surgically exteriorize these organs. The mesentery of mice, rats, and cats has been extensively used for intravital microscopy of microcirculation (House and Lipowsky 1988; Kubes et al. 1991; Kurose et al. 1993a,b; Higuchi et al. 1997; Milstone et al. 1998). Since only two thin membranes cover the microvessels in the mesentery, this model provides the best optical accessibility for in vivo microcirculation studies. Unfortunately, the preparation of intact mesenteric microcirculation requires extreme care because the mesentery is quite fragile. Furthermore, repeated or long-term observation is not possible. In addition to normal blood vessels, mesentery can be used to study peritoneal metastasis (Zeidman 1961; Yanagi and Ohsima 1996).

Our laboratory and others have shown that the host microenvironment influences tumor biology, affecting parameters such as gene expression, angiogenesis, physiological functions, tumor growth, invasion, metastasis, and responses to therapy. Consequently, the use of tumor models growing in the appropriate "orthotopic" (e.g., liver tumor growing in the liver) location is necessary to obtain a rigorous understanding in tumor pathophysiology and to correctly study antitumor treatments (Fidler 1995). Orthotopic tumor models include the liver (Morris et al. 1993; Fukumura et al. 1997b) and the mammary gland (Monsky et al. 2002). These preparations have provided remarkable insights into the effect of host-tumor interactions on tumor biology and therapeutic response. The protocols for their preparation are presented below.

Mesentery Preparation

The following procedure can be used for mice and rats (Jain et al. 1998a). Animals should fast for 24 hours prior to the procedure.

1. After putting the animal under anesthesia, remove abdominal hair, and open the abdomen via a midline incision.

2. Exteriorize the ileocecal portion of the intestine. Use a cotton swab premoistened with saline to gently nudge the intestinal loop onto a glass stage. Avoid direct contact with and tension in the mesentery.

3. Gently straighten the intestine and fix it in place with cotton sponges immersed in warmed saline, so that the mesentery does not unfold. Keep the mesentery moist and warm by superfusion with warm saline (37°C).

4. Observe the mesenteric microcirculation with transillumination, epifluorescence illumination, or the MPLSM in combination with appropriate tracers (Fig. 2A).

Liver Tumor Preparation

The liver is the most common site for distant metastasis of colorectal carcinomas. The mouse or rat liver tumor metastasis model is prepared by performing a splenic injection of tumor cells (Fukumura et al. 1997b).

1. Make a small incision in the left lateral flank, exteriorize the spleen, and inject a tumor cell suspension (1×10^6 to 5×10^6 cells in 100 µl) into the spleen just under the capsule. Replace the spleen in the peritoneal cavity. Close the two layers of incision (skin and abdominal wall) with metal wound clips.

2. One week later, remove the metal clips.

3. Three to four weeks after the tumor cell injection, open the abdominal wall via a midline incision. Functional parameters are typically measured on tumor foci of ~3–5 mm in diameter. Gently exteriorize the main liver lobe containing metastatic tumors and hold it in place with a liver support device (Fig. 2B).

4. Fix a circular glass coverslip onto the bottom surface of the liver lobe with cyanoacrylate adhesive, and fix this coverglass to a mechanical support with denture adhesive cream (e.g., CVS Drug Stores), allowing adjustment of the three-dimensional position and angle of the top surface of the liver to render it parallel to the microscope stage.

5. Gently place another circular glass coverslip, attached to a metal ring "lollipop"-shaped support, onto the top surface of the tumor or normal liver tissue.

Mammary Fat Pad Tumor Preparation

Breast cancer is a leading cause of death in women. The mouse mammary fat pad serves as an orthotopic site for breast cancer (Fig. 2D) (Monsky et al. 2002).

1. Inject breast carcinoma cells into the mammary fat pad just inferior to the nipple of female mice. Allow tumors to grow to ~5–8 mm in diameter, which typically takes 4–6 weeks.

2. When tumors are of the proper size, make a midline incision through the skin and fascia. Gently elevate a flap by blunt dissection, being careful to avoid disrupting the vasculature and irritating the tumor vessels.

3. Glue the flap to a stage that has been custom-designed for the liver preparation (see above).

4. Place a glass coverslip over the tumor using a lollipop-shaped metal ring support.

IN SITU PREPARATIONS

The anterior chamber of the eye is a natural site for observing tumor growth. Two assays used extensively for this purpose are implantation on the iris and implantation in a corneal pocket (Gimbrone et al. 1972, 1974, 1976; Ausprunk and Folkman 1977; Brem et al. 1977; Maiorana and Gullino 1978; Muthukkaruppan and Auerbach 1979; Sholley et al. 1984; Klintworth 1991). Of the two assays, the corneal pocket is more widely used (for review, see Klintworth 1991). Due to the three-dimensional nature of vessel growth, it is difficult to quantify the vascular response except in the early stages when the vessel length and number can be assessed. Some investigators have quantified the vascular response by perfusing the cornea with colloidal carbon and then measuring vascular length using computer-assisted image analysis (Klintworth 1991).

Originally developed for rabbits, the corneal micropocket assay (Fig. 3A) has been adapted to rats and mice (Muthukkaruppan and Auerbach 1979; Schlenger et al. 1994). Although it is less expensive to use rats/mice compared to rabbits, surgery becomes more difficult as the size of the eye gets smaller. Because the rat/mouse cornea is thinner than the rabbit's, the three-dimensional growth of vessels is more limited in these rodent models.

The chick chorioallantoic membrane (CAM) is also commonly used with the shell either intact or partially or completely removed (Leighton 1967; Auerbach et al. 1975; Nguyen et al. 1994; Friedlander et al. 1995). This is an inexpensive and hence widely used angiogenic assay. To eliminate the inflammatory response that develops in the 7–8-day-old CAM, the vitelline membrane of a 4-day-old chick embryo has also been used (Taylor and Weiss 1984). Due to the difficulty in precisely quantifying newly formed vessels, the CAM assay has been used primarily for screening purposes. However, Nguyen et al. (1994) have modified this assay for easy quantification. The CAM has also proved to be useful for analyzing the efficiency of metastatic cell extravasation and colonization (Koop et al. 1995) and for studying the kinetics of gene expression in metastasizing cells (Shioda et al. 1997).

The lymphatic system is a primary component of the immune system, and it is an important route for metastasis of cancer cells (Jain and Padera 2002). Lymphatic microvessels have been studied by adapting lymphangiography to the mouse ear or tail (Leu et al. 1994; Berk et al. 1996; Swartz et al. 1996; Jeltsch et al. 1997). In this technique, a high-molecular-weight fluorescently labeled tracer is injected into the interstitial compartment of the ear chamber or tail tissue (Figs. 3B and 7). The tracer migrates to the local lymphatic vessels and fills the network, allowing visualization. By implanting a tumor in the tail, the structure and function of the lymphatics at the tumor periphery can be monitored (Leu et al. 2000; Padera et al. 2002a,b). Similarly, primary and collecting lymphatics in various tissues, as well as lymphatic tissues such as lymph nodes or Peyer's patches with afferent and efferent lymphatics, can be visualized by injecting tracer molecules.

Corneal Pocket Assay

The following procedure for the corneal pocket assay in rabbits (Fig. 3A) (Gimbrone et al. 1974) can be used for mice with modification (Muthukkaruppan and Auerbach 1979).

1. After anesthesia and retrobulbar infiltration with 2% Xylocaine, move the eye forward and secure it in position by a fold clamped in the lower lid.

2. Make a superficial incision, 1.5 mm long, with a surgical blade in the corneal dome off-center. Reduce intraocular tension by draining a small amount of aqueous humor from the anterior chamber through a 27-gauge needle.

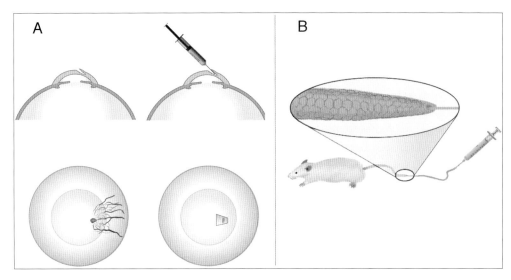

FIGURE 3. In situ preparations. Cornea pocket assay (*A*) and mouse tail lymphatic model (*B*) are in situ preparations to study angiogenesis and lymphangiogenesis. (Adapted, with permission, from Jain et al. 1997.)

3. Insert a malleable iris spatula (1.5-mm width) and fashion an oblong pocket within the corneal stroma. Peripheral pockets end 1–2 mm from the limbus.

4. Deposit a small piece of minced tumor or gel (5–10 µl) in the bottom of the pocket, which then seals spontaneously.

5. Use a stereomicroscope to observe the eyes with corneal implants. Measure tumor and new vessel growth *en face* with an ocular micrometer at 10×. A green filter allows clearer definition of fine vascular channels developing within the cornea.

Chick Chorioallantoic Membrane

1. Incubate fertilized White Leghorn chick eggs for 8–12 days at 37°C and ~60% relative humidity (Chambers et al. 1982; McDonald et al. 1992; Nguyen et al. 1994; Shioda et al. 1997).

2. To apply tumor cells or a matrix gel implant, identify large vessels in the CAM. Cut a window (~1 cm²) with an electric drill in the shell over the CAM and remove either a small section or the entire shell, leaving the CAM intact. In studies where the CAM is removed entirely from the shell, place it carefully in a Petri dish.

3. Place tumor cells (Chambers et al. 1982) or matrix implant (Nguyen et al. 1994) on top of the CAM or inject fluorescence-labeled tumor cells intravenously. (1×10^5 to 4×10^5 cells in 0.1 ml) (McDonald et al. 1992; Shioda et al. 1997).

4. Determine angiogenesis in the gel or tumor growth with a dissecting microscope (Chambers et al. 1982; Nguyen et al. 1994).

5. Visualize surviving tumor cells by intravital microscopy with epifluorescence illumination (McDonald et al. 1992; Shioda et al. 1997).

Lymphatics

Tail lymphatics, with or without tumor growth, provide an in situ model for lymphangiography (Fig. 3B) (Leu et al. 2000; Padera et al. 2002a,b). Similar methods can be used to visualize lymphatics in other tissue such as the ear (see Fig. 7). For tail studies, female Nu/Nu mice are preferable due to their reduced aggressiveness, which results in fewer tail wounds. In addition, hair removal is not necessary in females prior to observation. The animals should be housed separately prior to observation in order to prevent bite wounds to the tail.

1. For studies of tumor growth in the tail model, a single-cell suspension of tumor cells must be prepared as follows. Harvest tumor tissue, mince it, and digest it with trypsin until a uniform solution is formed.

2. To implant tumors, use a 26-gauge needle to inject ~0.2 ml of the single-cell suspension into the tail skin ~1 cm from the distal tip. Great care must be taken to avoid laceration of the tail veins.

3. Monitor the mouse for tumor growth in the tail, which typically takes ~4 weeks, depending on growth rate.

4. For microlymphangiography, place the mouse on a Lucite block adjacent to a Petri dish, and tape its tail to the bottom of the inside of the Petri dish using double-sided tape. Place a strip of tape over the mouse's hips to secure the mouse to theLucite block. Fill the dish with distilled H_2O to allow the use of long-working-distance water immersion lenses.

5. Connect a 30-gauge needle to a constant pressure source. This is used to inject fluorescently labeled dextran (2 million m.w.) into the interstitial space of the distal tail. Make a superficial injection into the skin above the tumor. Use a syringe to increase the pressure to encourage filling of peri-tumor lymphatics.

INTRAVITAL MICROSCOPY AND IMAGE ANALYSIS

Intravital Microscopy Workstations

Conventional Single-photon Microscopy

The standard microscopy workstation consists of an upright or inverted microscope equipped with transillumination and fluorescence epi-illumination, a flashlamp excitation device, a set of fluorescence filters, a motor-controlled filter wheel, a CCD (charge-coupled device) camera, a video monitor, a video recorder, and a frame-grabber board for image digitization (Fig. 4A). Advanced techniques require additional equipment such as a motorized x-y stage with ±1.0-μm lateral resolution, an intensified CCD camera, a photomultiplier tube, and a dual-trace digital oscilloscope (Berk et al. 1997; Fukumura et al. 1997b; Helmlinger et al. 1997b; Jain et al. 2001, 2002).

Multiphoton Laser-scanning Microscopy

The MPLSM consists of a mode-locked Titanium:Sapphire laser and a laser scanhead purchased either as part of a multiphoton system or as a confocal system with further modifications for good infrared transmission. The laser beam first passes through a Pockels Cell which allows rapid (~1 μsec) modulation of laser intensity and then is directed by the scanhead into the side- or top-entry port of an upright epifluorescence microscope. Non-descanned photomultiplier tubes are used for imaging through significant depths of scattering tissue and should be introduced into the beam path via a dichroic beam splitter located in the beam path between the scanhead and the objective lens (Fig. 4B) (Denk et al. 1990; Brown et al. 2001).

The size of a tumor is determined by systematically imaging it at low magnification. To quantify physiological parameters, randomly selected areas (3–6 locations/tumor or animal)

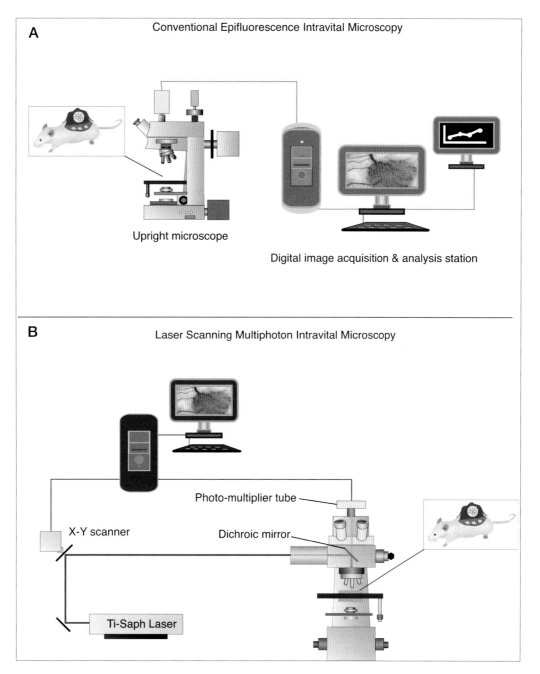

FIGURE 4. Intravital microscopy workstations. Mouse tumor models are observed using conventional intravital microscopy (*A*) or multiphoton laser-scanning microscopy (*B*). With appropriate tracer molecules and/or engineered vectors/cells and computer-assisted image analysis, one can monitor tumor size, vessel density, vessel diameter, red blood cell velocity, leukocyte endothelial interaction, vascular permeability, tissue pO_2, pH, gene promoter activity, enzyme activity, and delivery of drugs, including genes. (*B* adapted, with permission, from Brown et al. 2001.)

are investigated using long-working-distance objectives with appropriate magnification. The parameters that can be routinely measured include tumor size, angiogenesis (vascular density, length, diameter, etc.), hemodynamics (e.g., erythrocyte velocity), vascular permeability, leukocyte endothelial interactions, interstitial diffusion, convection and binding, gene promoter activity, and collagen structure and dynamics.

TABLE 2. Examples of noninvasive techniques for tumor studies

Technique	References
Vascular parameters	
Microvascular permeability of normal and neoplastic tissues	Gerlowski and Jain (1986)
Microvascular permeability of albumin, tumor vascular surface area, and vascular volume	Yuan et al. (1993)
Microvascular permeability of individual tumor vessel using MPLSM	Brown et al. (2001)
Pore cutoff size of tumor vessels	Yuan et al. (1995); Hobbs et al. (1998)
Perfusion of single-tumor microvessels; application to vascular permeability measurement	Lichtenbeld et al. (1996, 1999)
Effect of RBCs on leukocyte-endothelial interactions	Melder et al. (2000)
Extravascular parameters	
Fluorescence ratio imaging of pH gradients; calibration and application in normal and tumor tissues	Martin and Jain (1993, 1994)
Noninvasive measurement of microvascular and interstitial oxygen profiles in a human tumor xenograft	Torres Filho et al. (1994)
Simultaneous high-resolution measurements of interstitial pH and pO_2 gradients in solid tumors in vivo	Helmlinger et al. (1997b)
Direct measurement of interstitial diffusion and convection of albumin in normal and neoplastic tissues using fluorescence photobleaching	Chary and Jain (1989)
Fluorescence photobleaching with spatial Fourier analysis for measurement of diffusion and binding in tumors	Berk et al. (1993, 1997)
Flow velocity in the superficial lymphatic network of the mouse tail	Leu et al. (1994); Berk et al. (1996); Swartz et al. (1996)
Gene expression using an intravital reporter	Fukumura et al. (1998, 2001)
Cell identification using endogenous GFP	Chang et al. (2000); Brown et al. (2001)
Second harmonic signal imaging of type I collagen in tumors	Brown et al. (2003)

All of these techniques were developed in the authors' laboratory.

Tumor Growth and Regression

To measure two-dimensional tumor size, low-power transillumination or epi-illumination images are digitized and analyzed using an image processing system (Jain et al. 1998b). The tumor margin is identified either by the unusual shape of its angiogenic vessels, the expression of GFP in tumor cells, or by its protrusion above the surface of the surrounding tissue in the dorsal skin-fold chamber. If the tumor is sufficiently thick (i.e., it extends >400 μm from the tissue/window interface) or is optically dense (i.e., due to constitutive GFP expression or exceptionally high vascularization), the two-dimensional tumor size is calculated from the visible tumor surface area, and tumor volume is calculated using ex vivo measurements of thickness if available (Tsuzuki et al. 2000). For sufficiently small tumors (those that extend <400–500 μm from the window/tissue interface and are not optically dense), the tumor volume can be directly measured by MPLSM.

Vascular Parameters

Angiogenesis and Hemodynamics

To visualize microvessels within 150 μm (single photon) or ~600 μm (MPLSM) of the tumor/window interface, 100 μl of FITC-dextran (2 million m.w., 10 mg/ml) is injected into the tail vein of mice. During each observation period, FITC-fluorescence images are recorded

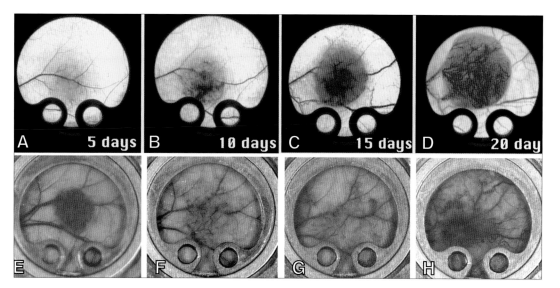

FIGURE 5. Angiogenesis, tumor growth, and regression in the dorsal skin chamber. Angiogenesis and tumor growth in a human colon carcinoma (*A–D*). At day 5 after tumor cell implantation, enlargement of host vessels is observed, and by day 10, occasional hemorrhage and sprout formation occurs. At day 15 tumor growth and further angiogenesis become apparent. By day 20, the tumor is fully vascularized. (Adapted with permission, from Leunig et al. 1992.) Tumor growth and regression in a Shionogi mouse mammary carcinoma (*E–H*). (*E*) 12-day tumor prior to orchiectomy; 3 days (*F*) and 9 days (*G*) after orchiectomy, the tumor vessels regress and the tumor shrinks. A second wave of angiogenesis is evident in *H*. (Adapted, with permission, from Jain et al. 1998.)

for 60 seconds, and the videotapes are analyzed off-line for single-photon microscopy (Fig. 5), or a three-dimensional image stack of the vessel network is generated and vessel properties are measured for MPLSM (Fig. 6 and Movie 24.1).

Single-photon Microscopy Procedure

1. Measure the vessel diameter (D) using an image-shearing device (Kaufman and Intaglietta 1985).

2. Measure the transverse red blood cell velocity (i.e., the red blood cell velocity in the plane of the image [V_{TRBC}]) by temporal correlation velocimetry using a four-slit apparatus connected to a personal computer (Intaglietta and Tompkins 1973).

3. Calculate the mean transverse blood flow rate of individual vessels (Q_{T}) using D and the mean V_{TRBC} ($V_{\mathrm{Tmean}} = V_{\mathrm{TRBC}}/a$). For blood vessels <10 μm, $a = 1.3$, and for those >15 μm, $a = 1.6$. By linear extrapolation, $1.3 < a < 1.6$ for blood vessels between 10 and 15 μm (Lipowsky and Zweifach 1978): $Q_{\mathrm{T}} = \pi / 4 \times V_{\mathrm{Tmean}} \times D^2$.

4. Use an image processing system to analyze the two-dimensional functional vessel density, which is a measure of angiogenesis, defined as the total length of vessels per unit area of the image (cm/cm^2), and the two-dimensional branching index, defined as the mean length of unbranched segments (μm) in the plane of the image (Fukumura et al. 1997a,b; Jain et al. 1997).

MPLSM Procedure

1. Measure the vessel diameters in micrometers (D) from the image stack using image analysis software (Abdul-Karim et al. 2003; Garkavtsev et al. 2004).

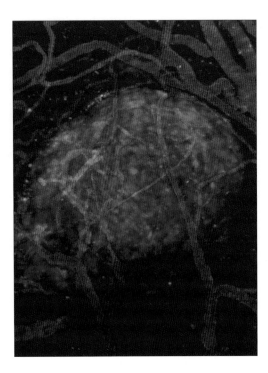

FIGURE 6. Tumor microcirculation imaged with MPLSM. A spontaneous metastasis of a U87 human glioblastoma growing in the cranial window of a SCID mouse. Vessels are highlighted with intravenous injection of 0.2 ml of 10 mg/ml, 2-million m.w. FITC-dextran. The image shown is a maximum intensity projection of 74 optical sections spaced 2 µm apart. Image is 500 µm across. (Image courtesy of E. diTomaso.)

2. Measure the transverse red blood cell velocity (V_{TRBC}) using the line scan technique (Brown et al. 2001). Since the majority of laser-scanning microscopes (LSPs) only scan in a single direction, this technique produces the red blood cell velocity tangential to the direcion of the line scan. Subsequently, calculate the absolute red blood cell velocity (V_{RBC}) by determining the angle θ between the line scan and the axis of the vessel using the three-dimensional image stack. The true red blood cell velocity is then $V_{RBC} = V_{TRBC} / \cos\theta$.

3. Calculate the mean absolute blood flow rate of individual vessels (Q) using D and the mean V_{RBC} ($V_{mean} = V_{RBC}/a$). For blood vessels <10 µm, $a = 1.3$, and for those >15 µm, $a = 1.6$. By linear extrapolation, $1.3 < a < 1.6$ for blood vessels between 10 and 15 µm (Lipowsky and Zweifach 1978): $Q = \pi / 4 \times V_{mean} \times D^2$.

4. Use an image processing system to analyze from the three-dimensional image stack, the three-dimensional functional vessel density, which is a measure of angiogenesis, defined as the total length of vessels per unit volume (cm/cm³), and the branching index, defined as the mean length of unbranched segments (µm) (Abdul-Karim et al. 2003).

Common Analysis

Fractal dimensions of vessels can be measured as described elsewhere (Gazit et al. 1995; Baish and Jain 2000).

Vascular Permeability

The effective average vascular permeability (P) of a region of vessels can be measured by single-photon microscopy using the following procedure (Yuan et al. 1993, 1994; Fukumura et al. 1997a).

Single-photon Microscopy Procedure

1. Inject a bolus of a fluorescent tracer (e.g., Rho-labeled or Cy5-labeled bovine serum albumin [BSA] or other molecule, 10 mg/ml; 0.1 ml/25 g body weight).

2. During the next 20 minutes, intermittently measure the fluorescence intensity of the tumor tissue.

3. Calculate the value of P as $P = (1 - HT)\ V/S\ \{1/(I_0 - I_b) \times dI/dt + 1/K\}$, where I is the average fluorescence intensity of the whole image, I_0 is the value of I immediately after the filling of all vessels by the fluorescent tracer, and I_b is the background fluorescence intensity. The average hematocrit (HT) of tumor vessels is estimated independently (Brizel et al. 1993). V and S are the total volume and surface area of vessels within the tissue volume covered by the surface image, respectively. The time constant of clearance of the tracer from plasma (K) is measured independently (Yuan et al. 1994).

MPLSM Procedure

The vascular permeability (P) of individual vessel segments within ~600 μm of the tumor-window interface can be measured by MPLSM using the following procedure (Brown et al. 2001).

1. Inject a bolus of a fluorescent tracer (e.g., TMR-labeled BSA or other molecule of interest, 10 mg/ml; 0.1 ml/25 g body weight).

2. Periodically record the fluorescence intensity of a single optical section containing the vessel segment of interest for up to 20 minutes using an MPLSM. TMR-labeled BSA is imaged with 840 nm excitation and a 550DF150 emission filter.

3. Measure the fluorescence intensity $F(r)$ along a line perpendicular to the vessel in each image using image processing software. The permeability P of the vessel segment is given by

$$P = \frac{\partial \int_{r=R}^{\infty} F(r)r\,dr}{\partial t (F_v - F_i) R}$$

where R is the radius of the vessel segment, F_v is the fluorescence signal from the plasma in the vessel, and F_i is the fluorescence signal immediately outside the vessel. The value of P is calculated from the $t = 0$ limit of the fluorescence distribution to minimize errors that can result from a finite integration length and due to fluorescence contribution from material leaking out of nearby vessels.

Leukocyte Endothelial Interactions

The flux of leukocytes, as well as the number of rolling and adhering leukocytes, is routinely measured by conventional single-photon microscopy (Fukumura et al. 1995).

Single-photon Microscopy Procedure

1. Inject mice with a bolus (20 μl) of 0.1% rhodamine-6G in 0.9% saline through the tail vein.

2. Visualize leukocytes with an intensified CCD camera and record their activity on S-VHS tape.

3. Count the numbers of rolling (Nr) and adhering (Na) leukocytes for 30 seconds along a 100-μm segment of a vessel. Also measure the total flux of cells for 30 seconds (Nt).

4. Calculate the ratio of rolling cells to total flux (rolling count), the density of adhering leukocytes (adhesion density), and the shear rate for each vessel, as follows:

$$\text{Rolling count (\%)} = 100 \times Nr/Nt$$
$$\text{Adhesion density (cells/mm}^2) = 10^6 \times Na/(\pi \times D \times 100\ \mu\text{m})$$
$$\text{Shear rate} = 8 \times V_{\text{mean}}/D$$

The majority of LSMs currently in use suffer from a limited temporal resolution, with frame rates on the order of ~1 per second for a 768×512 image. This temporal resolution can be insufficient for imaging selected freely flowing leukocytes multiple times, which is necessary to derive an accurate leukocyte velocity and overall particle count. Consequently, it is often necessary to reduce the spatial resolution of the image (by decreasing the number of pixels per image) in order to increase the temporal resolution. The exact procedures will depend on the LSM controlling software. Alternatively, high-speed video-rate LSMs can be used, usually with some frame averaging to build a sufficient signal-to-noise ratio in each pixel (Padera et al. 2002a). In typical (nonvideo-rate) LSMs, the flux of leukocytes, as well as the number of rolling and adhering leukocytes, is measured using the following procedure (Brown et al. 2001).

MPLSM Procedure

1. Inject mice with a bolus (20 μl) of 0.1% rhodamine-6G in 0.9% saline through the tail vein.

2. Visualize leukocytes via an MPLSM using 810 nm excitation and a 550DF150 emission filter.

3. Generate a series of 10–15 high-temporal-resolution images (i.e., ≥2 frames per second) of a given vessel, covering a total duration of 30 seconds. Count the numbers of rolling (*Nr*) and adhering (*Na*) leukocytes, as well as the total flux of cells (*Nt*).

4. Calculate the ratio of rolling cells to total flux (rolling count), the density of adhering leukocytes (adhesion density), and the shear rate for each vessel, as follows:

$$\text{Rolling count (\%)} = 100 \times Nr/Nt$$
$$\text{Adhesion density (cells/mm}^2) = 10^6 \times Na/(\pi \times D \times 100\ \mu\text{m})$$
$$\text{Shear rate} = 8 \times V_{\text{mean}}/D$$

Extravascular Parameters

Interstitial pH Measurements

Fluorescence ratio imaging microscopy (FRIM) for pH, its implementation, application to thick tissues, and calibration is performed as described previously (Dellian et al. 1996a; Helmlinger et al. 1997b). Key parameters are listed below.

1. Use the cell-impermeant form of the pH-sensitive fluorochrome 2′, 7′-bis-(2-carboxyethyl)-5,6-carboxyfluorescein (BCECF; 0.7 mg/kg intravenously).

2. Image the emission intensities (570 nm) through the CCD camera port following sequential excitations at 440 and 495 nm. To obtain a sampling depth of ≤25 μm and a lateral spatial resolution of 5 μm, use a 400-μm pinhole in the light excitation pathway and a 40× water immersion objective.

3. Cycle the *x-y* stage through the same locations for dynamic measurements.

4. Convert the ratio of intensities into pH using the calibration curves.

Interstitial and Microvascular pO$_2$ Measurements

This technique is based on the oxygen-dependent quenching of the phosphorescence of albumin-bound palladium meso-tetra (4-carboxyphenyl) porphyrin (Torres Filho and Intaglietta 1993; Torres Filho et al. 1994; Helmlinger et al. 1997b; Tsai et al. 1998).

1. Inject the porphyrin probe (60 mg/kg) into the mouse tail vein.

2. Expose the tissue to flashlamp excitation (540 nm). The resulting phosphorescence signal from the probe is detected at ≥630 nm using the photomultiplier tube and averaged on the oscilloscope prior to computer storage.

3. Cycle the *x-y* stage through the same tumor locations used for the pH measurements.

4. Reduce the illumination field to a 100-μm spot and place a 10 × 10-μm² slit in the light emission pathway. This reduces the sampling depth to ≤25 μm and gives a lateral spatial resolution of 10 μm, similar to the pH measurements.

5. Place a second eyepiece between the slit and the collecting tube, which allows refocusing on the region of interest prior to taking phosphorescence decay measurements. Phosphorescence measurements are valid within interstitial spaces as well as blood vessels.

6. Convert data to pO$_2$ values according to a standard calibration method (Lahiri et al. 1993).

Calibration tests indicate that there is a strong linear relationship (r^2 ≥0.99) between pO$_2$(0–60 mm Hg) and the inverse of the phosphorescence lifetime. Furthermore, the pO$_2$ calibration curves do not show any dependence on the pH of the solution (pH range: 6.60–7.40). Consequently, this porphyrin is an ideal probe for use in tumors, which can have variable pH values. We have also found that sequential measurements of pH and pO$_2$ in tissues in vivo are possible, since the presence of the pH probe (BCECF) does not affect lifetime measurements of the pO$_2$ probe (porphyrin) (Helmlinger et al. 1997b).

Interstitial Diffusion, Convection, and Binding

Fluorescence recovery after photobleaching (FRAP) with spatial Fourier analysis, its implementation, application to thick tissues, and calibration is performed as described by Berk et al. (1993, 1997) (see also Chapter 7). To measure interstitial diffusion coefficients, multiphoton fluorescence recovery after photobleaching (MPFRAP) (Brown et al. 1999) or multiphoton fluorescence correlation spectroscopy (MPFCS) (Alexandrakis et al. 2004) is performed as described previously.

Single-photon Microscopy Procedures

1. Infuse a fluorescently labeled molecule of interest into the tumor interstitium either via extravasation after intravenous injection or local low pressure microinfusion.

2. To bleach a subpopulation of the fluorescent molecules, expose the tissue to a brief (~msec) flash of focused laser light.

3. Generate consecutive images of the bleached region using epifluorescence, and capture them on the CCD camera as unbleached fluorophore diffuses back into the bleached region.

   IMPORTANT: During the bleaching flash, shutter the camera to avoid damage to the electronics.

4. Perform spatial Fourier analysis of the fluorescence recovery images as described previously (Chary and Jain 1989; Berk et al. 1993, 1997) to extract diffusion coefficients, convection velocity, and binding parameters.

MPLSM Procedures

1. Infuse an FITC-labeled molecule of interest into the tissue interstitium either via extravasation from the vasculature or low-pressure microinfusion. Highlight vessels with an injection of 100 µl of TMR-labeled dextran (2 million m.w., 10 mg/ml) into the tail veins of mice. Generate fluorescence with 840 nm excitation and collect it with 535DF40 (FITC) and 610DF75 (TMR) emission filters and a 570LP dichroic mirror.

2. Identify a location of interest from MPLSM images of blood vessels. Park the multiphoton focal volume at the location using the LSM control software. Use a brief (about a sub-millisecond) flash of high-intensity laser light to bleach out a subpopulation of the fluorescent molecules.

3. Monitor the bleached region using the same laser beam, but greatly attenuated. Use a multichannel scaler to record the recovery in fluorescence of the bleached region as the unbleached fluorophores diffuse back into the bleached region.

4. Perform mathematical analysis of the fluorescence recovery curve to extract the diffusion coefficient of the labeled molecules (Brown et al. 1999).

Lymphangiography

To visualize lymphatics and assess their function, 5 µl of 2.5% FITC-labeled or TMR-labeled dextran (2 million m.w.) is injected into tumor or dermis adjacent to the tumor at low pressure (Leu et al. 2000). The fluorescent dextran is taken up by lymphatics and serves as a marker of functional lymphatics (Fig. 7). Images up to 450-µm depth are obtained by using MPLSM (Padera et al. 2002a). Lymphatic network morphologies such as lymphatic vessel diameter and mesh geometry (maximum distance between parallel edges of each hexagon) are analyzed off-line using image analysis software such as NIH Image. Flow velocity is determined by monitoring the lymphatic network at various distances from the injection site (Leu et al. 1994) or by using FRAP techniques (Berk et al. 1996; Swartz et al. 1996).

Gene Expression: Promoter Activity via GFP Imaging

To monitor gene promoter activity in stromal and tumor cells, a fluorescent reporter gene driven by the promoter of interest is introduced into mice (Fukumura et al. 1998) and/or tumor cells (Fukumura et al. 1999). Once the gene is activated, the corresponding cells become fluorescent and the fluorescence intensity is detected. Currently, the most commonly used reporter gene is GFP. However, as other variants of GFP become available, the possibility of monitoring multiple genes simultaneously or gene activity in different cell populations will become practical. To image GFP fluorescence (emission: 509 nm) and vasculature simultaneously using the MPLSM, a bolus of red tracer (e.g., TMR-labeled 2-million-m.w. dextran, 10 mg/ml; 0.1 ml/25 g body weight) is injected into the tail vein, and an image stack is generated with 840–860 nm excitation, 525/100 and 610/75 emission filters, and a 550 LP dichroic (Fig. 7 and Movie 24.2) (Brown et al. 2001). To obtain quantitative data on promoter activity in vivo, a deconvolution algorithm must be used to adjust for the kinetics of GFP decay and the relationship between protein levels and the fluorescence emitted by GFP.

Collagen Dynamics Using Second Harmonic Generation

The structure of fibrillar collagen can be monitored using second harmonic generation by MPLSM as described in Brown et al. (2003) (Fig. 8). The procedure is summarized below.

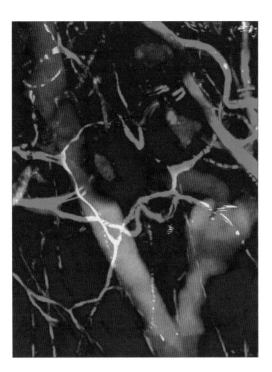

FIGURE 7. Noninvasive lymphangiography imaged using MPLSM. Lymphatics and blood vessels growing in the ear of a nude mouse. Vessels are highlighted with an intravenous injection of 0.2 ml of 10 mg/ml, 2-million-m.w. FITC-dextran, lymphatics are highlighted with direct injection of 1–2 μm of 2.5 mg/ml, 2-million-m.w. TMR-dextran. The image shown is a maximum intensity projection of 20 optical sections spaced 5 μm apart. Image is 250 μm across. (Image courtesy of T. Padera and R. Tong.)

1. To image fibrillar collagen simultanesously with blood vessels and GFP-labeled cells, choose an excitation wavelength based on the excitation cross section of fluorophores used to label the vasculature or cells of interest.

    For TMR-dextran-labeled vasculature and GFP-labeled cells, 850–860 nm excitation is appropriate.

2. Choose an emission filter and dichroic to transmit at exactly half the excitation wavelength and separate this wavelength from the emission of the fluorophores in the system.

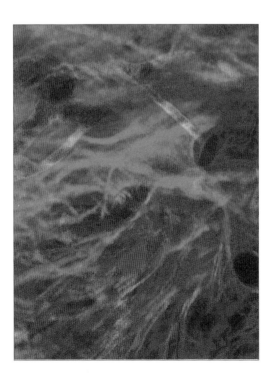

FIGURE 8. Collagen fibers in tumors imaged with second harmonic generation. An image of collagen in tumors imaged without extrinsic label using second harmonic generation (*green*); tumor vessels (*red*) are highlighted with an intravenous injection of 0.2 ml of 10 mg/ml TMR-dextran. Image is a maximum intensity projection of five optical sections spaced 5 μm apart, and is 250 μm across.

> In the example cited above, 860 nm excitation requires a 430DF30-nm emission filter (2HG) along with 525DF100 (GFP) and 610DF75 (TMR) emission filters and 460LP and 575LP dichroics.

3. Quantify the dynamics of acute collagen modification by generating repetitive images of a collagen matrix after addition of a collagen-degrading enzyme.

> For example, capture one image every minute after addition of 100 μl of 1% bacterial collagenase to a tumor growing in a dorsal skinfold chamber, or one image of a tumor every 2 days after implantation of an osmotic pump expressing the hormone relaxin (Brown et al. 2003).

NOVEL INSIGHTS

The ability to deliver therapeutic agents to all regions of a tumor is governed by the tumor blood supply. Using in vivo microscopy, we and other investigators have unequivocally demonstrated that the structure and function of tumor vessels are heterogeneous (Endrich et al. 1979a; Dudar and Jain 1984; Dewhirst et al. 1989; Leunig et al. 1992b) and have suggested the possibility that the presence of cancer cells of the vessel wall may contribute to this heterogeneity (Sasaki et al. 1991; Chang et al. 2000). Furthermore, our finding that cancer cells coerce the host cells into making vascular endothelial growth factor (VEGF), a potent angiogenic molecule, suggest that host cells are not passive bystanders but active participants in tumor angiogenesis, growth, and response to therapy (Fig. 9) (Fukumura et al. 1998; Brown et al. 2001). These concepts pull together a number of key observations made in our laboratory.

- Angiogenesis, pO_2, pH, permeability, and pore cutoff size in tumor vessels vary from one tumor to the next, from one region to the next within the same tumor, from one day to the next, and from one anatomical site to next (Yuan et al. 1994; Fukumura et al. 1997b; Helmlinger et al. 1997b).

- The production of angiogenic inhibitors, similar to angiogenic stimulators, is dependent on the site of primary tumor growth (Gohongi et al. 1999) and changes in response to treatment (Hartford et al. 2000).

- Surprisingly, in a hormone-dependent tumor, hormone withdrawal causes apoptosis of endothelial cells prior to that of cancer cells by down-regulating the production of VEGF by cancer cells. A second wave of angiogenesis is then driven by VEGF, presumably, from host cells (Jain et al. 1998b).

- One would expect antiangiogenic therapy to impair drug delivery by inducing vessel regression. However, in the initial phases, these therapies may prune immature vessels and induce a normal vascular network with more mature vessels, thus explaining the potential synergism between antiangiogenic and cytotoxic therapies (Jain et al. 1998b; Hansen-Algenstaedt et al. 2000; Lee et al. 2000; Willett et al. 2004).

Lymphangiography of a fibrosarcoma in the mouse tail model has shown that the lymphatics are impaired within the tumor mass and yet enlarged at the tumor periphery (Leu et al. 2000; Padera et al. 2002a,b). The former contributes to interstitial hypertension in tumors, a barrier to drug delivery (Jain 1994). The latter, presumably induced by VEGF type-C, facilitates lymphatic metastasis (Jeltsch et al. 1997). The impairment of intratumor lymphatics presumably results from solid stress generated by proliferating cancer cells (Helmlinger et al. 1997a; Padera et al. 2004). Releasing this solid stress should thus open lymphatics, lower pressure, and increase delivery of agents across tumor vessels (Griffon-Etienne et al. 1999).

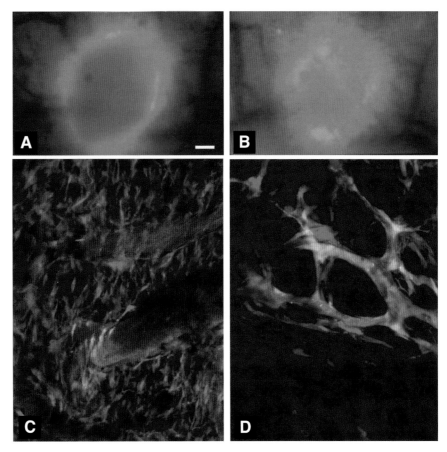

FIGURE 9. VEGF promoter activity during tumor growth. Epifluorescence images (*A* and *B*): Murine hepatoma *HcaI* grown in the dorsal skin chamber in the *VEGF^p-GFP* mice was visualized by intravital microscopy at 2 weeks (*A*) and 3 weeks (*B*) after the implantation. Implantation of solid tumors in the transgenic mice leads to an accumulation of green fluorescence, resulting from tumor induction of VEGF promoter activity in host stromal cells. With time, the fluorescent cells invade the tumor and can be seen throughout the tumor mass (Fukumura et al. 1998). MPLSM images (*C* and *D*): Murine mammary adenocarcinoma MCaIV grown in the dorsal skin chamber in the *VEGF^p-GFP* mice was imaged by MPLSM at 2 weeks (*C*) and 4 weeks (*D*) after the implantation. At the tumor periphery, a high density of GFP-positive host cells (*green*) that exhibit a fibroblast-like shape was observed. On the other hand, they often fortify angiogenic vessels (*yellow/red*) inside tumors (*D*). Bars: (A and B) 500 μm. (Panel *D* adapted, with permission, from Brown et al. 2001.)

Once a therapeutic agent has leaked from a blood vessel, it must migrate through the interstitial matrix to reach cancer cells (Jain 1987). Using FRAP, our lab has provided the first and to our knowledge the only measurements of interstitial diffusion, convection, and binding in vivo (Chary and Jain 1989; Berk et al. 1997). Furthermore, we showed that the anomalous assembly of collagen in tumors can prevent the penetration of therapeutic agents. This finding identified collagen synthesis as a potential target for improving the delivery of macromolecules with relaxin as a prototypic matrix modifier (Torres Filho et al. 1994).

In an attempt to understand heterogeneous localization of activated lymphocytes to tumor vessels, we discovered that angiogenic agents regulate adhesion molecules on the vasculature. This finding provided the first link between the disparate fields of angiogenesis and adhesion. For example, we showed that VEGF up-regulates while bFGF (basic fibroblast growth factor) down-regulates adhesion molecules on the vasculature (Melder et al. 1996; Detmar et al. 1998). This is one of the rare occasions where two molecules act synergistically for one function but antagonistically for another. This finding also provides a new mechanism of immune evasion by bFGF.

FUTURE PERSPECTIVES

Intravital microscopy has provided useful insight into angiogenesis and tumor biology (Ferrara and Alitaro 1999; Carmeliet and Jain 2000; Folkman 2000; Kerbel 2000). However, three key challenges remain. First, the most widely used microscopy techniques are surface-weighted. Given the heterogeneous nature of tumors, we need dynamic information about their internal milieu. Confocal microscopy can provide images up to a few hundred micrometers in depth (Dirnagl et al. 1992; Suzuki et al. 1998). The advent of multiphoton microscopy is likely to change this limit to greater than 500 μm, depending on the tissue and tracer used (Kleinfeld et al. 1998; Lendvai et al. 2000; Brown et al. 2001; Helmchen and Denk 2002). Other optical methods such as optical coherence tomography can image deeper regions, but they are still in the development phase and not yet widely available (Huang et al. 1991; Boppart et al. 1999). With more research in this area, we may some day be able to obtain dynamic images of the whole tumor with high spatial resolution.

The second challenge is the size and bulk of current devices, which limits their application. With advances in miniaturized multiphoton microscopes and high-speed scanning, this limitation may be overcome in the future (Helmchen and Denk 2002). The third challenge comes from the lack of molecular probes to interrogate various vascular processes taking place in vivo. Again, rapid development in this area will allow exciting opportunities for dissecting various molecular events in the future (Weissleder 2001; Weissleder and Ntziachristos 2003). With the availability of novel luorophores and computer analysis, we will be able to simultaneously image multiple events (visualized by distinct colors).

With these improvements in microscopy and probes, in vivo microscopy will continue to offer new opportunities for unexpected discoveries in tumor biology as well as cancer detection and treatment. Similarly, this approach can be used in other fields of biology and medicine, including neurobiology (Helmchen and Denk 2002) and immunology (Buosso et al. 2002; Cahalan et al. 2002).

ACKNOWLEDGMENTS

The work described here was supported by grants from the National Institutes of Health, National Science Foundation, American Cancer Society, United States Army, National Foundation for Cancer Research, and the Whitaker Foundation. This chapter is based on four previous related reviews (Jain et al. 1997, 1998a, 2001, 2002). We thank the publishers for allowing us to reproduce the relevant material.

MOVIE LEGENDS

MOVIE 24.1. Neovasculature in de novo adipose tissue. Images were captured by multiphoton laser-scanning microscopy with contrast enhancement of blood vessels by intravenous injection of tetramethylrhodamine-labeled dextran (MW 2,000,000). The movie presents a vertical stack of images over a depth of 300 μm, starting from the top (showing the vessels in the de novo adipose tissue) and continuing to the bottom (showing the typical pattern of the preexisting blood vessels in the underlining host subcutaneous tissue). Image is 1 μm across. (From Fukumura et al. 2004.)

MOVIE 24.2. Three-dimensional presentation of tissue-engineered blood vessels. Human umbilical vein endothelial cells (HUVECs) and 10T1/2 cells (mesenchymal precursor cells) were seeded in the fibronectin-collagen I 3-D constructs and implanted in the cranial window of SCID mice. Green fluorescent protein expressing HUVECs (*green*) and perfused blood vessels (*red*; contrast enhanced by tetramethylrhodamine-conjugated dextran) were visualized by multiphoton laser-scanning microscopy. At day 28, blood perfusion could be seen in all layers of the construct. Thickness of the construct is 165 μm. Image is 270 μm across. (From Koike et al. 2004 with permission from the Nature Publishing Group).

REFERENCES

Abdul-Karim A., Al-Kofahi K., Brown E., Jain R.K., and Roysam B. 2003. Automated tracing and change analysis of angiogenic vasculature from in vivo multiphoton image time series. *Microvasc. Res.* **66:** 113–125.

Alexandrakis G., Brown E., Tong R., McKee T., Campbell R., Boucher Y., and Jain R.K. 2004. Two-photon fluorescence correlation microscopy to quantify the transport of fluorescently labeled tracers in tumors. *Nat. Med.* **10:** 203–207.

Al-Mehdi A.B., Tozawa K., Fisher A.B., Shientag L., Lee A., and Muschel R.J. 2000. Intravascular origin of metastasis from the proliferation of endothelium-attached tumor cells: A new model for metastasis. *Nat. Med.* **6:** 100–102.

Algire G.H. 1943a. An adaptation of the transparent chamber technique to the mouse. *J. Natl. Cancer Inst.* **4:** 1–11.

———. 1943b. Microscopic studies of the early growth of a transplantable melanoma of the mouse, using the transparent-chamber technique. *J. Natl. Cancer Inst.* **4:** 13–20.

Algire G.H. and Chalkley H.W. 1945. Vascular reactions of normal and malignant tissues in vivo. I. Vascular reactions of mice to wounds and to normal and neoplastic transplants. *J. Natl. Cancer Inst.* **6:** 73–85.

Algire G.H. and Legallis F.Y. 1948. Growth and vascularization of transplanted mouse melanomas. *N.Y. Acad. Sci.* **4:** 159–175.

Asaishi K., Endrich B., Gotz A., and Messmer K. 1981. Quantitative analysis of microvascular structure and function in the amelanotic melanoma A-Mel-3. *Cancer Res.* **41:** 1898–1904.

Auerbach R., Arensman R., Kubai L., and Folkman J. 1975. Tumor-induced angiogenesis: Lack of inhibition by irradiation. *Int. J. Cancer* **15:** 241–245.

Ausprunk D.H. and Folkman J. 1977. Migration and proliferation of endothelial cells in preformed and newly formed blood vessels during tumor angiogenesis. *Microvasc. Res.* **14:** 53–65.

Baish J.W. and Jain R.K. 2000. Fractals and cancer. *Cancer Res.* **60:** 3683–3688.

Berk D.A., Swartz M.A., Leu A.J., and Jain R.K. 1996. Transport in lymphatic capillaries. II. Microscopic velocity measurement with fluorescence photobleaching. *Am. J. Physiol.* **270:** H330–H337.

Berk D.A., Yuan F., Leunig M., and Jain R.K. 1993. Fluorescence photobleaching with spatial Fourier analysis: Measurement of diffusion in light-scattering media. *Biophys. J.* **65:** 2428–2436.

———. 1997. Direct in vivo measurement of targeted binding in a human tumor xenograft. *Proc. Natl. Acad. Sci.* **94:** 1785–1790.

Boppart S., Herrmann J., Pitris C., Stamper D., Brezinski M., and Fujimoto J. 1999. High resolution optical coherence tomography-guided laser ablation of surgical tissue. *J. Surg. Res.* **82:** 275–284.

Brem S.S., Gullino P.M., and Medina D. 1977. Angiogenesis: A marker for neoplastic transformation of mammary papillary hyperplasia. *Science* **195:** 880–882.

Brizel D.M., Klitzman B., Cook J.M., Edwards J., Rosner G., and Dewhirst M.W. 1993. A comparison of tumor and normal tissue microvascular hematocrits and red cell fluxes in a rat window chamber model. *Int. J. Radiat. Oncol. Biol. Phys.* **25:** 269–276.

Brown E.B., Wu E.S., Zipfel W., and Webb W.W. 1999. Measurement of molecular diffusion in solution by multiphoton fluorescence photobleaching recovery. *Biophys. J.* **77:** 2837–2849.

Brown E., McKee T.D., di Tomaso E., Seed B., Boucher Y., and Jain R.K. 2003. Dynamic imaging of collagen and its modulation in tumors in vivo using second harmonic generation. *Nat. Med.* **9:** 796–801.

Brown E., Campbell R., Tsuzuki Y., Xu L., Carmeliet P., Fukumura D., and Jain R.K. 2001. In vivo measurement of gene expression, angiogenesis, and physiological function in tumors using multiphoton laser scanning microscopy. *Nat. Med.* **7:** 864–868.

Buosso P., Bhakta N.R., Lewis R.S., and Robey E. 2002. Synamics of thymocyte-stromal cell interactions visualized by two-photon microscopy. *Science* **296:** 1876–1880.

Cahalan M.D., Parker I., Wei S.H. and Miller M.J. 2002. Two-photon tissue imaging: seeing the immune system in a fresh light. *Nat. Rev. Immunol.* **2:** 872–880.

Carmeliet P. and Jain R.K. 2000. Angiogenesis in cancer and other diseases. *Nature* **407:** 249–257.

Chambers A.F., Shafir R., and Ling V. 1982. A model system for studying metastasis using the embryonic chick. *Cancer Res* **42:** 4018–4025.

Chang Y.S., diTomaso E., McDonald D.M., Jones R., Jain R.K., and Munn L.L. 2000. Mosaic blood vessels in tumors: Frequency of cancer cells in contact with flowing blood. *Proc. Natl. Acad. Sci.* **97:** 14608–14613.

Chary S.R. and Jain R.K. 1989. Direct measurement of interstitial convection and diffusion of albumin in normal and neoplastic tissues by fluorescence photobleaching. *Proc. Natl. Acad. Sci.* **86:** 5385–5389.

Chishima T., Miyagi Y., Wang X., Yamaoka H., Shimada H., Moossa A.R., and Hoffman R.M. 1997. Cancer invasion and micrometastasis visulaized in live tissue by green fluorescent protein expression. *Cancer Res.* **57:** 2042–2047.

Chute R.N., Sommers S.C., and Warren S. 1952. Heterotransplantation of human cancer. II. Hamster cheek pouch. *Cancer Res.* **12:** 912–914.

Clark E.R. and Clark E.L. 1932. Observations on living preformed blood vessels as seen in a transparent chamber inserted into the rabbit's ear. *Am. J. Anat.* **49:** 441–477.

Clark E.R., Kirby-Smith H.T., Rex R.O., and Williams R.G. 1930. Recent modifications in the method of studying living cells and tissues in transparent chambers inserted in the rabbit's ear. *Anat. Rec.* **47:** 187–211.

Dellian M., Helmlinger G., Yuan F., and Jain R.K. 1996. Fluorescence ratio imaging and optical sectioning: Effect of glucose on spatial and temporal gradients. *Br. J. Cancer* **74:** 1206–1215.

Dellian M., Witwer B.P., Salehi H.A., Yuan F., and Jain R.K. 1996. Quantitation and physiological characterization of angiogenic vessels in mice: Effect of basic fibroblast growth factor, vascular endothelial growth factor/vascular permeability factor, and host microenvironment. *Am. J. Pathol.* **149:** 59–72.

Denk W., Strickler J.H., and Webb W.W. 1990. Two-photon laser scanning fluorescence microscopy. *Science* **248:** 73–76.

Detmar M., Brown L.F., Schön M.P., Elicker B.M., Richard L., Velasco P., Fukumura D., Monsky W., Claffey K.P., and Jain R.K. 1998. Increased microvascular density and enhanced leukocyte rolling and adhesion in the skin of VEGF transgenic mice. *J. Invest. Dermatol.* **111:** 1–6.

Dewhirst M.W., Tso C.Y., Oliver R., Gustafson C.S., Secomb T.W., and Gross J.F. 1989. Morphologic and hemodynamic comparison of tumor and healing normal tissue microvasculature. *Int. J. Radiat. Onco. Biol. Phys.* **17:** 91–99.

Dirnagl U., Villringer A., and Einhuapl K. 1992. In vivo confocal laser scanning microscopy of the cerebral microcirculation. *J. Microsc.* **65:** 147–157.

Dudar T.E. and Jain R.K. 1984. Differential response of normal and tumor microcirculation to hyperthermia. *Cancer Res.* **44:** 605–612.

Eddy H.A. and Casarett G.W. 1973. Development of the vascular system in the hamster malignant neurilemmoma. *Microvasc. Res.* **6:** 63–82.

Endrich B., Intaglietta M., Reinhold H.S., and Gross J.F. 1979a. Hemodynamic characteristics in microcirculatory blood channels during early tumor growth. *Cancer Res.* **39:** 17–23.

Endrich B., Reinhold H.S., Gross J.F., and Intaglietta M. 1979b. Tissue perfusion inhomogeneity during early tumor growth in rats. *J. Natl. Cancer Inst.* **62:** 387–395.

Falkvoll K.H., Rofstad E.K., Brustad T., and Marton P. 1984. A transparent chamber for the dorsal skin fold of athymic mice. *Exp. Cell. Biol.* **52:** 260–268.

Ferrara N. and Alitaro K. 1999. Clinical application of angiogenic growth factors and their inhibitors. *Nat. Med.* **5:** 1359–1364.

Fidler I.J. 1995. Modulation of the organ microenvironment for treatment of cancer metastasis. *J. Natl. Cancer Inst.* **87:** 1588–1592.

Folkman J. 2000. Tumor angiogenesis. In *Cancer medicine,* 5th edition (ed. J.F. Holand et al.), pp. 132–152. Decker, Ontario, B.C.

Friedlander M., Brooks P.C., Shaffer R.W., Kincaid C.M., Varner J.A., and Cheresh D.A. 1995. Definition of two angiogenic pathways by distinct alpha v integrins. *Science* **270:** 1500–1502.

Fukumura D., Yuan F., Endo M., and Jain R.K. 1997a. Role of nitric oxide in tumor microcirculation: Blood flow, vascular permeability, and leukocyte-endothelial interactions. *Am. J. Pathol.* **150:** 713–725.

Fukumura D., Yuan F., Monsky W.L., Chen Y., and Jain R.K. 1997b. Effect of host microenvironment on the microcirculation of human colon adenocarcinoma. *Am. J. Pathol.* **151:** 679–688.

Fukumura D., Salehi H.A., Witwer B., Tuma R.F., Melder R.J., and Jain R.K. 1995. Tumor necrosis factor α-induced leukocyte adhesion in normal and tumor vessels: Effect of tumor type, transplantation site, and host strain. *Cancer Res.* **55:** 4824–4829.

Fukumura D., Xu L., Chen Y., Gohongi T., Seed B., and Jain R.K. 2001. Hypoxia and acidosis independently up-regulate vascular endothelial growth factor transcription in brain tumors *in vivo. Cancer Res.* **61:** 6020–6024.

Fukumura D., Gohongi T., Ohtaka K., Stoll B., Chen Y., Seed B., and Jain R.K. 1999. Regulation of VEGF promoter activity in tumors by tissue oxygen and pH levels. *Proc. Am. Assoc. Cancer Res.* **40:** 722.

Fukumura D., Ushiyama A., Duda D.G., Xu L., Tam J., Krishna V., Chatterjee K., Garkavtsev I., and Jain R.K. 2004. Paracrine regulation of angiogenesis and adipocyte differentiation during in vivo adipogenesis. *Circ. Res.* **93:** 88–97.

Fukumura D., Xavier R., Sugiura T., Chen Y., Parks E., Lu N., Selig M., Nielsen G., Taksir T., Jain R., and Seed B. 1998. Tumor induction of VEGF promoter activity in stromal cells. *Cell* **94:** 715–725.

Funakoshi N., Onizuka M., Yanagi K., Ohshima N., Tomoyasu M., Sato Y., Yamamoto T., Ishikawa S., and Mitsui T. 2000. A new model of lung metastasis for intravital studies. *Microvasc. Res.* **59:** 361–367.

Garkavtsev I., Kozin S., Chernova O., Xu L., Brown E., Barnett G., and Jain R.K. 2004. *ING4*: A novel regulator of brain tumor growth and angiogenesis. *Nature* **428:** 328–332.

Gazit Y., Berk D.A., Leunig M., Baxter L.T., and Jain R.K. 1995. Scale-invariant behavior and vascular network formation in normal and tumor tissue. *Phys. Rev. Lett.* **75:** 2428–2431.

Gerlowski L.E. and Jain R.K. 1986. Microvascular permeability of normal and neoplastic tissues. *Microvasc. Res.* **31:** 288–305.

Gimbrone M.A.J. and Gullino P.M. 1976. Angiogenic capacity of preneoplastic lesions of the murine mammary gland as a marker of neoplastic transformation. *Cancer Res.* **36:** 2611–2620.

Gimbrone M.A.J., Leapman S.B., Cotran R.S., and Folkman J. 1972. Tumor dormancy in vivo by prevention of neovascularization. *J. Exp. Med.* **136:** 261–276.

Gimbrone M.A.J., Cotran R.S., Leapman S.B., and Folkman J. 1974. Tumor growth and neovascularization: An experimental model using the rabbit cornea. *J. Natl. Cancer Inst.* **52:** 413–427.

Gohongi T., Fukumura D., Boucher Y., Yun C.-O., Soff G.A., Compton C., Todoroki T., and Jain R.K. 1999. Tumor-host interactions in the gallbladder suppress distal angiogenesis and tumor growth: Involvement of transforming growth factor β1. *Nat. Med.* **5:** 1203–1208.

Goodall C.M., Sanders A.G., and Shubik P. 1965. Studies of vascular patterns in living tumors with a transparent chamber inserted in hamster cheek pouch. *J. Natl. Cancer Inst.* **35:** 497–521.

Greene H.S.N. 1952. The significance of heterologous transplantability of human cancer. *Cancer* **5:** 24–44.

Griffon-Etienne G., Boucher Y., Brekken C., Suit H.D., and Jain R.K. 1999. Taxane-induced apoptosis decompress blood vessels and lowers interstitial fluid pressure in solid tumors: Clinical implications. *Cancer Res.* **54:** 3776–3782.

Hansen-Algenstaedt N., Stoll B.R., Padera T.P., Dolmans D.E. G.J., Hicklin D.J., Fukumura D., and Jain R.K. 2000. Tumor oxygeneation in hormone-dependent tumors during vascular endothelial growth factor receptor-2 blockage, hormone ablation, and chemotherapy. *Cancer Res.* **60:** 4556–4560.

Hartford A.C., Gohongi T., Fukumura D., and Jain R.K. 2000. Irradiation of a primary tumor, unlike surgical removal, enhances angiogenesis suppression at a distal site: Potential role of host-tumor interaction. *Cancer Res.* **60:** 2128–2131.

Helmchen F. and Denk W. 2002. New developments in multiphoton microscopy. *Curr. Opin. Neurobiol.* **12:** 593–601.

Helmlinger G., Yuan F., Dellian M., and Jain R.K. 1997a. Interstitial pH and pO_2 gradients in solid tumors *in vivo*: High-resolution measurements reveal a lack of correlation. *Nat. Med.* **3:** 177–182.

Helmlinger G., Netti P.A., Lichtenbeld H.C., Melder R.J., and Jain R.K. 1997b. Solid stress inhibits the growth of multicellular tumor spheroids. *Nat. Biotechnol.* **15:** 778–783.

Heuser L.S. and Miller F.N. 1986. Differential macromolecular leakage from the vasculature of tumors. *Cancer* **57:** 461–464.

Higuchi H., Kurose I., Fukumura D., Han J.Y., Saito H., Miura S., Hokari R., Watanabe N., Zeki S., Yoshida M., Kitajima M., Granger D.N., and Ishii H. 1997. Active oxidants mediate IFN-alpha-induced microvascular alterations in rat mesentery. *J. Immunol.* **158:** 4893–4900.

Hobbs S.K., Monsky W.L., Yuan F., Roberts W.G., Griffith L., Torchilin V.P., and Jain R.K. 1998. Regulation of transport pathways in tumor vessels: Role of tumor type and microenvironment. *Proc. Natl. Acad. Sci.* **95:** 4607–4612.

House S.D. and Lipowsky H.H. 1988. In vivo determination of the force of leukocyte endothelium adhesion in the mesentaric microvasculature of the cat. *Circ. Res.* **63:** 658–666.

Huang D., Swanson E.A., Lin C.P., Schuman J.S., Stinson W.G., Chang W., Hee M.R., Flottet T., Gregory K., Puliafito C.A., and Fujimoto J.G. 1991. Optical coherence tomography. *Science* **254:** 1178–1181.

Ide A.G., Baker N.H., and Warren S.L. 1939. Vascularization of the Brown-Pearce rabbit epithelioma transplant as seen in the transparent ear chamber. *Am. J. Roentgenol.* **42:** 891–899.

Intaglietta M. and Tompkins W.R. 1973. Microvascular measurements by video image shearing and splitting. *Microvasc. Res.* **5:** 309–312.

Jain R.K. 1987. Transport of molecules in the tumor interstitium: A review. *Cancer Res.* **47:** 3038–3050.
——— . 1994. Barriers to drug delivery in solid tumors. *Sci. Am.* **271:** 58–65.
——— . 1998. The next frontier of molecular medicine: Delivery of therapeutics. *Nat. Med.* **4:** 655–657.

Jain R.K. and Padera T.P. 2002. Prevention and treatment of lymphatic metastasis by antilymphangiogenic therapy. *J. Natl. Cancer Inst.* **94:** 785–787.

Jain R.K., Munn L.L., and Fukumura D. 2001. Transparent window models and intravital microscopy. In *Tumor models in cancer research* (ed. B.A. Teicher), pp. 647–671. Humana Press, Totowa, New Jersey.
——— . 2002. Dissecting tumor pathophysiology using intravital microscopy. *Nat. Rev. Cancer* **2:** 266–276.

Jain R.K., Munn L.L., Fukumura D., and Melder R.J. 1998a. In vitro and in vivo quantification of adhesion between leukocytes and vascular endothelium. *Methods Mol. Med.* **18:** 553–575.

Jain R.K., Schlenger K., Hockel M., and Yuan F. 1997. Quantitative angiogenesis assays: Progress and problems. *Nat. Med.* **3:** 1203–1208.

Jain R.K., Safabakhsh N., Sckell A., Chen Y., Jiang P., Benjamin L., Yuan F., and Keshet E. 1998b. Endothelial cell death, angiogenesis, and microvascular function after castration in an androgen-dependent tumor: Role of vascular endothelial growth factor. *Proc. Natl. Acad. Sci.* **95:** 10820–10825.

Jeltsch M., Kaipainen A., Joukov V., Meng X., Lakso M., Rauvala H., Swartz M., Fukumura D., Jain R.K., and Alitalo K. 1997. Hyperplasia of lymphatic vessels in VEGF-C transgenic mice. *Science* **276:** 1423–1425.

Kaufman A.G. and Intaglietta M. 1985. Automated diameter measurement of vasomotion by cross-correlation. *Int. J. Microcirc. Clin. Exp.* **4:** 45–53.

Kerbel R.S. 2000. Tumor angiogenesis: Past, present and the near future. *Carcinogenesis* **21:** 505–515.

Kleinfeld D., Mitra P.P., Helmchen F., and Denk W. 1998. Fluctuations and stimulus-induced changes in blood flow observed in individual capillaries in layers 2 through 4 of rat neocortex. *Proc. Natl. Acad. Sci.* **95:** 15741–15746.

Kleinman L.I. and Radford J., E.P. 1964. Ventilation standards for small mammals. *J. Appl. Physiol.* **19:** 360–362.

Kligerman M.M. and Henel D.K. 1961. Some aspects of the microcirculation of a transplantable experimental tumor. *Radiology* **76:** 810–817.

Klintworth G.K. 1991. *Corneal angiogenesis: A comprehensive critical review.* Springer Verlag, New York.

Koike N., Fukumura D., Gralla O., Au P., Schechner J.S., and Jain R.K. 2004. Tissue engineering: Creation of long-lasting blood vessels. *Nature* **428:** 138–139.

Koop S., MacDonald I., Luzzi K., Schmidt E., Morris V., Grattan M., Khokha T., Chambers A., and Groom A. 1995. Fate of melanoma cells entering the microcirculation: Over 80% survive and extravasate. *Cancer Res.* **55:** 2520–2523.

Kubes P., Suzuki M., and Granger D.N. 1991. Nitric oxide: An endogenous modulator of leukocyte adhesion. *Proc. Natl. Acad. Sci.* **88:** 4651–4655.

Kurose I., Miura S., Fukumura D., and Tsuchiya M. 1993a. Mechanisms of endothelin-induced macromolecular leakage in microvascular beds. *Eur. J. Pharmacol.* **250:** 85–94.

Kurose I., Fukumura D., Miura S., Suematsu M., Sekizuka E., Nagata H., and Tsuchiya M. 1993b. Nitric oxide mediates vasoactive effects of endothelin-3 on rat mesenteric microvascular beds *in vivo. Angiology* **44:** 483–490.

Lahiri S., Rumsey W.L., Wilson D.F., and Iturriaga R. 1993. Contribution of *in vivo* microvascular pO_2 in the cat carotid body chemotransduction. *J. Appl. Physiol.* **75:** 1035–1043.

Lee C.G., Heijn M., di Tomaso E., Griffon-Etienne G., Ancukiewicz M., Koike C., Park K.R., Ferrara N., Jain R.K., Suit H.D., and Boucher Y. 2000. Anti-vascular endothelial growth factor treatment augments tumor radiation response under normoxic or hypoxic conditions. *Cancer Res.* **60:** 5565–5570.

Leighton J. 1967. *The spread of cancer. Pathogenesis, experimental methods, interpretations.* Academic Press, New York.

Lendvai B., Stern E.A., Chen B., and Svoboda K. 2000. Experience-dependent plasticity of dendritic spines in the developing rat barrel cortex in vivo. *Nature* **404:** 876–881.

Leu A.J., Berk D.A., Yuan F., and Jain R.K. 1994. Flow velocity in the superficial lymphatic network of the mouse tail. *Am. J. Physiol.* **267:** H1507– H1513.

Leu A.J., Berk D.A., Lymboussaki A., Alitaro K., and Jain R.K. 2000. Absent of functional lymphatics within a murine sarcoma: A molecular functional evaluation. *Cancer Res.* **60:** 4324–4327.

Leunig M., Goetz A.E., Dellian M., Zetterer G., Gamarra F., Jain R.K., and Messmer K. 1992a. Interstitial fluid pressure in solid tumors following hyperthermia: Possible correlation with therapeutic response. *Cancer Res.* **52:** 487–490.

Leunig M., Yuan F., Menger M.D., Boucher Y., Goetz A. E., Messmer K., and Jain R.K. 1992b. Angiogenesis, microvascular architecture, microhemodynamics, and interstitial fluid pressure during early growth of human adenocarcinoma LS174T in SCID mice. *Cancer Res.* **52:** 6553–6560.

Li C.-Y., Shan S., Huang Q., Braun R.D., Lanzen J., Hu K., Lin P., and Dewhirst M.W. 2000. Initial stages of tumor cell-induced angiogenesis: Evaluation via skin window chambers in rodent models. *J. Natl. Cancer Inst.* **92:** 143–147.

Lichtenbeld H.C., Ferarra N., Jain R.K., and Munn L.L. 1999. Effect of local anti-VEGF antibody treatment on tumor microvessel permeability. *Microvasc. Res.* **57:** 357–362.

Lichtenbeld H.C., Yuan F., Michel C.C., and Jain R.K. 1996. Perfusion of single tumor microvessels: Application to vascular permeability measurement. *Microcirculation* **3:** 349–357.

Lipowsky H.H. and Zweifach B.W. 1978. Applications of the "two-slit" photometric technique to the measurement of microvascular volumetric flow rates. *Microvasc. Res.* **15:** 93–101.

Luckè B. and Schlumberger H. 1939. The manner of growth of frog carcinoma, studied by direct microscopic examination of living intraocular transplants. *J. Exp. Med.* **70:** 257–268.

Lutz B.R., Fulton G.P., Patt D.I., and Handler A.H. 1950. The growth rate of tumor transplants in the cheek pouch of the hamster (*Mesocricetus auratus*). *Cancer Res.* **10:** 231–232.

Maiorana A. and Gullino P.M. 1978. Acquisition of angiogenic capacity and neoplastic transformation in the rat mammary gland. *Cancer Res.* **38:** 4409–4414.

Martin G.R. and Jain R.K. 1993. Fluorescence ratio imaging measurement of pH gradients: Calibration and application in normal and tumor tissues. *Microvasc. Res.* **46:** 216–230.

———. 1994. Noninvasive measurement of interstitial pH profiles in normal and neoplastic tissue using fluorescence ratio imaging microscopy. *Cancer Res.* **54:** 5670–5674.

McDonald I.C., Schmidt E.E., Morris V.L., Chambers A.F., and Groom A.C. 1992. Intravital videomicroscopy of the chorioallantoic microcirculation: A model system for studying metastasis. *Microvasc. Res.* **44:** 185–199.

Melder R.J., Yuan J., Munn L.L., and Jain R.K. 2000. Erythrocytes enhance lymphocyte rolling and arrest in vivo. *Microvasc. Res.* **59:** 316–322.

Melder R.J., Koenig G.C., Witwer B.P., Safabakhsh N., Munn L.L., and Jain R.K. 1996. During angiogenesis, vascular endothelial growth factor and basic fibroblast growth factor regulate natural killer cell adhesion to tumor endothelium. *Nat. Med.* **2:** 992–997.

Milstone D.S., Fukumura D., Padget R.C., O'Donnell P.E., Davis V.M., Benavidez O.J., Monsky W.L., Melder R.J., Jain R.K., and Gimbrone J.M.A. 1998. Mice lacking E-selectin show normal numbers of rolling leukocytes but reduced leukocyte stable arrest on cytokine-activated microvascular endothelium. *Microcirculation* **5:** 153–171.

Monsky W.L., Carreira C.M., Tsuzuki Y., Gohongi T., Fukumura D., and Jain R.K. 2002. Role of host microenvironment in angiogenesis and microvascular functions in human breast cancer xenografts: Mammary fat pad vs. cranial tumors. *Clin. Cancer Res.* **8:** 1008–1013.

Monsky W.L., Fukumura D., Gohongi T., Chen Y., Yuan F., Kristensen C., and Jain R.K. 1998. Novel orthotopic models demonstrate effect of host microenvironment on human mammary carcinoma microcirculation. *Proc. Am. Assoc. Cancer Res.* **39:** 376–377.

Morris V.L., MacDonald I.C., Koop S., Schmidt E.E., Chambers A.F., and Groom A.C. 1993. Early interactions of cancer cells with the microvasculature in mouse liver and muscle during hematogenous metastasis: Videomicroscopic analysis. *Clin. Exp. Metast.* **11:** 377–390.

Muthukkaruppan V.R. and Auerbach R. 1979. Angiogenesis in the mouse cornea. *Science* **205:** 1416–1418.

Naumov G.N., Wilson S.M., MacDonald I.C., Schmidt E.E., Morris V.L., Groom A.C., Hoffman R.M., and Chambers A.F. 1999. Cellular expression of green fluorescent protein, couple with high-resolution in vivo videomicroscopy, to monitor steps in tumor metastasis. *J. Cell. Sci.* **112:** 1835–1842.

Netti P.A., Berk D.A., Swartz M.A., Grodzinsky A.J., and Jain R.K. 2000. Role of extracellular matrix assembly in interstitial transport in solid tumors. *Cancer Res.* **60:** 2497–2503.

Nguyen M., Shing Y., and Folkman J. 1994. Quantitation of angiogenesis and antiangiogenesis in the chick embryo chorioallantoic membrane. *Microvasc. Res.* **47:** 31–40.

Norrby K., Jakobsson A., and Sörbo J. 1990. Quantitative angiogenesis in spreads of intact rat mesenteric windows. *Microvasc. Res.* **39:** 341–348.

Nugent L.J. and Jain R.K. 1984. Extravascular diffusion in normal and neoplastic tissues. *Cancer Res.* **44:** 238–244.

Padera T.P., Stoll B.R., So P.T.C., and Jain R.K. 2002a. High-speed intravital multiphoton laser scanning microscopy of microvasculature, lymphatics, and leukocyte-endothelial interactions. *Mol. Imaging* **1:** 9–15.

Padera T.P., Stoll B.R., Tooredman J.B., Capen D., diTomaso E., and Jain R.K. 2004. Cancer cells compress intratumor vessels. *Nature* **427:** 695.

Padera T.P., Kadambi A., diTomaso E., Carreira C.M., Brown E.B., Munn L.L., and Jain R.K. 2002b. Lymphatic metastasis in the absence of functional intratumor lymphatics. *Science* **296:** 1883–1886.

Pluen A., Boucher Y., Ramanujan S., McKee T.D., Gohongi T., di Tomaso E., Brown E.B., Izumi Y., Campbell R.B., Berk D.A., and Jain R.K. 2001. Role of tumor-host interactions in interstitial diffusion of macromolecules: Cranial vs. subcutaneous tumors. *Proc. Natl. Acad. Sci.* **98:** 4628–4633.

Ramanujan S., Pluen A., McKee T.D., Brown E.B., Boucher Y., and Jain R.K. 2002. Diffusion and convection in collagen gels: Implications for transport in the tumor interstitium. *Biophys. J.* **83:** 1650–1660.

Reinhold H.S. 1971. Improved microcirculation in irradiated tumours. *Eur. J. Cancer* **7:** 273–280.

Reinhold H.S., Blachiwiecz B., and Blok A. 1977. Oxygenation and reoxygenation in "Sandwich" tumours. *Bibl. Anat.* **15:** 270–272.

Sandison J.C. 1924. A new method for the microscopic study of living growing tissues by the introduction of a transparent chamber in the rabbit's ear. *Anat. Rec.* **28:** 281–287.

Sasaki A., Melder R.J., Whiteside T.L., Herberman R.B., and Jain R.K. 1991. Preferential localization of human adherent lymphokine-activated killer cells in tumor microcirculation. *J. Natl. Cancer Inst.* **83:** 433–437.

Schlenger K., Höckel M., Schwab R., Frischmann-Berger R., and Vaupel P. 1994. How to improve the uterotomy healing. I. Effects of fibrin and tumor necrosis factor-α in the rat uterotomy model. *J. Surg. Res.* **56:** 235–241.

Schmidt J., Ryschich E., Maksan S., Werner J., Gebhard M., Herfarth C., and Klar E. 1999. Reduced basal and stimulated leukocyte adherence in tumor endothelium of experimental pancreatic cancer. *Int. J. Pancreatol.* **26:** 173–179.

Shioda T., Munn L., Fenner M., Jain R., and Isselbacher K. 1997. Early events of metastasis in the microcirculation involve changes in gene expression of cancer cells: Tracking mRNA levels of metastasizing cancer cells in the chick embryo chorioallantoic membrane. *Am. J. Pathol.* **150:** 2099–2112.

Sholley M.M., Ferguson G.P., Seibel H.R., Montour J.L., and Wilson J.D. 1984. Mechanisms of neovascularization. Vascular sprouting can occur without proliferation of endothelial cells. *Lab. Invest.* **51:** 624–634.

Suzuki T., Yanagi K., Ookawa K., Hatakeyama K., and Ohshima N. 1998. Blood flow and leukocyte adhesiveness are reduced in the mirocirculation of a peritoneal disseminated colon carcinoma. *Ann. Biomed. Eng.* **26:** 803–811.

Swartz M.A., Berk D.A., and Jain R.K. 1996. Transport in lymphatic capillaries. I. Macroscopic measurements using residence time distribution theory. *Am. J. Physiol.* **270:** H324–H329.

Taylor C.M. and Weiss J.B. 1984. The chick vitelline membrane as a test system for angiogenesis and antiangiogenesis. *Int. J. Microcirc. Clin Exp.* **3:** 337. (Abstr.)

Toolan H.W. 1954. Transplantable human neoplasms maintained in cortisone-treated laboratory animals: H.S. #1; H.Ep. #1; H.Ep. #2; H.Ep. #3; and H.Emb.Rh. #1. *Cancer Res.* **14:** 660–666.

Torres Filho I.P. and Intaglietta M. 1993. Microvessel PO2 measurements by phosphorescence decay method. *Am. J. Physiol.* **265:** H1434–H1438.

Torres Filho I.P., Leunig M., Yuan F., Intaglietta M., and Jain R.K. 1994. Noninvasive measurement of microvascular and interstitial oxygen profiles in a human tumor in SCID mice. *Proc. Natl. Acad. Sci.* **91:** 2081–2085.

Tsai A.G., Friesenecker B., Mazzoni M.C., Kerger H., Buerk D.G., Johnson P.C., and Intaglietta M. 1998. Microvascular and tissue oxygen gradients in the rat mesentery. *Proc. Natl. Acad. Sci.* **95:** 6590–6595.

Tsuzuki Y., Carreira C.M., Bockhorn M., Xu L., Jain R.K., and Fukumura D. 2001. Pancreas microenvironment promotes VEGF expression and tumor growth: Novel window models for pancreatic tumor angiogenesis and microcirculation. *Lab. Invest.* **81:** 1439–1452.

Tsuzuki Y., Fukumura D., Oosthuyse B., Koike C., Carmeliet P., and Jain R.K. 2000. Vascular endothelial growth factor (VEGF) modulation by targeting hypoxia inducible factor-1α → Hypoxia response element → VEGF cascade differentially regulates vascular response and growth rate in tumors. *Cancer Res.* **60:** 6248–6252.

Weissleder R. 2001. A clearer vision for in vivo imaging. *Nat. Biotechnol.* **19:** 316–317.

Weissleder R. and Ntziachristos V. 2003. Shedding light onto live molecular targets. *Nat. Med.* **9:** 123–128.

Willett C.G., Boucher Y., diTomaso E., Duda D.G., Munn L.L., Tong R., Chung D.C., Sahani D.V., Kalva S. P., Kozin S.V., Mino M., Cohen K.S., Scadden D.T., Hartford A.C., Fischman A.J., Clark J.W., Ryan D. P., Zhu A.X., Blaszkowsky L.S., Shellito P.C., Lauwers G.Y., and Jain R.K. 2004. Direct evidence that the anti-VEGF antibody Bevacizumab has anti-vascular effects in human rectal cancer. *Nat. Med.* **10:** 145–147.

Wood S. 1958. Pathogenesis of metastasis formation observed in vivo in the rabbit ear chamber. *Arch. Pathol.* **66:** 550–568.

Yamaura H., Suzuki M., and Sato H. 1971. Transparent chamber in the rat skin for studies on microcirculation in cancer tissue. *Gann* **62:** 177–185.

Yanagi K. and Ohsima N. 1996. Angiogenic vascular growth in the rat peritoneal disseminated tumor model. *Microvasc. Res.* **51:** 15–28.

Yuan F., Leunig M., Berk D.A., and Jain R.K. 1993. Microvascular permeability of albumin, vascular surface area, and vascular volume measured in human adenocarcinoma LS174T using dorsal chamber in SCID mice. *Microvasc. Res.* **45:** 269–289.

Yuan F., Salehi H.A., Boucher Y., Vasthare U.S., Tuma R.F., and Jain R.K. 1994. Vascular permeability and microcirculation of gliomas and mammary carcinomas transplanted in rat and mouse cranial windows. *Cancer Res.* **54:** 4564–4568.

Yuan F., Dellian M., Fukumura D., Leunig M., Berk D.A., Torchilin V.P., and Jain R.K. 1995. Vascular permeability in a human tumor xenograft: Molecular size dependence and cutoff size. *Cancer Res.* **55:** 3752–3756.

Zawicki D.F., Jain R.K., Schmid-Schoenbein G.W., and Chien S. 1981. Dynamics of neovascularization in normal tissue. *Microvasc. Res.* **21:** 27–47.

Zeidman I. 1961. The fate of circulating tumor cells. 1. Passage of cells through capillaries. *Cancer Res.* **21:** 38–39.

CHAPTER TWENTY FIVE

Development of Mammalian Cell Lines with *lac* Operator-tagged Chromosomes

Yuri G. Strukov and Andrew S. Belmont

Department of Cell and Structural Biology and Biophysics Program, University of Illinois, Urbana-Champaign, Illinois 61801

INTRODUCTION

THE DISCOVERY AND USE OF FLUORESCENT PROTEINS to label chromosomes has had a large impact in the fields of chromatin and chromosome structure and dynamics. Although the application of fluorescent proteins to the study of cytoplasmic proteins has provided invaluable insight into protein, cytoskeleton, and organelle dynamics, for several reasons the application of fluorescent proteins to tagging chromosomal proteins yields basic structural information, previously inaccessible, in additon to insight into dynamics.

During interphase, chromosomes generally have insufficient contrast to be visualized by differential interference contrast (DIC) or phase-contrast microscopy. Moreover, chromatin is exceptionally sensitive to small ionic variations in buffer conditions, such that dramatic changes in conformation, as visualized by light microscopy, may occur during the process of fixation and/or staining (Belmont et al. 1989). From a practical point of view, this has meant that different investigators studying the same cells from the same organism can come to very different conclusions regarding their chromatin structure on the basis of millimolar differences in the cation concentrations of the isolation buffers used. This has led the chromatin field to be very suspicious about conclusions and models based largely on cytological data from fixed cells. The integrity of chromatin structure is further compromised by the additional manipulations required for immunostaining or preparation of fixed samples for electron microscopy and escalates considerably with the use of in situ hybridization (requiring DNA denaturation procedures) to examine specific chromosomal regions. Finally, the extremely high density and complexity of higher-order chromatin folding is ill-suited to the resolution limits of light microscopy, and ultrastructural analysis by electron microscopy in well-fixed unperturbed samples, without detergent permeabilization of cells, is quite difficult due to the lack of good DNA-specific stains for electron microscopy (Belmont 1998).

The application of fluorescent proteins to image chromosomal proteins in live cells, combined with spectroscopic methods such as fluorescence recovery after photobleaching/fluorescence loss in photobleaching (FRAP/FLIP) and fluorescence resonance energy transfer (FRET), has provided an unprecedented view of nuclear and chromosomal structure and dynamics, summarized in a number of recent reviews (Misteli 2001; Reits and Neefjes 2001). This chapter focuses on methods for tagging specific chromosome sites using direct repeats of bacterial operator sequences coupled with green fluorescent protein (GFP)-tagged proteins that recognize these sequences. Specific protocols are provided for the isolation of cell lines with tagged chromosome loci, either as small transgene insertions or as large selectively labeled chromosome regions. The use of selectively labeled chromosome sites makes it possible to visualize large-scale chromatin structure in vivo. The method partially overcomes the complexity of chromatin folding by visualizing a small labeled region for which out-of-focus blur is minimized. This methodology facilitates the analysis in vivo of structural changes in chromosome structure as a function of cell cycle progression, transcriptional activation, and the spatial organization and timing of gene expression in tissue culture cells and tissues. Combined with the appropriate choice of activator-linked fusion protein and/or *cis*-linked transgenes, this method can also be used to observe in vivo the recruitment of transcription factors and coactivators (e.g., see Chapter 26).

OVERVIEW OF TECHNIQUES FOR THE CREATION AND OBSERVATION OF SPECIFICALLY LABELED CHROMATIN REGIONS IN LIVE CELLS

In principle, chromosome tagging can be accomplished through the use of any repetitive sequence to which a specific protein binds. The first application of this approach used the bacterial *lac* operator and Lac repressor combination (Robinett et al. 1996; Straight et al. 1996). The *lac* operator/Lac repressor-labeling system has been successfully adapted for tagging chromatin in live bacteria, yeast, and mammalian tissue culture cells (Fig. 1), as well as in multicellular organisms including *Caenorhabditis elegans*, *Drosophila*, and *Arabidopsis* (Belmont 2001). Subsequent approaches have used *tet* operator or glucocorticoid receptor element repeats (Michaelis et al. 1997; McNally et al. 2000; Becker et al. 2002). Recently, the *lac* operator system has been utilized in combination with the bacteriophage viral replicase translational operator and several color variants of GFP to label DNA, RNA, and protein in living cells (see Chapter 26) (Janicki et al. 2004).

Here, we focus on our experience using the *lac* operator/repressor system. Direct *lac* operator repeats were generated by a directional cloning method in which the number of direct repeats was doubled with each cloning cycle. Although the use of direct repeats, as opposed to inverted repeats, reduces recombination within the bacterial host, at high copy number, even the direct repeats are unstable, requiring the use of special bacterial hosts and low-copy-number plasmids for cloning. Introduction of the *lac* operator repeats into eukaryotic cells uses traditional transformation methods (e.g., see Spector et al. 1998; Sambrook and Russell 2001). Methods for the isolation of stable cell clones with varying transgene copy numbers are described later in this chapter. A key question is the number of repeats required for detection. Detection of the *lac* operator repeats is a question of both sensitivity and background and thus we discuss how this depends on the nature of the application. Finally, we describe several methods we have developed for visual screening of large numbers of stable cell clones to isolate rare clones containing labeled chromosomal regions with desired features.

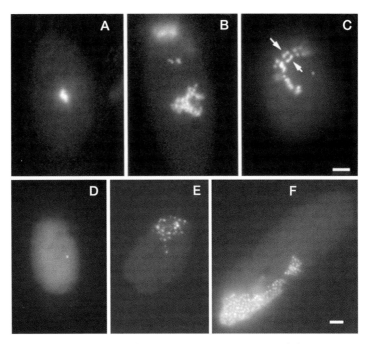

FIGURE 1. Examples of different stable cell lines with *lac* operator-tagged chromosome regions visualized by GFP-Lac repressor in vivo expression. (*A–C*) Three different gene-amplified cell lines, derived from multicopy plasmid insertions, each with different characteristic conformations of the *lac* operator-tagged chromosome region. (*A*) A03 cells with a heterochromatic-like amplified chromosome region. A 90-Mbp region forms an ~1-µm diameter spot during most of interphase, not much larger than its ~1.3µm metaphase chromosome length. (*B*) PDC cell line in which the amplified chromosome region appears as coiled, large-scale chromatin fibers by light and electron microscopy. (*C*) Bb cell line in which extended large-scale chromatin fibers are observed. In this example, what appear to be two sister chromatids are shown. (*D–E*) Different stages of gene amplification. (*D*) EP1-4 cell line containing a single-copy insertion of plasmid transgene containing *lac* operator direct repeat. (*E,F*) Two different stages of gene amplification using methotrexate selection. Chromosome regions consisting of clusters of individual dots are observed. These increase in number with increased amplification between *E* and *F*. Bars: (*A–C*), 1 µm; (*D–F*), 2 µm.

RECOMBINANT DNA CLONING OF DIRECT REPEATS

Details of the directional cloning scheme used to generate the *lac* operator direct repeats are described in detail elsewhere (Robinett et al. 1996; Belmont et al. 1999a). The approach requires the use of three restriction sites, two of which have compatible sticky ends. The stability of direct repeats appears to vary depending on the particular sequence, but in all cases, plasmids carrying a high number of direct repeats become unstable with respect to recombination. This instability appears to correlate with the copy number of the cloning vector, with very little instability present in single-copy bacterial artificial chromosome (BAC) vectors and the greatest instability in high-copy-number plasmids such as pUC18. The choice of a cloning vector therefore becomes a compromise between the repeat instability and the higher yield of plasmids with high-copy-number origins. Our experience with the *lac* operator has been that repeat instability was not a significant problem in a high-copy pUC18 plasmid derivative containing up to 64 *lac* operator copies, but for practical purposes, use of a lower-copy-number plasmid was required for greater than 128 copies. For typical cloning purposes, we now use a moderate-copy-number plasmid containing the pMB1 origin from pBR322, which is maintained at approximately 20 copies per cell. Bacterial strains that minimize recombination of direct repeats are used, including STBL2 cells from Invitrogen, which we use routinely, and EC100 from Epicentre Technologies (Madison, Wisconsin). Bacterial cell growth for plasmid isolation is achieved in small steps versus a large single preparation. Typically, DNA minipreps are used for multiple overnight cultures inoculated from individual bacterial colonies. Restriction endonuclease digestion is used to test each miniprep for correct repeat sizes, and DNA from cultures that did not undergo recombination are pooled. Alternatively, overnight cultures with intact repeats are used to inoculate larger volumes of LB medium for maxipreps (see the protocol, Preparing Vector DNA Containing Large Direct Repeats). At high repeat copy number (i.e., 256 *lac* operator copies), typically, one third to one half of overnight cultures prepared from colonies from a freshly streaked plate contain full-length repeats. This percentage is reduced after DNA transformation. Our experience is that electroporation-mediated transformation leads to significant repeat instability preventing cloning. Therefore, we use heat shock for transformation.

Detection Limits for Repeats

Using the directional cloning method, we started with a *lac* operator 8-mer, doubling this repeat length with each cloning cycle, producing repeat lengths of 16, 32, 64, 128, and 256 copies. The question then becomes: What number of repeats is required for detection? In vivo detection of the *lac* operator repeats uses an enhanced GFP (EGFP) fusion with a modified Lac repressor protein. The wild-type Lac repressor binds as a tetramer to two *lac* operator sites. To remove the capability for cross-linking between *lac* operator sites, we use an 11-amino-acid carboxy-terminal truncation, which prevents tetramerization, producing a dimer Lac repressor capable of binding only one *lac* operator site. In addition, the Lac repressor protein is fused to a nuclear localization signal which concentrates the EGFP-Lac repressor protein within the cell nucleus.

Detection of the *lac* operator repeats is therefore not a question of sensitivity per se, but of the ratio between signal strength and the nuclear background, which is a function of the expression level of the EGFP-Lac repressor construct. In mammalian Chinese hamster ovary (CHO) cells, a single copy of the 256-copy direct repeat is easily detected with a signal-to-noise ratio that would predict the ability to detect constructs containing far fewer *lac* operator repeats (severalfold lower) (Robinett et al. 1996). We stress, however, that this depends on the proper expression level of the EGFP-Lac repressor. Expression must be high enough to produce sufficient repressor binding, yet low enough to produce a nuclear background fluorescence intensity less than that over the *lac* operator repeat. Transient expression produces a high percentage of cells with a nuclear background level too bright for detection of single-copy direct-repeat insertions. Stable transformants typically express at tenfold or lower levels, allowing selection of cell clones with suitable expression levels. In practice, we have found optimal EGFP-repressor expression in cells that showed a low-level nuclear fluorescence to be just sufficient for detection with our imaging system. To obtain such expression levels, it is useful to avoid expression vectors with strong cytomegalovirus (CMV) promoters. The best results have been obtained using the p3′SS vector, originally from Stratagene, containing the weaker F9-1 polyoma promoter (Fieck et al. 1992). In theory, use of a tight-binding Lac repressor mutant might be predicted to improve detection sensitivity by producing the same signal against a lower nuclear background. Our experience with such a mutant (the proline 3 → tyrosine mutant of the Lac repressor), however, was that the nuclear background became less homogeneous, producing artifactual point-like GFP staining, which interfered with detection of single-copy *lac* operator direct repeats (M. Lakonishok and A. Belmont, unpubl.).

The above results suggest a lower limit for repeat detection, but they do not necessarily indicate the optimal level. This depends to a great extent on the actual experimental goals and design. For instance, on our imaging system, a 1–2-second exposure of an optimally focused optical section was required to obtain a good signal-to-noise ratio for a single copy of the 256-copy direct repeat (due largely to charge-coupled device [CCD] readout noise). Analysis of interphase chromosome dynamics might therefore be better accomplished using several copies of the 256-copy *lac* operator repeat (Chubb et al. 2002). For other research objectives, we have focused instead on cell lines containing tens to hundreds of copies of a *lac* operator repeat-tagged transgene. In this case, the use of 64-copy direct *lac* operator repeats still yielded strong, easily detected signals (S. Dietzel and A. Belmont, unpubl.).

It is impossible at this point to extrapolate our experience with *lac* operator repeats to other repeats. A GFP-glucocorticoid receptor (GR) has been used to detect a transgene array containing hundreds of copies of a 7-kb construct containing a *ras* gene driven by a mouse-mammary tumor virus (MMTV) promoter containing four to six glucocorticoid receptor elements (GRE) (McNally et al. 2000). The occupancy of these GREs is near 100% (Dundr et al. 2002). In contrast, we estimate that the typical occupancy in our mammalian cells for the *lac* operator repeats, at least for the A03 cell line (Li et al. 1998), is less than 10%. This estimate is based on an upper limit for occupancy established by comparing the maximum observed GFP fluorescence, after microinjection of very high EGFP-Lac repressor protein concentrations, with the fluorescence observed after EGFP-Lac repressor transient or stably transfected cells (T. Tumbar and A. Belmont, unpubl.). It is very likely that there are also position effects which determine occupancy levels of the *lac* operator repeats in a given chromosomal context.

Establishing Stable Cell Lines Containing *lac* Operator Repeats and Expressing the GFP-Lac Repressor

The *lac* operator/repressor tagging system requires stable cell lines containing the *lac* operator repeats and either transient or stable expression of the EGFP-Lac repressor. Investigators can either first establish a stable cell line expressing the EGFP-Lac repressor and then introduce the construct containing the *lac* operator repeats or instead first create stable cell lines containing the *lac* operator repeat. There are advantages to both approaches. The first approach allows rapid visual screening of cell clones containing *lac* operator insertions. This is particularly useful for selection of rare insertion events, for example, cell clones with transgene insertions of unusual size, conformation, or nuclear location. The second approach provides control cell lines that contain the *lac* operator repeats but do not express GFP-Lac repressor. These can be used for the generation of derivative lines expressing Lac repressor fused with different color GFP variants or newer fluorescent proteins and/or other proteins, such as transcriptional activators or repressors (Tumbar et al. 1999; Tsukamoto et al. 2000; Tumbar and Belmont 2001; Ye et al. 2001; Nye et al. 2002; Li et al. 2003).

A number of methods are used for introducing vector DNA into target eukaryotic cells, although there is little or no understanding of the actual mechanisms and pathways of DNA uptake. It is known that cells have several intrinsic mechanisms by which they can spontaneously take up large macromolecules from ambient medium irrespective of the cell type; endocytosis and pinocytosis are two examples. The backbone of DNA is negatively charged, as are most plasma membranes. To facilitate absorption of DNA by cells during a transfection procedure, the effect of electrostatic repulsion forces between DNA and membrane must be avoided or neutralized.

During calcium phosphate transfection, DNA forms an insoluble precipitate with hydroxyapatite, which is taken into cells by endocytosis. This method has been widely used for stable integration of vector DNA into host genomes for more than two decades. Lipofection (e.g., transfection with Lipofectamine and Fugene) takes advantage of the property of lipids to assemble into micelle-like structures. Micelles made of cationic lipids can spontaneously interact with vector DNA, forming macromolecular complexes. Positively charged lipid micelles shield vector DNA from the membrane charges and are able to fuse with cell membranes. Positively charged proteins and polypeptides have also been used to coat vector DNA, enabling the coated DNA to effectively penetrate cell membranes. Electroporation is based on the phenomenon of transient and reversible permeabilization of cell membranes with short pulses of high-strength electric fields. DNA reaches the nucleus by diffusion through breaks in membranes caused by electric pulses. Finally, microinjection is a method in which a solution of vector DNA is directly introduced into target cells or nuclei.

We have used electroporation to introduce single-copy insertions of our *lac* operator-tagged transgenes (Robinett et al. 1996). In this case, linear DNA was used to force integration of the vector with a specific linear-sequence orientation. We have used calcium-phosphate-mediated transfection and both Lipofectamine (GIBCO-BRL) and Fugene (Roche) for introduction of multiple vector copies into chromosomes. Typically, tens of copies are introduced with these methods, with a smaller fraction of stable transformants containing hundreds to thousands of copies. Transformation can be done with either circular or linear templates, but the investigator must keep in mind that the circular template will be integrated into the host-cell chromosome by breaking the circular plasmid at a random site. Vectors integrated as multiple copies appear to have similar breakpoints, based on underlying recombination methods, being integrated as direct or inverted repeats (Barsoum 1990). Our experience, however, is that these multiple copies are usually interspersed by large amounts of unknown genomic DNA (Y. Strukov and A. Belmont, unpubl.). The advantage of transforming with linear DNA is that vector sequences can be removed and a fixed linear layout of the template DNA can be chosen by the appropriate choice of restriction enzyme.

Selectable markers for the selection of stable transformants can be added directly to the GFP-repressor expression vector or the vector containing the *lac* operator repeats. Alternatively, DNA containing the selectable marker can be cotransfected using methods that integrate multiple vector copies. By adjusting the ratio of selectable marker to vector containing the *lac* operator repeats, this contransfection method could be used to bias selection of stable clones with high copy numbers of the vector containing the *lac* operator.

For eukaryotic cells, delivery of vector DNA into the cytoplasm is not sufficient for transfection, because the nuclear membrane is a significant barrier for exogenous DNA on its way into the nucleus. There is evidence that vector DNA uptake occurs in mitosis and hence transfection efficiency correlates with the percentage of mitotic cells in the culture (Wilke et al. 1996; Pellegrin et al. 2002). This is why it is important that cells be in their log phase of growth and of optimum confluency at the moment of transfection. Fresh tissue culture medium should be added to the cells the night before and again a few hours before transfection. Sample transfection protocols are provided later in this chapter.

Further increases in the size of the tagged chromosome region can be accomplished using gene amplification. We have used a dihydrofolate reductase (DHFR) cDNA transgene in *cis* with the *lac* operator repeat. Using CHO DG44 cells, containing a double X-ray-induced deletion of the endogenous *DHFR* gene (Urlaub et al. 1986), the *DHFR* transgene can be utilized directly as the selectable marker using medium prepared without thymidine or hypoxanthine, but containing dialyzed fetal bovine serum. Gene amplification is accomplished by progressive selection with increasing concentrations of methotrexate, an inhibitor of *DHFR*, as described elsewhere (Belmont et al. 1999b). Typically, gene-amplified chromosome regions are highly unstable in their size and chromosome location. This contrasts with the generally stable behavior of insertions after direct transformation without gene amplification. Interestingly, though, the general conformational properties of the amplified chromosomal region, whether the region is unusually condensed or decondensed, appear to be preserved even with chromosome rearrangements. In a few cases, more stable, labeled chromosome regions have been isolated (Li et al. 1998). In addition, it has been possible in some cases to select a subclone with a stable amplified chromosome region after prolonged passage (i.e., for a number of months) in medium without methotrexate (Dietzel and Belmont 2001).

SUBCLONING STABLE TRANSFORMANTS

Typically, approximately one half of the stable clones transfected with our expression vector for EGFP-Lac repressor show expression. Presumably, the remaining clones represent cases in which linearization of the transfected circular plasmid during transformation resulted in disruption of the EGFP-lac repressor transgene. In this case, a small number of clones must be screened in order to obtain one with an appropriate expression level. In cases in which the Lac repressor is used to tether other protein domains to a chromosome site via a Lac repressor fusion protein, however, there is frequently a strong selective bias against expression of the fusion protein. One example from our work is an EGFP–Lac repressor–VP16 acidic activation domain fusion protein (Tumbar and Belmont 2001). In this case, less than a few percent of stable clones show expression. Similarly, the isolation of clones with high-copy inducible transgene insertions, or gene-amplified chromosome regions with a particular size, chromatin conformation, and/or nuclear location, may require screening through large numbers of clones.

In general, screening clones is performed using either a visual, microscope-based screen (if possible) or a molecular or biochemical screen. In some cases, both might be combined, either as two sequential screens or one parallel screen. An example from our work is a screen of a set of 96 subclones transfected with a nonfluorescent VP16 transcriptional activator fusion protein. A visual immunofluorescence assay was used to test directly for expression of the activator, and a transient transfection expression assay was run to test the capability of each clone to activate a luciferase reporter gene (C. Chuang and A. Belmont, unpubl.). The intersection of these two screens yielded a clone that induced chromosome repositioning after induction of VP16 targeting.

A starting point for subcloning is to determine roughly what fraction of stable transformants show the desired properties. If this fraction is large, then we typically carry out the initial transformation in a single flask, followed by subsequent subcloning of the mixed stable transformant pool by serial dilution cloning into 96-well plates. A range of dilutions should be used to ensure success. The advantage of serial dilution is that the investigator can more easily verify whether the cells in a given well actually arise from a single clone.

If the fraction of desired cells is small, and if a visual screen is possible, we prepare a single coverslip from the mixed pool of stable transformants and obtain an initial estimate of exactly how many clones need to be screened. We have tried various approaches when the fraction of desired cells is very small, i.e., a few percent or less. One approach is to initially subclone in pools of clones (e.g., 5–10 per well). In theory, this should allow cloning of even rare transformants through several stages of subcloning. Our experience, however, is that often the desired subclones are rare because of a selective bias against growth of these clones. In this case, if the pool size is too great, the percentage of desired cells can drop in the culture prior to the next stage of subcloning, producing diminishing returns. A better approach, where possible, is cell sorting with flow cytometry to enrich the cell population with cells containing the desired expression levels (see the protocol, Cell Cloning with Enrichment of Stable Cells with Large-copy-number Inserts Using Fluorescence-activated Cell Sorting).

In cases where large numbers of clones are required, we turn to methods other than serial dilution. If visual screening of the clones is unnecessary, then we simply pick large numbers of colonies from a 150-mm-diameter Petri dish. Two simple methods for harvesting these colonies work well. The first uses small, trypsin-soaked pieces of filter paper to pick up each colony. The second method uses micropipette tips to directly pick each colony. Both are described below in the protocol, Subcloning Using Filter Paper or Micropipette Tips. These are used to deposit clonal cells into 96-well plates. Both methods are very rapid; however, both have a high risk of clone contamination, due to the use of a single Petri dish. Any clones selected, based on a secondary screen, must go through an additional round of clone "purification" using serial dilution.

In cases where live visual screening is desired, things become a little more complicated. An easy, low-cost method is to cover a large Petri dish (i.e., 60-mm diameter) with a layer of small, round coverslips (12-mm diameter). Cells are plated at low density such that one to five colonies grow per coverslip. Under sterile conditions, the coverslips are transferred to a Petri dish containing a hole over which a large glass coverslip is attached with vacuum grease and transported to the microscope for visual screening. Each colony on a given coverslip is screened. If a positive clone is found, then the coverslip is recovered under sterile conditions, and the region of the coverslip containing the desired colony is broken off with a diamond pen, and passaged. A more convenient, but more expensive, method is to pick up a large number of colonies, as described above, and place them into a 96-well, glass-covered microtiter dish (Whatman, Polyfiltronics). These dishes are available with number 1-1/2 thickness coverslips suitable for high-resolution imaging. After the cells in each well grow to a sufficient density, the 96-well plate can be directly screened using an inverted microscope. These plates, however, are expensive (about $25 per plate at the time of publication) and in high demand, so they may be back-ordered for 3–6 months at a time. We have also found that significant variability exists between batches of plates in their ability to support cloning at high efficiency. This appears to be related to contamination of the plates with residual glue. Our experience is that many lots of these plates support cell growth at high density but not at the limiting dilutions required for cloning.

IN VIVO MICROSCOPY

The sections above describe how stable cell lines suitable for imaging labeled chromosome regions are isolated. Other chapters in this book describe technical requirements for live cell imaging (see Chapters 1–3, and 17). Our experience, however, is that there remains one important consideration beyond the generation of cell lines for imaging and acquisition of imaging hardware and software. This involves the formulation of a general approach that will allow the investigator to determine whether the biological phenomenon under study can be successfully recapitulated during biological imaging. In our experience, in particular, phototoxicity remains a serious problem. This is especially true for experiments that exploit the labeling of small, selected chromosome regions, where high magnification, low signal intensity, and correct focus require significant photoexposure. Different biological phenomena show very different thresholds for phototoxicity. Whereas chromosome decondensation after mitosis or rapid constrained chromosome motions appear to be relatively photoinsensitive, long-range chromosome movements induced by targeting of VP16 are extremely photosensitive, and VP16-induced large-scale chromatin decondensation is somewhere in between.

It is thus important that most live cell imaging studies begin with a careful statistical survey using a fixed-cell approach. The goal here is to define a biological phenomenon using standard cytological approaches. Ideally, this description can be used to generate statistics, which can be used to monitor the reproducibility of the phenomenon in a corresponding live cell imaging experiment. The first goal of the live cell imaging approach is to determine whether these statistics can be reproduced under the imaging conditions used. If they are not reproducible, then there is an experimental problem, which can be traced either to the conditions used to grow the cells on the microscope stage or to phototoxicity. In the latter case, the statistical approach can be used to determine the maximum photoexposure permitted that still allows "the biology" to proceed normally.

If the phenomenon cannot be studied with fixed samples, for example, characterization of chromatin dynamics, then different approaches must be taken. At a minimum, we would like to demonstrate that the phenomenon being measured does not vary as a function of light exposure or time the cells spend being grown and imaged on the microscope stage, assuming that no time dependence is expected.

SAMPLE PROTOCOLS

It is assumed here that investigators are familiar with standard cloning and transfection procedures. For additional details on these procedures, see Sambrook and Russell (2001); for details of cell culture and manipulation, as well as transfection procedures, see Spector et al. (1998).

Preparing Vector DNA Containing Large Direct Repeats

This protocol is based on preparations from our laboratory using the pSV2-DHFR-8.32 plasmid containing a 256-copy *lac* operator repeat. This is a low-copy-number plasmid, so expect low yields at every purification step. It is important that once starting the protocol, it must be completed without delays, because leaving plates or cultures overnight at 4°C allows recombination to occur. Therefore, it should take no more than 3 days to complete (4 days if the optional second minipreps are performed [Step 8]). However, the plasmid DNA, once prepared from the cells, is stable.

1. From a stab culture, streak bacteria (STBL2 cells [GIBCO-BRL]) onto an LB plate containing 100 µg/ml ampicillin.

    The stab culture is stored at room temperature in the dark.

2. Incubate the inoculated LB plate for 24 hours at 30°C. Even after this length of time the colonies will be small.

3. Inoculate at least ten 3-ml cultures of LB containing 100 µg/ml ampicillin with well-isolated colonies. Incubate the cultures on a shaker overnight at 30°C.

4. Prepare DNA using a miniprep procedure (e.g., Qiagen) from 1.5 µl of each culture. Store the remainder of each culture at 4°C.

5. Digest aliquots of the DNA with *Sal*I and *Xho*I. Separate the DNA fragments on a 0.7% agarose gel.

    At this step, it is critical to be very discerning. Any bands present other than those expected could indicate recombination of the plasmid (in the case of the pSV2-DHFR-8.32 plasmid containing a 256-copy *lac* operator repeat, expect 10-kb and 5-kb fragments).

6. Prepare and freeze a glycerol stock or prepare a stab culture of each positive clone (from the stored cultures in Step 4).

7. Inoculate 500 ml of LB containing 100 µg/ml ampicillin with the remaining culture from one of the positive clones. Incubate the culture overnight at 30°C with shaking.

   It may be preferable to prepare two 500-ml cultures, each from a different positive clone, in the event that one clone is lost during large-scale growth.

8. *Optional:* Isolate DNA from an aliquot of the 500-ml culture.

   a. Repeat Steps 4 and 5 before proceeding with the maxiprep DNA isolation procedure.

   b. While preparing the miniprep DNA, centrifuge the 500-ml culture.

   c. Discard the supernatant and freeze the bacterial pellet until ready to use.

   d. Proceed with Step 9 using positive cultures.

9. Isolate DNA from the 500-ml culture using a maxiprep procedure.

   The yield of DNA will be low. A good yield is ~125 µg of DNA per 500-ml culture.

   If using Qiagen maxipreps with Qiafilters, double the volume of P1, P2, and P3 added to each pellet, and then split the prep into two Qiafilters. This prevents the Qiafilter from clogging.

10. Digest aliquots of the DNA with *Sal*I and *Xho*I to confirm that only DNA fragments of the expected size are present.

Cloning with the psv2-DHFR-8.32 Plasmid Containing a 256-copy *lac* Operator Repeat

- Use STBL2 competent cells (GIBCO-BRL) grown at 30°C to help maintain full-length *lac* operator repeats.
- Clone the *lac* operator repeats into low-copy-number vectors. Having too many copies in a cell increases the possibility of recombination.
- After transformation of the plasmid into cells, perform all steps as quickly as possible. Do not leave plates or cultures at 4°C for any longer than necessary.
- Use only chemical transformation with competent cells, not electroporation. Increased recombination following electroporation has been observed.

Making Stable Cell Lines of CHO Cells Containing *lac* Operator Repeats

Growth of Cells for Transfection

CHO DG44 cells are derivatives of CHO K1 cells and are doubly deleted for the *DHFR* locus. DG44 cells (Urlaub et al. 1986) are grown in F-12 (HAM) medium (GIBCO-BRL) supplemented with 10% defined fetal bovine serum (FBS, HyClone Lab). F-12 medium, lacking thymidine and hypoxanthine, with 10% dialyzed FBS (HyClone Lab) was used for selection and further passaging of stably transfected cells. It is important to note that the product of the *DHFR* gene is necessary for intermediary metabolism of thymidylates and purines. An imbalance of nucleotides in cells may be the cause of both insert instability due to gene amplification and an abnormal cell cycle due to interference with replication timing. Growth of the cells takes about 1 week.

Purification of Plasmid DNA for Transfection

1. Transfer 300 µl of vector DNA solution (~100 µg/ml) into a 1.5-ml microfuge tube.

2. Add 33 µl of 3 M sodium acetate (pH 5.5) to the DNA and thoroughly mix the solution with a pipette.

3. Add 660 μl of ice-cold 100% ethanol to the tube and mix the contents by inverting the tube several times. Let the tube stand on dry ice for 5 minutes. At this step, white fibers of DNA precipitate should be visible.

4. Centrifuge the reaction in a microfuge at 14,000 rpm for 15 minutes at 4°C. For convenience, label the location of the pellet inside the tube with a permanent marker.

5. Remove the supernatant, and wash the pellet with 1 ml of 70% ethanol (at 4°C) by inverting the tube several times. Centrifuge the tube in a microfuge at 14,000 rpm for 10 minutes at 4°C.

> Carry out all subsequent steps in a sterile environment.

6. Remove the supernatant and immerse the entire open tube in 70% ethanol for 10 seconds. Remove the supernatant. Air-dry the DNA pellet for 10–15 minutes.

7. Dissolve the dried DNA in 200 μl of sterile distilled H_2O or TE buffer.

> Isolated DNA should be aliquoted and frozen at −20°C for short-term storage if dissolved in water. DNA dissolved in TE buffer is stable for at least several weeks at 4°C.

8. Remove an aliquot of the DNA, and estimate the concentration and purity of the DNA.

Calcium Phosphate Transformation

Many transfection protocols use calcium-phosphate-precipitated DNA, with variations in the DNA concentration, incubation time, and precipitation reaction composition. However, all agree that the efficiency of transfection and the yield of stable colonies depend on many factors. The efficiency of transfection and possibly the copy number of inserts strongly depend on the growth state of recipient cells, concentration of calcium and phosphates, concentration of DNA, pH, temperature, and duration of the precipitation reaction. These factors influence initiation and growth of DNA–calcium phosphate coprecipitate complexes (Kjer and Fallon 1991; Jordan et al. 1996; Batard et al. 2001). DNA, if added in large excess to the precipitation mixture, can reduce the rate and efficiency of complex formation. Measurements of DNA concentration using absorption spectra of several dilutions are necessary to obtain a reasonable assessment of the amounts of DNA to be added to the reaction mixture. The pH of the reaction mixture should be adjusted to the desired value as accurately as possible. Transformation efficiency has been shown to be sensitive to pH variations as small as 0.05 pH units. Such accuracy is beyond the sensitivity of typical lab pH meters, taking into account temperature variations and calibration accuracy. Therefore, it is advisable to prepare several buffers with slightly different pH values (see Step 3 below). For example, if the pH of the transfection mixture is 7.05, prepare buffers with pH 6.95, 7.00, 7.05, 7.10, and 7.15 (Chen and Okayama 1988).

The solubility of calcium phosphate drops as the temperature decreases. Thus, for reproducible transfections, changes in the reaction temperature should be avoided. Transient transfection efficiency drops from 60% to 3% when the precipitation reaction incubation time increases from 1 minute to 20 minutes. Shorter incubation times also yield significantly smaller DNA–calcium phosphate precipitate particles as compared to the precipitate after longer incubation times. The growth of the precipitate is terminated with addition of the precipitation reaction to the culture medium. After calcium phosphate transfection, a large number of vector plasmids enter the cytoplasm of target cells and reside in multiple speckles; however, only a small fraction penetrates the nucleus. It is not known if the average size of stably integrated vector DNA is proportional to the average size of DNA–calcium phosphate precipitate particles.

1. Plate CHO DG44 cells in three 75-cm² flasks the day before transfection.

2. Replace the medium in the flasks with fresh medium 4 hours before transfection.

     Cells should be 50–80% confluent and be in the log phase of growth for efficient transfection.

3. Prepare three reaction tubes. In each one, dilute 7 μg of supercoiled plasmid DNA (from Step 7 of the previous protocol, Isolation of Plasmid DNA for Transfection) in 0.5 ml of 250 mM CaCl₂, mixed with 0.5 ml of TR buffer, varying only in its pH value (50 mM HEPES, 1.5 mM Na₂HPO₄, 0.28 M NaCl, 5 mM KCl, 5.5 mM dextrose [pH 6.95, 7.05, or 7.10]). Incubate the reactions for 30 minutes at room temperature.

4. Add each DNA mixture dropwise to a flask of cells (from Step 1). Incubate the flasks for 24 hours.

5. Replace the transfection medium with F-12 (HAM) medium supplemented with 10% defined FBS and incubate the cells for an additional 24 hours.

6. Replace the medium with selective medium (F-12 medium lacking thymidine and hypoxanthine, but containing 10% dialyzed FBS). Stable transformants are obtained after 2 weeks of selection.

Cell Cloning with Enrichment of Stable Cells with Large-copy-number Inserts Using Fluorescence-activated Cell Sorting

The goal of a typical cloning procedure is the isolation of genetically identical cells. Screening through large numbers of cells and colonies is the major rate-limiting step in cloning. This is why the time required for cloning is mostly defined by the screening design. Cloning procedures are roughly divided into two groups. Some cloning methods allow direct screening of the original cell pools or colonies followed by selection of the desired cells, whereas the others require initial duplication of cells before screening. The later cloning procedures are slower because of the extra steps involved in colony duplication.

Typically, staining mixed stable transformants with purified *Lac* repressor followed by immunofluorescence detection reveals only a small percentage of cells containing large chromosome insertions. The following protocol describes the selection of CHO cells expressing high amounts of exogenous DHFR from a pool of stably transformed cells using cell sorting, a strategy to enrich for cells harboring large chromosome insertions of the DHFR-containing transgene. Selection is based on binding a fluorescent cell-permeable DHFR inhibitor to the pool of cellular DHFR. Cells with high fluorescence show a high total expression of DHFR, which correlates roughly with copy number (as well as chromosome position effects).

1. Grow cells in 25-cm² tissue culture flasks until they are in log phase.

2. Replace the growth medium with fresh F-12 (HAM) medium supplemented with 10% FBS and 20 μM fluorescein-labeled methotrexate (FMTX) (Molecular Probes). Add 100 μM glycine, 30 μM hypoxanthine, and 30 μM thymine to the medium to relieve the toxic effects of FMTX (Sherwood et al. 1990). Incubate the cells for 8 hours.

3. Harvest cells by trypsinization and keep them in sterile PBS on ice.

     As FMTX is retained in cells at concentrations high enough to cause damage, cells are viable in PBS on ice only for several hours.

4. Sort the cells into flasks or 96-well plates.

5. Grow enough cells for freezing and freeze them after screening.

Preparing a Large Number of Clones for Screening or Subcloning

Cell cultures selected and sorted in the preceding protocol are plated, and individual colonies are isolated for visual screening or subcloning.

1. Plate cells onto a Petri dish at a low density, ideally, 20–100 cells per 150-mm dish. To minimize cross-contamination among neighboring colonies, make sure that the density of plating is low enough and that the colonies are spread out. Use several Petri dishes, titrating the cell concentration by factors of 2–3 between dishes, to ensure optimal cell density.

    The action of the trypsin (Step 3 of preceding protocol) must have progressed enough to ensure the absence of cell clumping and the predominance of single cells in suspension. This is aided by mechanical agitation or by vigorous pipetting of the culture.

2. Return the cells to the incubator, and if possible avoid moving the dishes (particularly after the first few days) for ~10 days to allow good size colonies to form.

    Movement of the dishes will cause cells to dislodge and form satellite colonies and/or a dispersed lawn of dispersed cells, increasing cross-contamination.

3. Allow the cells to grow until the colonies are well-defined and visible by eye while avoiding overgrowth to minimize colony cross-contamination. Mark individual isolated colonies on the bottom of the plate with a permanent marker. View the colonies under low magnification to check for the presence of undesirable smaller colonies growing in the vicinity of any colony of interest.

4. If visual screening of the clones is unnecessary, then pick and harvest a large number of colonies from the Petri dish using one of the two methods described in the following protocol. If live visual screening is desired, refer to the discussion above (preceding the protocols), Subcloning Stable Transformants.

Subcloning Using Filter Paper or Micropipette Tips

For subcloning, a large number of colonies are picked from a 150-mm-diameter Petri dish (from the preceding protocol). Either of two simple methods may be used for harvesting each colony. The first uses small pieces of filter paper soaked in trypsin, and the second uses micropipette tips to pick each colony. The clonal cells are then deposited into 96-well plates.

Trypsin Method for Picking Colonies

1. Cut filter paper into small squares each with an area of ~4 mm². Use sharp-tipped forceps to place the squares together in a glass Petri dish. Cover them and sterilize by autoclaving.

2. Aspirate the medium from the Petri dish containing the marked colonies (from Step 3 above).

3. Wash the cells with sterile calcium- and magnesium-free PBS (CMF-PBS), and remove excess liquid.

4. Dispense trypsin solution into a small sterile Petri dish. Using the sterile forceps, dip a sterile filter paper (from Step 1) in trypsin, blot it to remove excess trypsin, and place it over a colony.

5. Repeat Step 4 for each of the colonies. Return the Petri dish containing the cells to the incubator for ~5 minutes.

6. Transfer each filter paper into a separate well of a 96-well microtiter plate containing 0.1 ml of tissue culture medium per well. If visual screening will be performed, use a glass

bottom 96-well plate. Duplicate plating can be accomplished by dipping the filter paper into one well, to dislodge some cells, and then dropping the filter paper into a well of a second 96-well plate.

> The efficiency of colony duplication by this method is close to 100%. It is important that the cells not be left without buffer for more than 15–20 minutes or they will dry out. Also, CMF-PBS may induce rounding of cells and/or detachment after prolonged treatment (see Steps 1 and 2 in the protocol below).

7. Grow the collected cells in the 96-well plates for 1–2 days.

8. Remove the filter paper pieces from the wells, taking care to avoid cross-contamination. The following is one method for doing this.

    a. Attach a plastic hose to a vacuum system with an in-line trap. Insert the larger end of a 1-ml micropipette tip into the hose. Cut the other end of the tip to accommodate a sterile 200-μl micropipette tip.

    b. Insert the 200-μl tip into the bottom of a well. The vacuum will cause the filter paper to adhere to the tip. Withdraw the tip and discard both the filter paper and the 200-μl pipette tip.

    c. Repeat with the remaining wells, using a fresh sterile 200-μl tip each time.

9. Continue to grow the cells until their numbers are sufficient for screening. Screen the cells from each well and freeze the positive clones. Examine the positive clones by microscopy for the biological activity of interest.

Micropipette Tip Method for Picking Colonies

This method works well with more densely packed colonies, and it is faster than the trypsin method—a few hundred colonies can be easily transferred within a couple of hours. Once this method is optimized, efficiency is near 100%, although the number of cells transferred per colony into the well may be somewhat lower.

1. Mark the colonies that are less likely to be cross-contaminated with cells from other colonies. Rinse the cells twice with CMF-PBS (10 ml per 150-mm dish). Add a small amount of CMF-PBS (~4–6 ml), just sufficient to cover the surface of the dish, and return the dish to the incubator.

2. After ~5 minutes, look at the cells under a microscope to see if cells in colonies have rounded. If not, incubate further. Once the cells in the colony have rounded, but are still attached, they can be picked.

3. Use a permanent marker to circle small groups of colonies on the bottom of the dish. This helps to keep track of which colonies have already been harvested.

4. Using a fresh 200-μl pipette tip per colony, scrape 2–3 times over the colony while aspirating 10 μl into the pipette tip. Insert the pipette tip into the well of a 96-well plate, pipetting medium from the well several times into the tip to wash the cells off into the well.

5. Grow cells until their numbers are sufficient for screening. Screen the cells from each well and freeze the positive clones. Examine the positive clones by microscopy for the biological activity of interest.

ACKNOWLEDGMENTS

This work was supported by National Institutes of Health grants GM-42516 and GM-58460 to A.B.

REFERENCES

Barsoum J. 1990. Introduction of stable, high-copy-number DNA into Chinese hamster ovary cells by electroporation. *DNA Cell Biol.* **9:** 293–300.

Batard P., Jordan M., and Wurm F. 2001. Transfer of high copy number plasmid into mammalian cells by calcium phosphate transfection. *Gene* **270:** 61–68.

Becker M., Baumann C., John S., Walker D.A., Vigneron M., McNally J.G., and Hager G.L. 2002. Dynamic behavior of transcription factors on a natural promoter in living cells. *EMBO Rep.* **3:** 1188–1194.

Belmont A.S. 1998. Nuclear ultrastructure: Transmission electron microscopy and image analysis. *Methods Cell Biol.* **53:** 99–124.

———. 2001. Visualizing chromosome dynamics with GFP. *Trends Cell Biol.* **11:** 250–257.

Belmont A.S., Braunfeld M.B., Sedat J.W., and Agard D.A. 1989. Large-scale chromatin structural domains within mitotic and interphase chromosomes in vivo and in vitro. *Chromosoma* **98:** 129–143.

Belmont A.S., Li G., Sudlow G., and Robinett C. 1999a. Visualization of large-scale chromatin structure and dynamics using the lac operator/lac repressor reporter system. *Methods Cell Biol.* **58:** 99–124.

———. 1999b. Visualization of large-scale chromatin structure and dynamics using the lac operator/lac repressor reporter system. *Methods Cell Biol* **58:** 203–222.

Chen C.A. and Okayama H. 1988. Calcium phosphate-mediated gene transfer: A highly efficient transfection system for stably transforming cells with plasmid DNA. *BioTechniques* **6:** 632–638.

Chubb J.R., Boyle S., Perry P., and Bickmore W.A. 2002. Chromatin motion is constrained by association with nuclear compartments in human cells. *Curr. Biol.* **12:** 439–445.

Dietzel S. and Belmont A.S. 2001. Reproducible but dynamic positioning of DNA in chromosomes during mitosis. *Nat. Cell Biol.* **3:** 767–770.

Dundr M., McNally J.G., Cohen J., and Misteli T. 2002. Quantitation of GFP-fusion proteins in single living cells. *J. Struct. Biol.* **140:** 92–99.

Fieck A., Wyborski D.L., and Short J.M. 1992. Modifications of the *E. coli* Lac repressor for expression in eukaryotic cells: Effects of nuclear signal sequences on protein activity and nuclear accumulation. *Nucleic Acids Res.* **20:** 1785–1791.

Janicki S.M., Tsukamoto T., Salghetti S.E., Tansey W.P., Sachidanandam R., Prasanth K.V., Ried T., Shav-Tal Y., Bertrand E., Singer R.H., and Spector D.L. 2004. From silencing to gene expression: Real-time analysis in single cells. *Cell* **116:** 683–698.

Jordan M., Schallhorn A., and Wurm F.M. 1996. Transfecting mammalian cells: Optimization of critical parameters affecting calcium-phosphate precipitate formation. *Nucleic Acids Res.* **24:** 596–601.

Kjer K.M. and Fallon A.M. 1991. Efficient transfection of mosquito cells is influenced by the temperature at which DNA-calcium phosphate coprecipitates are prepared. *Arch. Insect Biochem. Physiol.* **16:** 189–200.

Li G., Sudlow G., and Belmont A.S. 1998. Interphase cell cycle dynamics of a late-replicating, heterochromatic homogeneously staining region: Precise choreography of condensation/decondensation and nuclear positioning. *J. Cell Biol.* **140:** 975–989.

Li Y., Danzer J.R., Alvarez P., Belmont A.S., and Wallrath L.L. 2003. Effects of tethering HP1 to euchromatic regions of the *Drosophila* genome. *Development* **130:** 1817–1824.

McNally J.G., Muller W.G., Walker D., Wolford R., and Hager G.L. 2000. The glucocorticoid receptor: Rapid exchange with regulatory sites in living cells. *Science* **287:** 1262–1265.

Michaelis C., Ciosk R., and Nasmyth K. 1997. Cohesins: Chromosomal proteins that prevent premature separation of sister chromatids. *Cell* **91:** 35–45.

Misteli T. 2001. Protein dynamics: Implications for nuclear architecture and gene expression. *Science* **291:** 843–847.

Nye A.C., Rajendran R.R., Stenoien D.L., Mancini M.A., Katzenellenbogen B.S., and Belmont A.S. 2002. Alterations of large-scale chromatin structure by the estrogen receptor. *Mol. Cell. Biol.* **22:** 3437–3449.

Pellegrin P., Fernandez A., Lamb N.J., and Bennes R. 2002. Macromolecular uptake is a spontaneous event during mitosis in cultured fibroblasts: Implications for vector-dependent plasmid transfection. *Mol. Biol. Cell* **13:** 570–578.

Reits E.A. and Neefjes J.J. 2001. From fixed to FRAP: Measuring protein mobility and activity in living cells. *Nat. Cell Biol.* **3:** E145–E147.

Robinett C.C., Straight A., Li G., Willhelm C., Sudlow G., Murray A., and Belmont A.S. 1996. In vivo localization of DNA sequences and visualization of large-scale chromatin organization using lac operator/repressor recognition. *J. Cell Biol.* **135:** 1685–1700.

Sambrook J. and Russell D.W. 2001. *Molecular cloning: A laboratory manual,* 3rd edition. Cold Spring Harbor Laboratory Press, Cold Spring Harbor, New York.

Spector D.L., Goldman R.G., and Leimwand L.L. 1998. *Cells: A laboratory manual.* Cold Spring Harbor Laboratory Press, Cold Spring Harbor, New York.

Sherwood S.W., Assaraf Y.G., Molina A., and Schimke R.T. 1990. Flow cytometric characterization of antifolate resistance in cultured mammalian cells using fluoresceinated methotrexate and daunorubicin. *Cancer Res.* **50:** 4946–4950.

Straight A.F., Belmont A.S., Robinett C.C., and Murray A.W. 1996. GFP tagging of budding yeast chromosomes reveals that protein-protein interactions can mediate sister chromatid cohesion. *Curr. Biol.* **6:** 1599–1608.

Tsukamoto T., Hashiguchi N., Janicki S.M., Tumbar T., Belmont A.S., and Spector D.L. 2000. Visualization of gene activity in living cells. *Nat. Cell Biol.* **2:** 871–878.

Tumbar T. and Belmont A.S. 2001. Interphase movements of a DNA chromosome region modulated by VP16 transcriptional activator. *Nat. Cell Biol.* **3:** 134–139.

Tumbar T., Sudlow G., and Belmont A.S. 1999. Large-scale chromatin unfolding and remodeling induced by VP16 acidic activation domain. *J. Cell Biol.* **145:** 1341–1354.

Urlaub G., Mitchell P.G., Kas E., Chasin L.A., Funanage V.L., Myoda T.T., and Hamlin J. 1986. Effect of gamma rays at the dihydrofolate reductase locus: Deletions and inversions. *Som. Cell Mol. Genet.* **12:** 555–566.

Wilke M., Fortunati E., van den Broek M., Hoogeveen A.T., and Scholte B.J. 1996. Efficacy of a peptide-based gene delivery system depends on mitotic activity. *Gene Ther.* **3:** 1133–1142.

Ye Q., Hu Y.F., Zhong H., Nye A.C., Belmont A.S., and Li R. 2001. BRCA1-induced large-scale chromatin unfolding and allele-specific effects of cancer-predisposing mutations. *J. Cell Biol.* **155:** 911–921.

Imaging Gene Expression in Living Cells

Susan M. Janicki and David L. Spector

Cold Spring Harbor Laboratory, Cold Spring Harbor, New York 11724

INTRODUCTION

THE PROGRESSION OF GENE EXPRESSION CAN EXTEND from the extracellular matrix to the cell nucleus, where input signals confer positive effects on the genome. Each step of a particular gene expression paradigm involves critical interactions among a large number of proteins or protein complexes (Lemon and Tjian 2000; Rappsilber et al. 2002). The individual steps must be orchestrated in a defined sequence such that the ultimate protein product is synthesized in a timely manner in the correct cell type (for reviews, see Maniatis and Reed 2002; Orphanides and Reinberg 2002). Due to its overall complexity, each aspect of gene expression has been studied primarily as an isolated process in vitro. Although this reductionist approach has contributed extensively to the identification of critical players, complexes, and functions of many of the protein constituents involved in gene expression, it has not afforded the ability to address the spatial and temporal aspects of gene expression that can only be studied within the context of the living cell.

Several recently developed approaches have been used to visualize one or more aspects of gene expression in living cells. In general, these efforts have utilized interactions between DNA-binding proteins and their target sequences. In a significant breakthrough, Belmont and co-workers developed an approach, based on the use of the *lac* operator/repressor system, to directly visualize chromatin organization and dynamics in living cells (Robinett et al. 1996). The introduction of bacterial *lac* operator repeats into the genomes of eukaryotic cells and expression of a green fluorescent protein (GFP) Lac repressor fusion protein is a noninvasive means of identifying and studying specific regions of chromatin (for a review, see Belmont 2001; Janicki and Spector 2003) (also see Chapter 25). By stably introducing a *lac* operator array into Chinese hamster ovary cells with the dihydrofolate reductase gene and amplifying it through methotrexate selection (Robinett et al. 1996), a stable cell line was selected that contained an approximately 90-Mb chromosomal array that can be visualized using a GFP-Lac repressor fusion protein. The array formed a late-replicating homogeneously staining region (HSR) (Li et al. 1998). Cell cycle analysis indicated that the integration site was peripherally localized throughout most of interphase. However, during several hours in mid-to-late S phase, the HSR decondensed and moved toward the nuclear interior, which correlated with its DNA replication (Li et al. 1998). Using this system, the Belmont group has been able to directly visualize activator binding in living cells, and they found that chromatin decondensation occurs upon activator binding and in the absence of transcription (Tumbar et al. 1999). However, subsequent studies using a tandem array ($\geq$2 Mb) of the mouse mammary tumor virus (MMTV) promoter driving a *ras* reporter (see discussion below) have shown that this array does require transcription for chromatin decondensation to occur (Muller et al. 2001b). Therefore, different loci may respond to different signals for chromatin decondensation. In a separate study, involving the use of a cell line with a smaller integration sequence (150–300 kb), VP16 targeting to the locus was also shown to induce its movement from the nuclear periphery to a more internal nuclear region (Tumbar and Belmont 2001), suggesting that internal nuclear regions may be more amenable to potentially active loci. Furthermore, recruitment of the endogenous histone acetyltransferase (HAT) and SWI/SNF components of the chromatin remodeling complex (Memedula and Belmont 2003), as well as of several histone acetyltransferases (including GCN5, P/CAF, and p300/CBP), and hyperacetylation of all core histones were observed (Tumbar et al. 1999). Examination of the extended chromosome fibers by light and electron microscopy supports the existence of a folded chromonema model based on approximately 100-nm chromonema fibers formed by compaction of 10-nm and 30-nm chromatin fibers (Robinett et al. 1996; Tumbar et al. 1999). The use of the *lac* operator/repressor system has also made it possible to observe the induction of a tetracycline-regulatable array of transcription units in living cells (Tsukamoto et al. 2000). The transcription units encode a cyan fluorescent protein (CFP) fused to a peroxisome targeting signal, thereby providing a direct read out of gene expression at the protein level in living cells.

A second approach to visualize gene expression relied on the use of a cell line that contains a stable tandem array of an MMTV/Harvey viral *ras* (MMTV/v-Ha-*ras*) reporter element (Kramer et al. 1999). Ligand activation of a GFP-tagged glucocorticoid receptor (GR) targeted it from the cytoplasm to the repeated regulatory elements in the array and allowed the site of transcription to be visualized in living cells (McNally et al. 2000; Muller et al. 2001a). Interestingly, with the use of this system and fluorescence recovery after photobleaching (FRAP), GFP-GR was shown to undergo rapid exchange between chromatin and the nucleoplasm, challenging the more traditional view of promoter-bound complexes as highly stable multiprotein complexes (Becker et al. 1984; Thanos and Maniatis 1995; Cosma et al. 1999; Shang et al. 2000) and substantiating a "hit-and-run" model (McNally et al. 2000). Although the GR and the GR coactivator GRIP/Tif-2 exchange rapidly on the locus ($t_{1/2}$ ~5 seconds), initial recovery of RNA polymerase II was found to occur rapidly, but was followed by a much slower overall exchange. This was interpreted to indicate the existence of two RNA polymerase II populations, a fast one undergoing multiple unsuccessful initiation events, and a slower one representative of RNA polymerase II molecules engaged in transcription elongation that must be completed prior to fluorescence recovery (Becker et al. 2002). The discovery that activator/chromatin interactions are highly dynamic suggests that transcription factors do not form stable holo-complexes at promoters and that their activity may be regulated in a similar manner.

In summary, the majority of studies investigating chromatin remodeling and gene expression have been performed in vitro or by chromatin immunoprecipitation, whereby the data represent an average of DNA/protein associations from many cells (for reviews, see Fry and Peterson 2001; Featherstone 2002; Narlikar et al. 2002). Approaches such as those described above have the advantage of assessing chromatin remodeling and transcription at the single-cell level, providing a higher degree of temporal resolution while at the same time correlating this with spatial information.

VISUALIZING THE CENTRAL DOGMA IN LIVING CELLS: DNA → RNA → PROTEIN

We have recently extended the studies described above by developing a system that allows us to directly visualize a stably integrated, regulatable genetic locus, its RNA product, and the protein encoded by this RNA in living cells (Janicki et al. 2004). This system allows us for the first time to access every level of gene expression directly in the living cell.

This live cell gene expression system is the second generation of a system that was previously developed in the authors' laboratory (D.L.S.) to study a specific transcription site and its protein product in living cells (Tsukamoto et al. 2000). The new system is composed of a human U2OS cell line (U2OS 2-6-3) that contains a stably integrated approximately 200-copy array of a 20-kb plasmid composed of a series of modules which have the potential to be exchanged (Fig. 1). The plasmid contains at its 5′ end 256 copies of *lac* operator repeats that, when bound by a CFP-Lac repressor fusion protein, allows the integration site to be visualized. Ninety-six copies of a tetracycline-responsive element (TRE) enable the minimal cytomegalovirus (CMV) promoter to be regulated using the pTet-ON or pTet-OFF systems, which utilize an activator composed of a tetracycline-responsive DNA-binding domain and a VP16 activation domain. Addition of doxycycline results in pTet-ON binding to the TRE, followed by transcription of an mRNA that encodes CFP with a peroxisome targeting signal-1, the Ser-Lys-Leu (SKL) tripeptide (Miyazawa et al. 1989), to allow the protein product of this transcription unit to be visualized as it accumulates in peroxisomes. In addition, this system provides direct proof that transcription, pre-mRNA processing, mRNP transport, translation, and correct protein targeting have taken place.

To visualize the mRNA encoded by the transcription unit, we incorporated a series of 24 MS2 bacteriophage viral replicase translational operators (a 19-nucleotide RNA stem loop) that are specifically recognized by the bacteriophage MS2 coat protein, which binds as a

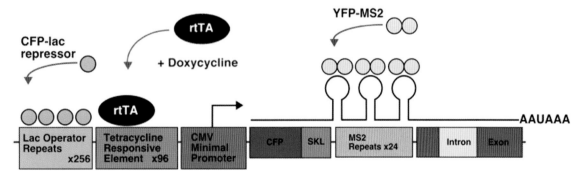

FIGURE 1. Schematic representation of the gene expression plasmid, p3216PECMS2β. The plasmid is composed of 256 copies of the *lac* operator, 96 tetracycline response elements, a minimal CMV promoter, CFP fused to the peroxisomal targeting signal SKL, 24 MS2 translational operators (MS2 repeats), a rabbit β-globin intron/exon module, and a cleavage/polyadenylation signal. Expression of the CFP-Lac repressor allows the DNA to be visualized and expression of pTet-ON (rtTA) in the presence of doxycycline (dox) drives expression from the CMV minimal promoter. When MS2-YFP (YFP fused to the MS2 coat protein) dimerizes and interacts with the stem-loop structure of the translational operator, it allows the transcribed RNA to be visualized.

dimer (Beckett and Uhlenbeck 1988; Valegard et al. 1994). Expression of MS2 coat protein fused to yellow fluorescent protein (YFP) containing a nuclear localization signal allows us to see the RNA as it is being transcribed and as it transits within the nucleoplasm and cytoplasm. This labeling approach was previously used to track the asymmetrical movement of *ASH1* mRNA in dividing yeast cells (Bertrand et al. 1998), the movement of RNA particles in the cytoplasm of hippocampal neurons and COS cells (Rook et al. 2000; Fusco et al. 2003), and the dynamic localization of endogenous *nanos* RNA in *Drosophila* oocytes (Forrest and Gavis 2003), although the kinetics of mRNA synthesis and movement had not been studied at a specific transcription site. Finally, the 3′ end of the transcription unit contains a rabbit β-globin intron flanked by two exons and a poly(A) signal, so that one can assess the recruitment of the pre-mRNA processing machinery to the site of transcription upon activation and its loss upon transcriptional shutdown. By combining all of these elements into a single plasmid, this system allows multiple aspects of the gene expression paradigm to be simultaneously assayed in living cells with high temporal and spatial resolution.

To visualize gene expression, U2OS cells (U2OS 2-6-3) stably expressing this plasmid are cotransfected by electroporation with three plasmids: pTet-ON (2 μg), pSV2-CFP-Lac repressor (2 μg), and pYFP-MS2 coat protein (0.5 μg), along with sheared salmon sperm DNA (40 μg) as a carrier; electroporation is carried out using a Gene Pulser II (Bio-Rad) (170 V, 950 μF). Electroporated cells are allowed to attach for 2.5 hours to Cell-Tak (BD Biosciences, Palo Alto, California)-coated coverslips, which allows them to flatten quickly and prevents movement during imaging. This short expression time ensures a low but easily detectable level of protein in the cells. The coverslip is then mounted into an FCS2 live cell chamber (Bioptechs Inc., Butler, Pennsylvania) that is placed on the stage of an Olympus IX-70 inverted fluorescence microscope (Olympus America, Melville, New York) equipped with a 100×/1.4 NA (numerical aperture) objective lens and an objective heater. The chamber maintains the temperature at 37°C, and pH is maintained by perfusion of HEPES-containing, phenol-red-free Leibovitz's L15 medium (Invitrogen). Fluorescent light is directed to the microscope by a fiber optic cable from a Polychrome II monochromator (TILL Photonics, Grafelfing, Germany) using a computer-controlled shutter. Images (80–100 msec) are acquired with a Peltier cooled IMAGO CCD (charge-coupled device) camera containing an SVGA interline chip (1280 × 1024) with a pixel size of 6.7 μm² using TILLvisION software. The objective lens is mounted with a piezoelectric driver that provides backlash-free movement in 75-nm steps to allow for image stacks to be obtained at each time point. Other imaging systems optimized for live cell imaging can also be used for observation and image acquisition. We have performed numerous time-course experiments with this cell line, and nuclear rotation does not seem to occur in these flat cells (Z = ~7 μm) and cell movement is minimal.

In the absence of doxycycline (dox), the transcriptionally inactive locus is visualized with the CFP-Lac repressor protein as a tightly condensed dot in the nucleoplasm (Fig. 2A). In the same cell, the YFP-MS2 fusion protein is diffusely distributed throughout the nucleoplasm with no apparent enrichment at the locus (Fig. 2B). Two hours and 30 minutes after the addition of dox (1 μg/ml), the chromatin at the locus is significantly decondensed (Fig. 2D), and there is an enrichment of the YFP-MS2 coat protein at the transcription site (Fig. 2E). Interestingly, as compared to the relatively smooth distribution of YFP-MS2 coat protein in the nucleoplasm prior to the addition of dox, after turn-on of transcription, the YFP-MS2 coat protein exhibits a granular appearance (Fig. 2E) indicative of packaging of the mRNA into mRNP particles. At this time point, we also observe the CFP-SKL protein in cytoplasmic peroxisomes (Fig. 2F), confirming that the mRNP has been transported out of the nucleus and that the mRNA has been translated. This change in chromatin structure and

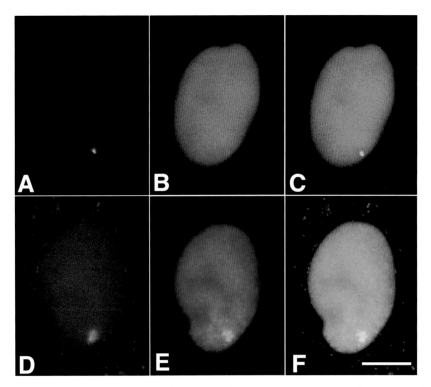

FIGURE 2. Visualization of DNA, RNA, and protein in living cells. U2OS 2-6-3 cells were transiently transfected with the pSV2-CFP-Lac repressor, pTet-ON (rtTA), and MS2-YFP, and imaging was begun 2.5 hours following transfection. (*A–C*) In the "off" state, the chromatin at the locus is condensed (*A*); the MS2 coat protein is diffusely distributed throughout the nucleoplasm (*B*); and no cytoplasmic protein product is observed (*A,C*). (*D–F*) 2.5 hours after induction of the transcription units, the chromatin is decondensed (*D*); the RNA coat protein is present at the transcription site as well as in particles in the nucleoplasm (*E*); and the protein product is observed in cytoplasmic peroxisomes (*D,F*). Bar, 10 μm.

the localization of the YFP-MS2 coat protein is seen in approximately 70% of the transfected cells. We can reproducibly detect the VP16 transcriptional activator and nascent RNA at the transcription site within 5 minutes after the addition of dox, and protein expression is observed 30–60 minutes later.

A major advantage of live cell imaging is that the kinetics of dynamic cellular processes can be tracked with high spatial and temporal resolution as they happen. To more closely examine the kinetics of RNA synthesis in individual cells during transcriptional activation, cells were imaged every 2.5 minutes for 4–6 hours in both the YFP and CFP channels (see Movie 26.1). An increase in RNA levels was detected immediately after the addition of dox. The highly condensed state of the chromatin at the early time points suggests that only a subset of the repeats are being transcribed, which is supported by the limited amount of the activator detected at the locus 5 minutes post dox. By 17.5 minutes, when noticeable changes in higher-order chromatin structure begin to be detected, the RNA coats the entire region of the locus. By about 130 minutes, when chromatin is maximally decondensed, RNA levels at the transcription site are at their peak (Janicki et al. 2004).

SUMMARY

The system described in this chapter thus provides an approach to directly visualize gene expression at the levels of DNA, RNA, and protein in living cells. The system is modular and can therefore be modified to incorporate different promoters and transcription units, thereby providing a more global assay for assessing the spatial and temporal aspects of the expression of genes associated with different chromosomal regions and/or functions. The ability to assess the different levels of gene expression simultaneously in single living cells provides a powerful approach for correlative analyses of many cellular processes, including transcription, DNA replication, and mRNA packaging and transport.

MOVIE LEGEND

MOVIE 26.1. Real-time analysis of gene expression. We have developed a cell line in which we can observe a stably integrated genetic locus as well as its mRNA and protein products using live cell microscopy and various FP fusion proteins. This locus can be induced to initiate transcription. The movie shows the fluorescent signal from the chromatin at the genetic locus as well as the protein product in the left panel. The right panel shows the mRNA synthesized when the genetic locus is turned on. Both movies show the same cell. When the genetic locus is not transcriptionally active, the chromatin is condensed (*tight blue dot in left panel*) and no protein product can be seen in the cytoplasm of the cell. At this time, the mRNA binding protein is diffusely distributed throughout the nucleus (*right panel*), because it has absolutely no substrate with which to bind. At the beginning of the movie, the chromatin in the left panel decondenses over time and eventually the protein product is encoded for by the genetic locus, distributed in peroxisomes in the cytoplasm (*blue dots in the cytoplasm*). Encoded within the protein is a peroxisome-targeting signal. In the right panel, the mRNA binding protein concentrates at the genetic locus (transcription site) as the chromatin decondenses. Since mRNA is being synthesized, it has a binding site for the fluorescently tagged mRNA binding protein, and the nucleoplasmic pool now looks granular. The granular structures represent mRNPs moving away from the genetic locus in all directions by diffusion. They will end up in the cytoplasm where the mRNA will be translated into the protein that is targeted to peroxisomes. (From Janicki et al. 2004 with permission of Elsevier.)

REFERENCES

Becker M., Baumann C., John S., Walker D.A., Vigneron M., McNally J.G., and Hager G.L. 2002. Dynamic behavior of transcription factors on a natural promoter in living cells. *EMBO Rep.* **3:** 1188–1194.

Becker P., Renkawitz R., and Schutz G. 1984. Tissue-specific DNaseI hypersensitive sites in the 5′-flanking sequences of the tryptophan oxygenase and the tyrosine aminotransferase genes. *EMBO J.* **3:** 2015–2020.

Beckett D. and Uhlenbeck O.C. 1988. Ribonucleoprotein complexes of R17 coat protein and a translational operator analog. *J. Mol. Biol.* **204:** 927–938.

Belmont A.S. (2001). Visualizing chromosome dynamics with GFP. *Trends Cell. Biol.* **11:** 250–257.

Bertrand E., Chartrand P., Schaefer M., Shenoy S.M., Singer R.H., and Long R.M. (1998). Localization of ASH1 mRNA particles in living yeast. *Mol. Cell* **2:** 437–445.

Cosma M.P., Tanaka T., and Nasmyth K. 1999. Ordered recruitment of transcription and chromatin remodeling factors to a cell cycle- and developmentally regulated promoter. *Cell* **97:** 299–311.

Featherstone M. 2002. Coactivators in transcription initiation: Here are your orders. *Curr. Opin. Genet. Dev.* **12:** 149–155.

Forrest K.M. and Gavis E.R. 2003. Live imaging of endogenous RNA reveals a diffusion and entrapment mechanism for *nanos* mRNA localization in *Drosophila. Curr. Biol.* **13:** 1159–1168.

Fry C.J. and Peterson C.L. 2001. Chromatin remodeling enzymes: Who's on first? *Curr. Biol.* **11:** R185–197.

Fusco D., Accornero N., Lavoie B., Shenoy S.M., Blanchard J.M., Singer R.H., and Bertrand E. 2003. Single mRNA molecules demonstrate probabilistic movement in living mammalian cells. *Curr. Biol.* **13:** 161–167.

Janicki S.M. and Spector D.L. 2003. Nuclear choreography: Interpretations from living cells. *Curr. Opin. Cell Biol.* **15:** 149–157.

Janicki S.M., Tsukamoto T., Salghetti S.E., Tansey W.P., Sachidanandam R., Prasanth K.V., Ried T., Shav-Tal Y., Bertrand E., Singer R.H., and Spector D.L. 2004. From silencing to gene expression: Real-time analysis in single cells. *Cell* **116:** 683–698.

Kramer P.R., Fragoso G., Pennie W., Htun H., Hager G.L., and Sinden R.R. 1999. Transcriptional state of the mouse mammary tumor virus promoter can affect topological domain size in vivo. *J. Biol. Chem.* **274:** 28590–28597.

Lemon B. and Tjian R. 2000. Orchestrated response: A symphony of transcription factors for gene control. *Genes Dev.* **14:** 2551–2569.

Li G., Sudlow G., and Belmont A.S. 1998. Interphase cell cycle dynamics of a late-replicating, heterochromatic homogeneously staining region: Precise choreography of condensation/decondensation and nuclear positioning. *J. Cell. Biol.* **140:** 975–989.

Maniatis T. and Reed R. 2002. An extensive network of coupling among gene expression machines. *Nature* **416:** 499–506.

McNally J.G., Muller W.G., Walker D., Wolford R., and Hager G.L. 2000. The glucocorticoid receptor: Rapid exchange with regulatory sites in living cells. *Science* **287:** 1262–1265.

Memedula S. and Belmont A.S. 2003. Sequential recruitment of HAT and SWI/SNF components to condensed chromatin by VP16. *Curr. Biol.* **13:** 241–246.

Miyazawa S., Osumi T., Hashimoto T., Ohno K., Miura S., and Fujiki Y. 1989. Peroxisome targeting signal of rat liver acyl-coenzyme A oxidase resides at the carboxy terminus. *Mol. Cell. Biol.* **9:** 83–91.

Muller W.G., Walker D., Hager G.L., and McNally J.G. 2001a. Large-scale chromatin decondensation and recondensation regulated by transcription from a natural promoter. *J. Cell. Biol.* **154:** 33–48.

———. 2001b. Large-scale chromatin decondensation and recondensation regulated by transcription from a natural promoter. *J. Cell. Biol.* **154:** 33–48.

Narlikar G.J., Fan H-Y., and Kingston R.E. 2002. Cooperation between complexes that regulate chromatin structure and transcription. *Cell* **108:** 475–487.

Orphanides G. and Reinberg, D. 2002. A unified theory of gene expression. *Cell* **108:** 439–451.

Rappsilber J., Ryder U., Lamond A.I., and Mann M. 2002. Large-scale proteomic analysis of the human spliceosome. *Genome Res.* **12:** 1231–1245.

Robinett C.C., Straight A., Li G., Willhelm C., Sudlow G., Murray A., and Belmont A.S. 1996. In vivo localization of DNA sequences and visualization of large-scale chromatin organization using *lac* operator/repressor recognition. *J. Cell Biol.* **135:** 1685–1700.

Rook M.S., Lu M., and Kosik K.S. 2000. CaMKIIalpha 3′ untranslated region-directed mRNA translocation in living neurons: Visualization by GFP linkage. *J. Neurosci.* **20:** 6385–6393.

Shang Y., Hu X., DiRenzo J., Lazar M.A., and Brown M. 2000. Cofactor dynamics and sufficiency in estrogen receptor-regulated transcription. *Cell* **103:** 843–852.

Thanos D. and Maniatis T. 1995. Virus induction of human IFN beta gene expression requires the assembly of an enhanceosome. *Cell* **83:** 1091–1100.

Tsukamoto T., Hashiguchi N., Janicki S.M., Tumbar T., Belmont A.S., and Spector D.L. 2000. Visualization of gene activity in living cells. *Nat. Cell. Biol.* **2:** 871–878.

Tumbar T. and Belmont A.S. 2001. Interphase movements of a DNA chromosome region modulated by VP16 transcriptional activator. *Nat. Cell. Biol.* **3:** 134–139.

Tumbar T., Sudlow G., and Belmont A.S. 1999. Large-scale chromatin unfolding and remodeling induced by VP16 acidic activation domain. *J. Cell Biol.* **145:** 1341–1354.

Valegard K., Murray J.B., Stockley P.G., Stonehouse N.J., and Liljas L. 1994. Crystal structure of an RNA bacteriophage coat protein-operator complex. *Nature* **371:** 623–626.

CHAPTER TWENTY SEVEN

Studying Mitosis in Cultured Mammalian Cells

Patricia Wadsworth

Department of Biology and Program in Molecular and Cellular Biology,
University of Massachusetts, Amherst, Massachusetts 01003

INTRODUCTION

THE PROCESS OF MITOSIS HAS FASCINATED BIOLOGISTS for more than 100 years. During this time, diverse cell types have been used to study various aspects of mitosis. Early investigators focused primarily on cells that were well suited for morphological studies; more recently, workers have developed experimental systems that can be used to study both morphology and the molecular basis of chromosome motion and cell cycle regulation. This chapter briefly discusses cell types that have been used to study mitosis in live cells and outlines methods that are used to examine mitosis in cultured mammalian cells.

Studying Mitosis in Living Cells

Several systems have historically provided a wealth of information regarding the events of mitosis in living cells. Endosperm cells of the African blood lily, *Haemanthus katherinia*, are among the most aesthetically appealing mitotic cells, and observations of these cells have provided detailed information about spindle assembly and chromosome motion throughout mitosis (Inoue and Bajer 1961; Bajer 1990). Endosperm cells, which lack a cell wall, are isolated directly from endosperm fluid and are allowed to spread on a slide; the preparation is mitotically active for many hours (Vos et al. 1999). These cells are suitable for all types of microscopy. A disadvantage of using these cells is that plants must be grown so that flowering occurs throughout the year, and the preparation is unsuitable for microinjection or micromanipulation. Newt lung epithelial cells, which are grown as primary cultures from lung explants, have also been used to examine chromosome behavior. These cells and their chromosomes are large and the cells very flat, making them easy to observe and microinject (Rieder and Hard 1990). Spermatocytes of various insects have large meiotic spindles, and spindle fibers can be clearly visualized using polarized light microscopy; these cells have proven to be the system of choice for chromosome micromanipulation experiments (Nicklas

1997). Insect colonies can be maintained in the laboratory, and recent work has demonstrated that these cells can be successfully microinjected (Zhang and Nicklas 1999; D. Zhang, pers. comm.). Despite their excellent morphology, however, genetic and molecular approaches are limited in these cell types.

Cell division can also be studied in the eggs and embryos of various species. Extracts of frog eggs have become a very popular system for studying cell division. These meiotic extracts support spindle assembly and can be easily manipulated—for example, one can add antibodies to remove specific components and add fluorescent probes for real-time imaging (Desai et al. 1999). In addition, the extract can be maintained at particular cell cycle states (Lohka and Maller 1985; Murray 1991). The eggs and early embryos of various marine organisms can also be used to study mitosis. For example, sea urchin embryos develop synchronously following fertilization, are available in biochemical quantities, and are easy to microinject and micromanipulate (Sluder et al. 1999). The genome of at least one species of sea urchin is planned for sequencing, so this system may undergo a renaissance regarding its use for mitotic studies. Several developmental systems, notably, *Caenorhabditis elegans* and *Drosophila melanogaster*, are also very useful for analysis of living cells, including cells in mitosis (see Chapters 20 and 21). Finally, various fungi, including *Aspergillis nidulans* and yeast (*Saccharomyces cerevisiae* and *Schizosaccharomyces pombe*), have provided a wealth of molecular information about mitosis, in large part due to the ease with which genetic manipulations can be performed in these organisms as well as the ability to image cell division in these cells (Bloom et al. 1999) (see Chapter 19).

Cultured cells have also been a popular and useful system for studying mitosis. Several cultured cell lines that remain flat during mitosis are available and are excellent for microscopy and microinjection. Although cultured cells are not suitable for genetic approaches, they are easily transfected with dominant negative or constitutively active constructs, and RNA interference can be used to selectively reduce or eliminate the expression of specific genes. Because the cultured cells can be microinjected, antibody inhibition experiments can also be performed. Cultured cells are easily grown and maintained in the laboratory. It is important to note that cultured cells are somatic rather than embryonic and have an extended G1 phase of the cell cycle, but for researchers interested in the regulation of the somatic cell cycle, cultured mammalian cells are an ideal choice.

Studying Mitosis in Cultured Mammalian Cells

Experimental design will dictate the cell line chosen and often, more than one cell line may be suitable for a given study. Rat kangaroo kidney (PtK) cells are a frequent choice for studies of mitosis in cultured cells. (Note that two different lines, PtK_1, with 12 chromosomes, and PtK_2, with 13 chromosomes, are available. This chapter uses the term PtK to collectively refer to both of these cell lines). PtK cells have the advantages that they remain especially flat throughout mitosis and they have a small number of chromosomes, facilitating observations of chromosome behavior (Fig. 1). Another advantage of studying mitosis in PtK cells is that a considerable body of information regarding mitosis in these cells, including structural analysis, is already available (McDonald et al. 1992; Mastronarde et al. 1993; McEwen et al. 1997). PtK cells can be transfected using standard procedures, are easily microinjected, and are excellent for both light and electron microscopy.

Pig kidney cells (LLC-PK-1) are another epithelial cell line that remains flat throughout mitosis. These cells have more chromosomes than PtKs and a slightly thicker morphology. However, the cells are very hardy, easy to microinject (Wadsworth 1999), and transfect at high efficiency (see Fig. 1 and Movie 27.1 on the accompanying DVD). African green mon-

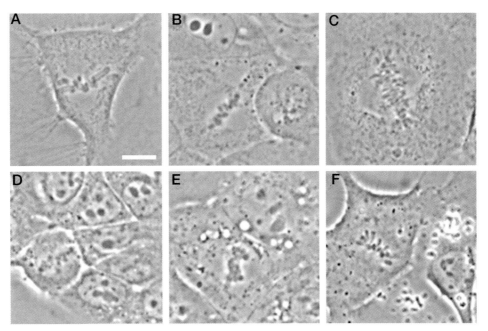

FIGURE 1. Phase contrast micrographs of living, cultured cells at the metaphase stage of mitosis. (*A*) PtK, (*B*) LLC-PK-1, (*C*) BS-C-1, (*D*) MDCK, (*E*) CFPAC-1, and (*F*) NRK-52E. Note the differences in spindle morphology. Bar, 10 μm.

key kidney cells (BS-C-1) are epithelial-like, remain flat throughout mitosis, and have a large, broad spindle. These cells are also characterized by a prominent centrosome and have been used for observations of centrosome behavior in live cells (Schliwa 1999) and for removal of the centrosome by cell cutting (Hinchcliffe et al. 2001). BS-C-1 cells can be transfected, microinjected, and used for microsurgery.

Madin Darby canine kidney (MDCK) epithelial cells are characterized by a polarized morphology in culture, especially when grown on filters. During cell division, the spindle moves to the apical region of the cell and orients parallel to the plane of the coverslip. Thus, these cells can be used to study not only the process of mitosis, but also the regulation of spindle orientation in mammalian cells (Busson et al. 1998). Polarized MDCK cells can measure up to 18 μm in the Z axis, much thicker than other cultured cells (Bacallao et al. 1989).

A disadvantage of all these cell lines (PtK, LLC-PK-1, BS-C-1, MDCK) is that they are derived from species for which genome projects are not anticipated in the near future, thus limiting the ease with which molecular manipulations can be performed. Given this limitation, cultured cells derived from human tissue or other organisms for which sequence information is, or soon to become, available, will be advantageous for studies of the molecular basis of chromosome motion. HeLa cells, which are derived from a human epithelial carcinoma, have been used to study many aspects of mitosis (see Haraguchi and Hiraoka, this volume). However, as these cells round up during mitosis, they can be difficult to microinject and are less suited for live cell imaging. An alternative to HeLa cells is CFPAC-1 cells, which are derived from a metastatic pancreatic adenocarcinoma. These cells are somewhat cuboidal in morphology and have numerous small chromosomes, but do not round up like HeLa cells. CFPAC-1 cells can be more easily microinjected, transfected, and used for live cell imaging studies (Compton 2000). Unfortunately, both HeLa and CFPAC-1 are carcinoma cells and likely to have genetic abnormalities. For example, CFPAC-1 cells are hyperdiploid.

Recently, human cell lines that express human telomerase reverse transcriptase have become commercially available (BD Biosciences Clontech). These cells are nontransformed, so genetic abnormalities are not an issue; such cell lines may be useful for mitotic research. Three lines are currently available, including human fibroblastic, mammary epithelial, and retinal pigment epithelial cells.

In addition to human cells, NRK-52E cells, derived from normal rat kidney, have also been used for analysis of mitosis and cytokinesis. These cells remain flat throughout cell division, can be readily microinjected, and can be used for live cell imaging experiments (Fishkind and Wang 1995). Because the rat genome is likely to be sequenced in the near future, these cells may be particularly useful.

Table 1 compares some key features of mitosis in various cultured cell lines. With the exception of PtK cells, most cells have a relatively large number of small chromosomes, making observations of individual chromosomes difficult, but not impossible. Most cultured cells also have a similar mitotic index, with ~2–5% mitotic cells in an exponentially growing culture. Interestingly, most of the cultured cells that remain flat during mitosis are derived from kidney and are of epithelial origin. It would be of considerable interest to determine if normal human kidney cell lines have an appropriate morphology for studies of mitosis.

Mitotic morphology in several cultured cell lines is shown in Figures 1 and 2. In Figure 1, images of living cells in phase contrast are shown; the spindle is recognized as the clear area that excludes cytoplasmic organelles and particles. Spindle morphology is clearly visible when cells are fixed and stained with antibodies to tubulin, the protein subunit of microtubules that make up the mitotic spindle. In all cases, the spindle microtubules converge at the two poles and in some cases, kinetochore fibers can be seen (Fig. 2); chromosomes are stained with propidium iodide. In all cases, a single confocal section is shown. The relative size of the mitotic spindle in various cultured cells can be seen in Figures 1 and 2, with PtK and LLC-PK-1 spindles showing a more elongated morphology and other spindles displaying a shorter and, in some cases, broader, spindle. Note that the degree of cell flattening, and thus spindle morphology, can be influenced by cell density and by coating the coverslips with various extracellular matrix molecules (Fishkind and Wang 1995).

A more detailed view of spindle structure can be seen in Figure 3, which shows projections of a complete Z series of images of an LLC-PK-1 spindle. The cells were fixed and

TABLE 1. Characteristics of cultured cell lines used to study mitosis

Cell line	Species	Tissue of origin	Tissue type, morphology	Mitotic index	Chromosome number
PtK$_1$ and PtK$_2$	rat kangaroo *Potorous tridactylis*	kidney	epithelial; flat	3.8	$2n = 12$ $2n = 13$
MDCK	dog *Canis familiaris*	kidney	epithelial; columnar	2.2	$2n = 78$
LLC-PK-1	*Sus scrofa*	kidney	epithelial; flat	3.6	$2n = 38$
CFPAC-1	human *Homo sapiens*	liver	epithelial-like; cuboidal	4.8	hyperdiploid
NRK-52E	rat *Rattus norvegicus*	kidney	epithelial; flat	3.7	$2n = 42$
BS-C-1	African green monkey *Cercopithecus aethiops*	kidney	epithelial-like; flat	2.6	$2n = 60$
HeLa	human *Homo sapiens*	cervix	epithelial-like; rounded	5.3	hyperdiploid

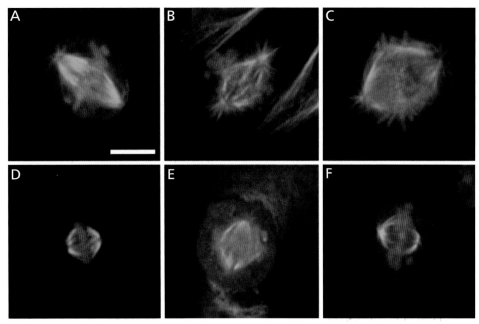

FIGURE 2. Spindle morphology in various cultured cells. Cells were fixed and stained with antibodies to tubulin; chromosomes were stained with propidium iodide. Images were collected using a Perkin-Elmer spinning disc confocal microscope; a single focal plane is shown. (*A*) PtK, (*B*) LLC-PK-1, (*C*) BS-C-1, (*D*) MDCK, (*E*) CFPAC-1, and (*F*) NRK-52E. Bar, 10 μm.

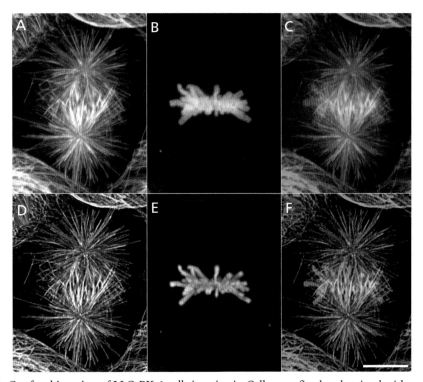

FIGURE 3. Confocal imaging of LLC-PK-1 cells in mitosis. Cells were fixed and stained with antibodies to tubulin (*A* and *D*); chromosomes were stained with propidium iodide (*B* and *E*); merged images are shown in *C* and *F*. The top row shows a projection of a Z series of images through a metaphase cell; these images were obtained using a Perkin-Elmer spinning disc confocal attached to a Nikon microscope. The bottom row shows the same images after deconvolution with AutoQuant software. Bar, 10 μm.

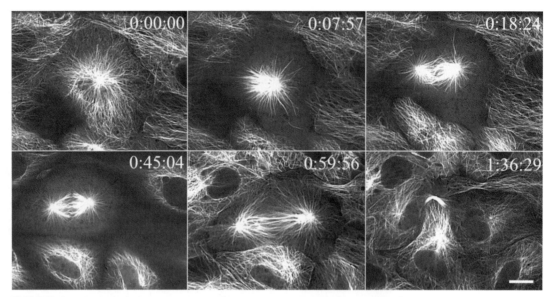

FIGURE 4. Mitosis in living LLC-PK-1 cells expressing GFP-tubulin. Images from a complete sequence of mitosis at the following time points (hours:minutes:seconds) are shown: (0:00:00) prophase, (0:07:57) prometaphase, (0:18:24) metaphase, (0:45:04) anaphase, (0:59:56) telophase, and (1:36:29) cytokinesis. Images were acquired using a Perkin-Elmer spinning disc confocal attached to a Nikon microscope. The complete movie is provided on the accompanying DVD. Bar, 10 μm.

stained with antibodies to tubulin (Fig. 3A) and with propidium iodide to stain the chromosomes (Fig. 3B; merged image shown in Fig. 3C). The same stack of images following deconvolution to remove out-of-focus information is shown in Figure 3D–F.

The process of mitosis is perhaps best appreciated from movie sequences illustrating the dynamic nature of the mitotic process. Several frames from a movie sequence of an LLC-PK-1α cell progressing through mitosis are shown in Figure 4. This type of cell constitutively expresses green fluorescent protein (GFP)-tubulin, and the dynamic changes in microtubule behavior throughout mitosis can be observed (Rusan et al. 2001). The complete movie sequence is provided on the accompanying DVD (Movie 27.1).

PROTOCOLS

Growing Cultured Cells

A major challenge for those who study mitosis is the need to keep cells alive and actively mitotic during observations. This protocol and those that follow describe methods for maintaining healthy, dividing cells in culture and during imaging. General procedures for tissue culture including basic equipment needed in a culture laboratory, sterile technique, media preparation, and procedures for subculturing, freezing, and thawing cells can be found in many laboratory methods books (e.g., Barker 1998).

1. Grow cells in 5–10% CO_2 at 37°C in a humid atmosphere. Use a medium recommended by the American Type Culture Collection, buffered with bicarbonate, and supplemented with 10% fetal calf serum and antibiotics.

> Note: In some cases, medium that has been mixed 1:1 with OptiMEM (Invitrogen) prepared with 5% serum improves cellular morphology.

2. Plate cells 1–3 days before use. Maximize the number of cells in mitosis at the time of observation by plating the cells densely enough so that they are in exponential growth at the time of the experiment. Determine the plating density empirically for each cell line.

3. For observation of living cells in mitosis, remove the bicarbonate-buffered medium and replace it with medium that lacks bicarbonate but instead maintains the pH with HEPES buffer (20 mM, pH 7.2).

> Note: Although cells proliferate faster in CO_2-containing medium, maintaining the cells in HEPES-buffered medium for short-term observations (several hours) causes no detectable adverse effects. For fluorescence observations, use medium that lacks the pH indicator dye, which is autofluorescent. To reduce photobleaching during fluorescence observations, add the oxygen scavenging reagent Oxyrase at a dilution of 1:100 in the medium.

4. For long-term storage, freeze cultured cells in the bicarbonate-buffered medium used for growing cells in Step 1. Supplement the medium with 15% dimethyl sulfoxide <!> and 20% serum. Store the frozen cells in liquid nitrogen tanks.

Mounting Coverslips for Imaging

> Note: <!> indicates hazardous material; see Cautions Appendix.

For use in experiments, cells are plated on sterile, glass coverslips that have been biologically cleaned. Each coverslip is hand washed in Alconox detergent, rinsed exhaustively in running hot water, and then rinsed in distilled water. Washed coverslips are stored in 95% ethanol and flame sterilized before use. See Lutz and Inoue (1986) for a detailed description of this cleaning method. An alternative, although less preferred, method for cleaning coverslips is to simply rinse them in 95% ethanol <!> and flame sterilize before use. For microinjection experiments or for experiments in which cells are to be fixed and relocalized after imaging, cells are grown on gridded coverslips (Bellco Glass).

Presented below are three ways to mount coverslips for imaging. Although the use of Rose chambers (Rose et al. 1958) for observation and microinjection of living mitotic cells is preferable (Wadsworth 1999), slide/coverslip preparations are easy to make and do not require any special equipment. Another inexpensive and easy-to-use alternative to using specially designed chambers is to culture cells in a culture dish with a glass bottom.

Slide/Coverslips

1. Plate cells on clean glass coverslips and allow them to grow to the desired density.

2. Remove a coverslip from the culture dish and blot excess medium onto a Kimwipe.

3. Invert the coverslip onto parafilm spacers on a clean glass slide.

> Note: Parafilm spacers are used to hold the coverslip above the slide surface and prevent the cells from getting crushed under the weight of the coverslip. To make the spacers, cut two thin strips of parafilm (~2 x 20 mm). Place them parallel to each other, about a coverslip's width apart, on the slide. Place a drop of medium between the spacers and carefully lower the coverslip so that it rests on the spacers.

4. Use the tip of a Kimwipe to remove excess medium from the top of the coverslip and the surface of the slide.

5. Seal the coverslip to the slide with Valap to prevent evaporation of medium during observation on the microscope.

SEALING MOUNTED COVERSLIPS WITH VALAP

Valap consists of equal parts of Vaseline petroleum jelly, lanolin, and paraffin (Lutz and Inoue 1986). Place Vaseline petroleum jelly, lanolin, and paraffin using a 1:1:1 ratio in a glass or ceramic vessel. Heat at a low setting on a hot plate until thoroughly blended. Apply the molten Valap with a small paintbrush to seal the edges of the coverslip to the slide's surface.

Rose Chambers

A Rose chamber (Fig. 5) consists of a lower metal plate with a circular cutout. The metal plate is covered with a layer of parafilm to prevent the coverslip from cracking when tightened against the metal plate. The coverslip is centered over the circular cutout and covered with a rubber gasket, a second coverslip, and the top metal plate. The entire assembly is held together by four screws. Medium can be added using a syringe and needle inserted into the gasket. Alternatively, the chamber can be constructed without the second coverslip; medium can then be added from above and the top opening of the chamber subsequently covered with a circular coverslip of 25-mm diameter.

The Rose chamber has many advantages:

- The rubber gasket makes a very tight seal and is thus virtually leakproof. Occasionally, a coverslip will crack and leak, but leakage almost never occurs around the gasket.

- The Rose chamber holds a larger volume of medium than a slide/coverslip preparation (the volume depends on the thickness of the rubber gasket), and this extra medium seems to keep the cells healthy for longer periods.

- The volume of the Rose chamber and the metal plates may help to reduce fluctuations in heating during observation (see below). Because the chambers are sealed, they can be returned to the 37°C incubator between observations or experiments.

- Coverslips mounted in a Rose chamber can be used for microinjection and subsequent high-magnification observations (the top coverslip is simply removed during injection and replaced when the cells are ready for imaging).

- Cells can be rapidly fixed after live cell observations by removing the top coverslip and adding fixative.

Although Rose chambers are not commercially available, several manufacturers make chambers similar in design, and some of these chambers also can be used for perfusion experiments (Bioptechs; for additional examples, see McKenna and Wang 1989). Rose chambers can be used on an inverted microscope or constructed with a thinner gasket and used on an upright microscope; alternatively, the chamber can be inverted (or constructed with the cell-containing coverslip on the top) for use with an upright microscope. Rose chambers can be held on the stage of most microscopes, or specially designed stage inserts can be made (Wadsworth 1999).

Glass-bottomed Culture Dishes

To culture cells in a culture dish with a glass (coverslip) bottom, the cells are seeded directly onto the bottom of the dish and allowed to grow. The medium can then be changed to non-

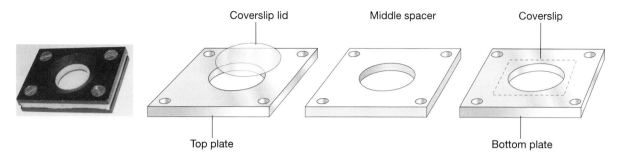

FIGURE 5. Photograph and diagram of a Rose chamber used for imaging living cells. The photograph shows the assembled chamber; the coverslip with cells is on the bottom, and the chamber is used on an inverted microscope. The diagram shows the parts of the chamber. The upper and lower plates are aluminum, and the inner spacer is made by polymerizing a sheet of silastic elastomer of the desired thickness and cutting the openings using a cork borer. To prevent cracking the coverslip when the chamber is assembled, a piece of parafilm (with a central opening) can be placed over the lower plate before adding the coverslip. The entire assembly is held together with four screws, and a round coverslip is added as a lid on the top.

CO_2 medium and the preparation simply moved to the stage of an inverted microscope. These dishes can also be used with an upright microscope with a water-immersion objective lens. Glass-bottomed dishes are available with a gridded bottom (e.g., MatTek) for following individual cells over long periods. The plastic lid of the dish will interfere with polarized light microscopy, but it can be removed and replaced with a round, glass coverslip of suitable diameter. Glass-bottomed dishes are inexpensive, easy to use, and suitable for microinjection experiments.

Maintaining Temperature

For observations of mitosis, the cells must be kept warm, ideally at 37ºC, although some cultured cells continue to divide at lower temperatures (Rieder 1981). For short-term studies, temperature can be maintained using an air-curtain-type incubator, which blows warm air at the sealed preparation. A thermister probe inserted directly into the Rose chamber can be used to monitor the temperature of the preparation during heating. This approach is relatively inexpensive and easy to assemble, but the microscope stand and lens (especially if using oil immersion) act as heat sinks, and the focus may fluctuate as the microscope warms. This fluctuation can be avoided by heating the microscope area before beginning the experiment. For long-term studies or for experiments that require more precise temperature control, the microscope can be enclosed in a chamber that surrounds the microscope or the stage region of the microscope. Such chambers are commercially available (e.g., Buck Scientific) or can be built from plexiglass or plywood. Also available are chambers that control the CO_2 as well as the temperature (McKenna and Wang 1989). It is not essential to regulate CO_2 for relatively short-term studies of mammalian cells in mitosis.

In addition to air curtain incubators, heated stages are available for some microscopes. These can be effective, especially if the observations are not made with oil-immersion objectives. If oil immersion is to be used, a heating collar that surrounds the objective lens can also be purchased (Bioptechs) or made, to prevent the lens from acting as a heat sink. (A lens heater can also be used in conjunction with the air-curtain-type incubator.) A heating block that holds a Rose chamber has also been designed (Rieder and Cole 1998b). A disadvantage of all of these methods is the introduction of strain into the lenses by the repeated cycles of heating and cooling. This problem can be circumvented by keeping the microscope lens in a dry incubator at 37°C (Bioptechs) when not in use. A final alternative is to place the entire microscope in an environmental room of constant temperature. (For long-term maintenance of temperature, see Chapter 18).

Maintenance of Mitotic Phase

In some cases, cells in early mitosis revert to interphase, a situation that must be avoided for studies of mitosis. For example, it has been noted that microinjection of cells in early mitosis can cause a return to interphase. This is thought to occur when calcium in the medium enters the cell during the process of microinjection (Vos et al. 1999). Removal of calcium from the culture medium can eliminate this problem, but the cells eventually detach from the surface in the absence of calcium ions. Prophase cells may also revert to interphase if they are damaged by light. In this situation, light-induced damage to DNA activates the G2 checkpoint, and the cells revert to interphase (Rieder and Cole 1998a). In general, progression through mitosis and overall cell health can be achieved by limiting the exposure of the cells to light, especially light at shorter wavelengths. In practice, when viewing cells with tungsten light for phase contrast or differential interference contrast microscopy, using green light and/or adding a heat cut filter is sufficient to keep cells actively mitotic. For fluorescence observations, expose cells to the least amount of fluorescent light as possible (see Chapter 28).

Cell Synchronization

For studies in which a large number of mitotic cells needs to be followed or a particular mitotic stage is desired, synchronization can be useful. For example, prophase cells constitute a minority of the mitotic cells on a coverslip, but incubation of coverslips of cells for 90 minutes in nocodazole (10 μM) followed by 90 minutes of washout results in a transient increase in prophase cells in the population (Rieder and Cole 2000). This procedure takes advantage of the observation that microtubule disassembly delays the G2–M transition in vertebrate cells (Rieder and Cole 2000).

Mitotic shake-off is an effective method to obtain relatively large numbers of cells in mitosis, but this method is limited to cells that round up in mitosis, such as HeLa. For cells that remain adherent during mitosis, thymidine can be used to block cells at the entry into S phase; following removal of the thymidine, the cells progress into G2 and M. The thymidine procedure is most effective when it is repeated (a double thymidine block) (Rao and Johnson 1970; Telzer and Rosenbaum 1979). Colcemid <!> can be used after the removal of thymidine to obtain cells synchronously entering but blocked in mitosis (Telzer and Rosenbaum 1979). A disadvantage of these methods is that, depending on the length of the cell cycle, it can take several days to obtain a synchronized population of cells.

MOVIE LEGEND

MOVIE 27.1. Mitosis in pig kidney cells expressing GFP-tubulin. The sequence shows a cell from prophase through cytokinesis.

REFERENCES

Bacallao R., Antony C., Dotti C., Karsenti E., Stelzer E.H.K., and Simons K. 1989. The subcellular organization of madin-darby canine kidney cells during the formation of a polarized epithelium. *J. Cell Biol.* **109**: 2817–2832.

Bajer A.S. 1990. The elusive organization of the spindle and the kinetochore fiber: A conceptual retrospect. *Adv. Cell Biol.* **3**: 65–93.

Barker K. 1998. *At the bench: A laboratory navigator.* Cold Spring Harbor Laboratory Press, Cold Spring Harbor, New York.

Bloom K., Beach D.L., Maddox P., Shaw S.L., Yeh E., and Salmon E.D. 1999. Using green fluorescent protein fusion proteins to quantitate microtubule and spindle dynamics in budding yeast. *Methods Cell Biol.* **61**: 369–383.

Busson S., Dujardin D., Moreau A., Dompierre J., and DeMay J.R. 1998. Dynein and dynactin are localized to astral microtubules and at cortical sites in mitotic epithelial cells. *Curr. Biol.* **8**: 541–544.

Compton D.A. 2000. Spindle assembly in animal cells. *Annu. Rev. Biochem.* **69**: 95–114.

Desai A., Murray A.W., Mitchison T.J., and Walczak C.E. 1999. The use of *Xenopus* egg extracts to study mitotic spindle assembly and function in vitro. *Methods Cell Biol.* **61**: 386–411.

Fishkind D.J. and Wang Y.-L. 1995. New horizons for cytokinesis. *Curr. Biol.* **7**: 23–31.

Hinchcliffe E.H., Cham M., Khodjakov A., and Sluder G. 2001. Requirement of a centrosomal activity for cell cycle progression through G1 into S phase. *Science* **291**: 1547–1550.

Inoue S. and Bajer A.S. 1961. Birefringence in endosperm mitosis. *Chromosoma* **12**: 48–63.

Lohka M.J. and Maller J.L. 1985. Induction of nuclear envelope breakdown, chromosome condensation, and spindle formation in cell-free extracts. *J. Cell Biol.* **101**: 518–523.

Lutz D.A. and Inoue S. 1986. Techniques for observing living gametes and embryos. *Methods Cell Biol.* **27**: 89–110.

Mastronarde D.N., McDonald K.L., Ding R., and McIntosh J.R. 1993. Interpolar spindle microtubules in PtK cells. *J. Cell Biol.* **123**: 1475–1489.

McDonald K.L., O'Toole E., Ding R., and McIntosh J.R. 1992. Kinetochore microtubules in PtK cells. *J. Cell Biol.* **118**: 369–383.

McEwen B.F., Heagle A.B., Cassels G.O., Buttle K.F., and Rieder C.L. 1997. Kinetochore fiber maturation in PtK1 cells and its implications for the mechanisms of chromosome congression and anaphase onset. *J. Cell Biol.* **137**: 1567–1580.

McKenna N.M. and Wang Y.-L. 1989. Culturing cells on the microscope stage. In *Fluorescence microscopy of living cells in culture* (ed. Y.-L. Wang and D.L. Taylor), vol. 29, pp. 195–205. Academic Press, New York.

Murray A.W. 1991. Cell cycle extracts. *Methods Cell Biol.* **36**: 581–605.

Nicklas R.B. 1997. How cells get the right chromosomes. *Science* **275**: 632–637.

Rao P.N. and Johnson R.T. 1970. Mammalian cell fusion: Studies on the regulation of DNA synthesis and mitosis. *Nature* **225**: 159–164.

Rieder C.L. 1981. Effect of hypothermia (20–25ºC) on mitosis in PtK1 cells. *Cell Biol. Int. Rep.* **5**: 563–573.

Rieder C.L. and Cole R.W. 1998a. Entry into mitosis in vertebrate somatic cells is guarded by a chromosome damage checkpoint that reverses the cell cycle when triggered during early but not late prophase. *J. Cell Biol.* **142**: 1013–1022.

———. 1998b. Perfusion chambers for high resolution video light microscopic studies of vertebrate cell monolayers: Some considerations and a design. *Methods Cell Biol.* **56**: 253–277.

———. 2000. Microtubule disassembly delays the G2–M transition in vertebrates. *Curr. Biol.* **10**: 1067–1070.

Rieder C.L. and Hard R. 1990. Newt lung epithelial cells: Cultivation, use, and advantages for biomedical research. *Int. Rev. Cytol.* **122**: 153–220.

Rose G.G., Pomerat C.M., Shindler T.O., and Trunnel J.B. 1958. A cellophane strip technique for culturing tissue in multipurpose culture chambers. *J. Biophys. Biochem. Cytol.* **4**: 761–764.

Rusan N.M., Fagerstrom C., Yvon A.C., and Wadsworth P. 2001. Cell-cycle-dependent changes in microtubule dynamics in living cells expressing GFP-alpha tubulin. *Mol. Biol. Cell* **12**: 971–980.

Schliwa M. 1999. Centrosomes, microtubules, and cell migration. *Biochem. Soc. Symp.* **65**: 223–231.

Sluder G., Miller F.J., and Hinchcliffe E.H. 1999. Using sea urchin gametes for the study of mitosis. *Methods Cell Biol.* **61**: 440–473.

Telzer B.R. and Rosenbaum J.L. 1979. Cell-cycle-dependent, *in vitro* assembly of microtubules onto the pericentriolar material of HeLa cells. *J. Cell Biol.* **81:** 484–497.

Vos J.W., Valster A.H., and Hepler P.K. 1999. Methods for studying cell division in higher plants. *Methods Cell Biol.* **61:** 413–434.

Wadsworth P. 1999. Microinjection of mitotic cells. *Methods Cell Biol.* **61:** 219–231.

Zhang D. and Nicklas R.B. 1999. Micromanipulation of chromosomes and spindles in insect spermatocytes. *Methods Cell Biol.* **61:** 209–217.

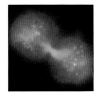

CHAPTER TWENTY EIGHT

Imaging Hoechst-33342-labeled Chromosomes and Fluorescent Proteins During the Cell Cycle

Tokuko Haraguchi and Yasushi Hiraoka

Kansai Advanced Research Center, National Institute of Information and Communications Technology, 588-2 Iwaoka, Iwaoka-cho, Nishi-ku, Kobe 651-2492, Japan

INTRODUCTION

FLUORESCENCE MICROSCOPY PROVIDES A POWERFUL TOOL for imaging specific molecular components in living cells. This is especially so today, as the construction of fusion proteins with green fluorescent protein (GFP) and its color variants has made the expression of fluorescent proteins within living cells a relatively simple and straightforward procedure: For example, the nuclear envelope (Haraguchi et al. 2000), centromere (Hori et al. 2003), microtubules (Nishihashi et al. 2002), microtubule motor proteins (Matuliene et al. 1999), and septin (Kinoshita et al. 1997) were visualized by using GFP fusions, and their cell cycle behaviors were followed in living vertebrate cells with Hoechst-33342-labeled chromosomes. This chapter includes protocols for imaging fluorescent cellular proteins in combination with Hoechst-33342-labeled chromosomes in living cells during the cell cycle.

Fluorescent Reagents

Several reagents can be used to fluorescently label specific molecular components in living cells, including fluorescent dyes, DNA constructs that express GFP-fusion proteins within the cell and fluorescently conjugated proteins that have been microinjected into the cell.

Low-molecular-weight fluorescent dyes that stain specific intracellular molecular components are widely used, including $DiOC_6(3)$ for staining lipid membranes and DNA-specific fluorescent dyes for staining chromosomes (Haraguchi et al. 1999; Swedlow and Platani 2002). The well-known DNA-specific fluorescent dyes DAPI, Hoechst 33258, and Hoechst 33342 all stain chromosomes equally well in fixed specimens, but they have different properties when used for staining chromosomes in living cells. Among these dyes, Hoechst 33342 is the best at permeating living cells and is best retained on DNA (Haraguchi et al. 1999). Chromosomes in living cells can be labeled simply by adding Hoechst 33342 to the culture medium.

Specific proteins within living cells can be made fluorescent by expressing them in vivo as GFP-fusion proteins. These fusion proteins are typically introduced into cells by transfection with the corresponding plasmid (see Chapter 2). Microinjection of the GFP-fusion plasmid into the nucleus generally gives better results than chemical transfection. Microinjection produces sufficient expression of the GFP-fusion protein with less disruption of the cells' normal physiology, but it is technically difficult and can encompass only a small number of cells. Fortunately, if sufficient care is taken, currently available chemical transfection methods provide relatively high transfection efficiencies, induce relatively little physiological damage to the cells, and are generally suitable for live analysis.

Another method used to visualize specific proteins within cells is microinjection of fluorescently conjugated proteins (see Chapter 5). These fluorescent proteins can be produced by chemically labeling the protein with a fluorescent dye or by purifying recombinant GFP-fusion proteins. There are several important advantages to microinjecting purified proteins rather than expressing GFP-fusion proteins in vivo: (1) Purified proteins can be tested for their native biochemical functions, (2) a small proportion of fluorescent proteins relative to endogenous proteins can be introduced into the cells, and (3) purified, fluorescent proteins can be introduced into cells immediately before observation, causing fewer adverse effects on cell physiology.

Instrumentation

Several microscope systems based on wide-field or confocal microscopy are now commercially available for recording live cell images. Which system to choose depends on the specimen and staining method. Confocal fluorescence microscopy typically requires a higher intensity of excitation light to obtain the same level of fluorescent light obtained with a wide-field fluorescence microscope. If fluorescent staining is sufficiently bright and of low toxicity, however, a confocal fluorescence microscope can be used. The phototoxic effects of fluorescently staining living cells depend largely on the fluorescent dye and the labeled molecule. Direct staining of DNA with a DNA-binding fluorescent dye is especially toxic. Currently, it is impractical to use confocal fluorescence microscope systems to image Hoechst-33342-labeled DNA in living cells.

The use of a high-sensitivity camera is important for live cell imaging. It is the most important factor for reducing the intensity level of excitation light, and thus for minimizing phototoxicity. Typically, a cooled, charge-coupled device (CCD) is used as an image detector with a wide-field fluorescence microscope, and a photomultiplier tube (PMT) is used with a laser-scanning confocal fluorescence microscope (see Chapters 6 and 14)

A shutter for the excitation light is another essential aspect of the microscope system for live cell imaging. Hoechst 33342 has an excitation maximum of 360 nm, which is in the UV spectrum and can cause considerable DNA damage. Continuous exposure of Hoechst-33342-labeled living cells to intense UV light causes metaphase arrest, probably due to checkpoint controls for DNA damage. Thus, time-lapse images are taken using a short exposure time for each time point, and the total exposure time should be limited. In addition, to reduce the damage induced by the short-wavelength spectra generated by the high-pressure mercury arc lamp, it is advisable to use an excitation filter with a narrow peak at 380 nm (Chroma Technology) to excite Hoechst 33342 molecules, instead of a standard UV filter with a peak at 360 nm. The use of neutral density (ND) filters is also recommended to reduce the level of excitation light when necessary.

Temperature control during microscopic observation is important not only for cell physiology, but also for stability of microscope focus. A change in temperature can easily cause a

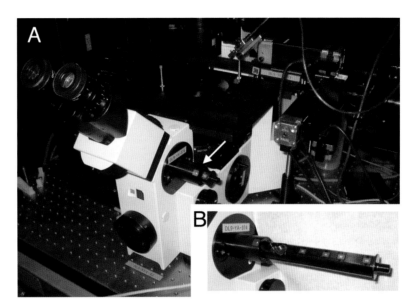

FIGURE 1. Microscope system setup. (*A*) A DeltaVision microscope system based on an Olympus inverted microscope IX70. A pseudo phase-contrast IX-PPC (Olympus Optical) is inserted beneath the binocular (indicated by the arrow) to generate phase-contrast images. The PPC pulled out from the microscope is shown in the inset (*B*).

focus shift due to thermal expansion of the microscope. The author's microscope system is located in a custom-made room in which the temperature can be controlled in a range from 10°C to 50°C. The computer and other control units are placed outside the room and control the microscope remotely (Haraguchi et al. 1997, 1999). When such a room is not practical, the combination of a heated stage and a temperature-controlled hood will suffice. The use of a heated stage alone cannot maintain the specimen at the desired temperature, because the objective lens acts as a heat sink. On the other hand, with a hood alone, the temperature may change due to a feedback circuit for temperature control. Combining the stage with the hood compensates for such a temperature oscillation, and thus provides stable and precise control. Details of these temperature-controlled devices are described in Spector et al. (1997).

To obtain high-resolution live cell images, the choice of the objective lens is critical. An oil-immersion objective lens 40× 1.35/NA (Olympus UApo40) is recommended for imaging HeLa cells and an oil-immersion objective lens 60× 1.4/NA (Olympus PlanApo60) for viewing fission yeast cells. These high-numerical-aperture (NA) objective lenses have no phase ring, however, and thus do not generate phase-contrast images, which are important for the visual screening of cells under transmitted light, especially during microinjection. To remedy this problem, we use a custom-made phase ring plate (designated pseudo phase contrast [PPC]) that is inserted beneath the binocular (Fig. 1). With this PPC insert, phase-contrast images may be observed through the binocular without affecting the images recorded by the camera.

PROTOCOLS

Specific molecular components in living cells can be efficiently labeled by a combination of three methods: (1) chemical transfection of GFP-fusion constructs, (2) staining of chromosomes with the DNA-specific, fluorescent dye Hoechst 33342, and (3) microinjection of fluorescently conjugated proteins. Figure 2 shows a flowchart of a procedure that includes all

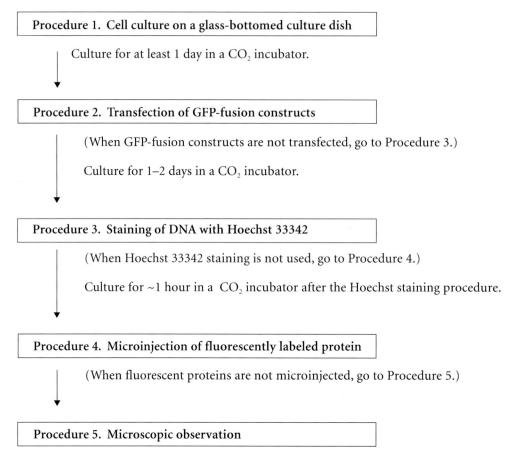

FIGURE 2. Flowchart of the Protocols.

three labeling methods. These methods should be followed in the order presented in the flowchart, but any of them can be omitted when not needed.

Note: <!> indicates hazardous material. See Cautions Appendix.

Cell Culture

Cell culture is the most important requirement for the success of live cell imaging. Whether the methodology used is Hoechst 33342 staining or expression of fluorescent proteins, fluorescent labeling can be toxic to cells. Culture conditions, including cell density and culture medium (serum, pH, temperature, etc.), must be carefully controlled to keep the cells healthy and optimally maintained during all procedures. Because high-resolution objective lenses are generally designed for viewing specimens through a glass coverslip, cells are cultured in a glass-bottomed culture dish instead of a plastic dish. Alternatively, cells can be grown on a glass coverslip fitted for a live cell culture chamber (e.g., Rose chamber [Rose et al. 1958], Focht FCS2 chamber [Bioptechs]). The cells should be checked for growth rate, morphology, adherence, etc., because some cell types may not adhere well to glass surfaces. If the cells exhibit a slow population growth rate due to loss of adherence to the dish, coating the glass surface with collagen, poly-L-lysine, or gelatin may help. The following steps describe the culture of HeLa cells as an example of a typical cell culture protocol.

1. Plate 1×10^{-5} to 2×10^{-5} HeLa cells on a 35-mm glass-bottomed culture dish (e.g., MatTek Corp., Ashland, Massachusetts).

The cells should be growing in 2 ml of DMEM (Dulbecco's modified Eagle's medium) supplemented with 10% calf serum.

2. Culture the cells in a CO_2 incubator for 1–2 days at 37ºC.

Transfection of the GFP-fused Gene

Cells should be transfected with a GFP-fusion construct 1–2 days before observation under the microscope. The following transfection protocol is based on the use of the LipofectaminePLUS transfection kit (Invitrogen), with a few minor modifications to the manufacturer's protocol. Transfection of eukaryotic cells with expression plasmids is described in detail in Chapter 2. Use standard sterile technique when handling cell cultures and reagents.

1. Prepare a DNA expression plasmid that encodes the GFP-fusion protein, using a phenol:chloroform <!> extraction method (e.g., see Sambrook and Russell 2001).

    Do not use DNA prepared with column-based plasmid purification kits for live cell imaging, because transfection with DNA prepared in this manner can cause an unacceptably high level of cell damage.

2. Label two microfuge tubes as Tube 1 and Tubes 2, and deliver 50 μl of serum-free medium into each. Prepare a set of tubes for each dish of cells (35-mm culture dish) to be transfected.

3. Add 0.2–0.5 μg of purified plasmid DNA to Tube 1, and mix gently.

4. Add 3 μl of Plus Reagent to the DNA solution (Tube 1) and mix gently. Incubate the mixture for 15 minutes at room temperature.

5. Add 4 μl of Lipofectamine Reagent to Tube 2 and mix gently.

6. Combine the contents of Tube 1 (DNA) and Tube 2 (Lipofectamine), and mix by vortexing. Incubate the mixture for 15 minutes at room temperature.

7. Remove a cell culture dish from the incubator, and replace the culture medium with 1 ml of serum-free medium. Add the DNA + Lipofectamine solution prepared in Step 6 to the cells, mix by swirling, and incubate in a CO_2 incubator for 1–2 hours at 37°C.

    Although the manufacturer's typical protocol states that cells should be incubated in the DNA + Lipofectamine solution for 3–5 hours, we recommend making the incubation as short as possible, because incubation of cells in serum-free medium can disturb the cells' physiology.

8. Replace the serum-free medium containing the DNA + Lipofectamine with complete cell culture medium (e.g., DMEM + 10% calf serum), and incubate in a CO_2 incubator for 1–2 days at 37°C.

Staining DNA with Hoechst 33342

As noted above, chromosomes in living cells are stained with Hoechst 33342 simply by adding the dye to the culture medium.

1. Prepare stock and working solutions of Hoechst 33342 <!>.

STOCK SOLUTION OF HOECHST 33342

FROZEN STOCK
10 mg/ml Hoechst 33342 in double-distilled H_2O. Store at –30°C.

WORKING SOLUTION
100 μg/ml Hoechst 33342 in double-distilled H_2O. Store in the dark at 4°C.

This Hoechst 33342 solution can be stored for 6 months at 4°C.

---

2. Add 2 μl of 100 μg/ml Hoechst 33342 to cells that are cultured in 2 ml of complete medium (final concentration 100 ng/ml), and mix by swirling.

3. Incubate the cell culture in a CO_2 incubator for 5–30 minutes at 37°C.

4. Wash the cells three times with culture medium (e.g., DMEM + 10% calf serum) that has been prewarmed to 37°C, and keep the culture in a CO_2 incubator for ~1 hour at 37°C.

Notes

Do not stain chromosomes too heavily with Hoechst 33342. To reduce phototoxicity, use a minimum detectable level of fluorescence intensity. Fluorescence intensity can be controlled by changing the duration of the Hoechst 33342 treatment. Conditions for Hoechst 33342 staining may differ depending on cell type. The concentration and duration of the Hoechst 33342 treatment must be optimized for each cell type being used. A less toxic method for staining chromosomes in living cells is the use of GFP-fused chromatin proteins such as histone (Kanda et al. 1998; Kimura and Cook 2001).

Microinjection of Fluorescently Labeled Proteins

The following method for microinjection of rhodamine-labeled tubulin is provided as an example protocol (see Haraguchi et al. 1997). Rhodamine-conjugated tubulin is prepared using tetramethylrhodamine succinimidylester (Molecular Probes). After rhodamine labeling, functional tubulin proteins are purified by cycling the assembly and disassembly of tubulin filaments (Hyman et al. 1991).

1. Prepare a solution of 1 mg/ml rhodamine-labeled tubulin in phosphate-buffered saline (PBS). Centrifuge 2 μl of this solution through a spin filter (Millipore 4-GV; pore size 0.2 μm) at 10,000 rpm for 2 minutes at 4°C to remove aggregates.

    If the available volume of fluorescently labeled protein is low (0.5–1 μl), apply 1 μl of PBS to the spin filter before applying the labeled protein solution. This method allows full recovery of the applied volume of protein.

2. Replace the medium in the cell culture (e.g., DMEM +10% calf serum) with observation culture medium prewarmed to 37°C, and place the culture dish on a microscope stage.

OBSERVATION CULTURE MEDIUM

DME medium without phenol red
10% fetal bovine serum (FBS)
20 mM HEPES buffer (pH 7.3)
80 μg/ml kanamycin sulfate <!>

DME medium can be replaced with any medium, but when observing red fluorescence (e.g., from rhodamine), phenol red can cause background and should be removed or reduced.

---

3. Inject the rhodamine-tubulin solution prepared in Step 1 into cells with a microinjector (0.1 second at an injection pressure of 300–500 hPa). Replace the observation culture medium with fresh, prewarmed (37°C) medium to remove residual fluorescence.

    When microinjecting fluorescent tubulins, load the tubulin solution into the microinjection needle at room temperature. It is important to microinject promptly once the needle is taken into the 37°C room, because tubulins polymerize to form microtubules at 37°C and will obstruct the needle.

    Residual fluorescent protein leaking into the medium during microinjection generates significant background fluorescence.

4. Leave the cell culture on a temperature-controlled microscope stage for ~2 hours before observation to allow the cells to recover from microinjection.

> Microinjected cells can be moved to a CO_2 incubator until microscopic observation, but the cells should be placed on a microscope stage at least 30 minutes before observation to avoid microscope focus movements.

Microscopic Observation

For microscopic observation, cells are cultured in the observation culture medium in a CO_2 incubator for ≥30 minutes at 37ºC before observation. HEPES buffer is used to maintain the pH in the absence of CO_2 gas.

1. If the microinjection step in the preceding protocol was skipped, replace the medium in the cell culture (e.g., DMEM) with observation culture medium (see preceding protocol for recipe) and place the culture dish on the microscope stage. If microinjection was performed, proceed to Step 2 (the cells should be in the observation culture medium on a microscope stage).

2. Overlay ~1–2 ml of mineral oil prewarmed to 37ºC on the culture medium to avoid evaporation of the medium during observation.

> For convenience, apply an excess amount of mineral oil, and then remove the excess so that just enough oil remains to cover the entire surface of the medium.

3. Allow the cell culture to sit on the microscope stage for ~1 hour before observation.

> If the cell culture is observed immediately, the microscope focus can change due to thermal expansion of the culture dish.

4. Record images with a computer-controlled microscope system.

Notes

Do not view Hoechst-labeled cells through the binocular when searching for cells to image; instead, use images taken by the computer controlled camera. Viewing through the binocular causes unnecessary exposure of cells to the excitation light. In addition, use a lamp shutter to minimize exposure time. If necessary, reduce the level of excitation light by using a neutral density (ND) filter. This is especially important for observing cells labeled with Hoechst 33342, which can cause DNA damage. The DNA damage checkpoint of the cell itself is probably the most sensitive sensor for DNA damage. Monitor the cells during or after data collection to verify that the cells proceed with metaphase normally. If the imaged cells remain at metaphase too long (more than 60–90 minutes in the case of HeLa cells), they probably have damaged DNA. In such cases, the total dose of excitation light should be reduced by taking images with shorter exposure times and longer time-lapse intervals.

PRACTICAL APPLICATION

Figure 3 shows an example of in vivo observation of chromosomes, centromeres, and microtubules in living HeLa cells. Movie 28.1 presents these observations on the accompanying DVD. In this example, all three staining methods described above were used: Chromosomes were labeled with the DNA-specific fluorescent dye Hoechst 33342, centromeres were labeled with CENP-B fused with GFP, and microtubules were labeled with rhodamine-conjugated tubulin. Time-lapse images were recorded at 5-minute intervals from interphase to the next interphase with the DeltaVision microscope system (Applied Precision); selected images during mitosis are shown. Note that Hoechst 33342 can also be used to stain chromosomes in combination with GFP staining in living fission yeast cells (Chikashige et al. 1994; Ding et al. 1998; Haraguchi et al. 1999).

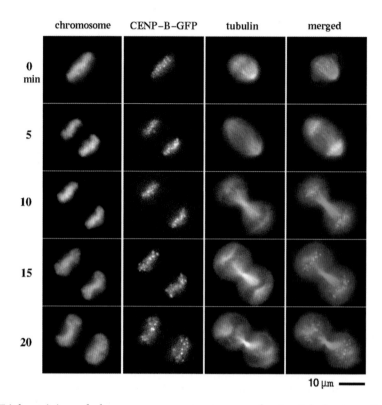

FIGURE 3. Triple staining of chromosomes, centromeres, and microtubules in a living HeLa cell. Chromosomes were labeled with Hoechst 33342 (*first column*), centromeres were labeled with GFP-fused CENP-B protein (*second column*), and microtubules were labeled with rhodamine-conjugated tubulins (*third column*). In the merged images (fourth column), chromosomes are shown in blue, centromeres are in red, and microtubules are in green. Images were obtained using an oil-immersion objective lens (Olympus UApo40× /NA = 1.35) on an Olympus inverted microscope IX70. See also Movie 28.1 on the accompanying DVD. (Modified, with permission, from Haraguchi and Hiraoka 2000.)

MOVIE LEGEND

MOVIE 28.1. Cell cycle behaviors of chromosomes and microtubules labeled with Hoechst 33342 and rhodamine-conjugated tubulin, respectively. More examples of live cell imaging can be seen on our Web site at http://www-karc.nict.go.jp/d332/CellMagic/index.html.

REFERENCES

Chikashige Y., Ding D.Q., Funabiki H., Haraguchi T., Mashiko S., Yanagida M., and Hiraoka Y. 1994. Telomere-led premeiotic chromosome movement in fission yeast *Schizosaccharomyces pombe*. *Science* **264:** 270–273.

Ding D.-Q., Chikashige Y., Haraguchi T., and Hiraoka Y. 1998. Oscillatory nuclear movement in fission yeast meiotic prophase is driven by astral microtubules as revealed by continuous observation of chromosomes and microtubules in living cells. *J. Cell Sci.* **111:** 701–712.

Hariguchi T. and Hiraoka Y. 2000. *Illustrated protocol of microscope*, pp. 142–146. Shujunsha, Tokyo. (Printed in Japanese.)

Haraguchi T., Kaneda T., and Hiraoka Y. 1997. Dynamics of chromosomes and microtubules visualized by multiple-wavelength fluorescence imaging in living mammalian cells: Effects of mitotic inhibitors on cell cycle progression. *Genes Cells* **2:** 369–380.

Haraguchi T., Ding D.-Q., Yamamoto A., Kaneda T., Koujin T., and Hiraoka Y. 1999. Multiple-color fluorescence imaging of chromosomes and microtubules in living cells. *Cell Struct. Funct.* **24:** 291–298.

Haraguchi T., Koujin T., Hayakawa H., Kaneda T., Tsutsumi C., Imamoto N., Akazawa C., Sukegawa J., Yoneda Y., and Hiraoka Y. 2000. Live fluorescence imaging reveals early recruitment of emerin, LBR, RanBP2, and Nup153 to reforming functional nuclear envelopes. *J. Cell Sci.* **113:** 779–794.

Hori T., Haraguchi T., Hiraoka Y., Kimura H., and Fukagawa T. 2003. Dynamic behavior of Nuf2-Hec1 complex that localizes to the centrosome and centromere and is essential for mitotic progression in vertebrate cells. *J. Cell Sci.* **116:** 3347–3362.

Hyman A., Drechsel D., Kellogg D., Salser S., Sawin K., Steffen P., Wordeman L., and Mitchison T. 1991. Preparation of modified tubulins. *Methods Enzymol.* **196:** 478–485.

Kanda T., Sullivan K.F., and Wahl G.M. 1998. Histone-GFP fusion protein enables sensitive analysis of chromosome dynamics in living mammalian cells. *Curr. Biol.* **8:** 377–385.

Kimura H. and Cook P.R. 2001. Kinetics of core histones in living human cells: Little exchange of H3 and H4 and some rapid exchange of H2B. *J. Cell Biol.* **153:** 1341–1353.

Kinoshita M., Kumar S., Mizoguchi A., Ide C., Kinoshita A., Haraguchi T., Hiraoka H., and Noda M. 1997. Nedd5, a mammalian septin, is a novel cytoskeletal component interacting with actin-based structures. *Genes Dev.* **11:** 1535–1547.

Matuliene J., Essneer R., Ryu J.-H., Hamaguchi Y., Baas P.W., Haraguchi T., Hiraoka Y., and Kuriyama R. 1999. Function of a minus-end-directed kinesin-like motor protein in mammalian cells. *J. Cell Sci.* **112:** 4041–4050.

Nishihashi A., Haraguchi T., Hiraoka Y., Ikemura T., Regnier V., Dodson H., Earnshaw W. C., and Fukagawa T. 2002. CENP-I is essential for centromere function in vertebrate cells. *Dev. Cell* **2:** 463–476.

Rose G. G., Pomerat C.M., Shindler T.O., and Trunnel J.B. 1958. A cellophane strip technique for culturing tissue in multipurpose culture chambers. *J. Biophys. Biochem. Cytol.* **4:** 761–764.

Sambrook J. and Russell D.W. 2001. *Molecular cloning: A laboratory manual,* 3rd edition. Cold Spring Harbor Laboratory Press, Cold Spring Harbor, New York.

Spector D.L., Goldman R.D., and Leinwand L.A., eds. 1997. *Cells: A laboratory manual,* Volume 2: *Light microscopy and cell structure,* Chapter 75. Cold Spring Harbor Laboratory Press, Cold Spring Harbor, New York.

Swedlow J.R. and Platani M. 2002. Live cell imaging using wide-field microscopy and deconvolution. *Cell Struct. Funct.* **27:** 335–341.

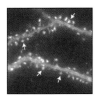

CHAPTER TWENTY NINE

Imaging the Actin Cytoskeleton

Andrew Matus, Virginie Biou, Heike Brinkhaus, and Martijn Roelandse
Friedrich Miescher Institute, 4002 Basel, Switzerland

INTRODUCTION

THROUGH ITS CAPACITY FOR DYNAMIC REARRANGEMENT, the actin cytoskeleton supports a wide variety of intracellular structures ranging from the rigid actomyosin arrays of skeletal muscle to the motile lamellipodia of crawling fibroblasts. The most convenient way of studying these transitions is to use living cells expressing a fusion construct of actin linked to green fluorescent protein (GFP) (see Chapter 2). Actin-GFP constructs can be introduced into cells both by transfection and by raising transgenic mice in which the construct is expressed intrinsically. The availability of transgenic mice opens up the possibility of studying changes in the actin cytoskeleton in tissue slices maintained in vitro or, as more advanced imaging techniques become available, in intact animals (see Chapters 22 and 23). Nevertheless, for high-resolution imaging of actin dynamics in living cells, the best results are obtained with dissociated cell cultures. For this purpose, it is often more convenient to use cells from wild-type animals transfected in vitro, rather than cells derived from transgenic mice. This chapter focuses on studying actin dynamics in neurons of the central nervous system and describes the use of both transfected hippocampal pyramidal cell in low-density culture and cultured tissue slices derived from transgenic animals intrinsically expressing actin-GFP.

GENERAL CONSIDERATIONS

Before using a fusion construct, it is important to determine whether the addition of GFP has influenced the function of the tagged protein (Fig. 1A and Movie 29.1 on the accompanying DVD). A basic control consists of comparing constructs where GFP is attached to either the amino or carboxyl terminus of the target protein. Tests using transfected fibroblasts show that both configurations—actin-GFP and GFP-actin—effectively label actin-containing structures. Similarly, primary neurons from the brain expressing GFP-actin by transfection

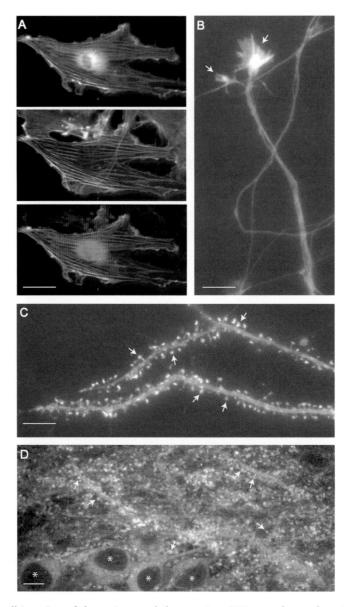

FIGURE 1. Live cell imaging of the actin cytoskeleton using GFP-tagged cytoplasmic actin. In all three examples, cells were transfected with a cDNA construct consisting of γ-cytoplasmic actin fused at the amino terminus to enhanced green fluorescent protein (eGFP). (*A*) Fluorescence micrographs of a culture of REF52 fibroblasts in which a single transfected cell expresses GFP-actin (*top panel*). (*Middle panel*) Corresponding image produced by staining the cells after fixation with rhodamine-phalloidin, which selectively labels actin filaments. (*Bottom panel*) Overlay of the two images in which the GFP fluorescence is shown in green and the rhodamine-phalloidin fluorescence is shown in red. Actin-rich stress fibers and surface accumulations of actin are labeled by both GFP-actin and rhodamine-phalloidin and appear yellow. Note the additional GFP-actin labeling of nonfilamentous actin in the cytoplasm (*green*). Motility associated with the cell surface actin accumulations is shown in Movie 29.1 (see below). (*B*) A developing neuron from embryonic rat brain transfected to express GFP-actin showing dynamic actin structures in growth cones (arrows). This frame is from the time-lapse recording shown in Movie 29.2. (*C*) Dendrites from a mature GFP-actin-expressing hippocampal neuron showing actin-rich dendritic spines, which appear as brightly fluorescent dots and some are marked with arrows (see also Movie 29.3). (*D*) Confocal image of a brain slice culture (area CA1 of hippocampus) from a transgenic mouse expressing actin-GFP showing actin-rich dendritic spines some of which are indicated by arrows. Asterisks indicate neuronal nuclei (see Movie 29.4). Bars: (*A*), 20 μm; (*B* and *C*), 10 μm; (*D*), 10 μm.

show the expected concentration of GFP-tagged actin in axonal growth cones, allowing actin dynamics to be readily observed by time-lapse recording (Fig. 1B and Movie 29.2 on the accompanying DVD).

Cells of the immortalized rat embryo fibroblast line REF52 are convenient for setting up time-lapse imaging of the actin cytoskeleton and evaluating GFP-tagged actin constructs. For general information on cell culture, transfecting fibroblasts with GFP constructs, and live cell imaging techniques, see Spector et al. (1998a, Chapters 75 and 78). In 10% serum, REF52 cells have motile lamellipodia suitable for imaging actin dynamics, whereas after 36 hours in reduced (0.25%) serum, the cells become quiescent and the more stable stress fibers predominate. For fixation and staining of the actin cytoskeleton with rhodamine-phalloidin see Spector et al. (1998b, Chapter 10).

IMAGING GFP-TAGGED ACTIN IN TRANSFECTED HIPPOCAMPAL NEURONS

In general, the more intact the tissue preparation, the more technically demanding the imaging and the lower the obtainable resolution. Consequently, for high-resolution imaging of cytoskeletal dynamics in brain neurons, low-density cultures are the best choice. Two methods for low-density culture of rat hippocampal neurons are widely used. The original method developed by Banker and colleagues (Bartlett and Banker 1984a) yields excellent results and produces relatively homogeneous cultures of embryonic hippocampal neurons that develop and mature on glass coverslips over several weeks and are optimal for high-resolution imaging of the actin cytoskeleton dynamics (Fig. 1B,C and Movies 29.2 and 29.3 on the accompanying DVD). For a detailed description of the technique, see Spector et al. (1998a, Chapter 9) and for an expanded discussion, see Goslin et al. (1998). A significant advantage of this method for high-resolution imaging is that the neurons are maintained on coverslips independently of the supporting astroglial cells necessary for their survival, which are grown as a separate layer attached to the culture dish. Although this is an excellent technique, it is time-consuming and technically demanding. Alternatively, neurons can be grown in a serum-free chemically defined medium, Neurobasal/B27 (Invitrogen-GIBCO), developed by Brewer and colleagues (Brewer 1995). This avoids the need for astroglial feeder layers but has the disadvantage that the composition of this medium is not known to the end-user. Additionally, hippocampal neurons grown in Neurobasal/B27 develop "twisted" dendrites and have a tendency to retain filopodia-like dendritic spines of immature appearance even after more than 3 weeks in culture. In contrast, the same cells grown in "Banker" cultures for 3 weeks are virtually free of filopodia and have dendritic spines equivalent in appearance to those of mature brain (see Bartlett and Banker 1984b). These same considerations regarding culture conditions also apply to low-density cultures of hippocampal neurons established from transgenic mice (V. Biou, I. Brünig, and A. Matus, unpubl.).

Transfecting Cultured Hippocampal Neurons

DOTAP Method

Various techniques have been described for expressing GFP-tagged constructs in primary neurons. The following transfection protocol, based on the method developed by Kaech et al. (1996), uses the lipid reagent DOTAP (Roche Diagnostics) and produces transfected actin-GFP-expressing hippocampal neurons that survive well during long periods in culture.

1. Resuspend neurons freshly prepared from E18 rat hippocampus in Hank's balanced salt solution (HBSS; Ca^{2+}-Mg^{2+}-free, GIBCO) and adjust to 10^6 cells per ml.

2. In a 15-ml polystyrene conical tube, add 15 μl of DOTAP to 5 ml of MEM (modified Eagle's medium) supplemented with 0.6% glucose (HC-MEM) and mix by vortexing. Add 1 ml of cell suspension (10^6 cells) and mix gently by swirling the tubes.

3. Incubate the tube for 15 minutes in a 37°C water bath. Add to the cells 3 μg of purified plasmid DNA diluted in 100 μl of HBSS and mix gently. Incubate for 30–40 minutes at 37°C.

4. Gently resuspend the cells and plate them onto prepared coverslips with paraffin "feet" (Spector et al. 1998a) in a 10-cm-diameter bacterial culture dish containing 15 coverslips in HC-MEM with 10% horse serum.

   > The density of the culture can be adjusted by using between 200,000 and 1,000,000 cells per dish. In either case, it is important to draw the cells gently up and down in the culture pipette immediately before plating to ensure that they are evenly distributed over the dish.

5. After 2–3 hours, transfer the coverslips with the attached neurons to 12-well culture dishes containing medium conditioned by preprepared glial feeder layers (Spector et al. 1998a), "flipping" the coverslips so that the neurons are underneath facing the glia. Return the culture dishes to the incubator immediately as the pH of neurobasal medium is not stable for long outside the incubator.

   > This method should yield 50 or more actin-GFP expressing cells per 18-mm coverslip.

Calcium Phosphate Method

This protocol, based on the method described by Kohrmann et al. (1999), avoids using proprietary commercial reagents. An advantage of the method is that it can be used to transfect neurons that have already been growing on coverslips in vitro, which allows the effects of genes that would otherwise interfere with neuronal development to be studied in relatively mature neurons. Transfected cells suitable for imaging can be obtained in cultures up to 15 days in vitro, but after this time, transfection efficiency declines dramatically. We have not been able to reproduce the level of transfected cells obtained in the original study Kohrmann et al. (1999); 1–2% transfected cells is a typical result in our hands. A disadvantage of the method compared to the DOTAP procedure is that hippocampal neurons become "fragile" after $CaPO_4$ treatment and tend to die following pharmacological and handling procedures.

1. Prepare a 12-well culture plate with 2 ml of glial conditioned medium in each well and equilibrate by placing it in an incubator with an atmosphere of 5% CO_2 for at least 30 minutes at 37°C.

   > Having the correct pH is crucial to obtaining the fine $CaPO_4$ precipitate necessary for optimum transfection efficiency: Check that the phenol red indicator in the medium is pink before using it. If the medium is too alkaline (phenol red indicator is red), the precipitate will form too quickly, whereas if the medium is too acid (phenol red indicator is yellow), the precipitate will not form or will form too slowly. Note that the procedure requires a separate incubator set to 37°C with an atmosphere of 2.5% CO_2 (see below).

GLIAL-CONDITIONED MEDIUM

Collect the medium from 2-week-old hippocampal neuron cultures (this is done when making the routine changes of medium during culturing). Store in aliquots at –20°C or –80°C, thaw, and prewarm in a 37°C water bath before use.

---

2. Transfer the coverslips with the neurons from the original well in which they have been growing to the equilibrated glial-conditioned medium in the new plate. If using the Banker culture method, turn the coverslips over so that the neurons are on top.

3. Store the original plate with the neuronal culture medium in the incubator for reuse after the transfection procedure.

4. Perform Steps 4–6 rapidly. For each coverslip, prepare a sterile 1.5-ml Eppendorf tube with 5 μg of actin-GFP plasmid DNA in 60 μl of 250 mM $CaCl_2$.

5. Add 60 μl of 2× BBS to the DNA vortex, mixing several times during the addition.

2× BBS (BES BUFFERED SALINE)

280 mM NaCl
1.5 mM Na_2HPO_4
50 mM BES buffer (pH 7.1) (Sigma)

It is convenient to make up 100 ml of this solution. Dissolve the components in 80 ml of distilled H_2O, adjust the pH to 7.1 with NaOH, and filter-sterilize.

---

6. Immediately add the mixture to the cells (120 μl/well) and gently mix by swirling the plate.

7. Incubate the cells in this transfection medium for 30–90 minutes at 37°C and 2.5% CO_2. After 30 minutes, begin checking for precipitate formation using an inverted microscope. It should appear as fine black speckles on the coverslip. If precipitate has not formed, continue the incubation, and reexamine at 30-minute intervals.

8. As soon as the precipitate has formed, gently wash the coverslips twice with prewarmed HBS (1ml/wash). Transfer them to the original plate that has been stored at 37°C and 5% CO_2, and return the cells to standard culture conditions.

HBS (HEPES-BUFFERED SALINE)

135 mM NaCl
1 mM Na_2HPO_4
4 mM KCl
2 mM $CaCl_2$
20 mM HEPES buffer (Sigma)

Make up 100 ml, adjust the pH to 7.5, and filter-sterilize.

Note

Good results are highly dependent on the quality of the calcium-phosphate precipitate, whose production requires some practice. We have tried several other transfection methods with varying success. Effectene (Qiagen), used as an alternative to calcium phosphate, is technically easier, but results with hippocampal neurons are generally inferior to $CaPO_4$. A recently introduced electroporation device, the Nucleofector, produced by Amaxa (Koeln, Germany; http://www.amaxa.com), gives superior results compared to previous transfection techniques, yielding ~30% GFP-actin expressing cells for rat hippocampal neurons.

MICROSCOPY

Successful high-resolution imaging of the neuronal cytoskeleton requires a precision-machined observation chamber, a suitable medium for maintaining neurons in good condition during long periods of viewing (minutes to hours), and careful control of temperature, which is essential for maintaining stable focus. Precision imaging chambers developed for this purpose are available from Life Imaging Services (Olten, Switzerland, http://www.lis.ch/thechamber.html). This firm also supplies custom chambers for the rectangular coverslips used in growing hippocampal slice cultures. The best means of maintaining effective temperature control is to mount the microscope together with ancillary equipment in a warm-air environment box such as "The Cube" from Life Imaging Services. A simple but effective medium for maintaining cells during viewing is HEPES-buffered Tyrode's solution.

HEPES-BUFFERED TYRODE'S SOLUTION

119 mM NaCl
5 mM KCl
25 mM HEPES buffer
2 mM $CaCl_2$
2 mM $MgCl_2$
6 g/liter glucose

Adjust pH to 7.4 with NaOH.

IMAGING ACTIN IN TISSUE SLICES FROM TRANSGENIC MOUSE BRAIN

The same GFP-tagged actin construct used in cell transfection experiments has been used to produce transgenic mice (Fischer et al. 2000), which were raised, screened, and maintained using standard procedures (Hogan et al. 1994). An advantage of using transgenic animals is that they offer the possibility of imaging brain tissue in the intact animal, as acutely cut slices, or as organotypic slice cultures. In addition, these animals serve as a source of cells for imaging neurons at high resolution in dispersed low-density cell culture. In contrast to cells transfected in culture, where the level of actin-GFP expression in neurons varies considerably, transgenic mice provide reproducibly labeled cells and tissues (M. Roelandse et al., in prep.)

Imaging GFP-tagged proteins in brain tissue requires confocal microscopy for which several techniques are available. For imaging tissue in the brains of living animals, a two-photon microscope is essential since other forms of illumination do not penetrate deeply enough for imaging neurons. However, for many experiments, organotypic slice cultures, which are established from neonatal brain and develop and mature in vitro, are advantageous, allowing ready access for physiological and pharmacological procedures. Two methods, using either semipermeable membrane supports (Stoppini et al. 1991) or rectangular glass coverslips (Gahwiler 1981), have been used. The latter technique has the advantage that since the slices are maintained on optical-quality glass, they can be examined repeatedly during development and they can be readily transferred to a suitable viewing chamber (e.g., from Life Imaging Services, see above).

Careful attention must be given to controlling the temperature and medium during imaging of brain tissue slices since minute variations in temperature or the flow of medium produce dramatic effects on the plane of focus. To maintain cultures in good condition during long periods of imaging, both the pH and levels of O_2 and CO_2 in the medium must be controlled. HEPES-buffered Tyrode's solution can be used, but for slice cultures, artificial cerebrospinal fluid (ACSF) is generally preferable. The medium must be aerated by bubbling with 95% O_2, 5% CO_2 prior to entry into the observation chamber. This can be done using two peristaltic pumps, the first pump adjusted to deliver medium at a suitable rate (100 ml/hour for a 1.5-ml observation chamber) and the second pump adjusted to remove medium so as to maintain a steady level in the chamber. When used in an open chamber configuration, pulsing from the peristaltic pumps is not detectable in the images.

ARTIFICIAL CEREBROSPINAL FLUID (ACSF)

124 mM NaCl
2.5 mM KCl
2.0 mM $MgSO_4$
1.25 mM KH_2PO_4
26 mM $NaHCO_3$
10 mM glucose
4 mM sucrose
2.5 mM $CaCl_2$

When the components are first mixed, the ACSF solution may be slightly milky, but it should clear when aerated with 95% O_2, 5% CO_2, after which the pH of the medium should be 7.4.

For imaging the neuronal cytoskeletal in brain slice cultures a long-focal-length high-aperture 63× or 100× lens is essential. Best results are obtained with water- or glycerol-immersion lenses where the refractive index of the immersion medium is closer to that of the cell interior than is the case for traditional oil-immersion lenses. In addition, the lens should have a focus correction collar to compensate for coverslip thickness and temperature. For imaging intact tissue such as brain slices, confocal microscopy is absolutely necessary. This can be achieved using purpose-built laser-scanning microscopes sold by major microscope manufacturers or with a device manufactured by the Yokogawa company in which the sample is scanned using an array of 20,000 microlens-equipped pinholes mounted in a spinning-disk unit. Necessary ancillary equipment includes a laser for illumination, a suitably sensitive CCD (charge-coupled device) camera for capturing images, and appropriate computer systems for controlling the microscope and camera. The spinning-disk unit has two advantages compared to conventional laser-scanning microscopes. First, it is extremely compact (overall dimensions are ~17 × 17 × 14 cm) so that it can be placed within a temperature-controlled box together with the microscope, which aids significantly in maintaining tissue quality and stable optical conditions during imaging. Second, the Yokogawa unit includes an eyepiece through which confocal optical sections of tissue sections can be observed directly by eye, providing a rapid and efficient means of orienting tissue slices for imaging. Compared to laser-scanning systems, the spinning-disk approach applies far less light to the tissue sample, which is a significant advantage in avoiding phototoxicity and fluorophore bleaching. Much valuable information on the basic features of the different systems is discussed by Inoue and Spring (1997). It is also a good idea to obtain advice from laboratories already active in live cell fluorescence imaging before deciding on a particular system. Trade exhibitions at major conferences such as the Annual Meeting of the American Society for Cell Biology (see http://www.ascb.org) offer excellent opportunities for comparative assessment of systems. A document with information on the Yokogawa spinning disk-unit is available at http://www.yokogawa.com/TR/No33.htm. Figure 2 shows the components of the imaging unit used in our laboratory.

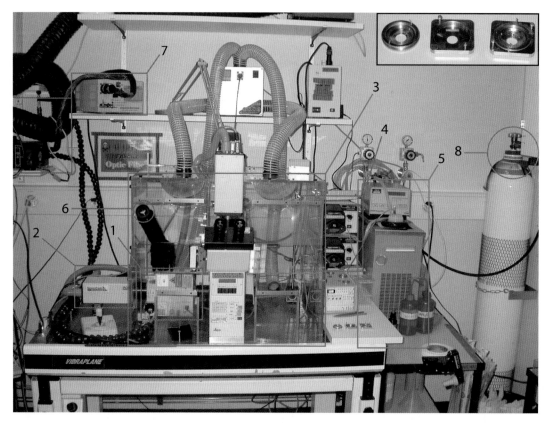

FIGURE 2. Confocal microscopy unit. (*1*) Laser-scanning device; (*2*) CCD camera; (*3*) temperature-controlled cabinet; (*4*) in- and efflux of the medium via two peristaltic pumps; (*5*) water bath; (*6*) monocular optical outlet of the laser-scanning device; (*7*) laser; (*8*) oxygen-carbon dioxide supply for the medium. (*Inset*): Viewing chambers for slice cultures.

MOVIE LEGENDS

MOVIE 29.1. Time-lapse recording of a rat embryo fibroblast transfected to express GFP-tagged cytoplasmic actin. This sequence, recorded from a living REF-52 fibroblast in cell culture, illustrates the difference in actin motility between more stable actin-rich stress fibers (longitudinal streaks across the cell interior) and surface ruffling produced by dynamic actin filaments at the cell surface.

MOVIE 29.2. Actin dynamics in the growth cone of a developing brain neuron. This time-lapse sequence shows the end of a process extending from an embryonic hippocampus neuron, transfected to express GFP-actin, after 5 days in vitro. Note the concentration of dynamic actin filaments in the growth cone showing typical actin-dependent filopodial and lamellipodial motility.

MOVIE 29.3. Actin dynamics in mature dendritic spines. Two dendrite branches from a hippocampal neuron transfected to express GFP-actin and maintained in vitro for 21 days until dendritic spines of mature morphology have formed. Actin-rich dendritic spines, appearing as bright fluorescent dots on the surfaces of dendrites, show actin-dependent surface ruffling in the region of synaptic contact, which is typical of mature dendritic spines following synapse formation.

MOVIE 29.4. Actin-based dendritic spine motility in brain slices. This time-lapse recording was made from a hippocampus slice culture prepared from a neonatal transgenic mouse expressing actin-GFP after one month in culture. Images collected using a spinning disk confocal microscope enable the high motility associated with actin-rich dendritic spines to be visualized in the dense neuropil of this brain slice.

REFERENCES

Bartlett W.P. and Banker G.A. 1984a. An electron microscopic study of the development of axons and dendrites by hippocampal neurons in culture. I. Cells which develop without intercellular contacts. *J. Neurosci.* **4:** 1944–1953.

———— 1984b. An electron microscopic study of the development of axons and dendrites by hippocampal neurons in culture. II. Synaptic relationships. *J. Neurosci.* **4:** 1954–1965.

Brewer G.J. 1995. Serum-free B27/neurobasal medium supports differentiated growth of neurons from the striatum, substantia nigra, septum, cerebral cortex, cerebellum, and dentate gyrus. *J. Neurosci. Res.* **42:** 674–683.

Fischer M., Kaech S., Wagner U., Brinkhaus H., and Matus A. 2000. Glutamate receptors regulate actin-based plasticity in dendritic spines. *Nat. Neurosci.* **3:** 887–894.

Gahwiler B.H. 1981. Organotypic monolayer cultures of nervous tissue. *J. Neurosci. Methods.* **4:** 329–342.

Goslin K., Asmussen H., and Banker G. 1998. Rat hippocampal neurons in low-density culture. In *Culturing nerve cells* (ed. G. Banker and K. Goslin), pp. 339–370. Bradford Books, MIT Press, Cambridge, Massachusetts.

Hogan B., Beddington R., Costantini F., and Lacy E. 1994. *Manipulating the mouse embryo: A laboratory manual.* Cold Spring Harbor Laboratory Press, Cold Spring Harbor, New York.

Inoue S. and Spring K.R. 1997. *Video microscopy: The fundamentals.* Plenum Press, New York.

Kaech S., Kim J.B., Cariola M., and Ralston E. 1996. Improved lipid-mediated gene transfer into primary cultures of hippocampal neurons. *Brain Res. Mol. Brain Res.* **35:** 344–348.

Kohrmann M., Haubensak W., Hemraj I., Kaether C., Lessmann V. J., and Kiebler M.A. 1999. Fast, convenient, and effective method to transiently transfect primary hippocampal neurons. *J. Neurosci. Res.* **58:** 831–835.

Roelandse M., Welman A., Wagner U., Hagmann J., and Matus A. 2003. Focal motility determines the geometry of dendritic spines. *Neurosci.* **121:** 39–49.

Spector D.L., Goldman R.D., and Leinwand L.A., 1998a. *Cells: A laboratory manual,* vol. I. *Culture and biochemical analysis of cells.* Cold Spring Harbor Laboratory Press, Cold Spring Harbor, New York.

———— 1998b. *Cells: A laboratory manual,* vol. III. *Subcellular localization of genes and their products.* Cold Spring Harbor Laboratory Press, Cold Spring Harbor, New York.

Stoppini L., Buchs P.A., and Muller D. 1991. A simple method for organotypic cultures of nervous tissue. *J. Neurosci. Methods.* **37:** 173–182.

Imaging Intermediate Filament Proteins in Living Cells

Edward R. Kuczmarski and Robert D. Goldman

Department of Cell and Molecular Biology, Feinberg School of Medicine, Northwestern University, Chicago, Illinois 60611

INTRODUCTION

INTERMEDIATE FILAMENTS (IF) ARE MAJOR CYTOSKELETAL SYSTEMS of vertebrate and many nonvertebrate cells. To date, approximately 70 genes encoding IF proteins have been identified, placing them in the 100 largest gene families in the human genome (Hesse et al. 2004). Based on sequence identity and patterns of expression, vertebrate IF can be organized into five different protein families (Herrmann and Aebi 2000). The type-I and -II homology groups comprise the keratins; type-III IF proteins include desmin, vimentin, GFAP, and peripherin; the type-IV homology group encodes α-internexin, syncoilin, nestin, synemin, and the neurofilament proteins (NF-L, NF-M, and NF-H); and the A- and B-type nuclear lamins make up type-V IF. The expression of this large family of proteins is cell-type specific and developmentally regulated. Based on these properties, IF have become important reporters for cell typing and tracking and identifying the developmental pathways of a variety of cell types including embryonic stem cells.

Until quite recently, the IF cytoskeletal system was assumed to be exceedingly stable in living cells. This assumption was based on the absence of drugs that could directly affect IF and biochemical studies revealing that after extracting cells with solutions containing detergents and high salt, well over 90% of IF protein could be sedimented by relatively low-speed centrifugation. The latter observation suggested that the "soluble" pools available for subunit exchange were very small and therefore IF were considered to be the most stable of the three major cytoskeletal systems in cells. Only after the development of live cell imaging methodologies utilizing fluorescently tagged proteins was it realized that IF form dynamic filamentous systems in vivo. Intriguingly, the dynamic properties of different IF systems vary in different cell types and during the various stages of the cell cycle. Studies to determine the dynamic properties of IF in living cells were initiated with the microinjection of purified IF proteins tagged with fluorochromes, whereas more recent studies have employed the transfection of cells with IF cDNA constructs fused to different forms of the green fluorescent protein (GFP). Using these approaches, it is now possible to study the properties of all five types of IF in living cells. The results to date have revealed an amazing array of dynamic and motile properties and have provided new insights into the unique mechanisms governing the assembly of IF networks. This chapter describes the methods that we have developed using both microinjection and transfection approaches to study the dynamic properties of IF networks in living cells.

SPECIFIC METHODS FOR THE DIRECT VISUALIZATION OF THE DYNAMIC AND MOTILE PROPERTIES OF INTERMEDIATE FILAMENTS IN LIVING CELLS

Microinjection of Fluorochrome-labeled IF Proteins

Before the establishment of modern cloning techniques, the only method available for visualizing IF dynamics in living cells involved the purification of milligram quantities of IF protein from animal tissues. Technically, this does not pose a major obstacle because many differentiated tissues such as the lens express large quantities of IF proteins. A major advantage of tissue-derived IF protein is its terminally modified state. The use of such proteins has the advantage that they more likely reflect the properties of IF especially at short time intervals following their injection into live cells. More recently, bacterial expression systems have been used as a source of large quantities of purified IF protein. Obvious advantages to this approach includes the avoidance of inconveniences incurred when large quantities of freshly derived tissues are used. Furthermore, bacterial expressed proteins can be purified in much less time than that required when starting with fresh tissue. In both cases, however, once properly purified protein is available, routine in vitro assembly assays can be carried out to demonstrate that the protein retains its competence to polymerize into 10-nm diameter IF in vitro (Vikstrom et al. 1992). These purified proteins can be easily derivitized with small fluorochromes such as rhodamine or fluorescein, purified, assayed for their polymerization properties in vitro, and then microinjected into the cytoplasm of living cells. A simple test to make certain that the exogenous protein is incorporated into the endogenous network is to fix and stain cells with antibody labeled with a different fluorophore. For example, several hours following microinjection with x-rhodamine-tagged vimentin, fixation and staining for indirect immunofluorescence using antivimentin and a fluorescein-tagged second antibody can be used to demonstrate whether there is a superimposition of the two probes, indicating that the derivitized protein is properly incorporated (Vikstrom et al. 1992). Microinjection of fluorescently tagged IF proteins into the cytoplasm of cells remains a useful technique, since it permits one to study pathways of incorporation into existing structures within minutes after they are introduced into the cytoplasm. Although we have isolated and purified several different types of IF proteins, we describe only methods for purifying and derivatizing bovine vimentin from lens tissue and human vimentin expressed in *Escherichia coli*. The basic methodology used for vimentin is also useful in the preparation of other IF protein subunits for microinjection studies.

Purification of Bovine Lens Vimentin

(See also Vikstrom et al. 1991 and methods adapted from the original procedures of Geisler and Weber 1981 and Lieska et al. 1980).

1. Obtain freshly excised bovine eyes at a slaughterhouse and immerse into crushed ice for transport to the laboratory. Remove the lenses within 1–2 hours and use immediately or store frozen at –70°C for subsequent use.

2. Homogenize approximately 15 lenses (22–25 g of tissue) in 200 ml of homogenization buffer at 4°C in a Sorvall (Newtown, Connecticut) Omni mixer at setting 2 for 1–2 minutes. Discard the small dense cores of the lenses that are not disrupted by this treatment.

> **HOMOGENIZATION BUFFER**
>
> 50 mM Tris-HCl (pH 7.4)
> 5 mM MgCl$_2$
> 0.1% 2-mercaptoethanol
> 1 mM phenylmethylsulfonyl fluoride (PMSF)

3. Centrifuge the homogenate at 32,000g for 20 minutes at 4°C and discard the supernatant.

4. Resuspend the pellets in approximately 200 ml of homogenization buffer, using a glass-teflon homogenizer. Centrifuge as in Step 3 and repeat once. After the second wash, the supernatant fraction should be clear.

5. Extract the washed pellets with approximately 3 ml of extraction buffer per gram of starting material. Stir at 4°C for at least 4 hours or overnight.

   Use ultrapure urea (e.g., Urea, ultrapure, from ICN Biomedicals, Inc.). Urea solutions should be freshly deionized using analytical-grade mixed bed resin, AG 501-X8 (D) from Bio-Rad (Richmond, California).

> **EXTRACTION BUFFER**
>
> 8 M urea
> 50 mM Tris-HCl (pH 7.4)
> 5 mM EGTA
> 0.2% 2-mercaptoethanol
> 1 mM PMSF

6. Clarify the urea extract by centrifugation at 100,000g for 30 minutes at 4°C.

7. Precipitate the protein by adding 36 g of ammonium sulfate/100 ml of solution. Stir for 30 minutes at 4°C and then collect the precipitate by centrifugation (25,000g, 20 min, 4°C).

8. Dissolve the pellet in ~ 0.5 ml of Column Buffer/g of starting material. After stirring for 30–60 minutes at room temperature, adjust the pH to 7.2.

> **COLUMN BUFFER**
>
> 6 M urea
> 8 mM sodium phosphate (pH 7.2)
> 0.14 M NaCl
> 1 mM dithiothreitol (DTT)
> 0.1 mM PMSF

9. Complete the exchange into Column Buffer by passing the sample over a column of Sephadex G-25 equilibrated in Column Buffer (80-ml bed volume).

10. Using a flow rate of 30–40 ml/hr, apply the sample to a column of hydroxyapatite (50-ml bed volume) equilibrated with Column Buffer.

    The hydroxyapatite column should be equilibrated until both the conductivity and the pH of the eluting solution are the same as that of the starting solution.

11. Wash the column with an amount of buffer equal to approximately one-half the column volume.

12. Elute the protein with a linear 8–35 mM phosphate gradient in Column Buffer.

13. Assay for vimentin-containing fractions by SDS-PAGE. Pool the most enriched fractions and dialyze overnight at room temperature against Assembly Buffer.

ASSEMBLY BUFFER

6 mM sodium/potassium phosphate (pH 7.4)
0.17 M NaCl
3 mM KCl
0.2% 2-mercaptoethanol
0.2 mM PMSF

---

14. The polymerized IF can be used immediately for x-rhodamine labeling as described below or stored by freezing dropwise in liquid N_2 and storing at $-70°C$.

Purification of Bacterially Expressed Vimentin

We have had success using the pET expression system for vimentin (see Sambrook and Russell 2001), as well as other types of IF such as peripherin, nuclear lamins, and keratins. The procedure for preparing human vimentin is as follows:

1. Prepare a 5-ml overnight culture (grown in 2YT or LB medium) of BL21 (DE3) bacteria carrying the appropriate pET plasmid and the pLysE or S plasmid if necessary. Include ampicillin and chloramphenicol to final concentrations of 100 and 50 μg/ml, respectively.

   To prepare vimentin, we used "pETvim," a plasmid made by cloning human vimentin cDNA into the NcoI-XbaI sites of pET-7 (Chou et al. 1996).

2. Add the entire 5-ml culture to 1l of culture medium (supplemented with ampicillin and chloramphenicol) that has been warmed to 37°C. Grow the culture at 37°C by shaking at ~ 275 rpm until the optical density at 595 nm reaches 0.6–1.0. This typically takes 3–5 hours.

3. Once the desired density is reached, add IPTG to 0.6 mM (from 100-mM stock in sterile water) to induce synthesis of the IF protein and continue growing for 3 hours.

   Bacteria expressing some types of IF proteins such as the nuclear lamins often grow very slowly. Therefore, it is sometimes more convenient to add a single transformed bacterial colony to 1 l of medium and grow this culture overnight at lower temperature (30–32°C). The next morning, increase the temperature to 37°C and continue growth until the desired density is achieved.

4. Chill the culture on ice and then collect the bacterial cells by centrifugation at 4000 rpm for 30 minutes at 4°C. Resuspend the pellet in 20 ml of ice cold TNE, transfer to a smaller tube (50 ml Falcon), and spin in a tabletop centrifuge for 10 minutes at 4000 rpm.

TNE

10 mM Tris-HCl (pH 8)
100 mM NaCl
1 mM EDTA

---

5. Freeze the cell pellet in liquid N_2. At this point, the pellet can be stored at $-20°C$, or the protein can be purified from inclusion bodies as described below.

Purification of IF Protein from Bacterial Inclusion Bodies

1. After freezing and thawing the cell pellets (as described in the preceding protocol), homogenize them with a glass-teflon homogenizer, on ice, in 20–30 ml of homogenization buffer per liter of cells. To inhibit proteolysis, add 1 mM PMSF and 5 μg/ml each of pepstatin A, leupeptin, and aprotinin to this, and all other buffers, just before use.

All solutions should be kept ice cold.

> **HOMOGENIZATION BUFFER**
>
> 50 mM Tris-HCl (pH 8)
> 25% sucrose
> 1 mM EDTA

2. Add two volumes of Detergent Buffer and homogenize once again. The release of DNA makes the solution very viscous, but continued homogenization should reduce the viscosity.

> **DETERGENT BUFFER**
>
> 25 mM Tris-HCl (pH 8)
> 0.2 M NaCl
> 1% NP-40
> 1% deoxycholate
> 1 mM EDTA

3. Centrifuge the homogenate at 12,000 rpm for 30 minutes at 4°C and wash the pellet of inclusion bodies by homogenizing it in 10–15 ml of Wash Buffer.

> **WASH BUFFER**
>
> TNE buffer
> 0.5% NP-40
> 3 mM MgCl$_2$

4. Add 1 mg DNase I and incubate for 20 minutes at 37°C.

5. Centrifuge as in Step 3 and repeat the wash once more with TNE buffer.

    At this stage, the inclusion bodies can be stored at −20°C for later use.

6. Homogenize the inclusion bodies in Column Buffer to dissolve the protein and centrifuge at 100,000g for 30 minutes at 4°C to remove any insoluble material. To determine if vimentin is present, analyze a sample of the supernatant fraction by SDS-PAGE.

> **COLUMN BUFFER**
>
> 10 mM Tris-HCl (pH 8)
> 8 M urea
> 2 mM EDTA
> 10 mM DTT

7. Fractionate the supernate by FPLC using a Mono-Q column, eluting the protein with a 0–0.6 M NaCl gradient.

8. Identify the vimentin-containing fractions by SDS-PAGE and dialyze pooled fractions against 2 mM phosphate buffer (pH 7.2) and 1 mM DTT, changing the dialysis buffer three times.

    For optimal storage, the soluble vimentin is first polymerized, then frozen dropwise in liquid nitrogen, and stored at −80°C (see above).

x-Rhodamine Labeling of Vimentin

1. Collect IF from a fresh or frozen/thawed aliquot of vimentin (either purified from lens or bacterially expressed) by centrifugation at 100,000g for 30 minutes at 4°C.

2. Resuspend the pellet in Disassembly Buffer and dialyze against the same for 30 minutes at room temperature.

DISASSEMBLY BUFFER

8 M urea
5 mM sodium phosphate (pH 7.2)
0.2% 2-mercaptoethanol
2 mM PMSF

---

3. Dialyze the disassembled IF protein against Dialysis Buffer at room temperature.

DIALYSIS BUFFER

5 mM sodium phosphate (pH 7.4)
0.2% 2-mercaptoethanol
0.2 mM PMSF

---

4. Remove the protein from the dialysis tubing and adjust the protein concentration to 2.0–2.5 mg/ml. Add 40:1 molar excess of the labeling agent 5- (and 6-) carboxy-x-rhodamine succimidyl ester (Molecular Probes), dimethylformamide to 10% (v/v), and NaCl to a final concentration of 0.170 M to assemble IF during the labeling process.

> Dissolve the labeling reagent in the dimethylformamide immediately before adding to the protein solution.
>
> We have tried conjugating to soluble nonfilamentous IF protein, assembling IF, and fully assembled IF and have found that conjugation during assembly works best for optimizing the amount of fluorophore and assembly properties of derivitized IF.

5. Following incubation for 60 minutes at room temperature, collect the x-rhodamine-labeled IF by centrifugation at 100,000*g* for 30 minutes. To remove labeling reagent, rinse the pellet and the sides of the tube several times with PBSa.

6. Solubilize the pellet in Disassembly Buffer (Step 2) and then exchange the protein into Sephadex Column Buffer by passing the protein over a Sephadex G-25 column equilibrated in this buffer.

SEPHADEX COLUMN BUFFER

5 mM sodium phosphate (pH 7.4)
0.2% 2-mercaptoethanol
0.2 mM PMSF

---

7. Adjust the protein concentration to 0.3–0.7 mg/ml. Add NaCl to a final concentration of 0.17 M to induce IF assembly and dialyze versus Assembly Buffer for several hours or overnight at room temperature.

ASSEMBLY BUFFER

5 mM sodium phosphate (pH 7.4)
0.17 M NaCl
0.2% 2-mercaptoethanol
0.2 mM PMSF

---

8. Collect the x-rhodamine-labeled vimentin IF by centrifugation at 100,000*g* for 30 minutes and then repeat Steps 6 and 7. A typical preparation contains 0.5 mole x-rhodamine

per mole of vimentin protein, calculated using an extinction coefficient of 5.3×10^4 /M/cm for x-rhodamine in 5 mM sodium phosphate and a molecular weight of 55,000 for vimentin.

> For long-term storage, assemble the derivitized vimentin (see above) and store at $-70°C$ following dropwise freezing in liquid N_2.

Microinjection of x-rhodamine Vimentin

1. Grow the cells on locator coverslips (Bellco Glass, Inc.) glued to 35-mm cell culture dishes to cover holes cut in their bottoms.

   > Make holes with a 3/4-inch drill bit and glue the coverslips to the outside of the bottom of the dish to seal the holes using Sylgard 184 (Dow Corning). Sterilize dishes by rinsing with 70% alcohol and exposing them to a germicidal UV lamp for 30 minutes. Seed cells into the dishes and when they reach 70% confluence, they are ready to be microinjected. This arrangement permits ready access to cells for microinjection on an inverted phase/fluoresence microscope. To permit observations at time intervals after microinjection without removing the culture dish from the microscope stage, maintain the cells in dishes in Leibovitz's L15 medium at 37°C with an airstream stage incubator.

2. Use an automated microinjection system (InjectMan NI 2, Ependorf Scientific, Inc.), mounted on an inverted microscope, to inject cells identified by their coordinates on the locator coverslips. Inject each cell with approximately 0.06 pl of x-rhodamine vimentin (1–2 mg/ml) in 5 mM sodium phosphate (pH 8.5) 0.05% 2-mercaptoethanol.

   > Best results are obtained by injecting at a shallow angle in the thicker region of the cell near the nucleus. The injected protein can be followed at time intervals after microinjection using fluorescence optics on the same inverted microscope equipped with a sensitive image-capturing system (see Chapter 6) to monitor the pathway of incorporation of the injected protein. Alternatively, the dishes can be transferred to a confocal microscope for higher-resolution time-lapse image capturing, and individual microinjected cells can be identified by their known position on the locator coverslips.

3. Once the x-rhodamine-coupled vimentin is injected, fluorescence recovery after photo-bleaching (FRAP) analyses can be undertaken. As a prerequisite, the exogenous x-rhodamine vimentin should be allowed to become completely incorporated into the endogenous IF network over a period of several hours or overnight (Vikstrom et al. 1992). This permits direct observation of vimentin fibrils in live fibroblasts.

   > The best subcellular region for carrying out FRAP analyses is in the peripheral regions in which individual fibrils can be readily observed. Now, we carry out photobleaching using conventional confocal techniques (see Chapter 7). Recovery of fluorescence can be monitored using time-lapse imaging of the bleach zone (Fig. 1).

The Use of GFP-tagged IF Proteins

Advances in cloning technology and development of the expression of green fluorescent protein (GFP)-tagged IF proteins have provided an extraordinary opportunity to visualize a variety of different types of IF in living cells. A major advantage of this approach is related to the fact that GFP fluorescence is not as sensitive to photobleaching, thereby permitting longer periods of observation during time-lapse analyses of the properties of IF. For example, we found that with the x-rhodamine microinjection approach, we never achieved complete photobleach recovery due to the loss of light attributable to generalized background photobleaching (Vikstrom et al. 1992). Under similar conditions, with FRAP analyses in cells expressing GFP-tagged vimentin, much less background photobleaching occurred, and we

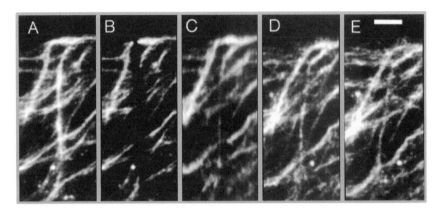

FIGURE 1. Recovery of vimentin fluorescence in a living cell. A 3T3 cell was microinjected with x-rhodamine vimentin, and after 12 hours of incorporation, a prominent vimentin fiber was photobleached along its length. Monitoring recovery at 15-minute intervals revealed that fluorescence returned throughout the fiber's length, with no apparent polarity to the recovery. (*A*) x-rhodamine-vimentin network in the living cell, prior to photobleaching, and recovery after (*B*) 0.5, (*C*) 15, (*D*) 30, and (*E*) 45 minutes. Bar, 5 μm. (Reprinted, with permission, from Vikstrom et al.1992.)

were able to achieve full recovery of fluorescence in specifically bleached areas (Yoon et al. 1998; also see Chapter 7). Moreover, by using various isoforms of GFP with distinct and separable emission spectra, several IF systems can be monitored in a single cell simultaneously (see below).

Constructing and Expressing GFP-IF Protein in Living Cells

Our first GFP-tagging studies employed cloned human vimentin (Yoon et al. 1998). A commercially available plasmid containing GFP (pEGFP-C1, CLONTECH Laboratories Inc., Palo Alto, California) was cut with BamH1. Then, a BamH1-BamH1 cDNA fragment containing the human vimentin sequence was subcloned into this plasmid (see Yoon et al. 1998). We found that it was necessary to place the GFP at the amino-terminal end of the vimentin, because when placed at the carboxyl terminus, the expressed protein did not incorporate into IF in vivo (Yoon et al. 1998). We have subsequently engineered GFP-, CFP-, and YFP-IF cDNA constructs for visualizing numerous types of IF in vivo, including keratin (Yoon et al. 2001), nuclear lamins (Moir et al. 2000b), and peripherin (Helfand et al. 2003). In all cases, the GFP-IF cDNA was engineered so that GFP was expressed at the amino terminus of the IF protein chain being studied. For expression, the construct was introduced into cells by one of the following transfection methods.

Electroporation

1. Trypsinize subconfluent cultures of cells in 100-mm petri dishes and suspend in 250 μl of culture medium containing 10 mM Hepes (pH 7.1).

2. Mix the suspension gently with 7 μg of target DNA and some carrier DNA (13 μg of sheared salmon sperm DNA, Amresco Inc.). To carry out cotransfection experiments, use 5 μg of each target DNA and reduce the amount of carrier DNA to 10 μg.

3. Electroporate the suspension at 0.25 V/960 uFD (Gene Pulser II; Bio-Rad Laboratories) and plate the transfected cells onto 22-mm number 1 glass coverslips or 35-mm petri dishes either with or without locator coverslips affixed to their bottoms as described in Step 1 of the protocol, "Microinjection of x-rhodamine Vimentin."

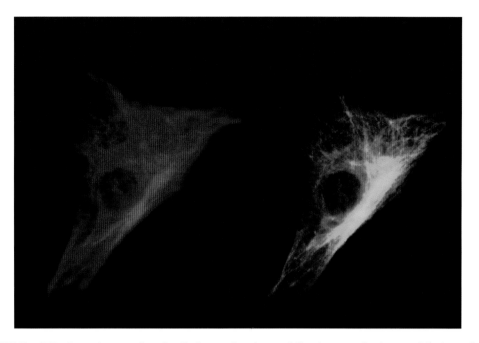

FIGURE 2. GFP-vimentin transfected cell observed 48 hours following transfection. At this time, the cell was fixed and processed for indirect immunofluorescence using rabbit antivimentin antibody and rhodamine goat antirabbit IgG. The GFP fluorescence is viewed directly (*green*). Note that the GFP-vimentin and antibody patterns show extensive colocalization.

Lipofection

1. For transient transfection, mix 1–2 μg of the appropriate plasmid with 6 μl Lipofectamine™ (Gibco-BRL) in a total volume of 200 μl of serum-free medium and incubate for 40 minutes at room temperature.

2. Incubate cells grown on coverslips with the DNA/Lipofectamine mixture (in a final volume of 1 ml of serum-free medium) for 3 hours at 37°C.

3. Following removal of the DNA/Lipofectamine, wash the cells with serum-free medium and allow them to recover overnight in complete growth medium.

Characterizing GFP-IF Proteins

Because added GFP protein can potentially interfere with the tagged protein's normal function, it is important to demonstrate that the GFP-IF fusion protein behaves normally (Yoon et al. 1998). Several criteria should be tested for IF derivatives. For example, BHK-21 cells transfected with GFP-vimentin display typical IF networks and the GFP-vimentin colocalizes with endogenous vimentin, as determined by double-label immunofluoresence (Fig. 2 [Yoon et al. 1998]). In vitro studies indicate that the GFP-tagged vimentin is coassembled with endogenous IF, as demonstrated by SDS-PAGE and immunoblotting analyses of IF-enriched cytoskeletal preparations (Yoon et al. 1998).

Dynamic Properties of IF Networks Revealed Using GFP Tagging

Type-III IF (Vimentin and Peripherin)

Studies of well-spread interphase fibroblasts expressing GFP vimentin show a wide range of dynamic and motile activities. Using time-lapse confocal microscopy, we have found that closely spaced vimentin fibrils can move in opposite directions and that they constantly display bending or straightening movements over relatively short time intervals. Moreover, these fibrils appear to be capable of lengthening and shortening, as evidenced by the movement of interconnected "foci" toward or away from each other (See Movie 30.1 on the accompanying DVD) (Yoon et al. 1998). GFP vimentin further revealed that short filamentous structures with two free ends (vimentin "squiggles") exhibited rapid movements. Studies of spreading cells revealed the presence of fluorescent nonfilamentous vimentin particles that moved at high speeds along microtubules (Prahlad et al. 1998). These particles are converted into short filaments or "squiggles" that are also motile (Movie 30.2) (Prahlad et al. 1998). Similar movements of another Type-III IF protein, peripherin, have also been detected in various subdomains of PC12 neurons including cell bodies, neurites and growth cones (Movie 30.3; Helfand et al. 2003) These studies revealed entirely new dimensions of IF assembly and motility. It should be noted that GFP vimentin has also provided important insights into changes in the cytoskeletal system in live endothelial cells during hemodynamic shear stress (Helmke et al. 2000, 2001) (see Chapter 3).

Types-I and -II Keratin IF

The keratins assemble as obligate heteropolymer IF in epithelial cells where they usually form wavy bundles known as tonofibrils. Tonofibrils have always been assumed to be the most stable of all IF systems because they are found abundantly expressed in epithelial tissues, such as skin, that must be able to resist large mechanical stresses. Using GFP-tagged keratin 8 or keratin 18, we have been able to directly monitor the motile properties of tonofibrils. One of the most intriguing aspects of their motility relates to their ability to form hairpin bends and propagate them as wave forms. Even closely spaced tonofibrils can be seen to change their shapes and move independently of one another (Yoon et al. 2001). Drugs known to inhibit both microtubules and actin/myosin cytoskeletal systems do not affect these intrinsic movements of tonofibrils (Movie 30.4; Yoon et al. 2001). We have also carried out FRAP analyses of tonofibrils expressing GFP-tagged keratins and found that their recovery rates are much slower than those recorded for Type-III IF, such as vimentin. This is seen most dramatically in studies of cells such as PtK2 that are known to express separate keratin and vimentin IF networks. This affords the opportunity to study the dynamic properties of two separate IF cytoskeletal networks within the same region of a single cell. For example, PtK2 cells can be doubly transfected to express YFP-keratin 8 and CFP vimentin. Bleach zones can be easily made simultaneously across keratin tonofibrils and vimentin fibrils. Fluorescence recovery can then be assayed by time-lapse observations. Interestingly, the half-time recovery rate for keratin is $t_{1/2}$ ~100 minutes and for vimentin, full recovery occurs in ~ 30 minutes (Fig. 3; Yoon et al. 2001). These observations suggest that different factors contribute to and regulate the dynamic properties of different IF systems within the same cytoplasmic region.

Dynamic Properties of Type-V Nuclear IF Revealed by GFP-lamin Expression

Similar to endogenous lamins, GFP lamins localize primarily to the nuclear lamina and the nucoleoplasmic veil (Moir et al. 2000b). By comparing the organization and distribution of GFP-A- and -B-type lamins, new information has been obtained about their dynamic properties during mitosis and reassembly of nuclei in daughter cells. These studies revealed that the A- and B-type lamins exhibit different pathways of assembly. The B-type lamins are the first to assemble onto chromosomes during the anaphase–telophase transition (Fig. 4; Moir

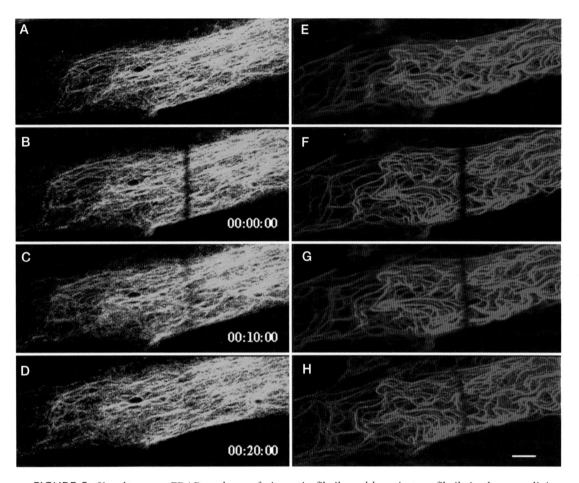

FIGURE 3. Simultaneous FRAP analyses of vimentin fibrils and keratin tonofibrils in the same living cell. A PtK2 epithelial cell was doubly transfected with CFP vimentin and YFP-keratin 8, and a bar-shaped region was photobleached. Recovery of fluoresence was monitored at 2-minute intervals. Complete recovery of fluoresence occurred within 30 minutes for bleached CFP vimentin (A–D). In contrast, no recovery of fluoresence was observed for bleached YFP-K8 during this same time interval (E–H). Elapsed time (h:m:s) is indicated at the lower left. Bar, 5 μm. (Reprinted, with permission, from Yoon et al. 2001.)

et al. 2000b), and by the time the chromosomes reach the poles, nearly all of this lamin has accumulated at the chromosome surface. In contrast, A-type lamin begins to associate with the reforming nucleus only after the major components of the nuclear envelope have assembled in daughter cells (Moir et al. 2000b).

Another important use of the live cell imaging method employing GFP lamins is related to our studies of the effects of mutant lamins on lamin assembly in live cells. This involves imaging a nuclear lamin network within a cell expressing GFP-lamin A while microinjecting bacterial expressed ΔNLA, a dominant negative mutant lamin designed to disrupt the head-to-tail assembly of dimers, the basic building blocks of lamin polymers (Moir et al. 2000a). Within minutes following injection into the cytoplasm (using the methods described above), the mutant lamin enters the nucleus and disrupts the endogenous network. In this experiment, the disrupted lamins are rapidly reorganized into nucleoplasmic aggregates as the lamina region disappears and the nucleus becomes misshapen (Fig. 5).

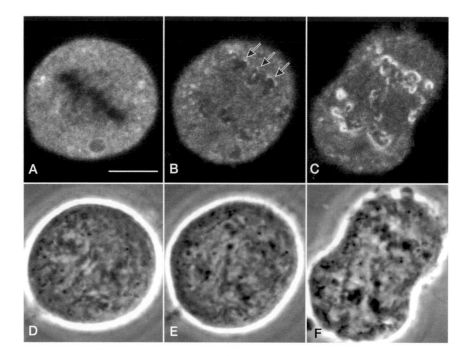

FIGURE 4. GFP-lamin B in a living BHK-21 cell as it progresses from metaphase to telophase. Confocal (*A–C*) and corresponding phase contrast (*D–F*) images reveal that at metaphase/early anaphase, the lamin B is diffusely distributed throughout the cell (*A* and *D*). Eight minutes later, at mid-to-late anaphase, the lamin B begins to associate with the surface of chromosomes in the region of the kinetochore that faces the spindle poles (*B, arrows*). Because the spindle is slightly tilted, localization at the corresponding set of chromosomes (*B, lower left*) is not as obvious. The cell reached mid-to-late telophase 4 minutes later, and as the cleavage furrow began to form, almost all detectable lamin B fluorescence was concentrated at the surface of the chromosomes (*C*). (Reprinted, with permission, from Moir et al. 2000b.)

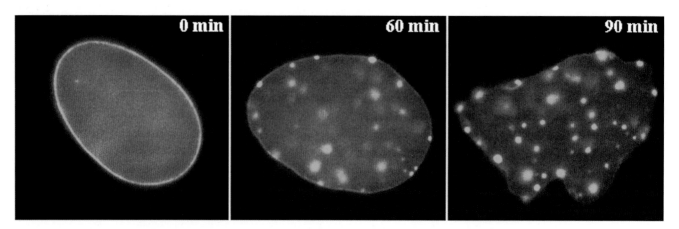

FIGURE 5. The nucleus of a live BHK-21 cell expressing GFP-lamin A before, and at time intervals following, the microinjection of the dominant negative mutant protein, ΔNLA. Note the disruption of the normal organization of the lamins in the lamina, the sequestration of the lamins (both from the lamina and the nucleoplasm) into aggregates, and the alteration in the shape of the nucleus.

MOVIE LEGENDS

MOVIE 30.1. A well-spread BHK-21 cell expressing GFP-vimentin. Note that many of the fibrils move, appear to shorten and lengthen, and focal points for fibrils also appear to move (see Yoon M. et al. 1998). Phase contrast images are used to show the outlines of the cell or regions of the cell under observation.

MOVIE 30.2. Spreading BHK-21 cells showing the motile properties of vimentin particles, the apparent fusion of some particles, and the formation of short filaments or squiggles. These cells were transfected with GFP-vimentin, trypsinized, replated, and observed as described in Prahlad et al. 1998.

MOVIE 30.3. A PC-12 cell expressing GFP-peripherin in the region of a newly emerging growth cone observed a short time following the addition of nerve growth factor. Note the peripherin particles and short filamentous structures (*green*) moving in the various domains of the growth cone. The growth cone is visualized with phase contrast optics (for details, see Helfand et al. 2003).

MOVIE 30.4. A region of a PtK2 cell transfected with GFP-keratin 14. Note that the tonofibrils exhibit bending movements and in some cases appear to propagate wave forms (for details, see Yoon et al. 2001).

REFERENCES

Chou Y.H., Opal P., Quinlan R.A., and Goldman R.D. 1996. The relative roles of specific N- and C-terminal phosphorylation sites in the disassembly of intermediate filament in mitotic BHK-21 cells. *J. Cell Sci.* **109:** 817–826.

Geisler N. and Weber K. 1981. Isolation of polymerization-competent vimentin from porcine eye lens tissue. *FEBS Lett.* **125:** 253–256.

Helfand B.T., Loomis P., Yoon M., and Goldman R.D. 2003. Rapid transport of neural intermediate filament protein. *J. Cell Sci.* **116:** 2345–2359.

Helmke B.P., Goldman R.D., and Davies P.F. 2000. Rapid displacement of vimentin intermediate filaments in living endothelial cells exposed to flow. *Circ. Res.* **86:** 745–752.

Helmke B.P., Thakker D.B., Goldman R.D., and Davies P.F. 2001. Spatiotemporal analysis of flow-induced intermediate filament displacement in living endothelial cells. *Biophys. J.* **80:** 184–194.

Herrmann H. and Aebi U. 2000. Intermediate filaments and their associates: Multitalented structural elements specifying cytoarchitecture and cytodynamics. *Curr. Opin. Cell Biol.* **12:** 79–90.

Hesse M., Zimek A., Weber K., and Magin T.M. 2004. Comprehensive analysis of keratin gene clusters in humans and rodents. *Eur. J. Cell Biol.* **83:** 19–26.

Lieska N., Chen J., Maisel H., and Romero-Herrera A.E. 1980. Subunit characterization of lens intermediate filaments. *Biochim. Biophys. Acta* **626:** 136–153.

Moir R.D., Spann T.P., Herrmann H., and Goldman R.D. 2000a. Disruption of nuclear lamin organization blocks the elongation phase of DNA replication. *J. Cell Biol.* **149:** 1179–1192.

Moir R.D., Yoon M., Khuon S., and Goldman R.D. 2000b. Nuclear lamins A and B1. Different pathways of assembly during nuclear envelope formation in living cells. *J. Cell Biol.* **151:** 1155–1168.

Prahlad V., Yoon M., Moir R.D., Vale R.D., and Goldman R.D. 1998. Rapid movements of vimentin on microtubule tracks: Kinesin-dependent assembly of intermediate filament networks. *J. Cell Biol.* **143:** 159–170.

Sambrook J. and Russell D. 2001. *Molecular cloning : A laboratory manual,* 3rd edition. Cold Spring Harbor Laboratory Press, Cold Spring Harbor, New York.

Vikstrom K.L., Miller R.K., and Goldman R.D. 1991. Analyzing dynamic properties of intermediate filaments. *Methods Enzymol.* **196:** 506–525.

Vikstrom K.L., Lim S.S., Goldman R.D., and Borisy G.G. 1992. Steady state dynamics of intermediate filament networks. *J. Cell Biol.* **118:** 121–129.

Yoon K.H., Yoon M., Moir R.D., Khuon S., Flitney F.W., and Goldman R.D. 2001. Insights into the dynamic properties of keratin intermediate filaments in living epithelial cells. *J. Cell Biol.* **153:** 503–516.

Yoon M., Moir R.D., Prahlad V., and Goldman R.D. 1998. Motile properties of vimentin intermediate filament networks in living cells. *J. Cell Biol.* **143:** 147–157.

CHAPTER THIRTY ONE

Methods for Expressing and Analyzing GFP-Tubulin and GFP-Microtubule-associated Proteins

Holly V. Goodson

Department of Chemistry and Biochemistry, University of Notre Dame, Notre Dame, Indiana 46556-5670

Patricia Wadsworth

Department of Biology and Program in Molecular and Cellular Biology, University of Massachusetts, Amherst, Massachusetts 01003

INTRODUCTION

IMPORTANT ADVANCES IN OUR UNDERSTANDING of the organization and dynamics of the cytoskeleton have been made by direct observations of fluorescently tagged cytoskeletal proteins in living cells. In early experiments, the cytoskeletal protein of interest was purified, covalently modified with a fluorescent dye, and microinjected into living cells (Kreis and Birchmeier 1982; Wang et al. 1982a). Direct labeling of proteins with fluorescent molecules can still be extremely useful, and the relevant methods are described elsewhere in this book (see Chapter 5). More recently, green fluorescent protein (GFP) technology has been used to prepare chimeric proteins that are expressed in cells following transfection with a vector containing the gene of interest fused to GFP. This chapter discusses the methods used to prepare GFP-tagged tubulin and microtubule-associated proteins (MAPs). Although some details may be system-specific, the methods and considerations outlined here can be adapted to a wide variety of proteins and organisms (e.g., see de Hostos 1999; Stephens and Pepperkok 2001; Reski 2002; Wehrle-Haller and Imhof 2002; Hynes et al. 2004; for reviews, see Ludin and Matus 1998; Yoon et al. 2002).

Background

Observations of living cells using differential interference contrast (DIC), polarized light, and phase-contrast microscopy clearly demonstrate that cytoskeletal elements are highly dynamic during cellular motility, mitosis, and cytokinesis. However, identification of the individual cytoskeletal components involved is not possible with these methods. In the early 1980s, Taylor and Wang introduced a method termed fluorescent analog cytochemistry

(Wang et al. 1982a) in which a cytoskeletal protein was isolated, tagged with a fluorescent dye, and microinjected into living cells. These pioneering studies provided novel information on the behavior of actin in living cells and embryos and demonstrated the potential of the technique for examination of the cytoskeleton in vivo (Wang and Taylor 1979; Wang et al. 1982b).

Although novel insights into the dynamics of the cytoskeleton continue to be made using fluorescent analog cytochemistry (see Chapter 5), the technique has several limitations. First, the protein of interest must be readily obtained in sufficient quantities for purification and covalent modification in vitro, and the protein must retain function after modification. Second, microinjection is required to deliver the fluorescent analog into the cells of interest, which limits the number and type of cells that can be studied. Finally, after microinjection, the cells must be incubated to allow incorporation of the fluorescent analog into cellular structures. This can be a limitation when particularly rapid events, such as the transition into mitosis, need to be imaged.

Recently, the ability to express cloned proteins containing a GFP tag has become an important approach for the analysis of the cytoskeleton in living cells (Chalfie et al. 1994). GFP technology offers several important advantages over fluorescent analog cytochemistry. First, GFP technology can be applied to any protein for which a cDNA clone is available; it is not necessary to biochemically purify and fluorescently modify the protein the investigator wishes to examine. The GFP tag can be added to the 5′ or 3′ end of the cDNA, so that the resulting chimeric protein has a GFP tag at the carboxyl or amino terminus. In practice, many proteins tolerate a tag at one, but not the other, end of the protein (see Chapter 2). Adding a GFP tag involves straightforward subcloning steps into commercially available vectors containing GFP or its spectral variants. Once a plasmid containing the gene of interest fused to GFP has been constructed, it can be transfected into cells using standard methods; constructs can also be injected into the nucleus, but this approach is optional. One significant advantage of performing experiments with GFP fusions instead of fluorescent analogs is that site-directed mutagenesis can be incorporated into the cloning steps to directly test the structure/function relationships involved in localization and dynamics (e.g., see Olson et al. 1995; Vaughan et al. 2002).

The major disadvantage of GFP technology is that the addition of GFP (238 amino acids) to the carboxyl or amino terminus of the target protein can potentially interfere with protein function. Thus, it is important to demonstrate that the chimeric protein retains its normal characteristics (see Chapter 2). In particular, the GFP-tagged protein may fail to assemble properly into macromolecular complexes. This problem can occur because of steric hindrance caused by the GFP, overproduction of the GFP fusion relative to the endogenous complex components (see Vaughan et al. 2002), or mismatch between the time course of the experiment and the assembly kinetics of the complex. Some assembly problems can be ameliorated by replacing the endogenous protein with the GFP-tagged version. This can be accomplished by expressing the GFP-tagged protein in model organisms or cells in which the endogenous copy of the gene has been deleted or mutated by homologous recombination. Alternatively, small interfering RNA (siRNA) methods combined with standard transfection techniques can be used to simultaneously deplete the endogenous protein and express the GFP-tagged version (accomplished by using siRNAs directed against a sequence not present in the GFP-tagged protein). Finally, overexpression of any protein can potentially interfere with cellular function. This latter limitation can be overcome by using an inducible promoter or by establishing permanent cell lines with the desired level of expression of the chimera.

GFP-Tubulin

The first application of GFP technology to the study of microtubule behavior was in yeast, when the yeast α-tubulin genes (*tub1* and *tub3*) were tagged with GFP (Carminati and Stearns 1997; Straight et al. 1997). Brief examination of the yeast GFP-tubulin literature illustrates several important considerations that have general relevance. First, constructs with GFP at the amino terminus of *tub3* complemented *tub3* null mutations, but a carboxy-terminal fusion to *tub3* did not complement the null, highlighting the importance of the position of the GFP (Carminati and Stearns 1997). On the other hand, GFP fusions at either end of *tub1* (the other yeast α-tubulin) failed to complement null mutations (Maddox et al. 1999; see also Carminati and Stearns 1997), demonstrating that similar molecules can apparently respond differently to the presence of the GFP tag. Although the GFP-*tub1* fusion cannot serve as the only source of *tub1*, indicating that the fusion is not fully functional, this fusion can still be useful as a "tracer" for the dynamics of tubulin polymerization: Yeast cells expressing a mixture of GFP-tagged and wild-type *tub1* have spindle microtubules with apparently normal dynamics (Straight et al. 1997) and grow at normal rates (Maddox et al. 1999). Finally, fusions to the yeast β-tubulin gene, *tub2*, have not been reported, in part because variation in the level of β-tubulin is highly deleterious to yeast (Katz et al. 1990; Weinstein and Solomon 1990). This observation provides a reminder that the study of protein complexes may be best pursued by tagging multiple subunits and comparing their behavior.

GFP-tubulin has now been successfully used to examine microtubule behavior in mammalian cultured cells (see Movie 31.1 on accompanying DVD), as well as a variety of organisms including *Aspergillis*, *Arabidopsis*, and *Drosophila* (see Table 1). Although some altered

TABLE 1. Fusions between tubulin and GFP or its spectral variants

Species/Cell Type	Protein Fused to GFP (gene name)	Type of Transfection	Tag Position	References (spectral variant if relevant)
A. nidulans	α-tubulin (TUBA)	integration into TUBA locus	N-term	Han et al. (2001)
A. thaliana	α-tubulin (TUA6); Tub1A and Tub3A (TUA1, TUA3?)	stable	~	Hasezawa et al. (2000); Shaw et al. (2003) (YFP)
C. albicans	β-tubulin (*TUB2*)	gene replacement	~	Hazan et al. (2002)
C. elegans	α tubulin (C47B2.3)	transgenic animal	~	Oegema et al. (2001)
Chicken/ Nuf23-63	human α-tubulin (Clontech)	stable	N-term	Hori et al. (2003)
C. lagenarium	α-tubulin (*tub1*)	~	~	Takano et al. (2001)
Dictyostelium	α-tubulin (TUBA?)	stable	N-term	Koonce et al. (1999); Kimble et al. (2000)
Drosophila	α-tubulin (*tub84B*)	transgenic animals and S2 cells	N-term	Grieder et al. (2000); Rogers et al. (2002)
Goldfish/fin fibroblasts	mouse β-tubulin~	transient and/ or stable	~	Kaverina et al. (2002); Krylyshkina et al. (2003)
Green monkey/ COS-1, COS-7, CV-1	Human α-tubulin (Clontech)	transient and/ or stable	N-term	Martin et al. (2002); Buster et al. (2002); Vaughan et al. (2002); Petrulis et al. (2003) (YFP)

(Continued)

TABLE 1. *(Continued)*

Species/Cell Type	Protein Fused to GFP (gene name)	Type of Transfection	Tag Position	References (spectral variant if relevant)
Human/SK-N-SH, U2-OS, RPMI 8226, MCF7, HeLa	α-tubulin (Clontech)	transient and/or stable	N-term	Popova et al. (2002); Seong et al. (2002); Kamath and Jordan (2003); Maxwell et al. (2003); Lauf et al. (2002) (YFP)
Mouse/B16 (melanoma)	mouse β-tubulin~	transient	~	Kaverina et al. (2002)
Pig/LLCPK1	human α-tubulin (Clontech)	stable	~	Rusan et al. (2001)
Rat/REF, NRK, REF 52	mouse β-tubulin, human α-tubulin (Clontech)	transient	C-term (β) N-term (α)	Heidemann et al. (1999); Chen et al. (2003); Schmoranzer and Simon (2003) (YFP)
S. cerevisiae	α-tubulin (*TUB1* and *TUB3*)	Cen plasmids, chromosomal integrations	N-term and C-term	Straight et al. (1997); Carminati and Stearns (1997); Jensen et al. (2001) (CFP); many additional strains are based on these fusions
S. pombe	α-tubulin (α2)	Multicopy plasmid	N-term	Ding et al. (1998)
Tobacco/BY-2	~	stable	~	Kumagai et al. (2001)
T. gondii	α-tubulin ~	transient	N-term and C-term	Striepen et al. (2000) (YFP)
U. maydis	α-tubulin (*tub1*)	stable	C-term	Straube et al. (2003)
Xenopus/ fibroblasts	mouse β-tubulin ~	transient	~	Krylyshkina et al. (2002)
E. coli	FtsZ	plasmid	C-term	Ma et al. (1996)

Fusions between tubulin and GFP or its spectral variants. This table is meant to provide some information about the range of tools available. References listed are in most cases the original descriptions of particular fusions and/or cell lines; a large amount of additional research using these tools can be identified by citation searches. When no species is indicated in the "Protein fused" column, the tubulin and recipient cell line/organism are conspecific. "~" indicates ambiguity in the primary source.

phenotypes have been reported upon expression of GFP-tubulin constructs, particularly in *Dictyostelium* (Kimble et al. 2000), the available data strongly support the view that expression of GFP-tubulin and its incorporation into microtubules does not detectably interfere with microtubule behavior in most systems.

GFP-MAPs

In addition to GFP-tagged tubulin, various MAPs have been fused to GFP. Traditional MAPs (MAP2 and Tau) were tagged first (Kaech et al. 1996), but the list now includes molecular motors and microtubule plus-end tracking proteins (Table 2). While providing the expected information about protein localization, GFP-MAP fusions have also been used to examine the kinetics of MAP-microtubule interactions (Bulinski et al. 2001), the behavior of single-microtubule motors (Pierce et al. 1999), and as a marker for axons and the mitotic machinery in transgenic animals (Pratt et al. 2000).

TABLE 2. Fusions between microtubule-binding proteins and GFP

Microtubule-Binding Protein (MBP) Family	MBP Species (individual gene or protein name)	Species/Cell Line Expressing GFP Fusion	Type of Transfection	References (spectral variant if relevant)
APC	Xenopus (APC)	*Xenopus* A6 cells	transient	Mimori-Kiyosue et al. (2000a)
CLIP-170	mammalian (CLIP-170)	Vero, CHO, NRK	transient	Perez et al. (1999); Komarova et al. (2002)
	mammalian (CLIP-170)	cowpea mesophyll protoplasts, tobacco BY-2	transient and or stable	Dhonukshe and Gadella (2003) (YFP)
	mammalian (CLIP-115)	COS-1	transient	Hoogenraad et al. (2000)
	S. cerevisiae (BIK1)	–	stable	Lin et al. (2001)
	S. pombe (TIP1)	–	FIX	Brunner and Nurse (2000); Niccoli and Nurse (2002) (YFP)
Clip R-59	mammalian (CLIPR-59)	HeLa	transient	Perez et al. (2002)
CLASPS/MAST/ ORBIT	mammalian (CLASP1/2)	COS-1	transient	Akhmanova et al. (2001)
DAM1	*S. cerevisiae (DAM1)*	–	plasmid, chromosomal integration	Hofmann et al. (1998); Jones et al. (2001)
Dynein heavy chain	*A. nidulans (NUDA)*	–	plasmid	Zhang et al. (2002)
	S. cerevisiae (DYN1)	–	plasmid, chromosomal integration	Shaw et al. (1997); Lee et al. (2003) (3× GFP) Yamamoto et al. (1999)
	S. pombe (dhc1)	–	chromosomal integration	
EB1	mammalian (EB1)	LLCPK	stable	Piehl and Cassimeris (2003)
	mammalian (EB1)	*Xenopus* A6 cells	transient	Mimori-Kiyosue et al. (2000b)
	mammalian (EB3)	Purkinje cells mouse hippocampal neurons	transient	Stepanova et al. (2003)
	Dictyostelium (DdEB1)	–	transient	Rehberg and Graf (2002)
	Arabidopsis (AtEB1a)	*Arabidopsis* tissue culture cells	~	Chan et al. (2003)
	S. cerevisiae (BIM1)	–	cen plasmid	Schwartz et al. (1997); Tirnauer et al. (1999)
Ensconsin (also called E-MAP-115, MAP7)	mammalian (Ensconsin)	African green monkey/TC-7, human/MCF-7	stable	Faire et al. (1999)
EF1-α	carrot (EF1-α)	fava bean leaf epidermal cells	~	Moore and Cyr (2000)
RHAMM/IHABP	mammalian (RHAMM/IHABP)	HeLa	transient	Assmann et al. (1999)

TABLE 2. *(Continued)*

Microtubule-Binding Protein (MBP) Family	MBP Species (individual gene or protein name)	Species/Cell Line Expressing GFP Fusion	Type of Transfection	References (spectral variant if relevant)
KAR9	*S. cerevisae (KAR9)*	–	cen plasmid; "wild-type levels"	Miller and Rose (1998); Lee et al. (2000)
Katanin*	*Arabidopsis* (AtKSS)	*Arabidopsis*	~	Bouquin et al. (2003)
	C. elegans (Mei-1)	HeLa	transient	Srayko et al. (2000)
	mammalian (katanin, spastin)	baculovirus, CV-1, COS-7, HeLa	transient	Hartman and Vale (1999) (GFP, YFP, CFP); Buster et al. (2002); Errico et al. (2002)
Kinesin*	mammalian (conventional kinesin)	bacterial expression	–	Pierce et al. (1997); Pierce and Vale (1998)
Lis1	*Neurospora (NUDF)*	–	stable	Han et al. (2001)
	S. cerevisiae (LIS1)	–	tagged PAC1 locus	Lee et al. (2003)
MAP2	mammalian (MAP2c)	HeLa	transient	Ludin et al. (1996); Ozer and Halpain (2000)
MAP4	mammalian (MAP4)	BHK, CHO, Ltk	stable	Olson et al. (1995); Chang et al. (2001)
	mammalian (MAP4)	fava bean epidermal cells, cowpea mesophyll protoplasts BY-2 cells	~	Marc et al. (1998); Dhonukshe and Gadella (2003)
	Xenopus (XMAP4/p220)	*Xenopus* A6 cells	stable	Shiina and Tsukita (1999)
MARK4 kinase	mammalian (MARK4)	CHO, Neuro2A	transient	Trinczek et al. (2004)
MIR1	mammalian (MIR1)	BHK	transient	Stein et al. (2002)
MID1	mammalian (MID1)	COS-7	transient	Schweiger et al. (1999)
NuMa	mammalian (NuMa)	HeLa	transient	Merdes (2000)
NuSAP	mammalian (NuSAP)	COS1, PtK2	transient	Raemaekers et al. (2003) (GFP, YFP, and CFP)
P150	mammalian (rat brain p150)	COS-7	transient	Vaughan et al. (2002) (GFP and RFP)
Plakins	*Drosophila* (Short Stop)	NIH-3T3, transgenic flies	transient	Lee and Kolodziej (2002)
PRC1	mammalian (PRC1)	HeLa	transient	Mollinari et al. (2002)
Tea1	*S. pombe (Tea1)*	–	chromosomal integration	Behrens and Nurse (2002) (GFP, YFP)
TOG/Dis1	*Dictyostelium* (CP224)	–	transient	Graf et al. (2000)
	S. cerevisiae (Stu2)	–	fusion to chromosomal *Stu2* locus	Kosco et al. (2001)

TABLE 2. *(Continued)*

Microtubule-Binding Protein (MBP) Family	MBP Species (individual gene or protein name)	Species/Cell Line Expressing GFP Fusion	Type of Transfection	References (spectral variant if relevant)
	Xenopus (XMAP-215)	XL177 cells	transient	Popov et al. (2001)
	S. pombe (Dis1, Mtc1/Alp14)	–	chromosomal integration	Nakaseko et al. (2001) (GFP and YFP)
Tau	mammalian (tau34, bovine tau)	B104, CHO, NIH-3T3, primary neurons, ES cells, transgenic mice	transient, adenoviral, chromosomal integration	Ludin et al. (1996); Illenberger et al. (1996); Lu and Kosik (2001); Stamer et al. (2002) (CFP); Pratt et al. (2000)
	mammalian (bovine tau)	zebrafish	embryo microinjection	Geldmacher-Voss et al. (2003)
	mammalian~	*Drosophila*	transgentic flies	Brand (1995)
Viral proteins	herpes simplex VP-22	COS-1	transient	Martin et al. (2002)

Fusions between microtubule-binding proteins (MBPs) and GFP. This table is meant to provide some information about the range of tools available. In many cases, the paper cited is only the first of a series using the listed fusion or its variants. Inclusion is biased toward proteins that are known to bind directly to microtubules, but some proteins with indirect interactions are listed. For reasons of space limitation, proteins with cellular localization limited to the centrosome are excluded. *Symbols*: (~) Ambiguous in the primary source; (–) fusion protein and expression system are conspecific; (*) the number of GFP fusions to kinesin and kinesin family members is too large to include here.

Microtubule plus-end tracking proteins (+TIPs) provide a particularly vivid example of the power of GFP fusions as experimental tools (see Movies 31.2 and 31.3 on the accompanying DVD). Immunofluorescent analyses of +TIPs had shown that they localize to microtubule plus ends in a comet-like pattern (Diamantopoulos et al. 1999; Vaughan et al. 1999), but no amount of static-image analysis prepared investigators for what they saw when they first examined movies of GFP-CLIP-170: The CLIP-170 comets dynamically track all growing microtubule plus ends, resulting in an appearance of "cellular fireworks" (Perez et al. 1999; Komarova et al. 2002). GFP fusions have revealed that a large number of proteins exhibit +TIP behavior, including dynactin p150 (Vaughan et al. 2002), EB1 (Mimori-Kiyosue et al. 2000b), NudF (Lis1) (Han et al. 2001), and even dynein itself (Han et al. 2001).

GFP fusions to +TIPs are useful not only for the study of these proteins themselves, but as markers for the plus ends of growing microtubules: Komarova and Borisy used GFP-CLIP-170 to demonstrate that microtubules in interphase cells grow processively in the interior of cells and exhibit classic dynamic instability only near the edges (Komarova et al. 2002). GFP-EB1 has been used to examine microtubule nucleation in both mammalian cells (Piehl et al. 2004) and *Arabidopsis* (Chan et al. 2003), and GFP-EB3 has been used to address questions about local microtubule dynamics and growth in cultured neurons (Stepanova et al. 2003).

The following sections outline methods used to prepare and express GFP-tubulin and MAPs in cultured mammalian cells. For methods used to express GFP-tagged proteins in other systems, see the reference columns in Tables 1 and 2.

PROTOCOLS

Note: <!> indicates hazardous material. See Cautions Appendix.

Tagging Tubulin and MAPs with GFP

Molecular Methods

To generate GFP-tagged proteins in mammalian cells, pEGFP vectors available from BD Biosciences/Clontech can be used. These vectors drive protein expression using the cytomegalovirus immediate early (CMV IE) promoter and contain the neomycin gene for selection. The cDNA is inserted into the multiple cloning site using standard molecular methods. Insertion of the gene upstream of the GFP tag results in a carboxy-terminal tag on the protein; alternatively, insertion downstream from the GFP tag results in a protein with an amino-terminal GFP tag. Human α-tubulin constructs tagged at the amino terminus with enhanced GFP (EGFP) and yellow fluorescent protein (YFP) are commercially available from BD Biosciences/Clontech. The behavior of these constructs is indistinguishable from the behavior of rhodamine-tagged tubulin injected into mammalian cells (Rusan et al. 2001). Mammalian β-tubulin has also been tagged at the carboxy terminus (Heidemann et al. 1999). Tubulin has been tagged with a photoactivatable variant of GFP (Tulu et al. 2003), but to date, there is no report of tubulin tagged with DsRed, a red variant of GFP.

Although most reported studies in mammalian cells describing fusions between GFP and proteins of the microtubule cytoskeleton use a constitutively active promoter like the CMV IE promoter, it is also possible to use an inducible promoter system, for example, the tetracycline <!> (Tet)-inducible or repressible promoters (see Chapter 2). In this case, the level of the chimeric protein can be regulated by the addition or removal of tetracycline from the medium (Zhu et al. 2002). Interestingly, the level of tubulin subunits in cells is autoregulated: Tubulin binds to its own mRNA, regulating the level of protein in the cytoplasm (Yen et al. 1988). In LLCPK cells expressing GFP-tagged tubulin, 17% of the total tubulin is GFP-tagged and the level of unlabeled tubulin is reduced to 81% of the level in parental cells, indicating that autoregulation of tubulin levels is maintained in the GFP-tubulin-expressing cells (Rusan et al. 2001).

Transfection

Once the DNA construct has been prepared, it must be introduced into the cells of interest. For mammalian cells, this is most easily accomplished by transient transfection. Numerous cell lines have been successfully transfected using a variety of transfection reagents. The optimum conditions for transfection using various reagents such as Lipofectamine 2000 (Invitrogen) can be found on product Web sites. For references describing transfections of tubulin and MAPs into particular cell types and organisms, see Table 1 and 2.

An advantage of transient transfection is that it is relatively fast—cells can usually be observed the day after transfection, although expression continues for several days. In some cases, cell health improves with time, but long-term expression of some proteins is deleterious (see Section below on Considerations Specific to Analysis of GFP-MAPs). The major disadvantage of transient transfection is that the expression level of the chimeric protein varies greatly from cell to cell. Some cells express very high levels of protein; these cells may show overexpression artifacts and are unlikely to survive for long periods. Other cells do not express the chimeric protein or express it at levels that are too low for imaging. A subset of cells expresses the protein at a level appropriate for imaging; it must be determined whether this level is sufficient to avoid overexpression artifacts. This variation in expression level can actually be an advantage in studying the cellular consequences of expressing the GFP fusions; transfected cells are generally surrounded by untransfected "control cells" that have been subjected to otherwise identical treatment. Clearly, it is important to determine that the expressed protein does not interfere with cell function and that it incorporates into the cellular structure of interest (see Chapter 2).

Stable Cell Lines

An alternative to transient transfection is to generate a clonal cell line permanently expressing the chimeric protein at an appropriate level. Although this procedure takes longer than transient transfection, it has the major advantage that all of the cells are expressing the chimeric protein at the same level and the cell line can be characterized.

In the following protocol, a bicistronic expression vector (pIRES; BD Biosciences/Clontech) is used to generate stable cell lines. This vector has been used to create cell lines that express GFP-tubulin and PA-GFP-tubulin (Rusan et al. 2001; Tulu et al. 2003). The vector uses an internal ribosome entry site (IRES) derived from the encephalomyocarditis virus and translates two open reading frames from one mRNA; thus, the gene of interest (e.g., GFP-tubulin) and the selection marker are expressed from the same promoter. The gene of interest is immediately downstream from the pCMV IE promoter and the selection marker is downstream from the IRES site. With this vector system, selection is more efficient, presumably because of the linked expression of both genes. IRES vectors with neomycin <!> and hygromycin <!> resistance markers have been used with this protocol.

Generating Stable Cell Lines

1. Subclone GFP-tubulin into the pIRES vector.

2. Transfect the cells using Lipofectamine 2000 as described above (see Section above on Transfection). Incubate the cells for 48 hours in normal medium to allow expression of the resistance gene.

3. Incubate the cells in antibiotic containing medium for 2 weeks. Change the medium frequently (every 1 or 2 days) to remove dead or dying cells.

   Use the lowest concentration of antibiotic that kills all control parental cells. Determine this concentration by generating a "death curve" for the parental cell line before transfection. Briefly, treat parental cells growing in a 24-well plate with various concentrations of antibiotic for 10 days; use the lowest level of antibiotic that kills all of the cells in the well.

4. Trypsinize <!> the surviving cells and plate them in 100-mm dishes at a sufficient dilution so that single cells will give rise to well-separated individual colonies (because this dilution depends on many variables, it must be determined empirically). Plate a dish or flask at higher cell density and keep it in reserve in case appropriate clones are not generated from the dilute plating.

5. Allow the cells to grow (usually for 1–2 weeks) until individual colonies of several hundred cells are present.

   If the cells are expressing a GFP-tagged cytoskeletal protein, colonies (of varying brightness) can be located using an inverted epifluorescence microscope and their location on the dish marked; alternatively, colonies can be selected randomly without assaying fluorescence (cloning blind).

6. Isolate and transfer individual colonies to a 24-well plate.

   a. Apply a thin layer of vacuum grease or petroleum jelly (i.e., Vaseline) to one face of a set of cloning rings (6 or 8-mm outer diameter size; Fisher Scientific), and set them aside in the tissue culture hood.

   b. Remove the medium from the cultured cells and wash them twice with calcium- and magnesium-free phosphate-buffered saline (PBS).

      > Because only a small amount of PBS remains on the plate during the following steps, care must be taken to avoid killing the cells by drying (with practice, ~20 colonies can be selected from a 100-mm plate before drying is a problem).

   c. Use sterile forceps to place cloning rings over the selected colonies, Vaseline side down, and press them onto the surface of the plate.

   d. Pipette 50–70 µl of trypsin into each cloning ring. After the cells in the ring have rounded up, usually after ~3 minutes, add medium to the ring, aspirate the cells, and place them in individual wells of a 24-well plate. Avoid pipetting up and down too many times.

      > After this point, consider each clone in the 24-well plate as a potential cell line; therefore, take care when splitting and growing to prevent intercontamination between different clones.

      > Alternatively, use cloning discs to select colonies on the 100-mm dish. If the cells of interest do not form tight colonies (i.e., fibroblasts), they can be cloned by limiting dilution in 96-well plates (Celis et al. 1994).

7. Trypsinize the cells in the 24-well plate when they have grown to sufficient density. Plate them in two wells of a 6-well plate, one of which contains a coverslip. Use the cells on the coverslip to check the level of GFP-tubulin expression with high-resolution fluorescence microscopy; propogate colonies expressing an appropriate level of the chimeric protein from the remaining well. Discard colonies that either do not express GFP-tubulin or are too dim for imaging.

Characterization of Clones Expressing GFP-Tubulin or GFP-MAPs

Selected clones are characterized to determine the level of expression of the chimeric protein and the dynamic behavior of the microtubules (or microtubule-binding proteins).

1. Determine the level of expression of the tagged protein by probing western blots of cell extracts with antibodies to the tagged protein and determining the percentage of the total protein that is tagged. The level of untagged protein can also be measured relative to the endogenous protein level in extracts of control cells.

2. Measure the mitotic index and doubling time of the cloned cell lines and compare with control cells to determine whether expression of the GFP fusion protein alters these parameters (Stein et al. 1994). In the case of GFP-MAP expression, examine transfected cells for possible effects on the organization of microtubules, membranes, and other cytoskeletal elements.

3. Measure the dynamic behavior of individual microtubules from time-lapse sequences of cells expressing GFP-tubulin and compare them with the behavior of microtubules in parental cells injected with rhodamine-labeled tubulin (Rusan et al. 2001).

   > In cases where expression of a full-length GFP fusion is deleterious to the cell, it may be difficult to obtain the desired stable cell line—either no stable transfectants are obtained or only mutated versions (e.g., truncations) of the GFP-tagged protein are recovered. Before proceeding with any in-depth analysis, confirm within reasonable limits (by western blot and/or PCR)

that observed transfectants do indeed express full-length protein. If problems are observed, changing to a weaker or regulated promoter (Zhu et al. 2002) may be necessary to obtain the desired stable cell lines (see Chapter 2).

Microscopy

Once cells stably expressing or transiently transfected with GFP-tubulin or GFP-MAPs have been generated, they are observed by using fluorescence microscopy. For a discussion of fluorescence microscopy, see Chapters 12, 27 and 28. The following is a brief summary of methods that we use to image cells that express GFP-tubulin and GFP-MAPs.

Cells can be imaged in a variety of situations. They are typically mounted on coverslips in Rose chambers or plated in dishes with a coverslip bottom (see Chapter 27). For observation, growth medium is replaced with medium lacking phenol red indicator dye and bicarbonate and containing HEPES buffer. Oxyrase (E.C. Oxyrase; Oxyrase, Inc. Mansfield, Ohio) is added to the medium before imaging to retard photobleaching. Changing to HEPES-buffered medium is important to prevent any change in pH upon exposure to atmospheric levels of CO_2. Alternatively, some investigators prevent pH changes by placing a layer of inert oil on top of bicarbonate-buffered medium while the plates are outside the incubator. For short-term observation, coverslips in normal growth medium can simply be inverted onto slides and sealed with VaLaP (Vaseline, lanolin, paraffin 1:1:1). Rogers et al. (2002), for example, invert coverslips onto fragments of broken coverslips.

Mounted cells are observed on an inverted microscope using a high-numerical-aperture (NA) objective lens ($100\times$ or 63×1.4 N.A.) and fluorescence filters optimized for EGFP excitation and emission such as the standard sets sold by Nikon (B-2E/C) and Chroma (Endow-GFP). Although wide-bandpass filters can give a more intense signal, they are also prone to problems with photobleaching and autofluorescence. The illuminating light should be electronically shuttered to minimize exposure of the cells to fluorescent illumination. A cooled CCD camera is used to collect images; exposure times of 400–700 msec are usually sufficient to image individual microtubules in the thin peripheral regions of cultured cells. For measurement of microtubule dynamic instability, images are normally collected at 1–2-second intervals. Lengthening the interval between successive images underestimates some values of dynamic instability because some transitions are not captured (Shelden and Wadsworth 1993). In addition, microtubule dynamics is temperature-dependent, so it is important to maintain temperature during the imaging process (see Chapter 27). For a detailed description of calculating dynamic instability parameters, see Dhamodharan and Wadsworth (1995).

To image individual microtubules, especially in thicker regions of cells, confocal microscopy is recommended. Spinning-disc confocal microscopy has the advantage that images can be acquired rapidly and photodamage to the sample is minimal (Maddox et al. 2003). Laser-scanning confocal microscopy can also be used. Both types of confocal microscopy reduce background fluorescence and permit observations of individual microtubules.

Considerations Specific to Analysis of GFP-MAPs

Most of the information described above can be and has been directly applied to characterization of GFP-MAP proteins. However, in the characterization of GFP-MAP fusions, a few additional issues must be considered. First, some MAPs, especially plus-end tracking proteins, are exquisitely sensitive to changes in the environment of the cell, especially changes in pH and temperature (Vaughan et al. 1999; H. Goodson and T.E. Kreis, unpubl.). Therefore, extra care must be taken to characterize how the behaviors observed in living cells relate to the particular conditions used. In addition, the feedback mechanisms regulating the total concentration of cellular tubulin (Yen et al. 1988; Rusan et al. 2001) do not generally exist for MAPs. Therefore, problems caused by variation in expression level and/or overexpression are likely to be more severe. For example, Perez and Kreis used inhibitors of protein synthesis after transfection to limit the expression level of CLIP-170 when it was expressed in Vero cells (Perez et al. 1999), but in other cell types (e.g., COS-7), these problems can be avoided simply by selecting cells with low levels of expression. Expression of some proteins (those that interfere with mitosis, etc.) may be deleterious to cell survival. If this problem occurs, it may be necessary to perform experiments less than 24 hours after transfection.

MOVIE LEGENDS

MOVIE 31.1. (GFP-MTs): Microtubule dynamic instability, visualized in the peripheral region of an LLCPK1 cell (pig kidney epithelial cells) expressing GFP-tubulin. Individual microtubules can be observed growing, shortening, and pausing. Images were acquired at 2-second intervals; movie is presented at 20× real time.

MOVIE 31.2. (GFP-EB1): Microtubule growth visualized in an LLCPK1 cell expressing GFP-EB1 (Piehl et al. 2004) to mark the growing microtubule plus ends. The central cell contains two centrosomes from which the GFP-EB1 dots emerge; individual growing microtubules can be followed as they approach the cell periphery. Images were acquired at 2-second intervals; movie is presented at 15× real time.

MOVIE 31.3. (GFP-CLIP-170): Microtubule growth visualized in a COS-7 cell expressing GFP-CLIP-170 (Perez et al. 1999) to study CLIP-170 plus-end tracking behavior. In particular, note the inhomogeneities (fluorescent speckling) in the CLIP-170 fluorescence (Danuser and Waterman-Storer 2003). Analysis of the patterns in these speckles as a function of time (kymograph analysis) shows that they are immobile and indicates that individual CLIP-170 molecules do not move in the process of plus-end tracking (Perez et al. 1999; E.S. Folker and H. Goodson, in prep.). Images were acquired by 0.5-second exposures at 1-second intervals; movie is presented at 30 frames/second. Microtubule growth is somewhat slower than in Movies 31.1 and 31.2 because the data were acquired at a lower temperature (28°C).

REFERENCES

Akhmanova A., Hoogenraad C.C., Drabek K., Stepanova T., Dortland B., Verkerk T., Vermeulen W., Burgering B.M., De Zeeuw C.I., Grosveld F., and Galjart N. 2001. Clasps are CLIP-115 and -170 associating proteins involved in the regional regulation of microtubule dynamics in motile fibroblasts. *Cell* **104:** 923–935.

Assmann V., Jenkinson D., Marshall J.F., and Hart. I.R. 1999. The intracellular hyaluronan receptor RHAMM/IHABP interacts with microtubules and actin filaments. *J. Cell Sci.* **112:** 3943–3954.

Behrens R. and Nurse P. 2002. Roles of fission yeast Tea1p in the localization of polarity factors and in organizing the microtubular cytoskeleton. *J. Cell Biol.* **157:** 783–793.

Bouquin T., Mattsson O., Naested H., Foster R., and Mundy J. 2003. The *Arabidopsis lue1* mutant defines a katanin p60 ortholog involved in hormonal control of microtubule orientation during cell growth. *J. Cell Sci.* **116:** 791–801.

Brand A. 1995. GFP in *Drosophila. Trends Genet.* **11:** 324–325.

Brunner D. and Nurse P. 2000. CLIP170-like Tip1p spatially organizes microtubular dynamics in fission yeast. *Cell* **102:** 695–704.

Bulinski J.C., Odde D.J., Howell B.J., Salmon T.D., and Waterman-Storer C.M. 2001. Rapid dynamics of the microtubule binding of ensconsin in vivo. *J. Cell Sci.* **114:** 3885–3897.

Buster D., McNally K., and McNally F.J. 2002. Katanin inhibition prevents the redistribution of gamma-tubulin at mitosis. *J. Cell Sci.* **115:** 1083–1092.

Carminati J.L. and Stearns T. 1997. Microtubules orient the mitotic spindle in yeast through dynein-dependent interactions with the cell cortex. *J. Cell Biol.* **138:** 629–641.

Celis A., Dejgaard K., and Celis J.E. 1994. Production of mouse monoclonal antibodies. In *Cell biology: A laboratory handbook* (ed. J.E. Celis), vol. 2, pp. 269–275. Academic Press, New York.

Chalfie M., Tu Y., Euskirchen G., Ward W.W., and Prasher D.C. 1994. Green fluorescent protein as a marker for gene expression. *Science* **263:** 802–805.

Chan J., Calder G.M., Doonan J.H., and Lloyd C.W. 2003. EB1 reveals mobile microtubule nucleation sites in *Arabidopsis*. *Nat. Cell. Biol.* **5:** 967–971.

Chang W., Gruber D., Chari S., Kitazawa H., Hamazumi Y., Hisanaga S., and Bulinski J.C. 2001. Phosphorylation of MAP4 affects microtubule properties and cell cycle progression. *J. Cell Sci.* **114:** 2879–2887.

Chen X., Kojima S., Borisy G.G., and Green K.J. 2003. p120 catenin associates with kinesin and facilitates the transport of cadherin-catenin complexes to intercellular junctions. *J. Cell Biol.* **163:** 547–557.

Danuser G. and Waterman-Storer C.M. 2003. Quantitative fluorescent speckle microscopy: Where it came from and where it is going. *J. Microsc.* **211:** 191–207.

de Hostos E.L. 1999. The coronin family of actin-associated proteins. *Trends Cell Biol.* **9:** 345–350.

Dhamodharan R. and Wadsworth P. 1995. Modulation of microtubule dynamic instability in vivo by brain microtubule associated proteins. *J. Cell Sci.* **108:** 1679–1689.

Dhonukshe P. and Gadella T.W., Jr. 2003. Alteration of microtubule dynamic instability during preprophase band formation revealed by yellow fluorescent protein-CLIP170 microtubule plus-end labeling. *Plant Cell* **15:** 597–611.

Diamantopoulos G.S., Perez F., Goodson H.V., Batelier G., Melki R., Kreis T.E., and Rickard J.E. 1999. Dynamic localization of CLIP-170 to microtubule plus ends is coupled to microtubule assembly. *J. Cell Biol.* **144:** 99–112.

Ding D.Q., Chikashige Y., Haraguchi T., and Hiraoka Y. 1998. Oscillatory nuclear movement in fission yeast meiotic prophase is driven by astral microtubules, as revealed by continuous observation of chromosomes and microtubules in living cells. *J. Cell Sci.* **111:** 701–712.

Errico A., Ballabio A., and Rugarli E.I. 2002. Spastin, the protein mutated in autosomal dominant hereditary spastic paraplegia, is involved in microtubule dynamics. *Hum. Mol. Genet.* **11:** 153–163.

Faire K., Waterman-Storer C.M., Gruber D., Masson D., Salmon E.D., and Bulinski J.C. 1999. E-MAP-115 (ensconsin) associates dynamically with microtubules in vivo and is not a physiological modulator of microtubule dynamics. *J. Cell Sci.* **112:** 4243–4255.

Geldmacher-Voss B., Reugels A.M., Pauls S., and Campos-Ortega J.A. 2003. A 90-degree rotation of the mitotic spindle changes the orientation of mitoses of zebrafish neuroepithelial cells. *Development* **130:** 3767–3780.

Graf R., Daunderer C., and Schliwa M. 2000. *Dictyostelium* DdCP224 is a microtubule-associated protein and a permanent centrosomal resident involved in centrosome duplication. *J. Cell Sci.* **113:** 1747–1758.

Grieder N.C., de Cuevas M., and Spradling A.C. 2000. The fusome organizes the microtubule network during oocyte differentiation in *Drosophila*. *Development* **127:** 4253–4264.

Han G., Liu B., Zhang J., Zuo W., Morris N.R., and Xiang X. 2001. The *Aspergillus* cytoplasmic dynein heavy chain and NUDF localize to microtubule ends and affect microtubule dynamics. *Curr. Biol.* **11:** 719–724.

Hartman J.J. and Vale R.D. 1999. Microtubule disassembly by ATP-dependent oligomerization of the AAA enzyme katanin. *Science* **286:** 782–785.

Hasezawa S., Ueda K., and Kumagai F. 2000. Time-sequence observations of microtubule dynamics throughout mitosis in living cell suspensions of stable transgenic *Arabidopsis*—Direct evidence for the origin of cortical microtubules at M/G1 interface. *Plant Cell Physiol.* **41:** 244–250.

Hazan I., Sepulveda-Becerra M., and Liu H. 2002. Hyphal elongation is regulated independently of cell cycle in *Candida albicans*. *Mol. Biol. Cell* **13:** 134–145.

Heidemann S.R., Kaech S., Buxbaum R.E., and Matus A. 1999. Direct observations of the mechanical behaviors of the cytoskeleton in living fibroblasts. *J. Cell Biol.* **145:** 109–122.

Hofmann C., Cheeseman I.M., Goode B.L., McDonald K.L., Barnes G., and Drubin D.G. 1998. *Saccharomyces cerevisiae* Duo1p and Dam1p, novel proteins involved in mitotic spindle function. *J. Cell Biol.* **143:** 1029–1040.

Hoogenraad C.C., Akhmanova A., Grosveld F., De Zeeuw C.I., and Galjart N. 2000. Functional analysis of CLIP-115 and its binding to microtubules. *J. Cell Sci.* **113:** 2285–2297.

Hori T., Haraguchi T., Hiraoka Y., Kimura H., and Fukagawa T. 2003. Dynamic behavior of Nuf2-Hec1 complex that localizes to the centrosome and centromere and is essential for mitotic progression in vertebrate cells. *J. Cell Sci.* **116:** 3347–3362.

Hynes T.R., Hughes T.E., and Berlot C.H. 2004. Cellular localization of GFP-tagged alpha subunits. *Methods Mol. Biol.* **237:** 233–246.

Illenberger S., Drewes G., Trinczek B., Biernat J., Meyer H.E., Olmsted J.B., Mandelkow E.M., and Mandelkow E. 1996. Phosphorylation of microtubule-associated proteins MAP2 and MAP4 by the protein kinase p110mark. Phosphorylation sites and regulation of microtubule dynamics. *J. Biol. Chem.* **271:** 10834–10843.

Jensen S., Segal M., Clarke D.J., and Reed S.I. 2001. A novel role of the budding yeast separin Esp1 in anaphase spindle elongation: Evidence that proper spindle association of Esp1 is regulated by Pds1. *J. Cell Biol.* **152:** 27–40.

Jones M.H., He X., Giddings T.H., and Winey M. 2001. Yeast Dam1p has a role at the kinetochore in assembly of the mitotic spindle. *Proc. Natl. Acad. Sci.* **98:** 13675–13680.

Kaech S., Ludin B., and Matus A. 1996. Cytoskeletal plasticity in cells expressing neuronal microtubule-associated proteins. *Neuron* **17:** 1189–1199.

Kamath K. and Jordan M.A. 2003. Suppression of microtubule dynamics by epothilone B is associated with mitotic arrest. *Cancer Res.* **63:** 6026–6031.

Katz W., Weinstein B., and Solomon F. 1990. Regulation of tubulin levels and microtubule assembly in *Saccharomyces cerevisiae:* Consequences of altered tubulin gene copy number. *Mol. Cell. Biol.* **10:** 5286–5294.

Kaverina I., Krylyshkina O., Beningo K., Anderson K., Wang Y.L., and Small J.V. 2002. Tensile stress stimulates microtubule outgrowth in living cells. *J. Cell Sci.* **115:** 2283–2291.

Kimble M., Kuzmiak C., McGovern K.N., and de Hostos E.L. 2000. Microtubule organization and the effects of GFP-tubulin expression in *Dictyostelium discoideum. Cell Motil. Cytoskel.* **47:** 48–62.

Komarova Y.A., Vorobjev I.A., and Borisy G.G. 2002. Life cycle of MTs: Persistent growth in the cell interior, asymmetric transition frequencies and effects of the cell boundary. *J. Cell Sci.* **115:** 3527–3539.

Koonce M.P., Kohler J., Neujahr R., Schwartz J.M., Tikhonenko I., and Gerisch G. 1999. Dynein motor regulation stabilizes interphase microtubule arrays and determines centrosome position. *EMBO J.* **18:** 6786–6792.

Kosco K.A., Pearson C.G., Maddox P.S., Wang P.J., Adams I.R., Salmon E.D., Bloom K., and Huffaker T.C. 2001. Control of microtubule dynamics by Stu2p is essential for spindle orientation and metaphase chromosome alignment in yeast. *Mol. Biol. Cell* **12:** 2870–2880.

Kreis T.E. and Birchmeier W. 1982. Microinjection of fluorescently labeled proteins into living cells with emphasis on cytoskeletal proteins. *Int. Rev. Cytol.* **75:** 209–214.

Krylyshkina O., Anderson K.I., Kaverina I., Upmann I., Manstein D.J., Small J.V., and Toomre D.K. 2003. Nanometer targeting of microtubules to focal adhesions. *J. Cell Biol.* **161:** 853–859.

Krylyshkina O., Kaverina I., Kranewitter W., Steffen W., Alonso M.C., Cross R.A., and Small J.V. 2002. Modulation of substrate adhesion dynamics via microtubule targeting requires kinesin-1. *J. Cell Biol.* **156:** 349–359.

Kumagai F., Yoneda A., Tomida T., Sano T., Nagata T., and Hasezawa S. 2001. Fate of nascent microtubules organized at the M/G1 interface, as visualized by synchronized tobacco BY-2 cells stably expressing GFP-tubulin: Time-sequence observations of the reorganization of cortical microtubules in living plant cells. *Plant Cell Physiol.* **42:** 723–732.

Lauf U., Giepmans B.N., Lopez P., Braconnot S., Chen S.C., and Falk M.M. 2002. Dynamic trafficking and delivery of connexons to the plasma membrane and accretion to gap junctions in living cells. *Proc. Natl. Acad. Sci.* **99:** 10446–10451.

Lee L., Tirnauer J.S., Li J., Schuyler S.C., Liu J.Y., and Pellman D. 2000. Positioning of the mitotic spindle by a cortical-microtubule capture mechanism. *Science* **287:** 2260–2262.

Lee S. and Kolodziej P.A. 2002. Short Stop provides an essential link between F-actin and microtubules during axon extension. *Development* **129:** 1195–1204.

Lee W.L., Oberle J.R., and Cooper J.A. 2003. The role of the lissencephaly protein Pac1 during nuclear migration in budding yeast. *J. Cell Biol.* **160:** 355–364.

Lin H., de Carvalho P., Kho D., Tai C.Y., Pierre P., Fink G.R., and Pellman D. 2001. Polyploids require Bik1 for kinetochore-microtubule attachment. *J. Cell Biol.* **155:** 1173–1184.

Lu M. and Kosik K.S. 2001. Competition for microtubule-binding with dual expression of tau missense and splice isoforms. *Mol. Biol. Cell.* **12:** 171–184.

Ludin B. and Matus A. 1998. GFP illuminates the cytoskeleton. *Trends Cell Biol.* **8:** 72–77.

Ludin B., Doll T., Meili R., Kaech S., and Matus A. 1996. Application of novel vectors for GFP-tagging of proteins to study microtubule-associated proteins. *Gene* **173**: 107–111.

Ma X., Ehrhardt D.W., and Margolin W. 1996. Colocalization of cell division proteins FtsZ and FtsA to cytoskeletal structures in living *Escherichia coli* cells by using green fluorescent protein. *Proc. Natl. Acad. Sci.* **93**: 12998–13003.

Maddox P., Chin E., Mallavarapu A., Yeh E., Salmon E.D., and Bloom K. 1999. Microtubule dynamics from mating through the first zygotic division in the budding yeast *Saccharomyces cerevisiae*. *J. Cell Biol.* **144**: 977–987.

Maddox P.S., Moree B., Canman J.C., and Salmon E.D. 2003. Spinning disk confocal microscope system for rapid high-resolution, multimode, fluorescence speckle microscopy and green fluorescent protein imaging in living cells. *Methods Enzymol.* **360**: 597–617.

Marc J., Granger C.L., Brincat J., Fisher D.D., Kao T., McCubbin A.G., and Cyr R.J. 1998. A GFP-MAP4 reporter gene for visualizing cortical microtubule rearrangements in living epidermal cells. *Plant Cell* **10**: 1927–1940.

Martin A., O'Hare P., McLauchlan J., and Elliott G. 2002. Herpes simplex virus tegument protein VP22 contains overlapping domains for cytoplasmic localization, microtubule interaction, and chromatin binding. *J. Virol.* **76**: 4961–4970.

Maxwell C.A., Keats J.J., Crainie M., Sun X., Yen T., Shibuya E., Hendzel M., Chan G., and Pilarski L.M. 2003. RHAMM is a centrosomal protein that interacts with dynein and maintains spindle pole stability. *Mol. Biol. Cell* **14**: 2262–2276.

Miller R.K. and Rose M.D. 1998. Kar9p is a novel cortical protein required for cytoplasmic microtubule orientation in yeast. *J. Cell Biol.* **140**: 377–390.

Mimori-Kiyosue Y., Shiina N., and Tsukita S. 2000a. Adenomatous polyposis coli (APC) protein moves along microtubules and concentrates at their growing ends in epithelial cells. *J. Cell Biol.* **148**: 505–518.

———. 2000b. The dynamic behavior of the APC-binding protein EB1 on the distal ends of microtubules. *Curr. Biol.* **10**: 865-868.

Mollinari C., Kleman J.P., Jiang W., Schoehn G., Hunter T., and Margolis R.L. 2002. PRC1 is a microtubule binding and bundling protein essential to maintain the mitotic spindle midzone. *J. Cell Biol.* **157**: 1175–1186.

Moore R.C. and Cyr R.J. 2000. Association between elongation factor-1alpha and microtubules in vivo is domain dependent and conditional. *Cell Motil. Cytoskel.* **45**: 279–292.

Nakaseko Y., Goshima G., Morishita J., and Yanagida M. 2001. M phase-specific kinetochore proteins in fission yeast: Microtubule-associating Dis1 and Mtc1 display rapid separation and segregation during anaphase. *Curr. Biol.* **11**: 537–549.

Niccoli T. and Nurse P. 2002. Different mechanisms of cell polarisation in vegetative and shmooing growth in fission yeast. *J. Cell Sci.* **115**: 1651–1662.

Oegema K., Desai A., Rybina S., Kirkham M., and Hyman A.A. 2001. Functional analysis of kinetochore assembly in *Caenorhabditis elegans*. *J. Cell Biol.* **153**: 1209–1226.

Olson K.R., McIntosh J.R., and Olmsted J.B. 1995. Analysis of MAP 4 function in living cells using green fluorescent protein (GFP) chimeras. *J. Cell Biol.* **130**: 639–650.

Ozer R.S. and Halpain S. 2000. Phosphorylation-dependent localization of microtubule-associated protein MAP2c to the actin cytoskeleton. *Mol. Biol. Cell* **11**: 3573–3587.

Perez F., Diamantopoulos G.S., Stalder R., and Kreis T.E. 1999. CLIP-170 highlights growing microtubule ends in vivo. *Cell* **96**: 517–527.

Perez F., Pernet-Gallay K., Nizak C., Goodson H.V., Kreis T.E., and Goud B. 2002. CLIPR-59, a new trans-Golgi/TGN cytoplasmic linker protein belonging to the CLIP-170 family. *J. Cell Biol.* **156**: 631–642.

Petrulis J.R., Kusnadi A., Ramadoss P., Hollingshead B., and Perdew G.H. 2003. The hsp90 co-chaperone XAP2 alters importin beta recognition of the bipartite nuclear localization signal of the Ah receptor and represses transcriptional activity. *J. Biol. Chem.* **278**: 2677–2685.

Piehl M. and Cassimeris L. 2003. Organization and dynamics of growing microtubule plus ends during early mitosis. *Mol. Biol. Cell* **14**: 916–925.

Piehl M., Tulu U.S., Wadsworth P., and Cassimeris L. 2004. Centrosome maturation: Measurement of microtubule nucleation throughout the cell cycle by using GFP-tagged EB1. *Proc. Natl. Acad. Sci.* **101**: 1584–1588.

Pierce D.W. and Vale R.D. 1998. Assaying processive movement of kinesin by fluorescence microscopy. *Methods Enzymol.* **298**: 154–171.

Pierce D.W., Hom-Booher N., and Vale R.D. 1997. Imaging individual green fluorescent proteins. *Nature* **388**: 338.

Pierce D.W., Hom-Booher N., Otsuka A.J., and Vale R.D. 1999. Single-molecule behavior of monomeric and heteromeric kinesins. *Biochemistry* **38:** 5412–5421.

Popov A.V., Pozniakovsky A., Arnal I., Antony C., Ashford A.J., Kinoshita K., Tournebize R., Hyman A.A., and Karsenti E. 2001. XMAP215 regulates microtubule dynamics through two distinct domains. *EMBO J.* **20:** 397–410.

Popova J.S., Greene A.K., Wang J., and Rasenick M.M. 2002. Phosphatidylinositol 4,5-bisphosphate modifies tubulin participation in phospholipase Cbeta1 signaling. *J. Neurosci.* **22:** 1668–1678.

Pratt T., Sharp L., Nichols J., Price D.J., and Mason J.O. 2000. Embryonic stem cells and transgenic mice ubiquitously expressing a tau-tagged green fluorescent protein. *Dev. Biol.* **228:** 19–28.

Raemaekers T., Ribbeck K., Beaudouin J., Annaert W., Van Camp M., Stockmans I., Smets N., Bouillon R., Ellenberg J., and Carmeliet G. 2003. NuSAP, a novel microtubule-associated protein involved in mitotic spindle organization. *J. Cell Biol.* **162:** 1017–1029.

Rehberg M. and Graf R. 2002. *Dictyostelium* EB1 is a genuine centrosomal component required for proper spindle formation. *Mol. Biol. Cell* **13:** 2301–2310.

Reski R. 2002. Rings and networks: The amazing complexity of FtsZ in chloroplasts. *Trends Plant Sci.* **7:** 103–105.

Rogers S.L., Rogers G.C., Sharp D.J., and Vale R.D. 2002. *Drosophila* EB1 is important for proper assembly, dynamics, and positioning of the mitotic spindle. *J. Cell Biol.* **158:** 873–884.

Rusan N.M., Fagerstrom C.J., Yvon A.M., and Wadsworth P. 2001. Cell cycle-dependent changes in microtubule dynamics in living cells expressing green fluorescent protein-alpha tubulin. *Mol. Biol. Cell* **12:** 971–980.

Schmoranzer J. and Simon S.M. 2003. Role of microtubules in fusion of post-Golgi vesicles to the plasma membrane. *Mol. Biol. Cell* **14:** 1558–1569.

Schwartz K., Richards K., and Botstein D. 1997. BIM1 encodes a microtubule-binding protein in yeast. *Mol. Biol. Cell* **8:** 2677–2691.

Schweiger S., Foerster J., Lehmann T., Suckow V., Muller Y.A., Walter G., Davies T., Porter H., van Bokhoven H., Lunt P.W., Traub P., and Ropers H.H. 1999. The Opitz syndrome gene product, MID1, associates with microtubules. *Proc. Natl. Acad. Sci.* **96:** 2794–2799.

Seong Y.S., Kamijo K., Lee J.S., Fernandez E., Kuriyama R., Miki T., and Lee K.S. 2002. A spindle checkpoint arrest and a cytokinesis failure by the dominant-negative polo-box domain of Plk1 in U-2 OS cells. *J. Biol. Chem.* **277:** 32282–32293.

Shaw S.L., Kamyar R., and Ehrhardt D.W. 2003. Sustained microtubule treadmilling in *Arabidopsis* cortical arrays. *Science* **300:** 1715–1718.

Shaw S.L., Yeh E., Maddox P., Salmon E.D., and Bloom K. 1997. Astral microtubule dynamics in yeast: A microtubule-based searching mechanism for spindle orientation and nuclear migration into the bud. *J. Cell Biol.* **139:** 985–994.

Shelden E. and Wadsworth P. 1993. Observation and quantification of individual microtubule behavior in vivo: Microtubule dynamics are cell-type specific. *J. Cell Biol.* **120:** 935–945.

Shiina N. and Tsukita S. 1999. Mutations at phosphorylation sites of *Xenopus* microtubule-associated protein 4 affect its microtubule-binding ability and chromosome movement during mitosis. *Mol. Biol. Cell* **10:** 597–608.

Srayko M., Buster D.W., Bazirgan O.A., McNally F.J., and Mains P.E. 2000. MEI-1/MEI-2 katanin-like microtubule severing activity is required for *Caenorhabditis elegans* meiosis. *Genes Dev.* **14:** 1072–1084.

Stamer K., Vogel R., Thies E., Mandelkow E., and Mandelkow E.M. 2002. Tau blocks traffic of organelles, neurofilaments, and APP vesicles in neurons and enhances oxidative stress. *J. Cell Biol.* **156:** 1051–1063.

Stein G.S., Stein J.L., Lian J.B., Last T.J., Owen T., and McCabe L. 1994. Synchronization of normal diploid and transformed mammalian cells. In *Cell biology: A laboratory handbook* (ed. J.E. Celis), vol. 1, pp. 282–293. Academic Press, New York.

Stein P.A., Toret C.P., Salic A.N., Rolls M.M., and Rapoport T.A. 2002. A novel centrosome-associated protein with affinity for microtubules. *J. Cell Sci.* **115:** 3389–3402.

Stepanova T., Slemmer J., Hoogenraad C.C., Lansbergen G., Dortland B., De Zeeuw C.I., Grosveld F., van Cappellen G., Akhmanova A., and Galjart N. 2003. Visualization of microtubule growth in cultured neurons via the use of EB3-GFP (end-binding protein 3-green fluorescent protein). *J. Neurosci.* **23:** 2655–2664.

Stephens D.J. and Pepperkok R. 2001. Illuminating the secretory pathway: When do we need vesicles? *J. Cell Sci.* **114:** 1053–1059.

Straight A.F., Marshall W.F., Sedat J.W., and Murray A.W. 1997. Mitosis in living budding yeast: Anaphase A but no metaphase plate. *Science* **277:** 574–578.

Straube A., Brill M., Oakley B.R., Horio T., and Steinberg G. 2003. Microtubule organization requires cell cycle-dependent nucleation at dispersed cytoplasmic sites: Polar and perinuclear microtubule organizing centers in the plant pathogen *Ustilago maydis*. *Mol. Biol. Cell* **14:** 642–657.

Striepen B., Crawford M.J., Shaw M.K., Tilney L.G., Seeber F., and Roos D.S. 2000. The plastid of *Toxoplasma gondii* is divided by association with the centrosomes. *J. Cell Biol.* **151:** 1423–1434.

Takano Y., Oshiro E., and Okuno T. 2001. Microtubule dynamics during infection-related morphogenesis of *Colletotrichum lagenarium*. *Fungal Genet. Biol.* **34:** 107–121.

Tirnauer J.S., O'Toole E., Berrueta L., Bierer B.E., and Pellman D. 1999. Yeast Bim1p promotes the G1-specific dynamics of microtubules. *J. Cell Biol.* **145:** 993–1007.

Trinczek B., Brajenovic M., Ebneth A., and Drewes G. 2004. MARK4 is a novel microtubule-associated proteins/microtubule affinity-regulating kinase that binds to the cellular microtubule network and to centrosomes. *J. Biol. Chem.* **279:** 5915–5923.

Tulu U.S., Rusan N.M., and Wadsworth P. 2003. Peripheral, non-centrosome-associated microtubules contribute to spindle formation in centrosome-containing cells. *Curr. Biol.* **13:** 1894–1899.

Vaughan K.T., Tynan S.H., Faulkner N.E., Echeverri C.J., and Vallee R.B. 1999. Colocalization of cytoplasmic dynein with dynactin and CLIP-170 at microtubule distal ends. *J. Cell Sci.* **112:** 1437–1447.

Vaughan P.S., Miura P., Henderson M., Byrne B., and Vaughan K.T. 2002. A role for regulated binding of p150(Glued) to microtubule plus ends in organelle transport. *J. Cell Biol.* **158:** 305–319.

Wang Y.L. and Taylor D.L. 1979. Distribution of fluorescently labeled actin in living sea urchin eggs during early development. *J. Cell Biol.* **81:** 672–679.

Wang Y.L., Heiple J.M., and Taylor D.L. 1982a. Fluorescent analog cytochemistry of contractile proteins. *Methods Cell. Biol.* **25:** 1–11.

Wang Y.L., Lanni F., McNeil P.L., Ware B.R., and Taylor D.L. 1982b. Mobility of cytoplasmic and membrane-associated actin in living cells. *Proc. Natl. Acad. Sci.* **79:** 4660–4664.

Wehrle-Haller B., and Imhof B. 2002. The inner lives of focal adhesions. *Trends Cell. Biol.* **12:** 382–389.

Weinstein B. and Solomon F. 1990. Phenotypic consequences of tubulin overproduction in *Saccharomyces cerevisiae*: Differences between alpha-tubulin and beta-tubulin. *Mol. Cell Biol.* **10:** 5295–5304.

Yamamoto A., West R.R., McIntosh J.R., and Hiraoka Y. 1999. A cytoplasmic dynein heavy chain is required for oscillatory nuclear movement of meiotic prophase and efficient meiotic recombination in fission yeast. *J. Cell Biol.* **145:** 1233–1249.

Yen T.J., Machlin P.S., and Cleveland D.W. 1988. Autoregulated instability of beta-tubulin mRNAs by recognition of the nascent amino terminus of beta-tubulin. *Nature* **334:** 580–585.

Yoon Y., Pitts K., and McNiven M. 2002. Studying cytoskeletal dynamics in living cells using green fluorescent protein. *Mol. Biotechnol.* **21:** 241–250.

Zhang J., Han G., and Xiang X. 2002. Cytoplasmic dynein intermediate chain and heavy chain are dependent upon each other for microtubule end localization in *Aspergillus nidulans*. *Mol. Microbiol.* **44:** 381–392.

Zhu Z., Zheng T., Lee C.G., Homer R.J., and Elias J.A. 2002. Tetracycline-controlled transcriptional regulation systems: Advances and application in transgenic animal modeling. *Semin. Cell Dev. Biol.* **13:** 121–128.

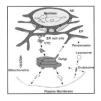

CHAPTER THIRTY TWO

Imaging of Organelle Membrane Systems and Membrane Traffic in Living Cells

Jennifer Lippincott-Schwartz and Erik Lee Snapp

Cell Biology and Metabolism Branch, National Institute of Child Health and Human Development, National Institutes of Health, Bethesda, Maryland 20892

INTRODUCTION

EUKARYOTIC CELLS ARE COMPOSED OF AN INTRICATE SYSTEM of internal membranes that are organized into different compartments, including the endoplasmic reticulum, the nucleus, the Golgi apparatus, lysosomes, endosomes, mitochondria, and peroxisomes (Figs. 1 and 2), with specialized roles within the cell. Only recently have the localization and dynamics of these organelles been studied in the context of the living cell due to the availability of green fluorescent protein (GFP) fusion proteins and recent advances in fluorescent microscope imaging systems. Results using these techniques are revealing how organelles maintain their steady-state organization and distributions, how they undergo growth and division, and how they transfer protein and lipid components to other organelles through the formation and trafficking of membrane transport intermediates.

This chapter describes methods using GFP fusion proteins to visualize the behavior of organelles and to track membrane-bound transport intermediates that bud off from organelles. Practical issues related to the construction and expression of GFP fusion proteins are first discussed. These are essential for optimizing the brightness and expression levels of GFP fusion proteins so that intracellular membrane-bound structures containing these fusion proteins can be readily visualized. Next, protocols for performing time-lapse imaging using a confocal laser-scanning microscope (CLSM) are detailed, including the use of photobleaching to highlight organelles and transport intermediates. Finally, methods for the acquisition and analysis of data and approaches for perturbing membrane traffic are outlined.

INSTRUMENTATION

To visualize organelles and the pathways by which they interconnect, it is desirable to use a laser scanning confocal microscope (CLSM) that permits imaging of multiple fluorescent

555

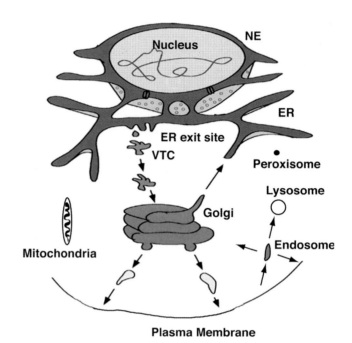

FIGURE 1. Illustration of the secretory pathway and associated organelles. Most membrane and secretory proteins are cotranslationally translocated into the endoplasmic reticulum (ER). In addition to protein processing and secretion, the ER forms the nuclear envelope. Newly synthesized proteins are sorted at ER exit sites and enter vesicles to traffic to intermediate compartments (VTCs), where proteins are sorted for movement forward to the Golgi apparatus or retrieval to the ER. Within the cisternal stacks of the Golgi apparatus, proteins are continually selectively sorted toward the TGN (*trans*-Golgi network) or retrieved back to earlier compartments. At the TGN, proteins are sorted for trafficking to the plasma membrane, endosomes, or lysosomes. From the plasma membrane, proteins can be recycled through endosomes back to the plasma membrane and to the TGN or sorted to late endosomes and potentially to lysosomes for protein degradation. Protein trafficking to mitochondria and peroxisomes appears to be mediated posttranslationally in the absence of vesicular traffic. Mitochondria are important for ATP production and lipid synthesis. Peroxisomes have a role in lipid metabolism and help reduce oxidative radicals.

markers and can aquire high resolution. In this system, a Plan-Neo 40× NA (numerical aperature) oil 1.3 or a Plan-Apo 63× NA 1.4 oil objective will provide excellent resolution and efficient light collection from a thin confocal slice of a sample such as a cell monolayer (see Chapter 14). For imaging of organelles in thicker samples, such as in tissues or embryos, a 63× NA 1.2 water objective can increase the available working distance (see Chapter 21).

Temperature control is essential for maintaining cellular processes and to stabilize microscope focus. In addition, some experiments require temperature blocks for maintaining nonpermissive temperatures. A simple approach for maintaining temperature on the microscope stage is use of a temperature-controlled airblower such as the Air Stream Stage Incubator (Model ASI 400, Nevtek, Burnsville, Virginia). A temperature probe, such as a Thermolyne Pyrometer (Carl Parmer, Vernon Hills, Illinois), can be used to confirm that the proper temperature is maintained at the coverslip or chamber. For tighter temperature control, thermal collars for objectives (available from Bioptechs, Butler, Pennsylvania) can be used (see Chapters 17 and 18).

SETTING UP THE IMAGING CHAMBER

Cells expressing GFP fusion proteins must be grown on or attached to a glass coverslip prior to imaging. It is important to use the correct glass coverslip thickness for imaging

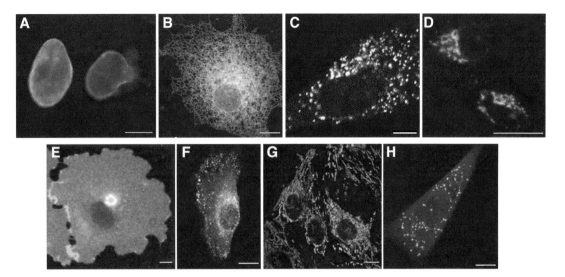

FIGURE 2. Fluorescence distributions of organelle markers in mammalian cells. (*A*) Nuclear envelopes highlighted in COS-7 cells expressing the Lamin B receptor GFP. (*B*) COS-7 cell expressing the ER marker GFP-cytochrome b(5). (*C*) NRK cell expressing the ER exit site marker Sec13-YFP. (*D*) Madin Darby canine kidney (MDCK) cells expressing the Golgi marker GalT-GFP. (*E*) MDCK cell expressing the plasma membrane marker GPI-YFP. (*F*) NRK cell expressing the endosome marker CFP-Rab5. (*G*) MDCK cells stained with the mitochondrial dye, MitoTracker. (*H*) HeLa cell expressing the peroxisome marker DsRed-SKL. Bars, 10 μm.

cells (i.e., number 1 glass coverslips for 63× and 100× objectives) to avoid significant spherical aberration of the image. The coverslip and its cells then must be mounted into a chamber that contains a buffered imaging medium. A simple inexpensive chamber can be built by taking a silicon rubber sheet and punching a hole through it (as described in Fig. 3). Commercially available chambers (Lab-Tek chambers, Nunc, Rochester, New York) can also be used. These have a glass coverslip for a base with individual plastic chambers adhered to the coverslip. These chambers allow the addition of drugs or other materials during the experiment. However, they are relatively expensive and can only be used on inverted microscopes. To prevent evaporation and alkalinization of the media, the top cover of the chamber must be sealed onto the chamber using petroleum jelly or silicon grease. Alternatively, the medium-filled well can be overlayed with a thin layer of mineral oil. Additional chambers are discussed in Chapters 17, 18, 26, and 27.

Cells grown in suspension can only be imaged after they have been first adhered to cov-

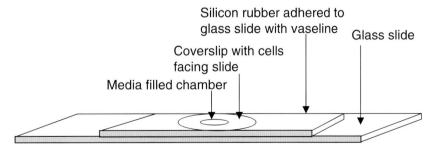

FIGURE 3. Rubber gasket imaging chamber construction. The rubber sheet with its hole is sealed to a glass slide with grease and the hole is filled with imaging medium. The coverslip is inverted (with the attached cells placed cell side down) and pressed onto the hole, allowing the cells to face the medium. The coverslip adheres to the gasket by capillary action. Excess liquid from the top of the coverslip is wicked away with absorbent tissue.

erslips or to Lab-tek chambers. This can be accomplished by coating the coverslip with poly-L-lysine, a cell-adhering agent. To coat, a concentrated poly-L-lysine (5–10 mg/ml in phosphate-buffered saline [PBS] diluted from a 10× stock [Sigma]) is layered on the coverslip surface for 15 minutes. After coating, the coverslip or chamber is washed twice in PBS. The coating lasts for up to 1 week. Suspension cells should be washed three times in PBS. Cells are than incubated on the poly-L-lysine-coated surface for 2–5 minutes and nonadhering cells are washed off by two rinses with PBS. Finally, the cells are immersed in imaging buffer.

CONSTRUCTING AND EXPRESSING ORGANELLE-TARGETED GFP FUSION PROTEINS

In constructing a fusion protein, use of the enhanced GFP variants, i.e., enhanced cyan fluorescent protein (ECFP), enhanced GFP (EGFP), enhanced yellow fluorescent protein (EYFP) (available from Clontech), or DsRed, is recommended. These are the brightest, most stable GFP variants and permit imaging with low light illumination, which minimizes problems such as photobleaching or photodamage. Because GFP variants (including EGFP, ECFP, and EYFP) can dimerize at sufficiently high concentrations (Zacharias et al. 2002), and this can perturb membrane structure (Snapp et al. 2003) or lead to false-positive interactions in fluorescence resonance energy transfer (FRET) experiments (Zacharias et al. 2002), monomeric variants of EGFP should be used, if possible (Snapp et al. 2003). In addition, because wild-type DsRed is an obligate tetramer, investigators should use the recently available monomeric red fluorescent protein 1 (mRFP1), described by Campbell et al. (2002). Even then, aberrant bright structures are sometimes observed when mRFP1 is attached to membrane proteins (J. Lippincott-Schwartz et al., unpubl.). Soluble DsRed proteins do not display this phenotype.

In addition to the intrinsic brightness of the GFP molecule, the expression level is another factor to consider in producing a bright intracellular signal from a GFP fusion protein. Use of vectors with strong promoters to enhance transcription levels and the use of proper codons to optimize translation are essential for increasing expression levels of the fusion protein and thereby enhancing its overall brightness in the cell. This is particularly important when cellular autofluorescence makes it difficult to distinguish GFP-fusion-protein-derived fluorescence from the background.

A variety of procedures can be used to enhance the expression level and brightness of the fusion protein. Addition of sodium butyrate (prepared as a 5 M solution in distilled H_2O and stored at –20°C) to the cell culture medium (1–5 mM for at least 12 hours) will increase the overall gene expression levels in stable cell lines expressing a GFP fusion protein (Gorman and Howard 1983). Use of transiently transfected cells is also possible, since they give much higher expression levels than stably transfected cell lines. However, transiently expressing cells can sometimes result in overexpression artifacts, such as protein aggregation or saturation of protein targeting machinery, which leads to inappropriate localization. Another method for increasing fusion protein brightness is to construct it with double or triple GFP molecules in tandem (Zaal et al. 1999).

Once a GFP fusion protein is optimally expressed within cells, the next goal is to determine whether it has targeted correctly. The simplest way to accomplish this is to compare the fusion protein's distribution to that of its parent molecule detected with fluorescently labeled antibodies in fixed and permeabilized nonexpressing cells. If the overall pattern of the two fluorescent signals is similar, then addition of the GFP tag to the parent protein does not perturb its targeting. In the event that the two patterns are dissimilar, then it is possible that the tag causes mistargeting by masking interactions with targeting machinery or by saturating targeting sites. In the latter case, lowering the expression level of the fusion protein by decreasing the amount of DNA used during transfection, or selecting stable cell lines or less active promoters may be helpful. It is also possible that the antibody staining does not properly reflect the parent protein distribution, because the epitope recognized by the antibody is masked in some way during fixation or is recognizing a modified form of the parent protein (i.e., a phospho-epitope).

The preservation of parental function by the GFP fusion protein may be demonstrated by showing that the fusion protein can rescue a phenotype in cells with mutations or deletions of the parent molecule or that the fusion protein can incorporate into a functional macromolecular structure. When a GFP fusion protein fails to either target or function in a manner similar to that of the parent molecule, one solution is to change the position where the GFP is attached to the parent molecule. For example, if the GFP is originally placed at the amino terminus, try it at the carboxyl terminus, and vice versa. For further discussion of fusion protein construction, see Chapter 2.

Performing a Time-lapse Imaging Experiment

The following protocol describes the use of the CLSM for time-lapse imaging of one or more fluorescent markers.

1. Set up the CLSM and its associated hardware. Grow cells in chambers or on coverslips as described in the section on Setting Up the Imaging Chamber.

   It is assumed that the investigator is familiar with the basic operation of a confocal microscope and understands both the concepts and the operation of pinholes, scan speed, zoom, detector gain, laser power, photobleach, and collection of a time series. Many CLSMs provide a number of options for data collection, including image size (512×512, 1024×1024 pixels, etc.), range of data collection (8-bit or 12-bit), and file formats. It is important to determine the requirements for the image analysis software in advance. For example, some image analysis programs cannot process 12-bit images or only process "PGM" (portable graymap) image files, instead of "tiff" (tagged-image file format) files.

2. Prewarm the imaging stage to 37°C or desired temperature, and warm up the microscope lasers for at least 5 minutes to avoid power fluctuations during imaging. Image living cells in imaging medium.

IMAGING MEDIUM

Phenol-red-free cell growth medium (e.g., RPMI or DMEM)
10% fetal bovine serum
2 mM glutamine
25 mM HEPES (pH 7.4)

---

3. Identify the cell of interest using the confocal microscope. Bring it to the desired focus. Scan an image of the whole cell at the desired excitation light intensity, line averaging, zoom, etc. Modify pinhole and detector gain for maximal fluorescence signal and minimal pixel saturation (pixel intensities that exceed the detector scale, i.e., >255 for an 8-bit image.

   It is essential to minimize saturated pixels, as they represent lost information. Saturated pixels only register as the maximum detector value, 255 (for an 8-bit image). Thus, information is lost. Detector gain and offset will vary depending on the concentration of the fluorophore, the laser power, and the width and thickness of the fluorescently labeled organelle or region. It is useful to record the detector gain settings to compare behaviors of cells expressing high versus low amounts of fluorescent protein.

4. Empirically determine conditions (i.e., scan speed, zoom, laser power, and microscope objective) that result in minimal photobleaching of the cell during the time course of imaging. Use imaging software to quantitate the fluorescence intensity of the whole cell over the time of the experiment.

> Quantitating the fluorescence intensity is necessary for determining the extent to which the whole cell undergoes photobleaching during the experiment. Significant continual fluorescence loss from the whole cell during the recovery adds an extra dimension to data interpretation and should be avoided as much as possible.

5. Use a 40-mW 488/514-nm Argon laser at 45–60% power with 1–5% transmission for imaging.

> The same conditions have been successfully used with a 25-mW 488-nm Argon laser. A 543-nm laser is typically used at 15–30% transmission and a 633-nm laser is typically used at 1–5% transmission. For rapid processes, cells usually are scanned at 0.798–3 seconds per 512 × 512 frame with either two-line averaging or no-line averaging. To enhance collection speed, reduce the imaging field by limiting imaging to a region of interest (ROI) around the structures of interest. For very rapid processes, use dual direction scanning on newer CLSMs (i.e., Zeiss 510 or Leica SP2) to effectively double the rate of data collection. For slower processes, the intervals at which fluorescence emission is collected need not be rapid (i.e., 3 seconds to several-minute intervals).

6. Since many trafficking processes are relatively rapid, it is necessary to optimize rapid imaging conditions. To image two different fluorescent markers simultaneously using the CLSM, select a chromophore pair with well-separated absorbance and emission spectra and then image both chromophores simultaneously onto separate detectors. (i.e., a photo multiplier tube).

> Note that the latest CLSMs (i.e., the Zeiss Meta and the Leica SP2) provide substantial control over fluorescence emission detection and separation. To avoid signal bleed through, image each fluorophore separately. To approximate simultaneous imaging, the newer CLSMs permit line imaging, which alternates laser excitation for each horizontal line of pixels. Alternating whole-image frames decreases the likelihood of colocalizing two different chromophores on dynamic structures.

7. Collect five to ten initial images to establish the baseline fluorescence intensity and distribution. These images can be used to confirm that the fluorescence intensities do not fluctuate significantly.

8. Collect multiple data sets for each experiment to confirm reproducibility and be able to perform statistical analyses.

> One way to increase data collection, with no loss in resolution, is to configure the CLSM for 2048 × 2048 image collection. At zoom 1, the resulting 2048 × 2048 image is equivalent to a zoom-4 512 × 512 image, but over an area four times the size of a normal zoom-4 image. Image collection time will increase. However, the increased area means that even more cells can be simultaneously imaged.

Troubleshooting Time-lapse Imaging Experiments

Some common imaging problems and their solutions are presented below.

Autofluorescence Noise

Lysosomes are notorious for autofluorescence when excited with light between 400 and 488 nm. To decrease autofluorescence, lower the intensity of the excitation beam and use narrower band pass emission filters. In addition, avoid phenol red and high serum concentrations (>20%) in the cell growth medium prior to imaging.

Fluorophore Photobleaches too Rapidly during Image Acquisition

Decrease the excitation light intensity using either neutral density filters or by lowering the voltage to the Acousto-Optical Modulator and increase the gain on the detector side to collect light more efficiently. If this does not decrease photobleaching, try adding 0.3 unit/ml Oxyrase (Oxyrase Inc., Mansfield, Ohio), an oxygen scavenger, to the medium.

Cells Are Excited with a Constant Light Intensity, but the Fluorescence Intensity Varies over Time

If this is not due to a biologically relevant process such as recruitment or degradation of fluorophore, then either the focus is shifting during acquisition or the laser power output is unstable. Maintain the focus either manually or by using autofocus software, which is available for some CLSMs. If laser output is a concern, then check it by exciting fluorescent beads and quantifying the emission over time. Fluctuating laser output may be due to the laser being operated at low power output. Increasing the power output to 50% may help. If this fails to correct the problem, then contact the microscope service representative.

Focus Drifts

Try using relatively flat cells (e.g., COS-7), which may alleviate the problem. Try imaging with the pinhole partly or entirely open and use lower-NA objectives. Make sure to pre equilibrate the sample to the temperature of the objective, as this will prevent expansion or contraction of the coverslip or chamber during imaging. In addition, ensure that the stage insert is mounted securely and that the sample is seated properly in the holder.

USE OF PHOTOBLEACHING TO HIGHLIGHT TRANSPORT INTERMEDIATES

When expressing GFP fusion proteins in cells, specific organelles or structures may appear substantially brighter than other labeled structures, or two fluorescently labeled organelles in close proximity may not be easily resolved. To visualize dimmer or closely associated structures, use of inverse fluorescence recovery after photobleaching (IFRAP) or selective photobleaching is recommended. A CLSM capable of photobleaching discrete regions of interest is a must for these methods. In IFRAP, the area surrounding a specific organelle is photobleached in order to visualize trafficking of molecules out of this structure. An example is shown in Figure 4A, in which IFRAP was used to visualize the trafficking of transport intermediates containing glycosyl phosphatidylinositol (GPI)-GFP molecules (Nichols et al. 2001). In IFRAP, the protected organelle fluorescence intensity must not be saturating if qualification of the fluorescent signals in the organelle and surrounding structures are to be analyzed later. For a more detailed description of photobleaching, see Chapter 7.

Selective photobleaching can reveal dim structures masked by bright organelles and can be used to visualize trafficking or flux through an organelle. For example, the Golgi complex (Nichols et al. 2001) expressing a GFP fusion protein can be photobleached and then fluorescence recovery into the photobleached organelle can be imaged for both qualitative and quantitative analyses (see Fig. 4B). For selective photobleaching, a region of interest appropriate to the organelle or structure of interest must be defined. Imaging conditions must be determined that lead to photobleaching only of the structure of interest without significant bleaching of the entire cell. To visualize dim structures, the conditions must be set such that the dim structure(s) is sufficiently bright, but not saturating. Any other bright structure that will be saturated can then be photobleached.

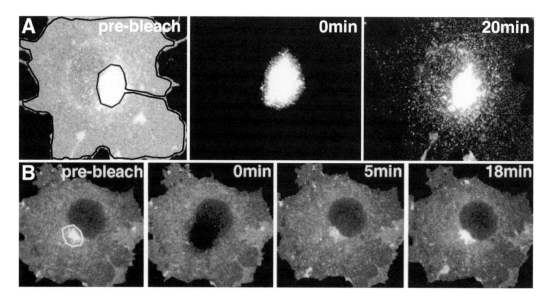

FIGURE 4. IFRAP and selective photobleaching. (*A*) A cell expressing a protein (GPI-GFP) that localizes and traffics between the Golgi apparatus and the plasma membrane is shown in the prebleach image. The fluorescence of the entire cell within the black outline is then photobleached (IFRAP) and trafficking of carriers out of the Golgi to the plasma membrane can be visualized. (*B*) Movement of fluorescent protein into the Golgi apparatus can be visualized when the Golgi apparatus fluorescence within the white outline is photobleached and the whole cell is monitored over time (Nichols et al. 2001).

Photobleaching with a CLSM

The following protocol describes selective photobleaching using a CLSM that can bleach discrete, selected regions of interest.

1. Follow the imaging protocol described above in the section Performing a Time-lapse Imaging Experiment. Identify the cell of interest on the confocal microscope, and bring it to the desired focus. Scan an image of the whole cell at the desired excitation light intensity, line averaging, zoom, etc. Modify pinhole and detector gain for maximal fluorescence signal with no pixel saturation.

2. Define a region of interest (ROI) for the photobleach and empirically determine photobleaching conditions (i.e., scan speed, zoom, laser power, microscope objective, and the minimal number of laser iterations required for photobleaching) so that after photobleaching, the fluorescent signal of the photobleach ROI decreases to within background intensity levels.

   Typical bleaching conditions require a 100–1000-fold increase in laser power (decrease in attenuation) for 1–10 bleach iterations (~0.1–2 seconds) for many organelles. Suggested conditions for photobleaching EGFP or EYFP with a 40-mW 488/514-nm Argon or 25-mW Argon laser are 45–60% power with 100% transmission. DsRed can be photobleached with 100% transmission of a 543-nm laser. ECFP can be photobleached with a 405- or 413-nm laser line. However, intense UV light can be phototoxic to cells and UV photobleaching is not encouraged.

Alternative Photobleaching Protocol

Photobleaching can also be performed with older CLSMs without the capacity for selective photobleaching. Photobleaching is accomplished by zooming in to a small region of the cell and scanning with full laser power.

1. Zoom in to the smallest possible area at the highest zoom possible (usually zoom 8–32, depending on the microscope). At high zooms, the laser dwells longer on an ROI per line scan and thus delivers more bleaching radiation.

2. Set laser power at maximum and remove all neutral density filters from the path of the laser beam.

3. Scan (photobleach) the zoomed region of interest.

USE OF PHOTOACTIVATABLE GFP

To follow a population of proteins over time, biochemists typically perform pulse-labeling experiments. Pulse-labeling in this fashion using the fluorescence is now available with the advent of a photoactivable variant of GFP (PA-GFP) (Patterson and Lippincott-Schwartz 2002). In stably or transiently transfected unactivated cells, the unactivated protein can be excited with low-intensity laser light (i.e., 1% transmission 413-nm Enterprise II ion laser or 5 μW) from a 405-nm or 413-nm laser and alternately excited with low-intensity 488-nm laser excitation (i.e., 1 μW or 1% transmission of a 40-mW argon laser) and visualized with an LP 505-nm filter for both excitation tracks. Fluorescence should be at background levels with 488-nm excitation. To photoactivate, the cell or ROI is rapidly irradiated with high-intensity UV laser light (i.e., 1 mW of 413-nm laser light), as if performing a photobleaching experiment. The number of photoactivating iterations and the amount of laser transmission must be determined experimentally for each microscope system. A typical protocol is one photobleach iteration of 5% transmission of 413-nm laser light at scan speed 8 and zoom 3 using a 63× oil NA 1.4 objective on a Zeiss LSM 510. After photoactivation, there should be up to a 100-fold increase in fluorescence when the sample is excited with the 488-nm laser line. Photoactivation is also possible with a mercury arc lamp, but it will lack specificity for particular cells or structures. In this approach, cells can be imaged prior to activation with excitation for GFP, then exposed to the arc lamp for an experimentally determined period of time with a CFP excitation filter set, and then imaged again with typical imaging conditions for GFP.

ANALYSIS OF DATA

Membrane Movement

Time-lapse imaging data allow the movement of fluorescent objects to be studied. To analyze the path and velocity of an object, the distance between two image locations in pixels must first be converted to actual distance. The calibration can be performed by imaging a calibration grid using bright-field illumination and then imaging with the same objective lens and camera system used for fluorescence imaging. The distance traversed by an object of interest between two successive frames can then be calculated. Dividing this distance by the elapsed time between the frames gives the velocity. The path followed by the object can be determined by plotting the X and Y coordinates at each time point.

This procedure can be simplified by writing a simple macro for NIH Image (NIH Image can be obtained free from the National Institutes of Health Web site, http://rsb.info.nih.gov/nih-image/). The aim of the macro is to allow the investigator to use a computer mouse to click on the object of interest in the first frame in a stack and then in subsequent frames. As output, it should produce a text with complete distance and velocity information in a form that is readable by a graphing program. It also should be able to produce a graph that describes the path followed by the object of interest. For additional information on analyzing live cell data and tracking dynamic movements, see Chapter 16.

Calculation of Changes in Protein Concentration

It is often useful to determine whether the number of GFP fusion proteins associated with an intracellular structure changes over time due to transport or sorting processes.

1. Identify the object of interest and a nearby region that gives an appropriate estimate of the background contribution of fluorescence intensity of the object.

2. Measure the average total intensity of the two regions in each frame of the time-lapse sequence. Make sure that the object of interest does not move out of the plane of focus during the experiment and that the background does not change in unexpected ways.

3. Subtract the background intensity from the intensity associated with the object of interest. If the total intensities are measured, the two regions should contain identical numbers of pixels or the background intensity should be appropriately normalized to account for the difference in the areas.

4. Plot the raw values, the background values, and the background-corrected values as a function of time on a graph.

5. It is also useful to plot total fluorescence associated with the images as a function of time on a log linear plot. This will reveal the extent of photobleaching, which should be a simple exponential decay. Any problems due to focal drift, which tend to show up as an abrupt or oscillating change in total fluorescence can easily be determined.

MEMBRANE TRAFFICKING AND ORGANELLE REAGENTS

Drugs

In this section, several common drugs, their targets, and protocols are described for studying organelle distribution and trafficking. The drugs are readily available from general suppliers including Sigma, Roche, and Calbiochem. All drugs should be aliquoted and stored for single use or made freshly, when noted.

CAUTION: Several of the listed drugs are toxic. Be sure to read the MSDS and follow safe-handling precautions.

Note: <!> indicates hazardous material; see Cautions Appendix.

Brefeldin A

Brefeldin A (BFA) <!> is a fungal metabolite that disrupts the function of the small GTPase, Arf1, whose active state is necessary for maintaining the Golgi apparatus. Adding BFA to cells causes the Golgi to disassemble and redistribute into the endoplasmic reticular (ER) (Sciaky et al. 1997). This can be visualized in living cells by treatment with a 5 µg/ml solution of BFA (prepared as a 5 mg/ml solution in ethanol and stored at −20°C) in imaging medium for 15–30 minutes at 37°C. BFA can be washed out by changing the imaging medium three times in rapid succession. The Golgi apparatus reforms over the course of 1 hour (Lippincott-Schwartz et al. 1989).

Microtubule Disruptors: Nocodazole and Colchicine

Many of the secretory and endocytic organelles (including the ER, Golgi apparatus, endosomes, and lysosomes) associate with microtubules to establish their spatial distribution within cells.

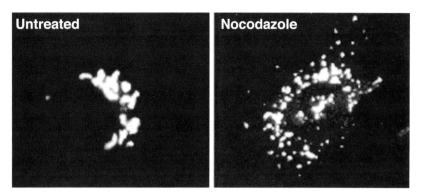

FIGURE 5. The distribution of the Golgi apparatus in formaldehyde-fixed HeLa cells is visualized with an antimannosidase II antibody and a fluorescent secondary antibody. In untreated cells, the Golgi apparatus is relatively compact and restricted to a perinuclear localization. After 1 hour of microtubule disruption with nocodazole treatment, the Golgi apparatus becomes redistributed throughout the cell.

Disrupting microtubules can therefore profoundly affect their distribution. Upon microtubule disruption, the Golgi apparatus, for example, redistributes into scattered sites localized adjacent to ER exit sites (Fig. 5) (Cole et al. 1996a). To acutely disrupt microtubules, cells are incubated for 10 minutes on ice and then warmed to 37°C in imaging medium containing nocodazole<!> or colchicine<!>, which are drugs that prevent microtubules from repolymerizing. Nocodazole<!> (prepared as a 5 mg/ml solution in dimethylsulfoxide (DMSO), stored at −20°C) is used at 5 µg/ml. Colchicine<!> (prepared as a 50 mM solution in distilled H_2O, and stored at −20°C) is used at 1 µM.

Actin Depolymerizing Drugs

The movement of some organelles, such as melanosomes (Wu et al. 1997), and the formation of phagosomes utilize actin-dependent processes. To depolymerize actin, treat cells with either cytochalasin B<!> or latrunculin A<!>. Cytochalasin B (prepared as a 10 mM stock solution in DMSO and stored at –20°C) is used at 1–20 µM for 15–60 minutes in imaging medium at 37°C. Latrunculin A (prepared as a 10 mg/ml stock solution in DMSO or ethanol and stored at –20°C) is used at 0.2–10 µg/ml in imaging medium for 1–12 hours at 37°C.

Aluminum Fluoride

Aluminum fluoride (AlF)<!> treatment causes persistent activation of heterotrimeric G proteins (Gilman 1987) and induces binding of peripheral coat proteins to Golgi membranes. The latter effect prevents trafficking of proteins through the secretory pathway. When applied directly to cells for 10 minutes, it will block ER-to-Golgi transport, as well as Golgi-to-plasma membrane trafficking (Hirschberg et al. 1998). AlF is prepared fresh as a mixture of 60 µM $AlCl_3$ and 20 mM NaF <!> in imaging medium. Cells are incubated with the drug for 0 minutes to 1 hour.

ATP Depletion

Membrane trafficking is sensitive to ATP depletion since this inhibits cytoskeletal motor protein activity and blocks vesicle budding and fusion steps. To minimize effects due to cytotoxicity, cells should not be depleted of ATP for longer than 45 minutes. To deplete ATP, cells are incubated in depletion medium (see below) for at least 15 minutes.

DEPLETION MEDIUM

Growth medium lacking glucose 10% dialyzed serum (available from suppliers of tissue culture reagents)

2 mM glutamine

50 mM 2-deoxyglucose (prepared as 1M stock in distilled H_2O and stored at 4°C),

0.02% sodium azide<!> (NaN_3, prepared as a 1 M solution in distilled H_2O and stored at room temperature).

Note: Prepare depletion medium fresh.

Protein Synthesis Inhibitors

To block new synthesis of proteins in eukaryotes, cycloheximide<!> (prepared as a 10 mg/ml solution in distilled H_2O and stored at −20°C) can be used at 10–150 µg/ml in imaging medium at 37°C depending on the cell type. Alternatively, puromycin (prepared as a 100 mM solution in distilled H_2O and stored at −20°C), a structural analog of aminoacyl-tRNA, can be used at 200–1000 µM and incubated in imaging medium for 10 minutes at 37°C.

N-ethylmaleimide

NEM (*N*-ethylmaleimide) <!> inhibits NSF and the related p97, a family of AAA ATPases that are important for membrane fusion reactions (Block et al. 1988; Beckers et al. 1989). Treatment with 0.5 mM NEM (prepared as a 10 mg/ml stock solution in ethanol and stored at −20°C) in imaging medium for 10 minutes at 37°C substantially inhibits most vesicular trafficking steps.

Temperature Blocks

Within the secretory pathway, it has been shown that distinct steps are differentially sensitive to temperature reduction. Incubation at 20°C leads, for example, to accumulation of secretory proteins in the trans-Golgi network (TGN), whereas incubation at 15°C leads to accumulation of these proteins in pre-Golgi transport intermediates (Griffiths et al. 1989).

Cells are incubated in imaging medium in an incubator with the appropriate level of CO_2 (i.e., 5%) for 1–3 hours at 15°C or 20°C to block protein traffic. When cells are removed from the incubator, the block can be released and trafficking can be monitored. The cells can be imaged and manipulated at different temperatures using an objective heater. Alternatively, they can be fixed for immunofluorescence or other imaging studies.

Dyes

A wide variety of organelle-specific fluorescent dyes are available for live cell imaging. For example, dyes are available for mitochondria (i.e., MitoTracker) (see Fig. 2G), lysosomes (i.e., LysoTracker), the Golgi apparatus (i.e., Bodipy ceramide), and endosomes (i.e., rhodamine-transferrin). For a more extensive list, see the Molecular Probes catalog or Web site (see http://www.probes.com). These dyes can often be imaged in conjunction with a GFP fusion protein. Unlike GFP, it is not recommended that the dyes be photobleached (i.e., a FRAP experiment), because photobleaching of the dyes can release cytotoxic free radicals. An additional caveat is that some dyes label multiple organelles. For example, $DiOC_6(3)$ preferentially labels mitochondria at low concentrations and the ER, as well, at higher concentrations. Finally, some of the dyes may be inherently toxic to cells and therefore should not be used for experiments lasting more than 1–2 hours.

REFERENCES

Bergmann J.E. 1989. Using temperature-sensitive mutants of VSV to study membrane protein biogenesis. *Methods Cell Biol.* **32:** 85–110.

Beckers C.J., Block M.R., Glick B.S., Rothman J.E., and Balch W.E. 1989. Vesicular transport between the endoplasmic reticulum and the Golgi stack requires the NEM-sensitive fusion protein. *Nature* **339:** 397–398.

Block M.R., Glick B.S., Wilcox C.A., Wieland F.T., and Rothman J.E. 1988. Purification of an *N*-ethylmaleimide-sensitive protein catalyzing vesicular transport. *Proc. Natl. Acad. Sci.* **85:** 7852–7856.

Campbell R.E., Tour O., Palmer A.E., Steinbach P.A., Baird G.S., Zacharias D.A., and Tsien R.Y. 2002. A monomeric red fluorescent protein. *Proc. Natl. Acad. Sci.* **99:** 7877–7882.

Cole N.B., Sciaky N., Marotta A., Song J., and Lippincott-Schwartz J. 1996a. Golgi dispersal during microtubule disruption: Regeneration of Golgi stacks at peripheral endoplasmic reticulum exit sites. *Mol. Biol. Cell.* **7:** 631–650.

Cole N.B., Smith C.L., Sciaky N., Terasaki M., Edidin M., and Lippincott-Schwartz J. 1996b. Diffusional mobility of Golgi proteins in membranes of living cells. *Science* **273:** 797–801.

Gilman A.G. 1987. G proteins: Transducers of receptor-generated signals. *Annu. Rev. Biochem.* **56:** 615–649.

Gorman C.M. and Howard B.H. 1983. Expression of recombinant plasmids in mammalian cells is enhanced by sodium butyrate. *Nucleic Acids Res.* **11:** 7631–7648.

Griffiths G., Fuller S.D., Hollinshead M., Pfeiffer S., and Simons K. 1989. The dynamic nature of the Golgi complex. *J. Cell Biol.* **108:** 277–297.

Lippincott-Schwartz J., Yuan L.C., Bonifacino J.S., and Klausner R.D. 1989. Rapid redistribution of Golgi proteins into the ER in cells treated with brefeldin A: Evidence for membrane cycling from Golgi to ER. *Cell* **56:** 801–813.

Miyawaki A., Sawano A., and Kogure T. 2003. Lighting up cells: Labelling proteins with fluorophores. *Nat. Cell Biol.* (Suppl): S1–S7.

Nichols B.J., Kenworthy A., Polishchuk R.S., Lodge R., Roberts T.H., Hirschberg K., Phair R.D., and Lippincott-Schwartz J. 2001. Rapid cycling of lipid raft markers between the cell surface and Golgi complex. *J. Cell Biol.* **153:** 529–542.

Patterson G.H. and Lippincott-Schwartz J. 2002. A photoactivatable GFP for selective photolabeling of proteins and cells. *Science* **297:** 1873–1877.

Presley J.F., Cole N.B., Schroer T.A., Hirschberg K., Zaal K.J.M., and Lippincott-Schwartz J. 1997. ER-to-Golgi transport visualized in living cells. *Nature* **389:** 81–85.

Sciaky N., Presley J., Smith C., Zaal K.J.M., Cole N., Moreira J.E., Terasaki M., Siggia E., and Lippincott-Schwartz J. 1997. Golgi tubule traffic and the effects of brefeldin A visualized in living cells. *J. Cell Biol.* **139:** 1137–1155.

Snapp E., Hegde R., Colombo S., Borgese N., Francolini M., and Lippincott-Schwartz J. 2003. Formation of stacked cisternae by low affinity protein interactions. *J. Cell Biol.* **163:** 257–269.

Wu X., Bowers B., Wei Q., Kocher B., and Hammer J.A. 3rd. 1997. Myosin V associates with melanosomes in mouse melanocytes: Evidence that myosin V is an organelle motor. *J. Cell Sci.* **110:** 847–859.

Zaal K.J.M., Smith C.L., Polishchuk R.S., Altan N., Cole N.B., Ellenberg J., Hirschberg K., Presley J.F., Roberts T.H., Siggia E., Phair R.D., and Lippincott-Schwartz J. 1999. Golgi membranes are absorbed into and reemerge from the ER during mitosis. *Cell.* **99:** 589–601.

Zacharias D.A., Violin J.D., Newton A.C., and Tsien R.Y. 2002. Partitioning of lipid-modified monomeric GFPs into membrane microdomains of live cells. *Science* **296:** 913–916.

Imaging Live Cells Under Mechanical Stress

Brian P. Helmke

Department of Biomedical Engineering and Cardiovascular Research Center, University of Virginia, Charlottesville, Virginia 22908

Peter F. Davies

Department of Pathology and Laboratory Medicine and Institute for Medicine and Engineering, University of Pennsylvania, Philadelphia, Pennsylvania 19104

INTRODUCTION

CELLULAR RESPONSES TO MECHANICAL STIMULI are implicated in the structural and functional adaptation of many tissues. For example, cellular mechanisms mediate bone and skeletal muscle remodeling during mechanical loading, lung function during ventilator-induced injury, hearing loss in the inner ear, and blood-flow-mediated cardiovascular pathophysiology. Because much of the authors' work investigates vascular biomechanics, the focus in this chapter is on the techniques used to study vascular endothelial cells in vitro; however, similar techniques can be used to study other cell types.

Why Image Vascular Cells Under Mechanical Stress In Vitro?

In the blood vessel wall, two primary mechanisms for arteriolar autoregulation of blood flow depend on vascular wall cell functions. First, flow-mediated vasodilation occurs in response to increased frictional shear stress acting on the endothelium at the blood-tissue interface. Mechanochemical signals from endothelial cells cause adjacent smooth muscle cells to relax, thereby increasing vessel diameter. This response reduces the pressure gradient driving flow through the vessel and returns volumetric flow rate back to baseline. Second, an increase in arteriolar pressure or the circumferential stretch of the vessel wall induces vascular smooth muscle cell contraction that results in vasoconstriction. This myogenic response limits volumetric flow through the vessel and returns the flow rate to baseline levels. Physiological control of blood flow to peripheral organs depends on a balance of these two acute vasoregulatory responses.

Although intravital microscopy allows access to cellular behavior in the microcirculation (see Chapter 24), direct measurements of intercellular interactions or intracellular dynamics in response to mechanical stimuli in larger arteries are challenging. In addition, precise measurement of the local hemodynamic force profile is often technically difficult, and physiological variation among subjects prevents control of fluid dynamics parameters. Furthermore, the complexity of the in vivo environment often prevents accurate elucidation of mechanisms. In vitro models of mechanical stimulation provide a means to precisely control the stimulus and monitor cellular responses. From a biomedical engineering perspective, this allows development of predictive models of cell behavior that will lead to new therapeutic approaches based on control of vascular function.

In vitro models for applying shear stress to endothelial cells rely on flow chambers that only recently have been optimized for visualization of living cells. Image acquisition methods with such chambers obtain the higher temporal and spatial resolution necessary for understanding cell-mediated mechanisms that direct vascular function. For example, hemodynamic shear stress rapidly induces transient intracellular events that have a role in the activation of multiple signaling networks (Davies 1995). The fastest of these events include heterotrimeric G-protein activation; opening of chloride, potassium, and volume-regulated anion channels; increased intracellular calcium concentration; and production of chemical mediators such as nitric oxide and prostaglandins. These events occur within seconds after a change in the local hemodynamic profile, requiring rapid measurement of biochemical functions during application of fluid shear stress. Although temporal changes of cellular responses can often be measured indirectly, detection of the spatial distribution of mechanical and chemical responses within the cell requires direct visualization. Examples include preferential locations of increased membrane lipid lateral mobility and heterogeneous displacement, deformation, and polymerization of cytoskeletal elements. Such events must be monitored on a subcellular length scale in order to elucidate mechanisms of mechanochemical sensing. Thus, live cell imaging during extracellular force application is required in order to capture mechanically induced responses with enough detail to investigate mechanochemical signaling mechanisms.

Relating Live Cell Imaging to Biomechanical Models

Phenomenological models have been derived based on generalized cell structures in an attempt to explain cell behavior in response to applied forces, and the mechanical properties of the cytoplasm have been proposed to depend on the cytoskeletal architecture. In one model, a tensegrity structure represents the cytoskeleton as an interconnected network of tension-bearing elastic filaments and compression-bearing struts that exists in a state of prestress (Wang et al. 1993). Indeed, experimental disruption of adhesion sites near the cell edge results in apparent elastic recoil toward the cell body. This model has been extended to the nucleus after experimental demonstration of nuclear deformation in response to pulling on the cell surface. An alternative model suggests that the cytoplasmic cytoskeleton is a viscoelastic polymer gel that undergoes phase transitions in response to applied stress or strain. In this model, locally applied forces are dissipated with distance away from the site of force application so that the applied force is not directly sensed at remote locations within the cell (Heidemann et al. 1999). Relatively local displacement of cytoskeleton-associated elements during poking with a micropipette or twisting of magnetic beads on the cell surface supports this model. It is clear that mechanical responses within the cytoplasm that mediate mechanochemical signal transduction are integrated in a complex manner that is not well understood. High-resolution live cell imaging of cells under mechanical stress provides experimental access to test these models.

Recent advances in image analysis of live cell images are revealing new details relevant to mechanical interactions predicted by these models at a subcellular length scale. For example, onset of shear stress induces heterogeneous displacement of intermediate filaments, implying that fluid shear stress induces spatial gradients of mechanical events in the cytoskeleton. Mechanical strain computed in the intermediate filament network is highly focused at discrete locations within the cell, suggesting that cytoskeletal tension is redistributed to sites that are putatively associated with signaling molecule scaffolds such as focal adhesion sites. Such strain focusing was only detectable after careful quantitative analysis of four-dimensional fluorescence data in living endothelial cells during onset of shear stress (see discussion below). These quantitative optical analyses have revealed new details of intracellular mechanical behavior that is relevant to the initiation of mechanochemical signal transduction.

CHAMBERS FOR OBSERVING LIVE CELLS UNDER MECHANICAL STRESS

Many of the considerations discussed in previous chapters (see Chapter 17) are critical for the successful imaging of live cell processes under mechanical stress. These include temperature control of both the chamber and the objective lens and maintenance of sterility for long-term observation of cellular adaptation to mechanical force. In addition, for accurate three-dimensional fluorescence reconstruction of intracellular processes, a no. 1.5 coverslip must be used, since the design of most microscope objectives is optimized for glass coverslips that are 170-μm thick. The fluid medium should be chosen to minimize its contribution to measured fluorescence intensity, as described in Chapter 17 and in Chapters 28–32. Special consideration is given here to imaging live cells during force application.

Coverslip Stability during Force Application

For accurate time-lapse imaging at subcellular resolution, several sources of error associated with coverslip movement must be minimized or eliminated. Air currents or temperature changes near the microscope stage often cause focus drift. To minimize room air currents, the microscope should be located in an area that has relatively constant airflow and temperature, or airflow from vents should be deflected away from the microscope. For closed-chamber experiments, it may be possible to block airflow locally around the chamber by isolating the microscope stage within an environmental chamber. This also reduces temperature variation during overnight experiments in locations where energy conservation requires adjusting thermostat settings during night hours. Fluid introduced into the observation chamber should be at the same temperature as that already in the chamber to prevent focus drift due to temperature gradients associated with convection. Furthermore, high-numerical-aperture (NA) objective lenses should be warmed to minimize heat loss through the immersion fluid.

Experimental interventions required to apply mechanical forces on cells may induce three-dimensional coverslip displacement. Chamber designs using thicker glass reduce the amount of coverslip displacement during force application, but the no. 1.5 coverslip required for high-resolution fluorescence imaging often flexes significantly with force application on the cells. Displacement of cellular components is subtracted from coverslip movement by including stationary fiducial markers on the coverslip during the time-lapse experiment. For example, fluorescent microspheres on the coverslip underneath the cells are tracked to measure coverslip displacement (see protocol below). Microspheres with a 100-nm diameter or less are easily detectable using high-NA optics, and a fluorescence wavelength is chosen that does not interfere with experimental measurements. In three-dimensional images, a coplanar set of microspheres indicates the plane of the coverslip surface. In two-dimensional images, a set of microspheres should be chosen that maintains constant geometry among points and is continuously in focus to maximize the probability that the microspheres are on the coverslip surface. The displacement of such a set of microspheres represents coverslip displacement during the experiment, and subtracting the change in microsphere positions as a function of time yields true cellular motions.

The following procedure yields a surface density of approximately 3×10^5 microspheres/cm^2 using 100-nm microspheres. This density typically gives a coplanar set of approximately. 3 microspheres/cell in a confluent monolayer.

Coating Coverslips with Microspheres

1. Sonicate a suspension (2% solids) of 100-nm fluorescent carboxylate-modified polystyrene microspheres for ~5 minutes to achieve a monodisperse suspension. A 2% suspension of 0.1-μm diameter polystyrene microspheres has a concentration of 3.6×10^{11} microspheres/ml. This is computed from the following equation:

$$\frac{\text{microspheres}}{\text{ml}} = \frac{6C \times 10^{12}}{\pi \rho \phi}$$

where C is the concentration of microspheres in grams per milliliter (e.g., 0.02 g/ml for a 2% solids suspension), ρ is the polymer density in grams per milliliter (1.05 for polystyrene), and ϕ is the microsphere diameter in micrometers.

2. Prepare Suspension A: Make a 1:1000 dilution of the microsphere suspension in 100% ethanol (e.g., add 1 μl of microsphere suspension to 999 μl of ethanol).

    It is important to use pure ethanol as the solvent since water will leave a residue on the coverslip after evaporation that will scatter light during fluorescence imaging.

3. Prepare Suspension B: Make a 1:100 dilution of Suspension A in 100% ethanol (e.g., add 1 μl of Suspension A to 99 μl of ethanol. The total dilution factor of Steps 2 and 3 is 10^5, yielding a final concentration of 3.6×10^6 microspheres/ml.

4. Compute the surface area (in cm^2) of the coverslip to be coated with microspheres.

5. Compute the required volume of Suspension B (in ml) from (0.08) × (surface area in cm^2). For example, a 4-cm diameter circular coverslip has a surface area of 12.6 cm^2 and requires ~1.0 ml of Suspension B.

6. Place a sterile dry coverslip into the center of a tissue culture dish in a laminar flow hood. Do not allow the coverslip to touch the edges of the dish. Carefully pipette the required volume of Suspension B onto the coverslip, creating a surface tension bubble. Allow the coverslip to dry completely by evaporation.

7. Follow usual procedures for plating cells onto dry microsphere-coated coverslips.

Timing of Transient Transfection Using GFP Fusion Proteins

For experiments imaging live cells under mechanical stress, green fluorescent protein (GFP) fusion proteins are often used to visualize structural proteins. In the case of endothelial cells, a confluent monolayer is desired to represent a physiological model. Transient transfection using liposomes induces only about 10–15% of endothelial cells to express GFP so that details of structural dynamics within a single cell in the monolayer can be measured. The timing of an experiment should be planned to simultaneously optimize expression levels of GFP fusion proteins and cell density. Maximum expression of EGFP (enhanced GFP) typically occurs at approximately 48 hours after transfection, but timing of expression varies slightly with the choice of GFP variant. However, because liposome-mediated transfection is performed at low cell density, confluence may not be achieved until at least 3–4 days after transfection, depending on the population doubling rate. The following procedure yields a compromise between GFP expression and obtaining a confluent monolayer on 4-cm-diameter circular coverslips. The timing and cell numbers should be adjusted as necessary for varying expression levels, coverslip surface areas, and cell proliferation rates.

Transfection and Plating Protocol for 4-cm Coverslips

1. Plate endothelial cells in a 6-cm dish at a density of 2.5×10^4 cells/cm². Allow the cells to spread for ~16 hours in complete growth medium.

2. Transfect the endothelial cells with 3–11 μg of DNA plasmid per dish in low-serum medium (e.g., OptiMEM I, GIBCO) using a liposomal method (e.g., Lipofectin, GIBCO).

   The amount of DNA required varies slightly with the plasmid used and should be optimized in dose-response studies.

3. Allow the cells to recover overnight in complete growth medium.

4. Replate the cells onto 4-cm-diameter sterile glass coverslips that have been precoated with fluorescent microspheres (see preceding protocol). The cell density should be at least 7×10^4 cells/cm².

5. *Optional*: If the cell density is too low, add nontransfected endothelial cells from the same line into the suspension before plating to achieve the desired density. This will reduce the proportion of cells expressing GFP but will help to achieve a confluent monolayer.

6. Allow the cells to grow to a confluent monolayer, usually about 2 days after plating.

Open Chambers for Access to the Cell Surface

Manual measurement and manipulation of the cell surface requires access to the cells, usually in an open chamber. As discussed above and in Chapter 17, temperature-controlled chambers or stage inserts are preferred for maintaining physiological activity during the experiment. For example, heated culture dishes with coverslip glass bottoms (Bioptechs) permit high-resolution fluorescence microscopy of living cells during force application. In addition, an objective lens heater should be used with high-NA lenses to minimize heat loss through the immersion fluid.

For open chambers, it is also important to maintain pH during experiments. One method is to use a basal medium containing an organic buffer such as HEPES instead of bicarbonate, since standard bicarbonate buffers will become basic when equilibrated with atmospheric levels of carbon dioxide. Alternatively, the microscope stage may be enclosed in an incubation chamber equilibrated with 5% CO_2 and 95% air to mimic incubator conditions. This second approach has the advantage that temperature can be controlled simultaneously by heating the gas flowing into the stage chamber. However, physical access to the cell chamber is often limited, and that may be a disadvantage when using larger instruments or if frequent intervention is needed. Two examples of force application using open chambers are discussed below.

Deforming the Endothelial Surface Using Microneedles

An example of an open-chamber application is deforming the cell surface with a glass microneedle. Microneedles are pulled from borosilicate glass capillary tubes using a micropipette puller. A tip diameter of approximately 1 μm allows application of force locally on the cell surface. Force can be applied nonspecifically to the surface, or specific adhesion receptors can be engaged by coating the microneedle with extracellular matrix ligands such as fibronectin, laminin, or collagen. The stiffness, or bending modulus, of the microneedle can be calibrated to compute the applied force from the amount of microneedle deflection while pushing on the cell surface. In one method for calibration, thin wires with known mass are hung from the microneedle, and the needle deflection distance at the tip is measured optically. A typical range of stiffness is on order of 10–30 nanoNewtons (nN)/μm (Davidson et al. 1999). By this method, known forces are applied to the cell surface.

Flow Chambers for Application of Shear Stress

Cell-mediated responses to fluid flows have recently generated interest in mechanobiology. Application of well-defined flow forces onto the cell surface is necessary to elucidate mechanisms of mechanochemical signaling in endothelium cells and other cell types. In addition, selection of chambers compatible with high-resolution fluorescence microscopy permits dynamic measurements in living cells during changes in extracellular fluid forces.

Fluid Dynamics Considerations

For most cell culture experiments, cells are exposed to a flow of an incompressible Newtonian fluid which is typically complete growth medium or physiological buffer. Incompressibility implies that the fluid has constant density, independent of flow rate. For a Newtonian fluid, the shear stress is linearly proportional to the shear rate, and the constant of proportionality is the viscosity. The motion of this fluid is predicted from the basic conservation equations of mass and momentum. These equations of motion are called the Navier-Stokes equations:

$$\rho \frac{D\vec{v}}{Dt} = \rho \frac{\partial \vec{v}}{\partial t} + \rho \vec{v} \cdot \nabla \vec{v} = -\nabla p + \mu \nabla^2 \vec{v} \tag{1}$$

where $\vec{v}$ is the velocity vector of a local fluid "particle," p is the hydrodynamic pressure, ρ is the fluid density, μ is the fluid kinematic viscosity, D/Dt is the material derivative, ∇ is the gradient operator, and ∇^2 is the Laplace operator. It is beyond the scope of this chapter to derive and solve these equations; however, well-characterized special cases will be discussed below.

Variations in biological response have been reported based on whether the flow profile is laminar or turbulent. In the laminar flow regime, the fluid can be described qualitatively as "sheets" or layers of fluid sliding past each other with continuous variation in speed from one layer to the next. In contrast, turbulent flow is characterized by random fluctuations in velocity of individual fluid "particles" superimposed on the mean bulk fluid motion. The tendency toward turbulence is determined by the ratio of magnitudes of inertial forces to viscous damping forces described by a dimensionless parameter called the Reynolds number:

$$\text{Re} = \frac{\rho v L}{\mu} = \frac{v L}{\nu} \tag{2}$$

where v is the mean fluid velocity, ν is the dynamic viscosity of the fluid, and L is a characteristic length associated with the flow. By convention, the characteristic length for a parallel plate flow chamber is its height, and L for a glass tube is its diameter. A higher Reynolds number indicates a higher probability of turbulent fluid structure as viscous damping becomes dominated by fluid inertia. The transition typically occurs for Reynolds numbers

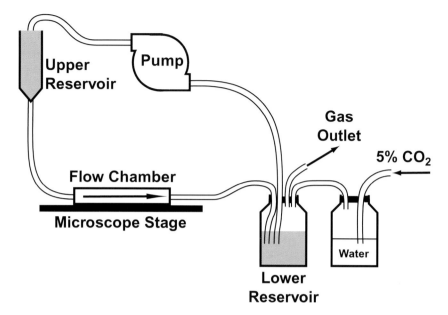

FIGURE 1. Continuous closed flow loop for long-term live cell microscopy. Fluid flows by gravity from an upper reservoir through the flow chamber positioned on the microscope stage and into a lower reservoir. A peristaltic pump continuously recirculates the fluid from the lower reservoir back into the upper reservoir. A gas mixture of 5% CO_2 and 95% air bubbled through sterile water in an adjacent humidifying chamber maintains pH in bicarbonate-buffered fluid. Both the lower reservoir and humidifying chamber are placed in a water bath warmed to 37°C to maintain temperature (not shown).

in the range of 1400–2300, depending on geometry. Most physiological flows fall in the laminar regime. However, a few examples of turbulent flows exist, such as airflow during a cough or sneeze, blood flow in the aortic arch during peak systole, and blood flow in the brachial artery during blood pressure measurement with a sphygmomanometer. In addition, turbulent flows often diagnose pathological states of blood flow, including heart valve insufficiency (heart murmur) or arterial stenosis.

The design of an experimental setup depends on the flow chamber geometry and desired flow profile. Perfusion with a syringe pump is acceptable for short experiments with low-volume flow rates. However, this may not be adequate for long-term experiments to study cell adaptation or for more complicated flow profiles. In this case, a continuous closed flow loop is used (Figure 1). A closed loop assembled aseptically in a biosafety cabinet prevents contamination during experimental observation on the microscope, and the perfusion method can be designed to produce the desired flow profile. For steady flow, a constant-volume flow rate is driven by gravity from an upstream reservoir positioned above the microscope stage. The hydrostatic driving pressure varies linearly with the height of the reservoir. Fluid is drained out of the chamber into a downstream reservoir, and a peristaltic pump recirculates the fluid back into the upstream reservoir. Bicarbonate-containing culture medium is maintained at proper pH by equilibrating it with humidified gas containing 5% CO_2 and 95% air, and the downstream reservoir is heated to maintain temperature.

Parallel Plate Flow Chambers

Parallel plate flow chambers are widely used in experimental studies of fluid dynamics effects on cultured cells. If the height-to-width and height-to-length ratios of the chamber dimensions are small (typically <1/50), then a fully developed parabolic velocity profile is sufficient to compute the shear stress acting on the cells. In this case, the only non-zero component of velocity is in the direction of flow and is a function of height in the chamber:

$$v(y) = \frac{3Q}{2wh}\left[1 - \left(\frac{2y}{h}\right)^2\right] \tag{3}$$

where v is velocity, Q is volume flow rate, w and h are the width and height of the chamber, and y is the vertical distance from the center line between the two parallel plates. From this equation, the shear stress τ is derived from the product of the velocity gradient and fluid viscosity μ as

$$\tau(y) = -\mu\frac{dv}{dy} = \frac{12\mu Q}{wh^3}y \tag{4}$$

For a fluid with constant viscosity, Equation (4) shows that the shear stress is zero at the center line between the parallel plates and increases linearly to a maximum on the plate surface, where the wall shear stress is given by

$$\tau_w = \frac{6\mu Q}{wh^2} \tag{5}$$

For imaging live cells under stress, two parallel plate flow chamber designs are commonly used. In the first (Frangos et al. 1988), cells are grown on a coverslip that is assembled onto the bottom of the flow chamber and held in place by a vacuum seal around the edge. This design is easy to assemble and disassemble quickly so that shear stress can be applied for only seconds, and cells can be recovered rapidly for biochemical assays such as protein phosphorylation or G-protein activation. In this design, temperature is maintained by perfusion of warmed fluid. A second design by Bioptechs (Model FCS2, Butler, Pennsylvania) is more convenient for reducing the shear stress to zero, since the coverslip and top plate of the chamber are maintained at the desired temperature (usually 37°C). In addition, the top plate in the Bioptechs chamber is optically clear to allow high-resolution transillumination microscopy in bright-field, phase-contrast, or differential interference contrast (DIC) modes. In both of these chamber designs, the Reynolds number typically lies in the range of 0–20 so that a fully developed parabolic velocity profile exists everywhere in the chamber except at the entrance; the entrance length necessary to achieve a fully developed velocity profile is approximately 1/100 of the chamber length.

Parallel Plate Flow Chamber with Physical Access to the Cells

To directly access cells for measurement and manipulation during exposure to hemodynamics forces, we have designed a parallel plate flow chamber that permits instrument access to adherent cells with minimal disturbance of the velocity field (Levitan et al. 2000). Based on the parallel plate flow chamber design, the "minimally invasive flow device" (MIF device) has 1-mm-wide longitudinal slits cut into the top plate of the chamber to allow insertion of a recording, measurement, or stimulating instrument. Surface tension forces at the slit openings are larger than the hydrodynamic pressure in the chamber, so overflow through the slits is prevented over a physiological range of steady shear stress (0–15 dyn/cm²). The components and assembled chamber are shown in Figure 2. A probe such as a micropipette is positioned near the cell surface, makes direct contact with the cell membrane, or enters the cell while inducing negligible deviations in laminar flow near the cell surface, as demonstrated by monitoring the trajectories of microspheres in the fluid. For example, the MIF device has

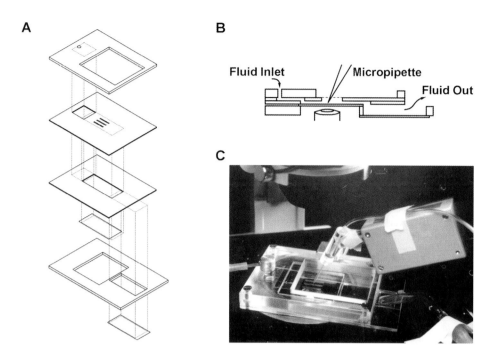

FIGURE 2. Components (*A*) and side view (*B*) of the MIF device that permits physical access to cells through slits in the top of a parallel plate flow chamber. (*C*) The MIF device mounted on a microscope stage with a typical electrophysiological micropipette lowered into the recording chamber. (Adapted, with permission, from Levitan et al. 2000.)

enabled measurement of flow-induced changes in membrane potential using patch-clamp electrophysiological techniques. The MIF device offers numerous possibilities for investigating real-time endothelial responses to well-defined flow conditions in vitro using electrophysiology and amperometric techniques, cell surface mechanical probing, local controlled chemical release or biosensing, and microinjection (Figure 3).

Glass Capillary Tubes

Endothelial cells have been cultured in glass capillary tubes with either circular or rectangular cross-section. Cylindrical tubes have the advantage of more closely modeling the in vivo geometry of blood vessels, especially if small tubes are chosen to match blood vessel diameter and radius of curvature. Rectilinear tubes provide a close approximation to vessel geometry with the additional advantage that cells are grown on an optically flat surface that allows high-resolution microscopy. Disadvantages include limited numbers of cells, which often prevents biochemical assays subsequent to imaging measurements.

For Reynolds numbers in the laminar flow regime, the Navier-Stokes equations can be solved for the velocity profile as a one-dimensional function of radius:

$$v(r) = \frac{2Q}{\pi a^2} \left[1 - \left(\frac{r}{a} \right)^2 \right] \tag{6}$$

where a is the tube radius and r is radial distance from the center line of the tube. For a Newtonian fluid with constant viscosity the shear stress is given by

$$\tau(r) = -\mu \frac{dv}{dr} = \frac{4\mu Q}{\pi a^4} r \tag{7}$$

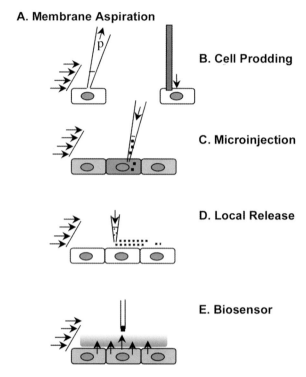

A. Membrane Aspiration

B. Cell Prodding

C. Microinjection

D. Local Release

E. Biosensor

FIGURE 3. Examples of MIF chamber uses include (*A*) membrane aspiration, (*B*) cell prodding, (*C*) microinjection, (*D*) local release of solutes, and (*E*) biosensing of locally released cell products. (Reprinted, with permission, from Levitan et al. 2000.)

and the shear stress at the wall is

$$\tau_w = \frac{4\mu Q}{\pi a^3} \tag{8}$$

The precision of manufacturing glass capillary tubes is important to accurately compute the shear stress profile. Reliable sources include Stoelting (Kiel, Wisconsin) and VitroCom (Mountain Lakes, New Jersey). Tubes with wall thickness of 0.17 mm, optimal for high-resolution microscopy, are available from VitroCom. A tight connection into the flow loop can be made directly from tubing with internal diameter slightly smaller than the outer diameter of the glass tube. For rectilinear tubes, connections can be sealed by carefully wrapping the end of the glass tube with Parafilm before connecting the flow loop tubing. To maintain sterility, Parafilm must not touch the fluid inside the glass tube.

ENGINEERING ANALYSIS OF RAPID INTRACELLULAR EVENTS INDUCED BY MECHANICAL STRESS

Analysis of Intermediate Filament Displacement

Using the methods discussed above, confluent monolayers of endothelial cells expressing GFP-vimentin were observed in the Bioptechs parallel plate flow chamber (see Movies 33.1 and 33.2). Stacks of optical sections were acquired at 90-second intervals and were deconvolved using a constrained iterative algorithm (Agard and Sedat 1983). GFP-vimentin distributed to the endogenous intermediate filament cytoskeleton (Helmke et al. 2000), which consisted of a prominent perinuclear ring and a mesh network of intermediate filaments radiating toward the cell periphery (Fig. 4A). In three-dimensional images, intermediate

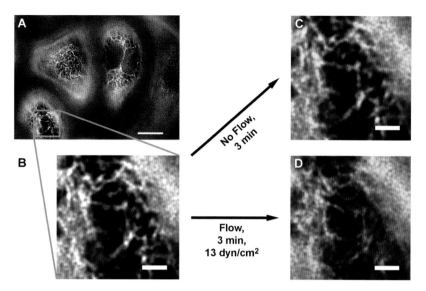

FIGURE 4. (*A*) Optical section GFP-vimentin intermediate filament network near the apex of living endothelial cells within a confluent monolayer. Bar, 10 μm. (*B*) Magnified view of inset indicated in *A*. Bar, 2 μm. (*C*) Small random displacement of GFP-vimentin illustrated by comparing false-colored images at the beginning (*red*) and end (*green*) of a 3-minute interval with no flow. Yellow indicates zero displacement. (*D*) Significant displacement of intermediate filaments shown by colored images at the beginning (*red*) and end (*green*) of a 3-minute interval just after onset of shear stress (13 dyn/cm², *left to right*). (Adapted, with permission, from Helmke et al. 2000.)

filaments were often visible reaching above and below the nucleus. In time-lapse movies, connections among individual filament segments within the intermediate filament network did not appear to change as filaments throughout the cytoplasm fluctuated or "wiggled" in apparently random directions (Fig. 4 B–C).

Onset of unidirectional laminar shear stress (τ_w = 12 dyn/cm²) induced significant directional displacement of intermediate filaments in regions of cells. Some filament segments were displaced by almost 1 μm within 3 minutes (Fig. 4B,D), although displacement was heterogeneous throughout the cytoplasm in three dimensions. Displacement continued with exposure to shear stress, and the rate of displacement varied with position in the cell.

A quantitative method for analyzing the structural dynamics in the cytoskeleton has been developed. The displacement of GFP-vimentin-labeled intermediate filaments is detected by analyzing the fluorescence distribution as a function of spatial location. For example, displacement projected onto the *x* axis can be detected by comparing the line-intensity profiles from two time points. Displacement of intensity peaks indicates filament movement in the *x* direction during the interval, whereas overlapping intensity peaks indicate zero displacement. Changes in peak intensity during the interval indicate displacement into or out of the focal plane.

Quantitative measurement of intermediate filament displacement has been extended to three dimensions by analyzing the four-dimensional fluorescence distribution function $f(x,y,z,t)$ from spatial regions in the image. Analysis of the entire image as a single spatial region yields a measure of degree of displacement throughout the cell, but is of limited use due to the constitutive movement of intermediate filaments throughout the cytoplasm. However, analysis of displacement at a subcellular length scale reveals regional heterogeneity of intermediate filament motion. An appropriate length scale is established by analyzing the spatial power spectrum from a three-dimensional image stack at the beginning of the experiment (see first protocol below). In analysis of GFP vimentin images, subimages were chosen that had dimensions $1.7 \times 1.7 \times 0.68$ µm. After dividing three-dimensional images into subimages, the displacement index (DI) is computed from the product moment cross-correlation coefficient, which measures the degree of overlap between fluorescence intensity distributions at two time points (see second protocol below Computation of Displacement Index). Increasing values of DI indicate increased displacement of intermediate filament segments within the subimage during an interval (Helmke et al. 2001).

Choosing a Length Scale for Subcellular Analysis

1. Read a series of tiff images into a three-dimensional matrix f_{xyz} in math software such as Matlab or IDL.

2. Compute the three-dimensional Fourier transform of the matrix with dimensions $X \times Y \times Z$ from

$$F_{uvw} = \sum_{x=0}^{X-1} \sum_{y=0}^{Y-1} \sum_{z=0}^{Z-1} f_{xyz} e^{-2\pi i \left(\frac{ux}{X} + \frac{vy}{Y} + \frac{wz}{Z}\right)} \qquad (9)$$

where $i = (-1)^{1/2}$. Note that (u,v,w) is the frequency domain coordinate position corresponding to the image coordinate (x,y,z).

3. Compute the power spectrum magnitude matrix from

$$S_{uvw} = F_{uvw} F_{uvw}^{*} \qquad (10)$$

where F^* is the complex conjugate of F obtained by replacing i with $-i$ in Equation (9).

4. Transform the matrix S_{uvw} to cylindrical coordinates $S_{r\theta w}$ where $r^2 = u^2 + v^2$ and $\tan \theta = v/u$.

5. Integrate the matrix S over all θ and w to yield the projection of the spectrum on the r axis:

$$S_r = \sum_{w=0}^{W-1} \sum_{\theta=0}^{2\pi} S_{r\theta w} \qquad (11)$$

6. Plot S_r vs. r and look for local peaks; these indicate dominant spatial frequencies in the optical plane. The spatial separation corresponding to these frequencies is given by $1/r$. Spatial regions for analysis of displacement should be at least twice this size so that all regions contain at least one filament segment.

7. Determine the minimum region size along the optical axis in a similar manner. Integrate S over all r and θ to yield the projection on the z axis:

$$S_w = \sum_{r=0}^{R-1} \sum_{\theta=0}^{2\pi} S_{r\theta w} \qquad (12)$$

Note that R is the dimension of the matrix on the r axis, computed from $R^2 = U^2 + V^2$. Plot S_w vs. w and look for local peaks, indicating dominant spatial frequencies along the optical axis. The spatial separation corresponding to these frequencies is given by $1/w$. Choose spatial regions with a z dimension of at least twice this size.

Computation of Displacement Index

1. The degree of overlap between fluorescence distributions is inversely correlated to the magnitude of filament displacement within a spatial region. Measure the degree of overlap

between images at times t and T by computing a spatial cross-product moment, or spatial covariance:

$$\mathrm{Cov}\left[f_{xyzt}, f_{xyzT}\right] = \frac{1}{XYZ} \sum_{z=0}^{Z-1} \sum_{y=0}^{Y-1} \sum_{x=0}^{X-1} \left[\left(f_{xyz} - \hat{f}_{xyzt}\right)\left(f_{xyzT} - \hat{f}_{xyzT}\right)\right] \qquad (13)$$

where

$$\hat{f}_{xyzt} = \frac{1}{XYZ} \sum_{z=0}^{Z-1} \sum_{y=0}^{Y-1} \sum_{x=0}^{X-1} f_{xyzt} \qquad (14)$$

is the mean intensity in the subregion at time t.

2. Compute the spatial variance at each time t from

$$\mathrm{Var}\left[f_{xyzt}\right] = \frac{1}{XYZ} \sum_{z=0}^{Z-1} \sum_{y=0}^{Y-1} \sum_{x=0}^{X-1} \left(f_{xyzt} - \hat{f}_{xyzt}\right)^2 \qquad (15)$$

3. To eliminate the dependence of spatial covariance on the number of nonzero data points and the absolute intensity scale, compute the product moment correlation coefficient using Equations (13) and (15):

$$\mathrm{PMCC}(t, T) = \frac{\mathrm{Cov}\left[f_{xyzt}, f_{xyzT}\right]}{\sqrt{\mathrm{Var}\left[f_{xyzt}\right] \mathrm{Var}\left[f_{xyzT}\right]}} \qquad (16)$$

Equation (16) is a measure of the degree of overlap of fluorescence intensity from time t to T in a spatial region.

4. To simplify interpretation, compute the displacement index as

$$\mathrm{DI}(t, T) = 1 - \mathrm{PMCC}(t,T) \qquad (17)$$

The DI has minimum value zero, indicating no displacement, and increases with increasing filament displacement within a spatial region during the time interval.

Analysis of Cytoplasmic Strain from Endogenous Intermediate Filament Networks

The relative displacement of adjacent intermediate filaments indicates mechanical strain in the cytoskeletal network. Since the shear stress acting at the cell surface may be transmitted throughout the cytoplasm via the cytoskeleton, analysis of intracellular strain using endogenous structures will reveal mechanisms that directly relate extracellular mechanical stimuli to initiation of biochemical signaling at discrete locations within the cell.

Computation of Cytoplasmic Strain Field in Two-dimensional Optical Sections

1. Select a time sequence of two-dimensional optical sections normalized to a constant height above the coverslip. Apply a median filter to minimize high frequency shot noise, and enhance contrast applying a high-pass filter. Apply a low-pass filter to smooth the resulting intensity distribution.

2. To optimize the extraction of filament positions, set all pixels with intensity below a threshold to value zero. Determine a threshold intensity value I_{thresh} from

$$I_{thresh} = \hat{I} - \frac{\sigma}{3} \qquad (18)$$

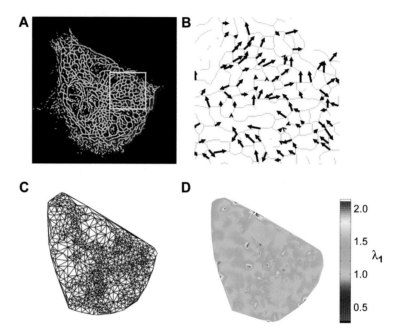

FIGURE 5. (*A*) Comparison of skeletons extracted from an optical section of GFP-vimentin just before (*red*) and 3 minutes after (*green*) onset of shear stress (12 dyn/cm², *left to right*). (*B*) Vertex displacement field from the inset box in *A* indicated by arrows on the skeleton. (*C*) Delaunay triangulation of vertices in the optical section in *A* to connect adjacent vertices on the skeleton for interpolation of the strain field. (*D*) Magnitude of principal stretch ratio λ_1 indicating distribution of maximum strain during the 3-minute interval after onset of shear stress. Note strain focusing corresponds to locations of dense interconnections among intermediate filament segments on the skeleton in *A*. (Adapted, with permission, from Helmke et al. 2003.)

where $\hat{I}$ is the global mean intensity and σ is the standard deviation of intensity in the optical section. Note that other methods for optimizing the threshold intensity are possible, and the choice depends to some extent on the signal-to-noise ratio in the image.

3. After setting background pixels to zero intensity, create a binary skeleton image (Fig. 5A) using built-in functions available in most image processing or math software (e.g., Scion Image, Matlab, IDL).

4. In each skeleton, vertices are defined as positions where three or more filament segments intersect. In the binary skeleton image (black pixels on a white background), identify coordinates of black pixels surrounded by at least three other black pixels. In areas of dense filament connections, this search may produce clusters of pixels rather than a single pixel coordinate. In this case, the position of the intersection is estimated from the geometric mean of the cluster.

5. Track the vertices from the current time point to the next to obtain the initial (X,Y,Z) and final (x,y,z) coordinates (Fig. 5B).

   Several automated tracking algorithms are readily available. For example, we have implemented a particle tracking algorithm based on a probability model for distance moved between time points (Crocker and Grier 1996).

6. To interpolate the strain field, connect adjacent vertices into triangles using a Delaunay triangulation function (Fig. 5C). This function, available in Matlab, creates a set of triangles for which no vertex is located inside another triangle, thereby providing maximum spatial resolution for estimating the strain field.

7. Compute the Lagrangian strain tensor from the shape change of each triangle derived from the initial and final coordinates of the vertices at the corner of the triangle. For each pair of vertices, the Lagrangian strain tensor E_{ij} is a 2 × 2 matrix computed from

$$ds^2 - ds_0^2 = 2E_{11}dX^2 + 4E_{12}dX \cdot dY + 2E_{22}dY^2 \tag{19}$$

In Equation (19), dX is the initial difference between x coordinates and dY is the initial difference between y coordinates of the two vertices, ds_0 is the initial distance between the vertices, and ds is the final distance between the vertices after deformation. Note that these distances are computed from

$$ds_0^2 = dX^2 + dY^2 \text{ and } ds^2 = dx^2 + dy^2 \tag{20}$$

In Equation (20), dx is the difference between final x coordinates and dy is the difference between final y coordinates of the two vertices after deformation.

8. Because Equation (19) contains three unknown components of the strain tensor E_{ij}, a unique solution is obtained by solving a simultaneous system of three equations derived from the vertex pairs determined by the three sides of each triangle. The resulting strain tensor describes the average deformation for the spatial region. The strain magnitude can then be mapped to the skeleton in the optical section (Fig. 5D).

9. The eigenvalues of E_{ij} are the principal values of strain oriented along the principal axes determined by the unit eigenvectors of the matrix. The principal values are the maximum and minimum values of strain in the spatial region for an equivalent strain field that contains zero shear components. For methods of computing the eigenvectors of a matrix, see linear algebra texts (e.g., see Boas 1983).

10. In some cases, it is desirable to express the principal values of strain E_1 and E_2 as stretch ratios λ_1 and λ_2, computed from

$$E_i = \frac{1}{2}\left(\lambda_i^2 - 1\right) \tag{21}$$

where the subscript i has values 1 and 2 corresponding to the principal axes. In particular, note that if the product of principal stretch ratios $\lambda_1\lambda_2$ is unity, then the area of the triangle is conserved during deformation.

ACKNOWLEDGMENTS

We gratefully acknowledge the assistance of David Thakker and Amy Rosen during the development of these techniques and valued discussions with Drs. John Murray, Paul Janmey, and Arjun Yodh of the University of Pennsylvania. Dr. Robert Goldman graciously provided a GFP-vimentin DNA construct that enabled the initial studies of intermediate filament dynamics.

MOVIE LEGENDS

MOVIE 33.1. Time-lapse movie of deconvolved optical section demonstrating intracellular GFP-vimentin filament movement in 90-second intervals (*green*) superimposed on stationary images (*red*) selected at the start of the no-flow and flow periods. Separation of green and red indicates intermediate filament movement; yellow represents zero displacement compared to the beginning of the period. Flow direction, left to right; shear stress, 12 dyn/cm²; scale bar, 2 μm. Elapsed time, mm:ss. (From Helmke et al. 2000 with permission.)

MOVIE 33.2. Time-lapse movie of deconvolved optical section showing GFP-vimentin filament movement near the apical cell surface, corresponding to the region shown in Figure 4B–D. Intermediate filament motion in 90-second intervals (*green*) is superimposed on stationary images (*red*) selected at the start of the no-flow and flow periods. Separation of green and red indicates intermediate filament movement; yellow represents zero displacement compared to the beginning of the period. Flow direction, left to right; shear stress, 12 dyn/cm²; scale bar, 2 μm. Elapsed time, mm:ss. (From Helmke et al. 2000 with permission.)

REFERENCES

Agard D.A. and Sedat J.W. 1983. Three-dimensional architecture of a polytene nucleus. *Nature* **302:** 676–681.

Boas M.L. 1983. *Mathematical methods in the physical sciences.* John Wiley, New York.

Crocker J.C. and Grier D.G. 1996. Methods of digital video microscopy for colloidal studies. *J. Colloid Interface Sci.* **179:** 298–310.

Davidson L.A., Oster G.F., Keller R.E., and Koehl M.A.R. 1999. Measurements of mechanical properties of the blastula wall reveal which hypothesized mechanisms of primary invagination are physically plausible in the sea urchin *Strongylocentrotus purpuratus. Dev. Biol.* **209:** 221–238.

Davies P.F. 1995. Flow-mediated endothelial mechanotransduction. *Physiol. Rev.* **75:** 519–560.

Frangos J.A., McIntire L.V., and Eskin S.G. 1988. Shear stress induced stimulation of mammalian cell metabolism. *Biotechnol. Bioeng.* **32:** 1053–1060.

Heidemann S.R., Kaech S., Buxbaum R.E., and Matus A. 1999. Direct observations of the mechanical behaviors of the cytoskeleton in living fibroblasts. *J. Cell Biol.* **145:** 109–122.

Helmke B.P., Goldman R.D., and Davies P.F. 2000. Rapid displacement of vimentin intermediate filaments in living endothelial cells exposed to flow. *Circ. Res.* **86:** 745–752.

Helmke B.P., Rosen A.B., and Davies P.F. 2003. Mapping mechanical strain of an endogenous cytoskeletal network in living endothelial cells. *Biophys. J.* **84:** 2691–2699.

Helmke B.P., Thakker D.B., Goldman R.D., and Davies P.F. 2001. Spatiotemporal analysis of flow-induced intermediate filament displacement in living endothelial cells. *Biophys. J.* **80:** 184–194.

Levitan I., Helmke B.P., and Davies P.F. 2000. A chamber to permit invasive manipulation of adherent cells in laminar flow with minimal disturbance of the flow field. *Ann. Biomed. Eng.* **28:** 1184–1193.

Wang N., Butler J.P., and Ingber D.E. 1993. Mechanotransduction across the cell surface and through the cytoskeleton. *Science* **260:** 1124–1127.

Imaging Single Molecules Using Total Internal Reflection Fluorescence Microscopy

Nico Stuurman and Ronald D. Vale

The Howard Hughes Medical Institute and the Department of Cellular and Molecular Pharmacology, University of California, San Francisco, California 94143–0450

INTRODUCTION

MANY QUESTIONS CONCERNING CELL BEHAVIOR will be answered eventually by a detailed description of the cell's molecular constituents. Numerous current investigations are directed toward defining gene expression and protein content through microarray and proteomic technologies, respectively. However, these measurements are made from populations of cells. In some cases, the properties of groups of cells (or even large collections of molecules) may mask biologically significant interactions that take place within a subset of the population. This is especially true in the case of signaling molecules, where a single binding event can be amplified into massive signals that change the cell's properties. Therefore, direct measurements of single molecules in living cells can potentially provide new information on cellular processes. Advances in fluorescent labeling techniques, as well as the optics and electronics needed for low-light detection, now make it possible to detect and analyze single molecules with relative ease.

The key to sensitive detection of fluorescent molecules is a high signal-to-noise ratio. Thus, although every effort should be made to excite the dye molecules efficiently and to detect as many emitted photons as possible, it is of prime importance to reduce the background signal to levels acceptable for single-molecule detection. Among the sources of background signal, the sample itself often proves to be the most difficult to control. In wide-field illumination, Raman scattering of water molecules and fluorescence of out-of-focus molecules contribute significantly to the background signal (Funatsu et al. 1995). Nevertheless, single molecules have been detected by wide-field fluorescence microscopy (Funatsu et al. 1995; Schmidt et al. 1996). A confocal pinhole enables a reduction in the background caused by out-of-focus fluorophores (see Chapter 35), but results in rejection of some of the in-focus photons as well. Moreover, the axial dimension of the confocal volume (typically ~500 nm) is relatively large, which results in a significant background signal. Single molecules can also be detected by near-field scanning microscopy (for a review, see de Lange et al. 2001).

585

However, this technique provides relatively slow imaging and because of the complexity of the instrumentation is less likely to be widely available in biologically oriented research laboratories in the near future.

A dramatic reduction in illumination volume, and hence background fluorescence, is made possible by total internal reflection fluorescence microscopy (TIRFM) (Axelrod 2001a). In TIRFM, only the interface between two media differing in their refractive indices is illuminated. When the excitation beam travels from a medium with a higher refractive index (n_t) into a medium with lower refractive index (n_i), under an incident angle greater than the critical angle (θ_c), it is completely reflected at the boundary (total internal reflection). However, an electromagnetic field (the evanescent wave) emanates from the interface and decays exponentially, resulting in fluorescence excitation near the interface only, with little or no background signal from elsewhere. Funatsu et al. (1995) first demonstrated single-molecule detection using TIRFM and laser illumination, and TIRFM has become the technique of choice for detecting single molecules, provided the fluorescent molecules can be illuminated at or near the interface between a glass slide and the aqueous medium.

GEOMETRIES OF TIRF ILLUMINATION

TIRFM can be achieved by two primary geometries that differ in their illumination paths. In one method, the laser illumination does not pass through the objective, but rather is guided directly onto the sample usually through a prism to achieve the correct angle (referred to as "prism-type" TIRFM). In the second method, the excitation light source light travels through the objective before illuminating the sample (referred to as "through-the-objective" TIRFM). These methods are described briefly below, but a more detailed explanation and variations on these basic illumination strategies are described elsewhere (Axelrod 2001a).

Prism-type

In the prism-type configuration (see Fig. 1, top), the sample is sandwiched between a coverslip and slide. The prism and slide (which should have the same refractive index) are optically coupled using a fluid with a refractive index higher than that of the cell, such as glycerol or immersion oil. The angle of incidence can be adjusted by translating the beam, and the size of the sample illumination area can be adjusted using a focusing lens. The fluorescence

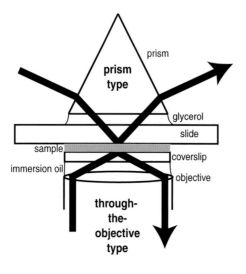

FIGURE 1. Schematic overview of the geometries used to achieve TIRF illumination. The top half of the figure shows the illumination path of a prism-type TIRF microscope. In this case, the evanescent wave is generated at the slide/sample interface, and the image is collected by the objective at the bottom. The bottom part shows the illumination path of a through-the-objective type TIRF microscope. Here, fluorescent dyes are excited at the coverslip/sample interface, and the image is collected through the same objective used for excitation.

emitted by the sample travels through the cell, buffer, coverslip, and immersion medium to reach the objective through which it is imaged on a camera. In practice, fused silica slides are used, since these have less autofluorescence than normal glass slides. For highly sensitive detection, it is critical to use an objective with a high numerical aperture (NA). Oil immersion objectives can be used as long as the sample thickness is small. The sample thickness is limited not only by the working distance of the objective, but also by spherical aberration caused by the mismatch of the sample medium refractive index (close to 1.33 for a typical aqueous buffer) and the immersion oil (1.52). Thus, when the sample thickness exceeds about 15 μm, the signal intensity will be higher with a water immersion objective compared with an oil immersion objective. This type of TIRF illumination has proven to be very fruitful for in vitro studies of single molecules (see section below on Applications). However, cell health is often impaired when the samples are sandwiched closely between two surfaces 15 μm apart for extended periods of time. Nevertheless, we have successfully imaged green fluorescent protein (GFP)-labeled proteins in *Schizosaccharomyces pombe*, *Dictyostelium discoideum*, and mammalian Jurkat cells using such a microscope setup.

Through-the-objective Type

Using high-NA objectives, it is possible to guide the exciting light through the objective such that it reaches the sample below the critical angle for total internal reflection (Fig. 1, bottom) (for a comprehensive overview, see Axelrod 2001b). In this method, the laser beam must be focused on the outer edge of the backfocal plane of the objective. The need for a high-NA objective is illustrated by the following calculation. The critical angle (θ_c) between two materials with refractive indices n_t (the higher refractive index) and n_i (the lower refractive index) is:

$$\theta_c = \sin^{-1}(n_i/n_t) \tag{1}$$

For a glass/water interface, this results in a critical angle of 61°. However, the cytoplasm of cells (adherent to the slide) has a refractive index of about 1.38, which translates into a critical angle of 65°. The angle of the incident light that can be generated with a 1.4-NA oil immersion objective is 67°, which makes it barely possible to achieve total internal reflection. Recently, several vendors (Olympus, Zeiss, and Nikon) have produced objectives with an NA of 1.45, resulting in a maximum illumination angle of about 72°. Aligning a microscope for TIRF illumination with these objectives is simpler than that with 1.4-NA objectives, since the area in the backfocal plane available for TIRF illumination is much larger. Olympus also sells a 100× 1.65-NA objective. This objective requires special (and costly) coverslips and volatile immersion oils with high refractive indices, making it a less practical choice for imaging living cells. Moreover, since the coverslip has a refractive index of 1.65, the critical angle for the coverslip/water interface in this system is about 54°, causing a significant loss in the detected signal in comparison with a normal glass coverslip (although this could possibly be offset by detection of near-field emission) (Axelrod 2001b).

Through-the-objective TIRF illumination is more difficult to set up than prism-type TIRF illumination (but see below), and it has more potential sources of unwanted background noise (due to light scattering within the objective, as well as fluorescence arising from internal components of the objective). On the other hand, because the emitted fluorescence does not have to travel through a layer of buffer, spherical aberration due to refractive index mismatch will be lower, resulting in a higher signal. Another clear advantage of this geometry is that samples are mounted on coverslips, which are commonly used for most types of microscopy. In the case of an inverted microscope, the upper surface can remain open, allowing for easy access of additional probes for stimuli, microinjection, electrophysiology, etc. Moreover, cells can be imaged in perfusion chambers, allowing temperature, gas regulation, and buffer exchange for improved cell viability or introducing experimental variations by buffer exchange. Therefore, through-the-objective type TIRFM is clearly the method of choice for live cell imaging.

Most commercial microscopes can be readily adapted for through-the-objective TIRF illumination. Some possible geometries have been described by Axelrod (2001b). Commercial solutions are available from Olympus, Nikon, and Till. In our laboratory, we use a Zeiss Axiovert 200m equipped with a Zeiss 100 × 1.45-NA objective. The Zeiss Axiovert has a fluorescence illumination pathway that provides a plane conjugate with the back focal plane (BFP; the aperture diaphragm is placed here) of the objective and a plane conjugate with the specimen (SP; the field diaphragm is placed here, see Fig. 2). We chose to utilize the existing fluorescence illumination path and adapt the beam path for TIRF illumination. A motorized mirror is used to select between a xenon or mercury lamp wide-field illumination source and laser TIRF illumination. To allow for continuous adjustment of the angle of the incident beam at the object plane, a planar convex lens (L2 in Fig. 2) is placed such that its focal point is conjugate with the first object plane within the microscope (Fig. 2). Another planar con-

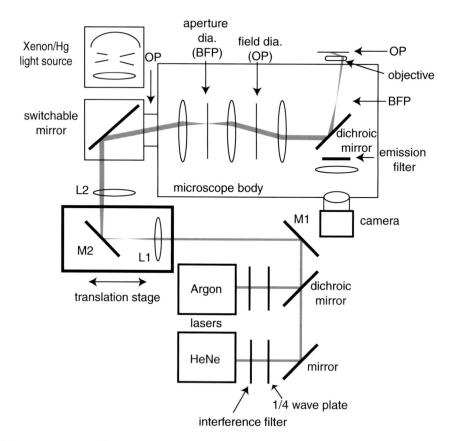

FIGURE 2. Arrangement of TIRF illumination with a Zeiss Axiovert 200. The laser light from the argon and HeNe lasers are combined with a dichroic mirror. The angle of incidence onto the sample is adjusted by moving the translation stage. This moves the beam in the plane of lens L2 and changes the angle of incidence on each of the planes confocal with the object plane (OP). The combination of lenses L1 and L2 functions as a beam expander and can be used to adjust the size of the illumination spot. We use a lens with a focal length of 76 mm for L1 and 150 mm for L2. The fluorescence light path of the Zeiss Axiovert 200 contains apertures confocal with the object plane (OP) and with the back focal plane (BFP) of the objective, which are useful for alignment. See text for further explanation.

vex lens (L1) and a mirror are placed together on a translation stage; movement of this stage changes the incident angle at the object plane, but the center of the beam still coincides with the center of the object plane. Lens L1 is placed such that its focal point coincides with the focal point of lens L2, creating a beam expander. The ratio of the focal lengths of lenses L1 and L2 determines the magnification of laser beam diameter, which translates directly into the size of the illumination spot. For alignment of such a system, the following points should be kept in mind:

1. The laser beam should first be centered along the optical path of the microscope. This can be achieved by removing lenses L1 and L2 and the objective. Adjust mirrors M1 and M2 such that the beam travels through the center of both the aperture and field diaphragms. It is helpful to project the beam onto a piece of paper (or the ceiling) for centering.

2. Place lens L2 back in position. Adjust its position such that the size of the projected beam spot is minimal. Maintain alignment through the aperture and field diaphragms.

3. Place lens L1 and the objective back in position. Adjust the position of lens L1 such that the beam is parallel. Maintain alignment through both diaphragms.

4. Place a coverslip (with immersion oil) on the objective. Place a hemispheric lens on the coverslip (the space in between should be filled with immersion oil). The purpose of this lens is to avoid total internal reflection at the glass/air interface. While projecting the beam on the ceiling, move the translation stage, which should result in movement of the beam. Continue until clipping on the BFP of the objective occurs and the beam begins to disappear. Now, the beam is focused on the outer rim of the BFP of the objective.

5. Replace the coverslip and hemispherical lens with a sample with a solution of fluorochrome-labeled protein. Make sure the right dichroic mirror and emission filter are in place (*Caution*: since a very significant proportion of the laser beam will travel back through the objective, a wrong choice of filter cube can result in a large amount of laser light traveling through the eye piece. As a precaution, *always* check for laser light exiting through the eyepiece by placing a piece of paper in front of the eyepiece). The illumination spot should now be visible through the eyepiece. If it is not completely centered, adjust mirror M1. Adjust the angle of the incident beam by changing the position of the translation stage. If the sample is dilute enough, an obvious transition between wide-field and TIRF illumination will be visible, characterized by diffuse fluorescence throughout the field of view in wide-field illumination and low background with visible diffusing single molecules in TIRF.

LIGHT SOURCES AND FILTERS

Although TIRF illumination is possible with nonlaser light sources such as mercury and xenon light bulbs, the light loss incurred in alignment for TIRFM with these light sources is usually prohibitive for single-molecule detection. Thus, in practice, laser light is used for TIRF illumination. One drawback of the use of coherent laser light is the relatively inhomogeneous illumination field caused by interference fringes due to slight differences in the optical path traveled by different rays in the beam. These interference patterns can be reduced by various techniques for reducing the coherence of the laser light. Another concern is the polarization of the laser light. Since the rotational movements of relatively immobilized molecules may take place on a time scale similar to that of the imaging, polarized light could cause variations in brightness of the single molecules depending on their orientation. It is

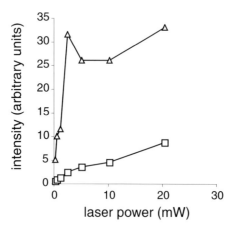

FIGURE 3. Saturation curve of Cy3. Kinesin labeled with Cy3 was attached to a coverslip, and single molecules were detected using the through-the-objective type TIRF microscope described in Figure 2. Images were acquired at varying intensity levels of the 514-nm laser line used for excitation, and the intensities of a number of spots (*triangles*) and their surrounding background (*squares*) were quantified and averaged. Under these conditions, the single Cy3 molecules saturate at about 2.5 mW.

therefore advisable to depolarize the exciting light with a diffuser or preferably a quarter-wave plate to generate circularly polarized light. Although solid-state lasers are becoming an increasingly attractive alternative due to their low (audible and visible) noise, power requirements, stability, and low maintenance, they are still significantly more expensive than argon and HeNe lasers. The power required to saturate fluorophores depends on the fluorophore used and the geometry of the microscope. In practice, an air-cooled argon laser with total output of at least 100 mW will generate sufficient power at the 457-, 488-, and 514-nm lines for single-molecule imaging. As an example, we find that the dye Cy3 is saturated in our through-the-objective type TIRF microscope using about 2.5 mW of 514-nm light (Fig. 3).

The choice of dichroic mirrors and excitation and emission filters is extremely critical when single molecules are imaged. The excitation (or interference) filter should completely block any light emitted from the laser that is not at the desired wavelength. The dichroic mirror and emission filters should have the highest possible light throughput, over as large a part of the emission spectrum of the dye to be imaged as possible. On the other hand, a somewhat narrower range should be chosen if this reduces background signal from the sample or the optics as this might improve the signal-to-noise ratio at the expense of a small percentage of the signal. Antireflection coatings will add a few percent to the light throughput and should be utilized whenever possible. The emission filter should completely block the wavelength used for excitation as well as Raman scattered light from the solution. Using carefully selected double dichroic mirrors and a method to simultaneously observe two different wavelengths (a commercial solution is provided by Optical Insights, Ltd), it is possible to image two dyes simultaneously or to observe fluorescence resonance energy transfer between two dyes.

FLUOROPHORES USEFUL FOR SINGLE-MOLECULE IMAGING

To detect emission from a single molecule, the fluorophore must have a high extinction coefficient at the excitation wavelength and a high quantum yield and should be able to emit a large number of photons before photobleaching. Moreover, to facilitate data analysis, it is desirable that the fluorophore does not have temporary dark states, i.e., not exhibit excessive blinking behavior (Dickson et al. 1997). When the fluorophore must be expressed in a living cell, the choice is obviously limited to the range of fluorescent proteins (or tags that can be made fluorescent in vivo) (Adams et al. 2002). Luckily, the properties of GFP and its derivates (see Chapter 1) are excellent in most respects, with the exception of their considerable blinking behavior, which can confound tracking and quantification. The cyan version of GFP (CFP) is less suited for single-molecule imaging due to its lower brightness. The yellow version (YFP) and the red variants (DsRed or RFP and its derivates) (Fradkow et al. 2000; Campbell et al. 2002) may be superior to GFP due to their excitation and emission spectra being further away from the autofluorescence of living cells (Harms et al. 2001). However, wc do not have direct experience with these variants for single-molecule observation. Interestingly, GFP and some of the DsRed derivates can be excited efficiently with the 488-nm line of an argon laser, with little overlap in their emission spectra. This makes it possible to detect simultaneously both dyes as single molecules without switching illumination sources.

When dyes are directly coupled to proteins and introduced to the cells by microinjection (or simply applied externally), the choice of dyes is much larger. In our laboratory (and in others'), the Cy3 and Cy5 dyes (Mujumdar et al. 1993) have been used with great success. They offer greater photostability and less blinking than GFP. Whenever possible, it is very important to reduce photobleaching rates by reducing oxygen radicals within the solution. Other dyes, such as the Alexa series (Molecular Probes) have been used successfully for single-molecule detection as well (M.Tomishige and R.D. Vale, pers. comm.), and new useful dyes are likely to become available (e.g., see Willets et al. 2003).

DETECTOR CHOICE

Diffusion coefficients of membrane-bound proteins are on the order of 1 μm^2/sec. Motor proteins generally move at rates of 0.1–2 μm/sec. Thus, the spatial and temporal resolution of the detector must be compatible with these values. To achieve the highest possible spatial and temporal resolution, the pixel size of the detector should match the Nyquest sampling rate (Pawley 1995); it should have a high quantum efficiency and low readout noise. In practice, investigators can choose between cooled charge-coupled device (CCD) and intensified CCD (ICCD) cameras. Cooled CCDs can have very high quantum efficiencies, but their readout speeds are limited due to the readout noise of the CCD chip. This can be overcome by imaging only a small part of the chip. The obvious drawback is that the field of view becomes limited, which is a handicap when imaging living cells. We use ICCD cameras for our studies. The quantum efficiency of current models of ICCD cameras can reach 50–60% in the green. By amplifying the signal before being detected by the CCD, the readout noise of the CCD chip becomes less of an issue. Modern ICCDs can be used to sample full frame at video rate or faster, making them ideal detectors for single-molecule imaging in living cells. Recently, on-chip multiplication gain systems have become commercially available (the Cascade from Ropers scientific, and iXon from Andor). With these chips, the signal is amplified inside the chip, before readout. This is an alternative technology to boost the signal/readout noise ratio and allow for high-speed full-frame detection. Cameras based on such technology might become viable alternatives for ICCD cameras, although their current performance is less satisfactory in our testing.

IMAGE ACQUISITION

Live cell imaging of cells by TIRFM is not different from imaging live cells by wide-field microscopy. For single-molecule imaging, however, the following considerations are crucial:

1. The density of labeled molecules should be low enough that single spots can be recognized and their trajectories between consecutive frames can be established (see below). The required minimum density depends on the diffusion coefficient of the molecules and the spatial and temporal resolution of the detection system used. In most cases, even cells that appear dimly labeled in wide-field fluorescence microscopy (e.g., in GFP-expressing

cells) can have too many molecules to allow single-molecule imaging. Selection for weakly expressing cells by fluorescence-activated cell sorting (FACS) and the use of weak or inducible promoters have been useful in our experience. Alternatively, but less desirable, it is possible to bleach a cell until the density of visible fluorochromes becomes amenable to single-molecule observation.

2. To obtain an optimal signal-to-noise ratio, the intensity of the exciting light source should be carefully calibrated. To do so, images of single molecules should be taken at varying light intensities. The intensity of fluorescence emitted by single molecules will increase until saturation is reached. The background will continue to increase linearly beyond this point (see Fig. 3). Thus, the highest signal-to-noise ratio can be obtained by illuminating with just enough power to saturate the fluorophore. Photobleaching rates can be reduced by lowering the excitation intensity at the expense of signal-to-noise ratio.

3. The image acquisition speed determines the upper limit of the diffusion coefficient of molecules that can be reliably analyzed. Molecules that are too fast for a given acquisition speed will appear "blurred" on the detector and are difficult to track. Molecules that moved very far between consecutive images also will be impossible to track. On the other hand, shorter exposure time will lower the number of photons that can be detected and lower the signal-to-noise ratio. Binning will help, but at the expense of spatial resolution. Conversely, very slowly moving molecules will photobleach if exposed continuously. For long-term tracking of such species, shuttering the light source is desirable.

TRACKING SINGLE MOLECULES

Microscopy of single molecules results in a sequence of images of diffraction-limited spots. The interesting parameters of these spots are their position and intensity. Since the intensity of the spots depends strongly on their position in the evanescent field (which will often be unknown), intensity data must be interpreted with caution. Nevertheless, changes in intensity from TIRFM experiments have been used to demonstrate receptor dimerization (Sako et al. 2000).

For motion tracking, the trajectories followed by the individual molecules need to be extracted from the image sequences. It is possible to track spots by hand; however, this is a laborious procedure and not very accurate. Since the signal-to-noise ratio of single-molecule images is rather low, a simple thresholding is not sufficient for single-molecule detection. Most algorithms used for automated spot recognition compare the available data to a two-dimensional point spread function (PSF) or theoretical approximation (like a Gaussian) by cross-correlation or fitting (Cheezum et al. 2001). Because the diffraction-limited "shape" of single molecules is taken into account in these approaches, it is possible to determine the position of the molecule with subpixel accuracy, substantially below the resolving power of the optical system used (Thomann et al. 2002). In a second step, the positions of the spots identified in the series of images are assembled in trajectories. During the assembly process, assumptions must be made about the expected (maximum) diffusion coefficient, and provisions should be made for possible blinking behavior of the dyes. There is also an upper limit to the density of molecules (determined by diffusion coefficient of the molecules and the image acquisition rate), beyond which it is impossible to accurately assign spot positions to individual tracks.

In our laboratory, we implemented the tracking method published by Gelles et al. (1988). First, the signal-to-noise ratio is improved by applying a median filter. Then, a two-dimensional PSF is generated by averaging a single spot over about 20 frames. Subsequently, the investigator assigns the approximate starting point of a spot, and the algorithm locates

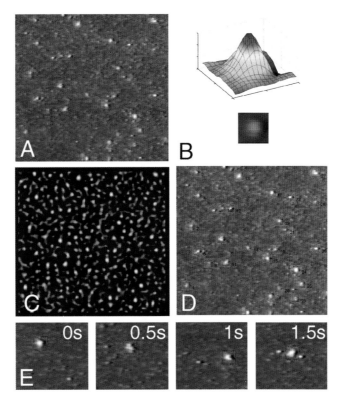

FIGURE 4. Spot detection and tracking algorithm. (*A*) Example of a digitized image displaying single molecules (Cy3-labeled Unc104 motor proteins on a supported phospholipid bilayer). This image was taken from a digitized video sequence and was median-filtered. (*B*) Averaged diffraction-limited spot (derived by averaging the image of single Cy3 molecules from images as the one shown in panel *A)* used for the tracking algorithm. Two representations are shown: A false-colored mesh plot on top and the normal gray-scale image at the bottom. (*C*) Image from panel *A* after cross-correlation with the point spread function shown in *B*. (*D*) Red brackets indicate spots identified in the image after segmentation of the image in *C* and rejection of those spots that had low intensities in the original image in *A*. (*E*) Examples of the tracking of an individual spot (single molecule) through a sequence of images. When tracking a diffraction-limited spot, the algorithm performs the cross-correlation and spot detection as described above only on a small field (limited by the expected maximum diffusion coefficient) around the last known location of the spot.

the spot by performing a cross-correlation with the PSF for every position in a field around the starting point whose size is determined by the expected maximum diffusion coefficient. The cross-correlation coefficients are assembled into an image, and the centroid of the cross-correlation image is determined by thresholding at a user-set value (Fig. 4). This process is repeated in the next image, starting at the position that was just determined. The intensity of the spot is calculated and background (as measured in the local environment of the spot) is subtracted. When no object with a cross-correlation coefficient higher than a specified cutoff is found (or the intensity of the object drops below a specified minimum), the track ends. Alternatively, to account for the blinking behavior of some dyes, the search can be continued on one or more consecutive frames. Finally, the track is visually inspected to ensure that the track did not move over to an unintended spot. The resulting centroid coordinates and intensity values are saved in a tabular format to be used for displacement analysis. Although still relatively labor intensive since every spot must be manually selected and the resulting analysis must be inspected, this approach allowed us to analyze many more spot tracks than would otherwise be possible, and yields subpixel precision. It should be possible, however, to improve spot tracking by more sophisticated prefiltering of the data, by using fitting methods instead of cross-correlation, and to improve speed by preselecting potential spots with line-scan-based methods and/or segmentation combined with curvature analysis.

DISPLACEMENT ANALYSIS

Once the location of individual molecules at different time points has been determined, it becomes possible to extract information concerning the average diffusion coefficient and the existence of multiple populations differing in diffusion coefficients. First, for every individual track, the average distance traveled by the molecule at every possible time interval is calculated. For freely diffusing molecules, the square displacement increases linearly with time. For molecules confined in a restricted area, the mean square displacement will soon reach a maximum (Fig. 5). In practice, these behaviors can readily be observed. However, since purely random walks will display all of these types of behaviors, the significance of a single

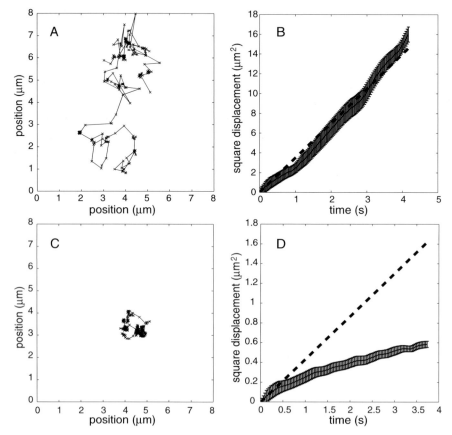

FIGURE 5. Mean square displacement analysis. (*A*) Trajectory of a single molecule of Cy3-labeled Unc104 motor protein moving on a glass-bound phospholipid bilayer. Each symbol (×) indicates the position of the spot in each frame of the sequence. Frames were 33 msec apart. (*B*) Mean square displacement plot of the trajectory shown in *A*. Mean square displacements were calculated as described in the text and are shown with error bars indicating the standard deviation. The dashed line shows the fit to the first four data points from which the (initial) diffusion coefficient can be calculated. Deviation from this line indicates anomalous diffusion. This particular object diffuses freely. (*C*) Trajectory of a single molecule of Cy3-labeled Unc104 motor protein moving on a glass-bound phospholipid bilayer that contained 40 mol% cholesterol and 20 mol% sphingomyelin. (*D*) Mean square displacement of the particle shown in *C*. The diffusion coefficient of this particle is much lower than that of the particle in *A*, and the particle's movements are restricted to a small area, which is reflected in the mean square displacement values.

"aberrant" diffusion measurement is not immediately obvious (Saxton and Jaoobson 1997). Nevertheless, at small time intervals (t), the square displacement in two dimensions (r^2) is related to the diffusion coefficient (D) as follows:

$$r^2 = 4Dt \tag{2}$$

Thus, diffusion coefficients for the tracked molecules can readily be calculated and presented as an average and/or as a histogram conveying information about the distribution of diffusion coefficients. However, such representations neglect differences in duration of the trajectories (which are in part caused by the stochastic nature of photobleaching) as well as the possibility that the molecules undergo changes in their diffusion states. To address this issue, we have used two approaches. First, it appeared useful to partition every available track into overlapping windows of the same duration and calculate the diffusion coefficient for each of these windows (Klopfenstein et al. 2002). This approach should uncover different diffusion states as long as their duration is longer than the time window chosen. However, the observation of multiple states in a histogram is highly dependent on the arbitrary choice of bin size. To reliably uncover multiple diffusion states, a better approach is to analyze the probability distribution function $P(r^2, t)$. For every trajectory being analyzed, the square displacement r^2 is calculated for a given time interval Δt (in practice, this should be the shortest time interval allowed by the experimental data), the square displacements are sorted by size, and for every square displacement, the number of square displacements (normalized by the total number of observations) is calculated (e.g., see Fig. 6). This probability distribution function can be fitted with one- or two-state models. Quality of fit will readily identify which of the models is more correct. This procedure provides both insight for which model describes the data better and information of the diffusion coefficients of the different populations and their relative contribution to the total pool of observations (Schutz et al. 1997).

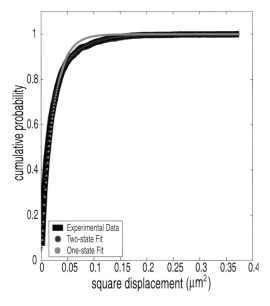

FIGURE 6. Probability distribution analysis of square displacements. The square displacement values derived from the tracking of single molecules of GFP-Lck moving on the surface of living Jurkat cells were used to construct a probability distribution, as described in the text. As can be seen from the fits, a two-state model describes the experimental data much better than a one-state model. In this particular case, 24% of the population was in a state with a diffusion coefficient of 0.039 μm^2/sec and 76% of the population was in a state with a diffusion coefficient of 0.306 μm^2/sec.

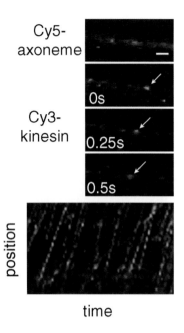

Cy5-axoneme

Cy3-kinesin

0s

0.25s

0.5s

position

time

FIGURE 7. Single kinesin motor proteins moving on an axoneme. The top panel shows the Cy5-labeled axoneme. The consecutive lower panels show a Cy-3 labeled Unc104 kinesin motor protein moving (leftward) along the filament (See Movie 34.1 on the accompanying DVD). Size bar in the top panel represents 1μm. The bottom panel shows a "kymograph." A line is drawn to trace the axoneme, and the lines of all consecutive time points are displayed along on the y axis of the kymograph. Moving particles are seen as a line in the kymograph. The slope of these lines is a measure for the velocity of the moving particle; their length indicates the run length. Moreover, it is immediately obvious (horizontal line) whether or not the motor temporarily stalls. Kymographs can be readily made using the freely available software package ImageJ (http://rsb.info.nih.gov/ij/).

APPLICATIONS

Imaging of Molecular Motor Proteins

The first application of single-molecule TIRFM was imaging molecular motor proteins in vitro. Funatsu et al. (1995) demonstrated binding and dissociation of fluorescently labeled nucleotides to single myosin motors, and Vale et al. (1996) visualized the motion of individual fluorescently labeled kinesin motors along microtubules. The latter assay has become an important tool for measuring and dissecting the mechanism of kinesin processivity (the ability of the motor to take many sequential steps along the microtubule without dissociating). This assay involves adding low concentrations of fluorescently labeled kinesin (0.5–20 nM, either chemically modified with Cy3 or genetically tagged with GFP and expressed/purified from bacteria) to axonemes (microtubule bundles from flagella) or microtubules attached to the coverslip surface (for details of the assay, see Pierce and Vale 1999). Single fluorescent "spots" can then be observed attaching to and moving linearly along the microtubule (Fig. 7 and Movie 34.1 on the accompanying DVD). The speed and distance traveled by more than 100 molecules can be readily recorded. The travel distances are distributed exponentially, reflecting a certain probability of release of motor per ATPase cycle, and the decay constant provides the average run length. Genetically modified motor proteins can be quickly produced and tested in this assay in order to dissect the structural determinants of motor processivity (e.g., see Romberg et al. 1998; Thorn et al. 2000). This assay also has been applied to other motors in the kinesin superfamily (Okada and Hirokawa 1999; Tomishige et al. 2002) and to the processive myosin V motor (Sakamoto et al. 2000). Measurements of fluorescence polarization (Adachi et al. 2000; Sosa et al. 2001) and fluorescent resonance energy transfer (our work in progress) can be used to measure conformational changes of motors in vitro.

Imaging Molecules in Membrane Bilayers

Membrane bilayers can be formed on glass surfaces (for a review, see Boxer 2000), allowing imaging of single lipid or protein molecules in the bilayer. As an example, we have imaged a motor protein that binds to phosphoinositol-containing lipids on such bilayers (Klopfenstein et al. 2002). Custom algorithms described earlier have allowed us to track these

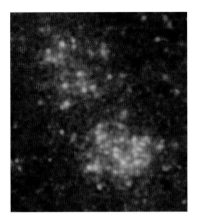

FIGURE 8. Imaging the interaction of phospholipid-bound single molecules with cells. Two membrane proteins, CD58-Cy3 (*green*) and ICAM-1-Cy5 (*red*), are present in a glass-bound planar phospholipid bilayer underneath two PMA/ionomycin-treated Jurkat cells. The image was obtained using the through-the-objective type TIRF microscope described in Figure 2. The dyes were excited with the 514-nm argon laser line (for Cy3) and the 633-nm HeNe laser line (for Cy5). A double dichroic mirror reflected both laser lines and transmitted the emitted fluorescence. The emission from Cy3 and Cy5 was separated using the Dual-View Micro-Imager (Optical Insights, Santa Fe, New Mexico) which projects the two colors side by side on the camera. (See Movie 34.2 on accompanying DVD.)

molecules in two dimensions and measure the diffusion coefficients of single molecules and even changes in diffusion coefficients over time. In this study, we demonstrated dramatic changes in single-molecule diffusion coefficients when agents (e.g., cholesterol) reported to cause lipid raft formation (Simons and Ikonen 1997) were introduced into the bilayers.

Glass-adsorbed membrane bilayers also can be used to interact with living cells. One example of such an application is the introduction of immune recognition proteins found in antigen-presenting cells (e.g., major histocompatibility complex [MHC] antigen and integrins such as ICAM-1) into membrane bilayers, which can then interact with T cells. Dustin and colleagues demonstrated that the bilayers can mimic antigen-presenting cells and create a complex signaling structure called the "immunological synapse," which displays a spatial segregation of various molecules involved in signaling and adhesion in an inner zone and outer zone pattern (Grakoui et al. 1999). Because the distances from the slide to the T cell are small, fluorescently labeled proteins can be simultaneously imaged on the bilayer and the T-cell membrane (Fig. 8 and Movie 34.2 on the accompanying DVD), allowing a single-molecule dissection of the formation of immunological synapse and the behavior of molecules in the two zones. Other applications of membrane bilayer-live cell interfaces are likely to emerge in the future.

Signaling Molecules at the Cell Surface

Membrane receptors of glass-adhered cells have been visualized using a fluorescently labeled receptor ligand, as well as GFP-labeled receptor protein. Yanagida and co-workers labeled epidermal growth factor (EGF) with Cy3 or Cy5 and demonstrated binding of the labeled compound to the cell surface of CHO cells (Sako et al. 2000). By varying the angle of incidence, they managed to generate the evanescent wave not only at the (basal) glass/cell interface,

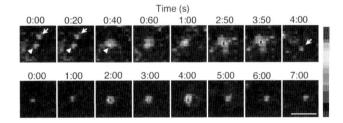

FIGURE 9. Dimerization of the EGF receptor. Diffusion of Cy3-labeled EGF on the surface of human carcinoma A431 cells was visualized by TIRFM. The upper part of the figure shows two spots coalescing and continuing as a single brighter spot, suggestive of receptor dimerization. The lower half shows a spot that suddenly doubles in intensity (presumably by binding Cy3-EGF from the medium) and subsequently drops back in intensity (probably by photobleaching of one of the two Cy3-EGF molecules). Bar, 5 μm. (Reprinted, with permission, from Sako et al. 2000 [©Nature Publishing Group].)

but also at the (apical) cell/medium interface, allowing visualization of single molecules at either side of the cell. Oligomerization of EGF receptors was suggested by fusion of spots containing Cy3-labeled EGF (Fig. 9), as well as by resonance energy transfer between Cy3-labeled and Cy5-labeled EGF. The same methodology was used to visualize cAMP labeled with Cy3 on the surface of *D. discoideum* in order to understand the molecular basis of chemotaxis (Ueda et al. 2001).

Kusumi and co-workers tracked GFP-labeled E-cadherin on mouse fibroblast cells using a through-the-objective type TIRF microscope, and they demonstrated that most of the cadherin was in a multimeric form (Iino et al. 2001). Likewise, we have visualized membrane-targeted GFP at the surface of T cells (A. Douglass and R. Vale, unpubl.).

Membrane Fusion and Endocytosis at the Plasma Membrane

Many important interactions occur between secretory vesicles and the plasma membrane, and these events are accessible for visualization by TIRFM. Almers and colleagues have applied TIRFM to measure vesicle fusion with the membrane (for review, see Steyer and Almers 2001). Fluorescently loaded vesicles appear bright when docked beneath the membrane, and when fusion occurs, the fluorescence disperses into the plasma membrane. This technique also has been valuable for studying endocytosis (Merrifield et al. 2002). In this study, the authors were able to observe the formation of clathrin-coated pits and the recruitment of dynamin to these structures. Transient actin assembly associated with endocytosis was visualized by this technique. Thus, TIRFM provides a powerful means of examining macromolecular assembly and disassembly reactions that occur at or near the plasma membrane.

Visualization of Intracellular Components

It is sometimes possible to visualize intracellular components by TIRFM. Because the refractive index of the cytoplasm (estimated to be 1.38) is higher than that of water, the evanescent wave propagates further into the cell than in water. Components not too far from the membrane can therefore be readily visualized by TIRFM. For instance, myofibrils formed by expression of GFP-labeled cytoplasmic myosin II in *D. discoideum* can be more clearly seen when using TIRF illumination than in wide-field due to the relatively high concentration of myofibrils at the cortex, and the higher concentration of unpolymerized myosin II–GFP in the cytoplasm. In fact, we have observed single-myosin-thick filaments in living cells, as evidenced by their characteristic bipolar morphology (Fig. 10 and Movie 34.3 on the accompanying DVD). Likewise, cytoplasmic vesicles can be readily observed by TIRFM (see above). These findings also raise the point that detection of a component by TIRFM is, by itself, no proof for localization of that component at the plasma membrane.

OUTLOOK

The relative ease with which single molecules can be visualized makes TIRFM an extremely powerful technique to peer into the life of cells at the molecular level. The combination of increased interest in the properties of single molecules in living cells and the commercial availability of microscope systems capable of TIRFM imaging suggest that TIRFM will become a widely used tool in cell biology laboratories in the near future.

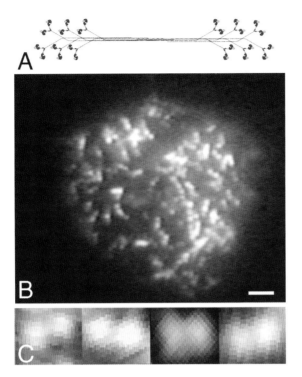

FIGURE 10. Imaging of single cytoplasmic myosin II-thick filaments in *Dictyostelium*. (*A*) A diagram showing the architecture of myosin molecules arranged into a thick filament. A GFP tag was placed on the myosin motor domains, which emerge at opposite ends of the thick filament (Moores et al. 1996). The GFP-myosin used in this imaging had a mutation in the coiled-coil domain (termed 3xAla myosin) that prevented phosphorylation and disassembly of the thick filament into myosin dimers (Egelhoff et al. 1993). The *Dictyostelium* strain expressing this myosin was kindly provided by Dr. J. Spudich (Stanford University). (*B*) A TIRFM image of GFP-myosin3xAla in a living *Dictyostelium* cell. Single-myosin-thick filaments are evident. Individual filaments are not evident by wide-field fluorescence microscopy. Bar, 1 μm. (*C*) Higher-magnification views showing individual myosin-thick filaments. Their bipolar architecture is evident from these images (GFP being localized to the ends of the thick filaments). (See Movie 34.3 on accompanying DVD.)

ACKNOWLEDGMENTS

We thank Mark J. Dayel for his help in devising the illumination scheme depicted in Figure 2. We are grateful to Adam Douglas and Michio Tomishige for providing unpublished data. Daniel Axelrod has greatly stimulated the use of TIRFM and his comments were very helpful during the construction of our through-the-objective type TIRF microscope.

MOVIE LEGENDS

MOVIE 34.1. Single kinesin motor proteins moving on an axoneme. The first frames show a Cy5-labeled axoneme. Subsequent frames show the movement of a Cy3-labeled unc104 kinesin motor protein. See the legend to Figure 7.

MOVIE 34.2. Imaging multiple phospholipid-bound single molecules. Two membrane proteins, CD58-Cy3 (*green*) and ICAM-1-Cy5 (*red*), are present in a glass-bound planar phospholipid bilayer under two PMA/ionomycin-treated Jurkat cells. The movie plays at a speed equivalent to real time. See the legend to Figure 8.

MOVIE 34.3. Imaging cytoplasmic myosin II in *Dictyostelium discoideum*. Wild-type GFP-myosin II was expressed in *Dictyostelium* cells and imaged by TIRFM. See the legend to Figure 10.

REFERENCES

Adachi K., Yasuda R., Noji H., Itoh H., Harada Y., Yoshida M., and Kinosita K., Jr. 2000. Stepping rotation of F1-ATPase visualized through angle-resolved single-fluorophore imaging. *Proc. Natl. Acad. Sci.* **97:** 7243–7247.

Adams S.R., Campbell R.E., Gross L.A., Martin B.R., Walkup G.K., Yao Y., Llopis J., and Tsien R.Y. 2002. New biarsenical ligands and tetracysteine motifs for protein labeling in vitro and in vivo: Synthesis and biological applications. *J. Am. Chem. Soc.* **124:** 6063–6076.

Axelrod D. 2001a. Total internal reflection fluorescence microscopy in cell biology. *Traffic* **2:** 764–774.

———. 2001b. Selective imaging of surface fluorescence with very high aperture microscope objectives. *J. Biomed. Opt.* **6:** 6–13.

Boxer S.G. 2000. Molecular transport and organization in supported lipid membranes. *Curr. Opin. Chem. Biol.* **4:** 704–709.

Campbell R.E., Tour O., Palmer A.E., Steinbach P.A., Baird G.S., Zacharias D.A., and Tsien R.Y. 2002. A monomeric red fluorescent protein. *Proc. Natl. Acad. Sci.* **99:** 7877–7882.

Cheezum M.K., Walker W.F., and Guilford W.H. 2001. Quantitative comparison of algorithms for tracking single fluorescent particles. *Biophys. J.* **81:** 2378–2388.

de Lange F., Cambi A., Huijbens R., de Bakker B., Rensen W., Garcia-Parajo M., van Hulst N., and Figdor C.G. 2001. Cell biology beyond the diffraction limit: Near-field scanning optical microscopy. *J. Cell Sci.* **114:** 4153–4160.

Dickson R.M., Cubitt A.B., Tsien R.Y., and Moerner W.E. 1997. On/off blinking and switching behaviour of single molecules of green fluorescent protein. *Nature* **388:** 355–358.

Egelhoff T.T., Lee R.J., and Spudich J.A. 1993. *Dictyostelium* myosin heavy chain phosphorylation sites regulate myosin filament assembly and localization in vivo. *Cell* **75:** 363–371.

Fradkov A.F., Chen Y., Ding L., Barsova E.V., Matz M.V., and Lukyanov S.A. 2000. Novel fluorescent protein from *Discosoma* coral and its mutants possesses a unique far-red fluorescence. *FEBS Lett.* **479:** 127–130.

Funatsu T., Harada Y., Tokunaga M., Saito K., and Yanagida T. 1995. Imaging of single fluorescent molecules and individual ATP turnovers by single myosin molecules in aqueous solution. *Nature* **374:** 555–559.

Gelles J., Schnapp B.J., and Sheetz M.P. 1988. Tracking kinesin-driven movements with nanometre-scale precision. *Nature* **331:** 450–453.

Grakoui A., Bromley S.K., Sumen C., Davis M.M., Shaw A.S., Allen P.M., and Dustin M.L. 1999. The immunological synapse: A molecular machine controlling T cell activation. *Science* **285:** 221–227.

Harms G.S., Cognet L., Lommerse P.H., Blab G.A., and Schmidt T. 2001. Autofluorescent proteins in single-molecule research: Applications to live cell imaging microscopy. *Biophys. J.* **80:** 2396–2408.

Iino R., Koyama I., and Kusumi A. 2001. Single molecule imaging of green fluorescent proteins in living cells: E-cadherin forms oligomers on the free cell surface. *Biophys. J.* **80:** 2667–2677.

Klopfenstein D.R., Tomishige M., Stuurman N., and Vale R.D. 2002. Role of phosphatidylinositol(4,5) bisphosphate organization in membrane transport by the Unc104 kinesin motor. *Cell* **109:** 347–358.

Merrifield C.J., Feldman M.E., Wan L., and Almers W. 2002. Imaging actin and dynamin recruitment during invagination of single clathrin-coated pits. *Nat. Cell Biol.* **4:** 691–698.

Moores S.L., Sabry J.H., and Spudich J.A. 1996. Myosin dynamics in live *Dictyostelium* cells. *Proc. Natl. Acad. Sci.* **93:** 443–446.

Mujumdar R.B., Ernst L.A., Mujumdar S.R., Lewis C.J., and Waggoner A.S. 1993. Cyanine dye labeling reagents: Sulfoindocyanine succinimidyl esters. *Bioconjug. Chem.* **4:** 105–111.

Okada Y. and Hirokawa N. 1999. A processive single-headed motor: Kinesin superfamily protein KIF1A. *Science* **283:** 1152–1157.

Pawley J.B. 1995. Fundamental limits in confocal microscopy. In *Handbook of biological confocal microscopy* (ed. J.B. Pawley), pp. 19–37. Plenum Press, New York.

Pierce D.W. and Vale R.D. 1998. Assaying processive movement of kinesin by fluorescence microscopy. *Methods Enzymol.* **298:** 154–171.

Romberg L., Pierce D.W., and Vale R.D. 1998. Role of the kinesin neck region in processive microtubule-based motility. *J. Cell Biol.* **140:** 1407–1416.

Sako Y., Minoghchi S., and Yanagida T. 2000. Single-molecule imaging of EGFR signalling on the surface of living cells. *Nat. Cell Biol.* **2:** 168–172.

Sakamoto T., Amitani I., Yokota E., and Ando T. 2000. Direct observation of processive movement by individual myosin V molecules. *Biochem. Biophys. Res. Commun.* **272:** 586–590.

Saxton M.J. and Jacobson K. 1997. Single-particle tracking: Applications to membrane dynamics. *Annu. Rev. Biophys. Biomol. Struct.* **26:** 373–399.

Schmidt T., Schutz G.J., Baumgartner W., Gruber H.J., and Schindler H. 1996. Imaging of single molecule diffusion. *Proc. Natl. Acad. Sci.* **93:** 2926–2929.

Schutz G.J., Schindler H., and Schmidt T. 1997. Single-molecule microscopy on model membranes reveals anomalous diffusion. *Biophys. J.* **73:** 1073–1080.

Simons K. and Ikonen E. 1997. Functional rafts in cell membranes. *Nature* **387:** 569–572.

Sosa H., Peterman E.J., Moerner W.E., and Goldstein L.S. 2001. ADP-induced rocking of the kinesin motor domain revealed by single-molecule fluorescence polarization microscopy. *Nat. Struct. Biol.* **8:** 540–544.

Steyer J.A. and Almers W. 2001. A real-time view of life within 100 nm of the plasma membrane. *Nat. Rev. Mol. Cell Biol.* **2:** 268–275.

Thomann D., Rines D.R., Sorger P.K., and Danuser G. 2002. Automatic fluorescent tag detection in 3D with super-resolution: Application to the analysis of chromosome movement. *J. Microsc.* **208:** 49–64.

Thorn K.S., Ubersax J.A., and Vale R.D. 2000. Engineering the processive run length of the kinesin motor. *J. Cell Biol.* **151:** 1093–1100.

Tomishige M., Klopfenstein D.R., and Vale R.D. 2002. Conversion of Unc104/KIF1A kinesin into a processive motor after dimerization. *Science* **297:** 2263–2267.

Ueda M., Sako Y., Tanaka T., Devreotes P., and Yanagida T. 2001. Single-molecule analysis of chemotactic signaling in *Dictyostelium* cells. *Science* **294:** 864–867.

Vale R.D., Funatsu T., Pierce D.W., Romberg L., Harada Y., and Yanagida T. 1996. Direct observation of single kinesin molecules moving along microtubules. *Nature* **380:** 451–453.

Willets K.A., Ostroverkhova O., He M., Twieg R.J., and Moerner W.E. 2003. Novel fluorophores for single-molecule imaging. *J. Am. Chem. Soc.* **125:** 1174–1175.

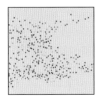

Visualization and Quantification of Single RNA Molecules in Living Cells

Yaron Shav-Tal, Shailesh M. Shenoy, and Robert H. Singer

Department of Anatomy and Structural Biology, Albert Einstein College of Medicine, Bronx, New York 10461

INTRODUCTION

THE ABILITY TO OBSERVE LIVING CELLS THROUGH THE MICROSCOPE as a result of technical advances made in the field of live cell imaging has opened a new dynamic subcellular world to the cell biologist. Identification of molecules in fixed cells can now be extended to the trafficking of these molecules within the cell. This makes it possible to follow changes in localization of molecules with changes in the environment and to describe the kinetics of these movements as well as the specific interactions of various molecules with different cellular structures. These data can be subjected to mathematical and statistical analyses and thus enhance our understanding of molecular dynamics in living cells. Moreover, it is now possible to quantify the number of molecules in different structures using fluorescence microscopy. Indeed, the power of studying an individual living cell brings a new perspective in biological studies typically performed on populations of cells representing an "average" understanding of the events taking place in one cell (Levsky and Singer 2003). This chapter describes some general approaches for detection of RNA, in particular, in living cells and presents methods for the detection and quantification of single RNA molecules.

Fluorescent fusion proteins are of common use in many laboratories and are excellent tools for following the movements of various proteins within a living cell. Such fluorescently tagged proteins are easily produced, efficiently introduced into cells, and, in most cases, display behaviors similar to those of their endogenous protein counterparts. However, the visualization of RNA dynamics in living cells is not as straightforward as in the case of proteins. One major obstacle in this field is the question of how to fluorescently tag RNA while retaining its normal properties for the use in living cells. Several technical approaches have been devised for detecting RNA in living cells, each having its advantages and drawbacks (for detailed reviews and references within, see Dirks et al. 2001; Fusco 2004).

VISUALIZATION OF RNA IN LIVING CELLS

A popular and powerful way to study RNA in fixed cells is the technique of fluorescent in situ hybridization (FISH). This concept has been modified for use in living cells (Politz et al. 1995) and is termed fluorescent in vivo hybridization (FIVH). In both cases, an oligonucleotide complementary to the RNA transcript under study is synthesized and covalently linked to fluorochromes. These fluorescent probes are introduced into the cell, hybridized with the RNA of interest, and subsequently followed in living cells. The obvious advantage of this approach is the ability to study an endogenous RNA species. However, the hybridized RNAs thus contain double-stranded regions. In addition, the translation machinery might encounter problems while scanning these RNAs. These features may affect some aspects of the metabolism of the respective RNA molecules and/or cells. Double-stranded RNA is regarded by the cell in many cases as an undesirable product that is therefore enzymatically degraded.

Variations on this technique of in vivo hybridization have been introduced in numerous studies. First, the means by which the fluorescent probes are introduced into cells has been approached in many ways (Dirks et al. 2001). The preferred method is microinjection of the fluorescent probes directly into living cells (Ainger et al. 1993; Theurkauf and Hazelrigg 1998; Wilkie and Davis 2001), although this allows only an exogenous RNA to be studied. It is possible to stain endogenous RNAs using the SYTO-14 dye, although this dye binds to all RNA species in the cell (Knowles et al. 1996; Knowles and Kosik 1997) Some protocols use permeabilization of the cell membrane either by enzymes (Barry et al. 1993) or with glass beads (McNeil and Warder 1987) in order to enhance the population of cells taking up the probe. Biological processes such as pinocytosis (Okada and Rechsteiner 1982) or receptor-mediated endocytosis (Loke et al. 1989) have also been applied for the introduction of probe.

The type of probes used in in vivo hybridization is another factor that has been tested and compared in several systems. Briefly, it is possible to choose between either DNA (oligodeoxynucleotides or ODNs) or RNA fluorescent probes (e.g., see Leonetti et al. 1991; Fisher et al. 1993; Politz et al. 1995, 1998). Factors to be taken into account when using these probes are probe stability, probe uptake, length of probe, preferred sequences to which the probe will hybridize, signal-to-noise ratio, and correct localization of the probe in the subcellular context. Some improvements have been applied successfully, such as the use of caged fluorophores (Politz et al. 1999), which fluoresce only when activated photolytically, thus improving the signal by avoiding the detection of nonhybridized sequences. Stability can be enhanced by the use of phosphorothioate oligodeoxynucleotides (PS-ODNs) (Fisher et al. 1993; Agrawal et al. 1997; Lorenz et al. 1998, 2000), $2'$-O-methyl-RNAs (Carmo-Fonseca et al. 1991; Molenaar et al. 2001), or peptide nucleic acid (PNA) probes (Dirks et al. 2001) in which the phosphodiester backbone of the nucleotide sequence is replaced by a glycine moiety and therefore structurally resembles a peptide. A relatively new tool used in RNA detection are molecular beacons (Tyagi and Kramer 1996; Matsuo 1998; Sokol et al. 1998). These DNA probes contain a stem-loop structure in their sequence and have fluorescent molecules attached to either end of the probe. While the stem-loop structure remains closed the fluorescent molecules are in close proximity so that fluorescent quenching occurs. Only when hybridization of the probe takes place and the stem-loop structure is melted do the fluorophores become distant from each other and fluoresce.

THE MS2-GFP SYSTEM

Obviously, it would be preferable if specific RNAs could be visualized directly in live cells without the need for hybridization. The MS2-GFP (green fluorescent protein) system

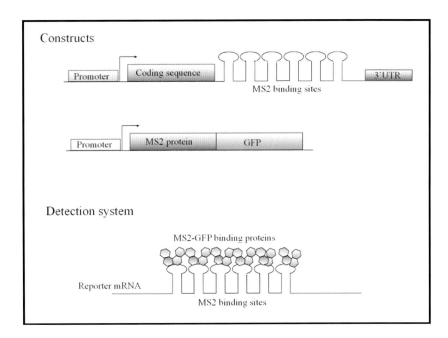

FIGURE 1. Two constructs are used for this RNA detection system in living cells. The top construct encoding the reporter mRNA contains stem-loop repeats of the MS2-binding site which in this case are inserted between the coding sequence and 3' UTR. The second plasmid encodes an MS2-GFP fusion protein. When both plasmids are coexpressed in a cell, the dimers of MS2-GFP proteins will specifically bind to the MS2-binding sites on the mRNA, and these RNAs can be detected via the fluorescence of the GFP moiety.

provides such a means to visualize RNA movement in a living cell that combines the simplicity of use of fluorescent proteins and the high specificity of a motif inserted into the RNA sequence under study (Bertrand et al. 1998). A phage sequence containing stem-loop-binding sites (6, 12, or 24 sites) for the phage capsid protein MS2 that binds single-stranded RNA sequences (Fig. 1) is added to the RNA sequence of interest. This plasmid, from which the RNA will be transcribed, is transfected into cells together with a plasmid encoding the MS2 protein fused to GFP. Since the binding of MS2 to the stem-loop-binding site is highly specific and stable (K_d = 39 nM in vitro; Lago et al. 1998) and binding occurs by the dimerization of the MS2-GFP protein giving a relatively strong signal, this provides a powerful system for following the travels of specific RNAs in living cells.

One should keep in mind that this method utilizes the overexpression of an exogenous RNA, although this can be minimized by stable expression. Another approach to overcome overexpression is to follow GFP-RNA proteins that bind to endogenous RNA under normal conditions; however, such an approach will also have some drawbacks. One study has used a fusion protein of GFP-exuperantia to follow *bicoid* mRNA (Theurkauf and Hazelrigg 1998), whereas another study has introduced general mRNA-binding proteins such as GFP-poly(A)-binding protein II (PABP2), which bind to all types of mRNA transcripts in the cell (Calapez et al. 2002). The first system is useful for only one type of RNA, whereas the latter is only specific for RNA polymerase II transcripts. Additionally, it is not possible to define what population of the GFP-RNA-binding protein is indeed bound to the RNA and what population is free of RNA. Therefore, this approach may prove problematic in the interpretation of data on RNA movement. The MS2-GFP system overrides these problems because of its high specificity for binding to the RNA. Moreover, the introduction of a nuclear localization sequence to the MS2 protein causes any unbound cytoplasmic MS2-GFP protein to be shuttled back into the nucleus, thereby providing excellent signal/noise for cytoplasmic mRNA localization.

OVERVIEW OF A METHOD FOR THE DETECTION AND QUANTIFICATION OF SINGLE RNAs IN LIVING CELLS

This chapter presents a method for detecting single RNA molecules in living cells using the MS2 system described above. The sensitivity of this system allows us to address questions relating to the very core of RNA movement in the cell.

- Visualization in real time of distinct RNA particles in motion throughout the nucleus and the cytoplasm.

- Tracking of the pathways in which these RNAs move.

- Analysis of the speeds of these movements.

- Characterization of the types of movements observed.

- Quantification of the number of RNA transcripts in the moving particles.

We have used this protocol for following the movement of two different mRNAs containing the coding sequence of LacZ, the MS2 sites and the 3′ UTR (untranslated region) of either the human growth hormone gene or SV40 (Fusco et al. 2003; see Movie 35.1 on the accompanying DVD). After establishing the system in COS-7 cells and observing discrete RNA particles in the cytoplasm, it was found that the number of GFP molecules in one particle corresponded to the number of particles expected to bind one mRNA transcript. This was further corroborated by in situ hybridization showing that indeed the GFP particles contained one RNA transcript each. Similarly, this system was used to follow the trafficking of single RNA particles in the nucleoplasm (Shav-Tal et al. 2004).

PROTOCOL FOR THE DETECTION OF SINGLE RNA MOLECULES IN LIVING CELLS

This discussion presents general considerations for setting up the system, followed by the protocols. The method is composed of several distinct steps, outlined as follows.

1. Construction of a reporter mRNA containing the repeats of MS2-binding sites.

2. Construction of an MS2-GFP plasmid with a suitable promoter.

3. Coexpression of the reporter RNA (with MS2 repeats) and MS2-GFP-binding protein and visualization of RNA particles in an area of interest in the cell.

4. Quantification of the number of RNA molecules in each mRNA particle.

Construction of a Reporter mRNA Containing the Repeats of MS2-binding Sites

Number of MS2-binding Sites

The MS2-binding sites are stem-loop repeats that bind dimers of the MS2 protein. The appropriate number of MS2-binding sites inserted into the mRNA sequence that will subsequently produce resolvable particles has been determined. We have found that RNA particles containing 24 MS2 repeats are required for detection of single molecules in mammalian cells. In this case, enough MS2-GFP fusion proteins accumulate on one molecule for an identifiable signal.

Where to Insert the MS2 Repeats

Typically, the MS2 sites should be inserted in regions that do not affect the coding sequence but not within intronic sequences, which will probably be excised during pre-mRNA-splicing reactions. A preferred region of insertion might be between the coding region and the 3′ UTR, thus preserving both the coding sequence and noncoding elements important for RNA stability, polyadenylation, and localization.

Construction of an MS2-GFP Plasmid with a Suitable Promoter

Expression Levels of MS2-GFP Protein

The levels of MS2-GFP protein should be regulated because overexpression of this protein might cause a high background of unbound fusion protein, which will interfere with the detection of RNA particles. Therefore, promoters of varying strengths should be tested for efficiency of the formation of resolvable RNA particles. We have used three types of promoters: the cytomegalovirus (CMV) promoter, the large subunit of RNA polymerase II promoter, and the ribosomal L30 promoter. In general, the CMV promoter overexpresses the fusion protein; the other two promoters are more suitable.

Fluorescent Tagging of the MS2 Protein

Although we refer to MS2-GFP throughout this chapter, it might be useful to consider tagging the MS2 protein with other available fluorescent proteins, especially when another fluorescent marker may be added to the system. For example, the use of a combination of yellow fluorescent protein (YFP) and cyan fluorescent protein (CFP) may be appropriate when another marker beside the RNA is required. We have used both MS2-GFP and MS2-YFP (the latter when another CFP protein was coexpressed) and have found both fusion proteins reliable, although the YFP fusion protein bleaches more rapidly compared to the GFP fusion protein.

Nuclear Localization Sequence

The MS2-GFP plasmid can have a nuclear localization sequence (NLS) within its coding region (Bertrand et al. 1998). This is especially useful when observing cytoplasmic RNA particles, since only MS2-GFP bound to the mRNA will be found in the cytoplasm, whereas the unbound fusion protein will be targeted to the nucleus. We have found that this NLS-containing MS2-GFP is also effective for visualizing nuclear RNA particles.

Coexpression of the Reporter RNA (with MS2 Repeats) and MS2-GFP-binding Protein and Visualization of RNA Particles in an Area of Interest in the Cell

Transfection of Reporter mRNA and MS2-GFP Plasmids

This system requires the simultaneous expression of the reporter mRNA harboring the MS2-binding sites and the expression of the fusion MS2-GFP protein which will bind to the transcribed reporter mRNA. Both plasmids can be transiently cotransfected into cells or a stable cell line can be produced that contains a stable integration of DNA encoding the reporter RNA. The expression of the latter might be under the control of an inducible promoter, therefore providing the means for controlling the levels of the reporter mRNA in the cell. In this case, only the MS2-GFP plasmid will be transiently transfected.

For transient transfections, we have performed both overnight transfection protocols using calcium phosphate precipitation or commercially available transfection reagents and electroporation of the plasmids, the latter allowing the visualization of the system on the same day of transfection. However, this entails the removal of cells from the culture plate, transfection by electroporation, and then replating of the transfected cells on coverslips appropriate for live cell imaging. For same-day observation, the coverslips should be coated to yield fast adherence rates (Janicki et al. 2004). An effective coating reagent that we have found to be useful is Cell-Tak (BD Biosciences).

Cell Type

Mammalian cells used for the visualization of moving RNA particles in living cells should be adherent cells, preferably a cell line which does not tend to move much on the coverslips (a factor that will make the imaging process easier). Other details to be considered are the thickness of the cytoplasm or nucleus that will provide the best conditions for imaging. Cells with lamellipodia that are extremely thin have the best optical properties for cytoplasmic studies.

The Experiment: Visualizing RNA Particles in Living Cells

After the constructs are ready and the cell type is chosen, it is now important to determine the conditions for transfection and visualization of the RNA. Transfection of the plasmids can be achieved either by calcium phosphate coprecipitation or by electroporation.

Transfection of Plasmids by Overnight Transient Transfection

1. Split cells to ~50–70% confluence one day prior to transfection on coverslips appropriate for live cell imaging (Bioptechs Inc., Butler, Pennsylvania).

2. On the following day, cotransfect the reporter plasmid together with the MS2-GFP plasmid.

   The ratio of plasmids must be tested specifically for best transfection efficiency (we have successfully used 5 μg of each plasmid). In the case of calcium phosphate coprecipitation, add sheared salmon sperm DNA to increase efficiency. Test the duration of transfection in order to find the optimal time for expression and detection.

3. On the following day, prior to visualization, change the medium to Leibovitz's L-15 medium (without phenol red) plus serum.

Transfection of Plasmids by Electroporation

1. Split cells to ~30–50% confluence one day prior to transfection.

   Cells should be suspended very well in order to obtain single-cell suspensions to be seeded in the plates.

2. On the following day, trypsinize the cells and suspend them in a small volume.

   Typically, cells from a 10-cm tissue culture plate will be suspended in 1 ml of medium containing serum.

3. Transfer 200 μl of the cell suspension into an electroporation cuvette.

4. Electroporate the reporter plasmid together with the MS2-GFP plasmid.

   The ratio of plasmids must be tested specifically for best transfection efficiency. Add sheared salmon sperm DNA (40 μg) to the transfection medium to increase efficiency. Conditions for electroporation are specific to the cell type, and parameters for commonly used cell lines can be found on the Internet site of Bio-Rad (http://www.biorad.com; search for "electroprotocols"). Otherwise, these conditions should be optimized.

5. After transfection, plate all the cells from the cuvette in the center of a coated coverslip suitable for live cell imaging. Allow the cells to settle down on the coverslip and then fill the plate with medium.

6. A few hours later, when cells are fully spread, change the medium to Leibovitz's L-15 medium (without phenol red) plus serum.

Live Cell Imaging

Mount the coverslip in an appropriate heated chamber (Bioptechs Inc.) for use in live cell imaging. Scan for positively transfected cells trying not to bleach them once found. In general, the presence of MS2-particles or granules in the cell should be obvious (see Movie 35.1). The movement of the labeled RNA particles is not always distinguishable by eye even with a 100× objective. Rapid imaging of the transfected cell (at least 3–9 frames per second) using a cooled CCD (charge-coupled device) camera should provide a movie sequence with moving RNA particles, which can be analyzed.

Troubleshooting the Visualization of RNA Experiments

No transfected cells observed

- Use a different transfection protocol. Vary the ratios of plasmid DNA used in the transfections.

- RNA particles are sometimes hard to observe in living cells. Test the procedure by fixing the cells (20 minutes in 4% formaldehyde followed by two 10-minute washes in phosphate-buffered saline). MS2-GFP-labeled RNA particles are easily identified in fixed cells, because they do not move.

- Cotransfect a known fluorescent protein (different from the MS2-fusion protein) to help in identifying positively transfected cells. The compound detection of the MS2-GFP reporter system can be more difficult to visualize than a simple transfection reporter.

No RNA particles observed

- It is vital to screen as many transfected cells as possible. Sometimes doubly transfected cells are rare.

- In cells with high expression levels of MS2-GFP, it is probably not possible to observe distinct particles. If a strong green background is observed, change the MS2-GFP plasmid concentration in the transfections or the promoter driving the expression. Decrease the amount of MS2-GFP plasmid transfected.

- To verify if the mRNA reporter is being expressed, try in situ hybridization. We have done this using probes designed to hybridize either with the coding region of the reporter mRNA or with the regions between the MS2-binding sites. The latter option will generate a stronger signal because there are multiple binding sites for this probe between the MS2 repeats.

Quantification of the Number of RNA Molecules in Each mRNA Particle

The number and spatial distribution of RNA molecules in fixed cells can be quantified by hybridizing the RNA to fluorescence probes, using RNA FISH, and then imaging the cells, using quantitative digital microscopy (Femino et al. 1998; Fusco et al. 2003). The number of RNAs can be quantified after the light output of each directly labeled FISH probe is measured and the optical blurring of the images acquired is corrected. Similarly, the number of RNA molecules can be estimated by quantifying GFP fluorescence.

An ideal fluorescent image of a cell, acquired by digital fluorescence microscopy, is composed of pixels whose intensities are proportional to the fluorescence emitted by the cell in the area represented by each pixel. As described by the Beer-Lambert Law, fluorescence emission is directly proportional to the concentration of fluorescent molecules observed. The number of molecules observed is calculated once the fluorescent output of each molecule is measured. By collecting an ideal three-dimensional digital volume of a cell, the cell can be digitally divided into discrete volume elements—called voxels, which are three-dimensional pixels—to measure the number of fluorescent molecules throughout the cell. (A complete technical review can be found in Femino et al. 2003).

The fluorescence output of a FISH probe is measured by imaging a known quantity of purified, labeled probe. By imaging a series of probe dilutions, and knowing the molecular weight of the labeled probe, a plot can be constructed of total fluorescence intensity (TFI) versus number of probe molecules. The slope of the best-fit line is the TFI per molecule.

No optical system can provide ideal images because the image formed by an optical system is uniquely blurred in a process mathematically defined as convolution. The unique blurring properties of an optical system are described by its point-spread function, or PSF, which can be measured. A PSF is a three-dimensional image of a point source of light used to describe the optical properties of an imaging system. Factors such as objective lens, numerical aperture, magnification, wavelength of emission, and index of refraction of the mounting medium and immersion oil influence the PSF. The effects of optical blurring in a digital image volume can be reversed by a constrained iterative deconvolution algorithm which uses the PSF of the fluorescence microscope to restore the image volume to its unblurred state (see Chapter 15). Mathematically, deconvolution can only approximate the unblurred, or ideal, image volume.

In Situ Hybridization

To quantify the number of RNA molecules in each particle, it is necessary to perform sensitive FISH with a probe that binds to one unique sequence in the mRNA studied. Many protocols exist for FISH, and we generally use the procedure described by Chartrand et al. (2000). Images of the FISH signal are then acquired in three dimensions for quantification. It is important for the quantification of these images to use constant magnification, CCD binning, exposure times, and z-step size while imaging. As the light output of each particle will be quantified, it is important to remove the blurred light from the image volume using deconvolution software with a constrained iterative deconvolution algorithm, such as Exhaustive Photon Reassignment (EPR, Scanalytics), DeltaVision (Applied Precision) derived from the original Agard and Sedat algorithm (for a review, see Agard et al. 1989), or Huygens Professional (Scientific Volume Imaging). Each of these programs performs quantitative calculations and contains algorithms that will reassign out-of-focus light to the point sources from which it originated. Deconvolution can be performed using either an actual or an assumed PSF. Using an acquired PSF for deconvolution will improve the accuracy of the data. This process cannot be done accurately using a confocal instrument, because the intensity of the laser illumination fluctuates as it scans the field and the photomultiplier tube (PMT) detectors generally do not provide quantitative measurements.

Probes for RNA FISH are usually 50 nucleotides long and are selected so that the probe contains at least four T positions for potential labeling and a G/C content of 50% for consistent hybridization kinetics among the probes. Fluorescent dyes are used to label the probe by attaching them to modified Ts that have an amine linker arm and are spaced no less than 8–10 bases apart. If necessary, a modified T can be added to the 5′ end.

Measuring Light Output of FISH Probes

Preparation of Probe Dilutions

The purpose of imaging dilutions of probes is to determine the light output of a single probe. This information is used to quantify the number of probes in each RNA particle in the cells imaged after FISH.

Visualization and Quantification of Single RNA Molecules in Living Cells
Yaron Shav-Tal, Shailesh M Shenoy and Robert H Singer

Calculation of Fluorescent Intensity of Single Probes

Instructions:
Enter values requested into yellow boxes.
Intermediate calculations are shown in green boxes.
Results are shown in purple boxes.

I) Calculate Probe Stock Concentration

a) Enter Sequence: (spaces are ignored, use any character to indicate modified T base)
cic cct ctg atg ltt alt tgt att taa atg ccc ala tgc cct tcc gag lg

b) Enter Absorbance (OD):
1.3

c) Enter Cuvette Path Length (cm):
1

Sequence Length (nucleotides):
50 [9:A,15:T,8:G,13:C,5:modified base]

Probe Molecular Weight (g/mol):
15314

Extinction Coefficient (cm⁻¹•M⁻¹)
500,800

Probe Stock Concentration (ng/µl):
39.8

2) Calculate Amount of Probe Imaged

a) Enter Pixel Size (adjust for binning) (µm/pixel):
0.0645

b) Enter ROI Width (pixels):
256

c) Enter ROI Height (pixels):
256

d) Complete Table Entries:

coverslip #	1	2	3	4	5	6	7	8	9	10	11
concentration (ng/µl)	0.4000000	0.4000000	0.4000000	0.4000000	0.0400000	0.0400000	0.0040000	0.0040000	0.0040000	0.0004000	0.0004000
distance between coverslip and slide (µm)	9.50	10.00	9.00	17.00	12.00	10.00	19.00	24.00	13.30	9.00	10.00
fluorescence intensity of ROI	73285160	71835640	70723856	35375076	27272086	27322336	25824320	29903504	21543098	24414720	24180936
concentration (ng/µm³)	4.0000E-10	4.0000E-10	4.0000E-10	4.0000E-11	4.0000E-11	4.0000E-11	4.0000E-12	4.0000E-12	4.0000E-12	4.0000E-13	4.0000E-13
concentration (probes/µm³)	15.72	15.72	15.72	1.57	1.57	1.57	0.16	0.16	0.16	0.02	0.02
volume of ROI (µm³)	2590.14	2726.46	2453.82	4634.98	3271.75	2726.46	5180.28	6543.51	3626.19	2453.82	2726.46
number of probe molecules in ROI	40728	42871	38584	7288	5145	4287	815	1029	570	39	43

3) Plot Fluorescence Intensity of ROI vs Number of Probe Molecules in ROI

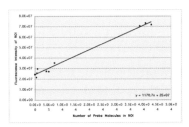

y = 1170.7x + 2E+07

FIGURE 2. Excel spreadsheet. An example spreadsheet reflecting real experimental data is provided to simplify the calculation process for single-molecule fluorescence intensity. In addition, an interactive Excel spreadsheet containing all the steps required for these calculations is included on the accompanying DVD, so that users can plug in their numbers to determine single-molecule fluorescence intensity.

1. Measure the concentration of the probe used in the FISH experiment in ng/µl (for calculation template, see the Excel spreadsheet on the accompanying DVD and for an example, scc Fig. 2).

   a. One way to calculate probe concentration is to use Beer's law: $A = \epsilon cd$, where A is absorbance (OD), ϵ is extinction coefficient (cm⁻¹ M⁻¹), c is concentration (M), and d is cuvette path length (cm). A simple way to calculate the extinction coefficient is to sum the individual coefficients for each base in the probe sequence: A = 15,400; T = 8,700; G = 11,500; C = 7,400 (cm⁻¹ M⁻¹).

   b. To convert concentration from moles/liter (M) to g/liter, multiply by the molecular weight of the probe (g/mole), which can be obtained using commercially available software such as Oligo (Molecular Biology Insights).

2. Prepare five serial dilutions of the probe in double-distilled H_2O. Prepare these gravimetrically to improve the accuracy of the dilutions. Suggested dilutions to begin with are as follows.

Cy3-labeled probes	10–20 ng/µl
Cy5	75 ng/µl
FITC	130 ng/µl

Further dilute each one of the above serial dilutions 1:10 in mounting medium.

3. Prepare a solution of diluted fluorescent beads (Molecular Probes).

> Apply the beads to many areas on the coverslip and on the slide surfaces and let them dry. We use 200-nm blue beads and dilute them 1:1000 to 1:2000 in H_2O for a dilute bead solution. The emission spectrum of the beads should differ from that of the fluorophore on the probe being measured.

4. To each slide, apply 5 µl of the appropriate dilution in mounting medium. Press the coverslip firmly onto the slide and let it settle for a few hours under some heavy books. Make sure that the surface of the coverslip containing the beads is applied face down onto the slide surface containing the beads.

Measuring the Fluorescence Intensity of a Single Probe

This procedure was obtained from Femino et al. (1998, 2003). The TFI per probe molecule is measured by first imaging the series of calibrated probe solutions and calculating the TFI in a specific volume on each slide. For imaging, use a wide-field microscope with cooled CCD camera. It is vital to image the probe dilutions at the same time as imaging the specimen. Use the identical optical configuration and acquisition parameters to image the probe dilutions and specimen. Make sure that the CCD gain, readout speed, and binning modes are the same. Perform the following procedure for each slide in the probe dilution series.

1. Search for stationary fluorescent beads. Some should be stuck to the coverslip and others to the slide. Exploit this difference in height to find the distance between the two glass surfaces, which will serve for calculating the imaged volume. To obtain a precise distance by which to determine an imaged volume, use a piezoelectric positioning device or a microscope with a precise internal focus motor and sum the increments to calculate the depth to multiply by the field area.

2. In the same field, move to a z, or axial, position that is halfway between the two glass surfaces (as measured in Step 1) and capture a single image at that position, with the same optical configuration and exposure time used to acquire images of the specimen. Crop the field (e.g., 256 × 256 pixels) to exclude regions that have beads. This represents the TFI emanating from the number of probe molecules in this cropped region. Record this value.

3. Since the distance between the glass surfaces will vary between fields, repeat the previous two steps to acquire images from multiple fields from each slide in the series.

4. Determine the dimension of each acquired pixel in micrometers. Divide the physical pixel size of the CCD detector by the total magnification used to image the cell. If CCD binning was used, multiply the calculated pixel size by the binning factor. To verify these calculations, use a stage micrometer to measure pixel size.

5. For each field of data collected:

   a. Determine the volume of probe solution between the glass surfaces that intersects with the cropped field.

   Volume (µm³) = (number of pixels in cropped field) × (x dimension of pixel) × (y dimension of pixel) × (distance between glass surfaces in acquired field)

   b. Calculate the number of probe molecules in the volume.

Number of probe molecules = Avogadro's number (molecules/mole) × probe concentration (g/μm³) × volume (μm³) ÷ probe molecular weight (mole/g).

c. Prepare a scatter plot with the number of probe molecules (collected in Step 5b) on the x axis and the TFIs (collected in Step 2) on the y axis. Perform a linear regression by fitting the points to a straight line. The slope of this line is the fluorescence intensity of a single probe (intensity/probe molecule) (see Fig. 2).

Measuring the Fluorescence Intensity of RNA Particles

Now that the fluorescence intensity of a single probe molecule is known, the number of probe molecules bound, and hence the number of RNA molecules, can be calculated in each particle identified by FISH. To do this, define RNA particles in the image volume, using commercially available image analysis software, such as Imaris (Bitplane), and calculate the total fluorescence intensity of each particle and divide by the fluorescence intensity of a single probe molecule. Since the image volumes were deconvolved, an additional correction factor must be applied to the data. When using Huygens Professional (SVI) or DeltaVision (Applied Precision), divide the total fluorescence intensity by the number of planes in the image volume. When using EPR (Scanalytics), divide the total fluorescence by the number of planes in the PSF used for deconvolution.

Measuring Light Output of GFP in MS2-GFP Particles

To quantify the number of GFP molecules in RNA particles labeled with MS2-GFP, use the

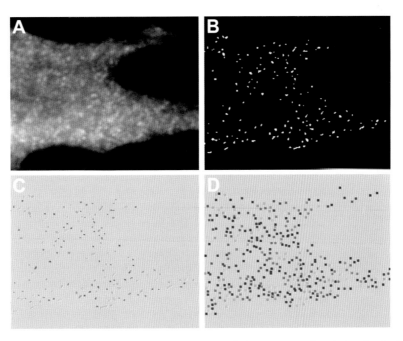

FIGURE 3. An example of the analysis of MS2-GFP-labeled RNA particles. (A) A focal plane from an image volume showing MS2-GFP particles, both in focus and out of focus, throughout the cytoplasm. (B) A single plane from a deconvolved version of the image volume from panel A. (C) A single plane from a color map indicating the objects detected in the deconvolved image volume. (D) A two-dimensional color map of the objects in the cell. Each object is marked with a square at the centroid of the object. The color code (C and D) indicates the number of GFP molecules in the particle. Blue particles contain the lowest number of molecules and red the highest.

same methods described to quantify the number of FISH probes. In this case, instead of using labeled probe, use purified GFP of a known concentration (Clontech) to make the concentration dilutions (Fig. 3).

ACKNOWLEDGMENTS

This work was supported by National Institutes of Health and Department of Energy grants GM-57071 and AR-41480 to R.H.S. The authors thank A. Femino for helpful advice on single-molecule detection and D. Fusco for providing the image used in Figure 3.

MOVIE LEGEND

MOVIE 35.1. Directed, corralled, static, and diffusive mRNA particles in a cell transfected with LacZ-hGH reporter mRNA and MS2-GFP. (From Fusco et al. 2003a.)

EXCEL SPREADSHEETS

Interactive Template: An interactive spreadsheet containing all of the steps required for the calculation process for single-molecule fluorescence intensity.

Example: The spreadsheet displayed in Figure 2, simplifying the calculation process for single-molecule fluorescence intensity reflecting real experimental data.

REFERENCES

Agard D.A., Hiraoka Y., Shaw P., and Sedat J.W. 1989. Fluorescence microscopy in three dimensions. *Methods Cell Biol.* **30**: 353–377.

Agrawal S., Jiang Z., Zhao Q., Shaw D., Cai Q., Roskey A., Channavajjala L., Saxinger C., and Zhang R. 1997. Mixed-backbone oligonucleotides as second generation antisense oligonucleotides: in vitro and in vivo studies. *Proc. Natl. Acad. Sci.* **94**: 2620–2625.

Ainger K., Avossa D., Morgan F., Hill S.J., Barry C., Barbarese E., and Carson J.H. 1993. Transport and localization of exogenous myelin basic protein mRNA microinjected into oligodendrocytes. *J. Cell Biol.* **123**: 431–441.

Barry E.L., Gesek F.A., and Friedman P.A. 1993. Introduction of antisense oligonucleotides into cells by permeabilization with streptolysin O. *BioTechniques* **15**: 1016–1018, 1020.

Bertrand E., Chartrand P., Schaefer M., Shenoy S.M., Singer R.H., and Long R.M. 1998. Localization of ASH1 mRNA particles in living yeast. *Mol. Cell.* **2**: 437–445.

Calapez A., Pereira H.M., Calado A., Braga J., Rino J., Carvalho C., Tavanez J.P., Wahle E., Rosa A.C., and Carmo-Fonseca M. 2002. The intranuclear mobility of messenger RNA binding proteins is ATP dependent and temperature sensitive. *J. Cell Biol.* **159**: 795–805.

Carmo-Fonseca M., Pepperkok R., Sproat B.S., Ansorge W., Swanson M.S., and Lamond A.I. 1991. In vivo detection of snRNP-rich organelles in the nuclei of mammalian cells. *EMBO J.* **10**: 1863–1873.

Chartrand P., Bertrand E., Singer R.H., and Long R.M. 2000. Sensitive and high-resolution detection of RNA in situ. *Methods Enzymol.* **318**: 493–506.

Dirks R.W., Molenaar C., and Tanke H.J. 2001. Methods for visualizing RNA processing and transport pathways in living cells. *Histochem. Cell. Biol.* **115**: 3–11.

Femino A.M., Fay F.S., Fogarty K., and Singer R.H. 1998. Visualization of single RNA transcripts in situ. *Science* **280**: 585–590.

Femino A.M., Fogarty K., Lifshitz L.M., Carrington W., and Singer R.H. 2003. Visualization of single molecules of mRNA in situ. *Methods Enzymol.* **361**: 245–304.

Fisher T.L., Terhorst T., Cao X., and Wagner R.W. 1993. Intracellular disposition and metabolism of fluorescently-labeled unmodified and modified oligonucleotides microinjected into mammalian cells. *Nucleic Acids Res.* **21**: 3857–3865.

Fusco D., Bertrand E., and Singer R.H. 2004. Imaging of single mRNAs in the cytoplasm of living cells. *Prog. Mol. Subcell. Biol.* **35:** 135–150.

Fusco D., Accornero N., Lavoie B., Shenoy S.M., Blanchard J.M., Singer R.H., and Bertrand E. 2003. Single mRNA molecules demonstrate probabilistic movement in living mammalian cells. *Curr. Biol.* **13:** 161–167.

Janicki S.M., Tsukamoto T., Salghetti S.E., Tansey W.P., Sachidanandam R., Prasanth K.V., Ried T., Shav-Tal Y., Bertrand E., Singer R.H., and Spector D.L. 2004. From silencing to gene expression: Real-time analysis in single cells. *Cell* **116:** 683–698.

Knowles R.B. and Kosik K.S. 1997. Neurotrophin-3 signals redistribute RNA in neurons. *Proc. Natl. Acad. Sci.* **94:** 14804–14808.

Knowles R.B., Sabry J.H., Martone M.E., Deerinck T.J., Ellisman M.H., Bassell G.J., and Kosik K.S. 1996. Translocation of RNA granules in living neurons. *J. Neurosci.* **16:** 7812–7820.

Lago H., Fonseca S.A., Murray J.B., Stonehouse N.J., and Stockley P.G. 1998. Dissecting the key recognition features of the MS2 bacteriophage translational repression complex. *Nucleic Acids Res.* **26:** 1337–1344.

Leonetti J.P., Mechti N., Degols G., Gagnor C., and Lebleu B. 1991. Intracellular distribution of microinjected antisense oligonucleotides. *Proc. Natl. Acad. Sci.* **88:** 2702–2706.

Levsky J.M. and Singer R.H. 2003. Gene expression and the myth of the average cell. *Trends Cell Biol.* **13:** 4–6.

Loke S.L., Stein C.A., Zhang X.H., Mori K., Nakanishi M., Subasinghe C., Cohen J.S., and Neckers L.M. 1989. Characterization of oligonucleotide transport into living cells. *Proc. Natl. Acad. Sci.* **86:** 3474–3478.

Lorenz P., Baker B.F., Bennett C.F., and Spector D.L. 1998. Phosphorothioate antisense oligonucleotides induce the formation of nuclear bodies. *Mol. Biol. Cell.* **9:** 1007–1023.

Lorenz P., Misteli T., Baker B.F., Bennett C.F., and Spector D.L. 2000. Nucleocytoplasmic shuttling: A novel in vivo property of antisense phosphorothioate oligodeoxynucleotides. *Nucleic Acids Res.* **28:** 582–592.

Matsuo T. 1998. In situ visualization of messenger RNA for basic fibroblast growth factor in living cells. *Biochim. Biophys. Acta* **1379:** 178–184.

McNeil P.L. and Warder E. 1987. Glass beads load macromolecules into living cells. *J. Cell Sci.* **88:** 669–678.

Molenaar C., Marras S.A., Slats J.C., Truffert J.C., Lemaitre M., Raap A.K., Dirks R.W., and Tanke H.J. 2001. Linear 2′ O-Methyl RNA probes for the visualization of RNA in living cells. *Nucleic Acids Res.* **29:** E89– E99.

Okada C.Y. and Rechsteiner M. 1982. Introduction of macromolecules into cultured mammalian cells by osmotic lysis of pinocytic vesicles. *Cell* **29:** 33–41.

Politz J.C., Taneja K.L., and Singer R.H. 1995. Characterization of hybridization between synthetic oligodeoxynucleotides and RNA in living cells. *Nucleic Acids Res.* **23:** 4946–4953.

Politz J.C., Browne E.S., Wolf D.E., and Pederson T. 1998. Intranuclear diffusion and hybridization state of oligonucleotides measured by fluorescence correlation spectroscopy in living cells. *Proc. Natl. Acad. Sci.* **95:** 6043–6048.

Politz J.C., Tuft R.A., Pederson T., and Singer R.H. 1999. Movement of nuclear poly(A) RNA throughout the interchromatin space in living cells. *Curr. Biol.* **9:** 285–291.

Shav-Tal Y., Darza X., Shenoy S.M., Fusco D., Janicki S.M., and Spector D.L. 2004. Dynamics of single mRNPs in nuclei of living cells. *Science* (in press).

Sokol D.L., Zhang X., Lu P., and Gewirtz A.M. 1998. Real time detection of DNA. RNA hybridization in living cells. *Proc. Natl. Acad. Sci.* **95:** 11538–11543.

Theurkauf W.E. and Hazelrigg T.I. 1998. In vivo analyses of cytoplasmic transport and cytoskeletal organization during *Drosophila* oogenesis: Characterization of a multi-step anterior localization pathway. *Development* **125:** 3655–3666.

Tyagi S. and Kramer F.R. 1996. Molecular beacons: Probes that fluoresce upon hybridization. *Nat. BioTechnology* **14:** 303–308.

Wilkie G.S. and Davis I. 2001. *Drosophila wingless* and *pair-rule* transcripts localize apically by dynein-mediated transport of RNA particles. *Cell* **105:** 209–219.

CAUTIONS

GENERAL CAUTIONS

The following general cautions should always be observed:

- Become **completely familiar** with the properties of substances used before beginning the procedure.

- **The absence of a warning** does not necessarily mean that the material is safe, since information may not always be complete or available.

- If **exposed** to toxic substances, contact your local safety office immediately for instructions.

- **Use proper disposal procedures** for all chemical, biological, and radioactive waste.

- For specific guidelines on **appropriate gloves**, consult your local safety office.

- Handle **concentrated acids and bases** with great care. Wear goggles and appropriate gloves, as well as a face shield if handling large quantities.

  Do not mix strong acids with organic solvents because they may react. Sulfuric acid and nitric acid especially may react highly exothermically and cause fires and explosions.

  Do not mix strong bases with halogenated solvent because they may form reactive carbenes that can lead to explosions.

- Handle and store **pressurized gas containers** with caution as they may contain flammable, toxic, or corrosive gases; asphyxiants; or oxidizers. For proper procedures, consult the Material Safety Data Sheet that must be provided by your vendor.

- Never **pipette** solutions using mouth suction. This method is not sterile and can be dangerous. Always use a pipette aid or bulb.

- Keep **halogenated and nonhalogenated solvents** separately (e.g., mixing chloroform and acetone can cause unexpected reactions in the presence of bases). Halogenated solvents are organic solvents such as chloroform, dichloromethane, trichlorotrifluoroethane, and dichloroethane. Some nonhalogenated solvents are pentane, heptane, ethanol, methanol, benzene, toluene, N,N-dimethylformamide (DMF), dimethyl sulfoxide (DMSO), and acetonitrile.

- **Laser radiation**, visible or invisible, can cause severe damage to the eyes and skin. Take proper precautions to prevent exposure to direct and reflected beams. Always follow manufacturer safety guidelines and consult your local safety office. For more detailed information, see caution below.

- Due to their light intensity, **flash lamps** can be harmful to the eyes and may explode on occasion. Wear appropriate eye protection and follow the manufacturer's guidelines.

- **Photographic fixatives and developers** also contain chemicals that can be harmful. Handle them with care and follow manufacturer's directions.

- **Power supplies and electrophoresis equipment** pose serious fire hazards and electrical shock hazards if not used properly.

- **Microwave ovens and autoclaves** in the lab require certain precautions. Accidents have occurred involving their use (e.g., when melting agar or bacto-agar stored in bottles or when sterilizing). If the screw top is not completely removedm and there is not enough space for the steam to vent, the bottles can explode and cause severe injury when the containers are removed from the microwave or autoclave. Always completely remove bottle caps before microwaving or autoclaving. An alternative method for routine agarose gels that do not require sterile agar is to weigh out the agar and place the solution in a flask.

- **Ultrasonicators** use high-frequency sound waves (16–100 kHz) for cell disruption and other purposes. This "ultrasound," conducted through air, does not pose a direct hazard to humans, but the associated high volumes of audible sound can cause a variety of effects, including headache, nausea, and tinnitus. Avoid direct contact of the body with high-intensity ultrasound (not medical imaging equipment). Use appropriate ear protection and display signs on the door(s) of laboratories in which the units are used.

- Use extreme caution when handling **cutting devices** such as microtome blades scalpels, razor blades, or needles. Microtome blades are extremely sharp! Use care when sectioning. If unfamiliar with their use, have someone demonstrate proper procedures. For proper disposal, use the "sharps" disposal container in your lab. Discard used needles *unshielded*, with the syringe still attached. This prevents injuries (and possible infections; see Biological Safety) while manipulating used needles since many accidents occur while trying to replace the needle shield. Injuries may also be caused by broken Pasteur pipettes, coverslips, or slides.

GENERAL PROPERTIES OF COMMON CHEMICALS

The hazardous materials list can be summarized in the following categories:

- Inorganic acids such as hydrochloric, sulfuric, nitric, or phosphoric are colorless liquids with stinging vapors. Avoid spills on skin or clothing. Dilute spills with large amounts of water. The concentrated forms of these acids can destroy paper, textiles, and skin as well as cause serious injury to the eyes.

- Inorganic bases such as sodium hydroxide are white solids that dissolve in water and under heat development. Concentrated solutions will slowly dissolve skin and even fingernails.

- Salts of heavy metals are usually colored powdered solids that dissolve in water. Many are potent enzyme inhibitors and therefore toxic to humans and to the environment (e.g., fish and algae).

- Most organic solvents are flammable volatile liquids. Breathing their vapors can cause nausea or dizziness. Also avoid skin contact.

- Other organic compounds, including organosulphur compounds such as mercaptoethanol or organic amines, have very unpleasant odors. Others are highly reactive and must be handled with appropriate care.

- If improperly handled, dyes and their solutions can stain not only the sample, but also skin and clothing. Some are also mutagenic (e.g., ethidium bromide), carcinogenic, and toxic.

- Nearly all names ending with "ase" (e.g., catalase, (-glucuronidase, or zymolase) refer to enzymes. There are also other enzymes with nonsystematic names such as pepsin. Many of them are provided by manufacturers in preparations containing buffering substances, etc. Be aware of the individual properties of materials contained in these substances.

- Toxic compounds often used to manipulate cells can be dangerous and must be handled appropriately.

- Be aware that several of the compounds listed have not been thoroughly studied with respect to their toxicological properties. Handle each chemical with the appropriate respect. Although the toxic effects of a compound can be quantified (e.g., LD_{50} values), this is not possible for carcinogens or mutagens for which one single exposure can have an effect. Also realize that dangers related to a given compound may also depend on its physical state (fine powder versus large crystals/diethylether versus glycerol/dry ice versus carbon dioxide under pressure in a gas bomb). Anticipate under which circumstances during an experiment exposure is most likely to occur and take steps to protect yourself and your environment.

HAZARDOUS MATERIALS

Note: In general, proprietary materials are not listed here. Kits and other commercial items as well as most dyes are also not included. Anesthetics also require special care. Follow manufacturer safety guidelines accompanying these products.

Acetone causes eye and skin irritation and is irritating to mucous membranes and the upper respiratory tract. Avoid breathing the vapors. Because it is also extremely flammable, wear appropriate gloves and safety glasses.

Acrylamide (unpolymerized) is a potent neurotoxin and is absorbed through the skin (effects are cumulative). Avoid breathing the dust. Wear appropriate gloves and a face mask when weighing powdered acrylamide and methylene-bisacrylamide. Use in a chemical fume hood. Polyacrylamide is considered nontoxic, but it must be handled with care because it may contain small quantities of unpolymerized acrylamide.

Ammonium sulfate, $(NH_4)_2SO_4$, may be harmful by inhalation, ingestion, or skin absorption. Wear appropriate gloves and safety glasses.

Ampicillin may be harmful by inhalation, ingestion, or skin absorption. Wear appropriate gloves and safety glasses. Use in a chemical fume hood.

Arc lamps are potentially explosive. Follow manufacturer guidelines. When turning on arc lamps, make sure nearby computers are turned off to avoid damage from electromagnetic wave components. Computers may be restarted once the arc lamps are in operation.

Azide, *see* **Sodium azide**

Bisbenzimide (Hoeschst No. 33342) may be harmful by inhalation, ingestion, or skin absorption. Wear appropriate gloves and safety glasses. Use in a chemical fume hood. Avoid breathing the dust.

Bleach (sodium hypochlorite), NaOCl, is poisonous, can be explosive, and may react with organic solvents. It may be fatal by inhalation, harmful by ingestion, and destructive to the skin. Wear appropriate gloves and safety glasses. Use in a chemical fume hood to minimize exposure and odor.

CaCl$_2$, *see* **Calcium chloride**

Calcium chloride, CaCl$_2$, is hygroscopic and may cause cardiac disturbances. It may be harmful by inhalation, ingestion, or skin absorption. Avoid breathing the dust. Wear appropriate gloves and safety goggles.

CHCl$_3$, *see* **Chloroform**

CH$_3$CH$_2$OH, *see* **Ethanol**

Chloramphenicol is a probable carcinogen and may be harmful by inhalation, ingestion, or skin absorption. Wear appropriate gloves and safety glasses. Use in a chemical fume hood.

Chloroform, CHCl$_3$, is irritating to the skin, eyes, mucous membranes, and respiratory tract. It is a carcinogen and may damage the liver and kidneys. It is also volatile. Avoid breathing the vapors. Wear appropriate gloves and safety glasses. Always use in a chemical fume hood.

Colcemid is highly toxic and may cause organ failure or death if inhaled or swallowed. It may cause reproductive or fetal effects and is harmful by inhalation, ingestion, and skin absorption. Wear appropriate gloves and safety goggles. Use only in a chemical fume hood. Avoid breathing the dust.

***N,N*-Dimethylformamide (DMF), HCON(CH$_3$)$_2$,** is a possible carcinogen and irritating to the eyes, skin, and mucous membranes. It can exert its toxic effects through inhalation, ingestion, or skin absorption. Chronic inhalation can cause liver and kidney damage. Wear appropriate gloves and safety glasses. Use in a chemical fume hood.

Dimethylsulfoxide (DMSO) may be harmful by inhalation or skin absorption. Wear appropriate gloves and safety glasses. Use in a chemical fume hood. DMSO is also combustible. Store in a tightly closed container. Keep away from heat, sparks, and open flame.

Dithiothreitol (DTT) is a strong reducing agent that emits a foul odor. It may be harmful by inhalation, ingestion, or skin absorption. When working with the solid form or highly concentrated stocks, wear appropriate gloves and safety glasses. Use in a chemical fume hood.

DMF, *see* **N,N-Dimethylformamide**

DMSO, *see* **Dimethylsulfoxide**

DTT, *see* **Dithiothreitol**

Ethanol (EtOH), CH_3CH_2OH, may be harmful by inhalation, ingestion, or skin absorption. Wear appropriate gloves and safety glasses.

EtOH, *see* **Ethanol**

Fluorescein may be harmful by inhalation, ingestion, or skin absorption. Wear appropriate gloves and safety glasses.

Heptane may be harmful by inhalation, ingestion, or skin absorption. Wear appropriate gloves and safety glasses. It is extremely flammable; keep away from heat, sparks, and open flame.

$HOCH_2CH_2SH$, *see* **β-Mercaptoethanol**

Hoechst No. 33342, *see* **Bisbenzimide**

Hygromycin is highly toxic and may be fatal if inhaled, ingested, or absorbed through the skin. Wear appropriate gloves and safety goggles. Use only in a chemical fume hood. Avoid breathing the dust.

IPTG, *see* **Isopropyl-β-D-thiogalactopyranoside**

Isopropyl-β-D-thiogalactopyranoside (IPTG) may be harmful by inhalation, ingestion, or skin absorption. Wear appropriate gloves and safety glasses.

Kanamycin may be harmful by inhalation, ingestion, or skin absorption. Wear appropriate gloves and safety glasses. Use only in a well-ventilated area.

KCl, *see* **Potassium chloride**

$KH_2PO_4/K_2HPO_4/K_3PO_4$, *see* **Potassium phosphate**

Laser radiation, both visible and invisible, can be seriously harmful to the eyes and skin and may generate airborne contaminants, depending on the class of laser used. High-power lasers produce permanent eye damage, can burn exposed skin, ignite flammable materials, and activate toxic chemicals that release hazardous by-products. Avoid eye or skin exposure to direct or scattered radiation. Do not stare at the laser and do not point the laser at someone else. Wear appropriate eye protection and use suitable shields that are designed to offer protection for the specific type of wavelength, mode of operation (continuous wave or pulsed), and power output (watts) of the laser being used. Avoid wearing jewelry or other objects that may reflect or scatter the beam. Some nonbeam hazards include electrocution, fire, and asphyxiation. Entry to the area in which the laser is being used must be controlled and posted with warning signs that indicate when the laser is in use. Always follow suggested safety guidelines that accompany the equipment and contact your local safety office for further information.

Ion lasers present a hazard due to high-voltage high-current power supplies. Always follow manufacturer's suggested safety guidelines.

Ultraviolet lasers present a hazard due to invisible beam, high-energy radiation. Always use beam traps, scattered light shields, and fluorescent beamfinder cards.

Blue-green lasers present a hazard due to photothermal coagulation. Blue and green wavelengths are readily absorbed by blood hemoglobin.

Magnesium chloride, $MgCl_2$, may be harmful by inhalation, ingestion, or skin absorption. Wear appropriate gloves and safety glasses. Use in a chemical fume hood.

Magnesium sulfate, $MgSO_4$, may be harmful by inhalation, ingestion, or skin absorption. Wear appropriate gloves and safety glasses. Use in a chemical fume hood.

β-Mercaptoethanol (2-Mercaptoethanol), $HOCH_2CH_2SH$, may be fatal if inhaled or absorbed through the skin and is harmful if ingested. High concentrations are extremely destructive to the mucous membranes, upper respiratory tract, skin, and eyes. β-Mercaptoethanol has a very foul odor. Wear appropriate gloves and safety glasses. Always use in a chemical fume hood.

$MgCl_2$, *see* **Magnesium chloride**

MgSO4, *see* **Magnesium sulfate**

Na_2HPO_4, *see* **Sodium hydrogen phosphate**

NaOCl, *see* **Bleach**

NaOH, *see* **Sodium hydroxide**

Neomycin may be harmful by inhalation, ingestion, or skin absorption. Wear appropriate gloves and safety glasses.

$(NH_4)_2SO_4$, *see* **Ammonium sulfate**

Phenol is extremely toxic, highly corrosive, and can cause severe burns. It may be harmful by inhalation, ingestion, or skin absorption. Wear appropriate gloves, goggles, and protective clothing. Always use in a chemical fume hood. Rinse any areas of skin that come in contact with phenol with a large volume of water and wash with soap and water; do not use ethanol!

Phenol red may be harmful by inhalation, ingestion, or skin absorption. Wear appropriate gloves and safety glasses. Use in a chemical fume hood.

Phenylmethylsulfonyl fluoride (PMSF), $C_7H_7FO_2S$ or $C_6H_5CH_2SO_2F$, is a highly toxic cholinesterase inhibitor. It is extremely destructive to the mucous membranes of the respiratory tract, eyes, and skin. It may be fatal by inhalation, ingestion, or skin absorption. Wear appropriate gloves and safety glasses. Always use in a chemical fume hood. In case of contact, immediately flush eyes or skin with copious amounts of water and discard contaminated clothing.

PMSF, *see* **Phenylmethylsulfonyl fluoride**

Polyacrylamide is considered nontoxic, but it must be treated with care because it may contain small quantities of unpolymerized material (*see* **Acrylamide**).

Potassium chloride, KCl, may be harmful by inhalation, ingestion, or skin absorption. Wear appropriate gloves and safety glasses.

Potassium phosphate, KH_2PO_4/K_2HPO_4/K_3PO_4, may be harmful by inhalation, ingestion, or skin absorption. Wear appropriate gloves and safety glasses. Avoid breathing the dust. K_2HPO_4 $3H_2O$ is *dibasic* and KH_2PO_4 is *monobasic*.

Sodium hypochlorite, NaOCl, *see* **Bleach**

Sodium hydrogen phosphate, Na_2HPO_4 (sodium phosphate, dibasic), may be harmful by inhalation, ingestion, or skin absorption. Wear appropriate gloves and safety glasses. Use in a chemical fume hood.

Sodium hydroxide, NaOH, and solutions containing **NaOH** are highly toxic and caustic and must be handled with great care. Wear appropriate gloves and a face mask. All concentrated bases must be handled in a similar manner.

Tetracycline may be harmful by inhalation, ingestion, or skin absorption. Wear appropriate gloves and safety glasses. Use in a chemical fume hood.

Trypsin may cause an allergic respiratory reaction. It may be harmful by inhalation, ingestion, or skin absorption. Avoid breathing the dust. Wear appropriate gloves and safety goggles. Use with adequate ventilation.

Urethane, a mutagen and suspected carcinogen, is also toxic and may be harmful by inhalation, ingestion, or skin absorption. Wear appropriate gloves and safety glasses. Avoid breathing the dust and use only in a chemical fume hood.

UV light and/or **UV radiation** is dangerous and can damage the retina. Never look at an unshielded UV light source with naked eyes. Examples of UV light sources that are common in the laboratory include hand-held lamps and transilluminators. View only through a filter or safety glasses that absorb harmful wavelengths. UV radiation is also mutagenic and carcinogenic. To minimize exposure, make sure that the UV light source is adequately shielded. Wear protective appropriate gloves when holding materials under the UV light source.

Index